High-Performance Computing Using FPGAs

Wim Vanderbauwhede • Khaled Benkrid
Editors

High-Performance Computing Using FPGAs

Springer

Editors
Wim Vanderbauwhede
School of Computing Science
University of Glasgow
Glasgow, United Kingdom

Khaled Benkrid
School of Engineering and Electronics
The University of Edinburgh
Edinburgh, United Kingdom

ISBN 978-1-4614-1790-3 ISBN 978-1-4614-1791-0 (eBook)
DOI 10.1007/978-1-4614-1791-0
Springer New York Heidelberg Dordrecht London

Library of Congress Control Number: 2013932707

Printed on acid-free paper

Springer is part of Springer Science+Business Media (www.springer.com)

Foreword

The field programmable gate array (FPGA) was developed in the middle of 1980s with the original intent to be a prototyping medium. The array of programmable logic blocks enabled it to be reconfigured to any of a variety of compute functions. As such it was an attractive vehicle for "in circuit hardware emulation" where designs could be prototyped and debugged before being committed to silicon. It was also an attractive teaching vehicle for students learning computer design. It was with this in mind that I was first introduced to the FPGA by one of the pioneers in the field, Ross Freeman, the founder of Xilinx (who tragically died just a few years after Xilinx's founding).

As the underlying silicon technology improved and the functional potential was better understood, the FPGA slowly permeated many aspects of computing. By the 1990s it was an accepted component in most communications technology, then consumer-electronics and automotive applications became apparent, and by the early 2000s the FPGA was well established in almost all areas of computing except high performance computing (HPC). It would seem that HPC is an unlikely target for FPGAs, as the FPGA with all of its flexibility in both routing and configuration has a clear disadvantage when compared to custom arithmetic design when measured in terms of an area time power product. Of course, even then it was understood that there were some small specialized compute applications for which the FPGA could offer some significant performance advantages mostly in areas such as cryptography and specialized arithmetic.

Around 2003 there was a seismic shift in the underlying silicon technology, and Moore's law of frequency scaling (processors double performance every 18–24 months) became inoperative because the power densities required to support higher frequencies could not be economically sustained. The future was parallel in one way or another. In HPC the obvious approach was to use multicore implementations, but there is a problem with attempting to scale performance by simply increasing the number of cores or processors to execute an application. The programming models that we have developed over the past decade have been oriented toward sequential processing and not parallel processing. The introduction of paradigms such as layers of abstractions that hide the underlying hardware thus makes it difficult to find the right form of parallelism to best express the execution of an application.

Still the notion of using FPGAs as a fabric to realize HPC for large classes of applications is surprising to many. It surely was unforeseen a decade ago. So what enables the FPGA to make its mark in HPC? There are at least three reasons:

1. The aforementioned difficulty in achieving scalable speed up with multicore implementations.
2. While Moore's law for frequency scaling ceased in 2003, Moore's law on transistor density scaling is still very much active so that over the intervening decade transistor densities have scaled up on more than an order of magnitude. These densities enable very large FPGA configurations. An enormous number of cells are available to realize complex compute engines. And because FPGAs necessarily operate at lower frequencies, they have not hit the power density limits of the CPU.
3. The flexibility of the FPGA enables the designer to realize almost any computer configuration that can be imagined and use any form of parallelism to suit the application. This flexibility provides the opportunity to create ideal machines for specific applications and unlike a decade ago where these applications would necessarily small they now can be of significant scope—really large, important, and interesting applications.

In a sense we have come full circle: the designer is again using the FPGA to do emulation but now that emulation is not of some established CPU but an emulation of an ideal machine for a particular application using techniques, representations, and processor forms unavailable to conventional processor designs. In effect the designer is emulating the future of computing high-speed computing.

This extraordinary book brings together the work of the leading technologists in this important field and points to the direction not only for high speed computing but also for the very future of computing itself.

Stanford, CA, USA
Palo Alto, CA, USA

Michael J. Flynn

Preface

The seamless exponential increase in computing power that scientists, engineers and computer users at large have enjoyed for decades has come to an end by the mid-2000s. Indeed, while until then, computer users could rely on computing power doubling every 18 months or so simply by means of increases in transistor integration levels and clock frequencies, with no major changes to software, physical limitations including voltage scaling and heat dissipation meant that this is no longer possible. Instead, the chip fabrication industry has turned to multicore chip technology to keep the "possibility" of doubling computer performance every 18 months alive. However, this is just a "potential" performance increase and not a seamless one as application software needs to be recoded to take full advantage of the performance potential of multicore technologies. Failing this, the computer industry would cease to become a growth industry as there would be no need for computer upgrades for performance sake. Instead, the industry would become a replacement industry where computers are only bought to replace faulty ones. This could have serious economic repercussions; hence the explosion of research activity in industry and academia in recent years aimed at bridging the semantic gap between applications, traditionally written in sequential code, and hardware, increasingly parallel in architecture.

The aforementioned semantic gap, however, is also opening a window of opportunity for niche parallel computer technologies such as field programmable gate array (FPGAs) and graphics processor units (GPUs) which have become more mainstream because the problem of parallel programming has to be tackled for general-purpose processors anyway. FPGAs in particular have the promise of custom-hardware performance and low power, with the software reprogrammability advantage of general purpose processors. This is precisely why this technology has attracted a great deal of attention within the high performance computing (HPC) community, giving rise to the new discipline of high performance reconfigurable computing (HPRC).

The aim of this book is to present a comprehensive view of the state of the art of HPRC to existing and aspiring researchers in the field. This book is split into three main parts: the first part deals with HPRC applications, the second with HPRC architectures, and the third with HPRC tools. Each part consists of a number of contributions from eminent researchers in the field. Throughout the book, emphasis is made on opportunities, challenges, and possible future developments, especially in relation to other technologies such as general-purpose multicore processors and GPUs. Overall, we hope that this book will serve as both a reference and a starting point for existing and future researchers in the field of HPRC.

Finally, we thank all contributors, reviewers, and Springer's staff for their efforts and perseverance in making this book project a reality.

Glasgow, UK Wim Vanderbauwhede
Edinburgh, UK Khaled Benkrid

Contents

Part I
Applications

The first part of the book covers research on applications in the emerging field of High-Performance Reconfigurable Computing. The first two chapters present work on FPGA-based financial computing, an application field which has grown considerably in the last decade in both research and industry. These are from de Schryver et al. of the University of Kaiserslautern, Germany, and Tian et al. from the University of Edinburgh, UK, respectively. These are followed by four chapter contributions on FPGA-based bioinformatics and computational biology (BCB), another application area which has attracted considerable attention in the last decade, mostly in academia but also industry. These are from Lars Wienbrandt of the Christian-Albrechts-University of Kiel, Germany, Herbordt et al. from Boston University, USA, Yamaguchi et al. from the Universities of Tsukuba, Ryukyus, Doshisha and Keio, in Japan, and Will Li et al. from the City University of Hong Kong, China. The following two contributions are on FPGA-based data search and processing, another interesting application in our information age characterised by an explosion of data. The two contributions are from Vanderbauwhede et al. of Glasgow University, UK, and the University of Massachussets, USA, and Sklyarov and Skliarova from the University of Aveiro, Portugal. The next two contributions are on FPGA-based stencil computations, a very important area with various applications in computational fluid dynamics, electromagnetic simulation based on the finite-difference time domain method, and iterative solvers e.g. for seismic modelling. The two contributions are from Kentaro Sano from Tohoku University, Japan, and Medeiros et al. from Universidad Federal de Pernambuco, Brazil. The following chapter from Gneysu et al. of Ruhr-University Bochum, Germany, Czech Technical University in Prague, Czech Republic, and the Christian-Albrechts-University of Kiel, Germany, presents dedicated FPGA-based cluster solutions for high performance efficient cryptanalysis. After this, Hamada and Shibata from Nagasaki University, Japan, present a contribution which deals with two floating point scientific applications, namely ocean model simulation with a particular emphasis on fast inter-task communications, and astronomical N-body simulations with a particular emphasis on performance per $ and performance per Watt measures of FPGAs compared to ASICs, GPUs and general purpose processors. Finally, Kapre and DeHon from Imperial College London, UK, and the University of Pennsylvania, USA, present an FPGA-accelerated solution for the SPICE simulator, a widely used open-source tool for the simulation and verification of analog circuits.

High-Performance Hardware Acceleration of Asset Simulations

Christian de Schryver, Henning Marxen, Stefan Weithoffer, and Norbert Wehn

Abstract State-of-the-art financial computations based on realistic market models like the Heston model require a high computational effort, since no closed-form solutions are available in general. Due to the fact that the underlying asset behavior predictions are mainly based on number crunching operations, FPGAs are promising target devices for this task. In this chapter, we give an overview about current problems and solutions in the finance and insurance domain and show how state-of-the-art market models and solution methods have increased the necessary computational power over time. For the reason of universality and robustness, we focus on Monte Carlo methods that require a huge amount of normally distributed random numbers. We summarize the state-of-the-art and present efficient hardware architectures to obtain these numbers, together with comprehensive quality investigations. Build on these high-quality random number generators, we present an efficient FPGA architecture for option pricing in the Heston model, tailored to FPGAs. For the problem *pricing European barrier options in the Heston model* we show that a Xilinx Virtex-5 device can save up to 97% of energy, providing the same simulation throughput as a Nvidia Tesla 2050 GPU.

1 The Need for High Performance Computing in Secure Economies

The happenings on the financial markets all around the world in the last years have clearly demonstrated the inherent risks prevailing in our current economic system.

C. de Schryver (✉) • S. Weithoffer • N. Wehn
Microelectronic Systems Design Research Group, University of Kaiserslautern, Germany
e-mail: schryver@eit.uni-kl.de; weithoffer@eit.uni-kl.de; wehn@eit.uni-kl.de

H. Marxen
Stochastic Control and Financial Mathematics Group, University of Kaiserslautern, Germany
e-mail: marxen@mathematik.uni-kl.de

W. Vanderbauwhede and K. Benkrid (eds.), *High-Performance Computing Using FPGAs*, DOI 10.1007/978-1-4614-1791-0_1,

Due to the permanent news, nowadays every citizen is sensitized to these problems, even if not everybody (not to say nearly nobody) understands what is really going on in the financial system right now.

One main reason for the financial crisis was the wrong assessment of financial products with respect to their values and risks. For example, *collateralized debt obligations* (CDOs) considered to be one of the major causes for the crisis [8] are challenging to evaluate. CDOs bundle other products together and split the resulting pool again into new tranches with different ratings. Determining realistic risks and values for these tranches is a highly compute-intensive task.

However, CDOs are just one example of demanding products. Financial institutes have to price complex product portfolios containing many different ingredients regularly. In addition to that, countermeasures taken by the governments after the crisis in 2007 have further increased the demand for a fast simulation environment. In the European Union, for example the *Basel III* and *Solvency II* regulations for the financial and insurance sector require frequent monitoring and analysis of the institutions' financial situation, in particular of the equity risks.

Besides that, the increasing mathematical complexity of the underlying stock market models and their calibration has already led to a tremendous increase of simulation effort in the past. For example, the Heston and jump-diffusion stochastic differential equations (SDEs) lacking closed-form solutions in general are currently state-of-the-art [11]. The construction of more and more complicated financial products has further contributed to this, and since those products are available right now, there is no perspective that the complexity will decrease again in the future.

The energy needed for portfolio pricing is immense and lies in the range of several megawatts for a single bigger institute nowadays. Already in 2008 the available power for the financial center of London had to be clipped to assure a reliable supply for the Olympic games in 2012 [35]. Therefore, there is an urgent need for bringing down the energy consumption on the one hand, and to allow even higher simulation speed in the future on the other hand. This gap can only be bridged by using optimized hardware accelerators for the simulations.

Most institutes are currently running their simulations on standard CPU clusters, exploiting the highest flexibility by using pure software models. We will see in Sect. 2 that a lot of simulation methods are based on basic number crunching operations. So, a standard CPU is certainly not the most efficient architecture for this task with respect to throughput and energy efficiency. GPUs are currently emerging in the financial business and are more and more used in productive environments, for example by JP Morgan Chase, Bloomberg, or BNP Paribas [21]. Optimized architectures based on FPGAs have a huge potential for saving energy and speed up the simulations at the same time. However, FPGAs have just been used for experimental studies in financial business [36, 37], and we are not aware of these devices being used in productive risk assessment environments today.

In this chapter we cover the following topics:

- We introduce the state-of-the-art Heston model and the Multi-Level Monte Carlo method to solve derivative pricing in this context in Sect. 2.

- For the application "pricing European double barrier options in the Heston model" we shortly present a comprehensive benchmark set that allows to compare implementations on different target architectures transparently.
- We present a hardware accelerator for European barrier option pricing with the Heston model in Sect. 3, together with throughput and energy measurement results. We show that hybrid FPGA-CPU systems can already today save far more than 60% of the energy consumed by a state-of-the-art Nvidia Tesla C2050 GPU.
- In Sect. 4 we show efficient hardware architectures to generate normally distributed high-quality random numbers. These random numbers are key for efficient Monte Carlo simulations.

2 Pricing Options: Model, Algorithm and Comparison

One *problem* in financial mathematics is the pricing of derivatives. In this chapter we focus on the valuation of barrier options in particular. In order to solve this real-world problem, we need a specific *model* to reflect the behavior of the underlying asset. In our case we employ the Heston model that is widely used nowadays and is a further development of the famous Black–Scholes model.

For the *solution* of the problem we need an *algorithm* and an *implementation* thereof. A detailed systematic methodology to clearly distinguish between these terms has been given by de Schryver et al. in 2011 [27].

In this section we give an overview about different Monte Carlo methods and why they fit well to the problem that we target. Section 3 shows the details of our hardware implementation.

Besides the implementation itself, evaluating and comparing it to different algorithms and architectures is a challenge. We suggest to rely on standardized application benchmarks for this task. In Sect. 2.3 we propose a meaningful benchmark set for European barrier option pricing in the Heston model.

2.1 The Heston Model

In 1973 Fisher Black and Myron Scholes have introduced the famous Black–Scholes model [4]. In the same year Robert C. Merton [19] expanded the mathematical understanding of the model. Therefore, the model is sometimes called Black–Scholes–Merton model.

Since prices for European vanilla options were easily calculated and for more complicated ones one could model the behavior of the asset prices, the Black–Scholes model has fundamentally changed the way how the financial industry works. In 1997, Merton and Scholes received the Nobel price for their work.

The Black–Scholes model consists of certain assumptions on the market behavior. The most important ones are the absence of arbitrage—which is needed to fairly evaluate prices—and the log normal characteristic of the asset price. The price of an asset under the risk-neutral measure follows the SDE

$$\mathrm{d}S(t) = S(t)r\mathrm{d}t + S(t)\sigma\mathrm{d}W(t). \tag{1}$$

S denotes the price process of the asset, r the risk-less interest rate, W a Brownian motion and σ the volatility. Furthermore, the process has some starting condition $S(0) = s_0$.

This SDE can be solved. Its solution is

$$S(t) = S(0)\exp\left(\left(r - \frac{\sigma^2}{2}\right)t + \sigma W(t)\right).$$

In order to price a derivative of an asset following the SDE above, the fundamental theorem of asset pricing states that the price is just the discounted expected payoff under the risk-neutral measure.

Even though the SDE of the Black–Scholes model can be solved, various derivatives can only be priced numerically in this setting.

Nevertheless, besides the huge impact on the financial world, the Black–Scholes model has some drawbacks. The main is that it assumes a constant volatility. From real market data of asset prices and options it is, however, known that the volatility is generally not constant.

The Heston model [9] tackles this problem by using a second SDE to describe the behavior of the volatility process. Under the risk-neutral measure the SDEs of the Heston model are as follows:

$$\mathrm{d}S(t) = S(t)r\mathrm{d}t + S(t)\sqrt{V(t)}\mathrm{d}W^S(t),$$
$$\mathrm{d}V(t) = \kappa(\theta - V(t))\,\mathrm{d}t + \sigma\sqrt{V(t)}\mathrm{d}W^V(t).$$

The asset price process is denoted by S, and V denotes the volatility process. The latter process has the important property that it is always non-negative. Under a certain condition, called the Feller condition, the origin cannot be obtained. This is important for several mathematical results. However, this condition is seldom satisfied in real-world applications. The Brownian motions W^S and W^V are correlated, typically in a negative way. This implies that if the stock price falls, the variance tends to increase and the market becomes more volatile.

The Heston model fits much better to the data observed in real markets and provides more realistic results compared to the Black–Scholes model. On the other hand analytic pricing formulas are known for simple European options. This is especially important to calibrate the model and one of the reasons why the model is so popular.

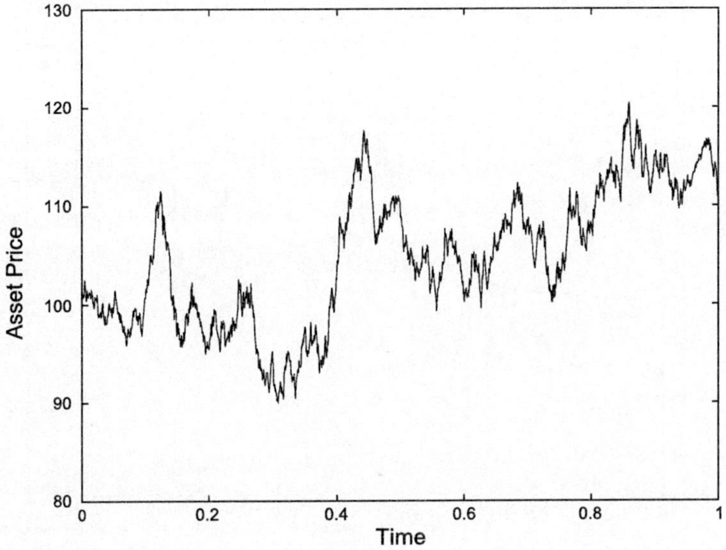

Fig. 1 A modeled asset price path in the Heston model

Figure 1 shows a realization of an asset price path following the Heston SDE. The erratic behavior is typical for most models and can be seen on the market.

2.2 *The Multi-Level Monte Carlo Method*

The price of an option is the discounted expected payoff of the option under the risk-neutral measure. One can analytically calculate the price of a plain European call or put option in the Heston model, i.e., $\mathbb{E}(e^{-rT} \cdot \max((S(T) - K), 0))$, where $\mathbb{E}$ means the expectation value, T is the maturity time, and K the strike price. However, for other options such as barrier options this is not the case. In these situations numerical methods have to be used to estimate the expectation. There are several methods available that fit best to different situations. To name the most popular ones, these are finite difference method, the quadrature scheme, tree-based methods such as binomial or trinomial trees, and the Monte Carlo method.

We will concentrate on the Monte Carlo method in this chapter. It is not always the fastest method but very flexible and applicable to a wide range of applications. The basic idea of the Monte Carlo method comes from the Law of Large Numbers. To calculate $\mathbb{E}X$ for some random variable X, one has to simulate independent realizations X^i of random variables with the same distribution as X. The mean value of all results is an estimator for the expectation. The variance of the estimator is depending on the variance of X and the number of simulations. The error introduced by this is called the *statistical error*.

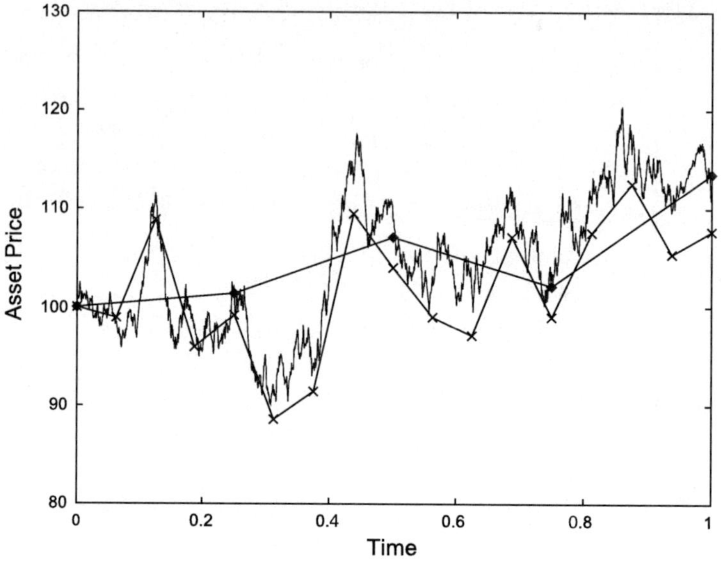

Fig. 2 A simulated Heston path and its discretizations on two different levels

Using Monte Carlo methods for asset simulations in the Heston model, however, leads to a problem: We cannot simulate $S(T)$ directly in the Heston setting. Therefore, the two SDEs are discretized and simulated. This introduces a second type of error called the *bias*. The bias is a systematic error and can be decreased by using more discretization steps. The plain Monte Carlo method now fixes the number of time steps and simulates many paths with these number of time steps. The chosen discretization has to be carefully selected, since it directly determines the bias.

It arises a second difficulty in the discretization of the volatility process in the Heston model. As we have seen, the variance process is always non-negative. The discretized version thereof, however, can become negative, if it is not adjusted. The obvious adjustment of setting a negative value to zero has turned out to be ineffective in general. More advanced schemes like the *full truncation* scheme that only set the volatility to zero when it is used as an argument of $sqrt()$ perform better [16].

Besides the discretization, the algorithm can be modified as well. The *Multi-Level Monte Carlo method*, for example, uses a slightly different approach than the plain Monte Carlo method. First, one simulates on a very coarse scale, that means with only a few time steps. These coarse scale simulations can be computed very fast. Then, iteratively, only the difference to the next finer *level* is simulated. Level in this context means a finer discretization (see Fig. 2). The variance of the difference is smaller and therefore less simulations on the finest level are needed compared to the plain Monte Carlo method. This gain can be a lot bigger than the cost of the additional simulations on the coarser levels. The benefits of the Multi-Level method increase with the required precision.

However, even though the Multi-Level Monte Carlo method is asymptotically better, the benefit is not always present in practical situations. Therefore, one has to be careful when to choose the method. In the Heston setting, a start level optimization that determines whether to use plain or Multi-Level Monte Carlo is mandatory. For more details about the Multi-Level Monte Carlo method and also the different discretization schemes in the Heston model, refer to Marxen et al. [16].

2.3 The Need for Fair Metrics: A Benchmark Proposal for Option Pricing with the Heston Model

Even though the Heston model is state-of-the-art and widely used in the financial industry, hardware accelerator publications are rare in that field (see Sect. 3.1). However, for the Black–Scholes model a lot of papers presenting sophisticated hardware architectures based on different methods exist.

The presented speedups look very impressive and the designs are likely well done. However, comparing the different implementations is a challenging task. A variety of attributes like speed, accuracy, and energy consumption can be considered. Furthermore, many different *solutions* are available in literature: not only the implementation and the architecture vary but also the algorithm. It is in many cases not clear by itself to which extent a speedup results from the employed algorithm and from the implementation. In addition to that, it is not possible to differentiate whether the presented algorithm or the implementation has the desired properties only for a special set of parameters, or if it performs well in a more general framework.

This challenge can only be bridged by using a unified benchmark set on application level, that means for a specific *problem* solved with a certain *model*. This application benchmark itself has to be independent of the algorithm and the implementation used. Morris and Aubury [20] already claimed the need for a benchmark for option pricing in 2007. By giving performance results for a benchmark set, authors allow their work to be compared fairly with respect to certain metrics without looking into details of the algorithm or the implementation.

In this section, we will describe our benchmark set for the application "pricing European double barrier options with the Heston model" presented in 2011 [26]. The benchmark was developed in a joint work with the financial mathematics group at the University of Kaiserslautern. It is freely available for download,[1] and we strongly encourage authors of future publications dealing with this problem to use it and provide application-specific metrics and therefore to make their work transparently comparable.

Twelve different settings for the Heston model, including parameter sets that have to be considered to be important in literature already, are used for the benchmark.

[1] http://www.uni-kl.de/benchmarking.

Table 1 One example of the benchmark parameters

Parameters for the Heston model	κ	θ	σ	r	S_0	V_0	ρ
	2.75	0.035	0.425	0	100	0.0384	−0.4644
Option specific parameters	Option type	Strike	Lower barrier	Upper barrier	Time to maturity (in years)		
	Double barrier call	90	80	120	1		
Price of the option	5.7538	Precision	0.0001				

They span a wide range of parameters observable on the markets. Our benchmark consists of three different components:

- The parameter sets defining the current *market situation*, such as the current volatility or the correlation between price and volatility
- The *option parameters* such as the type of option and the strike price
- The correct *reference price* or a good approximation thereof, together with a reference precision

To allow a comparison on application level, we recommend to provide the following metrics for all presented solutions:

- The consumed energy for pricing one option in *joule/option*
- The number of priced options per real time in *options/second*
- The numerical accuracy that is achieved by the proposed design, compared to the presented benchmark results
- The consumed area on chip for hardware architectures (slices, LUTs or mm^2 on silicon)

Table 1 exemplarily shows one of the twelve cases from the benchmark set [26]. The focus is not only on double barrier calls, but also on other types of options such as puts and digital calls are included.

In this section, we have briefly introduced our terminology, the Heston model, and the Multi-Level Monte Carlo method that we use in our hardware implementation described in Sect. 3, together with a benchmark set that allows to fairly compare different implementation on application level.

The key for Monte Carlo methods is a huge amount of high-quality random numbers. For hardware architectures, we therefore require efficient architectures for in our case non-uniform random numbers. We present suitable architectures for this task in Sect. 4. The next sections shows our proposed design for FPGA-based acceleration of option pricing in the Heston model.

3 Hardware Architectures for Asset Simulations

This section gives a short overview of the available FPGA implementations for option pricing. In the second part, we present an energy efficient FPGA architecture for this problem, together with detailed measured numbers for energy and throughput.

3.1 *Related Work*

Although the Heston model including its varieties (for example, the Heston–Hull–White model or the Heston model with additional jumps) is currently state-of-the-art in the financial industry [2, 11], the first GPU accelerators for solving this model have been presented just in 2010.

Zhang and Oosterlee have used the Fourier-Cosine Series Expansions (COS) method for multiple strike European and Bermudan option pricing in the Heston model on a NVIDIA GeForce 9800 GX2 GPU [40]. Compared to an Intel Core2Duo E6550@2.33 GHz CPU, they could achieve speedups between 10 and 100 for multiple strike European options, depending on the form of the characteristic function and on the number of strikes computed simultaneously.

Bernemann et al. have put the random number and path generation for Monte Carlo simulations on a Nvidia GPU, using a hybrid CPU-GPU option pricing system on top of the C++ *QuantLib* [23]. They could achieve up to 340 Gflops on a Nvidia Tesla C1060 GPU, compared to the maximum of about 11 Gflops given by a multi-threaded C++ implementation with SSE2 running on an Intel Xeon E5620@2.4 GHz [2]. Energy measurements are not provided in this work.

Based on this setup, investigations for exotic option pricing and Heston model calibration have been presented in 2011 [3]. Here Bernemann et al. have achieved a speedup between 10 and 50 for option pricing in the Heston model and 4–25 for simulations in the Heston–Hull–White model using a Hybrid Taus random number generator (RNG). The results are similar for a Mersenne Twister. For the Heston model calibration, they achieve a speedup between 15 and 50 with pseudo random numbers and 15–35 with quasi-random Sobol sequences, depending on the number of underlyings.

For option pricing in the Black–Scholes model, several FPGA architectures have been published in the last years [1,5,10,33,34,38]. These works show the wide range of potential speedups for FPGA-based accelerators, from 10 to more than 100.

In the last years, commercial FPGA systems have emerged for financial domain specific acceleration. Maxeler Technologies[2] offers hardware and software solution bundles for financial computing. They provide Xilinx Virtex-6 based platforms for professional server environments and desktop workstations. Their *MaxCompiler* for

[2] www.maxeler.com.

general purpose applications takes Java code and splits it into parts that remain on the host CPU and accelerated kernels executed on the FPGAs. The FPGA programming, including all the glue and interface logic, is done automatically.

Based on this system, the Maxeler CEO Oscar Mencer et al. presented speedup results for a single-asset Monte Carlo option pricer based on the Heston model with additional price jumps at the IEEE Workshop on High Performance Computational Finance (WHPCF) in November 2011. They have used a professional Maxeler MaxNode system with four MAX3 FPGA cards and could achieve a speedup of more than 100× over a 12 thread CPU version running on two Intel Xeon X5650@2.67 GHz CPUs [18]. Energy aspects have not been considered in this work.

Further available commercial systems are Wall Street FPGA,[3] Compaan Design,[4] and Impulse Accelerated Technologies.[5]

Wall Street FPGA uses National Instruments' LabView to bring a Monte Carlo-based European call option pricer on a Xilinx Virtex-5 FPGA [29, 30]. They state that their FPGA accelerated implementation is 131 times faster than the reference software running on an Alienware Area-51 7500 Dual Core CPU@3.0 GHz. Another application field for Maxeler is oil & gas exploration.

Impulse Accelerated Technologies and Compaan Design do not provide finance specific tools or benchmarks and cover a much wider application range.

3.2 *A Multi-Level Monte Carlo Accelerator for Option Pricing with the Heston Model*

In this section, we describe our dedicated FPGA accelerator architecture for pricing European double barrier options in the Heston model presented at ReConFig 2011 [25]. We give an overview about the architecture and provide detailed synthesis, performance, and energy results for a hybrid CPU-FPGA setup.

3.2.1 Architecture

By designing our FPGA-based accelerator, we wanted to achieve the maximum performance together with a minimal energy consumption. On the other hand, not all parts of the pricing process described in Sect. 2.1 are suitable for being implemented in hardware. For example, mathematical operations like $exp()$ or $/$ that are only needed for the final payoff computation would use up a lot of hardware resources, but could not contribute very much to increase the overall simulation speed.

[3] www.wallstreetfpga.com.

[4] www.compaandesign.com.

[5] www.impulseaccelerated.com.

Therefore we have decided to chose a hardware–software partitioning scheme that only brings those parts of the computation to hardware that are mainly data-flow oriented and use up most of the simulation time. We call these parts compute-intensive *kernels*. Complex mathematical operations or control driven parts remain on the host-CPU. This partitioning approach is also used by a number of authors proposing related accelerator designs [2, 3].

In particular, we have decided to chose the following partition for our implementation:

- The random number generation, the path simulation, and the barrier checking are ported to the FPGA. These kernels can be conveniently executed in parallel for different paths and return the final price for each path.
- The final path prices are transmitted to the host over USB. We have used the FTDI FT2232H interface module with a top average throughput of measured 6 MB/s.
- The reduction of all path results and the payoff computation remain on the host CPU.

For the random number generation, we have used a Tausworthe 88 uniform RNG together with our conversion unit described in Sect. 4.2.2. Since it provides a stream interface with handshaking, it can stall the rest of the design easily if no random number is present in the current clock cycle. However, any kind of uniform RNG may be used together with this converter. For example, interleaved parallel Mersenne Twisters as described in Sect. 4.1 that independent streams of random numbers from a single generator unit seem to be especially beneficial for high-quality multi-accelerator setups. Nevertheless, by simulating our benchmark introduced in Sect. 2.3 we have ensured that three Tausworthe 88 instances with independent seeds provide sufficient randomness for our application (see results in Sect. 3.2.2).

Our hardware has been implemented on a Xilinx ML-507 evaluation kit with a Virtex-5 XC5VFX70T FPGA. It uses single precision floating point units generated with the Xilinx CoreGen tool.

We have decided to use a similar approach to the automatically generated designs proposed by Thomas et al. [31]. However, we use our own host interface framework on top of the USB connection that allows to transparently read and write registers and data streams from a software application. Therefore we do not require a bus, but directly use a handshake-driven stream interface for the output prices and registers for the parametrization. Our protocol allows to dynamically reconfigure the accelerator parameters for the Monte Carlo simulation, the market and the option at runtime.

Our hardware design mainly consists of two parts: the control logic and the actual data path. In order to bring up the clock frequency to the maximum, our data path implementation is maximally pipelined. To get rid of additional control logic and to provide maximum scalability, we have decided to use a *packet-based* concept in our design:

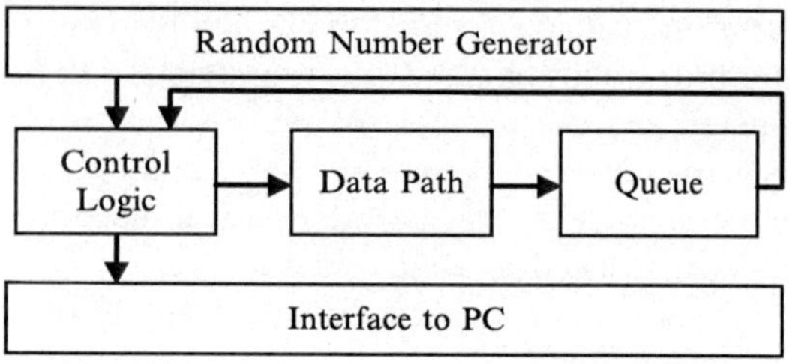

Fig. 3 High-level architecture of our hardware implementation

- Each packet describes the current state of a single path, including the price, volatility, step number, and a validity flag. Instead of having complex early termination strategies for paths that have hit a barrier, we change the status of those packets to *dummy packets* by clearing the validity flag. These packets remain in the processing pipeline, which decreases the throughput to some extent, but at the same time drastically reduces the hardware complexity.
- The data path is a pipeline that computes price and volatility for the next step and performs the barrier checking (see Sect. 2.1). It consumes one packet and produces another one in every clock cycle.
- The pipeline latency with 32-bit single precision floating point numbers is 60. This means that at every clock cycle, the pipeline outputs a packet that was sent to it 60 cycles earlier.
- When a packet goes through the pipeline, its contents are updated according to the selected algorithm for solving the Heston model, that is full truncation with antithetic variance reduction in our case (see Sect. 2.1).

Figure 3 shows the structure of our design and the interaction between the data path, a queue and the control logic. The queue buffers all packets coming out of the data path for future processing or final transmission to the host. This decision is made by the control logic. The depth of the queue has to be greater than the pipeline length of the data path, which is 60 in our case. We therefore have exploited the maximum depth of a BRAM36 slice from the target Virtex-5 device for the queue. It is important to note that the data path block is only made up of simple pipelined floating point cores, uses handshake-driven stream interfaces, and does not require support for any stall signals.

The role of the control logic is to act as a broker between the RNG, the data path and the host system. It follows the following set of rules:

- If the amount of created packets is less than the queue size, a new path is created.
- If enough packets are active, the control logic checks if a packet is available from the queue.
- If the queue contains a packet, its step number is checked. If the control logic sees that it was the last step, the final price is sent to the host, and a new packet is created. If not, the packet is resent to the pipeline along with a new pair of random numbers.

Table 2 Single precision floating point components in the data path

Component	Adders	Multipliers	Subtractors	sqrt()
Heston step generator	4	6	2	1
Barrier checker	1	1	1	0

The control logic has been implemented equivalently in a bit-true software model to allow easy testing of the design. Together with a bit-true model of the hardware RNG, each hardware component can be validated against the software reference independently. As the processing order of the packets does not depend on interface delays, this ensured bit-by-bit equivalence between software and hardware results.

The decomposition between the control logic and the data path further contributes to the reduction of the validation effort:

- The pipeline can be tested separately from the control logic, only considering the floating point operations.
- The control logic can be checked on its own by using a dummy pipeline that only counts the steps and has no floating point logic inside at all.

The internal structure of our pipeline is similar to the GARCH example presented by Thomas et al. [31], but includes the Heston specific modifications. Table 2 shows the number of floating point units in the Heston step generator part of the pipeline (that generates successive values for price and volatility) and the subsequent barrier checking.

We have used THDL++, a high-level approach for HDL design together with the free VisualHDL tool for the development.[6] For this task, the VisualHDL tool has been enhanced by a data path pipeline designer plugin that is shown in Fig. 4. It allows creating a data path by just dragging-and-dropping operations and connecting them from the inputs in the upper part of the screenshot to the output at the bottom.

3.2.2 Results

All synthesis results have been generated for a Xilinx Virtex-5 XC5VFX70T device (as on the ML-507 evaluation board) with the Xilinx ISE Design Suite 13.1. The results have been optimized for speed, are post place & route, and include the host interface logic. Although Xilinx is currently releasing the Virtex-7 family, no evaluation kits are available at the moment. Therefore we use the ML-507 kit in order to provide system level results for speed and energy (for all details refer to de Schryver et al. [25]).

Table 3 shows the number and percentage of resources used for two different corner scenarios: Using no DSP slices in the dataflow at all (the one remaining is

[6] visualhdl.sysprogs.org.

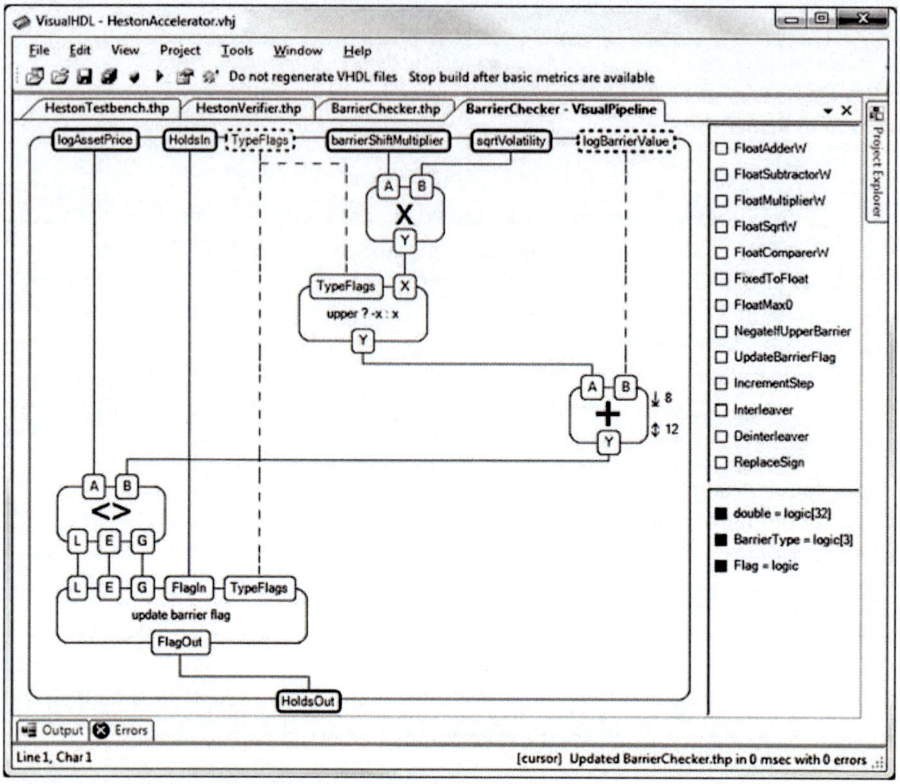

Fig. 4 VisualPipeline plugin editing the Heston barrier checker

Table 3 Synthesis results for one instance on a Virtex-5

	Minimum DSP usage		Maximum DSP usage	
	Number	Percentage (%)	Number	Percentage (%)
Slices	4,862	43	2,497	22
LUTs	11,382	25	5,481	12
Flip-flops	13,530	30	6,950	15
LUT-FF pairs	15,041	33	8,176	18
DSP48E slices	1	1	43	33
BRAM36 slices	5	3	5	3
Max. frequency	102 MHz		100 MHz	

occupied by the RNG from [24]) and using the maximum amount of DSP slices, depending on the Xilinx CoreGen settings.

From Table 3 we see that (without the triple interface logic) in total three instances can be put on a single XC5VFX70T device. We assume a three-instance FPGA accelerator for the following considerations. A Virtex-7 device would provide enough space for several hundreds of accelerator instances.

Table 4 Speed and energy results for the laptop-FPGA setup

	Laptop only		Laptop + FPGA		Factor (laptop/FPGA)	
Time steps	Real time (s)	Energy/step (J)	Real time (s)	Energy/step (J)	Real time	Energy
32	56	76.31	4	5.38	13.88	14.20
64	116	79.75	8	5.38	14.50	14.84
128	230	79.06	9	3.14	24.64	25.22
256	465	79.84	18	2.46	25.81	32.44
1,024	1,852	79.56	72	2.47	25.60	32.18
4,096	7,344	78.89	287	2.46	25.56	32.13
Average		78.90		3.55	21.66	25.17

Since the host CPU in the hybrid CPU-FPGA setup only computes the final payoff and performs the communication with the ML-507 board, we have chosen to use a low-power laptop as host: a Fujitsu Siemens Lifebook E8410 with an Intel Core 2 Duo T7250@2.0 GHz and 2 GB RAM, running Windows 7 Professional SP1 64 Bit. In the idle state, the laptop itself consumes around 20 W.

Detailed measured numbers for runtimes and energy consumptions in this setting are given in Table 4, with and without FPGA acceleration. In each case, ten millions of paths have been computed.

For the software-only simulations, it can be seen that the measured real time and consumed energy are linearly related to the number of time steps in the simulation. This is not surprising, since the power consumption of the laptop with the CPU fully loaded remains constantly 44 W. In this case, the idle power consumption of the FPGA board has not been included in the measurements.

Measuring the hybrid setup, the FPGA board with an idle power consumption of 9 W has been added to the 20 W of the laptop, so that we are talking about a 29 W idle load in total. In this setting Table 4 shows that the energy per step is much higher for small numbers of time steps (32–128). For 32 and 64 time steps, we have measured a power consumption of 40 W during the simulations for the whole system. For 256 and more steps, it remained constant 35 W.

The explanation for this observation is that the host-to-board interface provides a limited bandwidth. The host CPU also runs the tasks for communicating with the FPGA board, so that the amount of energy used for communication is very high for small step sizes, compared to the total energy consumed for one simulation. For more than 256 step sizes, the computations on the FPGA take enough time, so that the interface is no longer the limiting factor.

Table 4 also clearly shows that the average speedup of the hybrid system compared the the CPU-only scenario is 21 times in average, by only consuming 4% of energy per simulation.

Nowadays, financial simulations are performed on high-end CPU and GPU clusters. To give a fair comparison of our design to the state-of-the-art, we have implemented our Monte Carlo algorithm on a Nvidia Tesla C2050 graphics card. Preliminary work in our group has shown that the performance loss of using OpenCL is insignificant compared to CUDA. Thus, for the reason of higher flexibility, we have coded our accelerator in OpenCL.

Table 5 Speed and energy results for the server-GPU setup

	Server only		GPU accelerated		Factor (server/GPU)	
Time steps	Real time (s)	Energy/step (J)	Real time (s)	Energy/step (J)	Real time	Energy
32	5	29.06	0.95	9.22	5.25	3.15
64	10	29.06	1.88	9.09	5.33	3.20
128	21	30.88	3.74	9.05	5.69	3.41
256	41	29.97	7.43	9.00	5.55	3.33
1,024	166	30.20	29.68	8.99	5.60	3.36
4,096	660	29.97	118.46	8.97	5.57	3.34
Average		29.86		9.05	5.50	3.30

The Tesla GPU is hosted by a FluiDyna TWS 1xC2050-1xIQ-8 server workstation with an Intel Xeon CPU W3550@3.07 GHz and 8 GB RAM running OpenSuSE Linux 11.4 64 bit with Kernel 2.6.37.6-0.5-default (referred to as *server* in the following). The CPU provides four physical cores with hyperthreading, so that we can count them as eight cores. The idle power consumption for the server is 87 W without the GPU, and 148 W on average with the Tesla card plugged in. As in the laptop-FPGA setting, we have removed the GPU for all software-only measurements.

With the CPU fully loaded (but without the GPU), the server system consumes 186 W in average. If we run the simulations with full load on the GPU, the CPU still has to compute the payoff at the end of all Monte Carlo simulations. In this case, the overall power consumption of system is 310 W.

In Table 5 we see all measured runtime and energy results for the server-GPU setting. Again, we provide the numbers for a software only run on the virtual eight cores of the server and for the fully loaded GPU setup. We see that on average the simulations on the GPU run 5.5 times faster than the CPU-only simulations, by only requiring one third of the energy per simulation. Furthermore, Table 5 shows that the speedup and energy factors remain constant over different time steps. Therefore we conclude that in this setting with the fast PCIe connection of the GPU the interface is not a bottleneck, in contrast to the laptop-FPGA setup.

To provide a unified comparison of our four simulation setups including the GPU and FPGA accelerators, we have normalized the speedup and energy factors to the fully loaded 8-core server.

As mentioned above, the Virtex-5 device that we use is no longer state of the art, and the overhead of the ML-507 board for the idle energy consumption is immense. The complete board consumes 9 W in the idle mode, and not more than 10 W with the FPGA running. To obtain a power estimation for the FPGA itself, we used the Xilinx XPower Estimator [39] that gave an upper bound of less than 3 W for our design. The energy efficiency of the system could therefore be drastically increased by using optimized boards without peripherals, hosting several FPGAs with a tailored power supply.

To provide an estimation of potential energy savings, we have constructed the *FPGA chip only* scenario that assumes the 3 W from the XPower Estimator, with

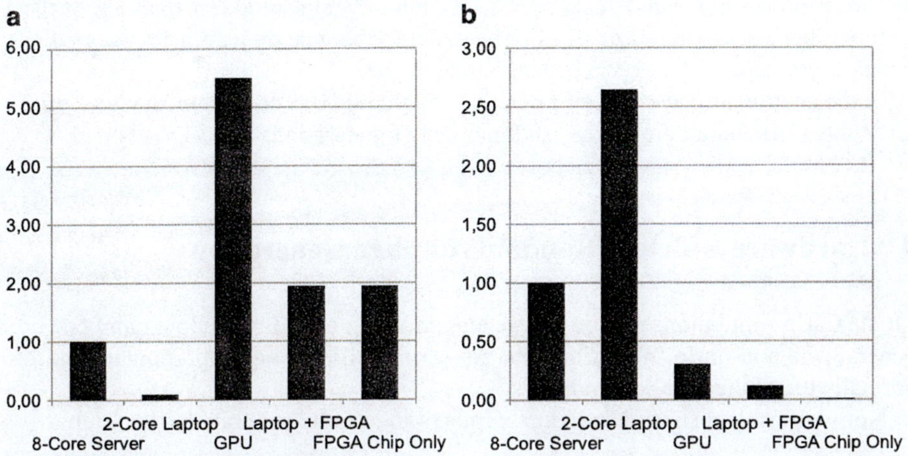

Fig. 5 Speedup and energy factors compared to the fully loaded 8-core server. **(a)** Average speedup factors **(b)** Average energy factors

the payoff computation that is currently performed on the host CPU implemented on the Virtex-5's hardwired PowerPC core. On the recent Xilinx Zynq platform, this part could be computed on the ARM cores.

Figure 5a illustrates the different throughputs of our implementations and the speedups compared to the eight-core software reference. It is obvious that the state-of-the-art Tesla C2050 outperforms all other implementations with respect to speed. However, even our (at least for HPC applications) obsolete Virtex-5 device running three accelerator instances achieves around 35% of the simulation speed of the Tesla C2050. With possible several hundreds of accelerators on a Virtex-7 device, apparently FPGAs have a huge potential to speed up Monte Carlo simulations for state-of-the-art option pricing.

Considering the consumed energy per simulation as shown in Fig. 5b, the FPGA setup clearly beats all other architectures. The measured laptop-FPGA setup only needs 12% of the energy compared to the eight-core server reference, and around 40% of the energy of the Tesla C2050 GPU. The laptop-only run consumes 2.5 times more energy than the server reference, which is not surprising due to the bad ratio of CPU performance contributing to the simulation and devices that consume energy but are not needed for the actual computations.

For both speed and energy comparison, we have used the average factors from the bottom line of Tables 4 and 5. The benefit of the FPGA accelerated setup is even higher if considering only 128 or more time steps per simulation and therefore going out of the interface bottleneck.

The FPGA chip only scenario highlights the enormous potential of FPGAs for energy efficient option pricing. It forecasts only 0.8% of energy per simulation, compared to the server reference, with a double throughput at the same time. A system with three FPGAs running three accelerators on each FPGA would achieve the same throughput as the Tesla 2050 GPU, but only consume less than 3% of the energy. This clearly shows that FPGAs can help to reduce the energy consumed in financial simulations by orders of magnitude.

In the next section, we will give detailed insight into one core element underlying all Monte Carlo simulations: the random number generation.

4 Hardware Efficient Random Number Generation

Monte Carlo simulations rely on a huge amount of high quality random numbers, in general with non-uniform distributions. Asset price simulations in particular require normally distributed random numbers.

Non-uniformly distributed random numbers are usually generated in two steps:

1. The *creation of uniformly distributed random numbers* with good statistical properties and
2. A *conversion step* that transforms this numbers into the desired target distribution.

Since these two steps are not linked to each other, they have been investigated rather independently in research up to now. A comprehensive overview of the available methods for Gaussian random number generation up to 2007 has been given by Thomas et al. [32].

To obtain meaningful simulation results, a high quality of the employed random numbers is absolutely crucial. For uniform RNGs, standardized and approved test suites exist, for example the TestU01 suite from L'Ecuyer and Simard [12]. Non-uniform RNGs require manual investigations to quantify the quality of the output distribution. In Sect. 4.2.2 we show which tests provide meaningful results for this task.

4.1 Uniform Random Number Generation

Pseudo random number generators (PRNGs) are widely used for simulation purposes. In contrast to true RNGs where the randomness comes from a physical process, for instance, radioactive decay, PRNGs are based on mathematical algorithms that yield deterministic number streams. This deterministic behavior makes it possible to obtain repeatable simulation results. In addition to that, common PRNGs are able to produce numbers much faster than true RNGs. The quality of PRNGs is described among others by the following attributes:

1. Quality of the desired uniform distribution (k-distribution [17])
2. Period length of the generator
3. Memory consumption

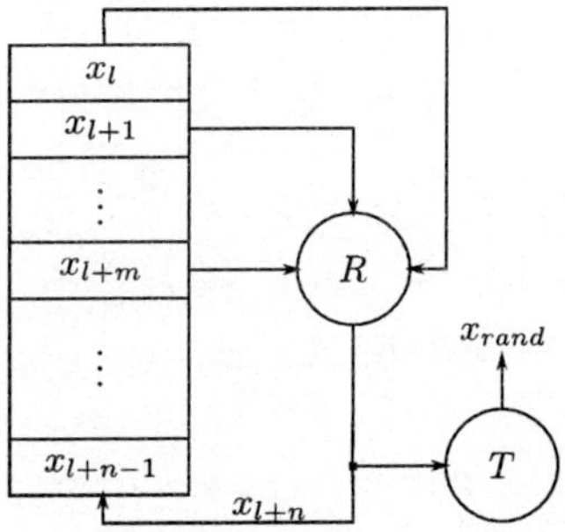

Fig. 6 Schematic view on the Mersenne Twister algorithm. Blocks *R* and *T* realize the linear recurrence and tempering, respectively

The "uniformness" of the generated number streams and the period length of the generator directly influence the simulation results. A highly uniform distribution indicates *good* randomness, as does a long period length.

State-of-the-art PRNGs used in software engineering offer extremely long period lengths, and, therefore, good statistical properties of the generated random numbers (RNs). Examples are the WELL [22] and Mersenne Twister [17] generators (see schematic in Fig. 6). They are based on the *generalized feedback shift register* (GFSR) concept proposed by Lewis and Pane [15]. Parallel Monte Carlo simulations require independent parallel random number streams. An approach to generate multiple random numbers in parallel with the Mersenne Twister generator is shown in the next section.

4.1.1 Interleaved Parallelized Mersenne Twister

Dalal and Stefan have described two methodologies for parallelizing GFSR-generators in 2008 [7]. The *Interleaved Parallelization* interleaves the state vector over a number of memory banks. With *Chunked Parallelization* the state vector is split into chunks of different sizes. Both methodologies allow for a high degree of flexibility and parallelism, we focus on the Interleaved Parallelization scheme here.

The recurrences of a GSFR occur at constant offsets (1 and m in the case of Mersenne Twister). Thus, if multiple memory banks holding the interleaved n-word state vector are utilized, multiple pseudo random numbers can be generated in parallel. Interleaving the state vector across β-memory banks gives the following possibilities:

1. β is a factor of n (i.e. $\beta \bmod n = 0$)
2. β is not a factor of n

In case 1 all memory banks would contain $\frac{n}{\beta}$ words, while in the latter case one memory bank would contain less words than the other memory banks. Any recurrence addresses $[(j+m_i) \bmod n]$ that initially pointed to such a memory bank for $j+M_i < N$ would point to a different memory bank for $j+M_i \geq N$. This leads to inefficient hardware architectures because additional conditional routing logic

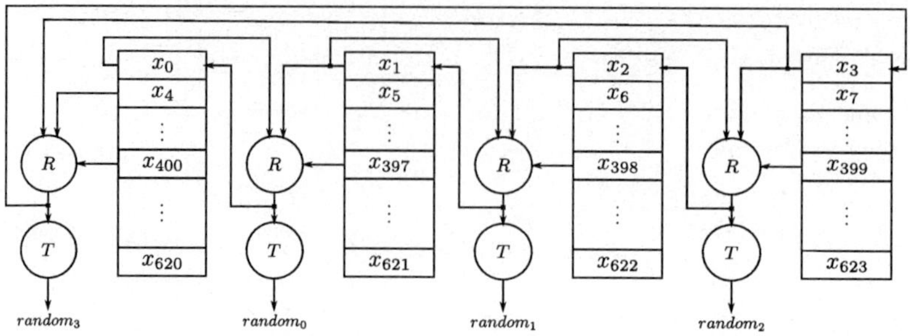

Fig. 7 Example of a 4-IP MT19337 ($n = 624$, $m = 397$). Blocks R and T realize the linear recurrence and tempering, respectively

(multiplexers) would be required. Thus, for efficient interleaved parallelizations, β should be a factor of n. This limitation is not too strict as, for example, for the Mersenne Twister MT19937 n factorizes as $624 = 13 * 3 * 2 * 2 * 2 * 2$. The state vector is interleaved across the β memory banks as follows: each bank b_i $(0 \leq i < \beta)$ contains the $\frac{N}{\beta}$ state vector word indices satisfying $j \bmod \beta = i$, while the corresponding recurrences are found in banks corresponding to $(j + M_i) \bmod \beta$. The memory banks can now supplement the inputs concurrently to multiple *Recurrence Units*, each implementing the recurrence equation. A Mersenne Twister generator with Interleaved Parallelization is enabled to produce β random numbers in parallel. Figure 7 illustrates the interleaved parallelization scheme on the example of a 4-IP Mersenne Twister MT19937.

4.1.2 Implementation Properties

We provide comparable synthesis results for several configurations of each implemented model in this section. Synthesis of the implementation models has been performed for the Virtex-5 FPGA (XC5VFX70T, package: FF1136, speed: -2) by Xilinx. The optimization goal for the synthesis process (xst) was set to the default value (speed), as well as the optimization effort switch (default: 1). Place & route (par) of the Mersenne Twister implementations was performed with the optimization strategy configured towards reducing the consumed area, with the effort level set to high.

Selected post place & route synthesis results for the interleaved parallelized Mersenen Twister implementations along with synthesis data provided in the paper by Dalal and Stefan are listed in Table 6. Taking into consideration that the internal structure was not yet optimized with regard to area and block RAM utilization, the results are satisfying. With future optimizations, like buffering all near recurrences, the throughput of our design can be doubled. Additional buffering can also save one block RAM, which would make the performance of the designs comparable to the reference.

Table 6 Post place and route synthesis results for various parameter configurations of the IP Mersenne Twister implementation model

Name	Used RAMs	Used slices	Max. freq [MHz]	Max. throughput $[10^6 \frac{\text{samples}}{\text{second}}]$	Target
2 IP	3	142	243	243	Virtex 5
3 IP	4	157	242	363	
4 IP	5	177	232	464	
2 IP[a]	2	159	349	698	Virtex II
3 IP[a]	3	222	265	795	
4 IP[a]	4	290	277	1,108	

[a]Taken from Dalal and Stefan 2008 [7]

4.2 Obtaining Non-uniform Distributions

Non-uniform distributions are, in general, generated out of uniformly distributed random numbers by application of appropriate conversion methods. State-of-the art conversion methods are based on one of the four mechanisms categorized by Thomas and Luk in 2007 [32]:

- *Transformation* (mathematical functions that provide a relation between the uniform and the desired target distribution),
- *Rejection sampling* (very high accuracy, but introduces unpredictable stalling by discarding several input numbers),
- *Recursion* (linear combinations of originally normally distributed random numbers), and
- *Inversion.*

The *inversion method* applies the *inverse cumulative distribution function* (ICDF) of the target distribution to uniformly distributed random numbers. It is the most genuine method to obtain non-uniform numbers, since it preserves the properties of the input sample sequence [11]. A piecewise approximation of the ICDF is the basis of hardware implementations of inversion-based converters, where the coefficients for various sampling points are stored in lookup tables (LUTs).

The Gaussian ICDF is symmetric at $x = 0.5$. Therefore, it is sufficient to implement a converter for only one half of the ICDF and to use one input bit as a *sign bit* to determine which half. The range (0,0.5), for example, is divided into non-equidistant segments with doubling segment sizes from the beginning of the interval to the end of the interval. Those segments are then subdivided into segments of equal size. Thus, the steep region of the ICDF (near zero) is covered by more segments than the more linear region close to 0.5. The inversion is performed by determining in which segment the input x is contained, retrieving the coefficients c_i of the polynomial for this segment from an LUT and evaluate the polynomial $y = \sum c_i \cdot x_i$.

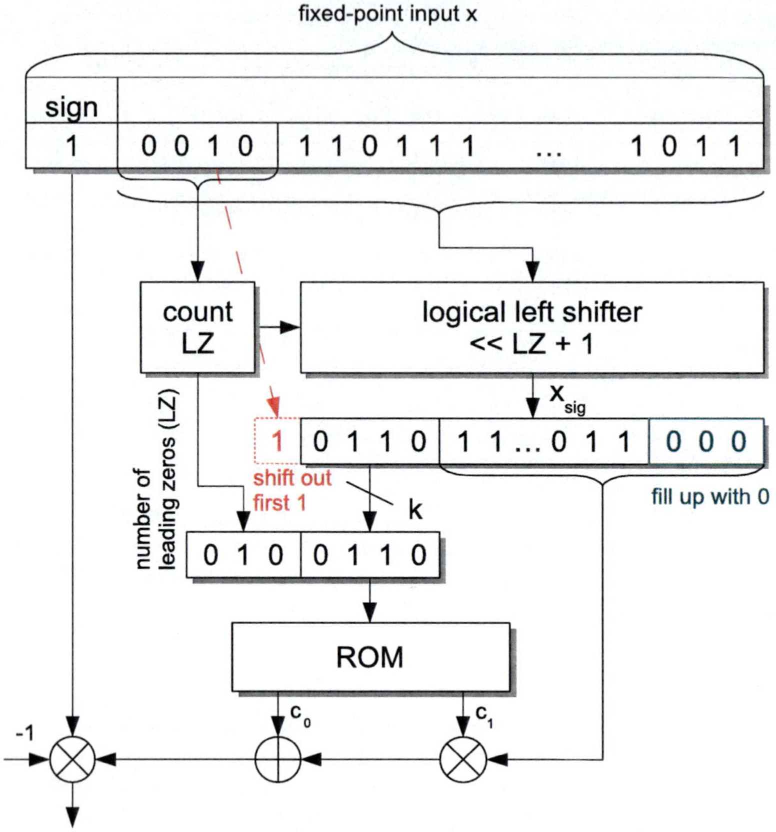

Fig. 8 ICDF lookup architecture presented by Cheung et al. [6]

4.2.1 Related Work

Hardware architectures for ICDF converters using hierarchical segmentation schemes have, for example, been presented by Cheung et al. [6] and Luk et al. [13, 14]. Figure 8 shows the architecture presented by Cheung et al. in 2007.

It illustrates how the number of the segment (i.e., the address for the LUT) in which a given input x in fixed point representation is located can be determined. First, the number LZ of leading zeros in the binary representation of x is counted. Numbers starting with a 1 lie in the segment [0.25, 0.5), numbers starting with the sequence 01 lie in the segment [0.125, 0.25) and so forth. The input x is shifted left by LZ + 1 bits, such that x_{sig} is the bit sequence following the most significant 1-bit in x. The equally spaced subsegments are determined by the k most significant bits (MSBs) of x_{sig}. Thus, the LUT address is the concatenation of LZ and $MSB_k\,(x_{\text{sig}})$. The remaining bits of x_{sig} are then used to evaluate the approximating polynomial for the ICDF in that segment.

However, this architecture has a number of drawbacks:

- *More than one uniform RNGs needed for a large output range.* Due to the fixed point implementation, the output range is limited by the available number of input bits.
- *Many input bits are wasted.* For example, a multiplier with a 53-bit input for the linear approximation requires a large amount of hardware resources. Therefore, the input is quantified to 20 significant bits before the polynomial evaluation. Thus, in the region close to the 0.5 a large amount of the generated input bits is just not used, but discarded by this architecture.
- *Low resolution in the tail region.* For the tail region (close to 0), there are no longer 20 significant bits available after shifting over the leading zeros (LZ). Thus, the tail resolution is limited. In fact, since there are no values between 2^{-53} and 2^{-52} in this fixed point representation, this architecture can not generate output samples between $icdf(2^{-52}) = 8.13\sigma$ and $icdf(2^{-53}) = 8.21\sigma$.

4.2.2 A Hardware Efficient Hardware Architecture for Non-uniform Distributions

To overcome the problems illustrated in Sect. 4.2.1, we propose to use a *floating point based* approach for the ICDF converter [24, 28].

The Floating Point Approach

In addition to the architecture shown in Fig. 8, we transform the uniformly distributed input RN sequences into floating point numbers. Figure 9 shows that for this step no floating point arithmetic units are used. The unit logically divides the input bit vector into the *sign_half* bit (that determines which half of the Gaussian ICDF to use), the exponent part and the mantissa part. *sign_half* and the mantissa are just mirrored at the output. The mantissa part is $mant_{bw}$ bits width and, therefore, (with a hidden bit) can have the values $1, 1+\frac{1}{2^{mant_{bw}}}, 1+\frac{2}{2^{mant_{bw}}}, \ldots, 2-\frac{1}{2^{mant_{bw}}}$. The output exponent part contains the number of leading zeros in its corresponding input section. If the input exponent section contains only zeros, another sample is taken and the number of zeros is accumulated, until a one occurs or the independently adjustable output range is exceeded. Thus, we can create arbitrary output precision with our approach, not relying on fixed uniform input bit vector sizes. Any uniform RNG can be used to generate inputs for this floating point converter unit.

We have carefully validated that the originally provided randomness and distribution of the input random numbers are preserved [24,28] by applying the standardized TestU01 test suite [12]. With a MT19337 Mersenne Twister RNG as input, the output of the uniform floating point generator passed all tests except those, that the MT19337 is known to fail itself. Thus, we conclude that our floating point converter unit maintains all the properties of the input RNs.

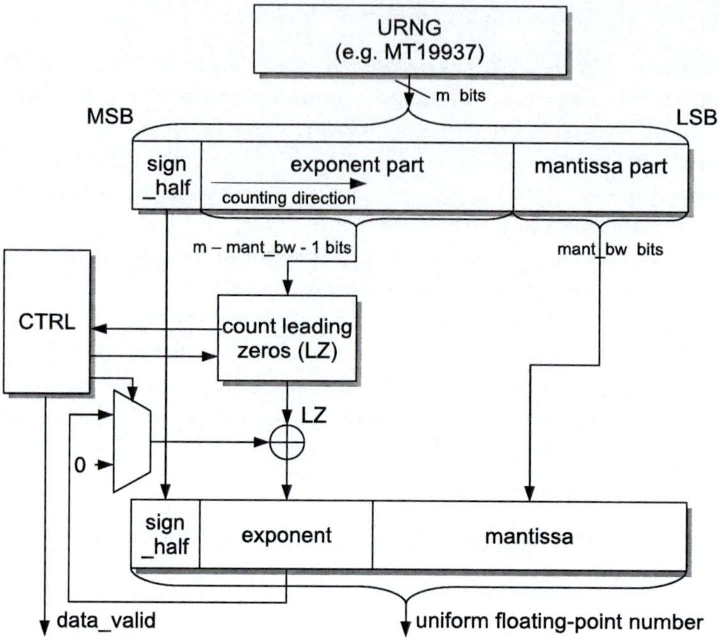

Fig. 9 Architecture of the proposed floating point converter unit

Table 7 Synthesis results for the proposed inversion based converter architecture

	Slices	FFs	LUTs	BRAMs	DSP48E
Floating point converter	13	11	26	–	–
LUT evaluator	18	47	7	1	1
Complete design	31	85	34	1	1

Figure 10 illustrates the corresponding ICDF lookup unit. In contrast to the proposed architecture by Cheung et al., we do no longer need the shifter, but directly rely on the provided exponent and mantissa values.

With optimized bit widths for the Virtex-5 DSP48E slice that supports a $18 * 25\,bit + 48\,bit$ MAC operation, the parameters are as follows: input bitwidth $m = 32$, $mant\_bw = 20$, $max\_exp = 54$, and $k = 3$ for subsegment addressing. The coefficient c_0 is quantized to 46 bits, c_1 has 23 bits.

Table 7 shows the resource consumption for the proposed architecture. We have used the Xilinx ISE 12.4 suite and the target device Xilinx Virtex-5 XC5FX70T-3. All provided results are post place & route. In total, our architecture saves more than 48% of the area compared to the design proposed by Cheung et al. [6], by providing a higher output resolution at the same time. It can run up to 286 MHz.

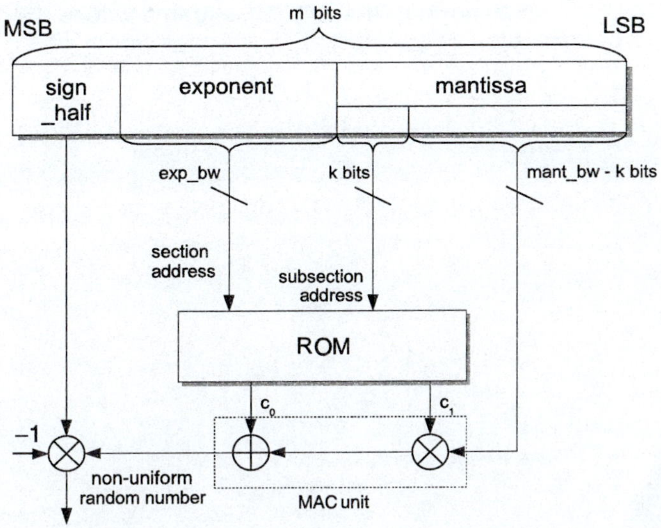

Fig. 10 The corresponding ICDF lookup unit for floating point inputs

For the convenience of the users who like to make use of our proposed architecture, we have developed a flexible C++ class package that creates the LUT entries for any desired distribution function. It is freely available for download.[7]

Quality Checking

For the normally distributed output of our unit, we have carried out intensive statistic analysis manually to verify the quality of the results. Several χ^2-tests have been applied that compare the empirical number of observations in several groups with the theoretical number of observations. The Kolmogorov–Smirnov test compares the empirical and the theoretical cumulative distribution function. We have also carried out this test on our results, and nearly all tests with different batch sizes were perfectly passed. Those that did not pass did not reveal an extraordinary *P* value.

Figures 11 and 12 illustrate the difference in the tail region between our proposed architecture and the standard *R* RNG. The random numbers of our generator seem to have the same distribution as the standard random numbers, with an exception of the reduced precision in the central region and an improved precision in the extreme values. It can be seen that our architecture achieves higher extreme values in the tail region compared to the *R* RNG. The smallest value from our floating point-based approach is $1 \cdot 2^{-54}$, compared to $1 \cdot 2^{-32}$ in standard RNGs. For that reason our

[7] http://ems.eit.uni-kl.de/fileadmin/downloads/icdf_lut_tool.tgz.

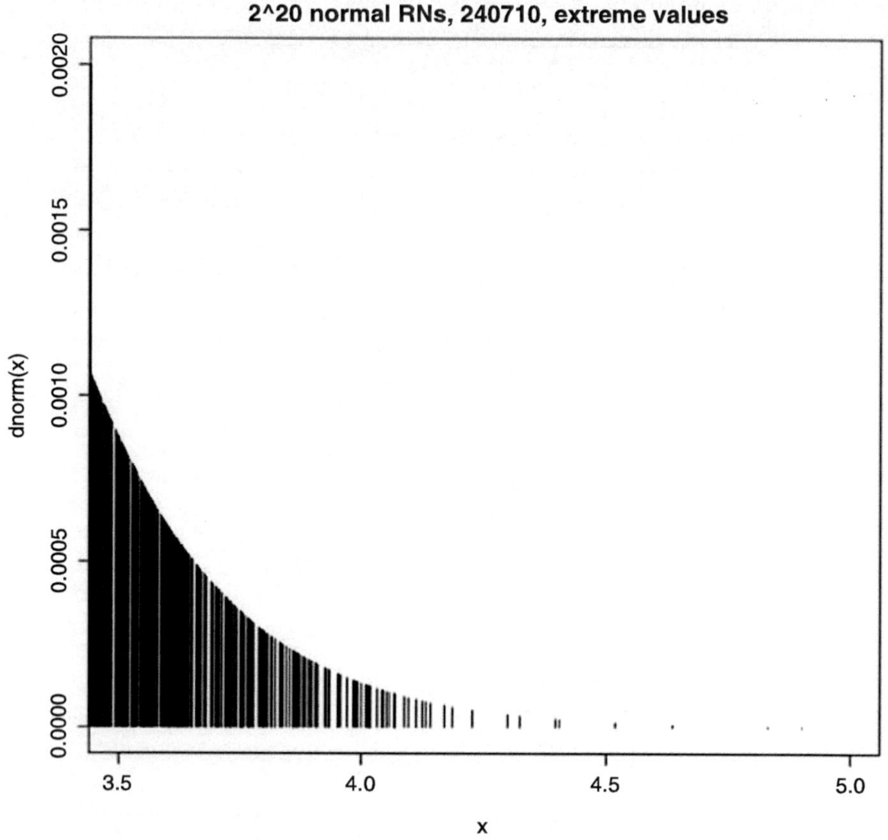

Fig. 11 Tail of the empirical distribution function produced by the proposed architecture

architecture can produce values of -8.37σ and 8.37σ. Therefore we expect our design to perform very well in the case of simulations rare extreme events can have a huge impact (consider risk simulations for insurances, for example).

In addition to the statistical investigations, a bit-true model of the Gaussian RNG has been validated in two application tests for practical scenarios: In an GNU Octave-based Monte Carlo simulation for option price modeling based on the Heston model, the Octave RNG *randn*() was replaced by the bit true model of the proposed hardware architecture. The same convergence behavior was observed and the same results were obtained, both for options with and without barriers.

Together with the accelerator structure shown in Sect. 3.2, our presented RNG architecture allows to build up very hardware efficient FPGA accelerators for asset simulations based Monte Carlo simulations.

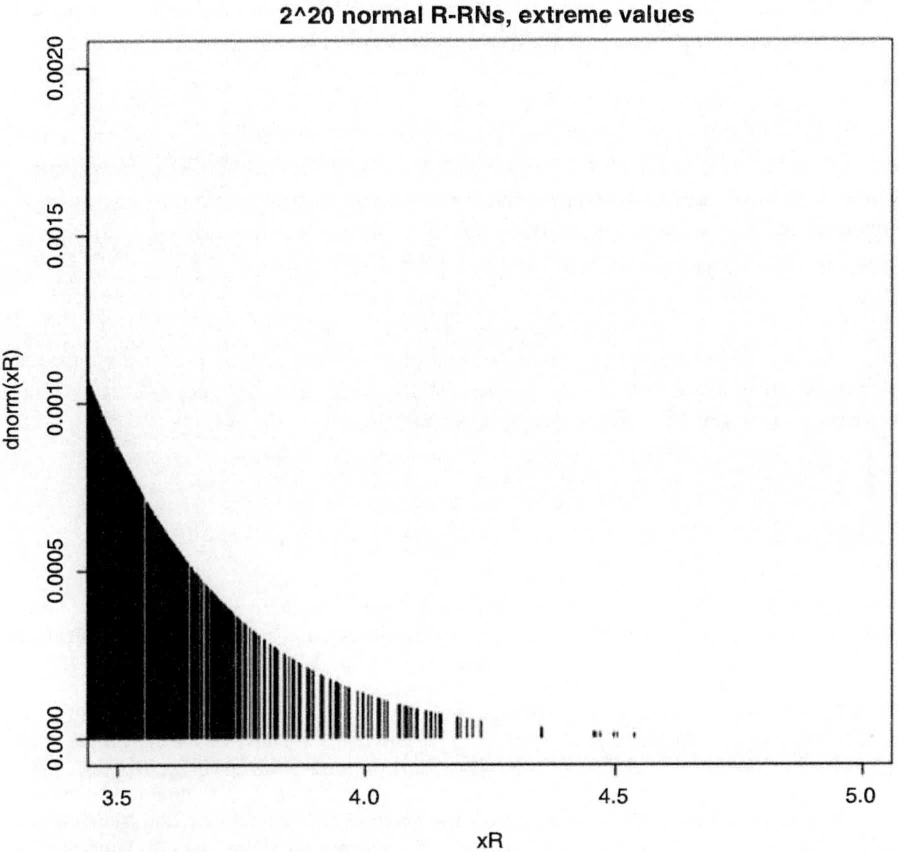

Fig. 12 Tail of the empirical distribution function for the R RNG

5 Conclusion

The increasing complexity of financial products and the need for more frequent simulation runs and higher accuracy have boosted the energy consumption continuously over time. Nowadays, inefficient standard CPUs and GPUs are still prevailing to perform very specific number crunching operations that fit much better to dedicated accelerators. In this chapter we show that optimized hardware architectures clearly outperform general purpose CPUs and GPUs with respect to energy efficiency. For the application "pricing European double barrier options in the Heston model" we present a dedicated FPGA architecture based on the advanced Multi-Level Monte Carlo method. This method (like all Monte Carlo method) strongly relies on high-quality random numbers with a Gaussian distribution. We illustrate how these numbers can be efficiently generated in hardware with an ICDF-based converter tool. This tool allows to produce random numbers with arbitrary output distributions and precisions. On top of this RNG, we have built a dedicated FPGA circuit for asset path simulations based on the Heston model. Together with a pricing engine running on the host CPU and the path simulation performed on a Xilinx Virtex-5 FPGA, we show that this hybrid CPU-FPGA system consumes only around 12% of the energy of an eight-core CPU-only system, providing twice the throughput. Compared to a state-of-the-art Nvidia Tesla C2050 GPU, this system achieves 35% of the simulation speed by consuming around 40% of the energy. However, an extrapolation with the path simulation and the pricing running on three FPGAs predicts that a pure FPGA accelerator can save incredible 97% of the energy compared to a Nvidia Tesla C2050 GPU, providing the same throughput. In addition, FPGAs are flexible devices that can be reconfigured for different pricing tasks quickly, allowing to dynamically instantiate different accelerator designs for the computation of various products. This clearly highlights the enormous potential for energy saving of FPGAs for financial simulations.

References

1. T. Becker, Q. Jin, W. Luk, S. Weston, Dynamic constant reconfiguration for explicit finite difference option pricing, in *2011 International Conference on Reconfigurable Computing and FPGAs (ReConFig)* (IEEE Computer Society, Los Alamitos, USA, 2011), pp. 176–181. ISBN-13: 978-0-7695-4551-6. doi:10.1109/ReConFig.2011.29
2. A. Bernemann, R. Schreyer, K. Spanderen, Pricing structured equity products on GPUs, in *2010 IEEE Workshop on High Performance Computational Finance (WHPCF)* (IEEE, Red Hook, USA, 2010), pp. 1–7. ISBN: 978-1-4244-9061-5. doi:10.1109/WHPCF.2010.5671821
3. A. Bernemann, R. Schreyer, K. Spanderen, Accelerating exotic option pricing and model calibration using GPUs (2011), http://ssrn.com/abstract=1753596. Accessed 28th January 2013
4. F. Black, M. Scholes, The pricing of options and corporate liabilities. J. Polit. Econ. **81**(3), 637–654 (1973)
5. G. Chatziparaskevas, A. Brokalakis, I. Papaefstathiou, An FPGA-based parallel processor for Black-Scholes option pricing using finite differences schemes, in *Proceedings of Design, Automation and Test in Europe, 2012 (DATE '12)*, EDAA (2012), ISBN: 978-3-9810801-6
6. R.C.C. Cheung, D.U. Lee, W. Luk, J.D. Villasenor, Hardware generation of arbitrary random number distributions from uniform distributions via the inversion method. IEEE Trans. Very Large Scale Integrat. (VLSI) Syst. **15**(8), 952–962 (2007). doi:10.1109/TVLSI.2007.900748, http://dx.doi.org/10.1109/TVLSI.2007.900748
7. I.L. Dalal, D. Stefan, A hardware framework for the fast generation of multiple long-period random number streams, in *Proceedings of the 16th International ACM/SIGDA Symposium on Field Programmable Gate Arrays, FPGA '08* (ACM, New York, 2008), pp. 245–254. doi:10.1145/1344671.1344707, http://doi.acm.org/10.1145/1344671.1344707
8. S. Gilani, The real reason for the global financial crisis...the story no one's talking about (2008), http://moneymorning.com/2008/09/18/credit-default-swaps/. Accessed 28th January 2013
9. S.L. Heston, A closed-form solution for options with stochastic volatility with applications to bond and currency options. Rev. Financ. Stud. **6**(2), 327 (1993). doi:10.1093/rfs/6.2.327
10. Q. Jin, W. Luk, D.B. Thomas, Unifying finite difference option-pricing for hardware acceleration, in *International Conference on Field Programmable Logic and Applications (FPL), 2011* (IEEE Computer Society, Los Alamitos, USA, 2011), pp. 6–9. ISBN: 978-0-7695-4529-5. doi:10.1109/FPL.2011.12

11. R. Korn, E. Korn, G. Kroisandt, *Monte Carlo Methods and Models in Finance and Insurance* (CRC Press, Boca Raton, 2010)
12. P. L'Ecuyer, R. Simard, TestU01: a C library for empirical testing of random number generators. ACM Trans. Math. Softw. **33**(4), 22 (2007). doi:http://doi.acm.org/10.1145/1268776.1268777
13. D.U. Lee, W. Luk, J. Villasenor, P.Y. Cheung, Hierarchical segmentation schemes for function evaluation, in *2003 IEEE International Conference on Field-Programmable Technology (FPT), 2003. Proceedings* (The University of Tokyo, Tokyo, Japan, 2003), pp. 92–99. ISBN: 0-7803-8320-6. doi:10.1109/FPT.2003.1275736
14. D.U. Lee, R. Cheung, W. Luk, J. Villasenor, Hierarchical segmentation for hardware function evaluation. IEEE Trans. Very Large Scale Integrat. (VLSI) Syst. **17**(1), 103–116 (2009). doi:10.1109/TVLSI.2008.2003165
15. T.G. Lewis, W.H. Payne, Generalized feedback shift register pseudorandom number algorithm. J. ACM **20**(3), 456–468 (1973). doi:10.1145/321765.321777, http://doi.acm.org/10.1145/321765.321777
16. H. Marxen, A. Kostiuk, R. Korn, C. de Schryver, S. Wurm, I. Shcherbakov, N. Wehn, Algorithmic complexity in the Heston model: an implementation view, in *2011 IEEE Workshop on High Performance Computational Finance (WHPCF)* (ACM, New York, USA, 2011), ISBN: 978-1-4244-9061-5
17. M. Matsumoto, T. Nishimura, Mersenne twister: a 623-dimensionally equidistributed uniform pseudo-random number generator. ACM Trans. Model. Comput. Simulat. **8**(1), 3–30 (1998). doi:http://doi.acm.org/10.1145/272991.272995
18. O. Mencer, E. Vynckier, J. Spooner, S. Girdlestone, O. Charlesworth, Finding the right level of abstraction for minimizing operational expenditure, in *2011 IEEE Workshop on High Performance Computational Finance (WHPCF)* (ACM, New York, USA, 2011), ISBN: 978-1-4244-9061-5
19. R.C. Merton, Theory of rational option pricing. Bell J. Econ. Manag. Sci. **4**(1), 141–183 (1973)
20. G.W. Morris, M. Aubury, Design space exploration of the European option benchmark using hyperstreams, in *International Conference on Field Programmable Logic and Applications, 2007. FPL 2007*, IEEE (2007), pp. 5–10. ISBN: 1-4244-1060-6 doi:10.1109/FPL.2007.4380617
21. NVIDIA Corporation: Computational finance website (2012), http://www.nvidia.com/object/computational_finance.html. Accessed 28th January 2013
22. F. Panneton, P. L'Ecuyer, M. Matsumoto, Improved long-period generators based on linear recurrences modulo 2. ACM Trans. Math. Softw. **32**(1), 1–16 (2006). doi:10.1145/1132973.1132974, http://doi.acm.org/10.1145/1132973.1132974
23. QuantLib - A free/open-source library for quantitative finance (2012), http://quantlib.org. Accessed 28th January 2013
24. C. de Schryver, D. Schmidt, N. Wehn, E. Korn, H. Marxen, R. Korn, A new hardware efficient inversion based random number generator for non-uniform distributions, in *2010 International Conference on Reconfigurable Computing and FPGAs (ReConFig)* (IEEE Computer Society, Los Alamitos, USA, 2010), pp. 190–195. ISBN: 978-0-7695-4314-7. doi:10.1109/ReConFig.2010.20
25. C. de Schryver, I. Shcherbakov, F. Kienle, N. Wehn, H. Marxen, A. Kostiuk, R. Korn, An energy efficient FPGA accelerator for Monte Carlo option pricing with the Heston model, in *2011 International Conference on Reconfigurable Computing and FPGAs (ReConFig)* (IEEE Computer Society, Los Alamitos, USA, 2011), pp. 468–474. ISBN-13: 978-0-7695-4551-6. doi:10.1109/ReConFig.2011.11
26. C. de Schryver, M. Jung, N. Wehn, H. Marxen, A. Kostiuk, R. Korn, Energy efficient acceleration and evaluation of financial computations towards real-time pricing, in *Knowledge-Based and Intelligent Information and Engineering Systems*, ed. by A. König, A. Dengel, K. Hinkelmann, K. Kise, R.J. Howlett, L.C. Jain. Lecture Notes in Computer Science, vol. 6884 (Springer, Berlin, 2011), pp. 177–186. Proceedings of 15th International Conference on Knowledge-Based and Intelligent Information & Engineering Systems (KES)

27. C. de Schryver, H. Marxen, D. Schmidt, Hardware accelerators for financial mathematics - methodology, results and benchmarking, in *Proceedings of 1st Young Researcher Symposium (YRS) 2011*, pp. 55–60 (Center for Mathematical and Computational Modelling $(CM)^2$, $(CM)^2$, Nachwuchsring, 2011). http://CEUR-WS.org/Vol-750/yrs08.pdf. ISSN: 1613-0073, urn:nbn:de:0074-750-0
28. C. de Schryver, D. Schmidt, N. Wehn, E. Korn, H. Marxen, A. Kostiuk, R. Korn, A hardware efficient random number generator for nonuniform distributions with arbitrary precision. Int. J. Reconfigurable Comput. (IJRC) **2012** (2012). doi:10.1155/2012/675130. Article ID 675130, 11 pages
29. J. Stratoudakis, Hardware acceleration of Monte Carlo simulation for option pricing (2012), http://wallstreetfpga.com. Accessed 28th January 2013
30. J. Stratoudakis, Hardware acceleration of Monte Carlo simulation for option pricing (2012), https://decibel.ni.com/content/docs/DOC-9984. Accessed 28th January 2013
31. D.B. Thomas, J.A. Bower, W. Luk, Automatic generation and optimisation of reconfigurable financial Monte-Carlo simulations, in *IEEE International Conference on Application-Specific Systems, Architectures and Processors, 2007. ASAP*, IEEE (2007), pp. 168–173. ISBN: 1-4244-1027-4. doi:10.1109/ASAP.2007.4429975
32. D.B. Thomas, W. Luk, P.H. Leong, J.D. Villasenor, Gaussian random number generators. ACM Comput. Surv. **39**(4), 11 (2007). doi:http://doi.acm.org/10.1145/1287620.1287622
33. X. Tian, K. Benkrid, American option pricing on reconfigurable hardware using least-squares Monte Carlo method, in *International Conference on Field-Programmable Technology, 2009. FPT 2009*, IEEE (2009), pp. 263–270. ISBN: 978-1-4244-4377-2. doi:10.1109/FPT.2009.5377662
34. X. Tian, K. Benkrid, X. Gu, High performance Monte-Carlo based option pricing on FPGAs. Eng. Lett. **16**(3), 434–442 (2008)
35. P. Warren, City business races the Games for power. The Guardian (2008), http://www.guardian.co.uk/technology/2008/may/29/energy.olympics2012. Accessed 28th January 2013
36. S. Weston, J.T. Marin, J. Spooner, O. Pell, O. Mencer, Accelerating the computation of portfolios of tranched credit derivatives, in *2010 IEEE Workshop on High Performance Computational Finance (WHPCF)* (IEEE, Red Hook, USA, 2010), pp. 1–8. ISBN: 978-1-4244-9061-5. doi:10.1109/WHPCF.2010.5671822
37. S. Weston, J. Spooner, J.T. Marin, O. Pell, O. Mencer, FPGAs speed the computation of complex credit derivatives. Xcell J. **74**, 18–25 (2011)
38. C. Wynnyk, M. Magdon-Ismail, Pricing the American option using reconfigurable hardware, in *International Conference on Computational Science and Engineering, 2009. CSE '09*, vol. 2 (IEEE Computer Society, Los Alamitos, USA, 2009), pp. 532–536. ISBN-13: 978-0-7695-3823-5. doi:10.1109/CSE.2009.496
39. Xilinx: XPower estimator (XPE) (2011), http://www.xilinx.com/products/technology/power/index.htm. Accessed 28th January 2013
40. B. Zhang, C.W. Oosterlee, Acceleration of option pricing technique on graphics processing units. Tech. Rep. 10-03, Delft University of Technology (2010)

Monte-Carlo Simulation-Based Financial Computing on the Maxwell FPGA Parallel Machine

Xiang Tian and Khaled Benkrid

Abstract Efficient computational solutions for scientific and engineering problems are a priority for many governments around the world, as they can offer major economic comparative advantages. Financial computing problems are a prime example of such problems where even the slightest improvements in execution times and latency can generate large amounts of extra profits. However, financial computing has not benefited relatively greatly from early developments in high performance computing, as the latter aimed mainly at engineering and weapon design applications. Besides, financial experts were initially focusing on developing mathematical models and computer simulations in order to comprehend the behavior of financial markets and develop risk-management tools. As this effort progressed, the complexity of financial computing applications grew up rapidly. Hence, high performance computing turned out to be very important in the field of finance.

Many financial models do not have a practical closed-form solution in which case numerical methods are the only alternative. Monte-Carlo simulation is one of the most commonly used numerical methods, in financial modeling and scientific computing in general, with huge computation benefits in solving problems where closed-form solutions are impossible to derive. As the Monte-Carlo method relies on the average result of thousands of independent stochastic paths, massive parallelism can be harnessed to accelerate the computation. For this, high performance computers, increasingly with off-the-shelf accelerator hardware, are being proposed as an economic high performance implementation platform for Monte-Carlo-based simulations. Field programmable gate arrays (FPGAs) in particular have been recently proposed as a high performance and relatively low power acceleration platform for such applications.

X. Tian (✉) • K. Benkrid
The University of Edinburgh, Institute of Integrated Systems, King's Buildings,
Mayfield Road, Edinburgh EH9 3JL, Scotland, UK
e-mail: X.Tian@ed.ac.uk; k.benkrid@gmail.com

W. Vanderbauwhede and K. Benkrid (eds.), *High-Performance Computing Using FPGAs*, DOI 10.1007/978-1-4614-1791-0_2, © Springer Science+Business Media, LLC 2013

In light of the above, the project presented in this chapter develops novel FPGA hardware architectures for Monte-Carlo simulations of different types of financial option pricing models, namely European, Asian, and American options, the stochastic volatility model (GARCH model), and Quasi-Monte Carlo simulation. These architectures have been implemented on an FPGA-based supercomputer, called Maxwell, developed at the University of Edinburgh, which is one of the few openly available FPGA parallel machines in the world. Maxwell is a 32-CPU cluster augmented with 64 Virtex-4 Xilinx FPGAs connected in a 2D torus. Our hardware implementations all show significant computing efficiency compared to traditional software-based implementations, which in turn shows that reconfigurable computing technology can be an efficacious and efficient platform for high performance computing applications, particularly financial computing.

1 Introduction

High performance computing (HPC) is a discipline concerned with the development and use of supercomputers or computer clusters, with applications in a variety of fields including bioinformatics, energy, climate modeling, and computational applications in engineering, of which typical computational demands exceed the TeraFlop/sec.[1] Supercomputers' development has been through several stages during the past decades starting with vector computers, and then symmetric multiprocessors (or SMPs[2]), to massively parallel processors (MPPs) which mostly use off-the-shelf commodity microprocessors nowadays [1]. Supercomputers' performance requirements, however, are increasing at a rate that exceeds the rate of chip-level improvements [2]. In the early days of the technology, mostly engineering and weapons' design applications benefited from the developments in high performance computing. In financial computing, for instance, financial experts were mostly focused on mathematical models and computer simulations in order to understand financial markets and develop risk-management tools. Generally, these models are stochastic process models. As the complexity of these models increased rapidly, personal computers were no longer able to perform the required computations in reasonable times; hence the adoption of high performance computing platforms became the mainstream. In 1999, for instance, a survey of high performance computing in finance and computer-aided design of financial products introduced the development of financial models and supercomputer-based high performance financial computing to a wider community [3].

[1]TeraFlop/sec is an acronym meaning 10^{12} floating point operations per second.

[2]An SMP is a computer system that has two or more processors connected in the same cabinet, managed by one operating system, sharing the same memory, and having equal access to input/output devices.

One widely used computational technique in financial computing is Monte-Carlo simulation. The latter is a numerical computational algorithm which is often used in simulating physical and mathematical systems. It relies on repeated random sampling to compute their result. This method is often used when it is impossible or impractical to get an analytical solution, or closed-form result, to system equations. The Monte-Carlo method is particularly important in physical chemistry, computational physics, and related applied fields. These are characterized by systems with a large number of coupled degrees of freedom, such as liquids, disordered materials, strongly coupled solids, and cellular structures. Monte-Carlo simulations are also used to forecast a wide range of events and scenarios, such as the weather, sales, and consumer demands. In financial computing, the Monte-Carlo technique is used to simulate the various sources of uncertainty that affect the value of the underlying instrument, portfolio, or investment in question. Many financial computing applications have no closed form solutions, as they depend on three or more stochastic variables. Here, Monte-Carlo simulation tends to be numerically more efficient than other procedures [4]. This is because the computational time of Monte-Carlo simulations increases approximately linearly with the number of variables, whereas in most other methods, computational time increases exponentially with the number of variables. One of the important characteristics of Monte-Carlo simulation is parallelism as multiple independent paths need to be computed. This makes it attractive to parallel implementation using multi-threading and/or multi-processing.

When evaluating a high performance computing platform, we have to consider several aspects. The cost of cluster computers and supercomputers can be prohibitive. Area and power consumption can also be a major problem with these computing platforms. For these reasons, various acceleration technologies are being considered. Field programmable gate arrays (FPGAs), for instance, offer the high performance of a dedicated hardware solution of a particular algorithm, with a fraction of the area and power consumption of equivalent microprocessor-based solutions. Moreover, the continuous developments in transistor integration levels mean that it is now possible to implement a considerable number of floating-point arithmetic units on modern FPGAs. If this trend is to continue, FPGA use is set to conquer new application domains, including financial computing.

The work presented in this chapter is mainly targeted on an FPGA parallel machine, called Maxwell. Maxwell was one of the first publicly accessible FPGA parallel machines and was built in Edinburgh, Scotland, by the FPGA High Performance Computing Alliance (FHPCA). Established in 2004, the FHPCA's aim was to explore the computing capability of a heterogeneous high performance computing platform, which combines general purpose processors (GPPs) and Xilinx FPGAs. Led by Edinburgh Parallel Computing Centre (EPCC) at The University of Edinburgh, the FHPCA was funded by Scottish Enterprise and built on the skills of Nallatech Ltd., Alpha Data Ltd., Xilinx Development Corporation, Algotronix and iSLI. The idea developed as a result of increasing FPGA hardware complexity, which makes it possible to execute relatively sophisticated general purpose numerical computing applications on modern FPGAs at a relatively low power budget. The work presented in this project focuses on financial computing with the aim of implementing a number of financial computing algorithms on Maxwell and evaluating the latter through a comprehensive strategy, whereby computation speed is not the only concern, but also accuracy, cost, and energy consumption. Indeed, the decision making procedure in a financial market always requires real-time processing of huge amounts of real time data. If the simulation results, which may need hours to get, can be achieved in few minutes, the benefit would definitely be remarkable. However, despite the importance of computation speed, we do not want to lose any considerable accuracy during the computation as an error of 0.001 in simulation, for instance, could bring losses in the thousands of pounds in trading if we are dealing with million-pound assets. Moreover, power efficiency is a very important issue nowadays. For instance, the cost of electricity consumed by modern supercomputers could be in the millions of pounds annually. Here lies the advantage of FPGAs as they achieve high speed performance not through high clock frequencies but mainly through massive data and instruction parallelism and deep pipelining. Indeed, typical clock frequencies of GPPs or graphic processing units (GPUs) are in the GHz range. However, FPGA chips are often clocked at few hundred MHzs, hence leading to considerable energy savings.

In light of the above, the aim of the work presented in this chapter is to rigorously assess the efficiency and efficacy of FPGAs in financial computing applications. This is done through the development of novel FPGA hardware architectures for a number of financial computing applications. Comparative evaluation against GPP and GPU technologies are made using the following criteria:

- *Speed performance*: a very important aim of our work is to maximize the computation speed of financial computing applications. This will need careful hardware design and optimization.
- *Arithmetic accuracy*: the accuracy requirement, which is critical in financial computing, needs to be guaranteed.
- *Power consumption*: since financial computing applications are often deployed in massively parallel computers, power consumption is a very important measure.
- *Cost of purchase and development*: the cost of the hardware and development effort also needs to be considered. Indeed, speed performance can always be increased arbitrarily if budgets were unlimited.
- *Productivity*: this is concerned with the return over time. Indeed, in business, time to market makes the difference between success and failure.

The remainder of this chapter is organized as follows. First, a brief overview of the state-of-the-art of high-performance financial computing is given. Then, an overview of the architecture and programming environment of the Maxwell Parallel FPGA machine is presented. After that, four case-study implementations of financial models on Maxwell are detailed, before an evaluation of the resulting implementations is presented. The chapter concludes with a general evaluation of FPGA-based high performance reconfigurable computing.

2 Brief Overview of the State-Of-The-Art of High Performance Financial Computing

In the last decade, researchers have started to use acceleration technology, e.g., in the form of FPGAs and GPUs in financial computing. In [5] for instance, an FPGA-based Monte-Carlo simulation core used for computing the BGM (Brace, Gatarek and Musiela) interest rate model for pricing derivatives was presented. The BGM interest rate model is commonly used to simulate the fluctuation of interest rates over time, something which has an influence on nearly all economic activity. Results show that ~25x speed-up can be obtained by using an FPGA, compared to an equivalent Pentium IV 1.5GHz-based software implementation. Other hardware architectures for Monte-Carlo-based financial simulations were published in [6]. In the latter, five different Monte-Carlo option pricing simulation algorithms were explored, including log-normal price movements, correlated asset value-at-risk calculation, and price movements under the GARCH model. Using a Xilinx Virtex-4 XC4VSX55 device, implementation results show that FPGA implementations run on-average 80x faster than equivalent software ones (running on a 2.66GHz PC). A comparison of different FPGA implementations of the European option[3] benchmark against other implementations using GPUs, Cell BE, and a traditional software implementation was presented in [7]. In this work, the FPGA implementation was produced using "HyperStreams," which is a high level abstraction for designing arithmetic pipelines built on the Handel-C programming language. An acceleration of 146x compared to a reference software implantation can be obtained using FPGAs. The implementation mapped onto a Xilinx XC5VSX50T chip is over 64 and 71 times faster than corresponding software running on a 3.4 GHz Intel Xeon processor, for one-factor and the multi-factor models, respectively.

The combination of cluster technology and reconfigurable hardware acceleration is a relatively new development in high performance computing, which promises to combine the relatively high performance and low power consumption of reconfigurable hardware with established design flows and consequent knowledge base in traditional microprocessor-based high performance computing. A simple Asian option[4] pricing core was designed as a demonstration application on Maxwell. The implementation results of this demonstrator application are shown in Fig. 1. Here, AlphaData (AD) and Nallatech (NT) refer to the two FPGA companies that donated the FPGA accelerator nodes on the Maxwell machine, 32 each.

The results show that the AlphaData nodes lead to ~320-time speed-up compared to an equivalent software implementation, whereas the Nallatech nodes lead to a 109-time speed-up. The discrepancy is due to the design language/flow used for each node type: VHDL for AlphaData and a proprietary C-based hardware language, called DIME-C, for Nallatech.

---

[3]A European option gives its holder the right to buy (a call option) or sell (a put option) an underlying asset at a particular fixed price (called Strike price) on a certain maturity date.

[4]Asian options are a special type of options where the strike price is the average price of the underlying asset over a period of time and not a fixed strike price as in European options.

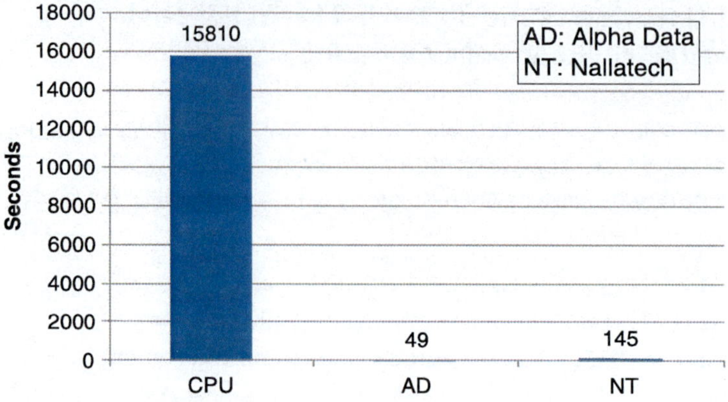

Fig. 1 Single node execution time of Asian option pricing simulation in a Maxwell-based demonstrator

Another important type of financial options is American options[5]. However there are relatively few corresponding FPGA-based implementations reported in the literature. One of these, the basic binomial-tree pricing model of American options, was implemented in [8]. The implementation of this model on a Virtex 4 FPGA achieved 250x speed-up compared to a 2.2GHz Core2 Duo CPU implementation. Another accelerated implementation used the LSMC algorithm [9] and implemented it on a 32 processor IBM BlueGene/P system to achieve 18x speed-up over a single processor implementation.

The Quasi-Monte Carlo method for financial option pricing has been researched for more than a decade. Nevertheless, high performance FPGA implementations of it have been rare in the literature. One such hardware implementation was reported in [10]. In it, a Quasi-Monte Carlo technique was applied to solve a 3-D IC partial inductance extraction problem. The number of dimensions of this design was reported to be 6. A Quasi-Monte Carlo simulator FPGA implementation was also reported in [11] where speed-ups in excess of 50x over a 3GHz multi-core processor were achieved.

As GPUs became more widely used for high performance computing, comparisons between GPUs, FPGAs, and other computing platforms became plentiful. For instance, a speed-up figure of 20x on FPGA compared to a CPU implementation for a European option pricing application was reported in [12], whereas an equivalent GPU implementation achieved a 2 order of magnitude speed-up. In another paper [7], the European option pricing application achieved 146x speed-ups on the FPGA compared to CPU.

As Monte-Carlo simulation relies on a stochastic procedure, random number generation is a key part of it. Software implementations of random number generators cannot meet the requirement of hardware Monte-Carlo simulation cores,

[5]Unlike European options, American options can be exercised at any date up to the maturity date.

and thus hardware random number generation is needed. There are many methods used for hardware random number generation including the Box–Muller method [13], Wallace method [14], and other methods [15, 16].

3 Overview of the Maxwell FPGA Parallel Machine

Maxwell was developed by the FPGA High Performance Computing Alliance in Scotland to demonstrate the feasibility of running computationally demanding applications efficiently on an array of FPGAs. In this section, we will introduce both the hardware architecture and the design flow on Maxwell.

3.1 Hardware Architectures

Maxwell comprises 32 blades housed in IBM Blade Centre. Each blade comprises one 2.8 GHz Xeon with 1 GB memory and 2 Xilinx Virtex4 FPGAs each on a PCI-X sub-assembly developed by Alpha Data or Nallatech. Each FPGA board has either 512 MB or 1 GB of off-chip memory. Whilst the Xeon CPUs and FPGAs on a particular blade can communicate with each other over the PCI bus (typical transfer bandwidths of 600 Mbytes/s), the principal communication infrastructure comprises a fast Ethernet network with a high-performance switch linking the Xeons together and RocketIO linking the FPGAs. Each FPGA has 4 RocketIO links enabling the 64 FPGAs to be connected together in an 8×8 toroidal mesh as shown in Fig. 2. The RocketIOs have a bandwidth of 2.5 Gbits/s per link [17].

Logically, Maxwell can be regarded as a collection of nodes where a node is defined as a software process running on a host machine, plus some FPGA acceleration hardware.

Figure 3 shows the structure of Maxwell, the interconnection, and the detailed architecture of a single FPGA node.

As we can see in the figure, there are three kinds of interconnection on Maxwell: FPGA-FPGA interconnection, which is formed by Rocket IO, the CPU-CPU interconnection, which is Ethernet-based, and the FPGA-CPU connection, which is the PCI-X interface.

There are totally 64 FPGAs on Maxwell as shown in Fig. 3: half of them are Nallatech's off-the-shelf H101-PCIXM with Xilinx XCV4100LX FPGA devices on them. The other 32 nodes are Alpha Data ADM-XRC-4FX cards using Xilinx XCV4FX100 FPGAs.

3.2 Design Flow on Maxwell

The design flow on Maxwell can be divided into four main steps:

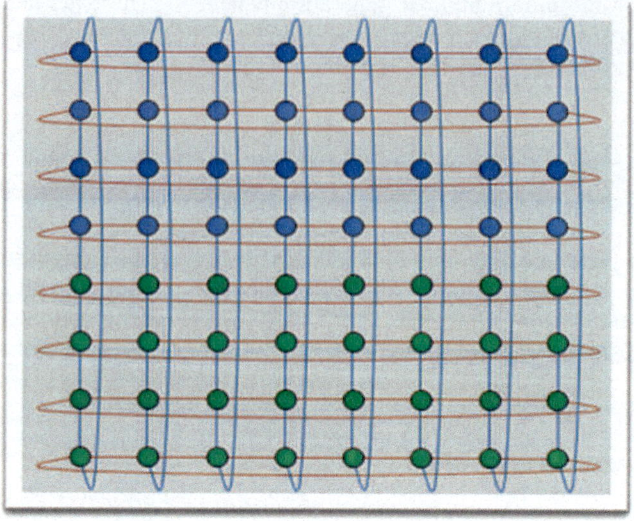

Fig. 2 FPGA links on Maxwell supercomputer

- Hardware design, including HDL coding, simulation, synthesis, and generating bitstream. This stage of work is the main part of the whole design flow.
- Software interface: This step mainly deals with the communication between the FPGA and CPU. On an Alpha Data board, for example, we use ADM-XRC-4FX Co-Processor Development Kit (CPDK) as shown in Fig. 4 to control all registers used to control the behavior of FPGAs.

At higher software level, an API will be used to deal with the standardization of high-level configuration. This tool is called Parallel Toolkit, developed by FHPCA and EPCC [18]. The aim is to configure the FPGA chip with target bitstream, as well as clock setting. The design flow also consists of:

- Message Passing Interface (MPI) [19] coding: communication between nodes is performed using MPI.
- Sun Grid Engine (SGE) job scheduler [20] scripts: allow for orderly safe job submission to the Maxwell machine.

4 Case Studies in High Performance Reconfigurable Computing

4.1 Hardware Random Number Generators

Monte-Carlo methods rely on random samples. Indeed, one random sample is needed for a single step of a Monte-Carlo simulation. For example, suppose

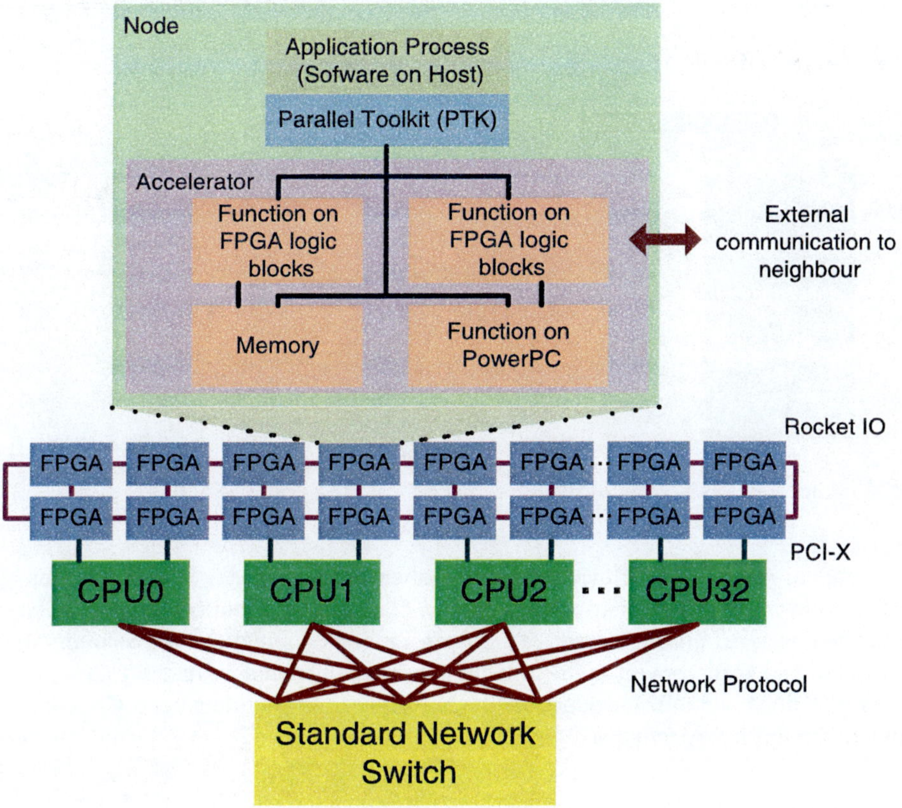

Fig. 3 Architecture of the Maxwell FPGA parallel machine

that a European option has a life length of 1 year, which is discretized to 100 time steps, and 10^6 paths are generated in the Monte-Carlo simulation, therefore, $100 \times 10^6 = 10^8$ random variables are needed for the simulation. Several considerations arise when constructing a random number generator [21]:

- Period length: any pseudo-random number generator will eventually repeat itself. Generally, we want generators with very longer periods.
- Reproducibility: it is often important to be able to re-run a simulation using exactly the same random samples as the previous simulation.
- Speed: as mentioned above, millions or even billions of samples are needed for a single simulation. It is very important to keep a very high throughput of random samples to feed a Monte-Carlo simulation engine.
- Portability: an algorithm for generating random numbers should produce the same sequence of values on all computing platforms.
- Randomness: this is the most important consideration. Theoretical properties and statistical tests could be used to evaluate the quality of the random samples.

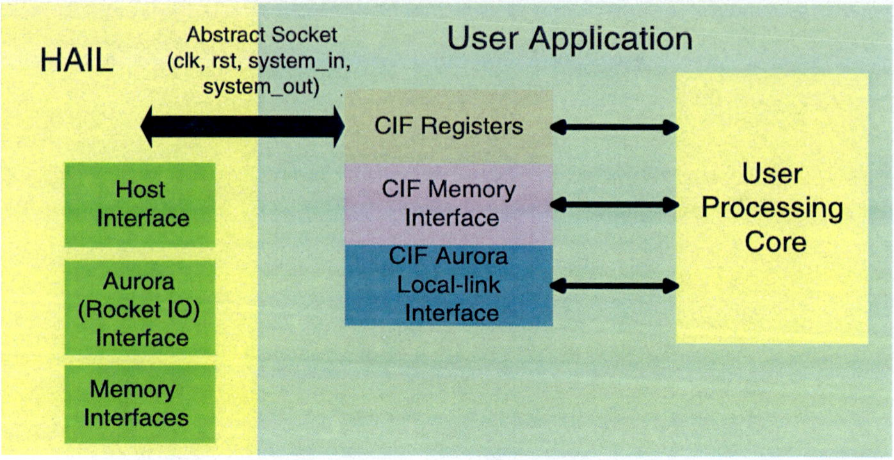

Fig. 4 Structure of CPDK application

The random samples in Brownian motion (often used to model financial option drifts) follow a Normal distribution, or say Gaussian distribution. A generally used method is to produce a set of uniform random samples (over the interval of (0, 1)), and then convert them to Gaussian random numbers. In the following two subsections, we will introduce methods for generating uniform and Gaussian random numbers, respectively.

4.1.1 Uniform Random Number Generator

Uniform random numbers are sampled from a distribution which has the following probability density function:

$$\begin{aligned} p(x) &= 1 \text{ if } 0 < x < 1 \\ &= 0 \text{ otherwise} \end{aligned} \tag{1}$$

It is a convenient distribution as there are many simple methods to transform uniform samples into samples from other distributions. In this section, three different categories of uniform random number generation methods will be introduced, namely: Linear Feedback Shift Registers (LFSR), Mersenne Twister, and Sobol.

LFSR

A generator introduced in [22] is based on a sequence of 0's and 1's generated by a recurrence of the form:

$$b_i = (a_p b_{i-p} + a_{p-1} b_{i-p+1} + \cdots + a_1 b_{i-1}) \bmod 2 \tag{2}$$

where all variables take on values of either 0 or 1. The bs are interpreted as bits and will be formed into binary representations of integers. Because the modulus is a prime, the generator can be related to a polynomial:

$$f(z) = z^p - (a_1 z^{p-1} + \cdots + a_{p-1} z + a_p) \quad (3)$$

over the Galois field GF(2) defined over the integers 0 and 1 with the addition and multiplication being defined in the usual way followed by a reduction modulo 2. An important result from the theory developed for such polynomials is that, as long as the initial vector of b's is not all 0's, the period of the recurrence in (2) is $2^p - 1$ if and only if the polynomial (3) is irreducible over GF(2).

For computational efficiency, most of the a's in Eq. (2) should be zero. The recurrence in Eq. (2) often has the form:

$$b_i = (b_{i-p} + b_{i-p+q}) \bmod 2 \quad (4)$$

Addition of 0's and 1's modulo 2 is the binary exclusive-or operation (represented as $\oplus$), and the recurrence can be written as:

$$b_i = (b_{i-p} \oplus b_{i-p+q}) \quad (5)$$

The recurrence can be performed in a feedback shift register, which is a vector of bits that is shifted, say, to the left, one bit at a time, and the bit shifted out is combined with other bits in the register to form the rightmost bit.

The uniform random number generator used in this design is called Tausworthe URNG [22], which is described by the pseudo-code shown in Fig. 5. Although traditional LFSRs are often sufficient as a uniform random number generator (URNG), Tausworthe URNGs are fast and occupy less area. Furthermore, they provide superior randomness when evaluated using the Diehard random number test suite.

Mersenne Twister

A very popular uniform random number generator, called the Mersenne Twister [23], is based on a recurrence [24] that has approximately 100 terms in the characteristic polynomial of a matrix A. Mersenne Twister has a period of $2^{19937} - 1$ and 623-variate uniformity. The Mersenne Twister algorithm generates a sequence of word vectors, which are considered to be uniform pseudo-random integers between 0 and $2^w - 1$. Dividing by $2^w - 1$, each word vector can be a real number in [0, 1].

A Mersenne prime is a number with the restriction of $2^{nw-r} - 1$. For a word x with w bit width, it is expressed as the recurrence relation:

$$x_{k+n} = x_{k+m} \oplus (x_k^u | x_{k+1}^l) A \; k = 0, 1, \ldots \quad (6)$$

```
unsigned int s0, s1, s2, b;

unsigned int taus()
{
    b = (((s0 << 13) ^ s0) >> 19);
    s0 = (((s0 & 0xFFFFFFFE) << 12) ^ b);
    b = (((s1 << 2) ^ s1) >> 25);
    s1 = (((s1 & 0xFFFFFFF8) << 4) ^ b);
    b = (((s2 << 3) ^ s2) >> 11);
    s2 = (((s2 & 0xFFFFFFF0) << 17) ^ b);
    return s0 ^ s1 ^ s2
}
```

Fig. 5 Tausworthe URNG algorithm

With $|$ as the bitwise or and $\oplus$ as the bitwise exclusive or (XOR), x_u, x_l being x with upper and lower bitmasks applied. We choose a form of the matrix A so that multiplication by A is very fast:

$$A = R = \begin{pmatrix} 0 & I_{w-1} \\ a_{w-1} & (a_{w-2}, \ldots a_0) \end{pmatrix} \tag{7}$$

with I_{n-1} as the $(n-1) \times (n-1)$ identity matrix (and in contrast to normal matrix multiplication, bitwise XOR replaces addition). The rational normal form has the benefit that it can be efficiently expressed as:

$$xA = \begin{cases} shiftright(x) & \text{if } x_0 = 0 \\ shiftright(x) \oplus a & \text{if } x_0 = 1 \end{cases} \tag{8}$$

where

$$x = (x_k^u | x_{k+1}^l) \quad k = 0, 1, \ldots \tag{9}$$

where $a = (a_{w-1}, a_{w-2}, \ldots, a_0)$, $x = (x_{w-1}, x_{w-2}, \ldots x_0)$.

Each generated word is multiplied by a suitable $w \times w$ invertible matrix T from the right to improve k-distribution to v-bit accuracy. For the tempering matrix $x \rightarrow z = xT$, we chose the following successive transformations:

$$y = x \oplus (x >> u) \tag{10}$$

$$y = x \oplus (y << s) \& b \tag{11}$$

$$y = x \oplus (y << t) \& c \tag{12}$$

$$y = x \oplus (x >> l) \tag{13}$$

In order to improve lower bit equi-distribution, we add the first and last transforms.

The coefficients for MT19937 (the commonly used variant of Mersenne Twister, which produces a sequence of 32-bit integers, and has a period of $2^{19937} - 1$) are:

$$(w, n, m, r) = (32, 624, 397, 31)$$

$$a = 9908B0DF_{16}$$

$$u = 11$$

$$(s, b) = (7, 9D2C5680_{16})$$

$$(t, c) = (15, EFC60000_{16})$$

$$l = 18$$

Figure 6 gives the pseudo code of MT19937. Mersenne Twister implementations cannot be parallelized across parallel computing cores simply through changing the initial seed for each core as this does not provide uncorrelated sequences on each generator sharing identical parameters. To solve this problem and enable Mersenne Twister parallel implementations, the authors of MT19937 developed a library for the dynamic creation of Mersenne Twister parameters. This library receives user's specification such as word length, period, size of working area, and a process ID, so that ID number is encoded in the characteristic polynomial of Mersenne Twister.

The process of generating Mersenne Twister numbers can be separated into the following 4 steps:

- Generating the tempering matrix for each computing core based on the given word length, size of working area, and process ID.
- Initializing the generator based on the given seed number.
- Generating the untempered numbers.
- Tempering.

After an initial feasibility analysis, we noticed that the first two steps consume the most of the hardware resources, but the least computing time: both of these two steps only run once at the beginning of the generation, and this time does not increase with the length of the random sequence. Moreover, the following two steps only require shift and XOR operations which consume relatively few logic resources on FPGA. The output of the first step is the 12 parameters needed by steps 3 and 4, and the output of the second step is 624 initialized numbers. However, since we target a parallel FPGA machine assuming the same bitstream on each FPGA node, each FPGA needs different initial numbers. There are two methods to achieve this: generate the initial numbers on each FPGA in the first stage using a different

```
// Create a length 624 array to store the state of the generator
 int[0..623] MT
 int index = 0

 // Initialize the generator from a seed
 function initializeGenerator(int seed) {
     MT[0] := seed
     for i from 1 to 623 { // loop over each other element
         MT[i] := last 32 bits of(1812433253 * (MT[i-1] xor (right shift by 30 bits(MT[i-1]))
+ i) // 0x6c078965
     }
 }
  /* Extract a tempered pseudorandom number based on the index-th value, calling
generateNumbers() every 624 numbers */
 function extractNumber() {
     if index == 0 {
         generateNumbers()
     }

     int y := MT[index]
     y := y xor (right shift by 11 bits(y))
     y := y xor (left shift by 7 bits(y) and (2636928640))
// 0x9d2c5680
     y := y xor (left shift by 15 bits(y) and (4022730752)) // 0xefc60000

     y := y xor (right shift by 18 bits(y))


     index := (index + 1) mod 624

     return y

 }

  // Generate an array of 624 untempered numbers
```

Fig. 6 Pseudo-code of MT19937

seed; or we can generate the initial numbers on CPU, and then transfer them to the FPGAs' BlockRAM. The latter needs 624 data to be transferred through the CPU-FPGA communication or even more if we instantiate more than one simulation core on one FPGA, which will consume considerable communication time. Hence, we chose to generate the initial numbers on FPGA at the expense of a slight resource

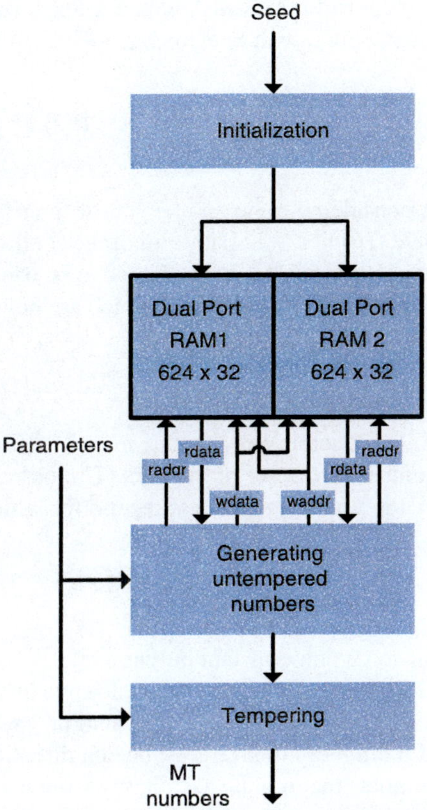

Fig. 7 Mersenne twister random number generator

overhead (mainly one multiplier for each core). The architecture of a Mersenne Twister random number generator core is hence given in Fig. 7.

Based on the Eqs. (8) and (9) we have to read two numbers from the BlockRAM at step 2 each clock cycle. To pipeline the random number generator, two block RAMs are needed: one is for x_l and the other is for x_u. Note that we can pre-compute the parameters and hardwire them to each Mersenne Twister core.

Sobol Random Number Generator

The theory of Sobol numbers starts with modular integer arithmetic. Two integers i and j are called congruent with respect to the modulus m, i.e.

$$i \overset{\Delta}{=} j \bmod m \tag{14}$$

if and only if the difference $i - j$ is divisible by m. For m being prime, the combination of addition and multiplication modulo m, plus a neutral element with respect to both, is also called a finite commutative ring which is isomorphic to a Galois Field with m elements, $GF[m]$. A polynomial $P(z)$ of degree g,

$$P(z) = \sum_{j=0}^{g} a_k z^{g-j}, \tag{15}$$

is considered to be an element of the ring $GF[m, z]$ of polynomials over the finite field $GF[m]$ if we assume all of the coefficients a_k to be $\in GF[m]$. A polynomial $P(z)$ of positive degree is considered to be irreducible modulo m if there are no other two polynomial $Q(z)$ and $R(z)$ which are not constant or equal to $P(z)$ itself such that:

$$P(z) \overset{\Delta}{=} Q(z)R(z) \bmod m, \tag{16}$$

An irreducible polynomial modulo m in $GF[m, z]$ is the equivalent to a prime number in the set of integers. The order of a polynomial $P(z)$ modulo m is given by the smallest positive integer q for which $P(z)$ divides $z_q - 1$, i.e.

$$q = \inf_{q'}\{q' | z^{q'} - 1 \overset{\Delta}{=} P(z)R(z) \bmod m\} \tag{17}$$

for some non-constant polynomial.

To construct a Sobol sequence, we initially construct a vector of numbers, known as direction numbers, of word length w which will serve as a base for the calculation of the Sobol numbers. We need a direction number for each digit, in base 2, of the numbers that will be used in the sequence. In our case, we used 24-bit fixed number representation for our implementation (i.e., $w = 24$). The dimension will be indexed by $k = 1, 2, \ldots D$. The construction of direction numbers is sketched below.

Given a series of integers $a_1, a_2, \ldots a_{d-1}$ that are zero or one, the primitive polynomial modulo 2 of degree d is defined as:

$$P = x^d + a_1 x^{d-1} + a_2 x^{d-2} + \cdots + a_{d-1} x + 1 \tag{18}$$

For each dimension k, a particular primitive polynomial is chosen and a series of integers, m_{ki}, $d_k < i < w$, is generated, starting from the following recursion with d_k terms, where d_k is the degree of the polynomial associated with the kth dimension:

$$\begin{aligned} m_{ki} &= 2a_{k,1}m_{k,i-1} \oplus 2^2 a_{k,2}m_{k,i-2} \oplus \cdots \oplus 2^{d_k-1} a_{k,d_k-1}m_{k,i-d_k+1} \\ &\oplus (2^{d_k} m_{k,i-d_k} \oplus m_{k,i-d_k}) \end{aligned} \tag{19}$$

for $k = 1, 2, \ldots D$, where $\oplus$ represents the bit to bit sum, XOR (exclusive OR), applied on the base 2 representation of the integer m_{ki}. It is necessary to supply the d_k initial values of m_{ki} in each dimension. We used the simple method described in [25] i.e. use a separate pseudo-random number generator to draw uniform variates from

(0, 1) and initialize as follows: draw u_{kl} from a separate uniform random number generator such that:

$$w_{kl} = \text{int}[u_{kl} \times 2^l] \tag{20}$$

is odd, and set:

$$m_{kl} = w_{kl} \times 2^{b-l} \quad for\, l = 1, \cdots d_k \tag{21}$$

The following step is to construct the Sobol numbers using the direction numbers. The nth Sobol number of kth dimension can be obtained from the $(n-1)$th using the following equation:

$$y_{nk} = \sum_{j=1}^{d \oplus 2} m_{kj} 1 \; \{j\text{th bit(counting from the right) of in teger is set}\} \tag{22}$$

However, if we realize it using Gray code instead of using the binary representation of the sequence counter n directly, (18) can be re-written as:

$$y_{nk} = y_{(n-1)k} \oplus m_{kj} \; \{j\text{th bit is the rightmost zero of } n-1\} \tag{23}$$

where m_{kj} is the direction number associated with the rightmost zero in the binary representation of $n-1$. Hence, the nth Sobol number for kth dimension can be obtained from the $(n-1)$th using just one direction number.

The architecture of our Sobol sequence generator is shown in Fig. 8. The black block represents a flip-flop. The architecture is a straightforward implementation of the algorithm above. As the generation of each Sobol number needs the number produced in the last cycle, we have to store the last Sobol number according to Eq. (23). Another point that needs to be clarified here is that since the XOR operation is applied between the direction number and the last Sobol number, we have to store the previously generated Sobol numbers for each dimension. In the case of a 100-day simulation, we have to keep 100 Sobol numbers (in a dual port RAM as shown in Fig. 8). The random number generator picks the Sobol number of the last path of the next day and pushes the current Sobol number in the RAM in each clock cycle.

After building the low-discrepancy numbers, we have to separate them into several sub-streams to implement the distributed Quasi-Monte Carlo simulation. There are two possible approaches of partitioning [26]:

Blocking: disjoint contiguous blocks of overall length l of the original sequence are generated by separate processing elements (PEs). This is achieved by simply using a different starting point on each PE (e.g. PE$_i$, $i = 0, \ldots, p-1$, generates the vectors $x_{il}, x_{il+1}, x_{il+2}, \ldots, x_{il+l-1}$).

Leaping: interleaved streams of the original sequence are generated by the PEs. Each PE skips those consumed by other PEs (*leap-frogging*) (e.g., PE$_i$, $i = 0, \ldots, p-1$, generates the vectors $x_i, x_{i+p}, x_{i+2p}, x_{i+3p}, \ldots$).

We adopt the *blocking* method to perform the partitioning in our implementations. In it, the separation of a Sobol stream can be realized by using a different initial number in each core as depicted in Fig. 9. Apart from the separation in a

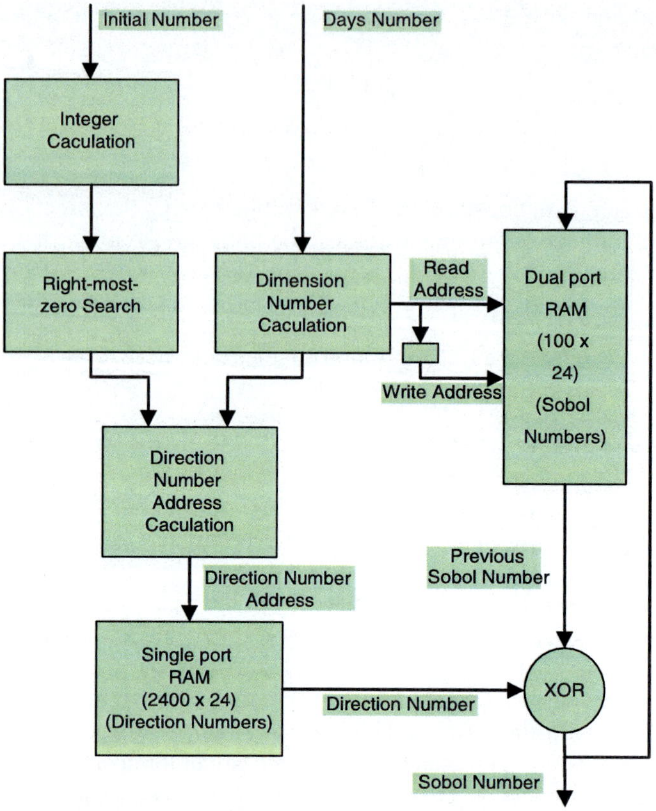

Fig. 8 Architecture of our Sobol sequence generator

single FPGA, or say, node, we have to separate the sequence twice, in the case of a multi-FPGA implementation. In that case, two levels of partitioning are necessary.

The architecture is different from the work mentioned in [10]. We parallelize the Sobol RNG by separating the whole sequence to several sub-sections. This is done by choosing a different initial number for each RNG, which can be seen in Fig. 9. This can be seen as blocking method.

Note finally that since Sobol sequence numbers are always integer numbers, the same module can be used for any subsequent processing arithmetic type, e.g., fixed or floating point.

4.1.2 Gaussian Random Number Generator

As the random variables required in Monte-Carlo simulations follow the Normal distribution, the uniform variables need to be converted to Normal random variables.

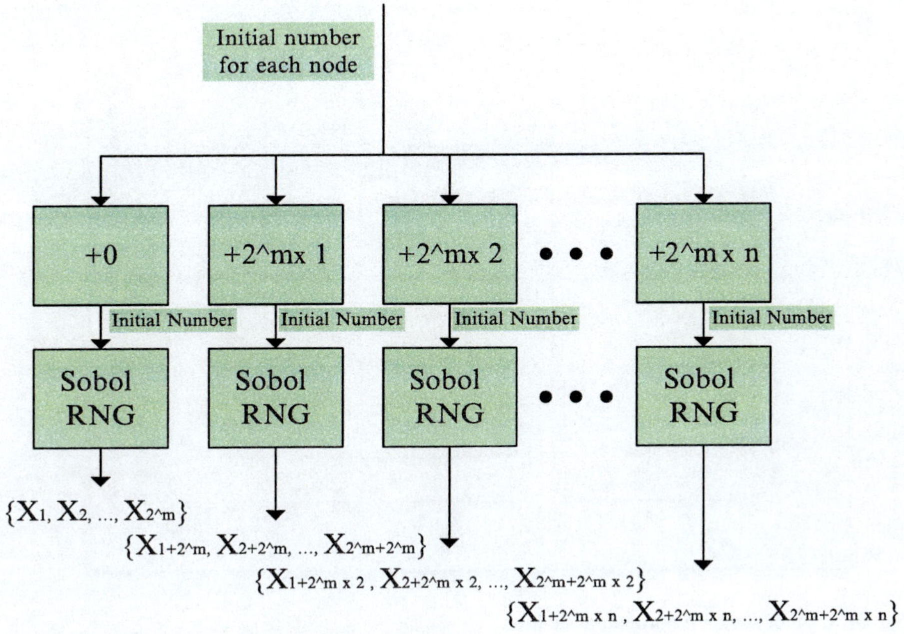

Fig. 9 Parallelism of random number generation

In this section, two methods of conversion will be introduced, namely the inverse cumulative distribution function method (ICDF) and the transformation method.

ICDF

For a random variable X, the cumulative distribution function (CDF) is the function P_X defined by:

$$P_X(x) = \Pr(X \leq x) \tag{24}$$

where $\Pr(A)$ represents the probability of the event A. Two important properties are: the CDF is non-decreasing, and it is continuous from the right. Particularly, the CDF and ICDF of Normal distribution are:

$$F(y) = \frac{1}{\sigma\sqrt{2\pi}} \int_{-\infty}^{y} \mathrm{e}^{(-(t-1)^2/2\sigma^2)} \mathrm{d}t \tag{25}$$

$$f(x) = \left| F^{-1}\left(\frac{x}{2}|\mu, \sigma\right) \right| \tag{26}$$

Figures 10 and 11 show the CDF and ICDF of the standard Normal distribution (with mean 0 and variance 1), respectively. As we can see in Fig. 11, if a uniform variable,

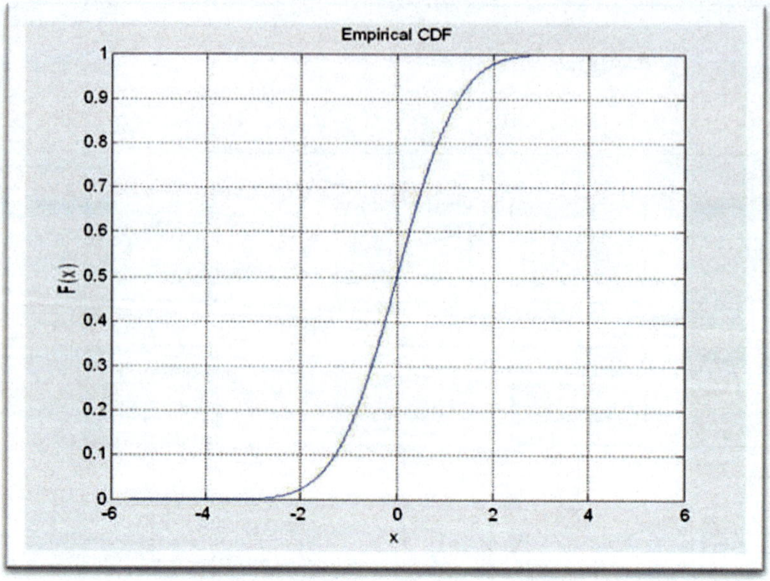

Fig. 10 CDF of standard normal distribution

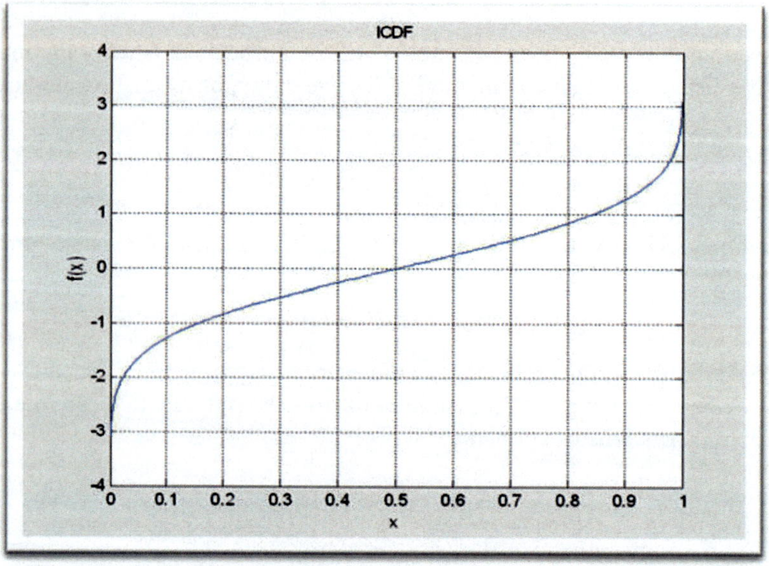

Fig. 11 ICDF of standard normal distribution

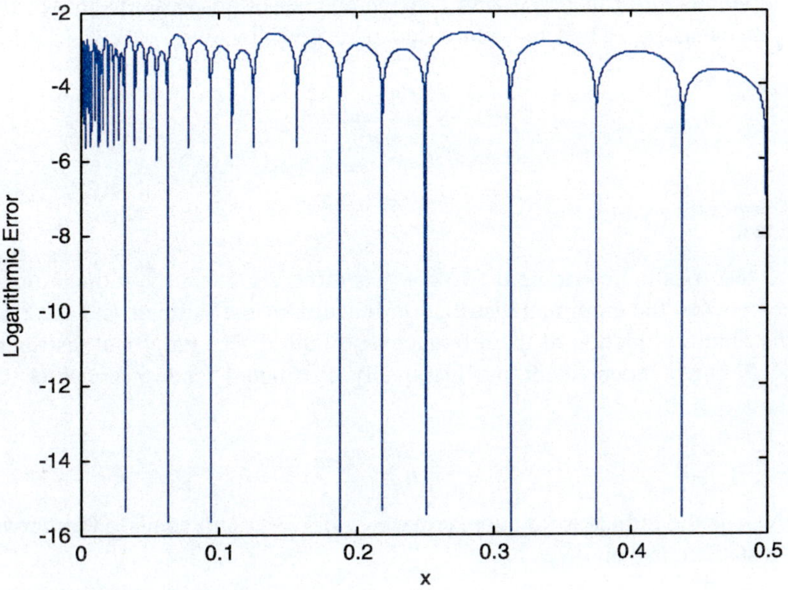

Fig. 12 Logarithmic error of ICDF

which is in the interval of 0 and 1, is the input of the ICDF, the output will follow the Normal distribution. As there is no closed-form for the ICDF, approximation must be adopted in the calculation. Piecewise polynomial approximation is used in this work.

We use a one degree piecewise linear approximation with 80 subsections used between 0 and 0.5 (since the ICDF is an odd function if shifted to the left by 0.5, we only need to calculate its values from 0 to 0.5). The coefficients of the function are pre-computed and stored in a single port RAM. Figure 12 gives the logarithmic error (The logarithmic error is defined as: $LogErr = log_{10}|LinearApproximationResult - GoldenReferenceValue|$) between our ICDF hardware core, using 26-bit fixed point arithmetic, and our golden reference from Matlab's ICDF function. The worst error is around $10^{-2.5}$.

We use two goodness-of-fit tests to check the normality of the Gaussian noise: the chi-square (χ^2) test and the Kolmogorov–Smirnov (K–S) test.

- Chi-Square test:

The Chi-Square test quantizes the x axis into k bins, and then calculates the actual number of samples appearing in each bin. Next, we compare this number with the number of samples which should appear in each bin based on a specific distribution and get a single number. This number can represent the overall quality metric. For example, if n is the number of observations, p_i is the probability that each observation falls into category I, and Y_i is the number of observations that actually do fall into category i. The Chi-Square statistic is given by

$$\chi^2 = \sum_{i=1}^{k} \frac{(Y_i - np_i)^2}{np_i}$$

- K–S test:

The K–S test tries to determine if two datasets differ significantly. It quantifies the distance between the empirical distribution function of the samples and the cumulative distribution function of the reference distribution. The empirical distribution function F_n for n independent and identically distributed random variables X_i is defined as:

$$F_n(x) = \frac{1}{n} \sum_{i=1}^{n} I_{X_i \leq x}$$

where $I_{Xi \leq x}$ is the indicator function, equal to 1 if $X_i \leq x$ and equal to 0 otherwise. The K–S statistic is then given by:

$$D_n = \sup_x |F_n(x) - F(x)|$$

where $\sup x$ is the supremum of the set of distances.

Both of the tests will give p-values for the outputs. The general convention is to reject the null hypothesis—that the samples are normally distributed if the p-value is less than 0.05. Results show that the p-value of samples from our design is more than 0.05.

Transformation

A very crude transformation method to construct normally distributed samples is to add up 12 uniform variates, and subtract 6. This method is the Central Limit Theorem applied to a sample of size 12. It is a poor approximation and always slower than other methods. Another way to transform the uniform variables is the Box–Muller transformation [27]. If u and v are independent standard uniform variables in (0, 1), a pair of independent Normal variables x and y can be generated using

$$x = \sqrt{-1 \ln u} \sin(2\pi v)$$
$$y = \sqrt{-1 \ln u} \cos(2\pi v) \qquad (27)$$

One problem of Box–Muller transformation is that it cannot be used for converting the low-discrepancy numbers such as Sobol numbers.

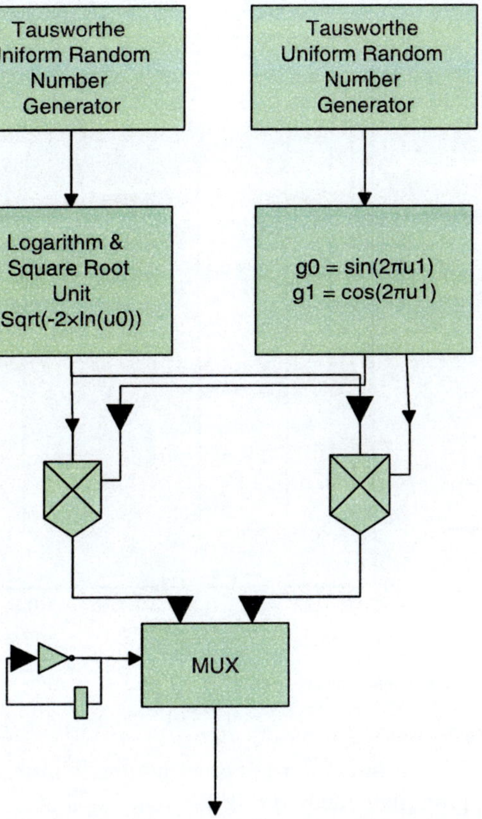

Fig. 13 Gaussian noise generator architecture

The Box–Muller method is conceptually straightforward. Given two independent realizations (u_1 and u_2) of a uniform random variable over the interval [0, 1), and a set of intermediate functions f, g_1 and g_2 so that:

$$f(u_1) = \sqrt{-2 \times \ln(u_1)} \tag{28}$$

$$g_1(u_2) = \sin(2\pi u_2) \tag{29}$$

$$g_2(u_2) = \cos(2\pi u_2) \tag{30}$$

$$x_1 = f(u_1)g_1(u_2) \tag{31}$$

$$x_2 = f(u_1)g_2(u_2) \tag{32}$$

Then providing two samples of a Gaussian distribution $N(0, 1)$, x_1 and x_2.

Based on this algorithm, the corresponding hardware architecture is given in Fig. 13.

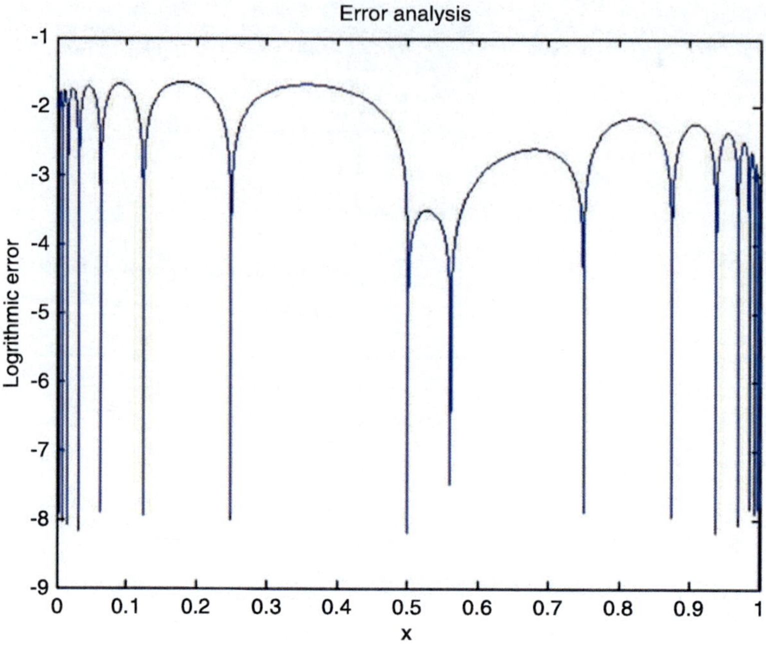

Fig. 14 Logarithmic error of $f(u_1) = \sqrt{-2 \times \ln(x)}$

Logarithm and trigonometric functions are computed using the piecewise linear approximate method [28, 29]. The logarithm errors of both functions are shown in Figs. 14 and 15. We generated 100,000 samples, and the PDF is shown in Fig. 16.

4.2 Financial Computing Models and Their Implementations on Maxwell

4.2.1 European Option Pricing

The Monte-Carlo method for European options pricing is based the Black–Scholes model of option price evolution:

$$S_{\Delta t} = S_0 \left(1 + \left(\left(\mu - \frac{\sigma^2}{2}\right) \delta t + \sigma \varepsilon \sqrt{\delta t}\right)\right) \tag{33}$$

where $S_{\Delta t}$ and S_0 are the stock prices at times Δt and zero, respectively, μ is the expected rate of return of the stock, σ is the volatility of the stock price, and ε is random variable with mean 0 and variance 1.

The simulation process can be described as the following algorithm:

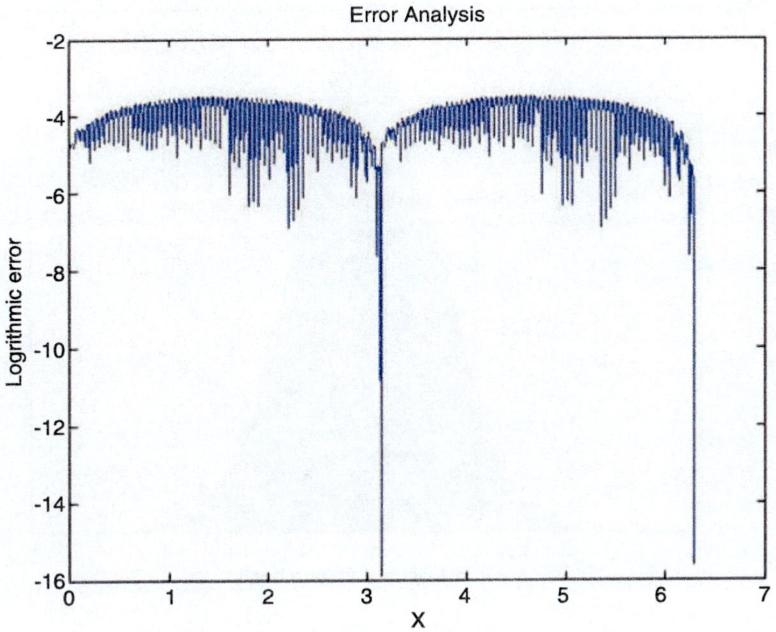

Fig. 15 Logarithmic error of $g_1(x) = \sin(2\pi x)$

For any $n \geq 1$, the estimator $\hat{C}_n$ of the option price is unbiased, in the sense that its expectation is the target quantity:

$$E[\hat{C}_n] = C \equiv E[\mathrm{e}^{-rT}(S(T) - K)^+]$$

The estimator is strongly consistent, meaning that as $n \to \infty$,

$$\hat{C}_n \to C \text{ with probability } 1$$

The algorithm is described by the pseudo-code in Fig. 17.

Note that the simple European option model does not need small time steps to build the paths. Single big time step can be used for generating the stock price path. However, as the Monte-Carlo simulation module is the basic part of all the other option pricing engines, in which small time steps must be used, the implementation of European option pricing model will use the small time step.

The European option pricing engine comprises an LSFR uniform random generator, a Box-Muller Gaussian random generator and the Monte-Carlo simulation core with implements Eq. (33). The architecture can be seen in Fig. 18.

The computing core was captured in Verilog and synthesized using Xilinx ISE 9.2i. We increased the number of computing cores until resources run out. Our user application processor implements 20 Monte-Carlo cores occupying 36144 slices on

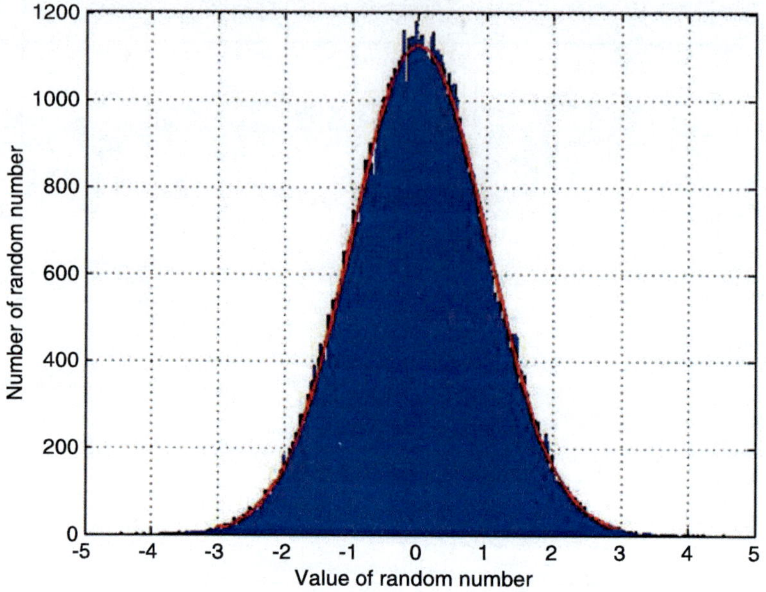

Fig. 16 PDF of the generated random variables

for $i = 1$ to n
  Generate m random samples
  for $t = 1$ to m

$$S_{t+1} = S_t(1 + ((\mu - \frac{\sigma^2}{2})\delta t + \sigma \varepsilon \sqrt{\delta t}))$$

$$C_i = e^{-rT}(S(T) - K)^+$$

$$\hat{C}_n = \text{mean}(C_i)$$

Fig. 17 Algorithm for path generation of European option prices

an XC4VFX100-10ff1517 FPGA, which has 42176 slices in all; all 160 DSP48s units are utilized. We set the clock frequency on Maxwell's FPGA node to 75 MHz.

We implemented our hardware Monte-Carlo simulation solution for European option pricing on Maxwell. We also ran an equivalent C++ based software solution on Maxwell and run it on the 2.8 GHz Xeon processors (each with 1 GB of memory). The execution time of both FPGA and CPU implementations is shown in Fig. 19.

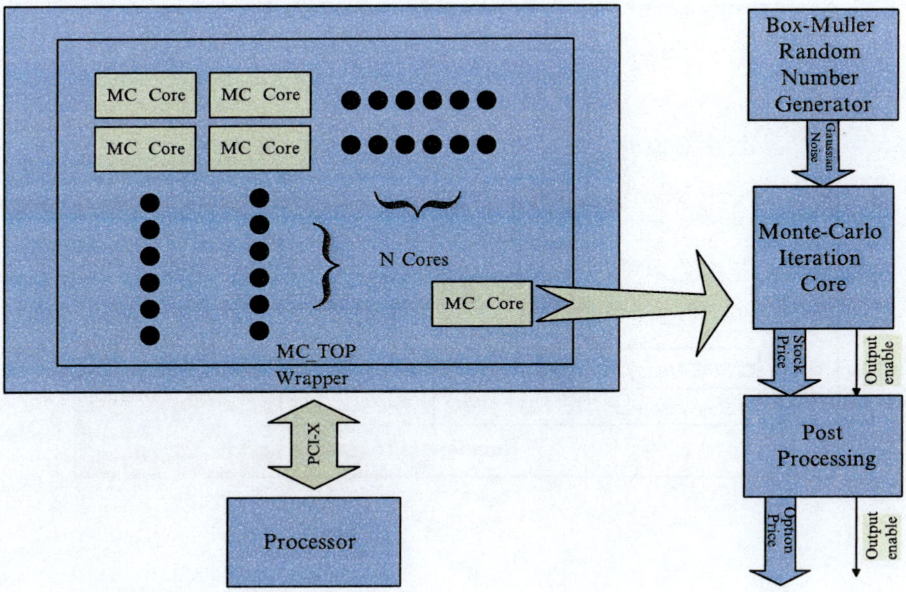

Fig. 18 Generic architecture of Monte-Carlo simulation engine4

From Fig. 19, we can see that the computing time decreases linearly as the number of nodes increases. The reason is that communication time is very limited during Monte-Carlo simulation: only broadcasting parameters at the beginning of simulation and gathering results at the end of simulation. As we pipelined the design and set clock frequency to 75 MHz, our Monte-Carlo computing core finishes one iteration in 13.3 ns. As the Input/Output and communication overheads are limited, we estimate the overall computing time to be:

$$ComputingTime = ClockPeriod \times (NumOfPaths \times NumOfDays) \\ -(NumOfCores \times NumOfNodes) \tag{34}$$

Take the result of a 32 nodes experiment as an example: the clock period is 13.3 ns; the number of paths is $2^{17} \times 10 = 1.31 \times 10^6$; the number of days is 100; the number of cores per FPGA is 20, and the number of nodes is 32. This gives us

$$ComputingTime = 13.3 \times (1.31 \times 10^6 \times 100) - (20 \times 32) \approx 2.731 \times 10^6 \text{ns} \tag{35}$$

Overall, the FPGA implementations are 750x faster than the equivalent software implementations. It is worth mentioning, however, that our software implementation was not optimized on the Xeon processors.

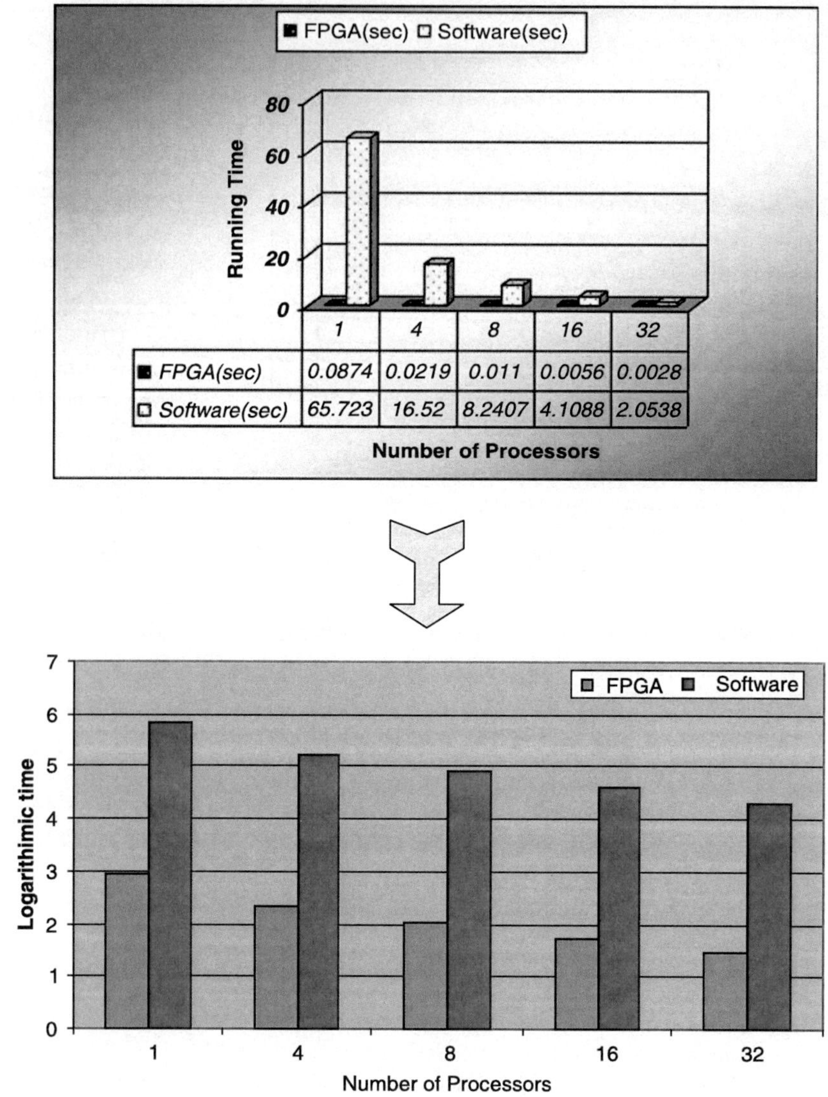

Fig. 19 Running time of C++ & FPGA implementation

4.2.2 Asian Option Pricing

In Asian options, the payoff is determined by the average underlying price over some pre-set period of time. Hence, an Asian option can be calculated as:

$$C_{\text{AsianCall}} = \max(0, S_{\text{avg}} - K)$$
$$C_{\text{AsianPut}} = \max(0, K - S_{\text{avg}})$$

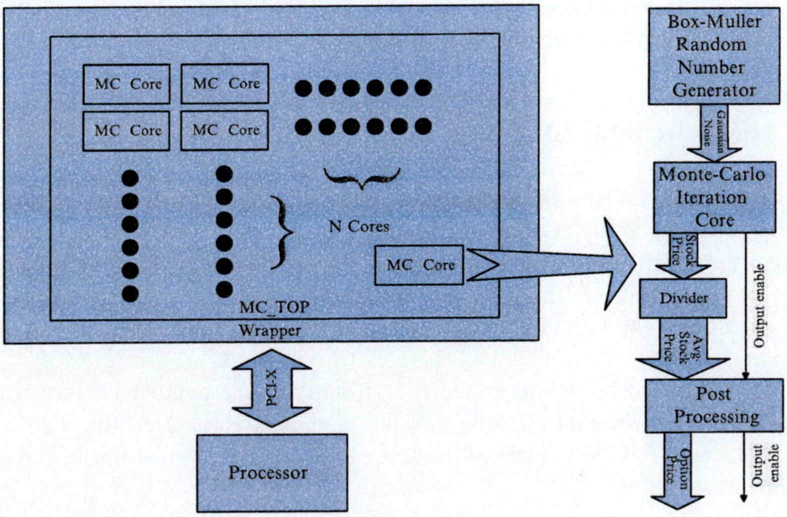

Fig. 20 Architecture of Asian option simulation engine

The average of S can be obtained in many ways. In the continuous case, this is calculated through an integral:

$$S_{\text{ave}} = \frac{1}{T}\int_0^T S(t)\mathrm{d}t$$

or in the discrete version:

$$S_{\text{ave}} = \frac{1}{N}\sum_{i=1}^{N} S(t_i)$$

There is also a type of Asian options with geometric average:

$$S_{ave} = \exp\left(\frac{1}{T}\int_0^T \ln(S(t))\mathrm{d}t\right)$$

From the characteristic of the Asian option, we can see that the Asian option has a lower volatility than the European option. Hence, it is less risky and cheaper than the European option. The Asian option is arguably more appropriate than regular options for meeting some of the needs of corporate treasurers.

In our work, we mainly dealt with the discrete version of the average. From the hardware point of view, the modification implies adding an accumulator (which is within the MC core) and divider after the Monte-Carlo simulation core as shown in Fig. 20.

With the extra divider, the Asian option hardware simulation engine was of 600x faster than the equivalent software implementation. This implementation is configured as the same wordlength, clock frequency, and the number of computation kernels as the European option pricing model.

4.2.3 The GARCH Model

One assumption in the Black–Scholes model that is not always true in practice is the assumption that volatility is constant. Indeed, practitioners often find it necessary to change the volatility parameter when using the Black–Scholes model to value options. In the case where the stock price and volatility are correlated, there is no simple solution to the model equations and Monte-Carlo-based simulations often become necessary.

One technique for modeling volatility that has become popular is Generalized Autoregressive Conditional Heteroskedasticity—GARCH model [30]. The most commonly used GARCH model is GARCH (1, 1) where the volatility is given by the following equation:

$$\sigma_i^2 = \sigma_0 + \alpha\sigma_{i-1}^2 + \beta\sigma_{i-1}^2\lambda^2 \tag{36}$$

Here α, and β are constants which can be estimated from historical data using maximum likelihood methods. σ_0 is the volatility of the stock price at time 0, σ_i and σ_{i-1} are the volatilities at time $i\Delta t$ and $(i-1)\Delta t$. λ is a random variable with a normal (Gaussian) distribution with a mean of zero and a standard deviation of 1.0. Notice that the random variable in the GARCH model is different from the one used in the describing the evolution of stock prices. The two random variables represent two independent stochastic processes.

For options that last less than 1 year, the pricing impact of a stochastic volatility is fairly small in absolute terms. It becomes progressively larger as the life of the option increases.

The GARCH model has only one extra module compared to the European pricing engine implementation, namely a stochastic volatility model implementation. The architecture is depicted in Fig. 21.

We captured our hardware architecture using Verilog-HDL and synthesized it using Xilinx ISE 9.2i. We could fit 11 Monte-Carlo cores on one single FPGA chip. These occupied 39,466 slices on an XCV4FX100-10ff1517 FPGA, which has 42,176 slices overall (the word length is configured as 26 bits). Besides, all 160 DSP48s units were utilized. The peak clock frequency of the core is 53 MHz. We set the clock frequency on the Maxwell's FPGA nodes to 50 MHz.

Figure 22 gives the execution time of the GARCH option pricing model on the Maxwell machine using an increasing number of nodes. This is shown for our FPGA implementation as well as for an equivalent software implementation running on the 2.8 GHz Xeon processors. In both cases, the execution time decreases linearly as the number of nodes increases. This is because inter-communication time is negligible compared to the computing time. Indeed, the only instances where communication between the host software and the Monte-Carlo cores (running on

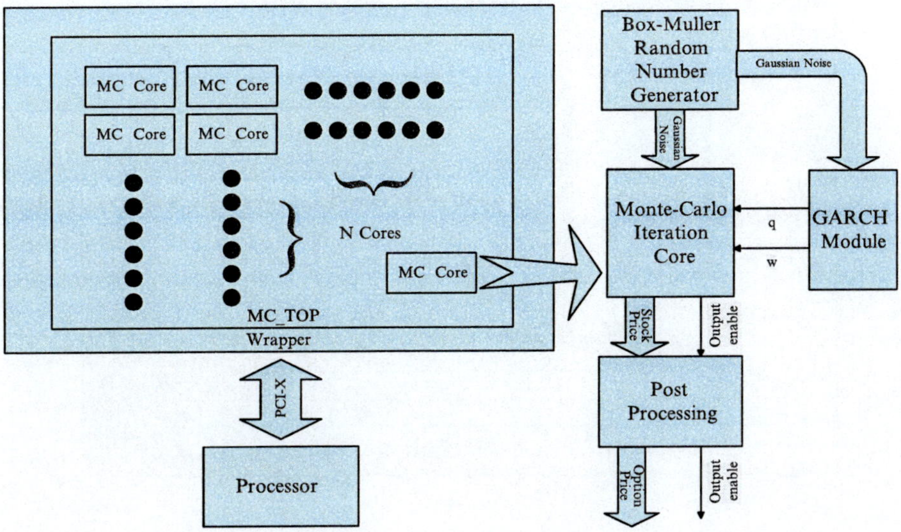

Fig. 21 Generic architecture of GARCH model simulation engine

FPGA or on the Xeon processors) is needed is when parameters are broadcasted to the cores at the beginning of the execution, and when results are gathered from the cores at the end of the simulation. Compared to software, our FPGA implementation results in a 340x speed-up. It is worth mentioning that this speed-up figure is independent of the number of nodes (FPGA/CPU) used.

The reason behind the high speed-up figure of the FPGA implementation, despite the huge difference in clock frequency (50 MHz for the FPGAs compared to 2.8 GHz for the Xeon's) is due to the high level of process parallelism (11 cores running in parallel on each FPGA device) as well as the high degree of pipelining used within each core.

4.2.4 American Option Pricing

As mentioned in Sect. 2, American options are call or put options that can be exercised at any time up to the expiration date. After generating the paths of the stock price, using the same approach as the simulation of European options, we need to go backward to find the best day to exercise the option. There are two different situations in the decision-making procedure: at the final exercise date, the optimal exercise strategy for an American option is to exercise the option if it is in the money; however, prior to the final date, the optimal strategy is to compare the immediate exercise value with the expected cash flows from continuing, and then exercise if immediate exercise is more valuable [31]. Thus, we can see that the strategy to optimally exercise an American option is to identify the conditional

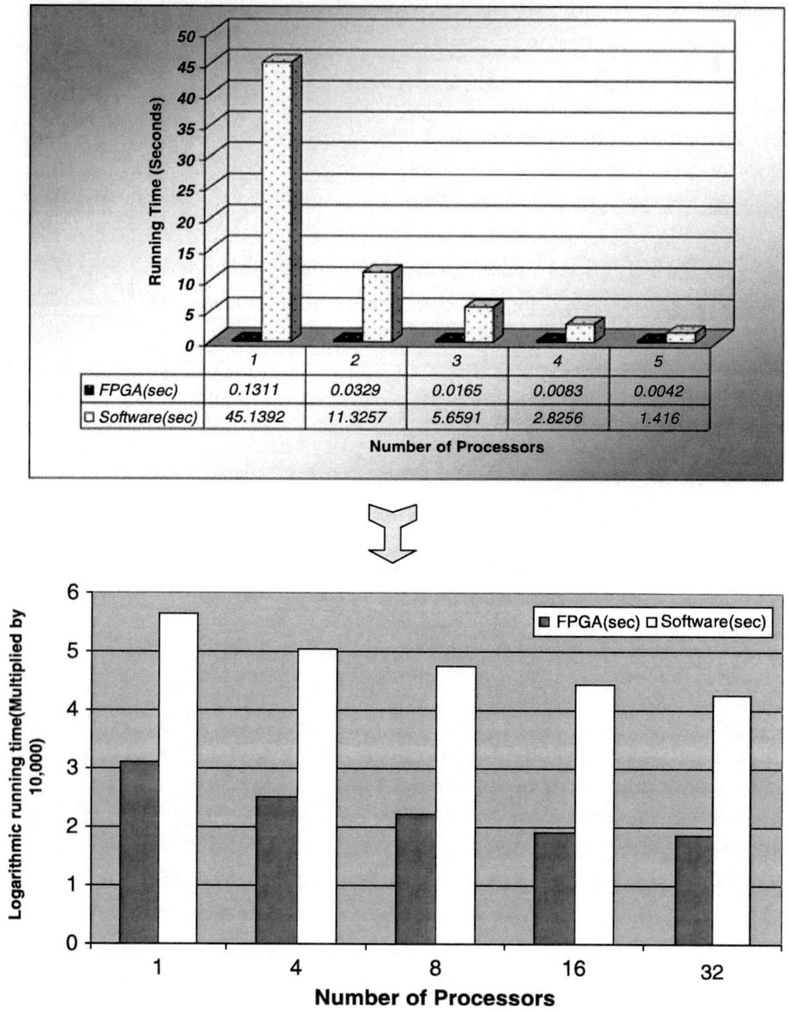

Fig. 22 Execution time of the GARCH option pricing model

expected value of continuation. We use the cross-sectional information in the simulated paths to identify the conditional expectation function. This is done by regressing the subsequent realized cash flows from continuation on a set of basis functions of the values of the relevant state variables. The fitted value of this regression is an efficient unbiased estimate of the conditional expectation function and allows us to accurately estimate the optimal stopping rule for the option.

Here we use a simple example to depict the least-squares regression. A more detailed description can be found in [31]. Consider an American put option on a share of non-dividend-paying stock. The put option is exercisable at a strike price of

Table 1 Stock price paths

Path	$t = 0$	$t = 1$	$t = 2$	$t = 3$
1	1.00	1.09	1.08	1.34
2	1.00	1.16	1.26	1.54
3	1.00	1.22	1.07	1.03
4	1.00	0.93	0.97	0.91
5	1.00	1.11	1.56	1.52
6	1.00	0.76	0.77	0.90
7	1.00	0.92	0.84	1.01
8	1.00	0.88	1.22	1.34

Table 2 Cash-flow matrix at time 3

Path	t=1	t=2	t=3
1	–	–	0
2	–	–	0
3	–	–	0.07
4	–	–	0.18
5	–	–	0
6	–	–	0.20
7	–	–	0.09
8	–	–	0

Table 3 Regression at time 2

Path	Y	X
1	0.00 × 0.94176	1.08
2	–	–
3	0.07 × 0.94176	1.07
4	0.18 × 0.94176	0.97
5	–	–
6	0.20 × 0.94176	0.77
7	0.09 × 0.94176	0.84
8	–	–

1.10 at times 1, 2, and 3, where time 3 is the final expiration date. The riskless rate is 6%. We use eight simulation paths for the price of the stock, which are shown in Table 1.

First, considering the situation of not exercising the option before the final expiration date at time 3, the cash flows realized by the option holder from following the optimal strategy at time 3 are given in Table 2.

If the put is in the money at time 2, the option holder must then decide whether to exercise the option immediately or continue the option's life until the final expiration date at time 3. From the stock-price matrix, there are five paths for which the option is the money at time 2. Let X denote the stock prices at time 2 for these five paths and Y denote the corresponding discounted cash flows received at time 3 if the put is not exercised at time 2, the regression at time 2 is shown in Table 3.

To estimate the expected cash flow from continuing the option's life conditional on the stock price at time 2, we regress Y on a constant, X, and X^2. This specification is one of the simplest possible; more general specifications are given

Table 4 Optimal early exercise decision at time 2

Path	Exercise	Continuation
1	0.02	0.0369
2	–	–
3	0.03	0.0461
4	0.13	0.1176
5	–	–
6	0.33	0.1520
7	0.26	0.1565
8	–	–

Table 5 Cash-flow matrix at time 2

Path	$t = 1$	$t = 2$	$t = 3$
1	–	0	0
2	–	0	0
3	–	0	0.07
4	–	0.13	0
5	–	0	0
6	–	0.33	0
7	–	0.26	0
8	–	–	0

Table 6 Regression at Time 1

Path	X	Y
1	1.09	0.00 × 0.94176
2	–	–
3	–	–
4	0.93	0.13 × 0.94176
5	–	–
6	0.76	0.33 × 0.94176
7	0.92	0.26 × 0.94176
8	0.88	0.00 × 0.94176

in [31]. Although we Although we only use this specification in the hardware implementation, a more general implementation can easily be adopted. The resulting conditional expectation function is $E[Y|X] = -1.070 + 2.983X - 1.813X^2$.

With this conditional expectation function, we now compare the value of immediate exercise at time 2, given the first column in Table 3, with the value from continuation, given in the second column in Table 4. This leads to the following matrix in Table 5, which shows the cash flows received by the option holder conditional on not exercising prior to time 2.

Proceeding recursively, we next examine whether the option should be exercised at time 1. Again we choose the paths where the option is in the money. Let X denote the stock prices at time 1 for these paths and Y denote the corresponding discounted cash flows received at time 2 if the put is not exercised at time 1. The regression data at time 1 is shown in Table 6.

The conditional expectation function at time 1 is estimated by again regressing Y on a constant, X, and X^2. Then we use the estimated conditional expectation

Table 7 Optimal early exercise decision at Time 1

Path	Exercise	Continuation
1	0.01	0.0139
2	–	–
3	–	–
4	0.17	0.1092
5	–	–
6	0.34	0.2866
7	0.18	0.1175
8	0.22	0.1533

Table 8 Option cash-flow matrix

Path	t=1	t=2	t=3
1	.00	.00	.00
2	.00	.00	.00
3	.00	.00	.07
4	.17	.00	.00
5	.00	.00	.00
6	.34	.00	.00
7	.18	.00	.00
8	.22	.00	.00

generated by the regression to calculate the estimated continuation values, as shown in Table 7.

After deciding the exercise strategy at times 1, 2, and 3, we can get the final option cash flow matrix as shown in Table 8. Then, the option can be valued by discounting each cash flow in the option cash flow matrix back to time zero, and averaging over all paths. This procedure results in a value of 0.1144 for the American put.

As the main operations in this step are matrix multiplication and inversion, the size of the matrix is very important to the overall architecture. If the number of simulation paths is 4,096, and the number of time steps is 100; the size of matrix X will be $4,096 \times 3$, and Y will be $4,096 \times 1$[6]. Figure 23 depicts the overall architecture of the regression step.

Our FPGA implementation targeted a Xilinx XC4VFX100-10 FPGA chip on an Alpha Data ADM-XRC-4FX card, which contains 42,176 slices, 160 DSP48s, and 376 BlockRAM units. We captured our hardware architectures in generic Verilog and synthesized them using Xilinx ISE 9.2i. To achieve the precision requirement of 10^{-4}, we needed 26 bits fixed point arithmetic for the Monte-Carlo simulation and 32 bits fixed point arithmetic for the regression part. The resource consumption breakdown is shown in Table 9. Moreover, 16 off-chip memory banks (4 physical

[6]Actually, the LSMC algorithm uses only the paths which are in the money. Hence, the number of row of X should be less than 4096. However, the memory size on FPGA is fixed, so we exclude the paths which are not in the money when doing the matrix multiplication. So the number of row of X can still be seemed as 4096.

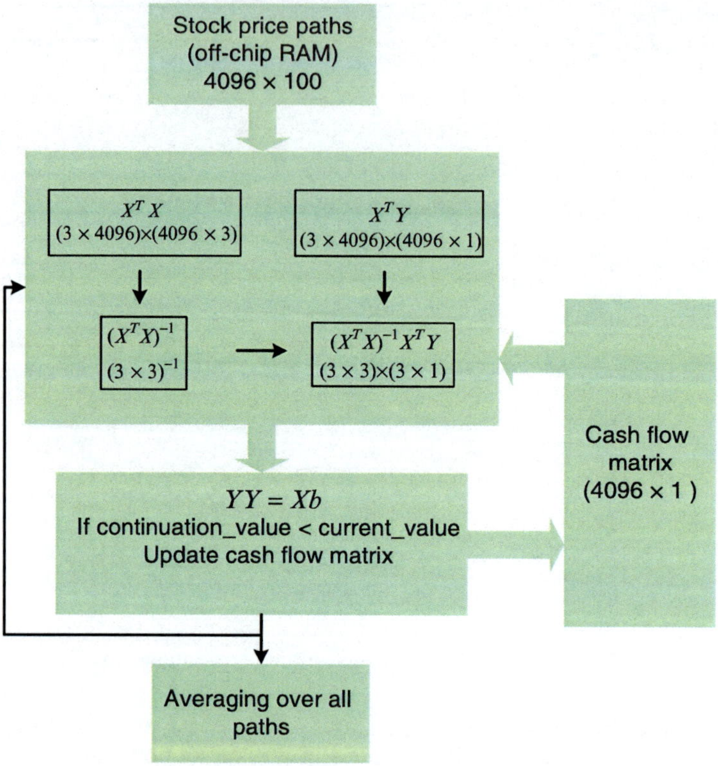

Fig. 23 Architecture of linear-squares regression

Table 9 Resource consumption breakdown

	Slices	FFs	LUTs	RAM	DSP48s
MC	2,655	2,996	3,656	24	44
Regression	37,667	18,385	66,384	53	116
Overall	40,322	21,381	70,040	77	160

banks are used, each bank has 52 bits word-length and we double the clock frequency to the memory) were used for storing the stock price paths, with each memory bank consisting of $512 \times 100 \times 26$ bits. The peak frequency achieved was 76 MHz, and the FPGA card was clocked at 75 MHz.

To compare the hardware implementation with an equivalent software implementation, we also wrote a C++ program for our Least-Squares Monte-Carlo engine and executed it on a 2.8 GHz Xeon processor-based machine with 1Gbyte memory. We used a fully optimized library, namely Intel Math Kernel Library (MKL), to generate the Sobol sequence and convert it to Gaussian noise using the ICDF method (also provided by the MKL). Single precision is used in the software implementation. The wall clock time of both FPGA and CPU-based implementations is shown in Table 10.

Table 10 Calculation time on FPGA and CPU (ms)

	FPGA	CPU
MC	0.683	16.778
Regression	1.368	24.608
Overall	2.051	41.386

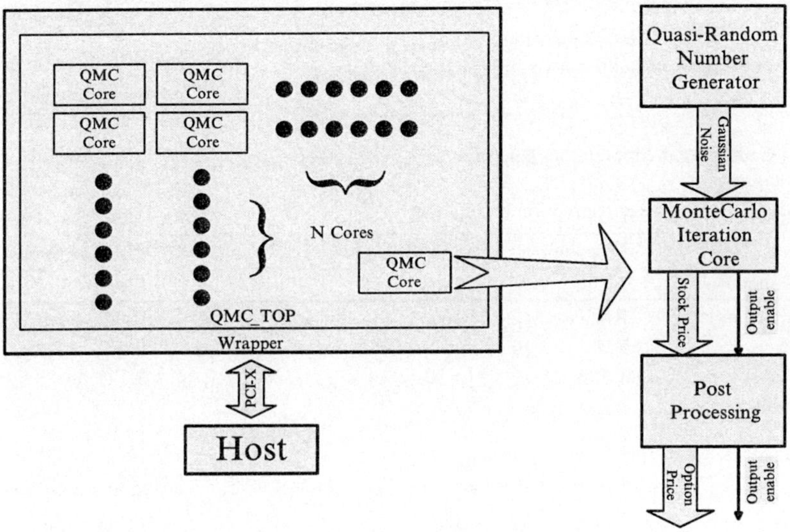

Fig. 24 Generic architecture of Quasi-Monte Carlo simulation engine

From Table 10, we can see that a 25x and 18x speed-up can be achieved in the Monte-Carlo simulation and regression steps, respectively, for an overall speed-up of the American option pricing calculation of 20x.

4.2.5 Quasi-Monte Carlo Simulation

The path generation part of the Quasi-Monte Carlo simulation core is the same as the one in the Monte-Carlo simulation core. However, the principle of the Quasi-Monte Carlo simulation is mainly based on the random number generation mechanism. Figure 24 gives the generic architecture of the Quasi-Monte Carlo simulation engine.

Hence, we plug a different random number generator (in Fig. 25) to the Monte Carlo simulation core. Since all of the modules in Fig. 24 have been described in the previous sections, we here only present the implementation results of the Quasi-Monte Carlo simulation core and the comparison with other implementations.

Our FPGA implementation targeted the XC4VFX100-10 FPGA chips on the Maxwell machine. We captured our hardware architectures in generic Verilog and synthesized them using Xilinx ISE 9.2i. We experimented with bit both fixed and floating point arithmetic as shown in Table 11, where the resources consumed by each module in a single Quasi-Monte Carlo computing core (excluding the PCI interface module) are given.

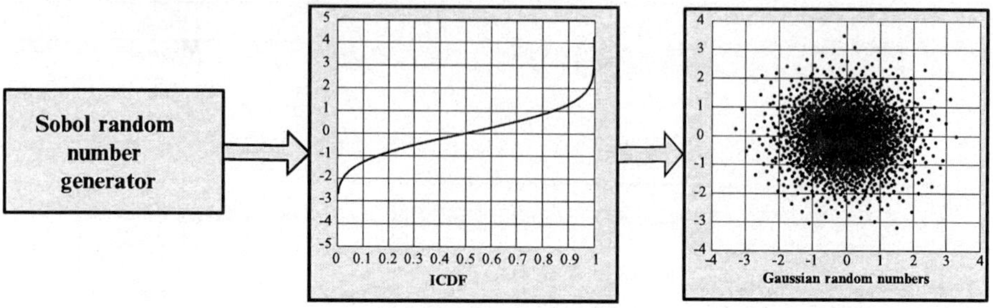

Fig. 25 Gaussian random number generator

Table 11 Resource consumption breakdown

	Fixed point						Floating point	
Arithmetic	26bits		29bits		32bits		Single precision	
Module name	RNG	MC	RNG	MC	RNG	MC	RNG	MC
Slices	944	614	945	675	951	777	951	2,790
	1,558		1,620		1,728		3,741	
DSP48s	1	8	1	9	1	10	1	13
	9		10		11		14	
RAM16s	6	0	6	0	6	0	6	0
	6		6		6		6	
LUTs	1,268	425	1,268	519	1,268	597	1,268	3,037
	1,693		1,787		1,865		4,305	
FFs	1,375	884	1,375	1087	1,375	1,273	1,375	4,392
	2,259		2,462		2,648		5,767	
Freq' (MHz)	211		211		194		180	
Precision	10-E4		10-E5		10-E6		10-E6	
Max No. of Cores	27		26		24		11	

Four RAM16s are used for storing 2,400 direction numbers and one is used for storing Sobol numbers from previous paths (in the Sobol number unit). We note that the precision of 32 bits fixed point implementation is the same as the single precision floating point one. This is largely due to the range of the input data. Another issue is that since we wanted to optimize the single precision floating point implementation, we used the most optimized pipeline stage for the floating point units, which results in a peak clock frequency of 180 MHz. The critical path for all the four implementation is the multiplier. The costs of addition, multiplication, and accumulation are 3, 3, and 1, respectively. In addition, the floating point arithmetic units are generated by Xilinx Core Generator.

Note that as we scaled the parameters of ICDF module, the range of Gaussian numbers in the 32 bits fixed-point implementation is the same as the single precision one. Hence we use the 32 bits fixed-point arithmetic for the RNG module in single precision implementation and converse the numbers to single precision before inputting to the MC module.

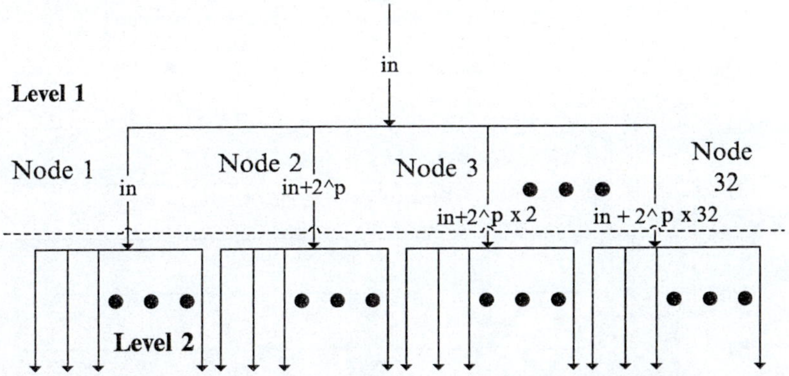

Fig. 26 Parallelism of Sobol sequence

Since we targeted a multi-FPGA platform, we use Message Passing Interface (MPI) [19] for process intercommunication (this is used for all the implementations in this chapter). Thirty-two FPGA nodes were used on the Maxwell machine, with each node loaded with the same bitstream and initial option price parameters, but with different initial numbers for the Sobol random number generator. Indeed, as explained above, we separate the Sobol sequence at two levels in order to allow for parallelism (see Fig. 26). As the initial number for each FPGA node is different, we calculate the initial number in one run of a recurrence and send it to the relative node using MPI. In each node, we calculate the initial number for each core using FPGA resources.

The pipelined design means that there is no need to store the random numbers as one Sobol number and one Gaussian variable are provided by the random number generator each cycle. Memory access time is hence reduced, and an option price estimate is generated every clock cycle in each core after the pipeline fills.

Figure 27 shows the execution time of our Quasi-Monte Carlo simulation engine on the Maxwell machine with different numbers of FPGA processing nodes and different arithmetic types, measured by the wall time function in MPI. As can be clearly seen, the execution time reduces linearly with the number of FPGA nodes used, which is to be expected since inter-process communication is negligible.

The second Quasi-Monte Carlo simulation engine was targeted at an NVIDIA 8800GTX GPU. This device has a core clock frequency of 575 MHz and a shader clock frequency of 1,350 MHz. The memory size is 768 MB, with 900 MHz clock frequency. We installed the newest Compute Unified Device Architecture (CUDA) 2.1 [32] development environment on a MacPro workstation with a 64-bit Linux system, and we implemented the exact same algorithm as in the FPGA implementation on the GPU. In the GPU, parallelism is mainly obtained through multi-threading. The thread hierarchy is as follows: the threads can be identified using a one-dimensional, two-dimensional, or three-dimensional index, forming a

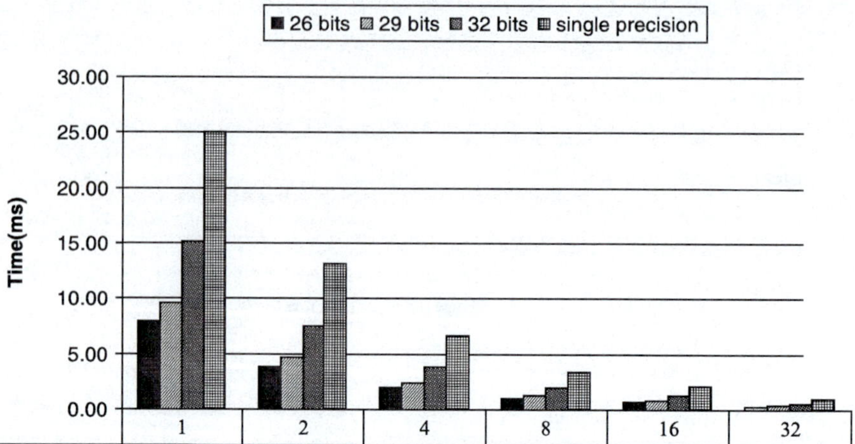

	1	2	4	8	16	32
26 bits	7.87	3.87	1.98	1.03	0.64	0.31
29 bits	9.56	4.66	2.41	1.25	0.78	0.37
32 bits	15.15	7.48	3.81	1.98	1.24	0.59
single precision	25.00	13.12	6.67	3.45	2.17	1.03

Number of FPGAs

Fig. 27 Running time of Quasi-Monte Carlo simulation engine on different number of FPGA processing nodes

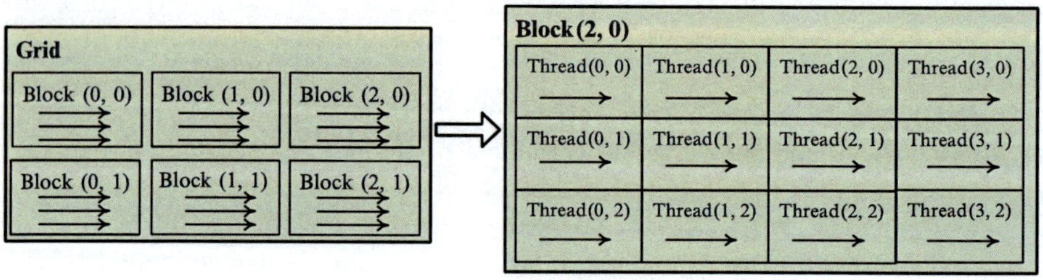

Fig. 28 Grid of thread blocks

one-dimensional, two-dimensional, or three-dimensional thread block, as illustrated in Fig. 28.

CUDA threads may access data from multiple memory spaces during their execution. Each thread has a private local memory. Each thread block has a shared memory visible to all threads of the block and with the same lifetime as the block. Finally, all threads have access to the same global memory. We used multiple threads per option to keep the GPU hardware efficiently occupied. We also used multiple thread blocks per option, in which case we have to get partial sums from each thread blocks, which in turn means that data transaction from shared memory to global memory is needed. Hence, we use a second kernel which uses a parallel reduction operation to compute the sums. A parallel reduction is a tree-based summation of

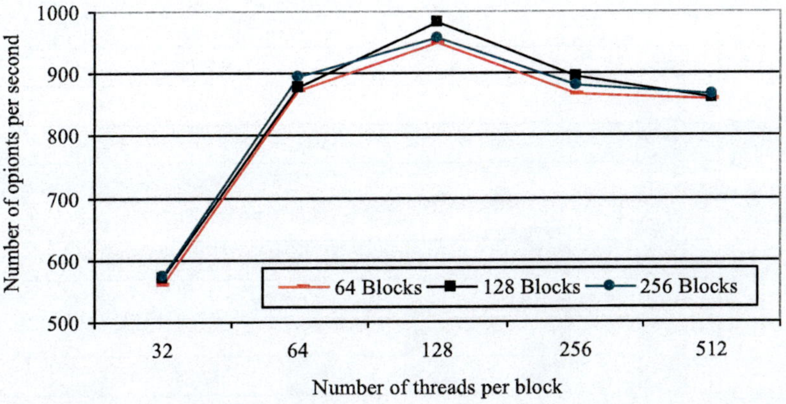

Fig. 29 Performance of our Quasi-Monte Carlo GPU implementation using different numbers of threads per block

values which takes $\log(n)$ parallel steps to sum n values. Parallel reduction is an efficient way to combine values on a data-parallel processor like a GPU; the larger the number of paths the better as this helps in hiding the latency of reading input values randomly.

Given a total number of threads per grid, the number of threads per block, or equivalently the number of blocks, should be chosen to maximize the utilization of the available computing resources. With a high enough number of blocks, the number of threads per block should be chosen as a multiple of 64, as the compiler and thread scheduler schedule the instructions as optimally as possible to avoid register memory bank conflicts, and the best results are achieved when the number of threads per block is a multiple of 64. After several experiments of multiple options pricing on our NVIDIA 8800GTX device, we found that using 128 thread blocks gives the best performance. Moreover, the performance is best when the number of threads per block is 128. Figure 29 shows the performance of our Quasi-Monte Carlo financial option pricing implementation when using different number of blocks and threads per block on the GPU.

We also wrote a C++ program of our Quasi-Monte Carlo simulation engine on the Maxwell machine and executed it on the Xeon processors. The most time consuming module of Quasi-Monte Carlo simulation core is the random number generator. Hence, we used a fully optimized library, namely Intel Math Kernel Library (MKL) [33, 34], to generate the Sobol sequence and transfer it to Gaussian noise using ICDF method (also provided by the MKL).

Figure 30 shows the execution time of our Quasi-Monte Carlo simulation engine on the Maxwell machine with different numbers of Xeon processors. As in the case of FPGAs, here also, the execution time scales linearly with the number of processors.

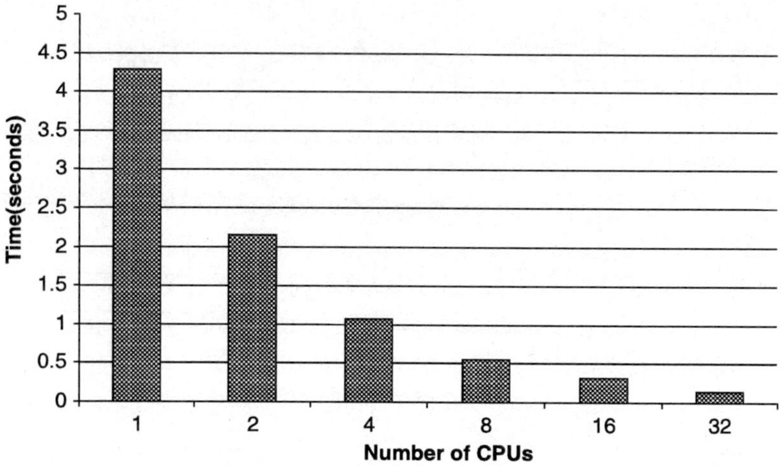

Fig. 30 Running time of our Quasi-Monte Carlo simulation engine on different number of Xeon processors

Table 12 Speed-ups of different platforms

	FPGA			GPU	
CPU	Fixed-26bits	Fixed-29bits	Fixed-32bits	Single Precision	Double Precision
1x	544x	448x	282x	162x	50x

To compare our three different implementation platforms, we ran our Quasi-Monte Carlo simulation engine to price a single option, using 524,288 simulation paths, on FPGA, GPU and GPP. This number of path is chosen as the precision can reach 10^{-4}. Moreover, it is a power of 2, which can benefit from the characteristic of Sobol numbers. We allot the same number of paths to each Quasi-Monte Carlo core on FPGA, Xeon CPU, or every thread and threads block on GPU. Although the optimized numbers of threads and threads blocks come from multiple options pricing, we still use the same parameters.

Table 12 shows comparative performance results between the FPGA, GPU and Xeon processor implementations, normalized to the Xeon CPU result. Here, the FPGA and Xeon implementations are both for a single node experiment, as we only use a workstation with a single GPU, and not a cluster of GPUs.

We can see significant speed-ups from both GPU and FPGA implementations. This is due to the high level of parallelism inherent in the Quasi-Monte Carlo simulation algorithm. Moreover, there is very limited conditional branching in the program, which is beneficial to both FPGAs and GPUs, especially the latter.

Apart from speed, we should also consider other factors when evaluating high performance computing implementations e.g. hardware and software cost, development time, power consumption, maintenance costs, technology maturity.

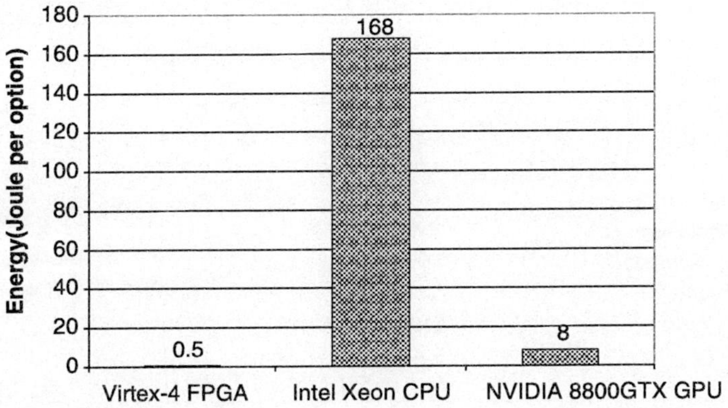

Fig. 31 Energy consumption in Joule

In the following, we will address one of these factors in our implementations, namely: power consumption.

We physically measured the power consumption of our FPGA, CPU and GPU implementations using a power meter, and deduced the energy consumption based on the execution times. Figure 31 gives the corresponding energy consumption results for each implementation. We can clearly see that FPGA offers the most energy efficient implementation, followed by the GPU, and then the CPU. Here FPGAs are 336x more energy efficient than CPUs, and 16x more energy efficient than GPUs.

5 Evaluation of Reconfigurable Hardware in High Performance Financial Computing

The evaluation of computational solutions in the literature is often focused on computation speed and accuracy. Nevertheless, when evaluating a high performance computing platform, several other metrics need to be considered. Those metrics include: the cost of equipment, development time, and power consumption. In fact, these metrics can all be valued in terms of only one metric: *Money*. Hence, to conclude this work in general, we consider the following aspects:

Equipment expense
Development expense
Energy expense

In this section, we perform a comprehensive comparison of the implementation of a Quasi-Monte Carlo simulation engine using three different computing platforms: FPGA, GPU, and GPP.

Table 13 Experimental parameters and results

	FPGA	GPU	GPP
Equipment cost	$10,000	$1,350	$1,000
Development time (days)	60	3	1
Development cost	$9600	$480	$160
Execution time	0.00787s	0.0858s	4.291s
Speed-up	545x	50x	1x
Dynamic power consumption	20W	95W	40W
Total power consumption	150W	225W	170W
Energy consumption	1.1805J	19.305J	729.47J
Annual energy cost	$197	$296	$223
Number of paths	524,288	524,288	524,288
Paths/second	66,618,551	6,110,583	122,183

5.1 *Evaluation of FPGA-Based Monte Carlo Simulation Engine*

In this comparison, we used the following device technologies: Xilinx Virtex 4 VFX100 FPGA, NVIDIA 8800GTX GPU, and Intel Xeon CPU 2.8GHz. Table 13 presents the experimental parameters and results. The equipment cost including both the expense for the host and the accelerator boards. The development cost is calculated using the following parameters: eight working hours per day and $20 per hour payment. The dynamic power consumption is the power measured at runtime, deducting the idle power. When calculating the energy consumed, we use the total power consumption. Annual energy costs are based on electricity price of $0.15 per kWh.

Several figures are plotted below to show the evaluation of the three implementations. The main metric used here is the number of paths per second, when normalized using development time, power consumption, and dollar expense. First, Fig. 32 shows the number of paths per second per development day using different platforms.

Results show that the GPU implementation gives the best result, followed by the FPGA implementation, and then the CPU one. Despite FPGA's high speed performance, its hardware description programming model is the most time consuming compared to GPU and CPU programming (which is essentially software programming), resulting in a lower performance per design effort compared to GPUs.

Considering power consumption, Fig. 33 presents the number of paths per second per Watt for each of the three implementations.

As we can see from Fig. 33, the normalized performance per Watt of the FPGA implementation outperforms the GPU and CPU implementations, respectively. Moreover, we note that although the power consumption of the GPU implementation is more than the CPU implementation, the former still beats the latter in terms of energy efficiency.

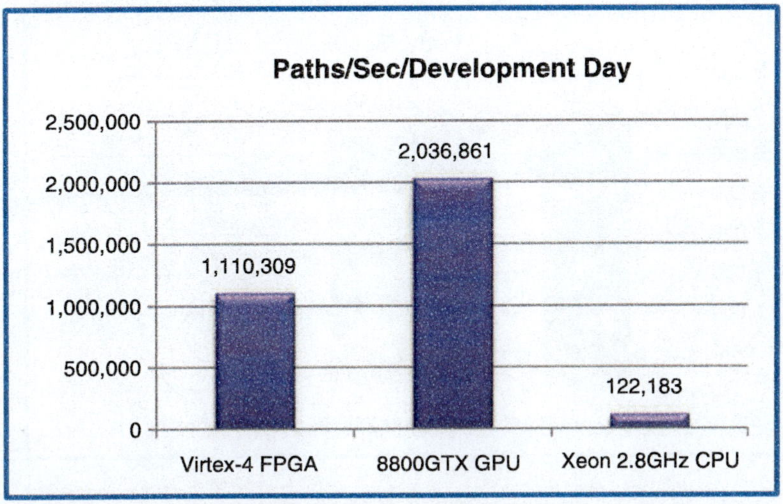

Fig. 32 Number of Paths/Sec/Development day

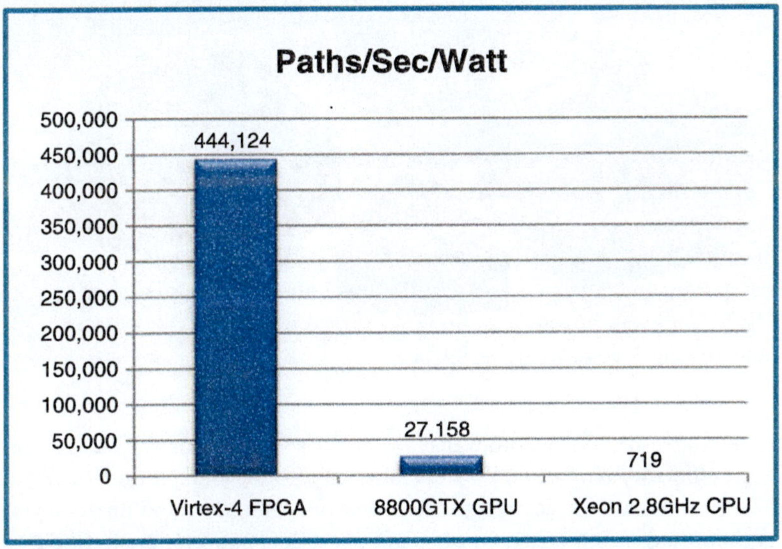

Fig. 33 Number of Paths/Sec/Watt

Considering the cost, Fig. 34 presents the performance normalized per total cost (purchase and development cost). As we can see, the normalized performance per cost of the FPGA and GPU implementations are very close. However, as in the above two figures, the normalized performance per cost of the CPU implementation is still much lower than the other two.

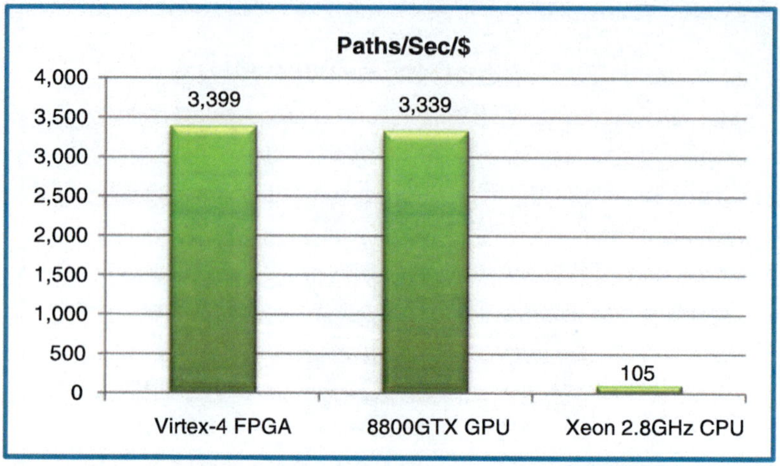

Fig. 34 Number of Paths/Sec/$ (purchase and development cost)

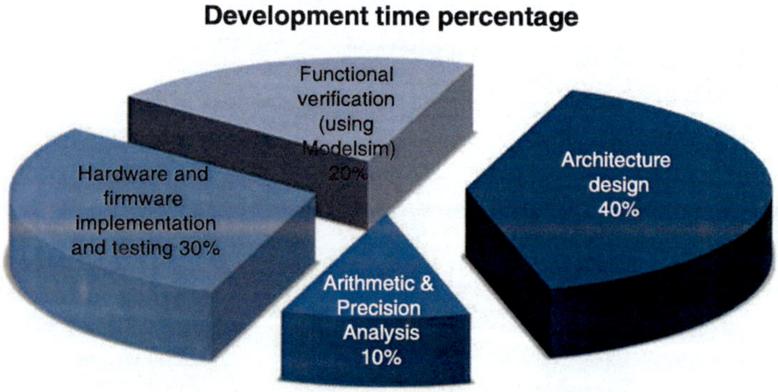

Fig. 35 Development time percentage on FPGAs

From the above results, we note that the main advantage of FPGAs resides in their energy efficiency. However, the development effort is a major drawback which impedes FPGAs' economic advantage. To have a deeper understanding of this issue, Fig. 35 presents a division of the whole FPGA solution's development time into its various steps.

From Fig. 35, we can see that the first three steps (namely arithmetic and precision analysis, architecture design, and verification) take 70% of the whole development time. This can be reduced in the future with high level design tools. Moreover, the hardware implementation and testing takes 30% of the whole development time. This could be reduced through the development of standard FPGA hardware boards with standard application programming interfaces (APIs).

6 Conclusion

Based on the work presented in this chapter, we can conclude that reconfigurable technology in the form of FPGAs has significant advantages compared to other technologies in high performance financial computing as it offers orders of magnitude speed-up compared to general purpose processors. Overall, FPGAs' main advantage lies in their high performance per watt, or energy efficiency. However, FPGAs' lack of high level programming tools and standard hardware and APIs is impeding the economic advantage of this technology, especially in comparison with GPU technology. Higher level programming tools and standard hardware and API platforms are necessary for further penetration of FPGA technology into high performance computing.

References

1. J. Dongarra, Trends in high performance computing: a historical overview and examination of future developments. IEEE Circ. Dev. Mag. 22, 22–27 (2006)
2. P. Marsh, High performance horizons. Comput. Contr. Eng. J. 42–48 (2004)
3. S.A. Zenios, High-performance computing in finance: the last 10 years and the next. Parallel Comput. 25, 2149–2175 (1999)
4. J.C. Hull, *Option, Futures, and Other Derivatives*, 4th edn. (Prentice Hall, Upper Saddle River, 2000)
5. G.L. Zhang, et al., Reconfigurable acceleration for Monte Carlo based financial simulation, in *Proceedings. 2005 IEEE International Conference on Field-Programmable Technology*, 2005, pp. 215–222
6. D. B. Thomas, et al., Hardware architectures for Monte-Carlo based financial simulations, in *FPT 2006. IEEE International Conference on Field Programmable Technology*, 2006, pp. 377–380
7. G.W. Morris, M. Aubury, Design space exploration of the european option benchmark using hyperstreams, in *FPL 2007. International Conference on Field Programmable Logic and Applications*, 2007, pp. 5–10
8. Q. Jin, et al., Exploring reconfigurable architectures for binomial-tree pricing models. Lect. Note Comput. Sci. 245–255 (2008)
9. A.R. Choudhury, et al., Optimizations in financial engineering: The Least-Squares Monte Carlo method of Longstaff and Schwartz, in *IPDPS 2008. IEEE International Symposium on Parallel and Distributed Processing*, 2008, pp. 1–11
10. I.L. Dalal, et al., Low discrepancy sequences for Monte Carlo simulations on reconfigurable platforms, in *International Conference on Application-Specific Systems, Architectures and Processors*, 2008, pp. 108–113
11. N.A. Woods, T. VanCourt, FPGA acceleration of quasi-Monte Carlo in finance, in *FPL 2008. International Conference on Field Programmable Logic and Applications*, 2008, pp. 335–340
12. J.H.C. Yeung, et al., Map-reduce as a programming model for custom computing machines, in *FCCM '08. 16th International Symposium on Field-Programmable Custom Computing Machines*, 2008, pp. 149–159
13. D.-U. Lee, et al., A hardware Gaussian noise generator using the Box-Muller method and its error analysis. IEEE Trans. Comput. **55**, 659–671 (2006)
14. D.-U. Lee, et al., A hardware Gaussian noise generator using the Wallace method, in *IEEE Transactions on Very Large Scale Integration (VLSI) Systems*, 2005, pp. 911–920

15. D.-U. Lee, et al., A hardware Gaussian noise generator for channel code evaluation, in *FCCM 2003. 11th Annual IEEE Symposium on Field-Programmable Custom Computing Machines*, 2003, pp. 69–78
16. D.-U. Lee, et al., A Gaussian noise generator for hardware-based simulations. IEEE Trans. Comput. 1523–1534 (2004)
17. R. Baxter, et al., Maxwell - a 64 FPGA Supercomputer, in *AHS 2007. Second NASA/ESA Conference on Adaptive Hardware and Systems*, 2007, pp. 287–294
18. R. Baxter, et al., The FPGA high-performance computing alliance parallel toolkit, in *AHS 2007. Second NASA/ESA Conference on Adaptive Hardware and Systems*, Edinburgh, 2007
19. M. Snir, S. Otto, *MPI-The Complete Reference*. (MIT Press, Cambridge, MA, 1998)
20. SUN. *Sun ONE Grid Engine Administration and User's Guide* [Online]. Available: http://www.sun.com
21. J.E. Gentle, *Random Number Generation and Monte Carlo Methods*, 2nd edn. (Springer, New York, 2003)
22. R.C. Tausworthe, Random numbers generated by linear recurrence modulo two. Math. Comput. **19**, 201–209 (1965)
23. M. Matsumoto, T. Nishimura, Mersenne twister: a 623-dimensionally equidistributed uniform pseudo-random number generator. ACM Trans. Model. Comput. Simulat. **8**, 3–30 (1998)
24. P. Bratley, et al., Implementation and tests of low-discrepancy sequences. ACM Trans. Model. Comput. Simulat. **2**, 195–213 (1992)
25. P. Jäckel, *Monte Carlo Methods in Finance* (Wiley, New York, 2002)
26. K. Entacher, et al., Defects in parallel Monte Carlo and quasi-Monte Carlo in tegration using the leap-frog technique. Int. J. Parallel Emergent Distributed **18**, 13–26 (2003)
27. G.E.P. Box, M.E. Muller, A note on the generation of random normal deviates. Ann. Math. Stat. **29**, 610–611 (1958)
28. O. Mencer, et al., Parameterized function evaluation for FPGAs, in *Proceedings of the 11th International Conference on Field-Programmable Logic and Applications*, 2001, pp. 544–554
29. O. Mencer, W. Luk, Parameterized high throughput function evaluation for FPGAs. J. VLSI Signal Process. **36**, 17–25 (2004)
30. J.-C. Duan, The garch option pricing model. Math. Finance **5**, 13–32 (1995)
31. F.A. Longstaff, E.S. Schwartz, Valuing American options by simulation: a simple least-squares approach. Rev. Financ. Stud. **14**, 113–147 (2001)
32. Nvidia, *NVIDIA CUDA Compute Unified Device Architecture Programming Guide*, 2.0 edn. (NVIDIA Corporation, 2008)
33. Intel. Intel® Math Kernel Library Reference Manual [Online]
34. Intel. Intel® Math Kernel Library Vector Statistical Library Notes [Online]. Available: http://developer.intel.com/

Bioinformatics Applications on the FPGA-Based High-Performance Computer RIVYERA

Lars Wienbrandt

Abstract Sequence alignment is one of the most popular application areas in bioinformatics. Nowadays, the exponential growth of biological sequence data becomes a severe problem if processed on standard general purpose PCs. Tackling this problem with large computing clusters is a widely accepted solution, although acquaintance and maintenance as well as space and energy requirements introduce significant costs. However, this chapter shows that this problem can be addressed by harnessing the high-performance computing platform RIVYERA, based on reconfigurable hardware (in particular FPGAs). The implementations of three examples of widely used applications in this area in bioinformatics are described: optimal sequence alignment with the Needleman–Wunsch and Smith–Waterman algorithm, protein database search with BLASTp, and short-read sequence alignment with a BWA-like algorithm. The results show a clear outperformance of standard PCs and GPU systems as well as energy savings of more than 90% compared to PC clusters, combined with the space requirements for one RIVYERA of only 3U–4U in a standard server rack.

1 Introduction

In bioinformatics computer scientists have to deal with the ever-growing amount of digital biological data stored in large sequence databases. Prominent examples are NCBI's Genbank database [22] and the UniprotKB/TrEMBL database [31], each observing an exponential growth in DNA sequence data and protein sequence data, respectively. Additionally, high-throughput sequencing data becomes available with the same growth in speed, strengthening the difficulties of a just-in-time data

L. Wienbrandt (✉)
Department of Computer Science, Christian-Albrechts-University of Kiel, Germany
e-mail: lwi@informatik.uni-kiel.de

W. Vanderbauwhede and K. Benkrid (eds.), *High-Performance Computing Using FPGAs*, 81
DOI 10.1007/978-1-4614-1791-0_3, © Springer Science+Business Media, LLC 2013

analysis. In fact, for many problems related to bioinformatics a single up-to-date standard PC already requires an unreasonable amount of processing time, although CPU technology has grown as well.

To keep up with this rising demand on computational power in bioinformatics, the focus is set to parallel processing. Many core CPUs already provide some degree of parallelism, which can be extended almost infinitely using computer clusters. However, with a linear growth in cluster size, the costs for acquisition, energy, and maintenance grow linearly as well.

This chapter is focused on the massively parallel utilization of FPGAs to address this problem. FPGAs provide an ASIC-like performance combined with an extensive degree of on-die parallelism according to its configuration and low energy requirements. However, the performance of bioinformatics applications is highly dependent on communication bandwidth and size of available memory as well. These problems have been addressed with the design of the RIVYERA architecture. RIVYERA provides the resources of 128 FPGAs (Xilinx Spartan6-LX150 in RIVYERA S6-LX150) connected by a high-throughput systolic bus system, and a significant amount of DRAM (512 MB up to 2.5 GB in RIVYERA S6-LX150) attached to each FPGA. These are ample resources to address most of the big problems in bioinformatics.

The RIVYERA architecture is shortly introduced in the next section (Sect. 2). In the following sections, three major applications are presented and implemented on the RIVYERA architecture: optimal sequence alignment using the Smith–Waterman algorithm (Sect. 3.1), database searches with BLAST (Sect. 3.2), and BWA-like short-read sequence alignment (Sect. 3.3).

2 RIVYERA S3-5000 and RIVYERA S6-LX150 Computing Platforms

The hardware platform RIVYERA was introduced in 2008 [25] and includes a completely reworked communication and memory infrastructure, compared to its predecessor COPACOBANA, introduced in 2006 for applications in cryptanalysis [12].

For the applications described here (see Sect. 3) two specific types of RIVYERA are used, the RIVYERA S3-5000 for the Smith–Waterman and BLAST applications (Sects. 3.1 and 3.2) and the more recent RIVYERA S6-LX150 for the BWA application (Sect. 3.3).

The RIVYERA architectures are now developed and distributed by SciEngines GmbH [26]. They have a common basic structure consisting of two elements, the in-built multiple FPGA-based supercomputer and a standard server grade mainboard, running a Linux operating system, e.g. equipped with an Intel Core i7-930 processor with 12 GB of RAM. The FPGA supercomputer is different for each RIVYERA type. The RIVYERA S3-5000 is equipped with up to 128 user configurable Xilinx Spartan3-5000 FPGAs, distributed over 16 FPGA cards, each

Fig. 1 The RIVYERA S6-LX150 system

containing 8 user FPGAs. Each user FPGA is directly attached to a DRAM module with a capacity of 32 MB. In contrast, the RIVYERA S6-LX150 consists of up to 128 Xilinx Spartan6-LX150 user FPGAs and attaches 512 MB DDR3-RAM to each user FPGA per default. However, the available memory can be extended with a memory extension module providing an additional amount of 2 GB DDR3-RAM per FPGA.

The high-performance bus system offered by RIVYERA is organized as a systolic chain, i.e. every FPGA on an FPGA card is connected by fast point-to-point connections to each neighbor forming a ring. All FPGA cards are connected in the same manner by an additional communication controller attached to each card. The PC mainboard in RIVYERA is connected via PCIe to at least one communication controller on one FPGA card forming a high-speed uplink to the host.

An intelligent routing scheme for this bus system has been implemented in an API to ensure usability. The API includes the communication interface for software and hardware to provide a transparent connection for the developer between host and FPGA or any two FPGAs by an automatic routing of data packets. The API includes broadcast facilities, methods for configuring the user FPGAs and a communication interface for the FPGA-attached DRAM.

RIVYERA allows a small packaging. RIVYERA S3-5000 is packed in a standard rack mountable $3U$ housing while the recent RIVYERA S6-LX150 is slightly larger, but still requires only $4U$. For more details on the RIVYERA architecture and its predecessor COPACOBANA, see Chap. 11. A picture of RIVYERA S6-LX150 is shown in Fig. 1.

3 Bioinformatics on the RIVYERA Architecture

The following sections address implementations of three major bioinformatics applications regarding sequence analysis. The well-known Needleman–Wunsch [24] and Smith–Waterman [27] algorithms to generate an optimal sequence alignment are presented in Sect. 3.1. RIVYERA S3-5000 is able to accelerate these applications to a speed of more than 3 TCUPS. Protein database searches can be accelerated significantly as well using the BLASTp implementation for the NCBI-BLAST [21] framework described in Sect. 3.2. The first bioinformatics application addressing the recent RIVYERA S6-LX150 architecture directly is the BWA-like short-read sequence alignment described in Sect. 3.3. This implementation, completely integrated into the BWA [14] framework, again outperforms common CPU and GPU implementations.

3.1 Optimal Sequence Alignment with Smith–Waterman and Needleman–Wunsch

Biological sequence alignment deals with the problem of finding the best way to align two nucleotide or protein sequences to each other. The more equal nucleotides or similar amino acids face each other in the alignment the higher is the achieved score. Additionally, *gaps* (insertions or deletions, sometimes referred to as *indels*) are allowed to be inserted with *linear* or *affine* costs.

An alignment is considered *optimal* if its score is the maximum achievable for the input sequences according to a previously selected scoring matrix. Figure 2 shows an example for an optimal local alignment of two short nucleotide sequences using the NUC44 scoring matrix and an affine gap penalty.

Algorithms generating alignments may be classified as heuristic or non-heuristic. Heuristic algorithms such as BLAST [1] or BWA [14] may produce a large amount of results but cannot guarantee to find the optimal alignment. Since they generally outperform non-heuristic types by far, they are commonly used for database searches (see Sect. 3.2) or short-read alignment (see Sect. 3.3). However, exactness is necessary for several applications as well. Especially many heuristic algorithms require an optimal alignment in their postprocessing.

Algorithms providing an optimal alignment are the Needleman–Wunsch [24] and the Smith–Waterman [27] algorithms. Both work similarly with the difference that Needleman–Wunsch finds optimal *global* alignments while Smith–Waterman finds optimal *local* alignments. In this section it is demonstrated how RIVYERA S3-5000 is used for the Smith–Waterman algorithm.

3.1.1 The Needleman–Wunsch and Smith–Waterman Algorithms

The Needleman–Wunsch [24] and Smith–Waterman [27] algorithms are capable to find the *optimal* alignments of two sequence to each other. In general, the major goal

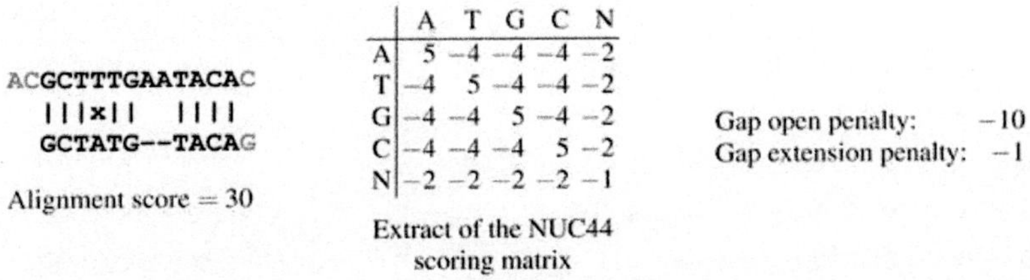

Fig. 2 Optimal local alignment of the two nucleotide sequences ACGCTTTGAATACAC and GCTATGTACAG using the NUC44 scoring matrix and affine gap penalty

is to find similar occurrences of one (shorter) sequence in another (long) sequence. Hence, for convenience, one sequence is referred to as *query* sequence q while the other one is called *subject* or *database* sequence s. Additionally, q_i denotes the symbol at position i in the query sequence q, likewise s_i for the subject sequence s.

In order to find an optimum out of every possible alignment, an alignment matrix $H_{n\times m}$ is generated, whereby n and m are the lengths of the query and subject sequence, respectively. The Smith–Waterman algorithm calculates the matrix cells according to the following simple scoring function (g denotes a linear gap penalty, S denotes the scoring matrix, e.g. NUC44):

$$H_{i,j} = \max \begin{cases} H_{i-1,j-1} + S(q_i, s_j) & \text{match/mismatch} \\ H_{i-1,j} + g & \text{insertion opening/extension} \\ H_{i,j-1} + g & \text{deletion opening/extension} \\ 0 & \text{do not allow negative values} \end{cases} \quad (1)$$

The Needleman–Wunsch algorithm performs likewise with the only difference that negative values are allowed in the alignment matrix. Hence, the scoring function simply misses the comparison with 0 from the maximum condition in (1). For an affine gap penalty, g becomes dependent on the condition of the gap, if it is either a gap opening or an extension.

After the calculation of the alignment matrix, a backtracking step is performed to generate the final alignment. Summarized, the backtracking starts at the lower right corner $H_{n,m}$ of the alignment matrix (Needleman–Wunsch) or at a matrix cell $H_{i,j}$ with the highest value (Smith–Waterman). The respective cell entries already state the score of the final alignment. The backtracking follows the path through the alignment matrix which reflects the chain of matrix cells whose values were taken for the maximum calculation in (1). For each chosen direction, either *up*, *left*, or *up-left*, the corresponding character or gap is inserted into the final alignment. The backtracking stops if either the upper left corner $H_{0,0}$ (Needleman–Wunsch) or a cell with $H_{x,y} = 0$ (Smith–Waterman) is reached. For a detailed description of the backtracking step, the original publications [24, 27] are referred to. Figure 3 illustrates the alignment matrix, the backtracking step and the final alignment of the example in Fig. 2.

i \ j	-	A	C	G	C	T	T	T	G	A	A	T	A	C	A	C
-	0	0	0	0	0	0	0	0	0	0	0	0	0	0	0	0
G	0	0	0	5	0	0	0	0	5	0	0	0	0	0	0	0
C	0	0	5	0	10	0	0	0	0	1	0	0	0	5	0	5
T	0	0	0	1	0	15	5	5	3	2	1	5	0	0	1	0
A	0	5	0	0	0	5	11	1	1	8	7	0	10	0	5	0
T	0	0	1	0	0	5	10	16	6	5	4	12	2	6	0	1
G	0	0	0	6	0	3	1	6	21	11	10	9	8	7	6	5
T	0	0	0	0	2	5	8	6	11	17	7	15	5	4	3	2
A	0	5	0	0	0	1	1	4	10	16	22	12	20	10	9	8
C	0	0	10	0	5	0	0	3	9	6	12	18	10	25	15	14
A	0	5	0	6	0	1	0	2	8	14	11	8	23	15	30	20
G	0	0	1	5	2	0	0	1	7	4	10	7	13	19	20	26

Fig. 3 Smith–Waterman alignment matrix and backtracking for the alignment of the sequences `ACGCTTTGAATACAC` and `GCTATGTACAG` (see example in Fig. 2)

3.1.2 Implementation of Smith–Waterman

Time and memory requirements for the calculation of the alignment matrix are clearly of complexity $\mathcal{O}(n \cdot m)$. This implies long runtimes for large datasets if processed subsequently, e.g. on a standard PC. For fine-grained parallel processing, possible on e.g. FPGAs, the runtime complexity can be reduced to $\mathcal{O}(n)$ if $n > m$, i.e. linear to the database sequence.

Considering a regular biologist's workflow, it is likely that a huge amount of alignments have to be processed, whereby the user is not interested in the alignment itself, but primarily in the quality of the alignment. Hence, it is not necessary to save the alignment matrix since the backtracking step is omitted if the alignment score is beneath a certain threshold. This reduces the memory requirements to $\mathcal{O}(m)$, i.e. linear to the length of the query sequence, if the database sequence is assumed not to be stored in memory.

However, if the final alignment is required anyway, valuable runtime and memory resources can still be saved. This presumes the matrix position of the alignment score to be saved as well. Then, the final alignment can be created in a postprocessing step considering only the necessary subsequences to perform the complete Smith–Waterman algorithm.

The parallelization is realized as follows, considering a query sequence q of length m. For every nucleotide or amino acid in the query sequence a Smith–Waterman cell *SWcell* is implemented on the FPGA. Hence, each cell has to be

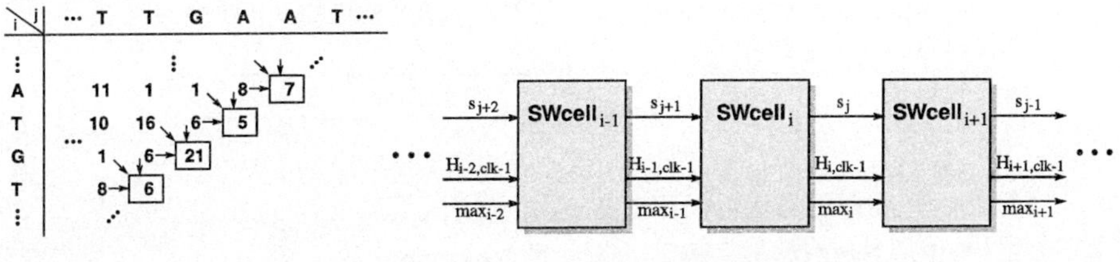

Fig. 4 Smith–Waterman chain structure and example for the calculation of an alignment matrix on an FPGA

initialized with the corresponding sequence character before an alignment starts. Its task is to calculate the values in the row of the alignment matrix corresponding to its assigned character, i.e. a direct implementation of (1), whereby i, the index of the row, will be fixed for each cell.

The aim is to calculate a matrix value in every SWcell in every clock cycle. Since for every calculation three already calculated values are required (up, left and up-left of the current matrix cell), all SWcells are able to compute an anti-diagonal of the alignment matrix in every clock cycle if all SWcells are connected in a chain. Then, the three values are accessed in the following way:

- The upper value $H_{i-1,j}$ is the calculated matrix entry from the previous clock cycle in the previous row (and therefore "left" neighboring cell).
- The left value $H_{i,j-1}$ is the matrix entry from the previous clock cycle in the same row (and therefore the same cell as well).
- The upper-left value $H_{i-1,j-1}$ is the matrix entry from two previous clock cycles in the previous row (left neighboring cell).

Since the processing is accomplished in anti-diagonals of the alignment matrix, the required database sequence can be streamed through the chain of SWcells character by character as well. Besides the calculation of (1) each cell has to store the current maximum matrix entry for the final Smith–Waterman score of the alignment, and its position to allow backtracking in a postprocessing step. The maximum is forwarded through the chain as well such that the final score can be determined at the end of the alignment from the last SWcell in the chain. Figure 4 shows a calculation step of the alignment matrix and a part of the chain structure.

3.1.3 Performance Evaluation

According to a particular alignment problem, the presented application can be implemented in several ways. For a short-read alignment experiment, the Smith–Waterman algorithm has been implemented with an SWcell chain of length $m = 100$. The configuration supports 5 nucleotide symbols (A, C, G, T, and N) and a scoring matrix of 5 bits per entry.

Table 1 Smith–Waterman performance on RIVYERA S3-5000 for DNA and protein sequence alignment

Architecture	DNA/prot.	Speed (GCUPS)
RIVYERA S3-5000	DNA	3,050
CLCbio Xeon X3210 @ 2.13 GHz (8 cores)	DNA	45
CLCbio Core2Duo @ 2.17 GHz (2 cores)	DNA	13
RIVYERA S3-5000	Protein	1,500
IBM QS20 blade (2x CellBE @ 3.2 GHz)	Protein	33
CUDASW++2.0 (GeForce GTX280)	Protein	17
Sony PS-3 (1x CellBE @ 3.2 GHz)	Protein	12

With this configuration, four chains fit onto a single Spartan3-5000 FPGA of the RIVYERA S3-5000. This sums up to 512 chains available on the whole RIVYERA, i.e. 512 queries to be processed concurrently. Hence, the alignment of a test set of 1 million Illumina 100 bp paired reads against the human genome (*hg19*, 3.2 Gbp) required only about 29 h. Since Smith–Waterman performance is often measured in *CUPS* (cell updates per second), i.e. how many cells of an alignment matrix are calculated per second, this leads to a speed of 3.05 TCUPS. Out of numerous available Smith–Waterman implementations, these results were compared to the speed of a commercial software solution for PCs and clusters provided by CLCbio [4] in Table 1.

The implementation can easily be adapted to generate protein sequence alignments, supporting 24 amino acid symbols and scoring matrix entries of up to 6 bits. This is sufficient for most available scoring matrices for protein alignments. With an SWcell chain length of $m = 100$ again, two chains can still be implemented on one Spartan3-5000 FPGA, leading to a concurrent processing of 256 queries on a whole RIVYERA S3-5000 machine. Of course, the performance is expected to be slower than for nucleotide sequence alignment now, but still 1.5 TCUPS are measured. Again, for comparison, numerous implementations are available. The RIVYERA performance was being compared to two other acceleration architectures. One is the Cell Broadband Engine processor (IBM QS20 blade with 2x CellBE @ 3.2 GHz and Sony PS-3 with 1x CellBE @ 3.2 GHz) [6], the other is a general purpose GPU (nVidia GTX280 running CUDASW++2.0) [19]. The results are stated in Table 1 as well.

Summarized, the massively parallel FPGA implementation of the Smith–Waterman algorithm using the RIVYERA S3-5000 outperforms other actual architectures by far.

3.2 *BLAST Database Search*

The Basic Local Alignment Search Tool (BLAST) [1] was originally developed to speed up biological database searches to find sequence similarities, which is primarily based on biological sequence alignment. Currently, NCBI provides one of the most commonly used versions [21], significantly improved by the *two-hit method* and a gapped alignment strategy [2]. BLAST generates alignments for either DNA (BLASTn) or protein (BLASTp) sequences. Several other variations like BLASTx, tBLASTn, tBLASTx or PSI-BLAST exist, but all with a similar core algorithm. This section mainly focuses on BLASTp for protein sequence alignments.

In contrast to the Needleman–Wunsch [24] or Smith–Waterman [27] algorithm described in the previous section (see Sect. 3.1), BLAST is heuristic. This leads to a significant runtime reduction with the drawback of losing alignment quality, since BLAST does not guarantee to find the optimum result. Several enhancements, such as the two-hit method and gapped BLAST [2], have further reduced computation time and improved result quality. However, with today's exponential growth of databases, BLAST reaches its limits on standard PC architectures especially for large query sets.

Recent development addresses alternative architectures, e.g. CUDA-BLASTp [18] utilizing general purpose GPUs with a speedup of up to 6 on an nVidia GeForce GTX 280 graphics card compared to a single CPU-thread of an Intel Core i7-920. Others provide single FPGA-based implementations such as Mahram and Herbordt [20], Kasap et al. [10] and Mercury BLASTp by Jacob et al. [9]. Two approaches are available for multiple FPGAs using the RIVYERA S3-5000 architecture [32, 33], whereby the latter has been developed to remove several bottlenecks detected in the dataflow of the former design, and will be described in the following.

3.2.1 BLAST Algorithm

The BLAST algorithm is organized in several steps, which will be explained shortly in the following. For details, it is refered to the original publications [1,2].

In the first step, the query sequence is preprocessed to identify its *neighborhood*. The neighborhood contains a list of short sequences of size k (k-mers) which are similar to k-mers of the query sequence, according to a scoring matrix (such as BLOSUM62) and a predefined threshold value. For BLASTp, k is fixed to $k = 3$. The value k is fixed, but different for either BLASTn ($k = 11$) or BLASTp ($k = 3$). A k-mer is declared similar to a k-mer of the query sequence if the score of a direct comparison, calculated according to a scoring matrix (such as BLOSUM62), exceeds a predefined threshold value.

The next step simply locates *hits*, i.e. exact matches of neighborhood words found in the database sequences. The hits are tested pairwise in the two-hit method if both hits of a pair hold the same distance to each other in the query sequence and in the subject sequence. The pair is then referred to as *two-hit*. To save runtime and memory the distance between the hits in a pair is bounded to a certain parameter A. Overlapping hits are omitted by applying the value k as lower bound. The following

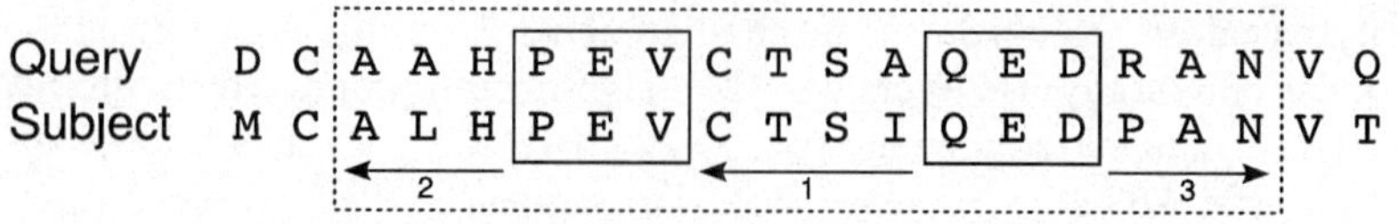

Fig. 5 Example for the ungapped extension of a two-hit in the NCBI BLAST implementation. The *solid rectangles* mark the hit pair, the *dashed* an extension. *Arrows* indicate the direction and the attached numbers, the order of the extensions

equation shows the condition for a two-hit whereby s_0 and s_1 state the location of two hits in the subject and q_0 and q_1 their locations in the query, respectively:

$$k \leq q_1 - q_0 = s_1 - s_0 < A \tag{2}$$

Each two-hit is further examined by an *ungapped extension* process. Both hits of a hit pair are extended forward and backward by calculating a similarity score of the current part of the subject and query sequence. In detail, the similarity score is firstly calculated for the hit pair itself and the gap between it. Then, the calculation of the score is extended residue by residue from the first hit of the pair to the left and afterward from the second hit to the right (in positional order). The calculation stops for each direction if the score declines a certain cutoff distance below the so far reached maximum. This method is referred to as *X-drop* mechanism. The *high-scoring pair* (HSP) of this extension, i.e. the two positions where the maximum score has been reached for each direction, states the result of this process. Figure 5 shows an example.

The last step in the BLAST algorithm states the *gapped extension*. To introduce gapped alignment, HSPs are analyzed by a slightly modified version of the Needleman–Wunsch algorithm [24]. First, the alignment is bound to the positional range of the ungapped extension, and second, in contrast to the original Needleman–Wunsch algorithm, the score of the alignment is stated by the maximum cell value rather than the value of the lower right corner of the alignment matrix. If a traceback is required to complete the final alignment (depending on the alignment score), it starts at the matrix cell with the calculated maximum as well. Runtime is reduced as well by using the X-drop mechanism again, i.e. omitting the calculation of matrix cell values where the score declines below a certain cutoff distance from the so far calculated maximum cell value.

3.2.2 Application Structure and Implementation

The implementation of BLASTp for the RIVYERA architecture is divided into two parts, the hardware and the software part. For transparency, the software is completely integrated in the original NCBI BLASTp v2.2.25+, including the user interface. The main and most compute intensive routines performing the core algorithm have been ported to the BLASTp pipeline implemented on the 128 Xilinx

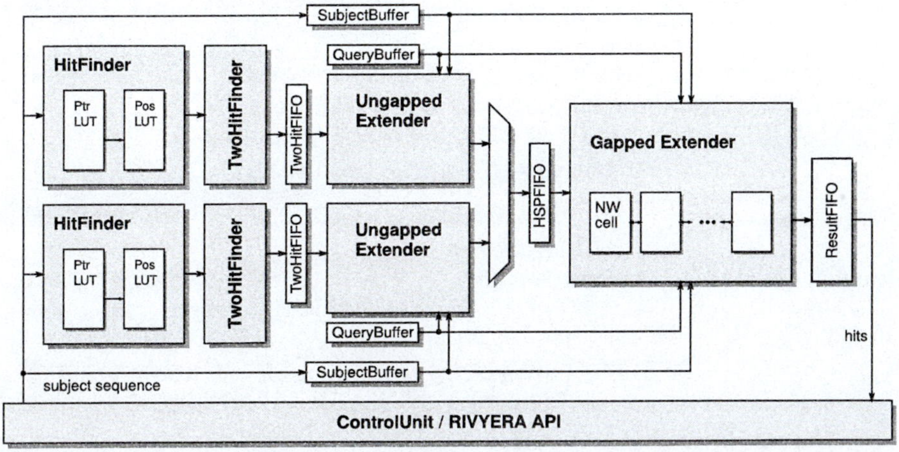

Fig. 6 Structure of two BLASTp hardware pipelines sharing one GappedExtender component

Spartan3-5000 FPGAs of the RIVYERA S3-5000 machine and thus replaced by a communication interface on the software side. The software is still responsible for pre- and postprocessing as well as controlling the BLASTp pipelines on the RIVYERA machine.

A BLASTp pipeline basically consists of four main components, the *HitFinder*, the *TwoHitFinder*, the *UngappedExtender*, and the *GappedExtender*, each representing one pipeline stage according to the processing steps in the BLAST algorithm (see Sect. 3.2.1). Figure 6 shows an overview of the FPGA implementation with two BLASTp pipelines.

The preprocessing of the queries includes the generation of the neighborhood and, if required, the splitting of long query sequences. For this implementation, a maximum query length of $1024 - A$ is supported directly (see (2) for definition of A, default $A = 40$), while longer queries are splitted automatically by the NCBI software routines. After the initialization and preprocessing phase, the database sequences are broadcasted to the FPGAs as stream while the BLASTp pipelines search for suitable alignments.

The HitFinder searches for occurences of k-mers of the subject sequence in the neighborhood, which is a simple look-up in a hashtable, organized in two separate tables.

Afterward, testing all possible pairs of hits to hold the condition for a two-hit (see (2)) results intuitively in a quadratic runtime complexity. With an easy strategy, basically consisting of a storage array for hit positions of a size corresponding to the query length, the runtime complexity can be reduced to linear.

First, an array of length $l = 1024$ is required. This corresponds to the maximum query length plus the parameter A for the bounds of (2). This array stores at position p the most recent subject position s_0 to the corresponding query position q_0. The position p is calculated from the following equation:

$$p = (s_0 - q_0) \bmod 1024 \tag{3}$$

Before inserting a new position, the content of the array cell is read. If this cell contains a valid subject position s_1, it holds $s_0 > s_1$ and:

$$s_0 - q_0 = s_1 - q_1 \pmod{1024} \tag{4}$$

$$\Leftrightarrow \quad s_0 - s_1 = q_0 - q_1 \pmod{1024} \tag{5}$$

Assuming $s_0 - s_1 < A$, it follows from (5):

$$s_0 - s_1 = \begin{cases} q_0 - q_1 & \text{if } q_0 \geq q_1 \\ 1024 - (q_1 - q_0) & \text{if } q_0 < q_1 \end{cases} \tag{6}$$

The second case (assuming $q_0 < q_1$) results in:

$$A > 1024 - (q_1 - q_0) \tag{7}$$

$$\Leftrightarrow \quad q_1 - q_0 > 1024 - A \tag{8}$$

This stays in contradiction to the bounds of the query length which is $l = 1024 - A$. Hence, if $s_0 - s_1 < A$ it directly follows $s_0 - s_1 = q_0 - q_1$, and if (2) holds, i.e. $k \leq s_0 - s_1$, this result is reported as a two-hit and buffered in a FIFO before processed further by the UngappedExtender component.

This method might be problematic if hits arrived unordered, i.e. if $s_1 > s_0$. However, this possibility can be counted out since the HitFinder provides hits only in ascending query positions, followed by an ascending order of the subject positions.

All resulting two-hits are buffered in a FIFO before processed further by the UngappedExtender. The ungapped extension process conforms to the order indicated in Fig. 5. In every clock cycle the score of a pair of residues from the query and the subject sequence is calculated using a scoring matrix, e.g. BLOSUM62, implemented in a dual-ported ROM. The process starts with the right hit of the hit pair and is directed left. The determined score is summed up continuously to the total score. The X-drop mechanism is implemented by checking the new calculated score in every clock cycle. If it drops a predefined cutoff distance below the so far calculated maximum score, the process stops, and the position of the current maximum score is stored. After finishing the left direction, the extension continues directed right from the right hit of the hit pair storing a new maximum position.

Finally, both stored maximum positions from each direction form a HSP, which is reported to the GappedExtender if its score exceeds another predefined threshold.

The UngappedExtender component contains a feedback path, controlling the elements stored in the preceding FIFO. According to the current progress, a pending two-hit may already be included in the running extension process. The UngappedExtender is able to remove such two-hits in advance to prevent the same extension with different starting points processed several times.

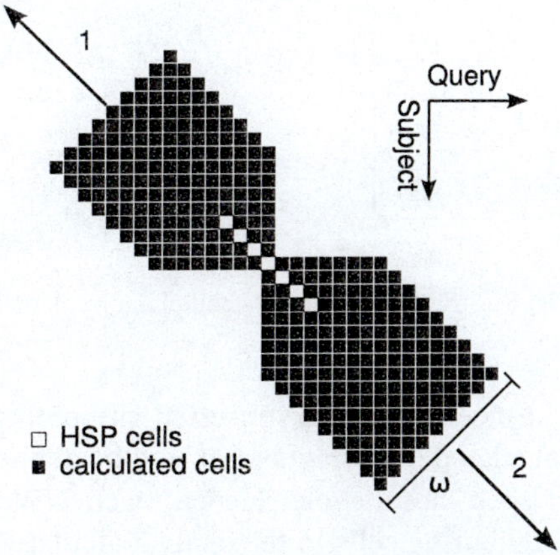

Fig. 7 Principle of the gapped extension of a high-scoring pair (HSP). *White cells* in the *middle* indicate the HSP, *black cells* the calculated cells in the Needleman–Wunsch alignment. *Arrows* with attached numbers indicate the direction and the order of the extension. The extension uses the X-drop mechanism to stop

The GappedExtender component basically performs a modified Needleman–Wunsch alignment with a banded matrix and a HSP at its center. In contrast to Mercury BLASTp [9], this implementation is kept close to the one in NCBI BLASTp using the X-drop mechanism to stop the extension process. However, the width of the matrix band is fixed to $\omega = 64$, but the length of the matrix band stays variable.

To create the alignment matrix with the HSP in the center it is necessary to do the calculation in two steps. The alignment starts at the center of the HSP and first, extends backward, using the reverse sequences for Needleman–Wunsch. Afterward, a forward directed alignment, again starting from the HSPs center, is performed in the same way. The original HSP is reported to the host software if the sum of both alignment scores exceed a predefined report threshold. The structure of this process is illustrated in Fig. 7.

Similar to the chain of SWcells for the Smith–Waterman algorithm (Fig. 4 in Sect. 3.1), the subcomponents of the GappedExtender component basically consist of ω *NWcells* connected in a chain. Since the calculation is restricted to a banded matrix now, a pre-initialization of the chain with the query sequence is impossible. Instead, the part of the query sequence, which is to be analyzed, is inserted from the one end of the chain, while the corresponding part of the subject sequence is inserted from the other end, alternating with every clock cycle.

The calculation of the score of a cell $H_{i,j}$ in the alignment matrix corresponds to the following equation, similar to (1) in Sect. 3.1. In the implementation an affine gap penalty is used whereby it is omitted here for simplicity (g denotes a linear gap penalty, S denotes the scoring matrix, e.g. BLOSUM62):

$$H_{i,j} = \max \begin{cases} H_{i-1,j-1} + S(q_i, s_j) & \text{match/mismatch} \\ H_{i-1,j} + g & \text{insertion opening/extension} \\ H_{i,j-1} + g & \text{deletion opening/extension} \end{cases} \quad (9)$$

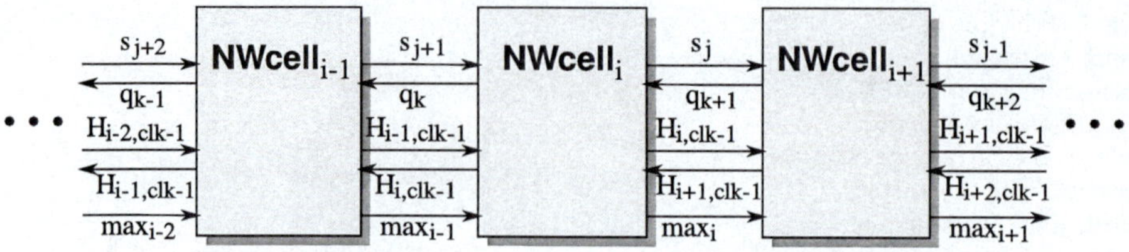

Fig. 8 Structure of the NW cell chain implemented in the GappedExtender component

Since residues are inserted alternating from both ends of the NWcell chain, the calculated anti-diagonal of width ω "moves" alternating rightward and downward in each clock cycle. Hence, each NWcell requires access to the scores of *both* neighboring cells in the chain, calculated in the previous clock cycle. The structure of an NWcell chain is depicted in Fig. 8.

The gapped extension step in hardware acts as an additional filter to keep the number of reports small. If a HSP passes the gapped extension filter, the exact alignment including the backtracking is generated on the host by the original NCBI routines. This way, valuable software runtime is saved by filtering nearly every HSP in advance in hardware, which would be omitted by the gapped extension of the host software anyway.

Before being fetched by the host software, the reports for each FPGA are collected in the attached DRAM. This way, the number of communication interruptions for the submission of reports during the core process can be kept small.

3.2.3 Performance Evaluation

Targeting the Xilinx Spartan3-5000 FPGAs of the RIVYERA 3-5000 machine, one FPGA provides the resources to hold two BLASTp pipelines as described above, but sharing one GappedExtender (s. overview in Fig. 6). This is easily possible, since the GappedExtender is the most resource occupying component but utilized the most infrequently as well. Therefore, the width of the matrix band is set to $\omega = 64$. Hence, a fully equipped RIVYERA S3-5000 is able to process 256 queries concurrently.

The reference system for the performance evaluation was a PC system equipped with two Intel Xeon E5520 CPUs, each containing 4 cores (8 threads) running at 2.26 GHz, 48 GB DDR3-RAM, and 64 bit Linux OS. The software is NCBI BLASTp v2.2.25+ with default parameters, BLOSUM62 scoring matrix and a varying number of threads (16 and 8, "`-num_threads`" switch). Three different query sets from *SUPERFAMILY* database [30] were tested (proteomes of *Arabidopsis thaliana*, *Populus trichocarpa*, and human (*Homo sapiens*)), each set randomly reduced to 2,335, 3,151, and 1,990 sequences, respectively, such that they contain about one million residues each. The reference database was the first part of the NCBI *RefSeq* BLAST database, release 50, containing 2,996,372 sequences ($\approx$1 billion residues).

Table 2 BLASTp runtimes (in seconds) of three randomly reduced query sets against part one of the NCBI *RefSeq* database

	RIVYERA (*n* FPGAs)					2x Xeon E5520		Mercury	CUDA
Query set	128	64	32	16	8	16 thr.	8 thr.	BLASTp	BLASTp
A. thaliana	353	648	1,106	1,934	3,531	8,301	9,995	3,780*	7,780*
P. trichocarpa	482	808	1,323	2,309	4,210	10,226	12,506	5,161*	9,615*
H. sapiens	561	987	1,723	2,817	4,409	9,464	11,602	6,007*	8,026*

The 2x Xeon E5520 reference system runs NCBI BLASTp v. 2.2.25+. The marked (*) runtimes are estimations calculated from published runtimes extrapolated to the changed database and query set

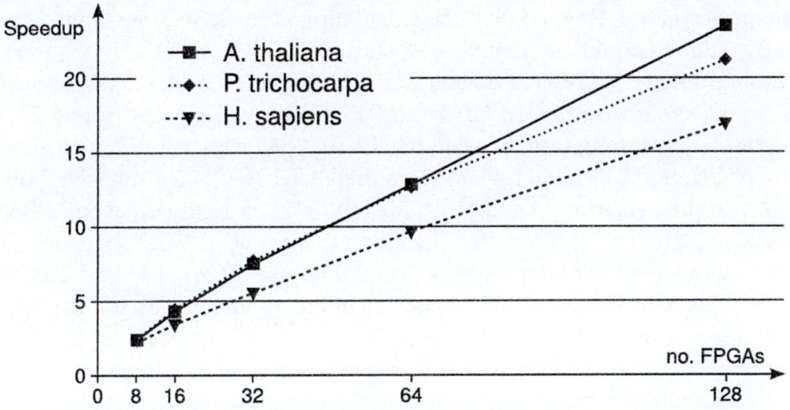

Fig. 9 BLASTp speedups of RIVYERA S3-5000 with different number of utilized FPGAs vs. 2x Xeon E5520 (16 threads)

All results are stated in Table 2. It shows that a fully equipped RIVYERA S3-5000 clearly outperforms the reference with a speedup of up to 23.5, i.e. about 376 against a single CPU thread. Hence, the runtime performance of one single FPGA conforms to about three CPU threads. Additionally, Fig. 9 illustrates the speedups of RIVYERA with a different number of utilized FPGAs versus the reference system, showing an approximately linear increase of speed with an increasing number of FPGAs.

The stated runtimes of Mercury BLASTp [9] and CUDA-BLASTp v2.0 [18] have been linearly extrapolated from the best results in the respective publications. Due to the lack of hardware for Mercury BLASTp and a non-functional CUDA-BLASTp on an nVidia GeForce GTX480 GPU, no real measurements could be made. Since the runtime is extremely dependent on the quality of the query, these results are only to be seen as a rough estimate. However, regarding these estimations, RIVYERA still outperforms these solutions as well.

Table 3 shows the energy consumption for the query sets measured with a customary power measurement device. The measured energy consumption of a fully equipped RIVYERA is only 590 W. Regarding the energy consumption of 290 W

Table 3 BLASTp energy consumption of three randomly reduced query sets against first part of the NCBI *RefSeq* database [23]

Query set	RIVYERA 128 FPGAs (525 W)	2x Xeon E5520 16 thr. (290 W)	RIVYERA / 2x Xeon
A. thaliana	51.5 Wh	668.7 Wh	7.7%
P. trichocarpa	70.3 Wh	823.8 Wh	8.5%
Homo sapiens	81.8 Wh	762.4 Wh	10.7%

The 2x Xeon E5520 reference system runs NCBI BLASTp v2.2.25+

by the reference system, up to 92.3% can be saved compared to a PC cluster with the same performance. Beyond that, the calculation is made without considering a potentially required cooling system for the cluster.

Regarding quality analysis, a detailed view on a query subset (109 sequences, 28,483 residues) showed 21,918 hits from RIVYERA while NCBI found 22,167 hits. A one-by-one comparison revealed 63 hits (0.29%) were additional results not found by NCBI, and 312 hits (1.41%) from the NCBI results are not found by the RIVYERA implementation. Another 24 hits (0.11%) in both sets were differing only in their alignment positions for the same query and subject sequence. This indicates that the alignment quality almost equals to the NCBI software. However, since BLAST is heuristic, small discrepancies in the alignments do not necessarily imply a difference in quality.

Summarized, due to the ability of an efficient processing of 256 queries at once, the massive parallelization of BLASTp benefits especially from large query sets. Regarding a permanent occupation of the machine more than 92% of the required energy can be saved while keeping almost the same alignment quality as in NCBI BLASTp.

3.3 Burrows–Wheeler Alignment

The sheer volume of short read data, produced by current high-throughput sequencing technologies, state more and more challenge for computers to align them in reasonable time. An optimal alignment using Smith–Waterman is unfeasible with standard computers resulting from its quadratic complexity. Several heuristic alignment algorithms emerged to speed up this process significantly. Many of them are based on hash tables and prefix or suffix tries, e.g. MAQ [15], SOAP [16] and BFAST [8]. However, the fastest aligners providing the best quality trade-off appear to be those based on Burrows–Wheeler transformation [3] and FM-indexing [7], e.g. Bowtie [13], SOAP2 [17] and BWA [14]. Still, their total runtime on complete read datasets is too slow to conform with a biologists workflow, although attempts exist to speed up the alignment using GPUs, e.g. CUSHAW [5] and BarraCUDA [11]. Harnessing the recent RIVYERA S6-LX150 architecture with 2 GB memory extension, ample resources are available to speed up this process.

$X =$ CATGTATGCC\$ ↓ rotate

0	CATGTATGCC$
1	ATGTATGCC$C
2	TGTATGCC$CA
3	GTATGCC$CAT
4	TATGCC$CATG
5	ATGCC$CATGT
6	TGCC$CATGTA
7	GCC$CATGTAT
8	CC$CATGTATG
9	C$CATGTATGC
10	$CATGTATGCC

sort ⟶

10	$CATGTATGCC
5	ATGCC$CATGT
1	ATGTATGCC$C
9	C$CATGTATGC
0	CATGTATGCC$
8	CC$CATGTATG
7	GCC$CATGTAT
3	GTATGCC$CAT
4	TATGCC$CATG
6	TGCC$CATGTA
2	TGTATGCC$CA

↑ last column

$B =$ CTCC\$GTTGAA

$S = (10, 5, 1, 9, 0, 8, 7, 3, 4, 6, 2)$

Fig. 10 Example for the Burrows–Wheeler transformation of sequence "CATGTATGCC"

3.3.1 BWA Algorithm

The Burrows–Wheeler alignment algorithm (BWA [14]) combines the Burrows–Wheeler transformation [3] and FM-indexing [7] with a method for inexact search in the index. It consists of three main steps. The first is the creation of the FM-index (`bwa index`), the second regards searching the alignment in the generated index (`bwa aln`). The last step generates the final alignments from the found positions in the SAM format (`bwa samse/sampe`).

Burrows–Wheeler Transformation

The Burrows–Wheeler transformation [3] of a reference sequence X can be explained in the following way. Let Σ be the alphabet of characters occurring in the reference sequence. A character $\$ \notin \Sigma$, defined to be lexicographically smaller than any other character in Σ, is added to the end of the reference sequence as "end-of-sequence" marker. Now, a table of all rotations of the resulting sequence $X\$$ is created. This table is sorted lexicographically, while the original position of each entry is stored in an array, which is called the suffix array S. The Burrows–Wheeler transformated (BWT) sequence B is now defined as $B[i] = \$$ when $S(i) = 0$ and $B[i] = X[S(i) - 1]$ otherwise. This can easily be created by concatenating the last characters of the rotated sequences in the sorted table. Figure 10 gives an example of the transformation of the sequence "CATGTATGCC." For details the original publication [3] is refered to.

Exact Search and FM-Indexing

Searching a subsequence of the reference means identifying an interval $[l,r]$ of positions in the sorted list where the entries begin with this subsequence. The original positions in the reference sequence can be obtained from the suffix array S in this interval. Exact searching equals to the identification of exactly one interval, while for an inexact search, there could be many. For example, searching the sequence "ATG" in the example in Fig. 10 leads to the inteval $[1,2]$ referring to positions 5 and 1 in the original sequence respectively.

The identification of the intervals is simplified with the help of FM-indexing [7]. Let $C(a)$ be the number of characters in the reference sequence X that are lexicographically smaller than a. Let $Occ(a,i)$ be the number of occurrences of a in the substring $b_0,\dots,b_i$ of the BWT sequence B. Let W be a substring of X. Now, to test if aW is a substring of X for any character a, let l and r be recursively defined as

$$l(aW) = C(a) + Occ(a, l(W) - 1) + 1 \tag{10}$$

$$r(aW) = C(a) + Occ(a, r(W)) \tag{11}$$

whereby $l(\varnothing) = 0$ and $r(\varnothing) = |X|$.

According to Ferragina and Manzini [7], it follows if and only if aW is a substring of X that $l(aW) \leq r(aW)$. The number of occurrences of aW in X can then be determined by the size of the interval $|[l(aW), r(aW)]| = r(aW) - l(aW) + 1$. For an efficient processing, the constants $C(a)$ have to be precalculated as well as at least parts of the occurrences function $Occ(a,i)$ whereby the latter is generally only precalculated in parts to save memory. Since a read is processed in reverse order, this procedure is called *backward search*.

Inexact Search: Backtracking

The inexact search algorithm of this implementation is based on the original BWA algorithm. The BWT of the reverse (not complemented) reference sequence and reverse reads are used to test for an occurrence in the reference X.

To introduce mismatches and gaps, several paths in a prefix trie of the reverse X have to be analyzed. BWA performs a breadth first search to find a suitable path in this trie. Against this behavior, the method described here uses recursion to perform a depth first search always starting with the most promising path, i.e. starting with a path only including matching characters until the first mismatch is found. Then, backtracking tests alternative paths until either a succesful path is found or the number of inserted errors (mismatches or gaps) exceeds a certain threshold. In the latter case, backtracking leads to a previous point where an alternative path is possible. This way, the required stack size can be bounded for the required recursions to $\mathcal{O}(n)$, whereby n is the length of a read.

```
INPUT:
- word W
- error threshold T(i), i ∈ [0,|W|−1]
OUTPUT:
- set of reports R containing BWT intervals [l,r]
PROCEDURE:
- search(W,i,k,l,r)
  if k > T(i) then
    return;
  if i < 0 then
    R = R ∪ {[l,r]};
    return;
  for each a ∈ {A,C,G,T} do
    l_a = C(a) + Occ(a,l−1) + 1;
    r_a = C(a) + Occ(a,r);
  // match
  if l_W[i] ≤ r_W[i] then
    search(W,i−1,k,l_W[i],r_W[i]);
  // mismatches
  for each a ∈ {A,C,G,T} and a ≠ W[i] do
    if l_a ≤ r_a then
      search(W,i−1,k+1,l_a,r_a);
  // deletion
  search(W,i−1,k+1,l,r);
  // insertions
  for each a ∈ {A,C,G,T} do
    if l_a ≤ r_a then
      search(W,i,k+1,l_a,r_a);
  return;
```

Fig. 11 BWA pseudo code for inexact search

To reduce the number of tested paths per read, an error threshold $T(i)$ is fixed for every position in a read. If during the search the number of inserted errors exceeds this threshold, the recursion stops and another path is tested. The distribution is per default optimized for Illumina reads, which are more likely to be erroneous at their ends.

Figure 11 states the pseudo code of the described version for the inexact search. The score, which is calculated from matches, mismatches, and gaps, is not shown in the code. As in BWA, different penalties are paid for mismatches, gap openings and extensions to be more realistic to biological data.

Since in general, there is not enough memory to store a second index in the available RAM, the original (not reversed) index is loaded for a second run to test reverse complement reads. This ensures the correct direction for processing the reads for processing the reads, e.g. according to the error probability of Illumina reads, which is higher at the read's end.

3.3.2 Implementation

The implementation of the BWA algorithm is divided into software and hardware part. While the software implements the interface to the original BWA software modules (BWA v0.59) and is responsible for preprocessing of reads and postprocessing of the results, the hardware performs the inexact search algorithm. The implementation is focused on the Spartan6-LX150 FPGAs of the RIVYERA S6-LX150 with 2 GB memory extension for each FPGA to handle large reference genomes (up to 4 Gbp) such as the human genome.

The main unit of the hardware implementation states the *ReadEntity*. It is responsible for performing the inexact search in the FM-index for a read. For this purpose, the DD3-RAM memory of the FPGA is initialized with this index. To implement the recursive behavior of a depth first search, a stack adjusted to the maximum recursion depth according to the read size is implemented in local FPGA memory.

The ReadEntity generates memory requests to look up the result of the *Occ*-function for calculating the new interval boundaries l and r in the index (s. (10) and (11)). The memory requests are buffered and sent to the memory controller. Since the memory contents are in a densely packed format, the memory replies have to be decoded in several clock cycles to get the final *Occ*-value. This process is completely pipelined to avoid blocking memory accesses.

Since the calculation of new interval boundaries is very fast, the total runtime solely depends on the memory access time. Therefore, it is important to maximize the load on all memory ports to avoid idle times. Hence, the memory contents are equally distributed over all memory ports, and all available FPGA resources are used to generate as many ReadEntity units as possible to maximize an equal distribution of memory requests.

For the current design the resources of a Spartan6-LX150 FPGA allow the utilization of 64 parallel ReadEntity units. Additionally, supplemental infrastructure is required, i.e. a switch to collect all memory requests, some distribution mechanism for the replies, and incoming and outgoing FIFOs for read data and results. An overview of the complete hardware structure is depicted in Fig. 12.

3.3.3 Performance Evaluation

For performance evaluation the alignments of two read datasets containing 11.3 million read pairs of length 2×36 bp (NCBI accession number ERR003014) and 14.5 million read pairs of length 2×76 bp (NCBI acc. no. SRR032215) against the human genome (*hg19*, 3.2 Gbp) have been generated. The runtime of the FPGA implementation has been measured on a fully and several partly equipped configurations of the RIVYERA S6-LX150 machine with memory extension, compared to the original BWA software v0.59 running on an Intel Xeon W3530 PC at 2.8 GHz with 24 GB of RAM. All parameters have been set to default (besides "`-t`" for the number of threads). The test runs have been repeated with a switched-off gapped

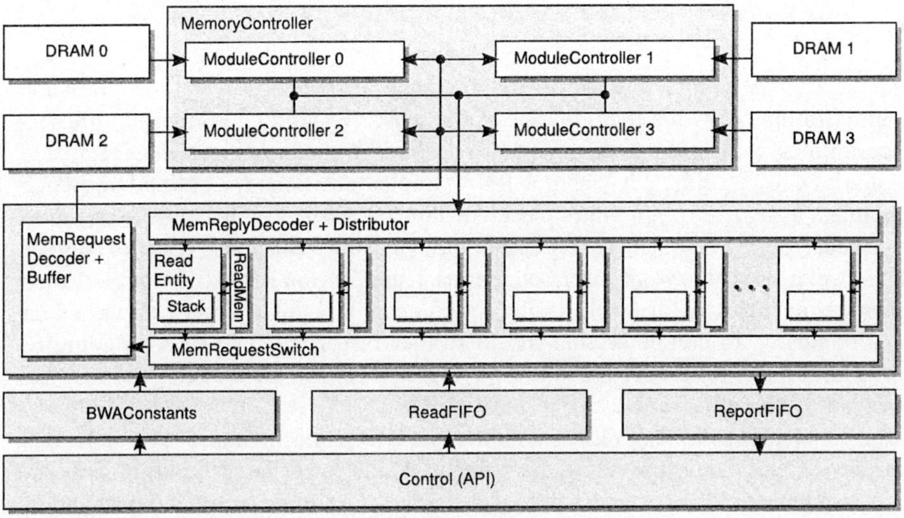

Fig. 12 Hardware structure of the BWA implementation

Table 4 Runtimes (in seconds) for the BWA alignment step (`bwa aln`) against the human genome (*hg19*, 3.2 Gbp)

	RIVYERA S6 (*n* FPGAs)					Intel Xeon W3530			
Query set	128	64	32	16	8	8 thr.	4 thr.	1 thr.	BarraCUDA
ERR003014 w/o gaps	88	109	155	243	418	1,019	1,481	5,455	346*
ERR003014	105	142	223	385	701	3,305	4,649	16,809	651*
SRR032215 w/o gaps	123	179	292	521	974	2,958	4,031	14,587	981*
SRR032215	202	339	611	1,154	2,247	14,801	19,503	70,976	2,401*

ERR003014 contains 11.3 million read pairs of length 2 × 36 bp, SRR032215 contains 14.5 million read pairs of length 2 × 76 bp. The marked (*) runtimes for BarraCUDA are directly taken from [11] (alignments against human genome NCBI build 36.54)

alignment (parameter "`-o 0`") as well. Since the same read datasets were used for the GPU-based aligner BarraCUDA [11], the results from the corresponding publication are included as well. According to the publication, BarraCUDA had been run on an nVidia Tesla M2090. The results are stated in Table 4. Since the postprocessing step to generate the final alignments (`bwa samse/sampe`) is always performed using the original software, the runtimes only include the processing time for the preceding alignment step (`bwa aln`).

The results show that a fully equipped RIVYERA S6-LX150 is able to outperform an Intel Xeon W3530 system with 24 GB of RAM and a full utilization with 8 threads easily by a factor of up to 73, or 351 compared to a single thread. Even the GPU accelerated variant BarraCUDA on an nVidia Tesla M2090 has been outperformed by a factor of about 12. Since the power consumption of RIVYERA S6 is less than three times the power consumption of a single Xeon system, energy savings of more than 95% can be achieved as well.

4 Summary

The RIVYERA architecture proves its efficiency for solving several big problems in bioinformatics, with a focus on one of the most frequently used areas in life sciences: biological sequence alignment, as required for the identification of sequence similarities in proteins, DNA and/or RNA. Implementations for optimal sequence alignment using the Smith–Waterman algorithm, database search using the common heuristical tool BLAST, and short-read alignment using a Burrows–Wheeler Aligner (BWA)-like algorithm were described. For each application RIVYERA outperforms other architectures by far, while saving a significant amount of energy of more than 90% compared to computer clusters. In further research, RIVYERA is going to prove its abilities in de novo assembly, phylogenetic tree analysis, and genome wide association studies (GWAS).

In conclusion, the RIVYERA architecture is capable to improve a biologists workflow significantly providing the computational power to speed up most common problems in bioinformatics at low costs due to its low energy and installation requirements. Due to its flexibility even other application fields, e.g. cryptanalysis (see Chap. 11 in this book) and stock market analysis [28, 29], can be addressed.

With the recent development of the RIVYERA S6-LX150, providing at least 2.5× more resources than RIVYERA S3-5000 at a roughly twofold increase of frequency, existing implementations may be easily ported with an expected speedup factor of about 4.

References

1. S.F. Altschul, W. Gish, W. Miller, E.W. Myers, D.J. Lipman, Basic local alignment search tool. J. Mol. Biol. **215**(3), 403–410 (1990)
2. S.F. Altschul, T.L. Madden, A.A. Schäffer, J. Zhang, Z. Zhang, W. Miller, D.J. Lipman, Gapped BLAST and PSI-BLAST: a new generation of protein database search programs. Nucleic Acids Res. **25**, 3389–3402 (1997)
3. M. Burrows, D.J. Wheeler, A block-sorting lossless data compression algorithm. Tech. rep., Digital Systems Research Center, Palo Alto, CA (1994)
4. CLCbio – High-Speed Smith–Waterman (2012), http://www.clcbio.com/index.php?id=1254. Accessed March 2012
5. CUSHAW: a CUDA compatible short read aligner to large genomes based on the Burrows–Wheeler transform (2011), http://cushaw.sourceforge.net/. Accessed March 2012
6. M.S. Farrar, Optimizing Smith–Waterman for the cell broadband engine (2010), http://sites.google.com/site/farrarmichael/smith-watermanfortheibmcellbe. Accessed March 2012
7. P. Ferragina, G. Manzini, Opportunistic data structures with applications, in *Proceedings of FOCS2000* (2000), IEEE Computer Society, Washington DC, USA, pp. 390–398
8. N. Homer, B. Merriman, S.F. Nelson, Bfast: an alignment tool for large scale genome resequencing. PLoS ONE **4**(11), 12 (2009). http://www.ncbi.nlm.nih.gov/pubmed/19907642
9. A. Jacob, J. Lancaster, J. Buhler, B. Harris, R.D. Chamberlain, Mercury BLASTp: accelerating protein sequence alignment. ACM Trans. Reconfigurable Tech. Syst. **1**, 9:1–9:44 (2008)
10. S. Kasap, K. Benkrid, Y. Liu, Design and implementation of an FPGA-based core for gapped BLAST sequence alignment with the two-hit method. Eng. Lett. **16**, 443–452 (2008)

11. P. Klus, S. Lam, D. Lyberg, M. Cheung, G. Pullan, I. McFarlane, G. Yeo, B. Lam, Barracuda - a fast short read sequence aligner using graphics processing units. BMC Res. Notes **5**(1), 27 (2012). doi:10.1186/1756-0500-5-27
12. S. Kumar, C. Paar, J. Pelzl, G. Pfeiffer, A. Rupp, M. Schimmler, How to break DES for €8,980, in *SHARCS2006*, Cologne, Germany (2006)
13. B. Langmead, C. Trapnell, M. Pop, S. Salzberg, Ultrafast and memory-efficient alignment of short dna sequences to the human genome. Genome Biol. **10**(3), R25 (2009). doi:10.1186/gb-2009-10-3-r25, http://genomebiology.com/2009/10/3/R25
14. H. Li, R. Durbin, Fast and accurate short read alignment with Burrows–Wheeler transform. Bioinformatics (Oxford, England) **25**(14), 1754–1760 (2009). doi:10.1093/bioinformatics/btp324, http://dx.doi.org/10.1093/bioinformatics/btp324
15. H. Li, J. Ruan, R. Durbin, Mapping short dna sequencing reads and calling variants using mapping quality scores. Genome Res. **18**(11), 1851–1858 (2008). doi:10.1101/gr.078212.108, http://dx.doi.org/10.1101/gr.078212.108
16. R. Li, Y. Li, K. Kristiansen, J. Wang, SOAP: short oligonucleotide alignment program. Bioinformatics (Oxford, England) **24**(5), 713–714 (2008). doi:10.1093/bioinformatics/btn025, http://dx.doi.org/10.1093/bioinformatics/btn025
17. R. Li, C. Yu, Y. Li, T.W.W. Lam, S.M.M. Yiu, K. Kristiansen, J. Wang, SOAP2: an improved ultrafast tool for short read alignment. Bioinformatics (Oxford, England) **25**(15), 1966–1967 (2009). doi:10.1093/bioinformatics/btp336, http://dx.doi.org/10.1093/bioinformatics/btp336
18. W. Liu, B. Schmidt, W. Müller-Wittig, CUDA-BLASTP: accelerating BLASTP on CUDA-enabled graphics hardware. IEEE/ACM Trans. Comput. Biol. Bioinformatics **8**, 1678–1684 (2011)
19. Y. Liu, B. Schmidt, D. Maskell, CUDASW++2.0: enhanced Smith–Waterman protein database search on CUDA-enabled GPUs based on SIMT and virtualized SIMD abstractions. BMC Res. Notes **3**(1), 93+ (2010). doi:10.1186/1756-0500-3-93
20. A. Mahram, M.C. Herbordt, Fast and accurate NCBI BLASTp: acceleration with multiphase FPGA-based prefiltering, in *Proceedings of ICS'10* (2010), ACM, New York, USA, pp. 73–28
21. NCBI BLAST, http://blast.ncbi.nlm.nih.gov/Blast.cgi. Accessed March 2012
22. NCBI GenBank database, http://www.ncbi.nlm.nih.gov/genbank/. Accessed March 2012
23. NCBI RefSeq database, http://www.ncbi.nlm.nih.gov/RefSeq/. Accessed March 2012
24. S.B. Needleman, C.D. Wunsch, A general method applicable to the search for similarities in the amino acid sequence of two proteins. J. Mol. Biol. **48**(3), 443–453 (1970)
25. G. Pfeiffer, S. Baumgart, J. Schröder, M. Schimmler, A massively parallel architecture for bioinformatics, in *ICCS2009*. Lecture Notes in Computer Science, vol. 5544 (Springer, Berlin, 2009), pp. 994–1003
26. SciEngines GmbH, http://www.sciengines.com. Accessed March 2012
27. T.F. Smith, M.S. Waterman, Identification of common molecular subsequences. J. Mol. Biol. **147**, 195–197 (1981)
28. C. Starke, V. Grossmann, L. Wienbrandt, M. Schimmler, An FPGA implementation of an investment strategy processor, in *ICCS2012*. Procedia Computer Science, vol. 9 (Elsevier, 2012), pp. 1880–1889
29. C. Starke, V. Grossmann, L. Wienbrandt, S. Koschnicke, J. Carstens, M. Schimmler, Optimizing investment strategies with the reconfigurable hardware platform RIVYERA. Int. J. Reconfigurable Comput. **2012**, 10 (2012). doi:10.1155/2012/646984
30. Superfamily HMM library and genome assignments server, http://supfam.cs.bris.ac.uk/SUPERFAMILY/. Accessed March 2012
31. UniProt Knowledgebase, http://www.ebi.ac.uk/uniprot/. Accessed March 2012
32. L. Wienbrandt, S. Baumgart, J. Bissel, F. Schatz, M. Schimmler, Massively parallel FPGA-based implementation of BLASTp with the two-hit method, in *ICCS2011*. Procedia Computer Science, vol. 1 (Elsevier, 2011), pp. 1967–1976
33. L. Wienbrandt, D. Siebert, M. Schimmler, Improvement of BLASTp on the FPGA-based high-performance computer RIVYERA, in *ISBRA2012*. Lecture Notes in Bioinformatics, vol. 7292 (Springer, Berlin, Heidelberg, 2012), pp. 275–286

FPGA-Accelerated Molecular Dynamics

M.A. Khan, M. Chiu, and M.C. Herbordt

Abstract Molecular dynamics simulation (MD) is one of the most important applications in computational science and engineering. Despite its widespread use, there exists a many order-of-magnitude gap between the demand and the performance currently achieved. Acceleration of MD has therefore received much attention. In this chapter, we discuss the progress made in accelerating MD using Field-Programmable Gate Arrays (FPGAs). We first introduce the algorithms and computational methods used in MD and describe the general issues in accelerating MD. In the core of this chapter, we show how to design an efficient force computation pipeline for the range-limited force computation, the most time-consuming part of MD and the most mature topic in FPGA acceleration of MD. We discuss computational techniques and simulation quality and present efficient filtering and mapping schemes. We also discuss overall design, host–accelerator interaction and other board-level issues. We conclude with future challenges and the potential of production FPGA-accelerated MD.

1 Introduction to Molecular Dynamics

Molecular dynamics simulations (MD) are based on the application of classical mechanics models to ensembles of particles and are used to study the behavior of physical systems at an atomic level of detail [39]. MD simulations act as virtual experiments and provide a projection of laboratory experiments with potentially greater detail. MD is one of the most widely used computational tools in biomedical research and industry and has so far provided many important insights into understanding the functionality of biological systems (see, e.g., [1, 25, 31]). MD models have been developed and refined over many years and are validated through

M.A. Khan (✉) • M. Chiu • M.C. Herbordt
Boston University, 8 Saint Mary's Street, Boston, MA 02215, USA
e-mail: azkhan@bu.edu; mattchiu@bu.edu; herbordt@bu.edu

W. Vanderbauwhede and K. Benkrid (eds.), *High-Performance Computing Using FPGAs*, DOI 10.1007/978-1-4614-1791-0_4,

fitting models to experimental and quantum data. Although classical MD simulation is inherently an approximation, it is dramatically faster than a direct solution to the full set of quantum mechanical equations.

But while the use of classical rather than quantum models results in orders-of-magnitude higher throughput, MD remains extremely time consuming. For example, the simulation of even a comparatively simple biological entity such as the STM virus (a million-atom system) for 100 ns would take 70 years if run on a single CPU core [14]. Fortunately MD scales well for simulations of this size or greater. The widely used MD packages, e.g., AMBER [6], CHARMM [5], Desmond [4], GROMACS [21], LAMMPS [37], NAMD [34], can take full advantage of scalability [27]. But typical MD executions still end up taking month-long runtime, even on supercomputers [45].

To make matters worse, many interesting biological phenomena occur only on far longer timescales. For example, protein folding, the process by which a linear chain of amino acids folds into a three-dimensional functional protein, is estimated to take at least a microsecond [12]. The exact mechanism of such phenomena remains beyond the reach of the current computational capabilities [44]. Longer simulations are also critical to facilitate comparison with physically observable processes, which (again) tend to be at least in the microsecond range. With stagnant CPU clock frequency and no remarkable breakthrough in the underlying algorithms for a decade, MD faces great challenges to meet the ever-increasing demand for larger and longer simulations.

Hardware acceleration of MD has therefore received much attention. ASIC-based systems such as Anton [43] and MD-Grape [32] have shown remarkable results, but their non-recurring cost remains high. GPU-based systems with their low cost and ease of use also show great potential. But GPUs are power hungry and, perhaps more significantly, are vulnerable to data communication bottlenecks [16, 48].

FPGAs, on the other hand, have a flexible architecture and are energy efficient. They bridge the programmability of CPUs and the custom design of ASICs. Although developing an FPGA-based design takes significantly longer than a GPU-based system, because it requires both software and hardware development, the effort should be cost-effective due to the relatively long life-cycle of MD packages. Moreover, improvements in fabrication process generally translate to performance increases for FPGA-based systems (mostly in the form of direct replication of additional computation units). And perhaps most significantly for emerging systems, FPGAs are fundamentally communication switches and so can avoid communication bottlenecks and form the basis of accelerator-centric high-performance computing systems.

This chapter discusses the current state of FPGA acceleration of MD-based primarily on the work done at Boston University [8, 9, 18]. The remainder of this section gives an extended introduction to MD. This is necessary because while MD is nearly trivial to define, there are a number of subtle issues which have a great impact on acceleration method. In the next section we present the issues universal to MD acceleration. After that we describe in depth the state-of-the-art in FPGA

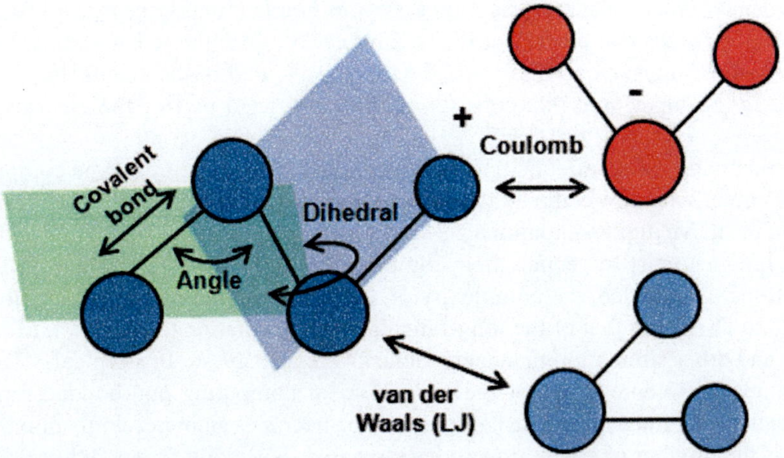

Fig. 1 MD Forces computed by MD include several bonded (covalent, angle, and dihedral) and nonbonded (van der Waals and Coulomb)

MD acceleration focusing on the range-limited force. Finally, we summarize future challenges and potential especially in the creation of parallel FPGA-based MD systems.

1.1 Overview of Molecular Dynamics Simulation

MD is an iterative process that models dynamics of molecules by applying classical mechanics [39]. The user provides the initial state (position, velocity, etc.), the force model, other properties of the physical system, and some simulation parameters such as simulation type and output frequency. Simulation advances by timestep where each timestep has two phases: force computation and motion update. The duration of the timesteps is determined by the vibration of particles and typically corresponds to one or a few femtoseconds (fs) of real time. In typical CPU implementations, executing a single timestep of a modest 100 K particle simulation (a protein in water) takes over a second on a single core. This means that the 10^6–10^9 timesteps needed to simulate reasonable timescales result in long runtimes.

There are many publicly available and widely used MD packages including NAMD [34], LAMMPS [37], AMBER [6], GROMACS [21], and Desmond [4]. They support various force fields (e.g., AMBER [38] and CHARMM [30]) and simulation types. But regardless of the specific package or force field model, force computation in MD involves computing contributions of van der Waals, electrostatic (Coulomb), and various bonded terms (see Fig. 1 and (1)).

$$F_{\text{total}} = F_{\text{bond}} + F_{\text{angle}} + F_{\text{dihedral}} + F_{\text{hydrogen}} + F_{\text{vanderWaals}} + F_{\text{electrostatic}}. \tag{1}$$

van der Waals and electrostatic forces are *non-bonded* forces, the others *bonded*. Non-bonded forces can be further divided into two types: the *range-limited* force that consists of the van der Waals and the short-range part of the electrostatic force and the *long-range* force that consists of the long-range part of the electrostatic force.

Since bonded forces affect only a few neighboring atoms, they can be computed in $O(N)$ time, where N is the total number of particles in the system. Non-bonded terms in the naive implementation have complexity of $O(N^2)$, but several algorithms and techniques exist to reduce their complexity; these will be described in later subsections. In practice, the complexity of the range-limited force computation is reduced to $O(N)$ and that of the long-range force computation to $N\log(N)$. Motion update and other simulation management tasks are also $O(N)$. In a typical MD run on a single CPU core, most of the time is spent computing non-bonded forces. For parallel MD, inter-node data communication becomes an increasingly dominant factor as the number of computing nodes increases, especially for small to medium sized physical systems. Sample timing profiles for both serial and parallel runs of MD are presented in Sect. 2.

The Van der Waals (VdW) force is approximated by the Lenard-Jones (LJ) potential as shown in (2):

$$\vec{F}_i(LJ) = \sum_{i \neq j} \frac{\varepsilon_{ab}}{\sigma_{ab}^2} \left\{ 12 \left(\frac{\sigma_{ab}}{|r_{ji}|} \right)^{14} - 6 \left(\frac{\sigma_{ab}}{|r_{ji}|} \right)^{8} \right\} \vec{r_{ij}}, \tag{2}$$

where ε_{ab} and σ_{ab} are parameters related to particle types and r_{ij} is the relative distance between particle i and particle j.

A complete evaluation of VdW or LJ force requires evaluation of interactions between all particle pairs in the system. The computational complexity is therefore $O(N^2)$, where N is the number of particles in the system. A common way to reduce this complexity is applying a cutoff. Since the LJ force vanishes quickly with the separation of a particle pair it is usually ignored when two particles are separated beyond 8–16$\AA$. To ensure a smooth transition at cutoff, an additional switching function is often used. Using a cutoff distance alone does not reduce the complexity of the LJ force computation because all particle pairs must still be checked to see if they are within the cutoff distance. The complexity is reduced to $O(N)$ by combining this with techniques like the cell-list and neighbor-list methods, which will be described in Sect. 1.2.

The electrostatic or Coulomb force works between two charged particles and is given by (3):

$$\vec{F}_i(CL) = q_i \sum_{i \neq j} \left(\frac{q_j}{|r_{ij}|^3} \right) \vec{r_{ij}}, \tag{3}$$

where q_i and q_j are the particle charges and r_{ij} is the separation distance between particles i and j.

Unlike the van der Waals force, the Coulomb force does not fall off sufficiently quickly to immediately allow the general application of a cutoff. The Coulomb force

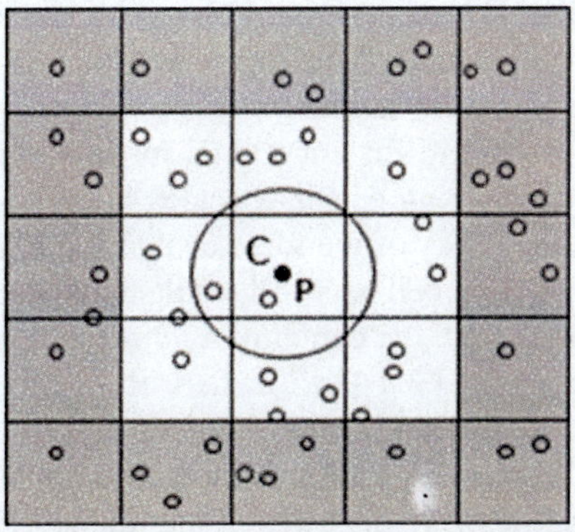

Fig. 2 2D Illustration of cell and neighbor lists. In the range-limited force, particles only interact with those in the cell neighborhood. Neighbor lists are constructed by including for each particle only those particles within the cutoff radius C (shown for P)

is therefore often split into two components: a range-limited part that goes to zero in the neighborhood of the LJ cutoff and a long-range part that can be computed using efficient electrostatic methods, the most popular being based on Ewald Sums [11] or Multigrid [46]. For example, one can split the original Coulomb force curve into two parts (with a smoothing function $g_a(r)$):

$$\frac{1}{r} = \left(\frac{1}{r} - g_a(r)\right) + g_a(r). \tag{4}$$

The short-range component can be computed together with the Lennard–Jones force using particle indexed lookup tables A_{ab}, B_{ab}, and QQ_{ab}. Then the entire short-range force to be computed is:

$$\frac{\mathbf{F}_{ji}^{\text{short}}}{\mathbf{r_{ji}}} = A_{ab}r_{ji}^{-14} + B_{ab}r_{ji}^{-8} + QQ_{ab}\left(r_{ji}^{-3} + \frac{g_a'(r)}{r}\right). \tag{5}$$

In addition to the non-bonded forces, bonded interactions (e.g., bond, angle, and dihedral in Fig. 1) must also be computed every timestep. They have $O(N)$ complexity and take a relatively small part of the total time. Bonded pairs are generally excluded from non-bonded force computation, but if for any reason (e.g., to avoid a branch instruction in an inner loop) a non-bonded force computation includes bonded pairs, then those forces must be subtracted accordingly. Because the long-range force varies less quickly than the other force components, it is often computed only every 2–4 timesteps.

1.2 Cell Lists and Neighbor Lists

We now present two methods of reducing the naive complexity of $O(N^2)$ to $O(N)$. In the cell-list method [22,40] a simulation box is first partitioned into several cells, often cubic in shape (see Fig. 2 for a 2D depiction). Each dimension is typically chosen to be slightly larger than the cutoff distance. This means, for a 3D system, that traversing through the particles of the home cell and 26 adjacent cells suffices, independent of the overall simulation size. If Newton's third law is used, then only half of the neighboring cells need to be checked. If the cell dimension is less than cutoff distance, then more number of cells need to be checked. The cost of constructing cell lists scales linearly with the number of particles but reduces the complexity of the force evaluation to $O(N)$.

Using cell lists still results in checking many more particles than necessary. For a particle in the center of a home cell, we only need to check its surrounding volume of $(4/3) * 3.14 * R_c^3$, where R_c is the cutoff radius. But in the cell-list method we end up checking a volume of $27 * R_c^3$, which is roughly 6 times larger than needed. This can be improved using neighbor lists [49]. In this method, a list of possible neighboring particles is maintained for each particle and only this list is checked for force evaluation. A particle is included in the neighbor list of another particle if the distance between them is less than $R_c + R_m$, where R_m is a small buffer margin. R_m is chosen such that the neighbor-list also contains the particles which are not yet within the cutoff range but might enter the cutoff range before the list is updated next. In every timestep, the validity of each pair in a neighbor list is checked before it is actually used in force evaluation. Neighbor lists are usually updated periodically in a fixed number of timesteps or when displacements of particles exceed a predetermined value.

Although neighbor lists can be constructed for all particles in $O(N)$ time (using cell-lists), it is far more costly as many particles must still be checked for each reference particle. But as long as the neighbor lists are not updated too frequently, which is the case generally, this method reduces the range-limited force evaluation time significantly. The savings in runtime comes at the cost of extra storage required to save the neighbor-list of each particle. For most high-end CPUs, this is not a major issue.

1.3 Direct Computation vs. Table Interpolation

The most time-consuming part of an MD simulation is typically the evaluation of range-limited forces. One of the major optimizations is the use of table lookup in place of direct computation. This avoids expensive square roots and $erfc$ evaluations. This method not only saves computation time but is also robust in incorporating small changes such as the incorporation of a switching function.

Typically the square of the inter-particle distance (r^2) is used as the index. The possible range of r^2 is divided into several sections or segments and each section is further divided into intervals or bins as shown in Fig. 3. For an M order interpolation, each interval needs $M+1$ coefficients and each section needs $N*(M+1)$ coefficients, where N is the number of bins in the section. Accuracy increases with both the number of intervals per section and the interpolation order. Generally the rapidly changing regions are assigned relatively higher number of bins, and relatively stable regions are assigned fewer bins. Equation (6) shows a third order interpolation.

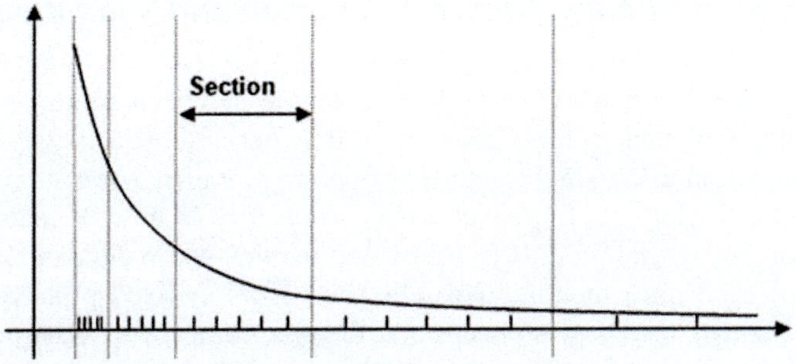

Fig. 3 In MD interpolation, function values are typically computed by *Section* with each having a constant number of bins, but varying in size with distance

$$F(x) = a_0 + a_1 x + a_2 x^2 + a_3 x^3 \tag{6}$$

For reference, here we present a sample of table interpolation parameters used in widely known MD packages and systems.

- NAMD (CPU)—[34] and Source code of NAMD2.7
  Order = 2 bins/segment = 64 Index: r^2
  Segments: 12—segment size increases exponentially, starting from 0.0625
- NAMD (GPU)—[48] and Source code of NAMD2.7
  Order = 0 bins/segment = 64 Index: $1/\sqrt{r^2}$
  Segments: 12—segment size increases exponentially
- CHARMM—[5]
  Order = 2 bins/segment = 10–25 Index: r^2
  Segments: Uniform segment size of $1\,\mathring{A}^2$ is used which results in relatively more precise values near cut-off
- ANTON—[28]
  Force Table Order = Says 3 but that may be for energy only. Value for force may be smaller.
  # of bins = 256 Index: r^2
  Segments: Segments are of different widths, but values not available, nor whether the number of bins is the total or per segment.
- GROMACS—[21] and GROMACS Manual 4.5.3, page 148
  Order = 2 bins = 500 (2000) per nm for single (double) precision
  Segments: 1 Index: r^2
  Comment: Allows user-defined tables.

Clearly there are a wide variety of parameter settings. These have been chosen with regard to cache size (CPU), routing and chip area (Anton), and the availability of special features (GPU texture memory). These parameters also have an effect on simulation quality, which we discuss next.

1.4 Simulation Quality: Numeric Precision and Validation

Although most widely used MD packages use double-precision floating point (DP) for force evaluation, studies have shown that it is possible to achieve acceptable quality of simulation using single-precision floating point (SP) or even using fixed point arithmetic, as long as the exact atomic trajectory is not the main concern [36, 41, 43]. Since floating point (FP) arithmetic requires more area and has longer latency, a hardware implementation would always prefer fixed point arithmetic. Care must be taken, however, to ensure that the quality of the simulation remains acceptable. Therefore a critical issue in all MD implementations is the trade-off between precision and simulation quality.

Quality measures can be classified as follows (see, e.g., [13, 33, 43]).

1. Arithmetic error here is the deviation from the ideal (direct) computation done at high precision (e.g., double-precision). A frequently used measure is the relative RMS force error, which is defined as follows [42]:

$$\Delta F = \sqrt{\left(\frac{\sum_i \sum_{\alpha \in x,y,z} [F_{i,\alpha} - F_{i,\alpha}^*]^2}{\sum_i \sum_{\alpha \in x,y,z} [F_{i,\alpha}^*]^2}\right)}. \tag{7}$$

2. Physical invariants should remain so in simulation. Energy can be monitored through fluctuation (e.g., in the relative RMS value) and drift. Total fluctuation of energy can be determined using the following expression (suggested by Shan et al. [42]):

$$\Delta E = \frac{1}{N_t} \sum_{i=1}^{N_t} \left| \frac{E_0 - E_i}{E_0} \right|, \tag{8}$$

where E_0 is the initial value, N_i is the total number of time steps in time t, and E_i is the total energy at step i. Acceptable numerical accuracy is achieved when $\Delta E \leq 0.003$.

2 Basic Issues with Acceleration and Parallelization

2.1 Profile

The maximum speed-up achievable by any hardware accelerator is limited by Amdahl's law. It is therefore important to profile the software to identify potential targets of acceleration. As discussed in Sect. 1.1, a timestep in MD consists of two parts, force computation and motion integration. The major tasks in force computation are computing range-limited forces, computing long-range forces, and computing bonded forces. Table 1 shows the timing profile of a timestep using the GROMACS MD package on a single CPU core [21]. These results are typical;

Table 1 Timing profile of an MD run from a GROMACS study [21]

Step	Task	% execution time
Force computation	Range-limited force	60
	FFT, Fourier-space computation, IFFT	17
	Charge spreading and force interpolation	13
	Other forces	5
Motion integration	Position and velocity updates	2
	Others	3

see, e.g., [43]. As we can see the range-limited force computation dominates and consumes 60% of the total runtime. The next major task is the long-range force computation, which can be further divided into two tasks, charge-spreading/force-interpolation and FFT-based computation. FFT, Fourier-space computation, and inverse FFT take 17% of the total runtime while charge spreading and force interpolation take 13% of the total runtime. Computing other forces takes only 5% of the total runtime. Unlike the force computation, motion integration is a straightforward process and takes only 2% of the total runtime. Other remaining computations take 3% of the total runtime. In addition to serial runtime, data communication becomes a limiting factor in parallel and accelerated version. We discuss this in Sect. 2.3.

2.2 *Handling Exclusion*

While combining various forces before computing acceleration is a straightforward process of linear summation, careful consideration is required for bonded pairs, especially when using hardware accelerators. In particular, covalently bonded pairs need to be excluded from non-bonded force computation. One way to ensure this is to check whether two particles are bonded before evaluating their non-bonded forces. This is expensive because it requires on-the-fly check for bonds. Another way is to use separate neighbor lists for bonded and non-bonded neighbors. Both of these methods are problematic for hardware acceleration: one requires implementing a branch instruction while the other forces the use of neighbor-lists, which may be impractical for hardware implementation (see Sect. 3.2).

A way that is often the preferred for accelerators is to compute non-bonded forces for all particle-pairs within the cutoff distance, but later subtract those for bonded pairs in a separate stage. This method does not need either on-the-fly bond checking or neighbor-lists. There is a different problem here though. The r^{14} term of the LJ force (2) can be very large for bonded particles because they tend to be much closer than non-bonded pairs. Adding and subtracting such large scale values can overwhelm real but small force values. Therefore, care needs to be taken so that the actual force values are not saturated. For example, an inner as well as an outer cutoff can be applied.

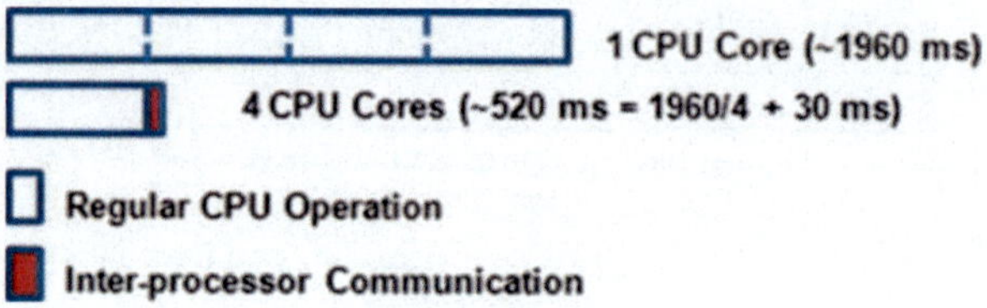

Fig. 4 Apoa1 benchmark runtime/timestep using NAMD showing overhead in a small-scale parallel simulation

2.3 Data Transfer and Communication Overhead

Accelerators are typically connected to the host CPU via some shared interface, e.g., the PCI or PCIe bus. For efficient computation on the accelerator, frequent data transfers between the main memory of the CPU and accelerator must be avoided. Input data need to be transferred to the accelerator before the computation starts and results need to be sent back to the host CPU. Although this is usually done using DMA, it may still consume a significant amount of time that was not required in a CPU-only version. It is preferred that the CPU remains engaged in other useful tasks while data transfer and accelerated computation take place, allowing efficient overlap of computation and communication, as well as parallel utilization of the CPU and the accelerator. Our studies show that host-accelerator data transfer takes around 5–10% of the accelerated runtime for MD (see Sect. 3.2).

In addition to intra-node (host-accelerator) data transfer, inter-node data communication may become a bottleneck, especially for accelerated versions of MD. MD is a highly parallel application and typically runs on multiple compute nodes. Parallelism is achieved in MD by first decomposing the simulation space spatially and assigning one or more of such decomposed sections to a compute node (see, e.g., [34]). Particles in different sections then need to compute their pairwise interaction forces (both non-bonded and bonded) which requires inter-node data communication between node-pairs. In addition to that, long-range force computation requires all-to-all communication [50]. Thus, in addition to the serial runtime, inter-node communication plays an important role in parallel MD. Figure 4 shows an example of inter-processor communication time as the number of processors increases from 1 to 4. We performed this experiment using Apoa1 benchmark and NAMD2.8 [34] on a quad-core Intel CPU (2 core2-duo) of 2.0 GHz each. For a CPU-only version the proportion may be acceptable. For accelerated versions, however, the proportion increases and becomes a major problem [35].

2.4 Newton's 3rd Law

Newton's 3rd law (N3L) allows computing forces between a pair of particles only once and uses the result to update both particles. This provides opportunities for certain optimizations. For example, when using the cell-list method, each cell now only needs to check half of its neighboring cells. Some ordering needs to be established to make sure that all required cell-pairs are considered, but this is a trivial problem.

The issue of whether to use N3L or not becomes more interesting in parallel and accelerated version of MD. It plays an important role in the amount and pattern of inter-node data communication for parallel runs, and successive accumulation of forces in multi-pipelined hardware accelerators (see discussion on accumulation in Sect. 3.1). For example, assume a parallel version of MD where particles x and y are assigned to compute nodes X and Y, respectively. If N3L is not used, we need to send particle data of y from Y to X and particle data of x from X to Y before the force computation of a timestep can take place. But no further inter-node communication will be required for that timestep as particle data will be updated locally. In contrast, if N3L is used, particle data of y need to be sent from Y to X before the computation and results need to be sent from X to Y. Depending on the available computation and communication capability, these two may result in different efficiency. Similar, but more fine-grained, issues exist for hardware accelerators too.

2.5 *Partitioning and Routing*

Parallel MD requires partitioning of the problem and routing data every timestep. Although there are various ways of partitioning computations in MD, practically all widely used MD packages use some variation of spatial decomposition (e.g., recursive bisection, neutral territory, half shell, or midpoint [4, 23]). In such a method, each compute node or process is responsible for updating particles in a certain region of the simulation space. In other words, it owns the particles in that region. Particle data such as position and charge need to be routed to the node that will compute forces for that particle. Depending on the partitioning scheme, computation may take place on a node that owns at least one of the particles involved in the force computation, or it may take place on a node that does not own any of the particles involved in the force computation. Computation results may also need to be routed back to the owner node. This also depends on several choices such as the partitioning scheme and the use of N3L.

For an accelerated version of MD, partitioning and routing may cause additional overhead. Because hardware accelerators typically require a chunk of data to work on at a time in order to avoid frequent data communication with the host CPU. This means fine-grained overlapping of computation and communication, which is possible in a CPU-only version, becomes challenging.

3 FPGA Acceleration Methods

Several papers have been published from CAAD Lab at Boston University describing a highly efficient FPGA kernel for the range-limited force computation [7–9]. The kernel was integrated into NAMD-lite [19], a serial MD package developed at UIUC to provide a simpler way to examine and validate new features before integrating them into NAMD [34]. The FPGA kernel itself was implemented on an Altera Stratix-III SE260 FPGA of Gidel ProcStar-III board [15]. The board contains four such FPGAs and is capable of running at a system speed of up to 300 MHz. The FPGAs communicate with the host CPU via a PCIe bus interface. Each FPGA is individually equipped with over 4GB of memory.

The runtime of the kernel was 26× faster over the end-to-end runtime of NAMD, for Apoa1, a benchmark consisting of 92,224 atoms [10]. The electrostatic force was computed every cycle using PME and both LJ and short-rage portion of electrostatic force were computed on the FPGAs. Particle data along with cell-lists and particle types are sent to the FPGA every timestep, while force data are received from the FPGA and then integrated on the host. A direct end-to-end comparison with the software-only version was not done since the host software itself (NAMD-lite) is not optimized for performance. In the next three subsections we discuss the key contributions of this work in depth. In the following two subsections we describe some preliminary work in the FPGA-acceleration of the long-range force and in mapping MD to multi-FPGA systems.

3.1 Force Pipeline

In Sect. 1.1 we described the general methods in computing the range-limited forces (see (5)). Here we present their actual implementation emphasizing compatibility with NAMD.

While the van der Waals term shown in (2) converges quickly, it must still be modified for effective MD simulations. In particular, a switching function is implemented to truncate van der Waals force smoothly at the cutoff distance (see (9)–(11)).

$$s = (cutoff^2 - r^2)^2 * (cutoff^2 + 2 * r^2 - 3 * switch\_dist^2) * denom \tag{9}$$

$$ds_r = 12 * (cutoff^2) * (switch\_dist^2 - r^2) * denom \tag{10}$$

$$denom = 1/(cutoff^2 - switch\_dist^2)^3. \tag{11}$$

Without a switching/smoothing function, the energy may not be conserved as the force would be truncated abruptly at the cutoff distance. The graph of van der Waals potential with the switching/smoothing function is illustrated in Fig. 5. The van der Waals force and energy can be computed directly as shown here:

IF ($r^2 \leq switch\_dist^2$) $U_{vdW} = U, F_{vdW} = F$
IF ($r^2 > switch\_dist^2$ && $r^2 < cutoff^2$) $U_{vdW} * s,\ F_{vdW} = F * s + U_{vdw} * ds_r$
IF ($r^2 \geq cutoff^2$) $U_{vdW} = 0, F_{vdW} = 0.$

For the Coulomb term the most flexible method used by NAMD for calculating the electrostatic force/energy is Particle Mesh Ewald (PME). The following is the pairwise component:

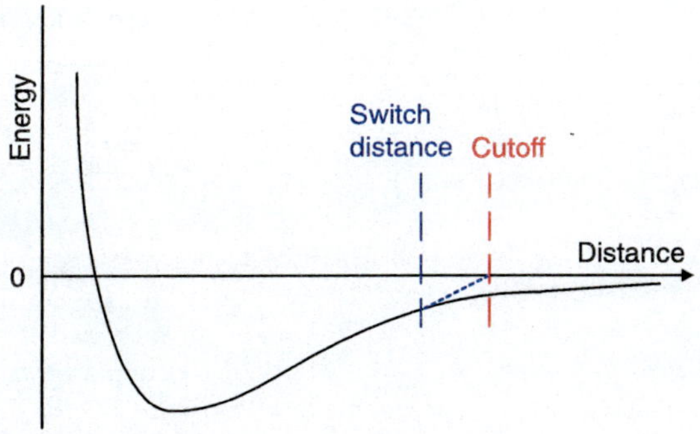

Fig. 5 Graph shows the van der Waals potential with switching/smoothing function (*dashed line*)

$$Es = \frac{1}{4\pi\varepsilon_0}\frac{1}{2}\sum_{n}\sum_{i=1}^{N}\sum_{i=0}^{n}\frac{q_i q_j}{|r_i - r_j + nL|} erfc\left(\frac{|r_i - r_j + nL|}{\sqrt{2}\sigma}\right). \quad (12)$$

To avoid computing these complex equations explicitly, software often employs table lookup with interpolation (Sect. 1.3). Equation (5) can be rewritten as follows:

$$\frac{\mathbf{F}_{ji}^{\text{short}}(|r_{ji}|^2(a,b))}{\mathbf{r_{ji}}} = A_{ab}R_{14}(|r_{ji}|^2) + B_{ab}R_8(|r_{ji}|^2) + QQ_{ab}R_3(|r_{ji}|^2), \quad (13)$$

where R_{14}, R_8, and R_3 are three tables indexed with $|r_{ji}|^2$ (rather than $|r_{ji}|$, to avoid the square root operation).

Designing a force computation pipeline on FPGA to accurately perform these tasks requires careful consideration of several issues. Figure 6 illustrates the major functional units of the force pipelines. The force function evaluators are the diamonds marked in red; these are the components which can be implemented with the various schemes. The other units remain mostly unchanged. The three function evaluators are for the R_{14}, R_8, and R_3 components of (13), respectively. In particular, *Vdw Function 1* and *Vdw Function 2* are the R_{14} and R_8 terms but also include the cutoff shown in (9)–(11). *Coulomb Function* is the R_3 term but also includes the correction shown in (12).

For the actual implementation we use a combination of fixed and floating point. Floating point has far more dynamic range, while fixed point is much more efficient and offers somewhat higher precision. Fixed point is especially advantageous for use as an index (r^2) and for accumulation. We therefore perform the following conversions: float to fixed as data arrive on the FPGA; to float for interpolation; to fixed for accumulation; and to float for transfer back to the host.

A significant issue is determining the minimum interpolation order, precision, and number of intervals without sacrificing simulation quality. For this we use two methods both of which use a modified NAMD-lite to generate the appropriate data. The first method uses (7) to compute the relative RMS error with respect to the reference code. The simulation was first run for 1,000 timesteps using direct computation. Then in the next timestep both direct computation and table lookup

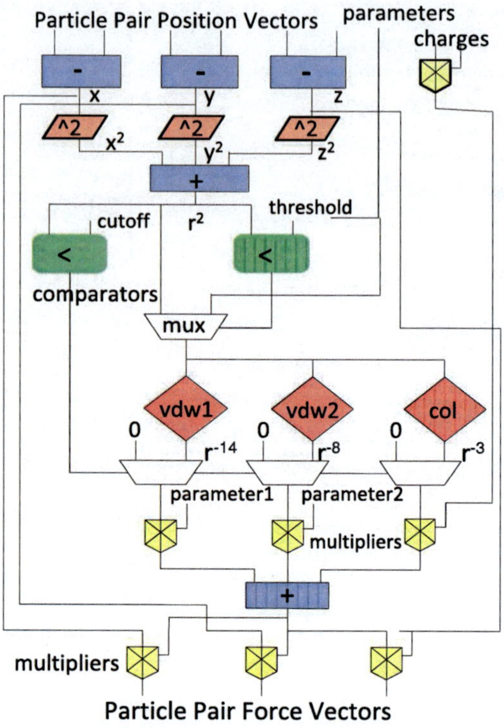

Fig. 6 Logic for computing the range-limited force. *Red diamonds* indicate respective table lookups for the two van der Waals force components and the Coulombic force

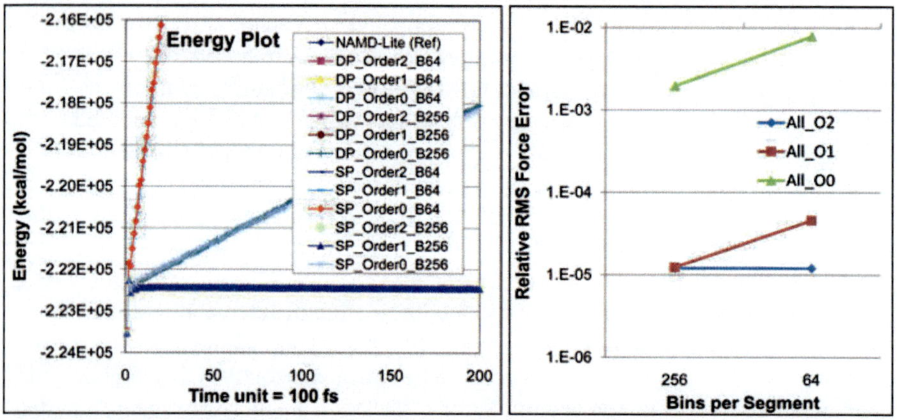

Fig. 7 *Right graph* shows relative RMS force error versus bin density for interpolation orders 0, 1, and 2. *Left graph* shows energy for various designs run for 20,000 timesteps. Except for 0-order, plots are indistinguishable from the reference code

with interpolation were used to find the relative RMS force error for the various lookup parameters. Only the range-limited forces (switched VdW and short-range part of PME) were considered. All computations were done in double-precision. Results are shown in the right half of Fig. 7. We note that 1st and 2nd order

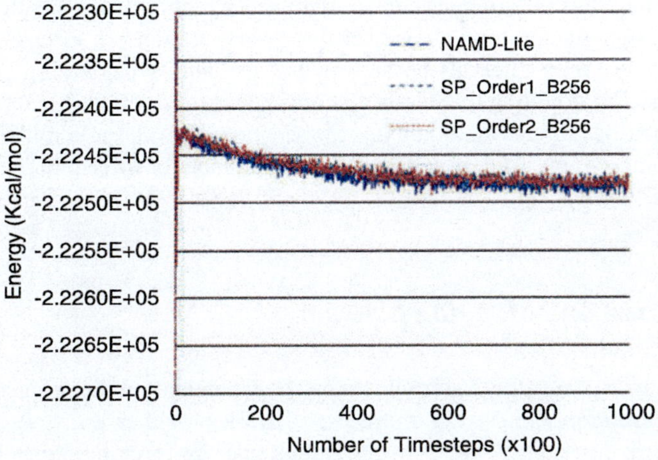

Fig. 8 Reference code and two designs run for 100,000 timesteps

interpolation have two orders of magnitude less error than 0th order. We also note that with 256 bins per section (and 12 sections), 1st and 2nd order are virtually identical.

The second method was to measure energy fluctuation and drift. Results are presented for the NAMD benchmark ApoA1. It has 92,224 particles, a bounding box of $108\AA \times 108\AA \times 78\AA$, and a cutoff radius of $12\AA$. The Coulomb force is evaluated with PME. A switching function is applied to smooth the LJ force when the intra-distance of particle pairs is between 10 and $12\AA$. Preliminary results are shown in the left side of Fig. 7. A number of design alternatives were examined, including the original code and all combinations of the following parameters: bin density (64 and 256 per section or segment), interpolation order (0th, 1st, and 2nd), and single and double-precision floating point. We note that all of the 0th order simulations are unacceptable, but that the others are all indistinguishable (in both energy fluctuation and drift) from the serial reference code running direct computation in double-precision floating point.

To validate the most promising candidate designs, longer runs were conducted. An energy plot for 100,000 timesteps is provided in Fig. 8. The graphs depict the original reference code and two FPGA designs. Both are single precision with 256 bins per interval; one is first order and the other second order. Good energy conservation is seen in the FPGA-accelerated versions. Only a small divergence of 0.02% was observed compared to the software-only version. The ΔE values, using (8), for all accelerated versions were found to be much smaller than 0.003.

One of the interesting contributions of this work was with respect to the utilization of Block RAM (BRAM) architecture of the FPGAs for interpolation. MD packages typically choose the interval such that the table is small enough to fit in L1 cache. This is compensated by the use of higher order of interpolation, second order being a common choice for force computation [9]. FPGAs, however, can afford having finer intervals because of the availability of on-chip BRAMs. It was found that, by doubling the number of bins per section, first order interpolation can achieve similar simulation quality as the second order interpolation (see Fig. 7). This saves logic and multipliers and increases the number of force pipelines that can fit in a single FPGA.

3.2 *Filtering and Mapping Scheme*

The performance of an FPGA kernel is directly dependent on the efficiency of the force computation pipelines. The more useful work pipelines do every cycle, the better the performance is. This in turn requires that the force pipelines be fed, as much as possible, with particle pairs that are within cutoff distance. Section 1.2 described two efficient methods for finding particle-pairs within cutoff distance. But for MD accelerators, this requires additional considerations. The cell list computation is very fast and the data generated are small, so it is generally done on the host. The results are downloaded to the FPGA every iteration. The neighbor-list method, on the other hand, is problematic if the lists are computed on the host. The size of the aggregate neighbor-lists is hundreds of times that of the cell lists, which makes their transfer to FPGA impractical. As a consequence, neighbor-list computation, if it is done at all, must be done on the FPGA.

This work first looks at MD with cell lists. For reference and without loss of generality, we examine the NAMD benchmark NAMD2.6 on ApoA1. It has 92,224 particles, a bounding box of $108\AA \times 108\AA \times 78\AA$, and a cutoff radius of $12\AA$. This roughly yields a simulation space of $9 \times 9 \times 7$ cells with an average of 175 particles per cell with a uniform distribution. On the FPGA, the working set is typically a single (home) cell and its cell neighborhood for a total of (naively) 27 cells and about 4,725 particles. Using Newton's third law (N3L), home cell particles are only matched with particles of part of the cell neighborhood, and with, on average, half of the particles in the home cell. For the 14- and 18-cell configurations (see later discussion on mapping scheme), the number of particles to be examined is 2,450 and 3,150, respectively. Given current FPGA technology, any of these cell neighborhoods (14, 18, or even 27) easily fits in the on-chip BRAMs.

On the other hand, neighbor-lists for a home cell do not fit on the FPGA. The aggregate neighbor-lists for 175 home cell particles is over 64,000 particles (one half of 732 per particle—732 rather than 4,725 because of increased efficiency).

The memory requirements are therefore very different. Cell-lists can be swapped back and forth between the FPGA and the DDR memory, as needed. Because of the high level of reuse, this is easily done in the background. In contrast, neighbor-list particles must be streamed from off-chip as they are needed. This has worked when there are one or two force pipelines operating at 100 MHz [26, 41], but is problematic for current and future high-end FPGAs. For example, the Stratix-III/Virtex-5 generation of FPGAs can support up to 8 force pipelines operating at 200 MHz leading to a bandwidth requirement of over 20 GB/s.

The solution proposed in this work is to use neighbor-lists, but to compute them every iteration, generating them continuously and consuming them almost immediately. There are three major issues that are addressed in this work, which we discuss next.

1. How should the filter be computed?
2. What cell neighborhood organization best takes advantage of N3L?
3. How should particle pairs be mapped to filter pipelines?

3.2.1 Filter Pipeline Design and Optimization

For a cell-list-based system where one home cell is processed at a time, with no filtering or other optimization, forces are computed between all pairs of particles i and j, where i must be in the home cell but j can be in any of the 27 cells of the cell neighborhood, including the home cell. *Filtering* here means the identification of particle pairs where the mutual short-range force is zero. A *perfect filter* successfully removes all such pairs. The *efficiency* of the filter is the ratio of undesirable particle pairs that were removed to the original number of undesirable particle pairs. The *extra work* due to imperfection is the ratio of undesirable pairs not removed to the desirable pairs.

Three methods are evaluated, two existing and one new, which trade off filter efficiency for hardware resources. As described in Sect. 3.1, particle positions are stored in three Cartesian dimensions, each in 32-bit integer. Filter designs have two parameters, precision and geometry.

1. *Full precision:* Precision = full, Geometry = sphere
   This filter computes $r^2 = x^2 + y^2 + z^2$ and compares whether $r^2 < r_c^2$ using full 32-bit precision. Filtering efficiency is nearly 100%. Except for the comparison operation, this is the same computation that is performed in the force pipeline.
2. *Reduced:* Precision = reduced, Geometry = sphere
   This filter, used by D.E. Shaw [28], also computes $r^2 = x^2 + y^2 + z^2, r^2 < r_c^2$ but uses fewer bits and so substantially reduces the hardware required. Lower precision, however, means that the cutoff radius must be increased (rounded up to the next bit) so filtering efficiency goes down: for 8 bits of precision, it is 99.5 for about 3% extra work.
3. *Planar:* Precision = reduced, Geometry = planes
   A disadvantage of the previous method is its use of multipliers, which are the critical resource in the force pipeline. This issue can be important because there are likely to be 6–10 filter pipelines per force pipeline. In this method we avoid multiplication by thresholding with planes rather than a sphere (see Fig. 9 for the 2D analog). The formulas are as follows:

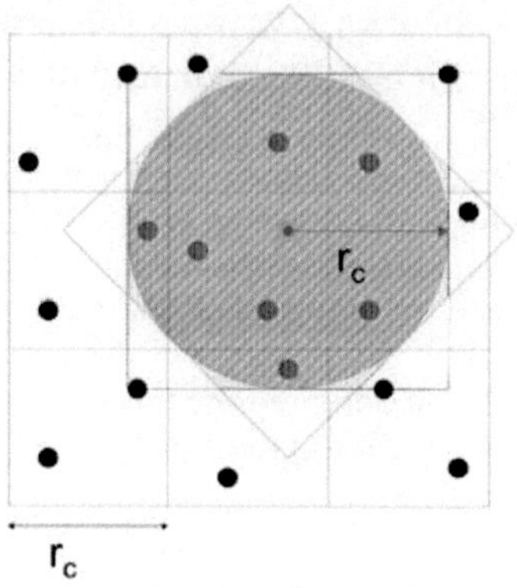

Fig. 9 Filtering with planes rather than a sphere—2D analogue

Table 2 Comparison of three filtering schemes with respect to quality and resource usage

Filtering Method	LUTs/registers		Multipliers		Filter eff.	Extra work
Full precision	341/881	0.43%	12	1.6%	100%	0%
Full prec.—logic only muls	2577/2696	1.3%	0	0.0%	100%	0%
Reduced precision	131/266	0.13%	3	0.4%	99.5%	3%
Reduced prec.—logic only muls	303/436	0.21%	0	0.0%	99.5%	3%
Planar	164/279	0.14%	0	0.0%	97.5%	13%
Force pipe	5695/7678	5.0%	70	9.1%	NA	NA

A force pipeline is shown for reference. Percent utilization is with respect to the Altera Stratix-III EP3SE260

- $|x| < r_c, |y| < r_c, |z| < r_c$
- $|x| + |y| < \sqrt{2}r_c, |x| + |z| < \sqrt{2}r_c, |y| + |z| < \sqrt{2}r_c$
- $|x| + |y| + |z| < \sqrt{3}r_c$

With 8 bits, this method achieves 97.5% efficiency for about 13% extra work.

Table 2 summarizes the cost (LUTs, registers, and multipliers) and quality (efficiency and extra work) of the three filtering methods. Since multipliers are a critical resource, we also show the two "sphere" filters implemented entirely with logic. The cost of a force pipeline (from Sect. 3.1) is shown for scale.

The most important result is the relative cost of the filters to the force pipeline. Depending on implementation and load balancing method (see later discussion on mapping scheme), each force pipeline needs between 6 and 9 filters to keep it running at full utilization. We refer to that set of filters as a *filter bank*. Table 2 shows that a *full precision* filter bank takes from 80 to 170% of the resources of its force pipeline. The *reduced (logic only)* and *planar* filter banks, however, require only a fraction: between 17 and 40% of the logic of the force pipeline and no multipliers at all. Since the latter is the critical resource, the conclusion is that the filtering logic itself (not including interfaces) has a minor effect on the number of force pipelines that can fit on the FPGA.

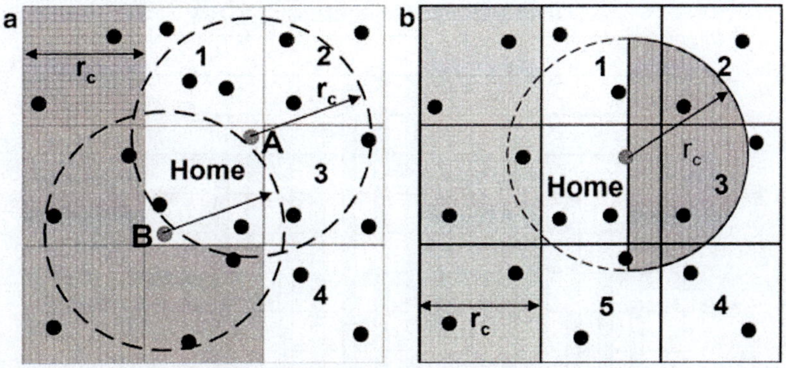

Fig. 10 Shown are two partitioning schemes for using Newton's 3rd law. In (**a**), 1–4 plus home are examined with a full sphere. In (**b**), 1–5 plus home are examined, but with a hemisphere (*shaded part of circle*)

We now compare the *reduced* and *planar* filters. The Extra Work column in Table 2 shows that for a *planar* filter bank to obtain the same performance as logic-only-*reduced*, the overall design must have 13% more throughput. This translates, e.g., to having 9 force pipelines when using *planar* rather than 8 for *reduced*. The total number of filters remains constant. The choice of filter therefore depends on the FPGA's resource mix.

3.2.2 Cell Neighborhood Organization

For efficient access of particle memory and control, and for smooth interaction between filter and force pipelines, it is preferred to have each force pipeline handle the interactions of a single reference particle (and its partner particles) at a time. This preference becomes critical when there are a large number of force pipelines and a much larger number of filter pipelines. Moreover, it is highly desirable for all of the neighbor-lists being created at any one time (by the filter banks) to be transferred to the force pipelines simultaneously. It follows that each reference particle should have a similar number of partner particles (neighbor-list size).

The problem addressed here is that the standard method of choosing a reference particle's partner particles leads to a severe imbalance in neighbor-list sizes. How this arises can be seen in Fig. 10a, which illustrates the standard method of optimizing for N3L. So that a force between a particle pair is computed only once, only a "half shell" of the surrounding cells is examined (in 2D, this is cells **1–4** plus **Home**). For forces between the reference particle and other particles in **Home**, the particle ID is used to break the tie, with, e.g., the force being computed only when the ID of the reference particle is the higher. In Fig. 10a, particle *B* has a much smaller neighbor-list than *A*, especially if *B* has a low ID and *A* a high.

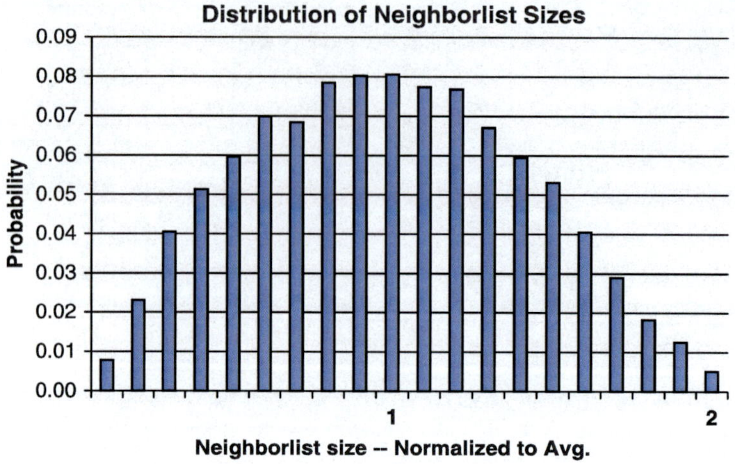

Fig. 11 Distribution of neighbor-list sizes for standard partition as derived from Monte Carlo simulations

In fact neighbor-list sizes vary from 0 to $2L$, where L is the average neighbor-list size. The significance is as follows. Let all force pipelines wait for the last pipeline to finish before starting work on a new reference particle. Then if that (last) pipeline's reference particle has a neighbor-list of size $2L$, then the latency will be double that if all neighbor-lists were size L. This distribution has high variance (see Fig. 11), meaning that neighbor-list sizes greater than, say, $\frac{3}{2}L$, are likely to occur. A similar situation also occurs in other MD implementations, with different architectures calling for different solutions [2, 47].

One way to deal with this load imbalance is to overlap the force pipelines so that they work independently. While viable, this leads to much more complex control.

An alternative is to change the partitioning scheme. Our new N3L partition is shown in Fig. 10b. There are three new features. The first is that the cell set has been augmented from a half shell to a prism. In 2D this increases the cell set from 5 cells to 6; in 3D the increase is from 14 to 18. The second is that, rather than forming a neighbor-list based on a cutoff sphere, a hemisphere is used instead (the "half-moons" in Fig. 10b). The third is that there is now no need to compare IDs of home cell particles.

We now compare the two partitioning schemes. There are two metrics: the effect on the load imbalance and the extra resources required to prevent it.

1. *Effect of load imbalance.* We assume that all of the force pipelines begin computing forces on their reference particles at the same time, and that each force pipeline waits until the last force pipeline has finished before continuing to the next reference particle. We call the set of neighbor-lists that are thus processed simultaneously a *cohort*. With perfect load balancing, all neighbor-lists in a cohort would have the same size, the average L. The effect of the

variation in neighbor-list size is in the number of *excess* cycles—before a new cohort of reference particles can begin processing—over the number of cycles if each neighbor-list were the same size. The performance cost is therefore the average number of excess cycles per cohort. This in turn is the average size of the biggest neighbor-list in a cohort minus the average neighbor-list size. It is found that, for the standard N3L method, the average excess is nearly 50%, while for the "half-moon" method it is less than 5%.

2. *Extra resources.* The extra work required to achieve load balance is proportional to the extra cells in the cell set: 18 versus 14, or an extra 29%. This drops the fraction of neighbor-list particles in the cell neighborhood from 15.5 to 11.6%, which in turns increases the number of filters needed to keep the force pipelines fully utilized (overprovisioned) from 7 to 9. For the reduced and planar filters, this is not likely to reduce the number of force pipelines.

3.2.3 Mapping Particle Pairs to Filter Pipelines

From the previous sections an efficient design for filtering and mapping particles follows.

- During execution, the input working set (data held on the FPGA) consists of the positions of particles in a single home cell and in its 17 neighbors;
- Particles in each cell are mapped to a set of BRAMs, currently one or two per dimension, depending on the cell size;
- The N3L algorithm specifies 8 filter pipelines per force pipeline; and
- FPGA resources indicate around 6–8 force pipelines.

The problem we address in this subsection is the mapping of particle pairs to filter pipelines. There are a (perhaps surprisingly) large number of ways to do this; finding the optimal mapping is in some ways analogous to optimizing loop interchanges with respect to a cost function. For example, one mapping maps one reference particle at a time to a bank of filter pipelines and relates each cell with one filter pipeline. The advantage of this method is that the outputs of this (8-way) filter bank can then be routed directly to a force pipeline. This mapping, however, leads to a number of load balancing, queuing, and combining problems.

A preferred mapping is shown in Fig. 12. The key idea is to associate each filter pipeline with a single reference particle (at a time) rather than a cell. Details are as follows. By "particle" we mean "position data of that particle during this iteration."

- A phase begins with a new and distinct reference particle being associated with each filter.
- Then, on each cycle, a single particle from the 18-cell neighborhood is broadcast to all of the filter.
- Each filters output goes to a single set of BRAMs.

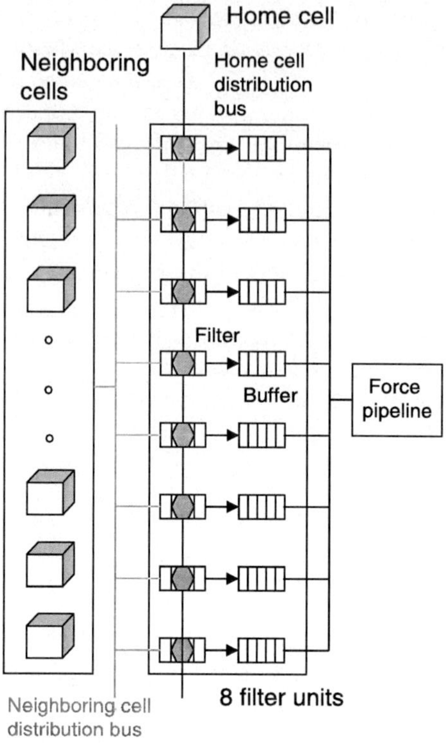

Fig. 12 A preferred mapping of particle pairs onto filter pipelines. Each filter is used to compute all interactions for a single reference particle for an entire cell neighborhood

- The output of each filter is exactly the neighbor-list for its associated reference particle.
- Double buffering enables neighbor-lists to be generated by the filters at the same time they are drained by the force pipelines.

Advantages of this method include:

- Perfect load balance among the filters;
- Little overhead: Each phase consists of 3,150 cycles before a new set of reference particles must be loaded;
- Nearly perfect load balancing among the force pipelines: Each operates successively on a single reference particle and its neighbor-list; and
- Simple queueing and control: Neighbor-list generation is decoupled from force computation.

This mapping does require larger queues than mappings where the filter outputs feed more directly into the force pipelines. But since there are BRAMs to spare, this is not likely to have an impact on performance.

A more substantial concern is the granularity of the processing phases. The number of phases necessary to process the particles in a single home cell is $\lceil |\text{particles-in-home-cell}| / |\text{filters}| \rceil$. For small cells the loss of efficiency can become significant. There are several possible solutions.

- Increase the number of filters and further decouple neighbor-list generation from consumption. The reasoning is that as long as the force pipelines are busy, some inefficiency in filtering is tolerable.
- Overlap processing of two home cells. This increases the working set from 18 to 27 cells for a modest increase in number of BRAMs required. One way to implement this is to add a second distribution bus.
- Another way to overlap processing of two home cells is to split the filters among them. This halves the phase granularity and so the expected inefficiency without significantly changing the amount of logic required for the distribution bus.

3.3 Overall Design and Board-Level Issues

In this subsection we describe the overall design (see Fig. 13), especially how data are transferred between host and accelerator and between off-chip and on-chip memory. The reference design assumes an implementation of 8 force and 72 filter pipelines.

1. *Host-Accelerator data transfers:* At the highest level, processing is built around the timestep iteration and its two phases: force calculation and motion update. During each iteration, the host transfers position data to, and acceleration data from, the coprocessor's on-board memory (POS SRAM and ACC SRAM, respectively). With 32-bit precision, 12 bytes are transferred per particle. While the phases are necessarily serial, the data transfers require only a small fraction of the processing time. For example, while the short-range force calculation takes about 55 ms for 100 K particles and increases linearly with particle count through the memory capacity of the board, the combined data transfers of 2.4 MB take only 2–3 ms. Moreover, since simulation proceeds by cell set, processing of the force calculation phase can begin almost immediately as the data begin to arrive.
2. *Board-level data transfers:* Force calculation is built around the processing of successive home cells. Position and acceleration data of the particles in the cell set are loaded from board memory into on-chip caches, POS and ACC, respectively. When the processing of a home cell has completed, ACC data are written back. Focus shifts and a neighboring cell becomes the new home cell. Its cell set is now loaded; in the current scheme this is usually nine cells per shift. The transfers are double buffered to hide latency. The time to process a home cell T_{proc} is generally greater than the time T_{trans} to swap cell sets with off-chip memory. Let a cell contain an average of N_{cell} particles. Then $T_{trans} = 324 \times N_{cell}/B$ (9 cells, 32-bit data, 3 dimensions, 2 reads and 1 write, and transfer bandwidth of B bytes per cycle). To compute T_{proc}, assume P pipelines and perfect efficiency. Then $T_{proc} = N_{cell}^2 \times 2\pi/3P$ cycles. This gives the following bandwidth requirement: $B > 155 * P/N_{cell}$. For $P = 8$ and $N_{cell} = 175$, $B > 7.1$ bytes per cycle. For many current FPGA processor boards $B \geq 24$.

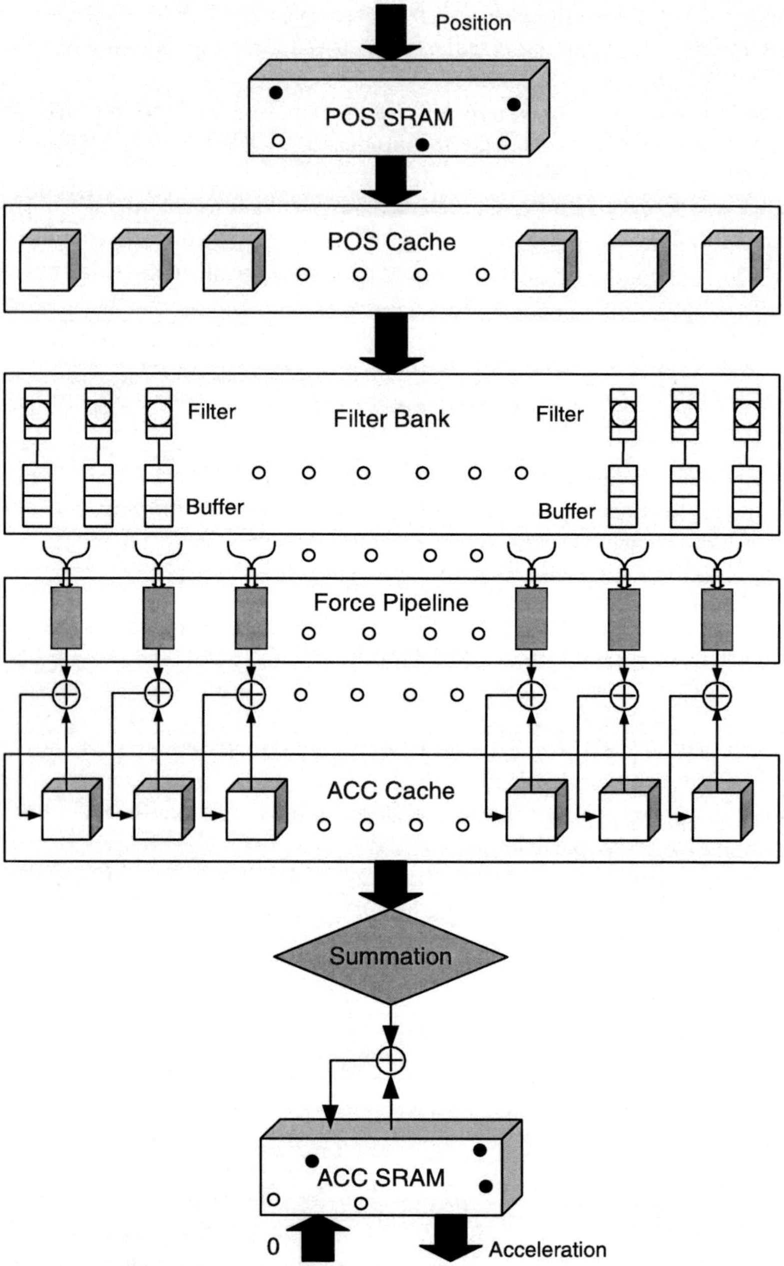

Fig. 13 Schematic of the HPRC MD system

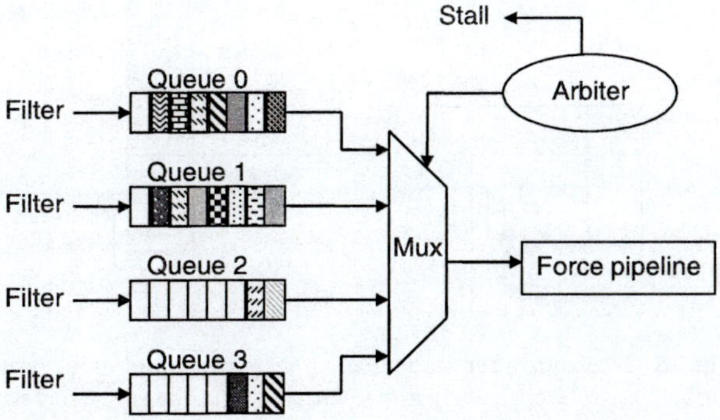

Fig. 14 Concentrator logic between filters and force pipeline

Some factors that increase the bandwidth requirement are faster processor speeds, more pipelines, and lower particle density. A factor that reduces the bandwidth requirement is better cell reuse.

3. *On-chip data transfers:* Force computation has three parts, filtering particle pairs, computing the forces themselves, and combining the accumulated accelerations. In the design of the on-chip data transfers, the goals are simplicity of control and minimization of memory and routing resources. Processing of a home cell proceeds in *cohorts* of reference particles that are processed simultaneously, either 8 or 72 at a time (either one per filter bank or one per force pipeline). This allows a control with a single state machine, minimizes memory contention, and simplifies accumulation. For this scheme to run at high efficiency, two types of load-balancing are required: (1) the work done by various filter banks must be similar and (2) filter banks must generate particle pairs having nontrivial interactions on nearly every cycle.
4. *POS cache to filter pipelines:* Cell set positions are stored in 54–108 BRAMS, i.e., 1–2 BRAMs per dimension per cell. This number depends on the BRAM size, cell size, and particle density. Reference particles are always from the home cell, partner particles can come from anywhere in the cell set.
5. *Filter pipelines to force pipelines:* A concentrator logic is used to feed the output of multiple filters to a pipeline (Fig. 14). Various strategies were discussed in [8].
6. *Force pipelines to ACC cache:* To support N3L, two copies are made of each computed force. One is accumulated with the current reference particle. The other is stored by index in one of the large BRAMs on the Stratix-III. Figure 15 shows the design of the accumulator.

3.4 Preliminary Work in Long-Range Force Computation

In 2005, Prof. Paul Chow's group at the University of Toronto made an effort to accelerate the reciprocal part of SPME on a Xilinx XC2V2000 FPGA [29]. The computation was performed with fixed-point arithmetic that has various precisions

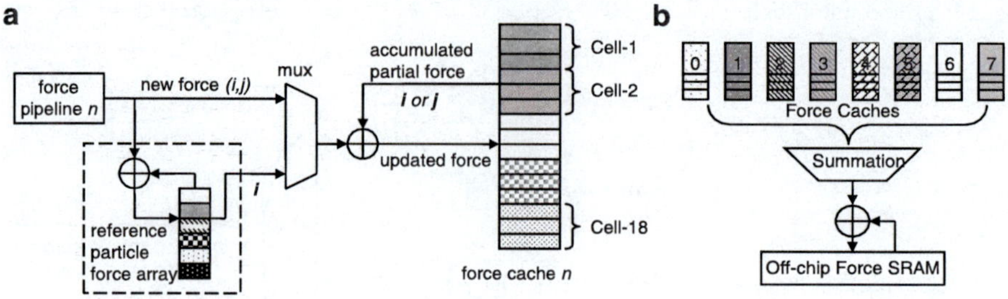

Fig. 15 Mechanism for accumulating per particle forces. (**a**) shows the logic for a single pipeline for both the reference and partner particles. (**b**) shows how forces are accumulated across multiple pipelines

to improve numerical accuracy. Due to the limited logic resources and slow speed grade, the performance was sacrificed by some design choices, such as the sequential executions of the reciprocal force calculation for *x*, *y*, and *z* directions and slow radix-2 FFT implementation. The performance was projected to be a factor of 3× to 14× over the software implementation running in an Intel 2.4GHz Pentium 4 processor. At Boston University the long-range electrostatic force was implemented using Multigrid [17] with a factor of 5× to 7× speed-up reported.

3.5 Preliminary Work in Parallel MD

Maxwell is an FPGA-based computing cluster developed by the FHPCA (FPGA High Performance Computing Alliance) project at EPCC (Edinburgh Parallel Computing Centre) at the University of Edinburgh [3]. The architecture of Maxwell comprises 32 blades housed in an IBM Blade Center. Each blade consists of one Xeon processor and 2 Virtex-4 FX-100 FPGAs. The FPGAs are connected by a fast communication subsystem which enables the total of 64 FPGAs to be connected together in an 8 × 8 torus. Each FPGA also has four 256 MB DDR2 SDRAMs. The FPGAs are connected with the host via a PCI bus.

In 2011, an FPGA-accelerated version of LAMMPS was reported to be implemented on Maxwell [24, 37]. Only range-limited non-bonded forces (including potential and virial) were computed on the FPGAs with 4 identical pipelines/FPGA. A speed-up of up to 14× was reported for the kernel (excluding data communication) on two or more nodes of the Maxwell machine, although the end-to-end performance was worse than the software-only version.

This work essentially implemented the inner-loop of a neighbor-list-based force computation as the FPGA kernel. Every time a particle and its neighbor-list would be sent to the FPGAs from the host and then corresponding forces would be computed on the FPGAs. This incurred tremendous amount of data communication which ultimately resulted in the slowdown of the FPGA-accelerated version.

They simulated a Rhodopsin protein in solvated lipid bilayer with LJ forces and PPPM method. The 32 K system was replicated to simulate larger systems. This work, however, to the best of our knowledge, is the first to integrate an FPGA MD kernel to a full-parallel MD package.

4 Future Challenges and Opportunities

The future of FPGA-accelerated MD vastly depends on the cooperation and collaboration among computational biologists, computer architects, and board/EDA tool vendors. In the face of the high bar set by GPU implementations, researchers and engineers from all of these three sectors must come together to make this a success. The bit-level programmability and fast data communication capability, together with their power efficiency, do make FPGAs seem like the best candidate for MD accelerator. But to realize the potential, computer architects will have to work with the computational biologists to understand the characteristics of the existing MD packages and develop FPGA kernels accordingly. The board and EDA tool vendors will have to make FPGA devices much easier to deploy. Currently FPGA kernels are mostly designed and managed by hardware engineers. A CUDA-like breakthrough here would make FPGAs accessible to a much broader audience.

Below, we discuss some of the specific challenges that need to be addressed in order to achieve the full potential of FPGAs in accelerating MD. These challenges provide researchers with great opportunities for inventions and advancements that are very likely to be applicable to other similar computational problems, e.g., N-body simulations.

4.1 Integration into Full-Parallel Production MD Packages

After a decade of research on FPGA-accelerated MD, with many individual pieces of work here and there, none of the widely used MD packages have an FPGA-accelerated version. Part of this is because FPGA developers have only focused on individual sections of the computation. But another significant reason is the lack of understanding of how these highly optimized MD packages work and what needs to be done to get the best out of FPGAs, without breaking the structure of the original packages. Researchers need to take a top-down approach and focus on the need of the software. Certain optimizations on the CPUs may need to be revisited, because we may have more efficient solutions on FPGAs, e.g. table-interpolation using BRAM as described in Sect. 3.1. Also, more effort must be given on overlapping computation and communication.

4.2 Use of FPGAs for Inter-Node Communication

While CPU-only MD remains compute-bound for at least a few hundred compute nodes, that is not the case for accelerated versions. It should be evident from the GPU experience that communication among compute nodes will become a bottleneck even for small systems. The need for fast data communication is especially crucial in evaluating the long-range portion of electrostatic force, which is often based on the 3D FFT and requires all-to-all communication during a timestep. Without substantial improvement in such inter-node communication, FPGA-acceleration will be limited to only a few times of speed-up. This presents a highly promising area of research where FPGAs can be used directly for communication between compute nodes. FPGAs are already used in network routers and seem like a natural fit for this purpose [20].

4.3 Building an Entirely FPGA-Centric MD Engine

As Moore's law continues, FPGAs are equipped with more functionality than ever. It is possible to have embedded processors on FPGAs, either soft or hard, which makes it feasible to create an entirely FPGA-centric MD engine. In such an engine, overall control and simple software tasks will be done on the embedded processors while the heavy work like the non-bonded force computations will be implemented on the remaining logic. Data communication can also be performed using the FPGAs, completely eliminating general purpose CPUs from the scene. Such a system is likely to be highly efficient, in terms of both computational performance and energy consumption.

4.4 Validating Simulation Quality

While MD packages typically use double-precision floating point for most of the computation, most FPGA work used fixed, semi-floating or a mixture of fixed and floating point for various stages of MD. Although some of these studies verified accuracy through various metrics, none of the FPGA-accelerated MD work presented results of significantly long (e.g., month-long) runs of MD. Thus it is important to address this issue of accuracy. This may mean revisiting precision and interpolation order in the force pipelines.

Acknowledgments This work was supported in part by the NIH through award #R01-RR023168-01A1 and by the MGHPCC.

References

1. S.A. Adcock, J.A. McCammon, Molecular dynamics: survey of methods for simulating the activity of proteins. Chem. Rev. **106**(5), 1589–1615 (2006)
2. J.A. Anderson, C.D. Lorenz, A. Travesset, General purpose molecular dynamics simulations fully implemented on graphics processing units. J. Comput. Phys. **227**(10), 5342–5359 (2008)
3. R. Baxter, S. Booth, M. Bull, G. Cawood, J. Perry, M. Parsons, A. Simpson, A. Trew, A. McCormick, G. Smart, R. Smart, A. Cantle, R. Chamberlain, G. Genest, Maxwell - a 64 FPGA supercomputer, in *Second NASA/ESA Conference on Adaptive Hardware and Systems (AHS)* (2007), IEEE Computer Society, Washington, DC, USA, pp. 287–294
4. K.J. Bowers, E. Chow, H. Xu, R.O. Dror, M.P. Eastwood, B.A. Gregersen, J.L. Klepeis, I. Kolossvary, M.A. Moraes, F.D. Sacerdoti, J.K. Salmon, Y. Shan, D.E. Shaw, Scalable algorithms for molecular dynamics simulations on commodity clusters, in *Proceedings of the 2006 ACM/IEEE Conference on Supercomputing (SC)* (2006), ACM New York, NY, USA, pp. 84:1–84:13
5. B.R. Brooks, C.L. Brooks III, A.D. Mackerell Jr., L. Nilsson, R.J. Petrella, B. Roux, Y. Won, G. Archontis, C. Bartels, S. Boresch, A. Caflisch, L. Caves, Q. Cui, A.R. Dinner, M. Feig, S. Fischer, J. Gao, M. Hodoscek, W. Im, K. Kuczera, T. Lazaridis, J. Ma, V. Ovchinnikov, E. Paci, R.W. Pastor, C.B. Post, J.Z. Pu, M. Schaefer, B. Tidor, R.M. Venable, H.L. Woodcock, X. Wu, W. Yang, D.M. York, M. Karplus, CHARMM: the biomolecular simulation program. J. Comput. Chem. **30**(10, Sp. Iss. SI), 1545–1614 (2009)
6. D.A. Case, T.E. Cheatham, T. Darden, H. Gohlke, R. Luo, K.M. Merz Jr., A. Onufriev, C. Simmerling, B. Wang, R.J. Woods, The Amber biomolecular simulation programs. J. Comput. Chem. **26**(16), 1668–1688 (2005)
7. M. Chiu, M.C. Herbordt, Efficient particle-pair filtering for acceleration of molecular dynamics simulation, in *International Conference on Field Programmable Logic and Applications (FPL)* (2009), ACM New York, NY, USA, pp. 345–352
8. M. Chiu, M.C. Herbordt, Molecular dynamics simulations on high-performance reconfigurable computing systems. ACM Trans. Reconfigurable Tech. Syst. (TRETS) **3**(4), 23:1–23:37 (2010)
9. M. Chiu, M.A. Khan, M.C. Herbordt, Efficient calculation of pairwise nonbonded forces, in *The 19th Annual International IEEE Symposium on Field-Programmable Custom Computing Machines (FCCM)* (2011), IEEE Computer Society Washington, DC, USA, pp. 73–76
10. S. Chiu, Accelerating molecular dynamics simulations with high-performance reconfigurable systems, PhD dissertation, Boston University, USA, 2011
11. T. Darden, D. York, L. Pedersen, Particle mesh Ewald: an N.log (N) method for Ewald sums in large systems. J. Chem. Phys. **98**(12), 10089–10092 (1993)
12. W.A. Eaton, V. Muñoz, P.A. Thompson, C.K. Chan, J. Hofrichter, Submillisecond kinetics of protein folding. Curr. Opin. Struct. Biol. **7**(1), 10–14 (1997)
13. R.D. Engle, R.D. Skeel, M. Drees, Monitoring energy drift with shadow Hamiltonians. J. Comput. Phys. **206**(2), 432–452 (2005)
14. P.L. Freddolino, A.S. Arkhipov, S.B. Larson, A. McPherson, K. Schulten, Molecular dynamics simulations of the complete satellite tobacco mosaic virus. Structure **14**(3), 437–449 (2006)
15. Gidel, Gidel website (2009), http://www.gidel.com. Accessed 17 April 2012
16. GROMACS, GROMACS installation instructions for GPUs (2012), http://www.gromacs.org/Downloads/Installation_Instructions/GPUs. Accessed 17 April 2012
17. Y. Gu, M.C. Herbordt, FPGA-based multigrid computation for molecular dynamics simulations, in *15th Annual IEEE Symposium on Field-Programmable Custom Computing Machines (FCCM)* (2007), pp. 117–126
18. Y. Gu, T. Vancourt, M.C. Herbordt, Explicit design of FPGA-based coprocessors for short-range force computations in molecular dynamics simulations. Parallel Comput. **34**(4–5), 261–277 (2008)
19. D.J. Hardy, NAMD-Lite (2007), http://www.ks.uiuc.edu/Development/MDTools/namdlite/. University of Illinois at Urbana-Champaign. Accessed 17 April 2012

20. M. Herbordt, M. Khan, Communication requirements of fpga-centric molecular dynamics, in *Proceedings of the Symposium on Application Accelerators for High Performance Computing* (2012)
21. B. Hess, C. Kutzner, D. van der Spoel, E. Lindahl, GROMACS 4: algorithms for highly efficient, load-balanced, and scalable molecular simulation. J. Chem. Theor. Comput. **4**(3), 435–447 (2008)
22. R. Hockney, S. Goel, J. Eastwood, Quiet high-resolution computer models of a plasma. J. Comput. Phys. **14**(2), 148–158 (1974)
23. L. Kalé, R. Skeel, M. Bhandarkar, R. Brunner, A. Gursoy, N. Krawetz, J. Phillips, A. Shinozaki, K. Varadarajan, K. Schulten, NAMD2: Greater scalability for parallel molecular dynamics. J. Comput. Phys. **151**, 283–312 (1999)
24. S. Kasap, K. Benkrid, A high performance implementation for molecular dynamics simulations on a FPGA supercomputer, in *2011 NASA/ESA Conference on Adaptive Hardware and Systems (AHS)* (2011), IEEE Computer Society Washington, DC, USA, pp. 375–382
25. F. Khalili-Araghi, E. Tajkhorshid, K. Schulten, Dynamics of K+ ion conduction through Kv1.2. Biophys. J. **91**(6), 72–76 (2006)
26. V. Kindratenko, D. Pointer, A case study in porting a production scientific supercomputing application to a reconfigurable computer, in *14th Annual IEEE Symposium on Field-Programmable Custom Computing Machines (FCCM)* (2006), IEEE Computer Society Washington, DC, USA, pp. 13–22
27. S. Kumar, C. Huang, G. Zheng, E. Bohm, A. Bhatele, J.C. Phillips, H. Yu, L.V. Kalé, Scalable molecular dynamics with NAMD on the IBM Blue Gene/L system. IBM J. Res. Dev. **52**(1–2), 177–188 (2008)
28. R. Larson, J. Salmon, R. Dror, M. Deneroff, C. Young, J. Grossman, Y. Shan, J. Klepeis, D. Shaw, High-throughput pairwise point interactions in Anton, a specialized machine for molecular dynamics simulation, in *IEEE 14th International Symposium on High Performance Computer Architecture (HPCA)* (2008), IEEE Computer Society Washington, DC, USA, pp. 331–342
29. S. Lee, An FPGA implementation of the Smooth Particle Mesh Ewald reciprocal sum compute engine, Master's thesis, The University of Toronto, Canada, 2005
30. A.D. MacKerell, N. Banavali, N. Foloppe, Development and current status of the CHARMM force field for nucleic acids. Biopolymers **56**(4), 257–265 (2000)
31. G. Moraitakis, A.G. Purkiss, J.M. Goodfellow, Simulated dynamics and biological macromolecules. Rep. Progr. Phys. **66**(3), 383 (2003)
32. T. Narumi, Y. Ohno, N. Futatsugi, N. Okimoto, A. Suenaga, R. Yanai, M. Taiji, A high-speed special-purpose computer for molecular dynamics simulations: MDGRAPE-3. NIC Workshop, From Computational Biophysics to Systems Biology, NIC Series, vol. 34 (2006), pp. 29–36
33. L. Nilsson, Efficient table lookup without inverse square roots for calculation of pair wise atomic interactions in classical simulations. J. Comput. Chem. **30**(9), 1490–1498 (2009)
34. J.C. Phillips, R. Braun, W. Wang, J. Gumbart, E. Tajkhorshid, E. Villa, C. Chipot, R.D. Skeel, L. Kalé, K. Schulten, Scalable molecular dynamics with NAMD. J. Comput. Chem. **26**(16), 1781–1802 (2005)
35. J.C. Phillips, J.E. Stone, K. Schulten, Adapting a message-driven parallel application to GPU-accelerated clusters, in *Proceedings of the ACM/IEEE Conference on Supercomputing (SC)* (2008), IEEE Press Piscataway, NJ, USA, pp. 8:1–8:9
36. L. Phillips, R.S. Sinkovits, E.S. Oran, J.P. Boris, The interaction of shocks and defects in Lennard-Jones crystals. J. Phys.: Condens. Matter **5**(35), 6357–6376 (1993)
37. S. Plimpton, Fast parallel algorithms for short-range molecular dynamics. J. Comput. Phys. **117**(1), 1–19 (1995)
38. J.W. Ponder, D.A. Case, Force fields for protein simulations. Adv. Protein Chem. **66**, 27–85 (2003)
39. D.C. Rapaport, *The Art of Molecular Dynamics Simulation*, 2nd edn. (Cambridge University Press, London, 2004)

40. P. Schofield, Computer simulation studies of the liquid state. Comp. Phys. Comm. **5**(1), 17–23 (1973)
41. R. Scrofano, M. Gokhale, F. Trouw, V.K. Prasanna, A hardware/software approach to molecular dynamics on reconfigurable computers, in *The 14th Annual IEEE Symposium on Field-Programmable Custom Computing Machines (FCCM)* (2006), IEEE Computer Society Washington, DC, USA, pp. 23–34
42. Y. Shan, J. Klepeis, M. Eastwood, R. Dror, D. Shaw, Gaussian split Ewald: a fast Ewald mesh method for molecular simulation. J. Chem. Phys. **122**(5), 54101:1–54101:13 (2005)
43. D.E. Shaw, M.M. Deneroff, R.O. Dror, J.S. Kuskin, R.H. Larson, J.K. Salmon, C. Young, B. Batson, K.J. Bowers, J.C. Chao, M.P. Eastwood, J. Gagliardo, J.P. Grossman, C.R. Ho, D.J. Ierardi, I. Kolossváry, J.L. Klepeis, T. Layman, C. McLeavey, M.A. Moraes, R. Mueller, E.C. Priest, Y. Shan, J. Spengler, M. Theobald, B. Towles, S.C. Wang, Anton, a special-purpose machine for molecular dynamics simulation, in *Proceedings of the 34th Annual International Symposium on Computer Architecture (ISCA)* (2007), ACM New York, NY, USA, pp. 1–12
44. D.E. Shaw, M.M. Deneroff, R.O. Dror, J.S. Kuskin, R.H. Larson, J.K. Salmon, C. Young, B. Batson, K.J. Bowers, J.C. Chao, M.P. Eastwood, J. Gagliardo, J.P. Grossman, C.R. Ho, D.J. Ierardi, I. Kolossváry, J.L. Klepeis, T. Layman, C. McLeavey, M.A. Moraes, R. Mueller, E.C. Priest, Y. Shan, J. Spengler, M. Theobald, B. Towles, S.C. Wang, Anton, a special-purpose machine for molecular dynamics simulation. Comm. ACM **51**(7), 91–97 (2008)
45. D.E. Shaw, R.O. Dror, J.K. Salmon, J.P. Grossman, K.M. Mackenzie, J.A. Bank, C. Young, M.M. Deneroff, B. Batson, K.J. Bowers, E. Chow, M.P. Eastwood, D.J. Ierardi, J.L. Klepeis, J.S. Kuskin, R.H. Larson, K. Lindorff-Larsen, P. Maragakis, M.A. Moraes, S. Piana, Y. Shan, B. Towles, Millisecond-scale molecular dynamics simulations on Anton, in *Proceedings of the Conference on High Performance Computing Networking, Storage and Analysis (SC)* (2009), ACM New York, NY, USA, pp. 39:1–39:11
46. R.D. Skeel, I. Tezcan, D.J. Hardy, Multiple grid methods for classical molecular dynamics. J. Comput. Chem. **23**(6), 673–684 (2002)
47. M. Snir, A note on N-body computations with cutoffs. Theor. Comput. Syst. **37**(2), 295–318 (2004)
48. J.E. Stone, J.C. Phillips, P.L. Freddolino, D.J. Hardy, L.G. Trabuco, K. Schulten, Accelerating molecular modeling applications with graphics processors. J. Comput. Chem. **28**(16), 2618–2640 (2007)
49. L. Verlet, Computer "Experiments" on classical fluids. I. Thermodynamical properties of Lennard-Jones molecules. Phys. Rev. **159**(1), 98–103 (1967)
50. C. Young, J.A. Bank, R.O. Dror, J.P. Grossman, J.K. Salmon, D.E. Shaw, A 32x32x32, spatially distributed 3D FFT in four microseconds on Anton, in *Proceedings of the Conference on High Performance Computing Networking, Storage and Analysis (SC)* (2009), ACM New York, NY, USA, pp. 23:1–23:11

FPGA-Based HPRC for Bioinformatics Applications

Yoshiki Yamaguchi, Yasunori Osana, Masato Yoshimi, and Hideharu Amano

Abstract Bioinformatics is one of the most frequently applied fields in FPGAs. Some applications in this field can be efficiently implemented by systolic arrays, which are intrinsically suited to FPGA implementations. Others can be expressed as numerical computations which can parallelize through pipelining, instruction-level and data-level parallelism. This chapter covers two sample applications encountered in bioinformatics, namely homology searches and biochemical molecular simulations, and shows how FPGAs can be effectively harnessed to achieve higher performances compared to off-the-shelf microprocessor technologies.

1 Introduction

Systems biology is a scientific domain to grasp the essentials of living organisms. In order to provide a profound insight for the research field, the performance expansion of analytical devices needs to be considered from the viewpoint of both

---

Y. Yamaguchi (✉)
University of Tsukuba, Tsukuba, Ibaraki, Japan
e-mail: yoshiki@cs.tsukuba.ac.jp

Y. Osana
University of the Ryukyus, Okinawa, Japan
e-mail: osana@eee.u-ryukyu.ac.jp

M. Yoshimi
University of Electro-Communications, Tokyo, Japan
e-mail: yoshimi@is.uec.ac.jp

H. Amano
Keio University, Tokyo, Japan
e-mail: hunga@am.ics.keio.ac.jp

W. Vanderbauwhede and K. Benkrid (eds.), *High-Performance Computing Using FPGAs*, DOI 10.1007/978-1-4614-1791-0_5,

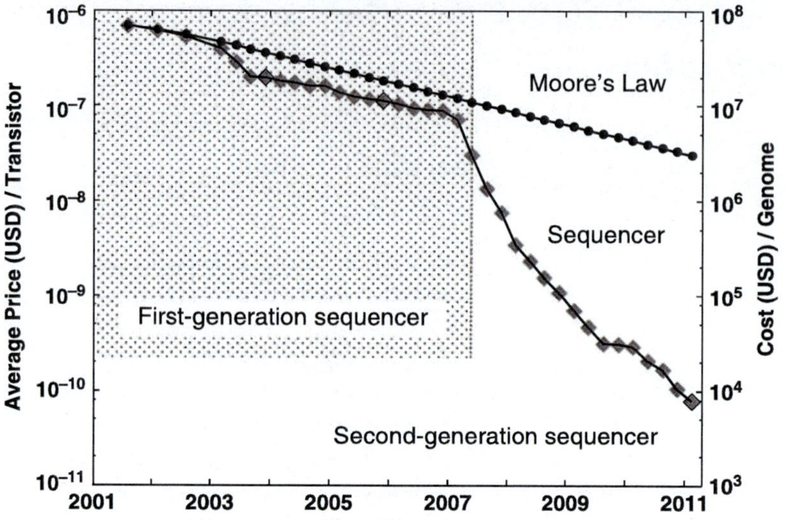

Fig. 1 Cost comparison of DNA sequencers and LSIs [1]

wet (experimental) and dry (computational) biological demands. Figure 1 shows the growth of DNA sequences in wet laboratories and LSIs in dry laboratories. The improved performance of DNA sequencers lets the cost reduced to 10^{-4} of that of a decade ago [1]. Technological innovations designed to LSIs, which is called the Moore's law, cannot follow the growth curve of DNA sequences in the second generation. It implies that software-level improvement on LSIs is currently having a hard time to continue computing big biological applications more than ever. Thus, hardware-level designs including computational hardware algorithm and architecture must be discussed for the expansion of the systems biology.

There is another problem in the systems biology. Figure 2 shows the Omic Space, which is one of the famous coordinate-based integration schemes for the systems biology [2, 3]. The computational system needs to have an analytical capability across these planes in Fig. 2. The results will provide us further novel biological insights when the system has multifaceted perspectives. The above factors will make us enter a new stage, which targets the entire cellular system.

This chapter offers a better solution for the bioinformatics system design. FPGAs can achieve high performance from highly paralleled and finely grained computation to numerical and sophisticated simulations. Moreover, the system with FPGAs proposes highly power-efficient performance compared with GPGPU and other architectures. Two application fields, homology search and biochemical reaction simulation are chosen. The following sections will describe the high potential capacity of FPGA systems.

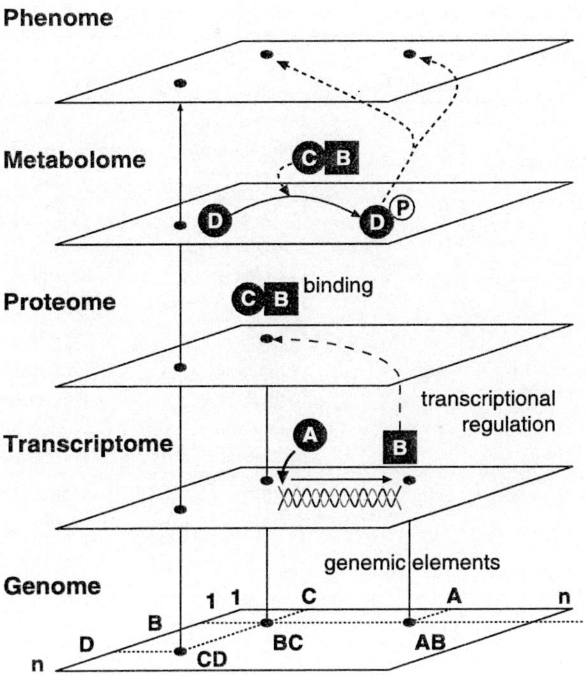

Fig. 2 Outline of the Omic space [2]

2 Homology Search

"The history of the earth is recorded in the layers of its crust;
The history of all organisms is inscribed in the chromosomes."

H. Kihara (1946)

There are ten million species of organisms on Earth and the diversity of them is derived from DNA sequences which are blueprints of the lives. If we decipher the meaning of them, we will achieve not only actual benefits that can accelerate medicine, pharmacy, agronomy and other fields, but also a theory that could possibly unite biology and physics. This section will specifically describe hardware-level acceleration with FPGAs.

2.1 Introduction

DNA sequences are composed of pairs of four nitrogenous nucleotide bases: thymine (T), cytosine (C), adenine (A), and guanine (G). A gene which is a portion of a DNA sequence is transcribed into an RNA and then translated to make proteins. In this process, each set of three nucleotide bases, namely codon, specifies a single amino acid as shown in the Table 1.

Table 1 Four nucleotide bases and 20 amino acids

		Third base			
First base	Second base	T(U)	C	A	G
T(U)	T(U)	Phenylalanine	Serine	Tyrosine	Cysteine
	C	Phenylalanine	Serine	Tyrosine	Cysteine
	A	Leucine	Serine	(*Stop*)	(*Stop*)
	G	Leucine	Serine	(*Stop*)	Tryptophan
C	T(U)	Leucine	Proline	Histidine	Arginine
	C	Leucine	Proline	Histidine	Arginine
	A	Leucine	Proline	Glutamine	Arginine
	G	Leucine	Proline	Glutamine	Arginine
A	T(U)	Isoleucine	Threonine	Asparagine	Serine
	C	Isoleucine	Threonine	Asparagine	Serine
	A	Isoleucine	Threonine	Lysine	Arginine
	G	Methionine (*Start*)	Threonine	Lysine	Arginine
G	T(U)	Valine	Alanine	Aspartate	Glycine
	C	Valine	Alanine	Aspartate	Glycine
	A	Valine	Alanine	Glutamate	Glycine
	G	Valine	Alanine	Glutamate	Glycine

U represents Uracil by which thymine is replaced

Each protein molecule develops a network of protein–protein interaction and signal transduction pathways which are central to a cell that regulates the cellular activites (the details are shown in Sect. 3). Thus, as shown in Fig. 2, a sophisticated network covers the entire life of the Omic Space production.

Here, comparative genomics is situated on the first layer in the Omic Space and it is the first step to understand the functions and evolutionary processes among different species. However, since it is not the exact string matching problem, it requires an approach which can assert the plausibility for some reason. The concept of gap offers one solution to this problem, "similarity computation", because it corresponds to genetic deletion and insertion. That is why we have to consider an intended sequence, a scoring system, and an effective algorithm before the hardware implementation.

It is preferable that the algorithm covers any comparisons: DNA–DNA, protein–protein, and DNA–protein comparisons. As shown in Table 1, DNA–DNA and protein–protein comparisons require at least 4-by-4 and 20-by-20 score matrices, respectively. The matrices will be bigger when the algorithm considers ambiguous character caused by laboratory equipments. The situation in DNA–protein comparison, namely translated nucleotide–protein comparison, is different from the preceding comparisons. The comparison requires not only nucleotide translation based on Table 1 but also flexible insertion and deletion of gap notions. This section will discuss about the first two comparisons aimed at a better understanding though the circuit can treat the last comparisons.

Secondly, a good scoring system is required to obtain a good alignment called similarity. For DNA–DNA comparisons, we can apply simple scoring matrices. But, protein–protein and DNA–protein comparisons have to treat conservative substitutions where some substitutions are more likely to occur than others because of the chemical property. The point accepted mutation (PAM) [4], block substitution matrices (BLOSUM) [5], WAG and WAG* matrices [6] are used for calculating the likelihood scores. The matrices used in applications vary depending on the intended use and are decided by the users. The design of the hardware should take into account of this situation. In this section, a matrix is stored in on-chip small memories and it will be reconfigured on demand.

Finally, basic local alignment search tool (BLAST) [7] is used extensively in comparative genomics; however, this method includes the risk which sequence similarity will be overlooked because of heuristic approach [8, 9]. The Needleman–Wunsch algorithm [10] and the Smith–Waterman algorithm [11] are based on dynamic programming specialized for comparative genomics. Dynamic programming can consider the best local alignment between a query sequence and database sequences. And therefore not BLAST but the Needleman–Wunsch and the Smith–Waterman algorithms are still used in bioinformatics. BLAST is a useful application and we acknowledge the meaningfulness. But, this section will introduce the hardware implementation for dynamic programming from the standpoint of understanding fundamental approach and widespread use.

2.2 Related Works

In dynamic programming, high-speed parallel approaches are always important. One reason is that their computational speed is often slow. Another reason is that the increase of genomic data volume, especially by the appearance of automated sequencers, has surpassed the growth of computational performance of MPUs. In recent years, researches in FPGAs [12–17] and other architectures [18–22] have started to focus on not only nucleotides but also amino-acid sequence comparison. While many studies have been reported on particular platforms, there is not much research on analytic treatment for parallelizing the Smith–Waterman algorithm, with novel systolic designs and experimental comparison of various FPGA and GPU implementations. Firstly, this section tries to evaluate how efficient FPGA is in comparative genomics through theoretical comparison of how to implement the application program. After that, this study aims to establish techniques for optimizing performance by organizing parallelism effectively and analyzing the resulting effects.

2.3 Smith–Waterman Algorithm

This section will define the fundamental nature of the Smith–Waterman algorithm. In 1982, the following recurrence formula was introduced to the Smith–Waterman algorithm [23].

$$s_{i,j} = \max \begin{cases} s_{i-1,j-1} + \delta(q_i, d_i) \\ s_{i,j,\downarrow} \\ s_{i,j,\rightarrow} \\ 0 \end{cases}, \tag{1}$$

where $s_{i,j}$ is the similarity score of a node at the (i,j) position, $\delta(x,y)$ is the similarity score by genetic-character comparison of x and y, q_i is the ith character of a query sequence, d_j is the jth character of a database sequence, $\downarrow$ denotes a gap insertion in query sequence direction, and $\rightarrow$ denotes a gap insertion in database sequence direction. $s_{i,j,\downarrow}$ and $s_{i,j,\rightarrow}$ are called affine gap cost functions and they are obtained by

$$s_{i,j,\downarrow} = \max \begin{cases} s_{i-1,j,\downarrow} + \beta \\ s_{i-1,j} + \alpha \end{cases} \tag{2}$$

$$s_{i,j,\rightarrow} = \max \begin{cases} s_{i,j-1,\rightarrow} + \beta \\ s_{i,j-1} + \alpha \end{cases} \tag{3}$$

α is the opening-gap-penalty cost and β is the continuous-gap-penalty cost, and generally $\alpha \le \beta \le 0$. These gap penalties were introduced for genetic character insertion or deletion. These genetic mutations are usually caused by the error during DNA replications. Considering that two or more DNA bases may be inserted or deleted at the same time, it is preferable to give a penalty score more gradually compared to the constant penalty [23].

Figure 3 illustrates the data movement for parallel processing involving a processing element called SWPE, Smith–Waterman Processing Element. The affine gap functions, (2) and (3), are also implemented as recursive functions in an SWPE.

The Smith–Waterman algorithm is known to be computable in parallel along the oblique line shown in Fig. 3. Thus, the number of SWPEs contributes to performance increase. In SWPE, it is necessary to store the maximum value used for judgement of a correlation between query and database sequences. Hence, SWPE includes circuits of not only (1) but also the maximum function. Additionally, the Smith–Waterman algorithm has a high degree of data locality. This is the reason why its acceleration has been tried in ASICs and FPGAs.

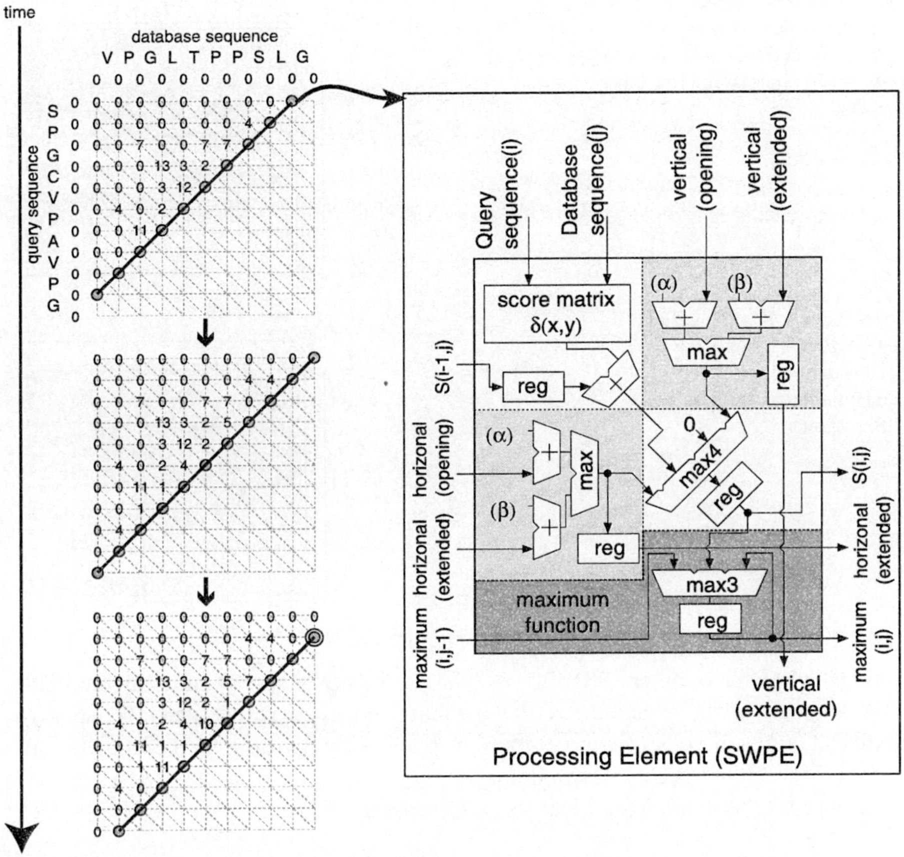

Fig. 3 Data movement for the parallel processing in the Smith–Waterman algorithm and SWPE

2.4 *Performance Analysis*

This section will provide an analytical estimation of the performance of our proposed design. In contrast to theoretical treatment based on parallelization of recurrences [24], our discussion focuses on a specific one-dimensional systolic array which includes: (a) multiple reconfigurable devices; (b) derivation of performance between line-based method and lattice-based method; (c) comparison of the achieved and theoretical peak performance of these methods.

Each SWPE can be regularly connected to form a systolic array. In the following discussion, we assume that the Smith–Waterman algorithm is implemented on SWPEs which are arranged as an one-dimensional systolic array as shown in Fig. 3. Then, the computational clock cycles becomes

$$f_{\text{ideal}}(Q,D) = Q + D - 1 \qquad (4)$$

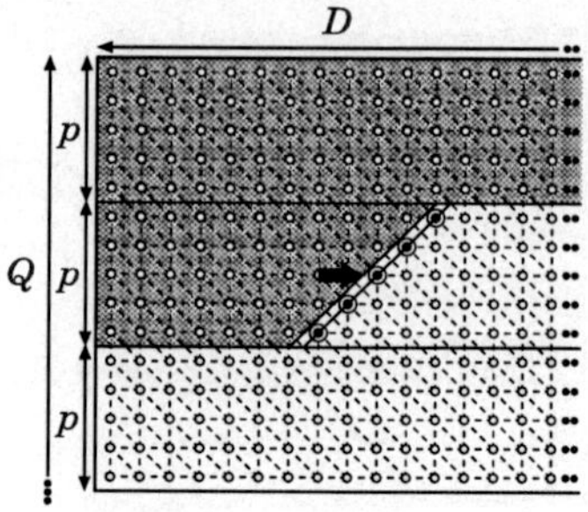

Fig. 4 Line-based parallel computation. The search space is divided into reed-shaped areas called line segment

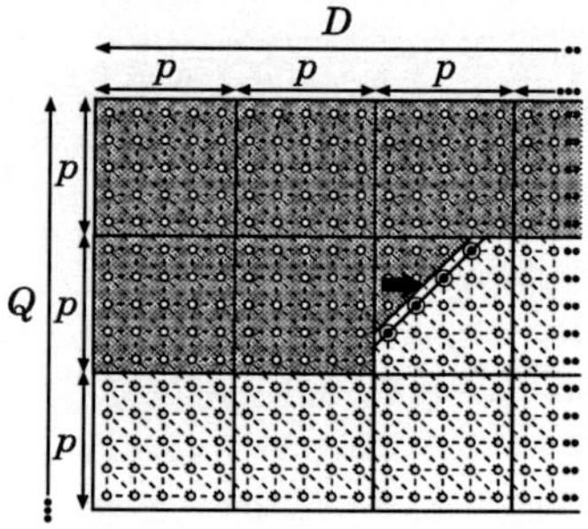

Fig. 5 Lattice-based parallel computation. The search space is divided into rectangular areas called lattice segment

from QD when the number of SWPE, p, is larger than Q, where Q and D are defined as the length of a query sequence and a database sequence, respectively. When p is smaller than Q, it is possible to compute in parallel by dividing the computational domain to multiple rows as shown in Fig. 4.

In this case, computational clock cycles become

$$\begin{aligned} f_{\mathrm{line}}(Q,D,p) &= f_{\mathrm{line_{segment}}}(p,D)\cdot N_{\mathrm{line_{segment}}} \\ &= (p+D-1)q \quad (\because f_{\mathrm{line_{segment}}}(p,D)=p+D-1), \end{aligned} \tag{5}$$

where $q = \lceil Q/p \rceil$. This computation can be extended to support multiple reconfigurable devices. Assuming that the number of devices is k, the computational time is obtained by the following expressions:

$$f_{\mathrm{line}_k}(Q,D,p,k) = \frac{q(pk+D-1)}{k} \tag{6}$$

$$\begin{aligned} \lim_{k\to q} f_{\mathrm{line}_k}(Q,D,p,k) &= pq+D-1 \\ &\approx Q+D-1, \end{aligned} \tag{7}$$

where $q \geq k \geq 1$. k should be sufficiently large for higher performance but the improvement stops when $kp > Q$.

Figure 5 is another possibility of multiple device acceleration. In this case, the rectangular area is delimited not by p but cache size on a single device, and the

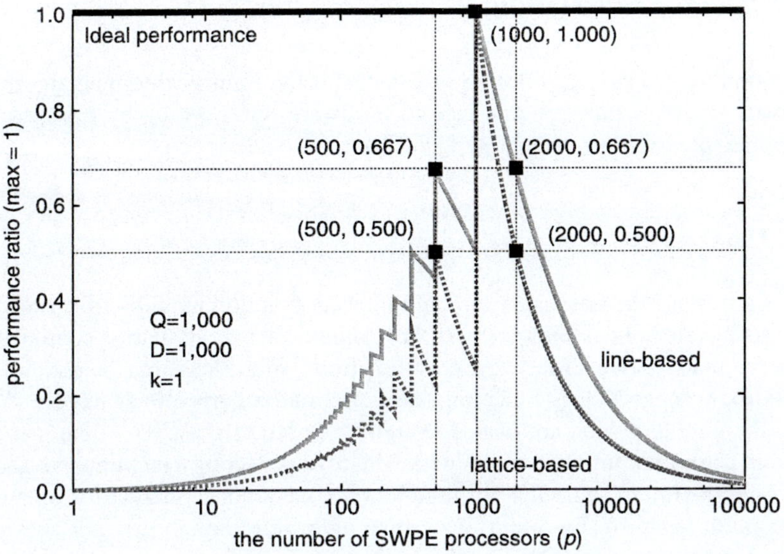

Fig. 6 Performance ratio of line-based and lattice-based performance to the theoretical peak performance: excessive parallelization induces the larger number of idle elements and it causes performance degradation

computation time is obtained by the following equation:

$$\begin{aligned} f_{\mathrm{lattice}}(Q,D,p) &= f_{\mathrm{lattice}_{\mathrm{segment}}}(p,p)\cdot N_{\mathrm{lattice}_{\mathrm{segment}}} \\ &= (2p-1)qd \quad (\because f_{\mathrm{lattice}_{\mathrm{segment}}}(p,p) = p+p-1), \end{aligned} \tag{8}$$

where $d = \lceil D/p \rceil$. The computational clock cycles with k devices are obtained by the following expression:

$$f_{\mathrm{lattice}_k}(Q,D,p,k) = (2p-1)\left(\frac{qd}{k} + k - 1\right) \tag{9}$$

$$\begin{aligned} \lim_{k\to q} f_{\mathrm{lattice}_k}(Q,D,p,k) &= (2p-1)(q+d-1) \\ &\approx 2(Q+D-1). \end{aligned} \tag{10}$$

Figure 6 shows the ratio of line-based and lattice-based performance to theoretical peak performance obtained by (4). It can be seen that line-based parallelism is always better than lattice-based parallelism since the former is closer to the ideal performance.

From Fig. 6, it can also be seen that the performance efficiency will decrease when the number of SWPEs and the length of a query sequence do not match. This result is important because it enables power-performance improvement.

2.5 Systolic Array Design

This section describes a new design of SWPE for the Smith–Waterman algorithm. The novel aspect of our design concerns hardware optimization by transforming numerical expressions in computing the affine gap cost function.

2.5.1 The Overview of an SWPE

Figure 3 describes the basic structure and internal function modules of the SWPE. A processing element is composed of five modules: one similarity computation with score matrix, two affine gap cost functions, one maximum detection, and one maximum score-history function. The score matrix corresponds to $\delta(q_i, d_i)$ of (1) and the circuits which are placed in light-grey boxes show the affine gap cost functions. The maximum detection function involves selecting a maximum as shown in (1). The maximum score-history function is an essential function for obtaining the maximum value of the Smith–Waterman algorithm though it is not shown in numerical expressions; it can be expressed by one single comparator.

2.5.2 Affine Gap Cost Function ($\alpha \leq \beta \leq 0$)

When the gene is copied, insertion and deletion of the characters might take place. Gap cost functions are quantification of the gap inserted or deleted at this time. The affine gap cost function [23] is well known and is obtained by (2) and (3). In the numerical equations, α is the cost of opening gap and β is the cost of continuous gap. In general, α is smaller than β.

When using this function, we must treat at least six values: $s_{i-1,j-1}$, $s_{i-1,j}$, $s_{i,j-1}$, $s_{i-1,j,\downarrow}$, $s_{i,j-1,\rightarrow}$, 0 for every SWPE. Therefore, at least five comparators, namely five branch instructions, are needed for getting the maximum of them in every SWPE. A single SWPE must require one or two additional comparators since the SWPE has a maximum value function as shown in Fig. 3. Based on this, the size of a single SWPE is estimated to be approximately 160 LUTs in this paper which is almost the same size as the design in [15] which is estimated to be 85 slices, i.e. 170 LUTs, on XILINX Virtex-II architecture. This section will discuss how to realize a better circuit for the SWPE.

The original numerical equations should be optimized for the hardware implementation; they use signed circuits in most operations. A signed comparator is larger than an unsigned one. In general, $\alpha \leq \beta \leq 0$. We can take advantage of this relationship and further reduce the number of LUTs. Here, (1)–(3) are revised to produce (11)–(13). Each underscored term in (11)–(13) means a signed number. $(\beta - \alpha)$ has to be larger than 0 as a precondition to this application.

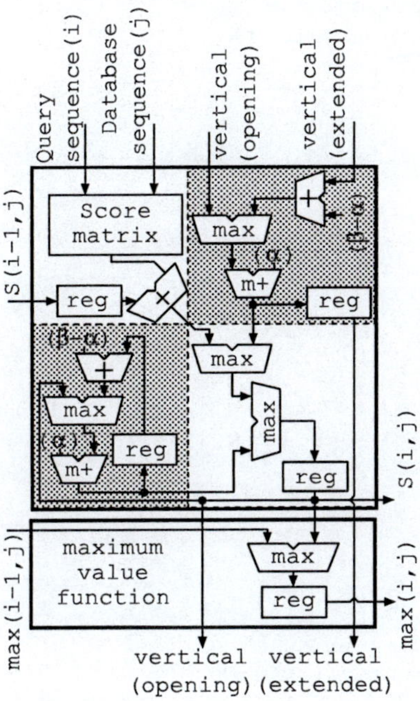

Fig. 7 An optimized SWPE by arranging Smith–Waterman equations

$$s_{i,j} = \max \begin{cases} s_{i-1,j-1} + \underline{\underline{\delta(q_i, d_i)}} \\ s_{i,j,\downarrow} = \max \begin{cases} s_{i,j,\downarrow,\alpha} + \underline{\underline{\alpha}} \\ 0 \end{cases} \\ s_{i,j,\rightarrow} = \max \begin{cases} s_{i,j,\rightarrow,\alpha} + \underline{\underline{\alpha}} \\ 0 \end{cases} \end{cases} \tag{11}$$

$$s_{i,j,\downarrow,\alpha} = \max \begin{cases} s_{i-1,j,\downarrow} + (\beta - \alpha) \\ s_{i-1,j} \end{cases} \tag{12}$$

$$s_{i,j,\rightarrow,\alpha} = \max \begin{cases} s_{i,j-1,\rightarrow} + (\beta - \alpha) \\ s_{i,j-1} \end{cases}. \tag{13}$$

In Fig. 7, each SWPE has four 15-bit comparators, one 16-bit comparator, one 16-bit adder–subtractor, two 15-bit adders, and two 15-bit positive-number discriminant adders, *m*+, whose details are illustrated in Fig. 8.

Compared to the previous approach, we can reduce 243 LUTs to 219 LUTs in case of 4-input LUTs for Virtex-4 FPGAs; and from 239 LUTs to 231 LUTs in case of 4-input LUTs for Virtex-II Pro and Virtex-II FPGAs. In case of 6-input LUTs for Virtex-5 FPGAs, the number of logic LUTs decreases even though the total does not decrease, as shown in Table 2. For these experiments, ISE12.2 (ver.M.63c) is used for Virtex 4 and 5, and ISE8.2i (ver.I.31) is also used for Virtex II and II pro.

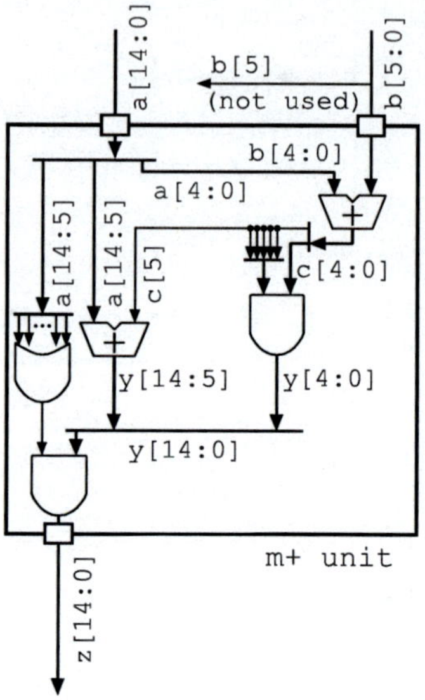

Fig. 8 A positive-number discriminated adder for SWPE

Table 2 Circuit resource requirement for a single SWPE

FPGA	LUT Total	Logic	Thru	Regs
Ideal	152			85
Virtex5	177	138	39	
Virtex4	219	191	28	
Virtex-II Pro	231	204	27	
Virtex-II	231	204	27	

Each SWPE includes a single embedded RAM

Here, a thru LUT means a "route-through" LUT in XILINX FPGAs and the number is counted when the circuits other than LUTs drives registers, carries, and some circuits because of the composition where a minimum unit contains a LUT and other circuits. As for the thru LUTs, there is a possibility of it being eliminated by the synthesis tools, which is confirmed by experiments in Sect. 2.6 for a larger circuit. Thus, this approach significantly improves the SWPE array performance.

Finally, this approach enables the efficient use of an embedded adder such as XILINX DSP48E. Three 15-bit adders are packed in one embedded adders because the proposed approach can reduce a signed bit. It reduces the use of LUTs and contributes to the high operating frequency. A compact saturation adder which keeps the maximum value when overflow occurs can also be realized by the use of 16-, 32-, 48-bit signal lines. The efficient use of embedded circuits is an important factor in the performance gain.

2.6 *Experimental Results*

SWPE arrays are also evaluated on current FPGA architectures. They are implemented on ADM-XRC-5T2 with XC5VLX330T produced by Alpha Data. We adopt Fig. 7 as an SWPE and optimize not only for an individual SWPE but also for a SWPE systolic array. In this section, billion cell update per second (GCUPS) is used for the performance measurement and it is obtained by $QD/T_{\text{comp}} \times 10^{-9}$, where T_{comp} is the computational time.

2.6.1 Performance Evaluation of FPGA

In this section, we evaluate the maximum performance obtained by applying the proposed techniques to the XILINX XC5VLX330T FPGA in the Alpha Data ADM-XRC-5T2 system. Our design for XILINX XC5VLX330 has 183,416 LUTs and 86,638 registers; it contains 1,000 SWPE elements and can reach a performance of up to 129 GCUPS. It is likely to be one of the fastest single-chip implementations of the Smith–Waterman algorithm. Moreover, since systolic architectures are regular and scalable, SWPE arrays and the associated optimizations would be applicable to next-generation FPGA devices as they become available.

In power-performance evaluation [25], the FPGA achieves approximately 16 GCUPS/W when we use two DDR-II Synchronous SRAM banks in the Alpha Data ADM-XRC-5T2 system. This power-performance is better than that of CPU and GPUs; CPU, such as Xeon E5420, is less than 0.5 GCUPS/W and GPUs, such as GTX295, are less than 0.08 GCUPS/W.

2.6.2 CPU and GPUs

There are recent researches on Smith–Waterman algorithm targeting GPUs [18, 20, 21, 26] but the effective performance of them is far inferior to the theoretical peak performance of the GPUs; for instance, the performance of four parallel GPUs hardly reaches one single XC4VLX200: approximately 40 GCUPS in our estimation. Moreover, a single Spartan produced by XILINX or one single Cyclone by ALTERA is comparable to the performance of a single GPU. Ligowski and Rudnicki [18] shows the bottleneck is not memory bandwidth and it implies current GPU architecture is not appropriate for this application when compared with FPGA architecture. We confirm this observation using NVIDIA GTX480; the memory bandwidth is not a bottleneck in this latest device. Recent other studies, [20, 21], also show the discrepancy between theoretical performance of GPU and real performance of this application. It seems that this application is unsuitable for GPU even though GPU has shown remarkable progress in accelerating various algorithms. On the other hand, appropriate CPUs can achieve a decent performance compared with their theoretical performance [22].

2.7 Summary of Homology Search

This section provides an analytical treatment of the SWPE systolic approach to the Smith–Waterman algorithm. The line-based and lattice-based methods for organizing parallelism are introduced; the effect of parallelism on the performance of these methods is analysed with respect to peak performance. The insights from this analysis are used in deriving a novel systolic design, which includes techniques for reducing affine gap cost functions.

The potential of our approach is demonstrated by the high performance of the resulting designs. For example, the XC5VLX330T FPGA can accommodate 1,000 SWPE cores operating at 130 MHz, resulting in a performance of up to 129 GCUPS, which is 3 times faster than the fastest quad-GPU processor, GTX295. Moreover, the FPGA is far more energy efficient than GPUs; XC5VLX330T achieves approximately 16 GCUPS/W, while the power-performance for CPU is less than 0.5 GCUPS/W, and for GPUs is less than 0.08 GCUPS/W.

3 Biochemical Kinetic Simulation with a Reconfigurable Platform

3.1 Introduction

Biochemical kinetic simulation, the so-called simulation of a cellular system, is one of the major applications in life science. Various biochemical kinetic simulators were developed since KINSIM [27], which was presented in 1983. Unlike small and simple targets of the simulators in the early days such as Gepasi [28] and DBsolve [29], recent simulators like E-Cell [30] and Virtual Cell [31] are targeted to large-scaled networks such as simple whole cells.

Since biochemical kinetics are basically modeled using ordinary differential equations (ODEs), the parameter fitting between experiments and simulations is essential in the modeling process. The parameter fitting is usually a time-consuming process because of its large number of model parameters. Today, many institutions and researchers have high-performance systems such as PC/WS clusters to solve this problem. Even when such systems provide enough performance, for a single researcher, it is difficult to occupy its full performance all time considering their high cost. In this chapter, we summarize the research result of the "ReCSiP" project, an FPGA-based simulation environment which aims to build a cost-effective alternatives of PC/WS clusters for personal use of biochemical kinetic simulations. The project started in 2004 in cooperation with life scientists and terminated in 2010.

Since life scientists are neither programmers nor hardware engineers who can write HDLs, the goal of ReCSiP is to develop an easy-to-use FPGA-based simulation environment for them. The front-end language is easy for them to use. We focus on a standard description form, Systems Biology Markup Language (SBML) to describe kinetic model in XML form. The hardware solvers are automatically generated from the description of SBML, and users don't have to touch its hardware description at all.

In the first 2 years of the project, we focused on the ODE solvers which simulate the model with numerical integration method. Although the solvers were approximately 80 times faster than that of the software run on the corresponding PCs, there is also a new trend in kinetic simulation using stochastic approach. So, we also tried to accelerate stochastic simulation. First, the simplest first reaction method (FRM) based on Gillespie's algorithm was implemented, and then more sophisticated next reaction method (NRM) was tried. The ODE solvers are introduced with a target model and SBML description and the stochastic solvers based on FRM and NRM are introduced.

3.2 *ODE Based Approach*

3.2.1 The Target Model and Its Description

Simulation of biochemical pathway kinetics is a numerical process to obtain the concentration change of the molecular species in time series. Figure 14a shows an example pathway, which has nine molecular species (S_1 to S_9) and five reactions (R_1 to R_5). The velocity of each reaction is determined by their rate-law function, corresponding to their reaction mechanism. For example, velocity of a simple second order reaction is:

$$v = k[S_1][S_2], \tag{14}$$

where $[S_1]$, $[S_2]$ are concentrations of substrates and k is the rate constant of the reaction. Rate-law functions are functions of concentrations, which have some constants (e.g., maximum velocity or rate constants) as their parameter and can be solved as ODEs. Each different reaction mechanism is expressed by the specific rate-law functions, as the examples in Table 3. ReCSiP runs simulations by calculating reaction rates with the modules called "solvers" which consist of several floating-point arithmetic units.

Here, a "model" to be simulated as an initial-value problem consists of:

- List of molecular species
- List of reactions
- Initial values of concentration for the molecular species
- And parameters of the reactions

And these contexts can be marked up as the standardized XML format, SBML [32]. ReCSiP framework described in the following sections automatically generates HDL modules corresponding to the given SBML model.

Table 3 Examples of rate-law functions defined in SBML level 1

Reaction mechanism	Rate-law function
Irreversible simple Michaelis–Menten	$v = \dfrac{V_m S}{K_m + S}$
Uni–Uni reversible simple Michaelis–Menten	$v = \dfrac{V_f S/K_{mS} - V_r P/K_{mP}}{1 + S/K_{mS} + P/K_{mP}}$
Uni–Uni reversible simple Michaelis–Menten with Haldane adjustment	$v = \dfrac{(V_f/K_{m1})(S - P/K_{eq})}{1 + S/K_{m1} + P/K_{m2}}$
Competitive inhibition (irreversible)	$v = \dfrac{VS/K_m}{1 + S/K_m + I/K_i}$
Competitive inhibition (reversible)	$v = \dfrac{V_f S/K_{mS} - V_r P/K_{mP}}{1 + S/K_{mS} + P/K_{mP} + I/K_i}$

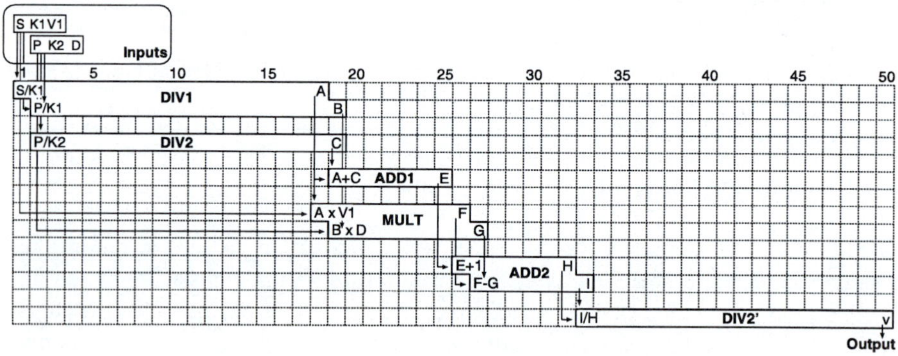

Fig. 9 An example of solver core: Uni–Uni reversible simple Michaelis–Menten with Haldane adjustment

3.2.2 The Solver Core Library

As described in the previous subsection, a biochemical kinetic simulation is driven by solving ODEs. The solver core library includes HDL implementation of basic rate-law functions was defined in SBML level 1 specification[33]. Each module in the library consists of some FP arithmetic units and calculates velocity of one or more reaction mechanisms. Some solvers core can process only specific reaction models, and others can process some similar reaction models.

Each solver core is an encapsulated, pipelined module to solve the specific rate-law function, as an example in Fig. 9. The pipeline schedule in the solver core is statically designed to maximize the effectiveness of pipelining operation.

Solver cores are not only designed manually but can also be generated automatically from equations described in MathML form[34] that can be embedded in an SBML file.

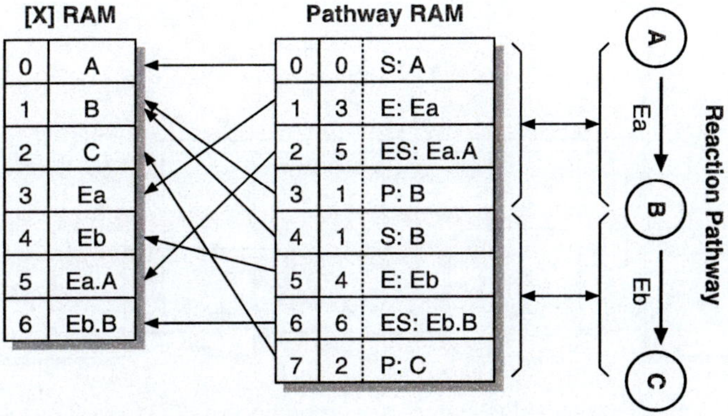

Fig. 10 Pathway description by pathway RAM

3.2.3 Mechanism for Integration and Pathway Mapping

Running a simulation task is not just to calculate velocities of reactions, but to track concentration changes of molecular species by numerical integration the velocities. ReCSiP has an external control mechanism around the Solver Core to enable this. The control mechanism plays two roles in the control mechanism: I/O stream management of the solver core and numerical integration. A module called "Integrator" provides these functions in ReCSiP.

The first role is I/O stream management. This is to supply and receive required datastream. ReCSiP controls solver cores' input and output by using array of pointers as shown in Fig. 10. "Pathway RAM" in the figure is the instruction memory, and others are data memory.

The instruction, each word in Pathway RAM is basically a set of pointers to read S RAM, k RAM and $[X]$ RAM beside a pointer to write $d[X]$ RAM. By scanning Pathway RAM serially from address 0, both I/O and memory access of a Solver Core are handled.

The second role is numerical integration. This operation is also controlled by Pathway RAM. $[X]$ RAM keeps concentrations of the molecular species, and $d[X]$ RAM is to accumulate its derivatives in a timestep. Contributions of each reaction on the concentration change come from the Solver Core are once accumulated in $d[X]$ RAM, but are not reflected to $[X]$ RAM immediately. This is because there are some molecular species appeared in two or more reactions like S_5, S_6, and S_7 in the example pathway in Fig. 14.

When velocities of all reactions are calculated, the derivatives of molecule concentrations are integrated on $[X]$ RAM. In the simplest implementation, one iteration of accumulation in $d[X]$ RAM and integration on $[X]$ RAM ticks one timestep dt in the simulated pathway as shown in Fig. 11. This simplest implementation is based on Euler's method. Methods of higher order can be implemented by adding several additional memory blocks and more iterations in a timestep [35].

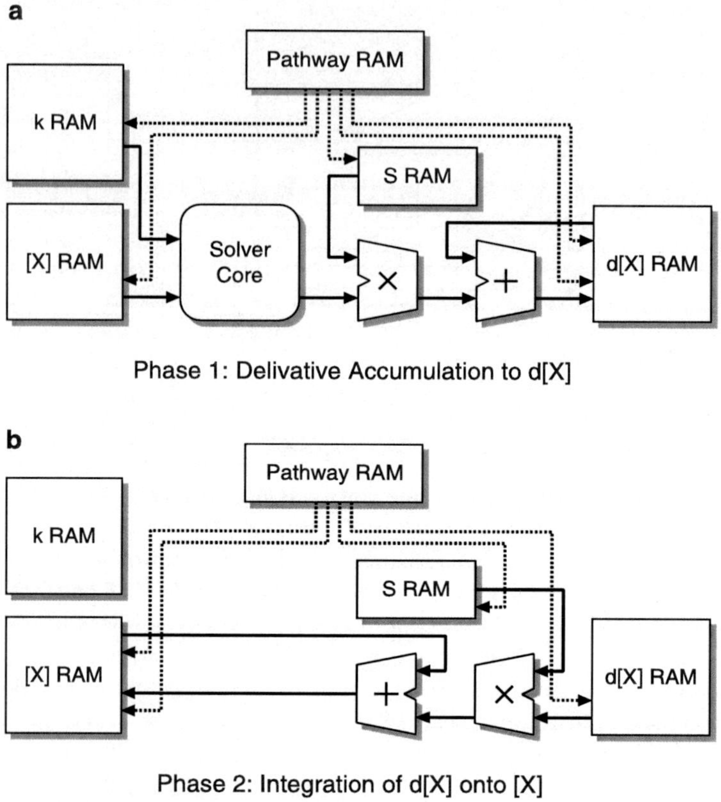

Fig. 11 Two-phase of integration mechanism

Since Integrator is isolated from Solver Core, users can choose arbitrary Solver Core and Integrator to meet their requirement. A module consists of one Integrator and one Solver Core called "Solver".

3.2.4 Heterogeneous Model Simulation Framework on ReCSiP

The previous section was about the basic mechanism of biochemical kinetic simulation for homogeneous model that contains only one reaction mechanism. Practically speaking, this is not the best method since the mechanism cannot run simulations containing two or more different reaction mechanisms.

In this section, the framework to enable cooperation of multiple solvers is presented.

Figure 12 shows the overview of this framework. The target system is described in SBML, which is generated by modeling tools such as CellDeginer [36]. The ReCSiP software extracts the list of reactions, molecular species and parameters

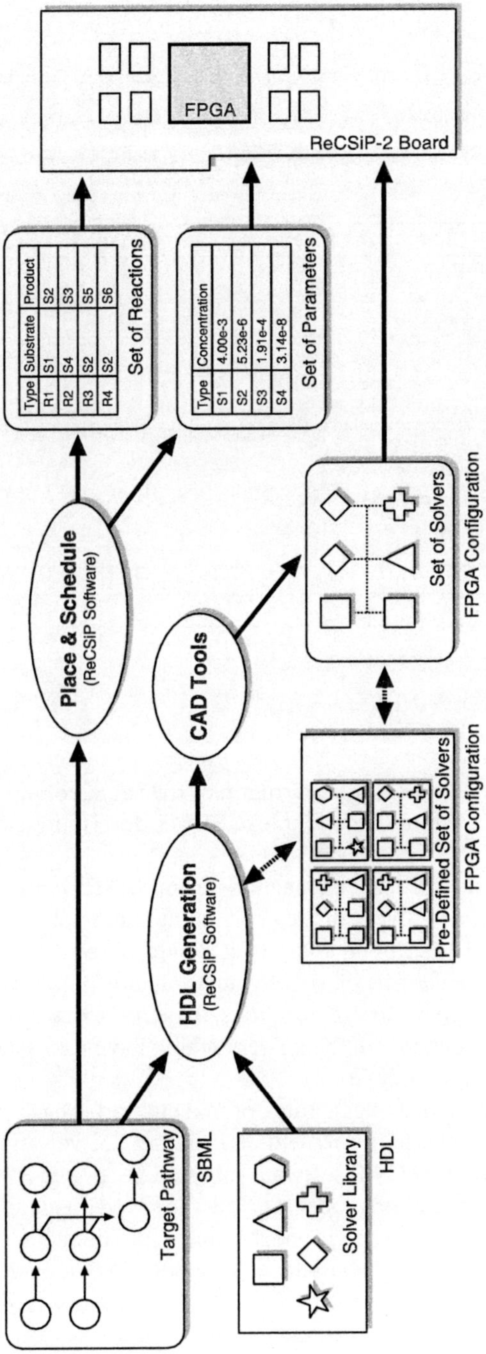

Fig. 12 Overview of the framework

```
<model name="sample_model">
  <listOfCompartments>
    <compartment name="cell"/>
  </listOfCompartments>
  <listOfSpecies>
    <specie name="S" initialAmount="0.3" compartment="cell"/>
    <specie name="P" initialAmount="0"   compartment="cell"/>
  </listOfSpecies>
  <listOfReactions>
    <reaction name="Reaction1" reversible="false">
      <listOfReactants>
        <specieReference specie="S" />
      </listOfReactants>
      <listOfProducts>
        <specieReference specie="P" />
      </listOfProducts>
      <kineticLaw formula="uui(S,P,km)">
        <listOfParameters>
          <parameter name="km" value="0.1" />
        </listOfParameters>
      </kineticLaw>
    </reaction>
  </listOfReactions>
</model>
```

Fig. 13 An example of SBML description

from SBML, then generates the RTL description of the set of solvers for simulation and its schedule. Figure 13 shows an example of SBML description, the input format of ReCSiP.

ReCSiP software extracts the list of reactions from SBML input, then generates RTL description of a set of solvers that can run simulations of the given reaction pathway. Generated RTL description consists of some solvers and solver-to-solver communication switches. The set of solver must contain the minimal set of the solvers which cover the given list of reactions. If some extra space is available after the minimum set is generated, the solvers which have excessive load will be duplicated to reduce simulation time.

However, it's not a good idea to generate optimal circuit for each model, because CAD tools to perform synthesis, placement and routing are very time consuming. To minimize CAD runtime, ReCSiP software saves all generated FPGA configuration bitstreams. When a simulation model is given, the saved bitstreams are scanned, and one of the bitstreams will be used if one that is compatible with the given model is found. In this case, ReCSiP software doesn't launch CAD tools to eliminate the runtime.

Then, the software generates the contents of the memory blocks (Pathway, $[X]$, k and S RAM) in the solver set that has been generated in the previous step.

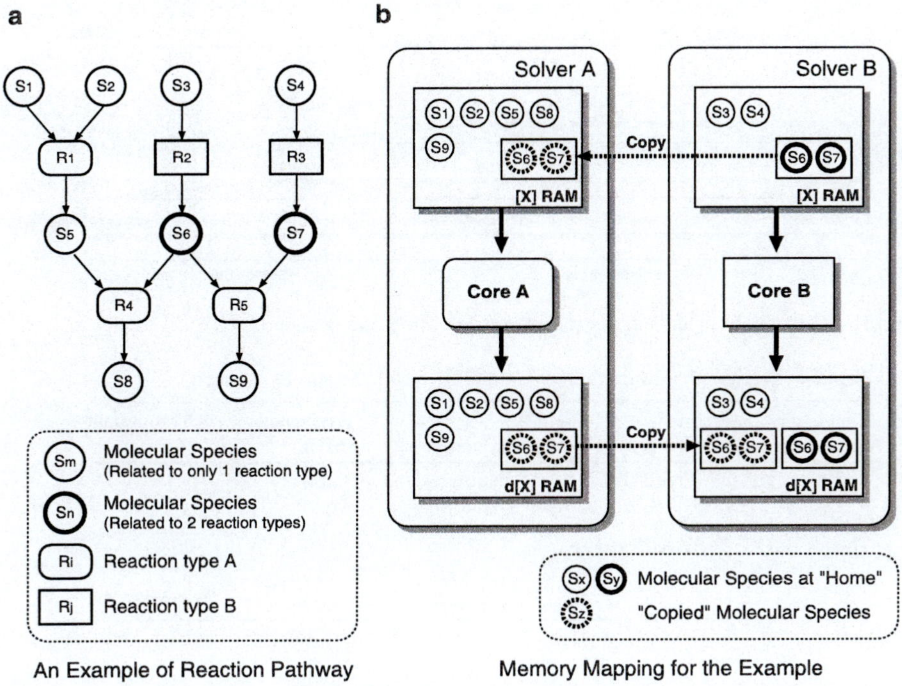

Fig. 14 An example pathway and its mapping on the solvers

This step is divided into two stages. First, the software determines the placement of reactions and molecular species on the solver set, as shown in Fig. 14. Each molecular species is assigned to one of the Solvers, where the master data of their concentration is stored in the $[X]$ RAM. This assigned Solver is "home" Solver of the molecular species.

After the placement of molecular species is determined, the reactions and transfers between the solvers are scheduled by the software. Schedule of solving the reactions is written on Pathway RAMs on each solvers, and the transfer schedule is written on Code RAM in the communication switch. Scheduling of reactions and their data transfers is important, since the solver can't start its operation until necessary concentration of molecular species is provided. If a molecular species is required at some other solver(s) than the home, its concentration has to be copied before used by the solver, as Fig. 14b shows. The figure also shows that the derivatives calculated from "copied" concentration have to be sent back to the home solver to perform integration.

3.2.5 Evaluation of ODE Based Approach

The whole system is written in Verilog-HDL and implemented with Xilinx's ISE-6.3i. Software benchmark was performed on FreeBSD 5.3-RELEASE environment, with gcc-3.4.2.

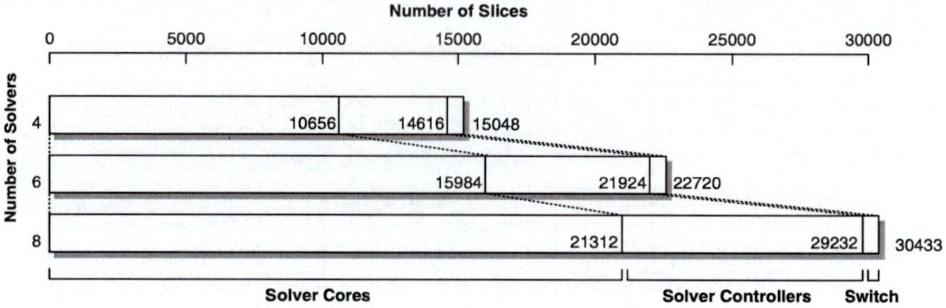

Fig. 15 Size of solvers and the switch (irreversible Michaelis–Menten)

Table 4 Software/hardware throughput on irreversible Michaelis–Menten

Implementation	System	Frequency (MHz)	Throughput (Mreaction/s)
Software (gcc-O3, FreeBSD 5.3)	Pentium4	3,200	5.75
	PentiumM	2,000	6.78
This work (XC2VP70-5FF1517)	4 solvers	119	238
	6 solvers	111	333
	8 solvers	107	428

3.2.6 Area Overhead of the Communication Mechanism

Figure 15 shows the number of slices occupied by solver cores (Irreversible Michaelis–Menten), integration modules (Euler), and the switch (4/6/8 port). Up to eight solvers can be implemented on an FPGA, and overall slice utilization is 30,433 in this case while the switch occupies approximately 1,200 slices. The area of switch is approximately 4 % of the whole design, a quite reasonable size.

3.2.7 Frequency and Throughput

Table 4 shows the result of throughput evaluation; 8 Irreversible Michaelis–Menten solvers and switch require 91 % of XC2VP70 in the area and achieved over 60-fold speedup compared to the Pentium M/Pentium 4 microprocessors. Although solver configuration in this table is homogeneous, heterogeneous configuration is also possible. In most cases, the FPGA can operate at 100 MHz+ in heterogeneous configuration, too.

3.3 Summary of ODE Approach

A biochemical kinetic simulator is designed and implemented on ReCSiP. This simulation framework can handle heterogeneous models, by using a set of customized

Fig. 16 Cascading model (D4S: $N = 4$)

$$
\begin{aligned}
R_1 &: S_1 + S_2 \xrightarrow{k_1} S_3 \\
R_2 &: S_2 + S_3 \xrightarrow{k_2} S_4 \\
&\vdots \\
R_N &: S_N + S_1 \xrightarrow{k_N} S_2
\end{aligned}
$$

rate-raw function modules called Solver Cores. Modular approach of the framework also provides flexibility to choose numerical integration method and scalability to fit larger FPGAs in the future.

To maximize the throughput of pipelined modules, optimized hardware and optimally scheduled data streams are put together on the FPGA. ReCSiP achieved over 60-fold speedup compared to latest Intel's microprocessors.

The custom Solver Cores, FPGA configuration, and input data are automatically generated by ReCSiP software, from an SBML description. This software significantly contributes to the usability of the framework, because users do not have to know about the FPGA designs.

3.4 *Stochastic Biochemical Simulation*

3.4.1 Stochastic Simulation Algorithm

Gillespie proposed a stochastic approach to compute chemically reacting systems[37]. Stochastic simulation algorithm, abbreviated to SSA, calculates "time-series" of the model which defines a list of reactions consists of chemical species as shown in Fig. 16.

Figure 16 is the model which is defined N reactions R_j and related M chemical species S_i. The combinatorial numbers of S_i represent the state of the models. For example, a collision of reactants S_1 and S_2 sets off the reaction R_1 which produces the products S_3, according to event probability k_1.

Since the Gillespie's FRM and the Direct Method (DM) had been proposed [37], several improved versions of SSA were presented such as NRM [38] and optimized direct method (ODM) [39].

3.4.2 First Reaction Method

The idea of SSA is to obtain the state-change of the model through a repetition of a process to select a next occurring reaction. The chosen reaction of FRM has the smallest τ_j, predicted time of occurrence among all reactions in the model [37]. Predicted time of occurrence for each reaction R_j is obtained by (15) per simulation cycle.

$$\tau_j = \ln(1/r)/a_j. \tag{15}$$

Value r is a uniform random number between 0 and 1. a_j is called "a propensity", which is a multiple of an event probability k_j and a combination number of all the reactants in R_j.

3.4.3 Next Reaction Method

NRM reduced the time complexity from $O(N)$ of FRM to $O(\log(N))$ maintaining the statistical equivalence by introducing two data structure called indexed priority queue (IPQ) and Dependency Graph (DG) [38]. NRM is applied to representative software biochemical simulators such as E-Cell3 [40].

The feature of the previous SSAs, that the number of reactions to be modified their predicted time is much less than the number of reactions defined in the simulation model, is utilized by Gibson and Bruck. At the first developing of NRM, the predicted times τ_j for each reaction are stored to a type of heap trees called IPQ for efficient determination of the smallest value τ_μ, which always located on the root node. Next innovation of NRM is DG, a list of reactions to be modified the predicted time by (16) when a reaction occurs.

$$\tau_{j,\text{new}} = a_{j,\text{old}}/a_{j,\text{new}}(\tau_{j,\text{old}} - \tau_\mu) + \tau_\mu. \tag{16}$$

For instance, DG(R_1) in Fig. 16 is given as (17). As the number of reactions listed in DG is much less than the whole number of reactions in the model, the ascendant factor determining time complexity becomes maintaining the order of IPQ rather than the number of reaction to recalculate the predicted time.

$$DG(R_1) = \{R_2, R_3, R_N\}. \tag{17}$$

3.4.4 Related Works

Several challenges have been made to design a stohastic biochemical simulator on FPGA since 2004. Keane et al. and Salwinski et al. both successfully achieved approximately 20 times speedup compared to microprocessors [41, 42]. However, both of their works are based on approximated stochastic algorithms, and calculation steps were also simplified. For example, they convert floating-point into integer values to perform high speed computation.

3.4.5 The FRM Implementation on an FPGA

We have been implementing and evaluating several SSA on FPGAs since 2004 [43–46]. In 2006, a circuit to compute FRM was designed with two simulation threads that time-shares one single-precision floating point computational unit calculating (15) and obtaining the smallest τ_j. Pipeline of the computational unit can receive consecutive input data to achieve high throughput [45]. And without

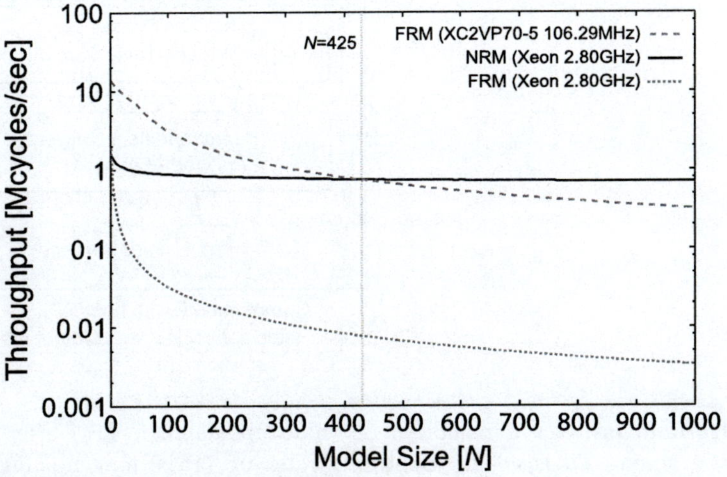

Fig. 17 Drop-off of the throughput versus model size

using approximated stochastic algorithm, this implementation achieved more than 80 times speedup compared to execution on Xeon 2.80 GHz by running six threads in parallel to simulate a model defined 1,000 reactions ($N = 1{,}000$).

For more detail, execution programs of FRM and NRM were written in C++, and their throughput versus model size was evaluated with the cascading model of Fig. 16. The results are shown in Fig. 17, together with the FPGA execution result of FRM. According to these results, throughput degradation of the FPGA implementation of FRM is more prominent than NRM execution on Xeon as the model size increases. Advantage of the two turns back at a point of $N = 425$, and NRM move out ahead by approximately three times at $N = 1{,}000$. This implies that calculation cost of FRM on an FPGA is purely disadvantageous compared with NRM on microprocessors, considering that NRM is proved to produce equivalent results with FRM.

3.4.6 Design Concept of NRM on FPGA

The hardware modules of SSA is expected not to perform any approximation or simplification of the original algorithm, even the computation is executed on FPGA. Therefore the strategy to design hardware to compute NRM is by high-throughput computation with multi-thread execution instead of some large floating-point arithmetic modules such as logarithmic unit which needs long computational cycle.

NRM procedures can be divided into two groups; commonly utilized computing sequences of NRM such as (15) and (16), and arrays to store variables and intermediate simulation data which should be prepared for each simulation thread. The former is called as "Thread Share Units (TSUs)" and the latter is called as "Thread Private Unit (TPU)," respectively.

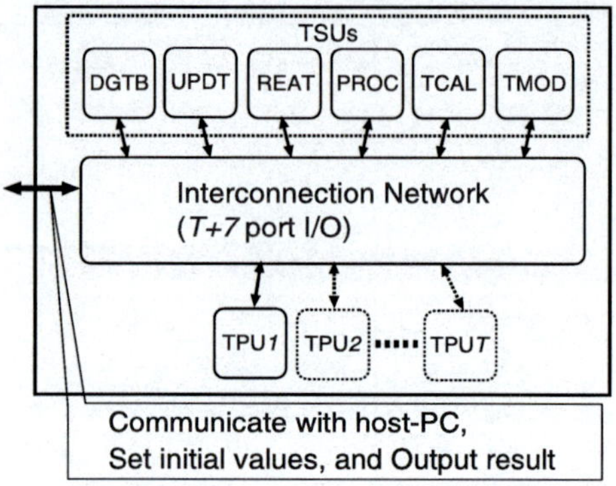

Fig. 18 Connection diagram

The algorithm involves a repetition of value update in a TPU after many data transfer among TSUs. Thus, low utility rates of TPUs, long floating-point operations, and ineffective restructure of tree-structured memories may degrade performance. Consequently, the proposed design connects several TPUs to TSUs by some interconnection network as shown in Fig. 18.

3.4.7 Implementation

Each module in the NRM circuit was written in Verilog-HDL, and synthesis, placement, and routing were done by Xilinx's ISE8.2i. Target device of the design is Virtex-4 (XC4VLX100-10FF1148), which is currently a middle-range FPGA. A single-precision floating-point arithmetic unit utilizes an IP core called Xilinx's LogiCORE floating-point. 32-bit$\times 1{,}024$ words BlockRAMs on the Virtex-4 were used as storage for variables in each unit. FIFOs are also implemented with 32-bit$\times$512 words BlockRAMs. The maximum number of biochemical reactions supported in this implementation is 1,023, which is sufficient for the existing stochastic models.

3.4.8 Interconnection Network

To provide a capability to flexibly connect various numbers of TPU and TSUs, the I/O of each module must have a common interface. Figure 19 illustrates a prototype design and protocols of a common interface for TPUs, TSUs, and routers within the interconnection network.

Figure 20 shows an example of the router structure when the number of I/O ports P is 4. The router module has output buffers implemented with P FIFOs, and output FIFO controllers send packets. A $P \times P$ crossbar arbitrates the order to send packets to the target output buffers, based on the predefined priority of the input. Data transfer adopts source routing protocol, and *forwarding information* in header flit is referred for the routing. Each TPU transfers data packets to each TSU via the network which consists of these routers.

Fig. 19 Signaling sequence

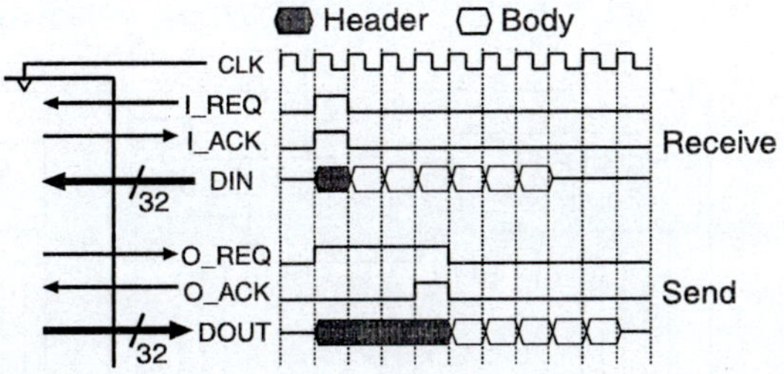

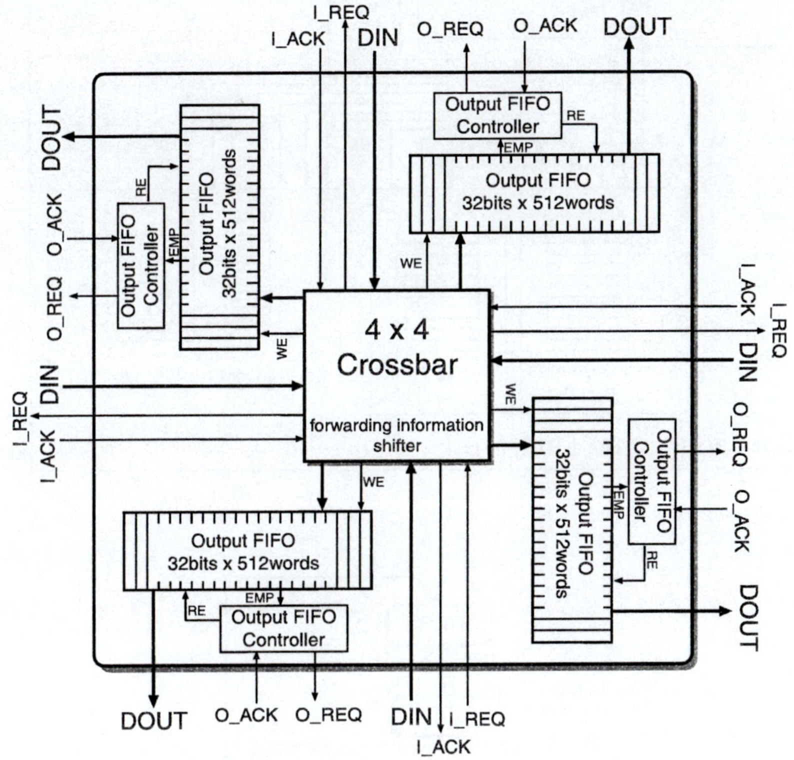

Fig. 20 An example of the router structure

3.4.9 Thread Private Unit

A TPU owns a data set about a status of one thread. A TPU has three arrays in BlockRAMs, as shown in Fig. 21. It also has a packet controller to communicate with each TSU based on the algorithm of NRM.

As shown in Fig. 21, the sequences of sending packets are the repetition of reading data from arrays in TPU and sending operations to TSUs. Figure 22 shows the dependency of each operation. The calculation of a reaction cycle begins from

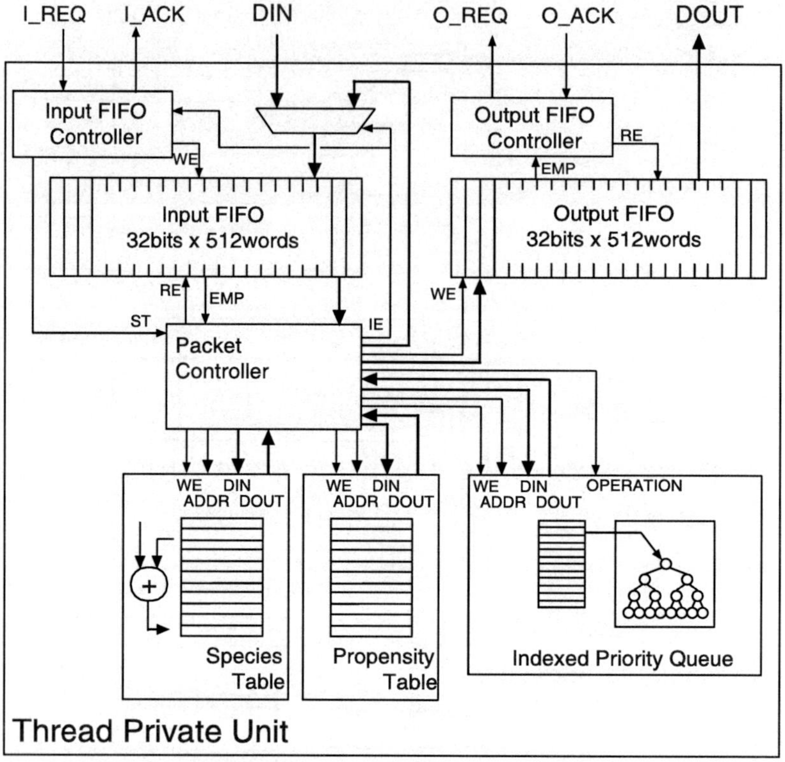

Fig. 21 Thread private unit (TPU)

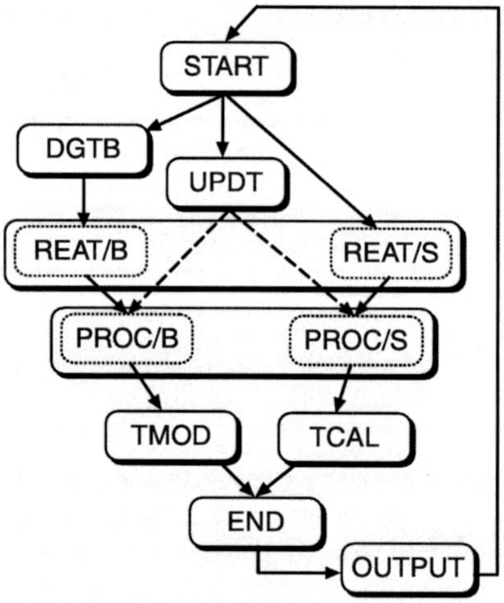

Fig. 22 Dependency per reaction cycle

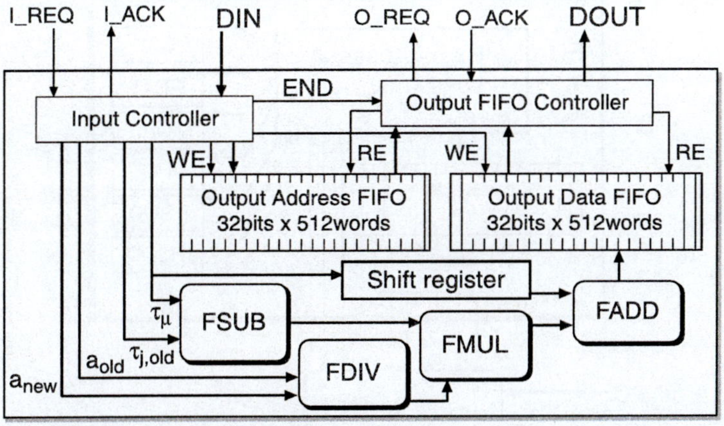

Fig. 23 An example of a TSU: TMOD

Table 5 Resource utilization and operating frequency of TPU and TSU

		TSU					
	TPU	REAT	TCAL	PRPC	UPDT	DGTB	TMOD
Slices	1,270	226	4,084	1,055	234	219	1,885
BRAMs	10	3	8	4	6	5	2
Mult.	0	0	13	8	0	0	4
Latency	–	1	21	11	2	2	27
Freq. [MHz]	113.234	157.324	130.733	123.362	165.696	161.409	127.585

a "`START`" node in Fig. 22, and the packet controller sends three packets to the next pointed nodes: `DGTB`, `UPDT`, and `REAT/S`. In the same figure, `/S` and `/B` packets are sent to the same TSU, but the number of their operation is different. The calculation of a simulation cycle ends when both `TCAL` and `TMOD` are completed.

3.4.10 Thread Share Unit

Each TSU calculates its own operation based on the received packet and sends back the result to the TPU. `TMOD` module as a typical example of TSU is shown in Fig. 23. The flits following the header flit are stored into the output address FIFO, as the header flit of the packet for return to the TPU. The result of arithmetic is stored into the output data FIFO. Each TSU is pipelined and can receive packet continuously and calculate its operation in fixed clock cycles.

Table 5 shows a rough estimation of the area and operating frequency of each unit. The area of `TCAL` in TSU is large, because it owns a logarithmic arithmetic unit that calculates (15). Random numbers required in the same equation are generated with M-sequence random number generator, and logarithmic values are obtained with second order interpolation. `PRPC` and `TMOD` are calculation units to obtain

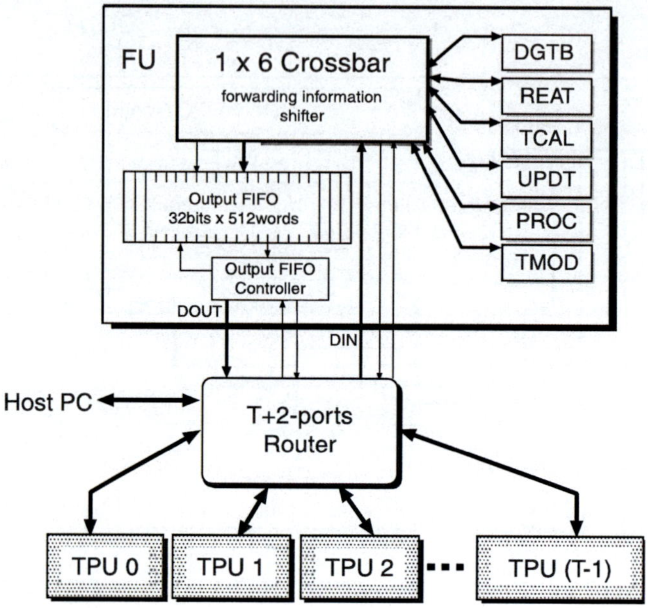

Fig. 24 Type AT: star structure

propensity a_j and τ_j modified with (16), respectively. DGTB are tables for storing constant values: species IDs of reactants in each reaction, state update vectors, and a dependency graph.

3.4.11 Evaluation

This section evaluates three types of network structures for NRM circuit implementation.

Type AT (Star structure): Type AT circuit (Fig. 24) has T TPUs which are connected to TSUs only by one router with $T+2$ ports. A group of TSUs is called a Functional Unit (FU), and it owns a 1×6 crossbar and a FIFO for output port.

Type B (Tree structure): This adopts a tree structure with 16 TPUs (Fig. 25). White round rectangles in Figs. 25 and 26 indicate the router, and numbers represent its number of ports. Dotted squares denote TPUs, and dark shaded squares represent FUs or TPUs.

Type C (Fat-Tree structure): In Type AT and B, many packets are concentrated on the gateway of FU. Therefore, Type C distributes TSUs in a circuit with more routers compared to the previous types (Fig. 26). Four REATs and four PROCs are shared by four TPUs. Each TSU accesses these units two times per simulation cycle. Only one set of TMOD and TCAL are placed in the circuit, because their access rate

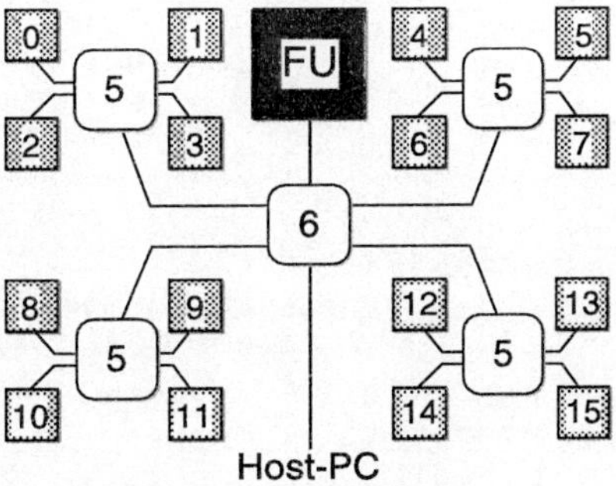

Fig. 25 Type B: cascading structure

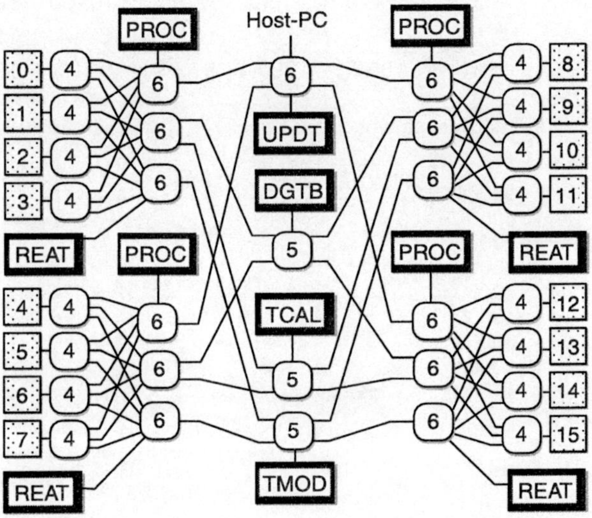

Fig. 26 Type C: network structure

is low in spite of their large area consumption. This topology is a composite structure of tree networks and can be categorized as a fat tree network.

3.4.12 Area and Operation Frequency Evaluation

Performance of the area and the operating frequency of Type AT was evaluated with different numbers of TPU for the router and the whole NRM circuit implemented. The results are shown in Fig. 27. Increase of the area of TPU is proportional to the numbers of TPU, whereas the interconnection network between TPUs and FU is larger than other components. This means that the router requires large resources according to its numbers of port. Similarly, the maximum operating frequency of the circuit degrades according to number of TPUs. There is a longer delay for arbitration and transition due to their complex operations. From Fig. 27, in case when $T \geq 4$, critical path of Type AT would be due to the router.

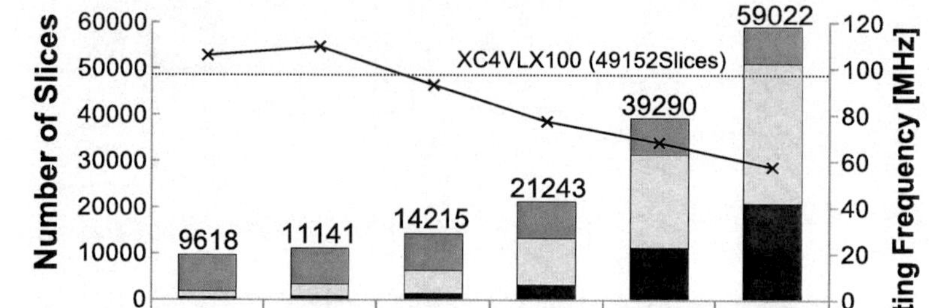

Number of TPU	1	2	4	8	16	24
FU	7819	7819	7819	7819	7819	7819
TPU	1270	2540	5080	10160	20320	30480
IC	529	782	1316	3264	11151	20723
* Freq.	105.82	109.47	92.899	77.101	68.161	57.657

Fig. 27 Evaluation of type AT

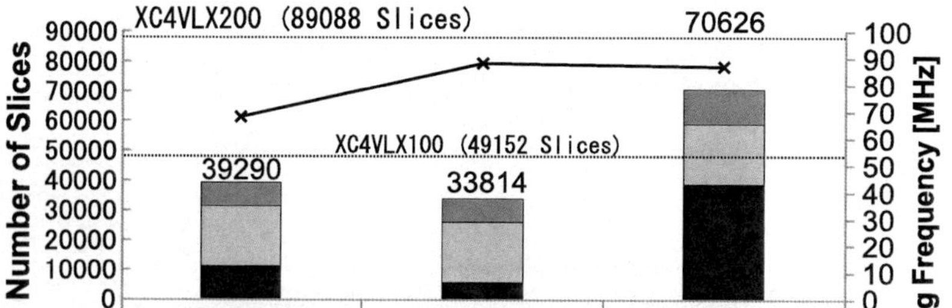

	TypeA16	TypeB	TypeC
FU	7819	7819	11546
TPU	20320	20320	20320
IC	11151	5675	38760
* Freq.	68.161	88.236	87.011

Fig. 28 Evaluations of type A16, B and C

Considering the resources of the target FPGA, 16 TPUs can be configured when implementing the Type A16. Numbers of TPU of Type B and Type C will also be fixed to 16 according to this result.

Figure 28 shows the area and the maximum operating frequency of Type A16, Type B, and Type C. Type B suppresses the area of the interconnection networks by 49 % compared to Type A16. Additionally, maximum operation frequency of Type B is also improved. Three additional modules of `REAT`s and three `PROC`s were added in the TPU of Type C, so its area is larger than the other types. Type C exceeds the area capability of the target FPGA(XC4VLX100), as the area of interconnect is 3.5 times larger than Type A16. Figure 28 shows a maximum capacity of the largest FPGA device(XC4VLX200: 89,088 slices) in Virtex-4 series.

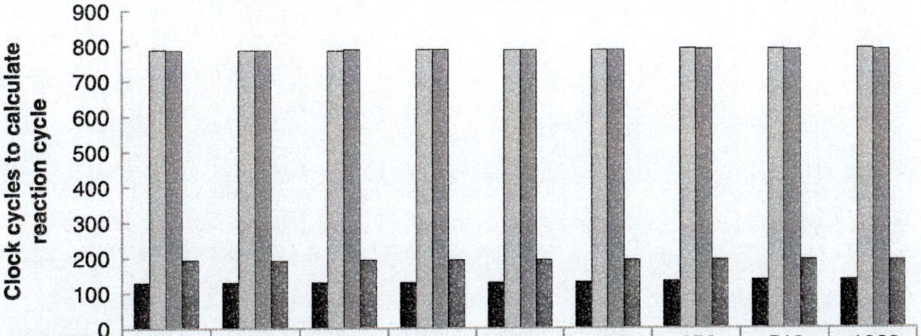

Model Size	4	8	16	32	64	128	256	512	1020
TypeA1	129.034	129.254	129.086	129.351	129.874	131.165	132.852	135.344	135.686
TypeA16	787.412	786.294	785.028	787.311	786.273	786.410	789.543	787.417	789.289
TypeB	786.322	786.224	787.476	787.410	786.693	785.332	787.476	785.874	786.191
TypeC	192.053	191.650	192.125	192.125	192.317	192.000	191.998	192.019	191.888

Fig. 29 Clock cycles per simulation cycle

3.4.13 Performance Evaluation

Stochastic biochemical models defined as $N/4$ sets of Lotka reactions [37] are used for the performance evaluation. In this section, we evaluated the throughput of each circuit type by the various model size according to the average clock cycles to calculate a simulation cycle and its operating frequency.

Average clock cycles to calculate a simulation cycle in Type A1, Type A16, Type B, and Type C are shown in Fig. 29. As Type A1 has a very simple network structure which has only a TPU, there is no jam with other TPUs in packet flow. According to Fig. 29, model size has small effect on the calculation time of a simulation cycle. TPU has an IPQ, which has a memory of binary tree structure, calculation time varies due to the number of rebuilding IPQ. Therefore, tree depth elongates calculation cycles, but it is within 5 % of whole clock cycles for a simulation cycle. However, Type A16 and Type B requires much longer time compared to Type A1, because six types of TSU are assembled into an FU which has only one I/O port. Packets are concentrated to the I/O port of the FU, which is a bottleneck to calculate a simulation cycle.

These performances are compared with a general-purpose processor. The program was written in C++, compiled with gcc-3.4.6 (-O3) on Linux 2.6.18, and executed on Xeon 3.20 GHz with 6.0 GB RAM. Even the program is run on a single thread, several techniques to increase the computational speed are introduced such as allocating tree structure on an array and address list of reactions which points the location on the tree.

Figure 30 shows throughput of RTL simulations with different model sizes between $N = 4$ and $N = 1{,}020$ of Lotka model. SW shows execution on general-purpose processor. Its throughput was calculated by a number of simulation cycles per second, which is approximately the execution time of 10^6 simulation cycles.

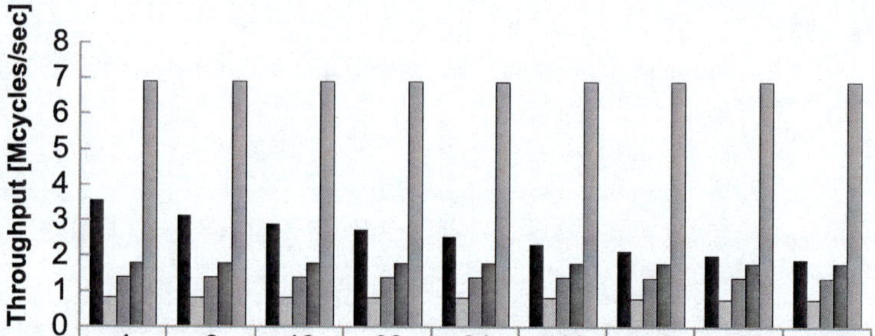

Model Size	4	8	16	32	64	128	256	512	1020
■SW	3.568	3.139	2.890	2.734	2.544	2.314	2.138	2.020	1.901
□TypeA1	0.820	0.819	0.820	0.818	0.815	0.807	0.797	0.782	0.780
■TypeA16	1.385	1.387	1.389	1.385	1.387	1.387	1.381	1.385	1.382
■TypeB	1.795	1.796	1.793	1.793	1.795	1.798	1.793	1.796	1.796
□TypeC	6.899	6.913	6.896	6.896	6.889	6.901	6.901	6.900	6.905

Fig. 30 Throughput

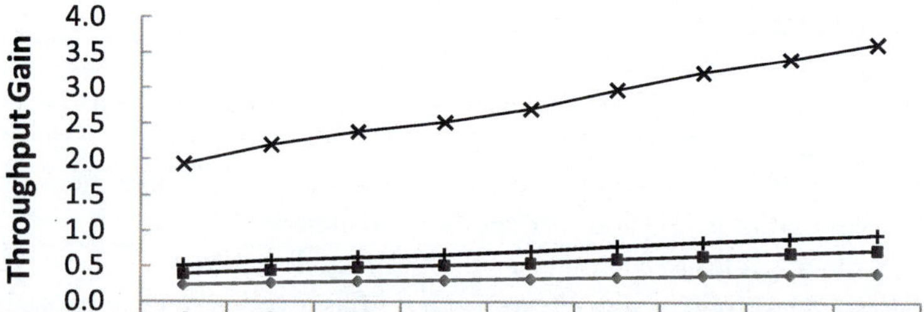

Model Size	4	8	16	32	64	128	256	512	1020
TypeA1	0.230	0.261	0.284	0.299	0.320	0.349	0.373	0.387	0.410
TypeA16	0.388	0.442	0.481	0.507	0.545	0.599	0.646	0.686	0.727
TypeB	0.503	0.572	0.620	0.656	0.706	0.777	0.838	0.889	0.945
TypeC	1.933	2.202	2.386	2.522	2.709	2.983	3.227	3.416	3.632

Fig. 31 Throughput gain

Throughput of FPGA was estimated by operating frequency in Fig. 27 or Fig. 28 divided by a number of clock cycles per simulation cycle in Fig. 29.

As shown in Fig. 30, throughput of SW degrades according to the model size due to the algorithmic property of NRM, whereas circuit on the FPGA maintains its throughput.

Figure 31 shows a throughput gain compared with the SW execution on Xeon 3.20 GHz. This indicates the advantage of Type C, which maintains high throughput even when model size increases. Although the throughput improvement on a chip is not so remarkable, use of FPGA for stochastic biochemical simulation may be advantageous for performance per watt, because of operating frequency is much lower than microprocessor dozens of times.

3.4.14 Discussion: For More Improvement

These evaluation results indicate several future approaches to improve performance. First, area of the current router consumes the vast majority of the whole circuit, for each router has large number of ports, and uses a BlockRAM. The router architecture can be minimized by using several distributed RAMs instead of using BlockRAMs. It is also worth considering interconnection networks that consist of routers with fewer ports. By saving the resource utilization of the Meanwhile, unnecessary transfer delay between the units may be prevented by utilizing small-port routers as repeater buffers. Although current throughput gain is 3.6 times higher than SW execution on Xeon 3.20 GHz at most, further enhancements on resource utilization of the interconnection network may bring out better performance, so that higher performance can be obtained even with lower-ranged FPGA chips.

3.5 Summary of Stochastic Approach

This section introduced a framework for designing and implementing a circuit on the FPGA that accelerates the execution of the NRM, one of the most recent stochastic biochemical simulation algorithms. By utilizing the on-chip interconnection network structure of the proposed circuit, calculation units can run multi-thread process for each independent simulation data of multiple data units. In this work, various types of network were configured depending on the capacity of target FPGA devices, and each performance was evaluated. Although the circuit achieved 3.6 times higher throughput on a high-end FPGA compared to that on Xeon 3.20 GHz at most, the area evaluation results indicated possibilities to improve throughput by reducing the area ratio of the interconnection network.

4 Conclusion

This chapter introduced three major examples in the field of bioinformatics. In recent years, various algorithms and methods, whether new or old, have been reconsidered since bioinformatics is broadening the field of applications. Furthermore, in order to ensure the smooth handling of huge biological databases and large-scale simulations, it is essential to utilize cluster computing technologies as accelerating them. Over the past 10 years, some studies have been done on FPGA cluster systems in bioinformatics [47–50]. These research achievements have also developed some new technical products. For example, DeCypher [51] is a famous parallel computing system designed to analyze genome databases from the early days of the FPGA acceleration history. It accelerates BLAST, Hidden Markov Model (HMM), and the Smith-Waterman algorithm. GeneMatcher2 [52] is another FPGA-based custom computer for HMM, the Smith-Waterman algorithm, and GeneWise.

HC-2 system [53] is applied not only in bioinformatics but also in the wide range of areas and it was the 44th place on the Graph500 list in 2012. Pico Computing [54] and SciEngines [55] are known as FPGA companies whose products are applied to dynamic programming algorithms. As stated above, FPGAs have long been contributed in the acceleration of bioinformatics application systems; however, we are now facing a new difficulty to build such a complicated heterogeneous cluster. Here, MaxCompiler [56] proposes one of the solutions. Its proposed system allows programmers to describe their target applications in a high-level language. The question now arises: how do we manage a number of bioinformatics applications as drawn up by Fig. 1 even if we can develop an independent application easily? Here, we may expand the application reconfigurability into user groups. Each group is expert in the field of one or a few biological disciplines and it means no one is able to put the whole picture together. Thus, Novo-G project which is composed of 11 academic groups is one reasonable approach. Novo-G system is a powerful cluster with 192 FPGAs and all team can use the system. This will be a valid method if there is a comprehensive framework which aims to integrate the computational results to one goal, such as system biology.

References

1. T.P. Niedringhaus, D. Milanova, M.B. Kerby, M.P. Snyder, A.E. Barron, Landscape of next-generation sequencing technologies. Anal. Chem. **83**(12), 4327–4341 (2011)
2. T. Toyoda, A. Wada, Omic space: coordinate-based integration and analysis of genomic phenomic interactions. Bioinformatics **20**(11), 1759–1765 (2004)
3. J. Lederberg, A.T. McCray, 'Ome sweet 'omics – a genealogical treasury of words. Scientist **15**(7), 8 (2001)
4. M.O. Dayhoff, R.M. Schwartz, B.C. Orcutt, A Model of Evolutionary Change in Proteins, in Atlas of protein sequence structure, Natl. Biomedical Research, **5**(3), 345–352 (1978)
5. S.F. Altschul, Amino acid substitution matrices from an information theoretic perspective. J. Mol. Biol. **219**(3), 555–565 (1991)
6. S. Whelan, N. Goldman, A general empirical model of protein evolution derived from multiple protein families using a maximum-likelihood approach. J. Mol. Biol. Evol. **18**, 691–699 (2001)
7. S.F. Altschul, W. Gish, W. Miller, E.W. Myers, D.J. Lipman, Basic Local Alignment Search Tool. Mol. Biol. **215**(3), 403–410 (1990)
8. W.R. Pearson, Comparison of methods for searching protein sequence databases. Profein Sci. **4**(6), 1145–1160 (1995)
9. E.G. Shpaer, M. Robinson, D. Yee, J.D. Candlin, R. Mines, T. Hunkapiller, Sensitivity and selectivity in protein similarity searches: a comparison of Smith-Waterman in hardware to BLAST and FASTA. Genomics **38**, 179–191 (1996)
10. S.B. Needleman, C.D. Wunsch, A general method applicable to the search for similarities in the amino acid sequence of two proteins. J. Mol. Biol. **48**(3), 443–453 (1970)
11. T.F. Smith, M.S. Waterman, Identification of common molecular subsequences. J. Mol. Biol. **147**(1), 195–197 (1981)
12. T.V. Court, M.C. Herbordt, Families of FPGA-based accelerators for approximate string matching. Microprocess. Microsyst. **31**, 135–145 (2007)
13. T.F. Oliver, B. Schmidt, D.L. Maskell, Reconfigurable architectures for bio-sequence database scanning on fpgas. IEEE Trans. Circ. Syst. II **52**(12), 851–855 (2005)

14. P. Zhang, G. Tan, G.R. Gao, Implementation of the Smith-Waterman algorithm on a reconfigurable supercomputing platform, in *Proceedings of the 1st International Workshop on High-Performance Reconfigurable Computing Technology and Applications: Held in Conjunction with SC07* (ACM, New York, 2007), pp. 39–48
15. K. Benkrid, Y. Liu, A. Benkrid, A highly parameterised and efficient FPGA-based skeleton for pairwise biological sequence alignment. IEEE Trans. Very Large Scale Integr. (VLSI Syst.) **17**(4), 561–570 (2009)
16. S. Lloyd, Q.O. Snell, Hardware accelerated sequence alignment with traceback. Int. J. Reconfigurable Comput. Article ID **762362**, 1–10 (2009)
17. M.N. Isa, K. Benkrid, T. Clayton, C. Ling, A.T. Erdogan, An FPGA-based parameterised and scalable optimal solutions for pairwise biological sequence analysis, in *Proceedings of 2011 NASA/ESA Conference on Adaptive Hardware and Systems* (IEEE, Piscataway, 2011), pp. 344–351
18. L. Ligowski, W.R. Rudnicki, An efficient implementation of Smith Waterman algorithm on GPU using CUDA, for massively parallel scanning of sequence databases, in *Proceedings of Eighth International Workshop on High Performance Computational Biology: Held in Conjunction with 2009 IEEE International Symposium on Parallel & Distributed Processing* (IEEE, 2009), pp. 1–8
19. Y. Liu, D.L. Maskell, B. Schmidt, CUDASW++: optimizing smith-waterman sequence database searches for CUDA-enabled graphics processing units. BMC Res. Notes **2**(1), 73–82 (2009)
20. L. Ligowski, W.R. Rudnicki, GPU-SW Sequence Alignment server, in *International Conference on Computational Science 2010* (2010), pp. 1–10, http://gpucomputing.net/?q=node/3149. Accessed February 2011
21. K. Dohi, K. Benkrid, K.C. Ling, T. Hamada, Y. Shibata, Highly efficient mapping of the Smith-Waterman algorithm on CUDA-compatible GPUs, in *Proceedings of the 21st IEEE International Conference on Application-Specific Systems Architectures and Processors* (IEEE, 2010), pp. 29-36
22. M. Aldinucci, M. Danelutto, M. Meneghin, P. Kilpatrick, M. Torquati, Efficient streaming applications on multi-core with FastFlow: the biosequence alignment test-bed, in *Proceedings of Parallel Computing: from Multicores and GPU's to Petascale* (IOS Press, Amsterdam, 2009), pp. 273–280
23. O. Gotoh, An improved algorithm for matching biological sequences. J. Mol. Biol. **162**(3), 705–708 (1982)
24. A.C. Jacob, J.D. Buhler, R.D. Chamberlain, Design of throughput-optimized arrays from recurrence abstractions, in *Proceedings of the 21st IEEE International Conference on Application-Specific Systems Architectures and Processors* (IEEE, 2010), pp. 133–140
25. Y. Yamaguchi, H. Tsoi, W. Luk, FPGA-based Smith-Waterman algorithm: analysis and novel design, in *Proceedings of the 7th International Conference on Reconfigurable Computing: Architectures, Tools and Applications* (Springer, Berlin, 2011), pp. 181–192
26. S.A. Manavski, G. Valle, CUDA compatible GPU cards as efficient hardware accelerators for smith-waterman sequence alignment. BMC Bioinformatics **9**(suppl 2), S10 (2008)
27. B.A. Barshop et al., Analysis of numerical methods for computer simulation of kinetic processes: development of kinsim — a flexible, portable system. Anal. Biochem. **130**, 134–145 (1983)
28. M. Pedro, Gepasi: a software package for modelling the dynamics, steady states and control of biochemical and other systems. Comput. Appl. Biosci. **9**(5), 563–571 (1993)
29. I. Goryanin et al., Mathematical simulation and analysis of cellular metabolism and regulation. Bioinformatics **15**(9), 749–758 (1999)
30. M. Tomita et al., E-cell: software environment for whole-cell simulation. Bioinformatics **15**(1), 72–84 (1999)
31. I.I. Moraru, J.C. Schaff, B.M. Slepchenko, L.M. Loew, The virtual cell: an integrated modeling environment for experimental and computational cell biology. Ann. N Y Acad. Sci. **971**, 595–596 (2002)

32. M. Hucka, A. Finney, B.J. Bornstein, S.M. Keating, B.E. Shapiro, J. Matthews, B.L. Kovitz, M.J. Schilstra, A. Funahashi, J.C. Doyle, H. Kitano, Evolving a lingua franca and associated software infrastructure for computational systems biology: the systems biology markup language (SBML) project. IEE Syst. Biol. **1**(1), 41–53 (2004)
33. M. Hucka, A. Finney, H. Sauro, H. Bolouri, in *Systems Biology Markup Language (SBML) Level 1: Structures and Facilities for Basic Model Definitions*. Systems Biology Workbench Development Group, ERATO Kitano Symbiotic Systems Project, version 2nd edn. (California Institute of Technology, Pasadena, 2003)
34. H. Yamada, Y. Ogawa, T. Ooya, T. Ishimori, Y. Osana, M. Yoshimi, Y. Nishikawa, A. Funahashi, N. Hiroi, H. Amano, Y. Shibata, K. Oguri, Automatic pipeline construction focused on similarity of rate law functions for an FPGA-based biochemical simulator. IPSJ Trans. Syst. LSI Des. Methodol. **3**, 244–256 (2010)
35. Y. Osana, M. Yoshimi, Y. Iwaoka, T. Kojima, Y. Nishikawa, A. Funahashi, N. Hiroi, Y. Shibata, N. Iwanaga, H. Kitano, H. Amano, An FPGA-based biochemical simulator recsip (to appear/english translation of Osana et al. on trans. IEICE j89-d). Syst. Comput. Jpn. **J89-D**(6), 1163–1172 (2007)
36. A. Funahashi, N. Tanimura, M. Morohashi, H. Kitano, Celldesigner: a process diagram editor for gene-regulatory and biochemical networks. BIOSILICO **1**(5), 159–162 (2003)
37. D.T. Gillespie, A general method for numerically simulating the stochastic time evolution of coupled chemical reactions. J. Comput. Phys. **22**, 403–434 (1976)
38. M.A. Gibson, J. Bruck, Efficient exact stochastic simulation of chemical systems with many species and many channels. J. Phys. Chem. A **104**(9), 1876–1889 (2000)
39. Y. Cao et al., Efficient formulation of the stochastic simulation algorithm for chemically reacting systems. J. Chem. Phys. **121**(9), 4059–4067 (2004)
40. K. Takahashi et al., A multi-algorithm, multi-timescale method for cell simulation. Bioinformatics **20**(4), 538–546 (2004)
41. J.F. Keane, C. Bradley, C. Ebeling, A compiled accelerator for biological cell signaling simulations, in *Proceedings of the 2004 ACM/SIGDA 12th International Symposium on FPGA* (ACM, New York, 2004), pp. 233–241
42. L. Salwinski et al., In silico simulation of biological network dynamics. Nat. Biotechnol. **22**(8), 1017–1019 (2004)
43. M. Yoshimi, Y. Osana, T. Fukushima, H. Amano, Stochastic simulation for biochemical reactions on FPGA, in *Proceedings of the 14th IEEE International Conference on FPL* (Springer, Berlin, 2004), pp. 105–114
44. M. Yoshimi, Y. Osana, Y. Iwaoka, A. Funahashi, N. Hiroi, Y. Shibata, N. Iwanaga, H. Kitano, H. Amano, The design of scalable stochastic biochemical simulator on FPGA, in *Proceedings of the 15th IEEE Field Programmable Technology* (IEEE, 2006), pp. 339–340
45. M. Yoshimi, Y. Osana, Y. Iwaoka, Y. Nishikawa, T. Kojima, A. Funahashi, N. Hiroi, Y. Shibata, N. Iwanaga, H. Kitano, H. Amano, An FPGA implementation of high throughput stochastic simulator for large-scale biochemical systems, in *Proceedings of the 16th IEEE International Conference on Field Programmable Logic and Applications* (IEEE, 2006), pp. 227–232
46. M. Yoshimi, Y. Iwaoka, Y. Nishikawa, T. Kojima, Y. Osana, A. Funahashi, N. Hiroi, Y. Shibata, N. Iwanaga, H. Yamada, H. Kitano, H. Amano, FPGA implementation of a data-driven stochastic biochemical simulator with the next reaction method, in *Proceedings of the 17th IEEE International Conference on Field Programmable Logic and Applications* (IEEE, 2007), pp. 254–259
47. B. Schmidt, H. Schroder, M. Schimmler, Massively parallel solutions for molecular sequence analysis, in *Proceedings of the 1st International Workshop on High Performance Computational Biology: Held in Conjunction with 2002 IEEE International Symposium on Parallel & Distributed Processing* (IEEE, 2002), pp. 186–193
B. Schmidt, H. Schroder, M. Schimmler, Massively parallel solutions for molecular sequence analysis, in Proceedings of the 1st International Workshop on High Performance Computational Biology: held in conjunction with 2002 IEEE International Symposium on Parallel & Distributed Processing, IEEE, pp. 186–193 (2002)

48. K. Regester, J.-H. Byun, A. Mukherjee, A. Ravindran, Implementing bioinformatics algorithms on nallatech-configurable multi-FPGA systems. Xcell J. Second Quarter, 100–103 (2005)
49. S. Masuno, T. Maruyama, Y. Yamaguchi, A. Konagaya, Multidimensional dynamic programming for homology search on distributed systems, in *Proceedings of Euro-Par 2006 Parallel Processing* (Springer, Berlin, 2006), pp. 1127–1137
50. A.G. Schmidt, S. Datta, A.A. Mendon, R. Sass, Investigation into scaling I/O bound streaming applications productively with an all-FPGA cluster. Parallel Comput. **38**, 344–364 (2012)
51. T. Mittler, M. Levy, C. Feller, K. Schlauch, Multblast: a web application for multiple blast searches. Bioinformation **5**, 224–226 (2010). http://www.timelogic.com/
52. M.A. Rieffel, T.G. Gill, W.R. White, *Bioinformatics Clusters in Action* (Paracel, Inc., Pasadena, 2004) 8 pp., http://www.paracel.com/pdfs/clusters-in-action.pdf
53. Convey to deliver FPGA cluster to virginia bioinformatics institute, HPC wire (2011), http://conveycomputer.com/. Accessed August 2011
54. FPGA cluster accelerates bioinformatics application by 5000X, Dr. Dobb's Journal (2009), http://www.picocomputing.com/. Accessed November 2009
55. Rivyera s3-5000 (white paper, v2.1), SciEngines (2012), http://www.sciengines.com/. Accessed April 2012
56. Maxcompiler (white paper), Maxeler Technologies (2011), http://www.maxeler.com/. Accessed February 2011

High-Performance Computing for Neuroinformatics Using FPGA

Will X.Y. Li, Rosa H.M. Chan, Wei Zhang, Chiwai Yu, Dong Song, Theodore W. Berger, and Ray C.C. Cheung

Abstract The brain represents information through the ensemble firing of neurons. These neural processes are difficult to study in vivo because they are highly non-linear, dynamical and often time-varying. Hardware systems, such as the FPGA-based platforms, are very efficient in doing such studies given their intrinsic parallelism, reconfigurability and real-time processing capability. We have successfully used the Xilinx Virtex-6 FPGA devices to prototype the generalized Laguerre–Volterra model (GLVM), which is a rigorous and well-functioning mathematical abstraction for the description of neural processes from a system input/output relationship standpoint. The hardware system first conducts GLVM parameters estimation using the neural firing data from experiments; then it is able to predict the neural firing outputs based on the field estimated model coefficients and the novel model inputs. The hardware system has been prototyped and is proved very efficient in this study compared to our previous software model running on the Intel Core i7-2620M CPU (with Turbo Boost to 3.4 GHz). It achieves up to a 2,662 times speedup in doing GLVM parameters estimation and a 699 times speedup in conducting neural firing outputs prediction. The calculation results are very precise with the NMSE being successfully controlled at the 10^{-11} scale compared to the software approach. This FPGA-based architecture is also significant to the future cognitive neural prostheses design.

W.X.Y. Li (✉) • R.C.C. Cheung
City University of Hong Kong, Hong Kong
e-mail: xyli@ee.cityu.edu.hk; rcheung@cityu.edu.hk

R.H.M. Chan • W. Zhang • C. Yu
Department of Electronic Engineering, City University of Hong Kong, Hong Kong

D. Song • T.W. Berger
Department of Biomedical Engineering, University of Southern California, Los Angeles, USA

W. Vanderbauwhede and K. Benkrid (eds.), *High-Performance Computing Using FPGAs*, 177
DOI 10.1007/978-1-4614-1791-0_6, © Springer Science+Business Media, LLC 2013

1 A Brief Introduction to Cognitive Neuroscience and Cognitive Neural Prosthesis

As an important branch of neuroscience, cognitive neuroscience, which focuses on the biological substrates underlying cognition, grew rapidly over the last several decades. The study of cognitive neuroscience is of multidisciplinary nature, for it requires knowledge of fields such as neurobiology, bioengineering, philosophy, and computer science. In this chapter, we explore how modern reconfigurable computing facilities can be effectively applied to the study of cognitive neuroscience, especially, to the research regarding the important brain functions of learning and memory.

1.1 The Hippocampus: Hub of Brain Network Communication for Memory

Learning is the process of the acquisition of enduring information, behavior patterns or skills through study, experience, or being taught. Memory refers to the capability of the retention of such information or abilities and the retrieval of them while being stored [1]. Learning and memory appear as two important aspects of animal cognition and are highly relevant in their biological basis to an important component of the brain—the hippocampus. It belongs to the limbic system and is responsible for the formation of long-term declarative or explicit memories, such as the specific personal experiences or factual information that can be recollected through later conscious brain activity. The hippocampus itself is comprised of several different sub-systems which form a closed feedback loop, as shown in Fig. 1. The flow of information within the hippocampus is largely unidirectional. Bioelectrical signals propagate through a series of tightly packed layers of nerve cells. Input signals from the dentate gyrus (DG) first go to the CA3 layer, and then the CA1 layer, after that the subiculum, finally out of the hippocampus to the entorhinal cortex (EC). After this process of signal processing, new interpretation of patterns of information

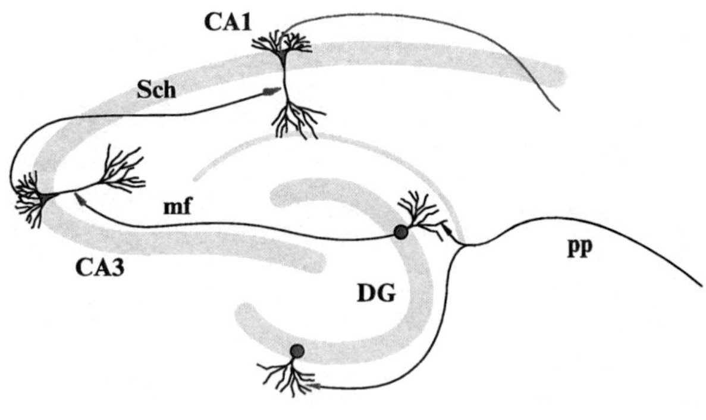

Fig. 1 The double-C shaped anatomical structure of the hippocampus. The *arrows* show the way of bioelectrical signal transmission within the hippocampal system. DG: dentate gyrus; CA3: Cornu Ammonis 3; CA1: Cornu Ammonis 1

are directed to other cortical areas for the purpose of long-term storage. Thus, hippocampus is not the brain region for storage of long-term declarative memories, but instead, a hub for the transmission and re-encoding of mnemonic information through it intrinsic biological circuitry.

Clinical practices reveal that the degeneration or malformation of hippocampal cells results in different kinds of pathological changes, some are highly related to the formation of long-term memory [2, 3]. The well-known Alzheimer's disease and dementia are in close association with the selective loss of hippocampal cells during their formation. Besides, symptoms such as the epileptic activity are highly susceptible to be the outcomes of dysfunction of the hippocampal CA3 region; while stroke is very possibly to be brought about by the preferential damage to CA1 pyramidal cells. Even the brain trauma is manifested to be of consequence to the selective loss of the hilar neurons of the hippocampal system. While medication can serve as a supplementary means for the treatment of these brain lesions, a more permanent cure may be the application of artificial substitutions to the pathological areas within the brain. These substitutions can also be termed neural prosthetic devices or neural prostheses.

1.2 Neural Prosthesis for Restoring Lost Cognitive Function

Neural prostheses are devices that can substitute modality that might have been damaged as a result of injury or disease. Based on their specific functions, the neural prosthetic devices fall into two major categories. Devices belonging to the first category can be adopted for making a compensation for the lost communications between the central nervous system (CNS) and the exosomatic environment. There are two subdivisions of such devices, also based on their functionalities. Devices such as artificial retina or cochlear implant pertain to the first subdivision, which works for the transduction of physical energy to bioelectrical simulation of sensory nerve fibers, bypassing the damaged primary sensory cells [4, 5]. Devices such as artificial limb belong to the second subdivision, which attempt to decode the bioelectrical signals coming from the CNS and generate functional electrical stimulation, thus compensating for the loss of motor control [6–13]. The other main category of the neural prostheses is developed to restore the lost communication between two individual brain areas, such as two different sub-regions within the animal hippocampal system. This kind of prostheses presents special challenges to the designers due to their dual roles in performing both signal decoding and encoding. By bridging the brain regions by an artificial, silicon-based means, the once fractured communication can be well reestablished.

We are now endeavoring to develop a hippocampal CA3 cognitive neural prosthesis. The prosthesis functions to transform the spatial temporal pattern of CA3 input spike trains to the spatial temporal pattern of CA1 output spike trains. The spike trains refer to a sequence of all-or-none, point process spiking events, with variations of intervals among individuals of them. Owing to the fact that underlying molecular mechanisms and synaptic connections are of highly dynamical and nonlinear nature [14–20], it is necessary for the hippocampal neural prosthesis, which mimics the short-term to long-term memory re-encoding process, to work in good adherence to a well-functioning mathematical model that can be constructed to represent the very complex brain process.

2 Modeling Techniques for Neural Systems

Mathematical model describing the process of the transmission of neural signals as they flow through neuronal ensembles can facilitate us to better understand the underlying mechanisms of the brain. However, this attempt is often hindered by the intrinsic complexity of these neural processes given their high nonlinearity, dynamic property, and potential variations contributed by internal and/or external factors. In order to study these nonlinear time-varying dynamics, various computational models have been proposed. Some of them have been successfully applied to describe different parts of the neural systems such as the retina [21–23], the auditory cortex [24], and the motor systems [25]. These computational models largely fall into two main categories: parametric models and nonparametric models. The former aims to investigate mechanism of the underlying biophysical and physiological processes while the latter seeks to find a solution to the problem by quantifying the functional interactions among the broad-range mechanisms found in the neurons. Table 1 draws a brief comparison between the parametric models and nonparametric models from the perspectives of their theoretical foundation and respective features.

2.1 *The Parametric Models*

The parametric models, which are also often termed mechanistic models, build their own structures starting from the mathematical equations delineating the molecular mechanisms of each neuron. A typical example for the mechanistic models is the Hodgkin and Huxley model, which is derived from experiments using squid giant axon and adopts a set of nonlinear ordinary differential equations to describe the initiation and propagation process of the action potentials (APs) [26]. Notable ongoing projects based on the mechanistic models include the IBM Blue Brain Project, which aims to reconstruct the brain piece by piece and build a virtual brain by reverse-engineering the real one down to the molecular level [27]; the elementary objects of neural systems (EONS), which provides an integrated synaptic platform for the study of the interactions between elements within the synapse [28]; and the Brains in Silicon project by Stanford, which is established to investigate how cognitive behavior arises from the brain's physiology such as the mechanism of the ion-channels [29]. Various simulation tools are also developed, such as NEURON and MCell [30, 31]. However, since the exact function of the neural circuit for cognition remains unknown, it would be difficult to verify the model's accuracy or generate new insights into how particular neural circuit works or what it

Table 1 Comparison between the parametric and nonparametric models

	Parametric models	Nonparametric models
Form of description	Utilizing the differential equations to design specific mechanisms or compartments	Utilizing Volterra type functional expansions
Prerequisite of application	A priori model postulates; Adjustment of unknown parameters in each application case	Broad-band stimulation trains
Advantages	Capable of being directly biophysically and physiologically interpreted; Some neuronal functions such as synaptic transmission, dendritic integration can be well described thereby	Exemption of model specification; Possessing the predictive capabilities for arbitrary inputs; Generality for a wild range of applications
Disadvantages	Becoming untrackable during scaling up due to the increasing model complexity with more inputs	Unable to be directly interpreted by low-level system mechanisms
Scope of application	Primarily individual neuron modeling (can be scaled up for neural ensembles modeling which takes much more computation)	Individual neuron or neural ensembles modeling

does while implementing these mechanistic models. Meanwhile, there are actually too many mechanisms underlying neural systems to be modeled parametrically. The effects of these mechanisms vary with ion channel densities, distributions in dendrites, and many other parameters. Given billions of inputs and outputs, a detailed but incomplete mechanistic model of the nervous system would still require supercomputing facilities.

2.2 The Nonparametric Models

The nonparametric models are also called data-driven models, which use engineering modeling techniques such as network analysis, information theory, and statistical methods to investigate or describe the behavior of biological neuron or neural networks. The data-driven models provide less biological information than the mechanistic models do, but reduce potential errors in the postulation of model structures required in the mechanistic modeling approach. Several approaches can be adopted when using the data-driven models to decode the neural firing signals. One simple example is the linear decoder, which postulates that the decoded neural firing rate is the linear combination of the neural activity which has been recorded. It uses linear least squares to fit the statistical model to data and the quality of decoding can be evaluated by calculating the root mean square errors [32]. Another common model estimation method is maximum likelihood, which is also convenient for assessment of model goodness-of-fit and for construction of the confidence intervals [33, 34]. Various versions of Kalman filter, which uses a series of measurements observed over time and produces a statistically optimal estimate, were also used to estimate model coefficients in a number of motor prosthesis projects [35]. In the next section, we will focus our discussion on using functional power series as an efficient way to describe the nonlinear input–output properties of neural circuits.

2.3 Theoretical Background of Our Model

The modeling algorithm we use in this work belongs to the category of nonparametric approaches.

Previous studies carried by Volterra [36], Wiener [37], and Marmarelis [38, 39] and other researchers have demonstrated that for any nonlinear and time-invariant system with finite memory length, the system output can be represented as a functional power series of the system input. For the single-input, single-output, discrete-time case, the input–output relationship can be expressed as:

$$y(t) = k_0 + \sum_{\tau=0}^{M} k_1(\tau)x(t-\tau) + \sum_{\tau_1=0}^{M}\sum_{\tau_2=0}^{M} k_2(\tau_1,\tau_2)x(t-\tau_1)x(t-\tau_2) + \cdots \quad (1)$$

In the above equation, the system dynamics is revealed through the temporal convolution between system input and the kernel functions k; while the system nonlinearity is suggested by multiple convolutions between the input and the higher order kernel functions. Theoretically, if the orders of the kernels employed are high enough, then arbitrary nonlinearities can be represented. For performing animals trained to achieve an asymptotic behavior, their brain activity can be viewed as in nonlinear and time-invariant pattern.

Our model can be extended to time-varying neural systems using similar model structure. In that case, the system input–output transfer function should be updated following a learning rule. And this learning rule can be eventually substantiated by conducting mathematical analysis. Thereby, the system parameters can be optimized and the changes/variances of the input–output transfer function can be replicated in neural decoding for generation of the stimuli.

3 Modeling the Brain Activity Using Mathematics

The neural model we use in our work is the generalized Laguerre–Volterra model (GLVM), which was developed earlier in our group by Song et al. and is a mathematical model for description of the highly complicated neural processes from a system input/output relationship standpoint [40–42]. The GLVM belongs to the category of nonparametric models. It utilizes the real-time Laguerre expansion of Volterra kernels and the point process filters for online study of the time-varying neural system using both natural spike inputs and outputs [41] with their values being recorded by means of the multi-electrode array technology. The GLVM is inspired by the electrophysiological properties of single spiking neurons and is first applied to the study of animal hippocampal region where it is proved to be very efficient by animal experiments [43].

In the GLVM, we propose the integration of (1) generalized Volterra model (GVM), (2) real-time Laguerre expansion, and (3) steepest descent point process filter (SDPPF) to track the time-varying neural system using both natural spike inputs and outputs.

3.1 Configuration of the Generalized Volterra Model

A MIMO system can be decomposed into a series of multiple-input, single-output (MISO) systems. The MISO models are identical in structure and each module projects to a separate output as shown in Fig. 2.

Each MISO model has physiologically plausible components which can be described by the following equations:

$$w = u(k,x) + a(h,y) + \varepsilon(\sigma) \quad (2)$$

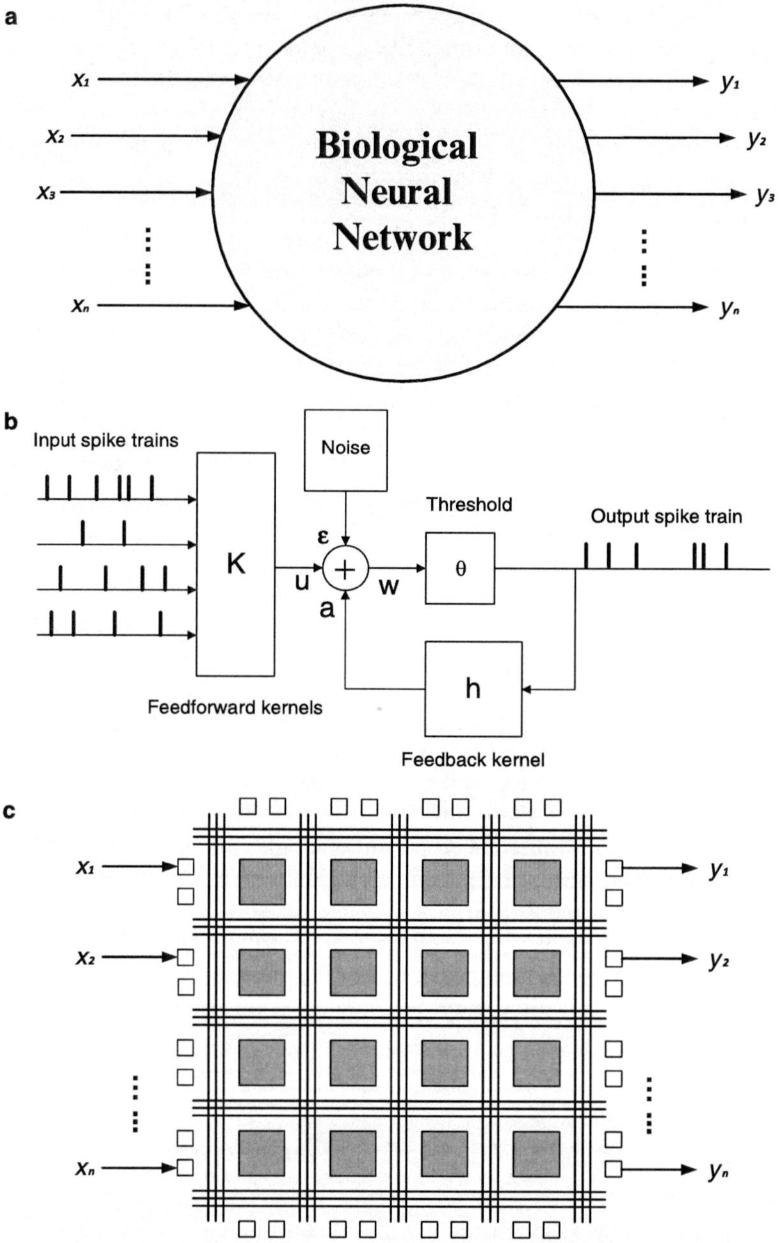

Fig. 2 FPGA implementation of a MIMO model for neural population activities. (**a**): brain region processes information by transforming input spike trains to output spike trains. (**b**): this input–output transformation can be described with a MIMO generalized Laguerre–Volterra model which can be further decomposed into a series of MISO models; each of them has physiological components, as shown. (**c**): the goal of this study is to implement such a MIMO model on FPGA (Subfigure (**c**) is only for illustration of concept, not representing the real circumstance of place and route)

and

$$y = \begin{cases} 0 & \text{where } w < \theta \\ 1 & \text{where } w \geq \theta \end{cases}. \tag{3}$$

The input and output spike trains are denoted by x and y, respectively. The hidden variable w represents the "pre-threshold membrane potential" of the output neuron. It is the summation of the "synaptic potential" u, the output spike-triggered "after-potential" a, and a Gaussian white noise input ε with standard deviation σ. The noise term models both the intrinsic noise of the output neuron and contributions from unobserved inputs. When w crosses the threshold θ, an output spike is generated and a feedback after-potential a is triggered and then added to w. Consider the first order Volterra kernel k_1 where N is the number of inputs. Then, the "synaptic potential" u can be expressed as

$$u(t) = k_0 + \sum_{n=1}^{N} (k_1^{(n)} * x_n)(t). \tag{4}$$

k_0 is the baseline potential when the inputs are absent. The first order kernels k_1 in (4) describes the linear transformation of input spike trains x into the hidden variable u, as functions of the time intervals between the present time and the previous spikes. The feedback variable a can be expressed as

$$a(t) = (h * y)(t), \tag{5}$$

where h is the linear feedback kernel. It is modeled by first order Volterra kernel. The feedback kernels captured spike-triggered processes that influence the firing behavior of hippocampal and also other cortical neurons [14, 44–48]

3.2 *Laguerre Expansion of GVM: Generalized Laguerre–Volterra Model*

The Laguerre expansion of the Volterra kernel (LEV) technique is used to reduce the number of open parameters to be estimated [41]. Using the LEV technique, both feedforward kernels k and feedback kernel h are expanded through L orthonormal Laguerre basis functions [49]. Input and output spike trains x and y are convolved with jth basis function b_j, such that the convolution products v are expressed as $v_j^{(n)} = b_j * x_n$ and $v_j^h = b_j * y$. Synaptic potential u and after potential a can be rewritten as

$$u(t) = c_0 + \sum_{n=1}^{N} \sum_{j=1}^{L} c_1^{(n)} v_j^{(n)}(t) \tag{6}$$

and

$$a(t) = \sum_{j=1}^{L} c_h v_j^h(t). \tag{7}$$

The convolved functions v include the temporal dynamics. Another advantage of the Laguerre expansions is that the convolutions are generated in real time. Let α_n $(0 < \alpha_n < 1)$ be the pole of the Laguerre basis functions of the nth input x_n. The Laguerre basis functions can be obtained by inverse Z-transform of transfer function of the Laguerre filter

$$b_j^{(n)} = Z^{-1}\left\{ \frac{\sqrt{1-\alpha_n^2}}{1-\alpha_n z^{-1}} \left(\frac{z^{-1}-\alpha_n}{1-\alpha_n z^{-1}} \right)^{j-1} \right\}. \tag{8}$$

A Laguerre basis function of jth order will have $j-1$ intercepts with the x-axis. The decay time of the built-in exponential of the Laguerre basis functions increases when the value of the Laguerre pole increases. The convolved product v can also be computed iteratively at each time t [50]. Let $V^{(n)}(t) = [v_1^{(n)}(t) \cdots v_L^{(n)}(t)]$,

$$V^{(n)}(t)A_1 = V^{(n)}(t-1)A_2 + \sqrt{1-\alpha_n^2}A_3 x_n(t), \tag{9}$$

where $A_1 = I + \alpha_n I_+$; $A_2 = \alpha_n I + I_+$; $A_3 = [1\ 0 \cdots 0]$; I is an $L \times L$ identity matrix and I_+ is an upper shift matrix.

3.3 Estimation of Parameters

Given the recorded input and output spike trains x and y, u and a can be readily calculated based on the present values of v and the model coefficients in real time. The estimated firing probability $P(t)$ is then calculated using the error function:

$$P(t) = 0.5 - 0.5\,\mathrm{erf}\left(\frac{\theta - u(t) - a(t)}{\sqrt{2}\sigma} \right). \tag{10}$$

Without the loss of generality, θ and σ can be set to 0 and 1, respectively. The model parameters to be estimated are the Laguerre coefficients C. Using the steepest descent point process filtering algorithm (SDPPF), the parameter vector $C(t)$ is updated iteratively at each time step t:

$$C(t) = C(t-1) + R\left[\left(\frac{\partial \log P(t)}{\partial C} \right)' (y(t) - P(t)) \right]_{C(t-1)}, \tag{11}$$

where R is learning rate. During adaptive parameter estimation, the gradient can also be generated in real time. The derivatives with respect to the Laguerre coefficients are given as the products of v calculated in (9), such as $\partial u(t)/\partial c_0 = 1$, $\partial u(t)/\partial c_1^{(n)}(j) = v_j^{(n)}(t)$ and $\partial a(t)/\partial c_h(j) = v_j^h(t)$. After observation of actual output spike train and prediction of firing probability in (10), R acts as the learning rate for the parameter estimations in (11). The estimated Laguerre coefficients C are used to reconstruct the feedforward and feedback kernels.

3.4 Model Selection

Model selection is a key stage in the study of the highly nonlinear and dynamic neural process using the GLVM. It is the procedure of efficiently reducing the model complexity by selecting only a subset of its parameters. We perform model selection when applying the generalized Laguerre–Volterra algorithm out of the following three considerations. First, the connections among neurons in a particular brain region are generally sparse, thus, a model output is usually not affected by all its inputs. The inputs which effectively contribute to the output under the MISO model can thereby be sifted out after the selection process. Second, as the number of inputs increases, the number of coefficients to be estimated increases rapidly. Without model selection, computation time rises steeply. Third, too many open parameters bring about the overfitting problem, i.e. the noise is more likely to be fitted than the signal which largely cripples the predicting ability of the GLVM when applying the novel data.

The model selection process consists of two stages as shown in Fig. 3. At the first stage, we conduct model parameters estimation using the training data (in-sample data) which are pre-recorded neural firing inputs and outputs by the multi-electrode array. At the second stage, we conduct model outputs prediction using the testing data (out-of-sample data) which are pre-recorded input spikes—the output spikes are used as objects of reference while verifying the correctness of the model outputs. The goodness-of-fit could be assessed by two methods, either of direct evaluation or quantification of the similarity between the two data sets after a certain smoothing process [42].

Our hardware architecture can be well applied to the model selection process. This platform consists of two parts, i.e. software part and hardware part. The software part includes the spike sorting module and the selection result analysis (SRA) module. The hardware part is designed for doing GLVM parameters estimation and outputs prediction. A neural spiking input recorded by a single electrode is often contributed by multiple neurons [51] and in principle, each neuron tends to fire the spiking in its own shape [52]. The spike sorting module is used for determination of the corresponding neurons to each of the observed spikes, i.e. for the disambiguation process between unique neurons. The hardware-based processing core performs the DSP which is consisted of two stages, namely, estimation and prediction which use the in-sample data and out-of-sample data, respectively. The prediction results are then sent back to the PC via the data interface. The SRA software compares

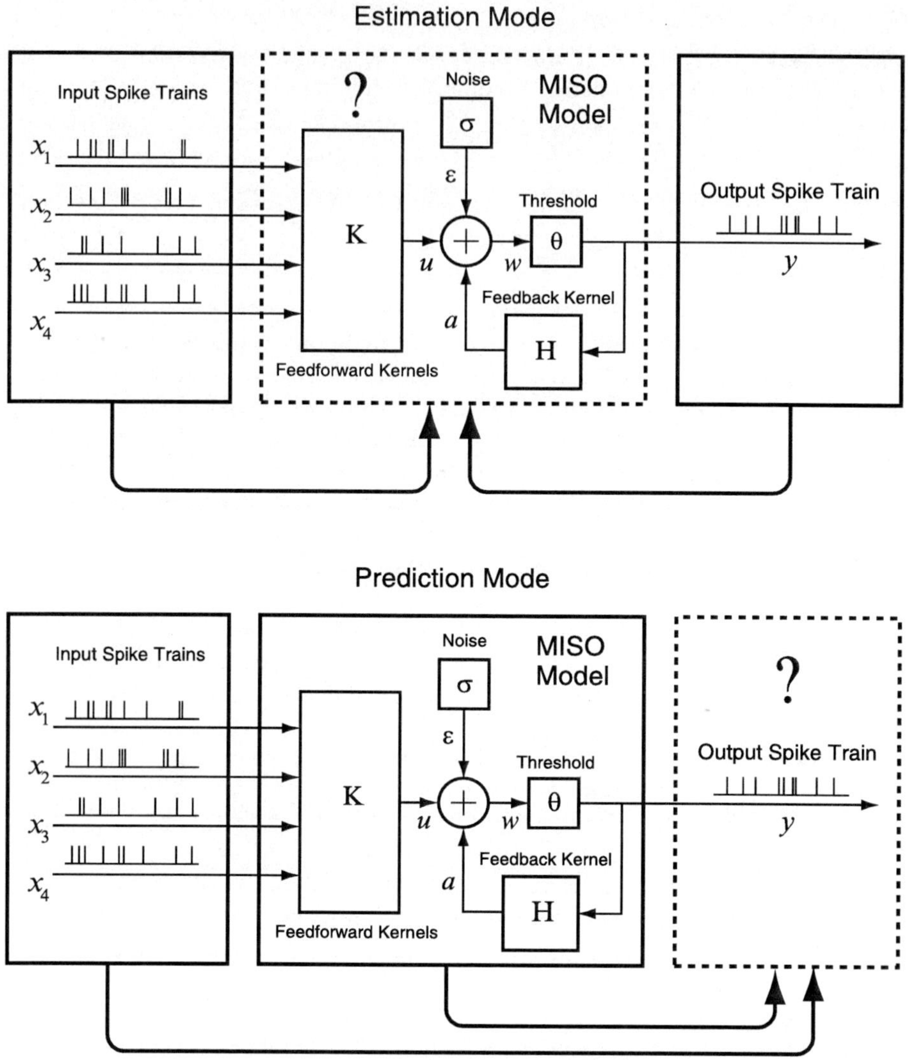

Fig. 3 The model selection process consists of two stages; thus, the system is required to work in dual modes

the predicted spiking activity with the spike activity and decides if the input(s) contribute(s) to the output in a previous causally linked inputs/outputs functionality. In our previous work, the key processing procedures of parameters estimation and outputs prediction were all conducted by software. However, with the aid of FPGA-based hardware platform, we can achieve a good speedup in doing such calculations with an enhancement in data throughput as to be elaborated in the following sections.

4 Using Reconfigurable Hardware to Predict Neural Activity

A major focus of this work is to utilize the hardware-level parallelism and on-chip resources of modern FPGA devices to model the generalized Laguerre–Volterra system which plays a vital role in neural spiking activity study, calculate the important coefficients, and predict neural firing output with desirable calculation precision.

The computational intensive parts of the hardware model include the convolution of the $N+1$ input channel signals with the basis function, the MAC (Multiplication and Accumulation) operation between the convolution products V and the Laguerre coefficients vector C, and the updating of firing probability.

4.1 Hardware Architecture

Figure 4 provides an overview of the hardware architecture for doing GLVM parameters estimation and outputs prediction. The circuit can work in different modes (estimation/prediction) and can switch between the two important system functions. In the figure, U1–U7 are important processing units which will be described in more detail in the following subsections. U1–U7 as a whole forms the hardware processing core. U1 and U2 are designed for the calculation of the pre-threshold membrane potential w. U3 and U4 function as the neuron firing probability P calculation unit and the Laguerre coefficients C updating unit, respectively. U5 is the component which generates the Gaussian white noise. U6 is the threshold trigger which conducts neural firing outputs prediction using a threshold function. U7 is the control unit which generates the control signals guiding the data flow under different modes of operation and also determining the cycles of execution in accordance with the number of valid model input(s) and processing elements within each component. H is the register array that stores the value of the augmented horizontal vector which is of the size of $1+(N+1)\times L$ (N is the number of inputs and L is the number of basis functions).

At the estimation stage, the training data are sent from PC via the data interface to the Ethernet IP which is embedded in the FGPA. The Ethernet IP then forwards the data to the hardware processing core. The processing core performs the DSP and sends back the calculation results of the pre-threshold membrane potential, the firing probability and the Laguerre coefficients to the PC via the Ethernet IP using the same data interface. At the prediction stage, the field estimated Laguerre coefficients appear as a constant vector and are stored in a register array. The testing data are transmitted to the processing core via the identical data path as at the estimation stage. After the value of the pre-threshold membrane potential is computed, the processing core conducts prediction of the model outputs measuring the potential, the random neuronal noise and the preset firing threshold. The predicted outputs are

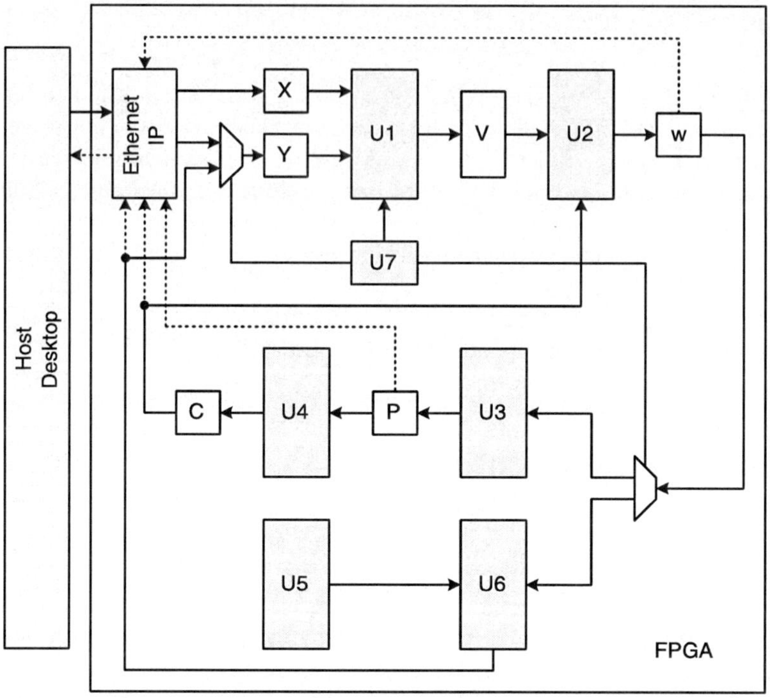

Fig. 4 Overview of the hardware architecture (U1 is the convolution unit; U2 is the multiplication and accumulation unit; U3 is the firing probability calculation unit; U4 is the Laguerre coefficients updating unit; U5 is Gaussian random number generation unit; U6 is the threshold triggering unit; U7 is the control unit; *X* and *Y* are the neural spike inputs (*Y* is the feedback of neural spike output into the input port); *V* is the convolution products; *w* is the pre-threshold membrane potential; *P* is the neural firing probability; *C* is the Laguerre coefficients)

then relayed back to the PC for further analysis employing the SRA software. They are also fed back to input ports of the processing core as after-potentials for iterative calculation.

The hardware architecture can be made very scalable with adoption of different number of processing elements (PEs) within each component. This can enhance the generality of our design to be applied to a broader range of FPGA devices on market. For example, supposing the number of valid inputs is 65 (which is also the maximum value the current architecture can accommodate), in the vector convolution and multiplication-and-accumulation components, the number of PEs can be set at 1–65, and accordingly, the convolution between the neural inputs and elements of the augmented horizontal vector is divided into different number of successive rounds. If the employed amount of PEs increases, the count of operational rounds decreases. We can also see that the fully paralleled architecture ($N_{\mathrm{PE}} = 65$) consumes much less time in execution than the compact architecture ($N_{\mathrm{PE}} = 1$) at both the estimation and the prediction stage. However, if the hardware is used for doing GLV model selection, the compact architecture is more suitable for the following two reasons:

Algorithm 1 Pre-threshold membrane potential updating algorithm (number of inputs N=64; number of basis functions L=3)

```
H(1)=1;
for n = 1: N
H(1+(n-1)*L:n*L) = Convolve(H(1+(n-1)*L:n*L),
InvA1, A2, A3, x(t,n));
end
H(2+n*L:1+(n+1)*L)= Convolve(H(2+n*L:1+(n+1)*L),
InvA1, A2,A3, y(t-1));
w(t) = H*C
```

Algorithm 2 Signal convolution algorithm

```
Convolve (V, InvA1, A2, A3, x)
V = V*A2
V(1) = V(1)+x*A3
V = V*A1
```

1. In the model selection process, the number of valid inputs to the processing core appears as a variable. Under the compact architecture, all the neural inputs are streamed into the processing core in a serial fashion. System timing will be made much easier in that scenario. We only need to alter the maximum accessible value of the counters which record the round number of processing. Additional circuitry for doing, such as inputs multiplexing and dynamic PE resource allocation is spared.
2. Even for processing of experimental data consisting of 65 inputs, the calculation efficiency is still satisfactory for current animal research.

4.2 *Calculating the Pre-threshold Membrane Potential*

The functionality of this unit is to calculate the pre-threshold membrane potential of the neuronal outputs w in real time. This unit appears as a combination of U1 and U2 in Fig. 4.

This unit is designed to implement the following algorithms (Algorithms 1 and Algorithm 2) in neural signal processing. Algorithms in this chapter are generally presented using the syntax of Matlab.

Pre-threshold membrane potential is calculated using Algorithm 1 at each time step. C is not a constant in Algorithm 1, it is updated using Algorithm 3 (which will be explained in Sect. 4.3) in hardware in sync with H. InvA1, A2 and A3 correspond to matrices A_1^{-1}, A_2 and A_3, respectively, in (9). They are used in the real-time convolution as shown in Algorithm 2. H is the augmented horizontal vector, which includes all the convolved products. This algorithm is designed to calculate the pre-

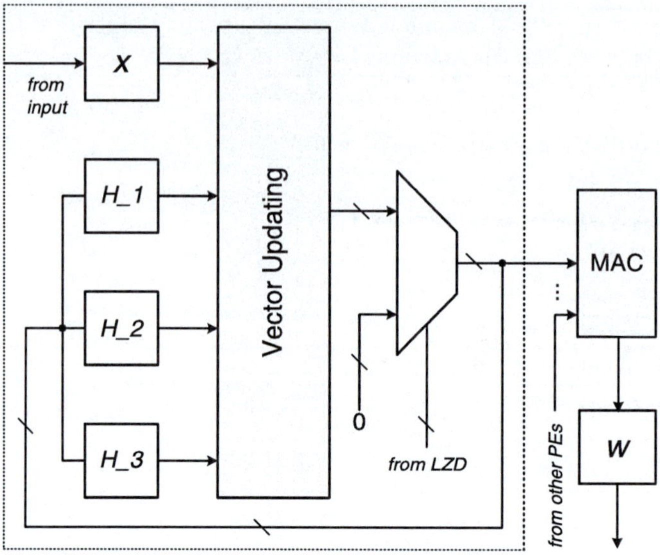

Fig. 5 Datapath of convolution and MAC units (PE is the processing element; LZD is the leading zeros detector component; MAC is the multiplication and accumulation component)

threshold membrane potential using C estimated from the previous time step and the convolved products, which account for the history of the neuronal inputs x and output y.

In Algorithm 2, V is a vector containing L elements; InvA1, A2 and A3 are $L \times L$ matrixes; and x is a scalar. This algorithm is designed to convolve each input with L basis functions as in (9).

U1 and U2 together contain $N+1$ leading zeros detector (LZD) components, $N+1$ vector updating (VU) components and one Multiplication and Accumulation (MAC) component. Figure 5 shows the configuration of a basic processing element of this part of circuit with the MAC component included. U1 deals with one sampling frame per processing cycle. Neurons in some parts of the brain, such as the hippocampus, have low firing frequency; so both X and Y could be sparse. The LZD component is designed to detect the zero elements in the input vectors. If zeros are detected, part of or the whole VU circuit is put on hold. Signals will bypass the VU component and the H registers will be reset directly. This saves power given that the frequent updating of the VU registers will be prevented. The VU component is designed to implement Algorithm 2. New values of H can be acquired after completing the combinational routine of VU. The MAC component is of tree structure consisting of stages of adders. It is designed to implement the 6th statement in Algorithm 1. Elements of H and C register arrays are first multiplied. The $1+(N+1)\times L$ products are added through stages of adder arrays. The size of

Algorithm 3 Calculating the firing probability, its gradient and the coefficients

```
P = f(th, w, σ)
dP = g(H, w, th, σ)
if P ~= 0
    dL = 1/P*dP
end
C = C+R*(dL′+(ŷ-P))
```

the adder array shrinks by half from the utmost leaves to the root stage by stage. The value of w is acquired at the root stage. The number of the PEs in the convolution units can be adaptive under the compact architectures and the number of adder stages will change accordingly.

4.3 Calculating the Firing Probability and Laguerre Coefficients

Firing probability P and Laguerre coefficients C are two other important parameters in the GLV algorithm. Their values need to be tracked during each round of calculation. The method for updating P and C are shown in Algorithm 3. In this algorithm, th, σ and R represent the threshold value, the noise strength and the learning rate, respectively. $\hat{y}$ is the instant (current round) value of the firing output. dP is the gradient of P with respect to the coefficients and dL is $\partial \log P(t)/\partial C$ as appeared in (11). f and $\mathbf{g}$ are the functions calculating P and its gradient. The P, C updating algorithm is intrinsically computationally intensive for it includes function series such as the exponential equation. These often demand long computing time using digital software based on serial instruction streams. The parallel processing capability of the FPGA hardware can greatly facilitate this process with its programmability on the circuit level.

For calculation of the exponential function, there are many methods proposed to date. Some methods use predetermined values for table look-up [53]; some methods are based on the CORDIC algorithm [54–58]. According to our observation of data from animal experiments, the values of the pre-threshold membrane potential in current recordings are restricted within the range of $[-4\sqrt{2}, 4\sqrt{2}]$. Based on this, we adopt direct Taylor series expansion in current implementation for the series can converge fast under this condition and the number of expansion terms is limited, hence resource cost is affordable for the FPGA hardware. However, we are also developing other computation methods to better accommodate any unpredictable variance in future applications. All these methods are being incorporated to the hardware IP library which is now under development. The hardware IP library is part of the higher-level self reconfiguring platform (SRP) [59].

For the calculation of the error function erf(x), we have also developed several computation methods. These methods have their respective pros and cons and can be chosen according to the specific application requirements, e.g. the precision level and the FPGA model availability. Some methods we have developed can be found in Table 2. The more detailed discussions can be found in Sect. V of [60].

4.4 Predicting the Output Spikes

This part of circuitry is activated during the prediction stage of the GLV system only. It is based on the summation of the pre-threshold membrane potential and the Gaussian random white noise, passing the threshold triggering component to generate the predicted output neural spiking signals. The noise term here is able to capture the system uncertainty which results from both the intrinsic neuronal noise and the unobserved inputs (neurons whose spiking activities essentially contribute to the model outputs but are not included into the model). In hardware, this noise source can be simulated by a Gaussian random number generator (GRNG). The circuit structure of the GRNG can be found in Fig. 6. In the figure, URNG is the uniform random number generator component which produces signals whose strengths are uniformly distributed in the range of $[0,1]$. It can be implemented by the bitwise XOR operations between the lower 32 bits of a 43-bit linear feedback shift register (LFSR) and the lower 32 bits of a 37-bit cellular automata shift register (CASR). This method was first proposed by Tkacik [61]. The URNG has a cycle length of $2^{80}-2^{43}-2^{37}+1$ with reduced bias to 2^{-80} and is very efficient to be implemented in our platform given its high operating frequency, low resource demand and non-repetition property in generating the random noise sequence during long animal experimental sessions. Each N uniformly distributed numbers generated by the URNG are added. According to the Central Limit Theorem, if N approaches infinity, the series composed by these summation products satisfies the Gaussian distribution. However, due to restrictions imposed by the execution time, we set N at 100 in the current design according to the analysis by Jeruchim et al. [62]. We use FIFO for data caching between different clock domains of the RNG and other circuit components. The post-processing (PP) unit in Fig. 6 is used to transform the non-standard Gaussian distribution generated by the RNG to a standard normal distribution by adjusting its mean and variance according to the value of N. The control unit is designed for functions such as URNG seed loading, FIFO r/w signal generation and component enabling. The Gaussian white noise generated after the post-processing stage is incorporated into the pre-threshold membrane potential and then passed to the threshold trigger, which produces the spiking outputs using the threshold function.

Table 2 Methods for the calculation of the error function

Design method	Description	Calculation precision	Storage demand	Logic demand
Direct LUT	Mapping the error function to look-up table entries with uniform size of data step in all range of the function curve	Best	Very high	Low
Indirect LUT	Mapping the error function to look-up table entries with nonuniform size of data step according to the slope of the curve in different regions	Good	High	Moderate
Converting to exponential	$\mathrm{erf}(z) \cong \sqrt{1 - \mathrm{e}^{-\frac{4}{\pi}z^2}}$	Region dependent	Very low	High
Polynomial expansion	$\mathrm{erf}(z) = \frac{2}{\sqrt{\pi}} \sum_{n=0}^{\infty} \frac{(-1)^n z^{2n+1}}{n!(2n+1)}$	Good	Very low	Very high

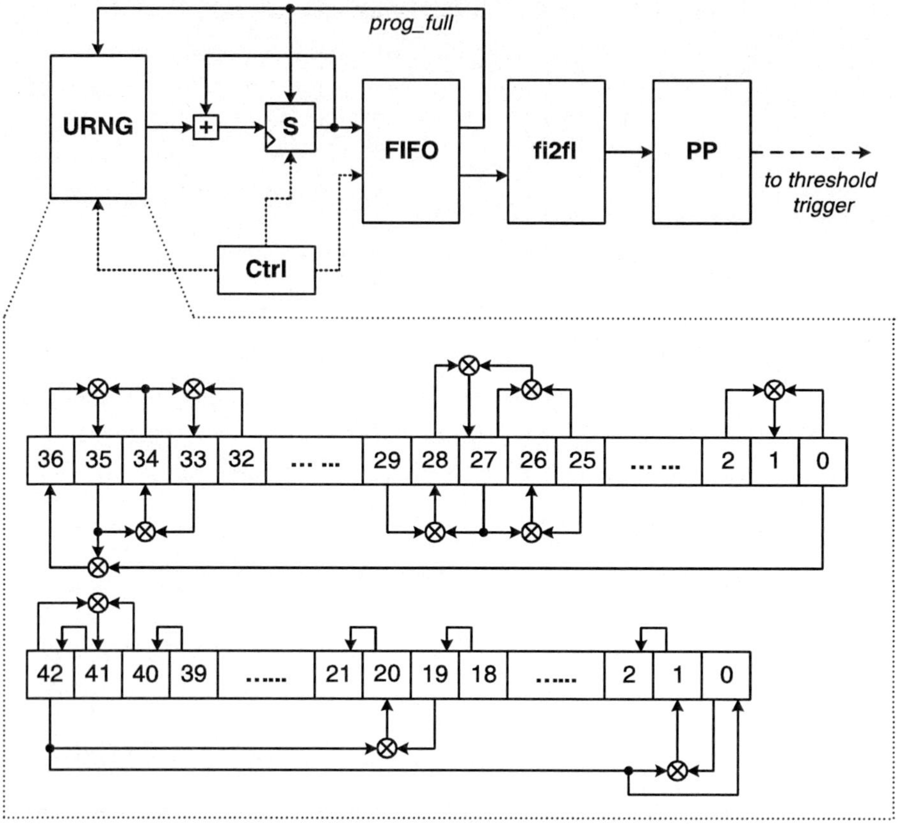

Fig. 6 Gaussian random noise generator (URNG is the uniform random number generator which includes a 43-bit Linear Feedback Shift Register and a 37-bit Cellular Automata Shift Register; S is the summation unit; fi2fl is the fixed point to floating point conversion unit; PP is the post processing unit)

4.5 System Scalability

One important feature of our hardware platform is its multifold scalability. The scalability of the system can be derived largely from two design considerations. One is module reuse and the other is MISO model extendibility.

When we design the hardware architecture, we want to make it fit into as many FPGA platforms available on market as possible. In order to do that, we have proposed both fully paralleled architectures, in which the number of system input is equal to the number of recording electrodes, and more compact architectures with the adoption of module reuse. Using these compact architectures, resource utilization can be reduced. For example, in the MAC units, the number of PEs which performs the operation of vector convolution can be reduced from 65 ($N = 64$) to

1 or $2^n + 1$ ($n \in Z$ and $1 \leq n \leq 5$), and the register that carries the value of pre-threshold membrane potential should then be updated after 65 or $2^{6-n} + 1$ rounds of convolution. The processing time is prolonged and more clock cycles are demanded. At the same time, when the system works at the estimation mode, in the Laguerre coefficients updating unit, the C register array is updated by 66 or $2^{6-n} + 2$ rounds of identical arithmetical routines. Due to the reduced parallelism, the new approach bears the clear defect of lower data throughput. However, it has several distinct advantages compared to the full parallel architecture such as lower system resource utilization and power consumption.

Another aspect revealing system scalability is that multi-FPGA extendable design can be easily implemented with our architecture. This is largely due to the data independence of the elements in the Y input (the feedback from model output as shown in Fig. 2) set under our proposed circuit architecture. In the convolution process, X inputs and Y inputs are data irrelevant. This means that one set (64 elements) of X can perform convolution operation with different Y elements concurrently. Parallelism can be greatly enhanced by adding additional FPGAs. A four FPGA network structure is shown in Fig. 4 of [60]. Input set Y is shifted to FPGA1-4 at each processing cycle. Neural network parameters which are stored in BRAMs are read simultaneously by the transceiver core and relayed to the host PC. By adding n extra FPGA stages, we will increase the data throughput $n + 1$ times compared to the mono FPGA structure. A maximum number of 64 FPGAs (element number in input set Y is also 64) could work together in a coordinated way.

4.6 Implementation Results: Hardware Versus Software

We use the proposed FPGA-based hardware and the original software platforms to process approximately 1,000 bins (discretized time units) of neural firing data (synthetic data). Both the hardware and software run the two stages of the generalized Laguerre–Volterra algorithm, i.e. estimation and prediction. During the estimation stage, the values of the three important system parameters: (1) the pre-threshold membrane potential, (2) the firing probability, and (3) the Laguerre coefficients are recorded. While at the prediction stage, the values of the membrane potential and the predicted firing output are recorded.

The calculation results of the three important parameters during the estimation stage are shown in Fig. 7. The differences between each of the three value sets are also plotted in Fig. 8.

If we define the NMSE as:

$$\mathrm{NMSE} = \sum_{t=1}^{T} (y(t) - \tilde{y}(t))^2 / \sum_{t=1}^{T} \tilde{y}(t)^2, \tag{12}$$

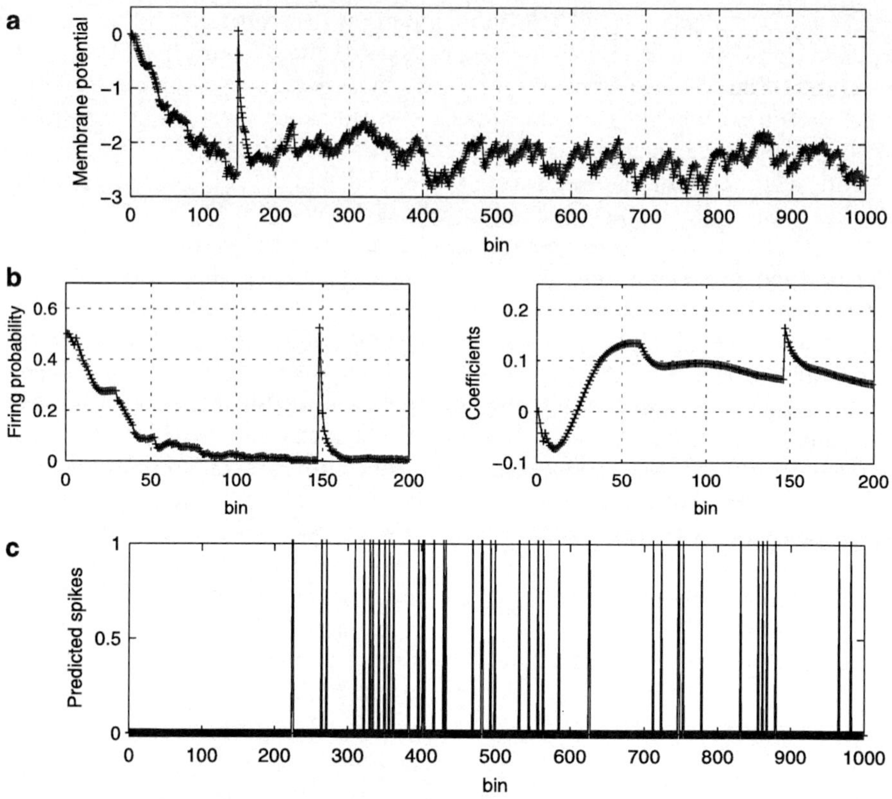

Fig. 7 The implementation results by the FPGA-based hardware platform. (**a**): result of the pre-threshold membrane potential; (**b**): result of the firing probability; (**c**): result of element of the Laguerre coefficients; (**d**): hardware predicted neural spike outputs. The hardware system changes from the model parameters estimation mode to the firing outputs prediction mode at the 200th bin. Threshold value is set at -0.600 when doing the firing outputs prediction

then we can calculate the NMSEs for the membrane potentials, the firing probability, the coefficients and the prediction results during each stage, respectively. The calculation results are shown in Table 3.

Also, we use both the hardware and software platforms to process a session of neural firing data, and we calculate the speed-up of the hardware platform versus the software platform.

For the software part, after the compilation of the C codes using the gcc compiler, the executable file is successively run for 10 rounds. The configuration of the software platform we use for running the generalized Laguerre–Volterra algorithm is shown in Table 4. Both the time consumed for conducting parameters estimation and model output prediction are recorded as shown in Fig. 9.

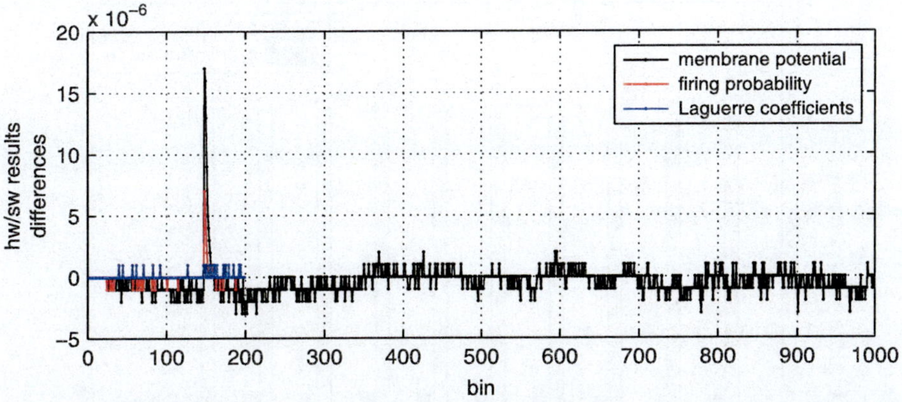

Fig. 8 The hardware and software calculation differences of the three important parameters: (1) the pre-threshold membrane potential (*black line*); (2) the firing probability (*red line*); (3) the Laguerre coefficients (*blue line*). In the figure, bin=200 is the time boundary between the estimation stage and the prediction stage

Table 3 Normalized mean square error of calculation

Variable name	Estimation mode	Prediction mode
Membrane potential	1.620×10^{-12}	1.678×10^{-13}
Firing probability	2.584×10^{-11}	–
Laguerre coefficient	1.481×10^{-11}	–
Firing outputs	–	0

Table 4 Host PC configuration

CPU	Intel Core i7-2620M (Turbo Boost to 3.40 GHz)
Memory size	8GB
C compiler	gcc 3.4.4-999 running on Cygwin 1.7 platform
Interface	Gigabit ethernet with jumbo frame enabled

The data throughput of the software platform is calculated by:

$$T_{\text{sw}} = \frac{D \times l}{\sum_{i=1}^{l} t_i} \tag{13}$$

In (13), D represents the number of data frames utilized for the test (in our test, it is set at 10,000); l indicates the times of iteration and t is the execution time of each round. Using the data gathered, we can calculate the data throughputs of the software platform are 500.90 data frames/s (for parameters estimation) and 1430.94 data frames/s (for model output prediction), respectively.

For the hardware part, the calculation speed can be estimated by the equation below:

$$T_{\text{hw}} = \frac{D}{\frac{D_{\text{e}} \times K_{\text{ce}}}{f_{clk}} + \frac{D_{\text{p}} \times K_{\text{cp}}}{f_{clk}}} \tag{14}$$

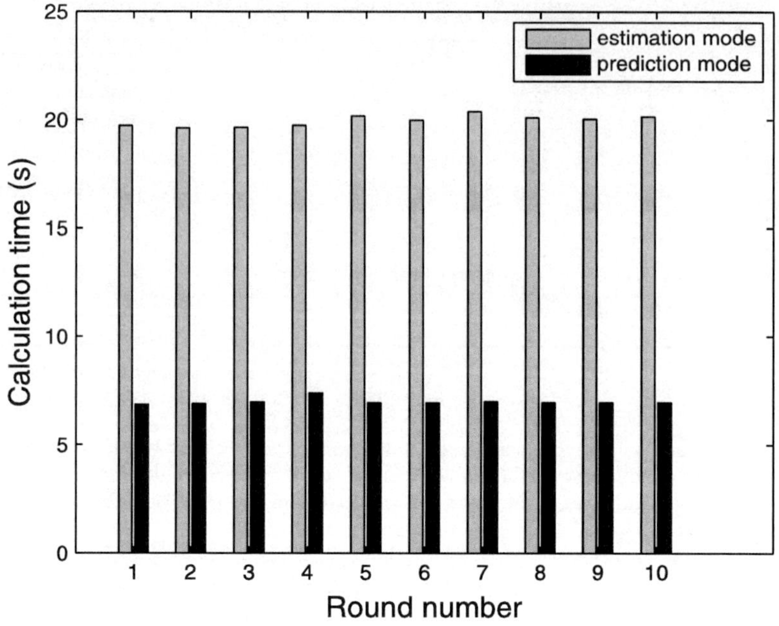

Fig. 9 Time consumption for conducting model parameters estimation and firing output prediction using software. (*Gray bar*: time records during the estimation mode; *black bar*: time records during the prediction mode)

In (14), D represents the total number of data frames utilized for test (in our test, it is set at 10,000); D_e is number of data frames used for parameters estimation and D_p is the number of data frames used for output prediction. K is cycles needed to doing one round of calculation. For parameters estimation, $K_{ce} = 67$; for firing output prediction, $K_{cp} = 68$. In our experiment, the hardware switches from the mode of estimation to prediction from the 2,000th input data frame. The overall data throughput for the hardware platform can be calculated as 3.83×10^4 data frames/s. For the stage of estimation, the data throughput can be calculated as 3.88×10^4 data frames/s (D=De, Dp=0). For the stage of prediction, the data throughput can be calculated as 3.82×10^4 data frames/s (D=Dp, De=0). The hardware to software speedups are 77.47x and 26.72x, respectively, at the two stages of calculation.

As mentioned previously, the hardware system is very scalable by implementing different number of processing elements within each design component. For the fully paralleled architecture, where the number of PEs equals the number of inputs, the data throughput can reach 1.33×10^6 data frames/s at the estimation stage and 1.00×10^6 data frames/s at the prediction stage. The speedups are 2.66×10^3x and 698.84x, respectively.

5 Discussions

In the above sections, we present our work of conducting the neural firing pattern prediction employing the reconfigurable hardware. The research objective of the current stage work has been successfully achieved. However, we are still in the process of upgrading the current hardware platform to better meet the future application requirements from the neural prosthetic device which will be implanted into be mammal brain.

5.1 The Ultra-Low Power Design Principle

For an implantable neural prosthesis, power consumption is of critical importance. The battery life should be ideally as long as possible, for the frequent recharge of the on-board power source would bring great inconvenience to the patients wearing such devices, such as sufferings from the medical surgeries. Although technologies of inductive coupling or energy conversion (electromagnetic to electrical) are emerging, the more fundamental solutions would lie in the optimization of the circuitry of the prosthetic device itself. For FPGA designers, one such solution is by resorting to more advanced, low-power FPGAs such as the newly released Xilinx Virtex-7 series devices, which adopt the 28 nm and high-K metal gate process optimized for low power applications, slashing the static power by 50% compared with their predecessors. However, there is a trade-off among power, development cost and risk of operational faults here due to the lower operating voltages introduced. Another solution is by adopting more tailored design modules. For instance, in our current FPGA architecture for the GLVM, we have extensively employed the off-the-shelf Intellectual Properties (IPs) like the DSP48E cores provided by the FPGA vendors. The IPs can perform floating point arithmetic operations with high precision. They also provided a fast-to-product solution by means of shortened design time. However, these IPs are more power hungry and area inefficient compared with more tailored units such as the ones employing the fixed point representations of model variables. So there is also a trade-off of design metrics in this aspect. A third solution for saving power can be well directed to the employment of run-time reconfiguration techniques as we will discuss in more detail in the next subsection. By reconfiguring FPGA devices on the fly, we can dynamically shutdown circuit blocks when they are not producing useful data; this, undoubtedly provides us with a new, intriguing approach for saving power.

5.2 The Dynamical Partial Reconfiguration Technique

The dynamical reconfiguration refers to the modification of the FPGA functions during its operation time. It is often achieved by altering portions of gate array while keeping other parts running. In our current research regarding the FPGA prototyping of the GLVM, we do not delve deep into this scope of research; however, it is indubitably a very promising technique given the huge advantages introduced such as reduction in power dissipation, decrease in space of the FPGA chip, avoidance of hardware obsolescence and more flexibility in implementation.

Two general approaches are often adopted when dynamically reconfiguring the device, each has applications where desirable. One approach is by external reconfiguration under which the compiled bitstream is transmitted to the device by JTAG boundary scan port or serial port. The other approach is by internal configuration where the internal configuration access port (ICAP) is utilized for transmission employing an embedded microcontroller or state machine. In our design, given the hard real-time requirement of the prosthetic application, it is much preferred that a hardware IP library of existing implementations tailored for different application scenarios be pre-stored in on-chip memory and be deployed by the microcontroller to a target module.

There are two scenarios of partial reconfiguration, as shown in Fig. 10. In the first scenario, a particular module of the design is ineffective and can be "blocked" during certain period of operating time. In our FPGA implementation of the GLVM, when the platform works in the mode of firing output prediction, the U5, U6, and U7 units shown in Fig. 4 are deemed as redundant and the related modules can be dynamically dropped off. This brings two distinct benefits. The first is reduced power dissipation and the second is spared chip area which can be utilized for other system modalities such as hardware redundancy for the fault-tolerance purpose. In the second scenario, a particular module of the design is substitutable and can be "replaced" during certain period of operating time. A good illustration would be the implementation of the error function of in the GLVM as stated above. The requirements of the prosthetic system may be time variant. For a specific application requirement, the most optimized method of implementation can be automatically chosen by the device in run-time with certain preset constraints and the compiled bitstream can be employed via the ICAP. This will ensure the hardware platform can always produce the results with desirable precision while keeping good trade-offs to other criteria such as power consumption. For our hardware system which is designed for doing neural firing patterns prediction, hard real-time is required. The reconfiguration time should be strictly limited within a certain range. Given the natural brain firing rate is very low, this requirement can be guaranteed by current technology.

5.3 *Fault-Tolerance Redundancy Design*

For a cognitive neural prosthesis targeting clinical applications, possessing fault-tolerance property is a fundamental requirement. It would result in disastrous consequences if the hardware fails in operation or produces erroneous prediction results. Traditional fault proof paradigms can be well adopted in current design such

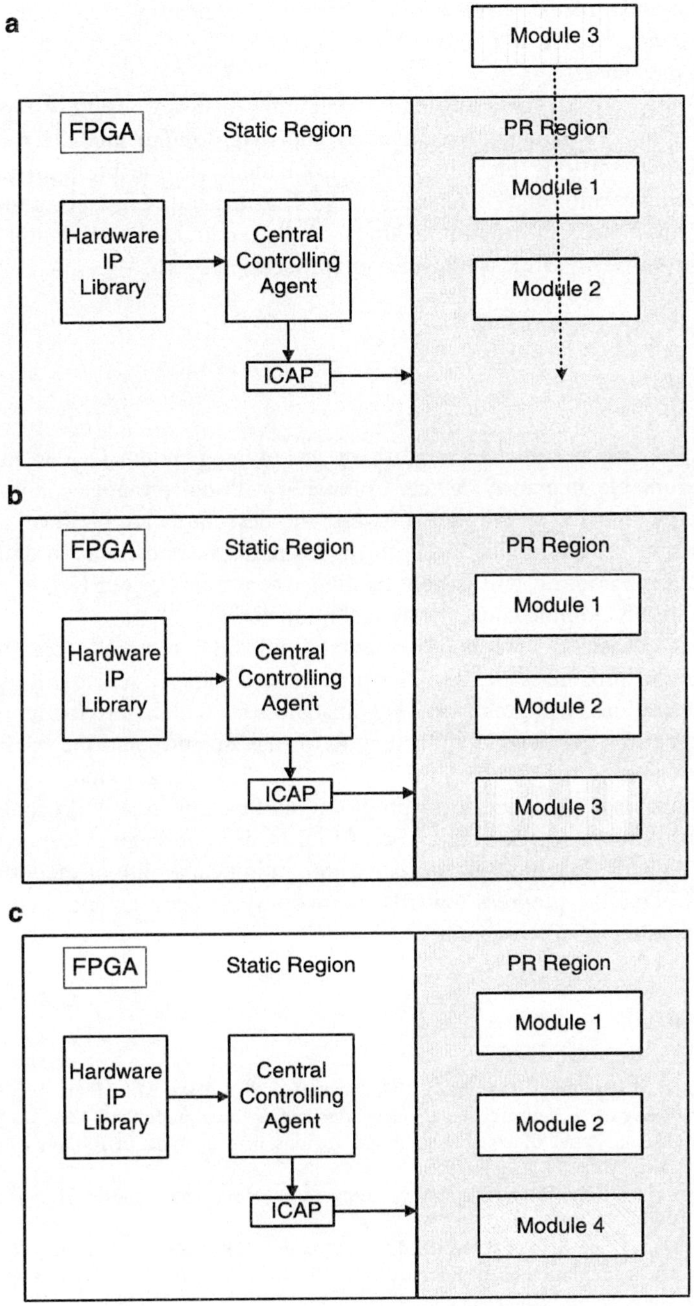

Fig. 10 Two scenarios of dynamical partial reconfiguration. (**a**)→(**b**): modules can be blocked/added on during device operation time; (**b**)→(**c**): modules can be substituted during device operation time

as the employment of hardware redundancy by the replication of critical circuitry, time redundancy by the multi-round execution and information redundancy by performing data checking.

Another way to implement the cognitive neural prosthesis design is to place the modalities of model parameters estimation and firing output prediction into two different chips with only the latter being implanted into the brain area. In that scenario, the integration and correctness of the transmission of model coefficients is of critical importance. Information redundancy is needed for the validation purpose during either wired or wireless transmission process.

6 Conclusions

In this chapter, we described our work of prototyping of the GLVM using the field-programmable gate array device. Compared with our previous computational platform employing PC and digital software, the new hardware platform achieves remarkable speedup in conducting both model parameters estimation and neural firing output prediction. It will largely facilitate the process of our further research towards the highly nonlinear and dynamical brain activity.

The work presented here is important to our final research objective—the cognitive neural prosthesis design. The prosthetic device, if successfully applied to clinical operations, will greatly avail against diseases with regard to hippocampal region dysfunction and degeneration such as stroke, seizure, and the Alzheimer's disease by bypassing the pathological regions. The field-programmable gate array, given its convenient hardware level programmability, low cost and capability of doing fast prototyping, serves as an ideal tool at current stage research of such biomedical devices. The high-performance computing capability of modern FPGA devices will boost the progress towards future study of learning and memory and the implementation of silicon brain.

References

1. D. Purves, E.M. Brannon, R. Cabeza, S.A. Huettel, K.S. LaBar, M.L. Platt, M. Woldorff, *Principles of Cognitive Neuroscience* (Sinauer Associates Inc., Sunderland, MA, USA, 2007)
2. B. Milner, Memory and the medial temporal regions of the brain, in *Biology of Memory* (Academic, New York, 1970), pp. 29–50
3. L.R. Squire, S.M. Zola, Episodic memory, semantic memory, and amnesia. Hippocampus **8**, 205–211 (1998)
4. M.S. Humayun, E. de Juan, J.D. Weiland, G. Dagnelie, S. Katona, R. Greenberg, S. Suzuki, Pattern electrical simulation of the human retina. Vis. Res. **39**, 2569–2576 (1999)
5. G.E. Loeb, Gochelear prosthetics. Annu. Rev. Neurosci. **13**, 357–371 (1990)
6. G.E. Loeb, R.A. Peck, W.H. Moore, K. Hood, Biontm system for distributed neural prosthetic interfaces. Med. Eng. Phys. **23**, 9–18 (2001)
7. K.H. Mauritz, H.P. Peckham, Restoration of grasping functions in quadriplegic patients by functional electrical stimulation (FES). Int. J. Rehabil. Res. **10**(4), 57–61 (1987)

8. J.P. Donoghue, Connecting cortex to machines: recent advances in brain interfaces. Nat. Neurosci. **5**, 1085–1088 (2002)
9. L.R. Hochberg, M.D. Serruya, G.M. Friehs, J.A. Mukand, M. Saleh, A.H. Caplan, A. Branner, D. Chen, R.D. Penn, J.P. Donoghue, Neuronal ensemble control of prosthetic devices by a human with tetraplegia. Nature **442**(7099), 164–171 (2006)
10. M.A.L. Nicolelis, Brain-machine interfaces to restore motor function and probe neural circuits. Nat. Neurosci. **4**, 417–422 (2003)
11. K.V. Shenoy, D. Meeker, S.Y. Cao, S.A. Kureshi, B. Pesaran, C.A. Buneo, A.R. Batista, P.P. Mitra, J.W. Burdick, R.A. Andersen, Neural prosthetic control signals from plan activity. Neuroreport **14**, 591–596 (2003)
12. D.M. Taylor, S.I.H. Tillery, A.B. Schwartz, Information conveyed through brain-control: cursor versus robot. IEEE Trans. Neural Syst. Rehabil. Eng. **11**(2), 195–199 (2003)
13. J.R. Wolpaw, D.J. McFarland, Control of a two-dimensionalmovement signal by a noninvasive brain-computer interfacein humans. Proc. Natl. Acad. Sci. USA **101**, 17849–17854 (2004)
14. T.W. Berger, G. Chauvet, R.J. Sclabassi, A biological based model of functional properties of the hippocampus. J. Physiol. **7**, 1031–1064 (1982)
15. J. Magee, D. Hoffman, C. Colbert, D. Johnston, Electrical and calcium signaling in dendrites of hippocampal pyramidal neurons. Annu. Rev. Physiol. **60**, 327–346 (1998)
16. S.S. Dalal, V.Z. Marmarelis, T.W. Berger, A nonlinear positive feedback model of glutamatergic synaptic transmission in dentate gyrus, in *Proceedings of the The 4th Joint Symposium on Neural Computation*, vol. 7 (Institute for Neural Computation, San Diego, CA, USA, 1997), pp. 68–75
17. D. Song, Z. Wang, V.Z. Marmarelis, T.W. Berger, Non-parametric interpretation and validation of parametric models of short-term plasticity, in *Proceedings of Annual International Conference of the IEEE EMBS* (Institute of Electrical and Electronics Engineers, New York, NY, USA, 2003), pp. 1901–1904
18. Z. Wang, X. Xie, D. Song, T.W. Berger, Probabilistic transformation of temporal information at individual synapses, in *Proceedings of Annual International Conference of the IEEE EMBS* (Institute of Electrical and Electronics Engineers, New York, NY, USA, 2003), pp. 1909–1912
19. G. Gholmieh, S.H. Courellis, D. Song, Z. Wang, V.Z. Marmarelis, T.W. Berger, Characterization of short-term plasticity of the dentate gyrus-ca3 system using nonlinear systems analysis, in *Proceedings of Annual International Conference of the IEEE EMBS* (Institute of Electrical and Electronics Engineers, New York, NY, USA, 2003), pp. 1929–1932
20. A. Dimoka, S.H. Courellis, D. Song, V. Marmarelis, T.W. Berger, Identification of lateral and medial perforant path using single- and dual-input random impulse train stimulation, in *Proceedings of Annual International Conference of the IEEE EMBS* (Institute of Electrical and Electronics Engineers, New York, NY, USA, 2003), pp. 1933–1936
21. M.C. Citron, J.P. Kroeker, G.D. McCann, Nonlinear interactions in ganglion cell receptive fields. J. Neurophysiol. **46**, 1161–1176 (1981)
22. M.C. Citron, R.C. Emerson, W.R. Levick, Nonlinear measurement and classification of receptive fields in cat retinal ganglion cells. Ann. Biomed. Eng. **16**, 65–77 (1988)
23. P.Z. Marmarelis, K.I. Naka, Nonlinear analysis and synthesis of receptive field responses in the catfish retina II: one-input white-noise analysis. J. Neurophysiol. **36**, 619–633 (1973)
24. D. McAlpine, Creating a sense of auditory space. J. Physiol. **566**, 21–28 (2005)
25. L. Paninski, M.R. Fellows, N.G. Hatsopoulos, J.P. Donoghue, Spatiotemporal tuning of motor neurons for hand position and velocity. J. Neurophysiol. **91**, 515–532 (2004)
26. A.L. Hodgkin, A.F. Huxley, A quantitative description of membrane current and its application to conduction and excitation in nerve. J. Physiol. **117**, 500–544 (1952)
27. H. Markram, The blue brain project. Nat. Rev. Neurosci. **7**, 153–160 (2006)
28. Elementary Objects of the Nervous System. [Online]. Available: http://synapticmodeling.com/ Accessed January 2013
29. Brain in Silicon. [Online]. Available: http://brainsinsilicon.stanford.edu Accessed January 2013
30. NEURON. [Online]. Available: http://www.neuron.yale.edu/neuron/ Accessed January 2013

31. Mcell: A Monte Carlo Simulator of Cellular Microphysiology. [Online]. Available: http://www.mcell.cnl.salk.edu/ Accessed January 2013
32. J.P. Cunningham, V. Gilja, S.I. Ryu, K.V. Shenoy, Methods for estimating neural firing rates, and their application to brain-machine interfaces. Neural Networks **22**(9), 1235–1246 (2009)
33. D.R. Brillinger, Nerve cell spike train data analysis: a progression of technique. J. Am. Stat. Assoc. **87**, 260–271 (1992)
34. E.N. Brown, R. Barbieri, U.T. Eden, L.M. Frank, Likelihood methods for neural data analysis, in *Computational Neuroscience: A Comprehensive Approach*, vol. 7 (Chapman & Hall/CRC, London, UK, 2003), pp. 253–286
35. W. Wu, Y. Gao, E. Bienenstock, J. Donoghue, M. Black, Bayesian population decoding of motor cortical activity using a Kalman filter. Neural Comput. **18**(1), 80–118 (2006)
36. V. Volterra, *Theory of Functionals and of Integral and Integro-Differential Equations* (Dover, New York, 1959)
37. N. Wiener, *Nonlinear Problems in Random Theory* (MIT, New York, 1958)
38. V.Z. Marmarelis, P.Z. Marmarelis, *Analysis of Physiological Systems: The White-Noise Approach* (Plenum, New York, 1978)
39. V.Z. Marmarelis, *Nonlinear Dynamic Modeling of Physiological Systems* (Wiley-IEEE Press, Hoboken, 2004)
40. D. Song, R.H.M. Chan, V.Z. Marmarelis, R.E. Hampson, S.A. Deadwyler, T.W. Berger, Nonlinear dynamic modeling of spike train transformations for hippocampal-cortical prostheses. IEEE Trans. Biomed. Eng. **54**, 1053–1066 (2007)
41. R.H.M. Chan, D. Song, T.W. Berger, Tracking temporal evolution of nonlinear dynamics in hippocampus using time-varying volterra kernels, in *Annual International Conference of the IEEE Engineering in Medicine and Biology Society*, vol. 54 (Institute of Electrical and Electronics Engineers, New York, NY, USA, 2008), pp. 4996–4999
42. D. Song, R.H.M. Chan, V.Z. Marmarelis, R.E. Hampson, S.A. Deadwyler, T.W. Berger, Nonlinear modeling of neural population dynamics for hippocampal prostheses. Neural Networks **22**, 1340–1351 (2009)
43. S.A. Deadwyler, T. Bunn, R.E. Hampson, Hippocampal ensemble activity during spatial delayed-nonmatch-to-sample performance in rats. J. Neurosci. **16**, 354–372 (1996)
44. B.E. Alger, R.A. Nicoll, Pharmacological evidence for two kinds of GABA receptor on rat hippocampal pyramidal cells studied *in vitro*. J. Physiol. **328**, 125–141 (1982)
45. J. Keat, P. Reinagel, R.C. Reid, M. Meister, Predicting every spike: a model for the responses of visual neurons. Neuron **30**, 803–817 (2001)
46. L. Paninski, J.W. Pillow, E.P. Simoncelli, Maximum likelihoodestimation of a stochastic integrate-and-fire neural encoding model. Neural Comput. **16**, 2533–2561 (2004)
47. D. Song, Z. Wang, T.W. Berger, Contribution of T-type VDCC to TEA-induced long-term synaptic modification in hippocampal CA1 and dentate gyrus. Hippocampus **12**, 689–697 (2002)
48. J.F. Storm, Action potential repolarization and a fast after-hyperpolarizationin rat hippocampal pyramidal cells. J. Physiol. **385**, 733–759 (2002)
49. V.Z. Marmarelis, Identification of nonlinear biological systems using Laguerre expansions of kernels. Ann. Biomed. Eng. **21**, 573–589 (1993)
50. C. Boukis, D.P. Mandic, A.G. Constantinides, L.C. Polymenakos, A novel algorithm for the adaptation of the pole of Laguerre filters. IEEE Signal Process. Lett. **13**, 429–432 (2006)
51. M.D. Linderman, G. Santhanam, C.T. Kemere, V. Gilja, S. O'Driscoll, B.M. Yu, A. Afshar, S.I. Ryu, K.V. Shenoy, T.H. Meng, Signal processing challenges for neural prostheses. IEEE Signal Process. Mag. **25**, 18–28 (2008)
52. C. Hansang, D. Corina, J.F. Brinkley, G.A. Ojemann, L.G. Shapiro, A new template matching method using variance estimation for spike sorting, in *Proceedings of the 2nd International IEEE EMBS Conference on Neural Engineering* (Institute of Electrical and Electronics Engineers, New York, NY, USA, 2005), pp. 225–228
53. W.F. Wong, E. Gogo, Fast hardware-based algorithms for elementary function computations using rectangular multipliers. IEEE Trans. Comp. **43**(3), 278–294 (1994)

54. J.C. Bajard, S. Kla, J.M. Muller, BKM: a new hardware algorithm for complex elementary functions. IEEE Trans. Comp. **43**(8), 955–963 (1994)
55. H. Bui, S. Tahar, Design and synthesis of an IEEE-754 exponential function, in *IEEE Canadian Conference on Electrical and Computer Engineering*, vol. 1 (Institute of Electrical and Electronics Engineers, New York, NY, USA, 1999), pp. 450–455
56. G. Even, P.M. Seidel, A comparison of three rounding algorithms for IEEE floating-point multiplication. IEEE Trans. Comp. **49**(7), 638–650 (2000)
57. M.A. Figueiredo, C. Gloster, Implementation of a probabilistic neural network for multispectral image classification on an FPGA based custom computing machine, in *Proceedings of the 5th Brazilian Symposium on Neural Networks* (IEEE Computer Society Press, Washington, D.C., USA, 1998), pp. 174–179
58. V. Kantabutra, On hardware for computing exponential and trigonometric functions. IEEE Trans. Comp. **45**(3), 328–339 (1996)
59. B. Blodget, P. James-Roxby, E. Keller, S. McMillan, P. Sundararajan, A self-reconfiguring platform, in *Proceedings of the 13th International Conference on Field Programmable Logic and Applications (FPL'03)* (Springer, New York, NY, USA, 2003), pp. 565–574
60. W.X.Y. Li, R.H.M. Chan, W. Zhang, R.C.C. Cheung, D. Song, T.W. Berger, High-performance and scalable system architecture for the real-time estimation of generalized Laguerre–Volterra MIMO model from neural population spiking activity. IEEE J. Emerg. Sel. Top. Circ. Syst. **1**(4), 489–501 (2011)
61. T.E. Tkacik, A hardware random number generator, in *Proceedings of the 4th International Workshop on Cryptographic Hardware and Embedded Systems* (Springer, New York, NY, USA, 2002), pp. 450–453
62. M.C. Jeruchim, P. Balaban, K.S. Shanmugan, *Simulation of Communication Systems: Modeling, Methodology, and Techniques* (Springer, Berlin, 2000). ISBN 978-0-306-46267-2

High-Performance FPGA-Accelerated Real-Time Search

Wim Vanderbauwhede, Sai. R. Chalamalasetti, and Martin Margala

Abstract This chapter presents our work on accelerating real-time search applications using FPGAs.

A real-time search (or document classification) application matches terms (words and groups of words) in a stream of documents against a static list of terms with associated weights. We discuss a novel FPGA design for the term matching and relevancy computation part of a high-throughput real-time search application. The design, implemented on the GiDEL PROCStar-IV board, is capable of processing streams of documents at a rate of 32 Gterms/s. We present a mathematical analysis of the throughput of the application, discussing in particular the problem of scaling the Bloom filter used to reduce the number of required lookups in external memory. Using the performance and power measurements obtained from our implementation and a high-performance multithreaded software reference implementation as inputs for an economic cost model, we show that using our technology can reduce the total cost of ownership of data centres for processing data-centric workloads with a factor of 10.

1 Introduction

The focus on real-time search is growing with the increasing adoption and spread of social networking applications. Real-time search is equally important in other areas such as analysing emails for spam or search web traffic for particular patterns.

W. Vanderbauwhede (✉)
School of Computing Science, University of Glasgow, G12 8QQ Glasgow, UK
e-mail: Wim.Vanderbauwhede@Glasgow.ac.uk

S.R. Chalamalasetti • M. Margala
Department of Electrical and Computer Engineering, University of Massachusetts Lowell, Lowell, MA, USA
e-mail: sairahul_chalamalasetti@student.uml.edu; martin_margala@uml.edu

W. Vanderbauwhede and K. Benkrid (eds.), *High-Performance Computing Using FPGAs*, DOI 10.1007/978-1-4614-1791-0_7, © Springer Science+Business Media, LLC 2013

Servicing millions of user requests and processing very large volumes of information requires massive amounts of computational resources. The data centres that process these requests consume huge amounts of energy for both computing and cooling [1]. According to a recent study by McKinsey, commissioned by the Uptime Institute [2], CO_2 emissions from data centres are projected to surpass those of the airline industry 2020. The same report shows that the energy cost associated with running a data centre is now the dominant cost. Thus the development of energy efficient solutions is motivated both from an economic perspective and an environmental perspective. As a result, there are currently many initiatives, both from academia and industry, to reduce the energy consumption of data centres. These efforts attempt to address this challenge by focusing on better layouts, more efficient cooling systems, power reduction in the servers by switching off unused cores, and more efficient software, e.g. through virtualisation. However, while all of these solutions contribute to a reduction in power consumption, none of them constitute a radical departure from current system architectures.

Our aim is to achieve dramatic energy savings through the use of Field Programmable Gate Arrays. FPGAs are eminently suited for acceleration of Information Retrieval tasks: this type of tasks is often inherently parallelisable because document models generally do not treat documents as a sequence of words, but rather as a collection of uncorrelated or loosely correlated terms.

In [3, 4] we presented our initial work on applying FPGAs for acceleration or search algorithms. In this chapter we describe the design, architecture and performance of our novel document filtering (real-time search) solution.

After introducing the application domain and workload, we present a mathematical analysis of the throughput of the system. This novel analysis is applicable to a much wider class of applications than the one discussed in the paper: any algorithm that performs non-deterministic concurrent accesses to a shared resource can be analysed using the model we present. In particular, the technology presented in this paper can also be used for "traditional", i.e. inverted index-based, web search. We present an FPGA implementation of our novel architecture on the GiDEL PROCStar-IV board and a performance comparison with a multicore CPU system. Finally, we present an analysis of the reduction in total cost of ownership of data centres that can be achieved by deploying our technology.

2 Real-Time Search Applications

2.1 Choice of Workload

There is a great variety in data-centric workloads and how the data is processed. Of the various operations that can be performed on data—collect and distribute, maintain and manage, organise, analyse and classify—in this work we focus on the last category. There are different classes of operations associated with organisation and analysis of data including ad hoc retrieval, classification, clustering, and filtering. However, fundamentally, all of these tasks require the matching between information items and information needs; for example, a user query or a predetermined profile matched against data content. Representative of these operations, in this paper, we focus on real-time unstructured search or information filtering where *given a collection of unstructured data sources (e.g., documents), we identify the match to a pre-determined weighted feature vector profile (e.g., topic signatures, keywords)*. With the growing increase in unstructured and semi-structured data, this class of workloads is likely to be more important in the future. Some example applications include searching patent repositories for related work comparison, searching emails and sharepoints for enterprise information management, detecting spam in incoming emails, monitoring communications for terrorist activity, news story topic detection and tracking, searching through books, images, and videos for matching profiles.

2.2 Algorithm Description

Real-time search, in Information Retrieval parlance called "document filtering", consists of matching a stream of documents against a fixed set of terms, called the "profile". Typically, the profile is large and must therefore be stored in external memory.

The algorithm implemented on the FPGA can be expressed as follows:

- A *document* is modelled as a "bag of words", i.e. a set D of pairs (t, f) where $f \triangleq n(t,d)$ is the *term frequency*, i.e. number of occurrences of the term t in the document d; $t \in \mathbb{N}$ is the *term identifier*.
- The profile M is a set of pairs $p = (t, w)$ where the *term weight* $w \triangleq \log\left(\frac{(1-\lambda)P(t|M)}{P(t)} + \lambda\right)$.

In this work we are concerned with the computation of the document score, which indicates how well a document matches the profile.

The document has been converted to the bag-of-words representation in a separate stage. (This approach is also followed by the Google n-gram project [5].) Note that a "term" can correspond to a single word, an n-gram, or a fixed sequence of n words. Inclusion of bigrams and trigrams, removal of stop words and stemming (reducing words to their root) lead to better search results.[1] We perform this stage on the host processor using the Open Source information retrieval toolkit Lemur [6]. We note that this stage could also be very effectively performed on FPGAs.

---

[1]Note that the inclusion of n-grams makes our vocabulary size much larger than expected. For example, the Oxford Dictionaries FAQ indicates about a quarter of a million distinct English terms, but the vocabulary size for our collection is 16 million terms.

The profile is precomputed offline based on specific user requirements using the relevance-based language model proposed by Lavrenko and Croft [7] or alternatively using a Bayesian algorithm.

Simplifying slightly, to determine if a document matches a given profile, we compute its score: the sum of the products of term frequency and term weight

$$\mathrm{score}(D,M) = \sum_{i \in D} f_i w_i \tag{1}$$

The weight is typically a high-precision word (64 bits) stored in a lookup table in the external memory. If the score is above a given threshold, we return the document identifier and the score by writing it into the external memory.

This function is the basis of most relevancy scoring algorithms, the main difference being the weighting of terms in profiles. For example, the same function can be used for Nave Bayes spam filtering, Support Vector Machine classification, Relevancy Feedback information filtering and even image recognition.

2.3 *Target Platform*

The target platform for this work is the Novo-G FPGA supercomputer [8] hosted by the NSF center for high-performance reconfigurable computing (CHREC).[2] This machine consists of 24 compute servers which each hosts a GiDEL PROCStar-IV board. The board contains four FPGAs with two banks of DDR SDRAM per FPGA used for the document collection and one for the profile. The data width is 64 bits, which means that the FPGA can read 128 bits per memory per clock cycle [9]. For more details on the platform, see Sect. 4.

2.4 *Term Scoring Algorithm*

To simplify the discussion, we first consider the case where terms are scored sequentially, and that, as in our original work, we use a Bloom filter to limit the number of external memory accesses.

For every term in the document, the application needs to look up the corresponding profile term to obtain the term weight. As the profile is stored in the external SDRAM, this is an expensive operation (typically 20 cycles per access). The purpose of document filtering is to identify a small amount of relevant documents from a very large document set. As most documents are not relevant, most of the lookups will fail (i.e. most terms in most documents will not occur in the profile).

---

[2] www.chrec.org.

Therefore, it is important to discard the negatives first. For that purpose we use a special Bloom filter implemented using the FPGA's on-chip memory.

2.4.1 Perfect Bloom Filter

A Bloom filter [10] is a datastructure used to determine membership of a set. False positives are possible, but false negatives are not. With this definition, the design we use to reject negatives is a Bloom filter. However, in most cases a Bloom filter uses a number (k) of hash functions to compute several keys for each element in the set and adds the element to the table (assigns a "1") if element is in the set. As a result, hash collisions can lead to false positives.

Our Bloom filter is a special case of this more general implementation: our hashing function is the identity function $key = elt$, and we only use a single hash function ($k = 1$) so every element in the set corresponds to exactly one entry in the Bloom filter table. As a result, the size of the Bloom filter is the same as the size of the set and there are no false positives, hence the name "perfect Bloom filter". Furthermore, no elements are added to the set at run time.

The dimensioning of the Bloom filter depends on the available memory on the FPGA and is discussed in Sect. 4.3.2.

2.4.2 Document Stream Format

The document stream is a list of *(document identifier, document term pair set)* pairs. Physically, the FPGA accepts a fixed number n of streams of words with fixed width w. The document stream must be encoded onto these word streams. As both elements in the document term pair $d_i = (t_i, f_i)$ are unsigned integers, m pairs can be encoded onto a word if w is larger than or equal to m times the sum of the magnitudes of the maximum values for t and f:

$$w \geq m(\lceil \log_2 t_{\max} \rceil + \lceil \log_2 f_{\max} \rceil) \tag{2}$$

To mark the start of a document we insert a header word (identified by $f = 0$) followed by the document ID.

2.4.3 Profile Lookup Table Implementation

In the current implementation, the lookup table that stores the profile is implemented in the most straightforward way: as the vocabulary size is 2^{24} and the weight for each term in the profile can be stored in 64 bits, a profile consisting of the entire vocabulary could be stored in the 512 MB SDRAM, which is less than the size of the fixed SDRAM on the PROCStar-IV board. Consequently, there is no need for hashing, the memory contains zero weights for all terms not present in the profile.

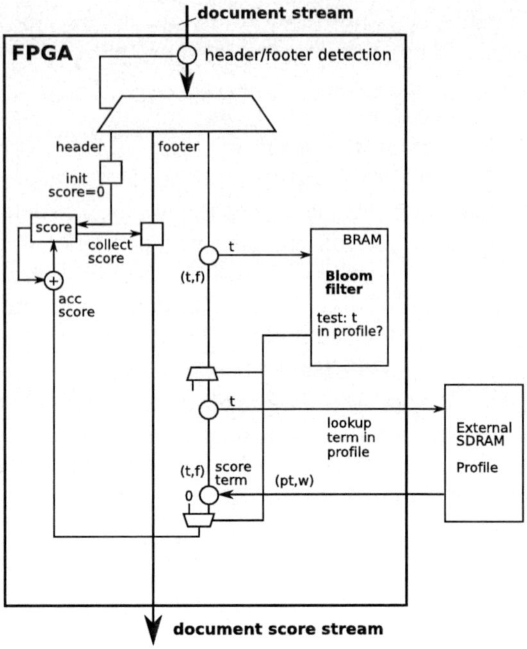

Fig. 1 Sequential document term scoring

2.4.4 Sequential Implementation

The diagram for the sequential implementation of the design is shown in Fig. 1.

Using the lookup table architecture and document stream format as described above, the actual lookup and scoring system is quite straightforward: the input stream is scanned for header and footer words. The header word action is to set the document score to 0; the footer word action is to collect and output the document score. For every term in the document, first the Bloom filter is used to discard negatives and then the profile term weight is read from the SDRAM. The score is computed and accumulated for all terms in the document and finally the score stream is filtered against a threshold before being output to the host memory. The threshold is chosen so that only a few tens or hundreds of documents in a million are returned.

If we would simply look up every term in the external memory, the maximum achievable throughput would be $1/\Delta t_S$, with Δt_S the number of cycles required to look up the term weight in the external memory and compute the term score. The use of a Bloom filter greatly improves the throughput as the Bloom filter access will typically be much faster than the external memory access and subsequent score computation. If the probability for a term to occur in the profile is P_P and the access time to the Bloom filter is Δt_B, the average access time will become $\Delta t_B + P_P \cdot \Delta t_S$. In practice P_P will be very low as most document terms will not occur in the profile (because otherwise the profile would match all documents). The more selective the profile, the fewer the number of document terms that match it.

2.5 *Parallelising Lookups*

The scoring process as described above is sequential. However, as in the bag-of-words representation all terms are independent, there is scope for parallelisation. In principle, all terms of a document could be scored in parallel, as they are independent and ordering is of no importance.

2.5.1 Parallel Document Streams

In practice, even without the bottleneck of the external memory access, the amount of parallelism is limited by the I/O width of the FPGA, in our case 64 bits per memory bank. A document term can be encoded in 32 bits (a 24-bit term identifier and an 8-bit term frequency). As it takes at least one clock cycle of the FPGA clock to read in two new 64-bit words (one per bank), the best case for throughput would be if four terms per document would be scored in parallel in a single cycle. However, in practice scoring requires more than one cycle; to account for this, the process can be further parallelised by demultiplexing the document stream into a number of parallel streams. If, for example, scoring would take four cycles, then by scoring four parallel document streams the application could reach the maximal throughput.

2.6 *Parallel Bloom Filter Design*

Obviously, the above solution would be of no use if there would be only a single, single-access Bloom filter. The key to parallelisation of the lookup is that, because the Bloom filter is stored in on-chip memory, accesses to it can be parallelised by partitioning the Bloom filter into a large number of small banks. The combined concepts of using parallel streams and a partitioned Bloom filter are illustrated in Fig. 2. To keep the diagram uncluttered, only the paths of the terms (Bloom filter addresses) have been shown.

Every stream is multiplexed to all m Bloom filter banks; every bank is accessed through an n-port arbiter. It is intuitively clear that for large numbers of banks, the probability of contention approaches zero, and hence the throughput will approach the I/O limit—or would if none of the lookups would result in an external memory access and score computation.

3 Throughput Analysis

In this section, we present the mathematical throughput analysis of the Bloom filter-based document scoring system. The analysis consists of four parts:

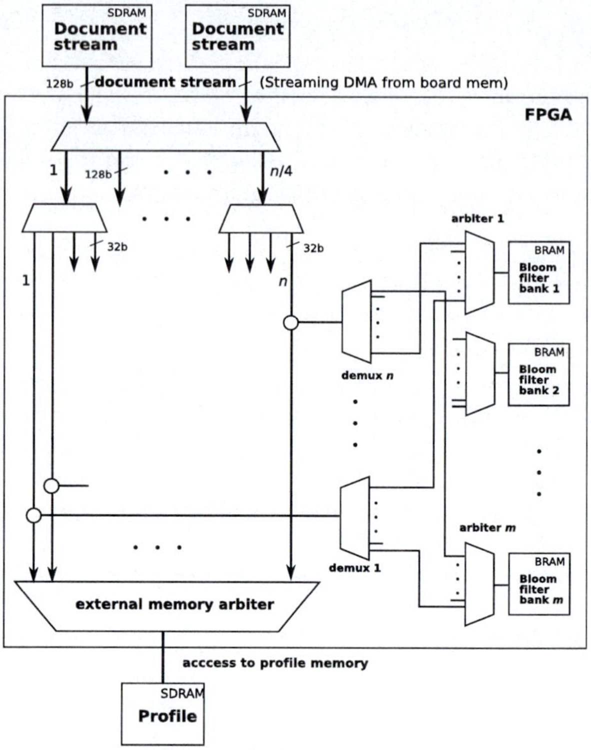

Fig. 2 Parallelising lookups using parallel streams and a multi-bank Bloom filter

- In Sect. 3.1 we derive an expression to enumerate all possible access patterns for n concurrent accesses to a Bloom filter built of m banks and use it to compute the probability for each pattern.
- In Sect. 3.2 we compute the average access time for each pattern, given that n_{H} accesses out of n will result in a lookup in the external memory. We consider in particular the cases of $n_{\mathrm{H}} = 0$ and $n_{\mathrm{H}} = 1$ and propose an approximation for higher values of n_{H}.
- In Sect. 3.3 we compute the probability that n_{H} accesses out of n will result in a lookup in the external memory.
- In Sect. 3.4, combining the results from Sects. 3.2 and 3.3, we compute the average access time over all n_{H} for a given access pattern; finally, we combine this with the results from Sect. 3.1 to compute the average access time over all access patterns.

3.1 Bloom Filter Access Patterns

We need to calculate the probability of contention between c accesses out of n, for a Bloom filter with m banks. Each bank has an arbiter which sequentialises the contending accesses, so c contending accesses to a given bank will take a time $c \cdot \Delta t_B$, with Δt_B the time required for a single lookup in the Bloom filter. We also account for a fixed cost of contention Δt_C. We use a combinatorial approach: we count all possible arrangements of n accesses to m banks. Then we count the arrangements that result in c concurrent accesses to a bank.

To do so, we need first to compute the *integer partitions* of n [11] as they constitute all possible arrangements of n accesses. For the remainder of the paper, we will refer to "all possible arrangements that result in x" as the *weight* of x. Each partition of n will result in a particular average access time over all accesses. If we know the probability that each partition will occur and its resulting average access time, we can compute the total average access time.

3.1.1 Integer Partitions

A *partition* $p(n,k)$ of a positive integer n is a non-increasing sequence of k positive integers $p_1, p_2, \ldots, p_k$ with n as their sum. Each integer p_i is called a *part*. Thus, with n in our case being the number of access ports to our Bloom filter, each partition is a possible access pattern for the Bloom filter. For example, if $n = 16$ and $k = 8$, the partition $(5\,3\,2\,2\,1\,1\,1\,1)$ means that the first bank in the Bloom filter gets 5 concurrent accesses, the next 3, and so on. For $n \leq m$, $k \in [1,n]$; if $n > m$, we must restrict k to $k \in [1,m]$ because we can't have more than m parts in the partition as m is the number of banks. In other words, $k \in [1, \min(n,m)]$. We denote this as $p(n,k)$.

3.1.2 Probability of Each Partition

For each partition, we can compute the probability of it occurring as follows: if there are n concurrent accesses to the Bloom filter's m banks, $n \leq m$, then each access pattern can be written as a sequence of numbers. We are not interested in the actual numbers but in the patterns, e.g. with $n = 8$ and $m = 16$, we could have a sequence $(aaa\ bb\ cc\ d)$, $a,b,c,d \in 0 \ldots m-1; a \neq b \neq c \neq d$ which results in a partition (3221). Consequently, given a partition we need to compute the probability for the sequence which it represents. The probability for each number occurring is the same, $1/m$. We can compute this probability as a product of three terms. First, we consider the probabilities for sequences of length n of events with probability α_i where each event occurs x_i times. These are given by the multinomial distribution:

$$n! \prod_{i=1}^{k} \frac{\alpha_i^{x_i}}{x_i!} \tag{3}$$

where $0 < x_i \leq n$ and $n = \sum_{i=1}^{k} x_i$.

In our case, each event has the same probability $1/m$ and the number of times each event occurs is the size of each part p_i in the partition, so

$$\frac{n!}{m^n}\prod_{i=1}^{k}\frac{1}{p_i!} \tag{4}$$

This gives the probability for a sequence of k groups of p_i events, n events in total.

The actual sequence will consist of numbers $1,\ldots,m$, so we must consider the total number of different sequences of numbers that result in a given partition. This is simply the number of possible combinations of k numbers out of m, C_m^k.

Finally, we must consider the permutations as well, for example, for $(2\,1\,1)$ we must also consider $(1\,2\,1)$ and $(1\,1\,2)$. This is a combinatorial problem in which the bins are distinguishable by the number of elements they contain; however, the actual number of elements is irrelevant, only the fact that the bins are distinguishable. The derivation is slightly more complicated. We proceed as follows: we transform the partition into a tuple with as many elements as the number of different integers in the partition, and the value for each element the number of times this integer occurs in the partition. For example, $(5\,5\,3\,3\,2\,1\,1) \rightarrow (2\,2\,1\,2)$ and $(4\,3\,2\,2\,1\,1\,1) \rightarrow (1\,1\,2\,3)$. We call the new set the frequencies of the partition p, $F(p(n,k))$. As partitions are non-increasing sequences, the transformation is quite straightforward:

First we create an ordered set $\mathcal{S} = \{S_1,\ldots,S_i,\ldots\}$ with $P = \bigcup S_i$ i.e. $\mathcal{S}$ is a set partition of P. The elements of $\mathcal{S}$ are defined recursively as

$$S_1 = \{\forall p_j \in P \,|\, p_j = p_1\} \tag{5}$$

$$S_i = \{\forall p_j \in P \backslash \bigcup_{k=1,\ldots,i-1} S_k \,|\, p_j = p_1\} \tag{6}$$

i.e. S_1 contains all parts of P identical to the first part of P; for S_2 we remove all elements of S_1 from P and repeat the process, and we continue recursively until the remaining set is empty. Finally, we create the (ordered) set of the cardinal numbers of all elements of $\mathcal{S}$:

$$F = \{f_i \triangleq \#S_i, \forall S_i \in S\} \tag{7}$$

We are looking for the permutations with repetition of $F(p(n,k))$, which is given by

$$n'! \prod_{\forall f_i \in F(p(n,k))} \frac{1}{f_i!} \tag{8}$$

where $n' = \sum f_i$.

Thus the final probability for each partition of n and a given m becomes:

$$\mathcal{P}(p(n,k),m) = \frac{C_m^k}{m^n}\cdot n! \prod_{\forall p_i \in p(n,k)} \frac{1}{p_i!} \cdot n'! \prod_{\forall f_i \in F(p(n,k))} \frac{1}{f_i!} \tag{9}$$

We observe that

$$\sum_{k=1}^{n} \mathcal{P}(p(n,k),m) = 1 \tag{10}$$

regardless of the value of m.

In the next section we derive an expression for the access time for a given partition, depending on the number of accesses that will result in an external memory lookup.

3.2 *Average Access Time per Pattern*

The time to perform n lookups in the Bloom filter is of course determined by the number of contending accesses. For c contending accesses, it will take a time $c\Delta t_{\mathrm{B}}$. However, not all Bloom filter lookups will result in a subsequent access to the external memory—in fact most of them will not, this is exactly the reason for having the Bloom filter. We will call a Bloom filter lookup that results in an access to the external memory a *hit*.

3.2.1 Case of No Hits

First, we will consider the case of 0 hits, i.e. the most common case. In this case, the average access time for a given partition $p(n,k)$ is the average of all the parts in the partition:

$$\overline{\Delta t_{\mathrm{H},p,0}} = \frac{k_{>1}}{k} \cdot \Delta t_{\mathrm{C}} + \frac{n}{k} \Delta t_{\mathrm{B}} \tag{11}$$

where $k_{>1}$ is the number of parts $p_i > 1$. For the case of $k = n$ (no contention), $k_{>1} = 0$ so there is no fixed cost of contention Δt_{C}. Note again that $k \leq \min(n,m)$.

In practice, a small number of Bloom filter lookups will result in a hit, and consequently there is a chance of having one or more hits for concurrent accesses.

3.2.2 Case of a Single Hit

Consider the case of a single hit (out of n lookups). The question we need to answer is, how long on average will it take to encounter a hit? Because as soon as we encounter a hit we can proceed to perform the external memory access, without having to wait for subsequent hits. This time depends on the particular integer partition. To visualise the partition, we use a so-called Ferrers diagram [12], in which every part is arranged vertically as a list of dots. For example, consider the Ferrers diagram for the partition $(8\,4\,1\,1\,1\,1)$, i.e. $n = 16, k = 6$ (Fig. 3). Each row can be interpreted as the number of concurrent accesses to different banks; each column represents the number of contending accesses to a particular bank.

Fig. 3 Ferrers diagram for the partition $(8\,4\,1\,1\,1\,1)$

Fig. 4 Ferrers diagram for the conjugate partition $(6\,2\,2\,2\,1\,1\,1\,1)$

From this graph it is clear that the probability for finding the hit on the first cycle is 6/16; on the second to fourth cycle 2/16, on the fifth to eighth cycle 1/16. Consequently, the average time to encounter a hit will in this case be

$$1\cdot\frac{6}{16}+(2+3+4)\cdot\frac{2}{16}+(5+6+7+8)\cdot\frac{1}{16}$$

To generalise this derivation, we observe first that the transposition of the Ferrers diagram of an integer partition p yields a new integer partition $p'(n,k')$ for the same integer called the *conjugate* partition. In our example $p' = (6\,2\,2\,2\,1\,1\,1\,1)$ with $k' = 8$ (Fig. 4).

We observe that the time it takes to reach a hit in part p'_i is $\Delta t_{\mathrm{B}} \cdot i$. Using the conjugate partition p', we can write the lower bound for average time it takes to reach a hit in partition p as

$$\overline{\Delta t_{\mathrm{H},p,1}} = \left(1-\frac{k-k_{>1}}{n}\right)\cdot\Delta t_{\mathrm{C}}+\Delta t_{\mathrm{B}}\cdot\frac{1}{n}\sum_{i=1}^{k'} i\cdot p'_i \qquad (12)$$

The term in Δt_{C} only occurs when the hit is in a bank with contention, i.e. in a part greater than 1. There are $k-k_{>1}$ parts of size 1, so the chance of a hit occurring in one of them (i.e. a hit on a bank without contention) is $\frac{k-k_{>1}}{n}$. Thus the probability for the term in Δt_{C} is

$$1-\frac{k-k_{>1}}{n} \qquad (13)$$

And of course, as the hit results in an external access, the average access time is

$$\overline{\Delta t_{\mathrm{A},p,1}} = \left(1-\frac{k-k_{>1}}{n}\right)\cdot\Delta t_{\mathrm{C}}+\Delta t_{\mathrm{B}}\cdot\frac{1}{n}\sum_{i=1}^{k'} i\cdot p'_i+\frac{1}{n}\cdot\Delta t_{\mathrm{S}} \qquad (14)$$

For the case of $k = n$, the equation reduces to

$$\overline{\Delta t_{\mathrm{A},p,1}} = \Delta t_{\mathrm{B}} + 1/n\cdot\Delta t_{\mathrm{S}} \qquad (15)$$

3.2.3 Case of Two or More Hits

If there are two or more hits, the exact derivation would require enumerating all possible ways of distributing n_H hits over a given partition; furthermore, simply enumerating them is not sufficient: we would have to consider the exact time of occurrence of each hit to be able to determine if a subsequent hit was encountered during or after the time to perform an external lookup and compute the score (Δt_S) from a given hit. It is easy to see that for large n_H, this problem is so complex as to be intractable in practice. However, we can make a simplifying assumption: in practice, Δt_S will be much larger than the time to perform a Bloom filter lookup.

If that is the case, a good approximation for the total elapsed time is the time until the *first* hit is encountered plus n_H times the time for external access. This approximation is exact as long as the time it takes to sequentially perform all external lookups is longer than the time between the best and worst case Bloom filter access time for n_H hits on a single bank, in other words as long as $\Delta t_S > p_i \Delta t_B$. The worst case is of course $p_1 = n$ but this case has a very low probability: for example, for $n = 16$, the average value of all parts is 2.5; even considering only the parts >1, the average is still <4. For $n = 32$, the numbers are resp. 3 and 5. In practice, if $\Delta t_S / \Delta t_B > 10$, the error will be negligible.

Conversely, we could consider the time until the *last* hit is encountered plus n_H times Δt_S. This approximation provides an upper bound for the access time.

Therefore, we are only interested in these two cases, i.e. the lowest resp. highest part of the partition with at least one hit. We need to compute the probability that the lowest (resp. highest) part will contain a hit, and the next but lowest (resp. highest) one, etc. For simplicity, we leave off Δt_C in the following derivation.

Lower Bound

The number of all possible cases is $N_{p'} = C(n, n_H)$, all possible arrangements of n_H elements in n bins. To compute the weight of a hit in the lowest part p'_1, we compute the complement: all possible arrangements without any hits in p'_1. That means that we remove p'_1 from n. Then, using the notation $\neg p_1$for "not a hit in p'_1", we compute

$$N_{\neg p_1} = C(n - p'_1, n_H) \tag{16}$$

These are all the possible cases for not having a hit in p'_1. Thus, $N_{p_1} = N_p - N_{\neg p_1}$ is the number of possible arrangements with $1, \ldots, n_H$ hits in p'_1.

We now do the same for p_2, etc. That gives us all possible cases for *not* having a hit in p_i:

$$N_{\neg p_i} = C\left(n - \sum_{j=1}^{i} p'_j, n_H\right) \tag{17}$$

Obviously, there must be enough space in the remaining parts to accommodate n_H hits, so i is restricted to values where

$$n - \sum p'_i \geq n_\mathrm{H} \tag{18}$$

We call the highest index for which (18) holds, k^*.

To obtain the weight of a hit in p'_i, we must of course subtract the weight of a hit in p'_{i-1}, because $N_p - N_{\neg p_i}$ would give the weight for having a hit in all parts up to p_i. It is easy to show [by substitution of (16)] that

$$N_{p_i} = N_{\neg p_{i-1}} - N_{\neg p_i} \tag{19}$$

Finally, the average time it takes to reach a part in a given p' with at least one hits out of n_H is

$$\overline{\Delta t_{\mathrm{H},p,n_\mathrm{H}}} = \Delta t_\mathrm{B} \cdot \frac{1}{N_{p'}} \sum_{i=1}^{k^*} i \cdot N_{p_i} \tag{20}$$

With the above assumption, the average access time for n_H hits can then be approximated as

$$\overline{\Delta t_{\mathrm{A},p,n_\mathrm{H}}} = \left(1 - \frac{k - k_{>1}}{n}\right) \cdot \Delta t_\mathrm{C} + \Delta t_\mathrm{B} \cdot \frac{1}{N_{p'}} \sum_{i=1}^{k^*} i \cdot N_{p_i} + n_\mathrm{H} \Delta t_\mathrm{S} \tag{21}$$

We observe that for $n_\mathrm{H} = 1$, (21) indeed reduces to (14) as $N_{p'} = n$ and $N_{p_i} = p'_i$. For $n = k$ the equation reduces to $\Delta t_\mathrm{B} + n_\mathrm{H} \Delta t_\mathrm{S}$.

Upper Bound

The upper bound is given by the probability that the highest part is occupied etc., so the formula is the same as (17) but starting from the highest part p_c, i.e.

$$N_{\neg p_{c-i}} = C\left(n - \sum_{j=c-i+1}^{c} p'_j, n_\mathrm{H}\right) \tag{22}$$

with the corresponding restriction on i that

$$\sum_{i=1}^{k^*} p'_i \geq n_\mathrm{H} \tag{23}$$

As we will see in Sect. 3.5, in practice the bounds are usually so close together that the difference is negligible.

3.3 *Probability of External Memory Access*

The chance that a term will occur in the profile depends on the size of the profile $N_{\mathcal{P}}$ and the size of the vocabulary $N_{\mathcal{V}}$:

$$P_{\mathcal{P}} = \frac{N_{\mathcal{P}}}{N_{\mathcal{V}}} \tag{24}$$

This is actually a simplified view: it assumes that the terms occurring in the profile and the documents are drawn from the vocabulary in a uniform random way. In reality, the probability depends on how discriminating the profile is. As the aim of a search is of course to retrieve only the relevant documents, we can assume that actual profiles will be more discriminating than the random case. In that case (24) provides a worst case estimate of contention.

The probability of n_{H} hits, i.e. contention between n_{H} accesses to the external memory is then

$$C(n, n_{\mathrm{H}}) \cdot P_{\mathcal{P}}^{n_{\mathrm{H}}} \cdot (1 - P_{\mathcal{P}})^{n-n_{\mathrm{H}}} \tag{25}$$

That is, there are $C(n, n_{\mathrm{H}})$ arrangements of n_{H} accesses out of n and for each of them, the probability that it occurs is $P_{\mathcal{P}}^{n_{\mathrm{H}}} \cdot (1 - P_{\mathcal{P}})^{n-n_{\mathrm{H}}}$. Furthermore, n_{H} contending accesses will take a time $n_{\mathrm{H}}\Delta t_{\mathrm{S}}$. Of course, if no external access is made, the external access time is 0.

3.4 *Average Overall Throughput*

3.4.1 Average Access Time Over All n_{H} for a Given Pattern

We can now compute the average access time over all n_{H} for a given access pattern p by combining (21) and (25):

$$\overline{\Delta t_{\mathrm{A},p}} = \sum_{n_{\mathrm{H}}=0}^{n} C(n, n_{\mathrm{H}}) \cdot P_{\mathcal{P}}^{n_{\mathrm{H}}} \cdot (1 - P_{\mathcal{P}})^{n-n_{\mathrm{H}}} \overline{\Delta t_{\mathrm{A},p,n_{\mathrm{H}}}} \tag{26}$$

3.4.2 Average Access Time Over All Patterns for Given n and m

Finally, using (9) and (26), we can compute the average access time over all patterns for given n and m, i.e. the average overall throughput of the application with n parallel threads and an m-bank Bloom filter.

$$\overline{\Delta t_{\mathrm{A}}(n,m)} = \sum_{\forall p(n)} \mathcal{P}(p(n,k), m) \cdot \overline{\Delta t_{\mathrm{A},p}} \tag{27}$$

Fig. 5 Accuracy of the approximation for $n_H \geq 2$

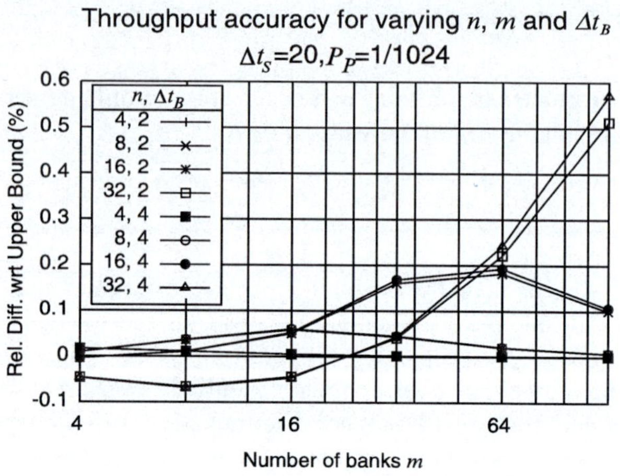

3.5 *Analysis*

In this section the expression obtained in Sect. 3.4 is used to investigate the performance of the system and the impact of the values of n, m, Δt_B, Δt_S, and P_P on the throughput.

3.5.1 Accuracy of Approximation

To evaluate the accuracy of the approximations introduced in Sect. 3.2.3, we compute the relative difference between the "first hit" approximation and the "upper bound" approximation. From Fig. 5, it can be seen that the difference is less than 1% of the throughput over all simulated cases. As the upper bound always overestimates the delay, and the "first hit" approximation will in most cases return the correct delay, this demonstrates that both approximations are very accurate. An interesting observation is that for $\Delta t_B = 4$ the error is almost the same as for $\Delta t_B = 2$, which illustrates that the condition $\Delta t_S > p_i \Delta t_B$ is sufficient but not necessary.

Next, we consider a more radical approximation: we assume that for $n_H > 1$, $P_P = 0$, in other words we ignore all cases with more than 1 hit.

From Fig. 6 we see that the relative difference between the throughput using this approximation and the "first hit" is very small, to such an extent that in almost all cases it is justified to ignore $n_H > 1$. This is a very useful result as this approximation speeds up the computations considerably.

3.5.2 Maximum Achievable Throughput

The throughput depends on the number of hits in the Bloom filter. Let us consider the case where the Bloom filter contains no hits at all. This is the maximum throughput

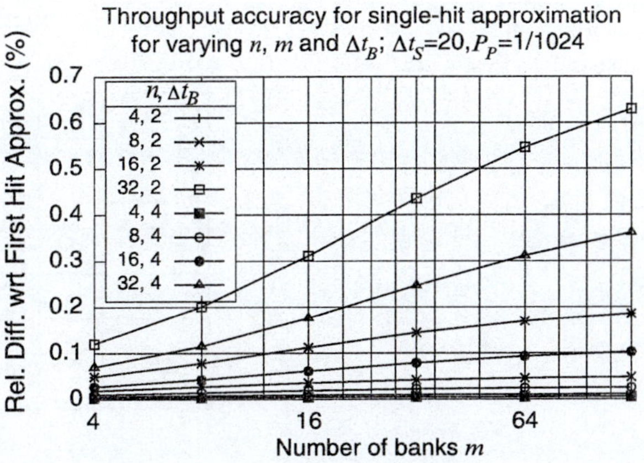

Fig. 6 Accuracy of the single-hit approximation for $n_{\mathrm{H}} \geq 2$

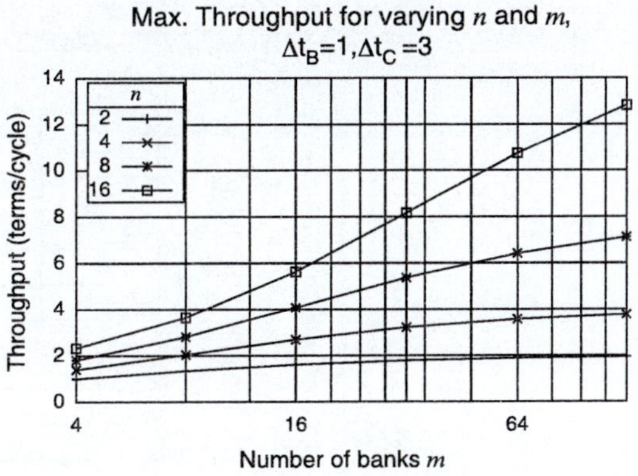

Fig. 7 Best case (0 hits) average access time for a Bloom filter with m banks and n access ports, $\Delta t_{\mathrm{B}} = 0.125$, $\Delta t_{\mathrm{C}} = 1.2$

the system could achieve, it corresponds to a profile for which no document in the stream has any matches. We can use (11) and (9) to calculate the best-case average access time for a Bloom filter with m banks and n access ports:

$$\overline{\Delta t_{\min}(n,m)} = \sum_{k=1}^{n} \sum_{\forall p(n,k)} \left(\left(\frac{k_{>1}}{k} \Delta t_{\mathrm{C}} + \frac{n}{k} \Delta t_{\mathrm{B}} \right) \cdot \mathcal{P}(p(n,k),m) \right) \tag{28}$$

Note that for $m < n$, $\mathcal{P}(p(n,n)) = 0$.

The results are shown in Fig. 7. The figure shows that for $\Delta t_{\mathrm{B}} = 0.125, \Delta t_{\mathrm{C}} = 1.2$ (the values for our current implementation), the I/O-limited throughput (8 terms/cycle for the PROCStar-IV board) is achieved with $n = 8$ and $m = 16$.

3.5.3 Throughput Including External Access

Figure 8 shows the effect of the external memory access and score computation. The important observation is that the performance degradation is quite small for

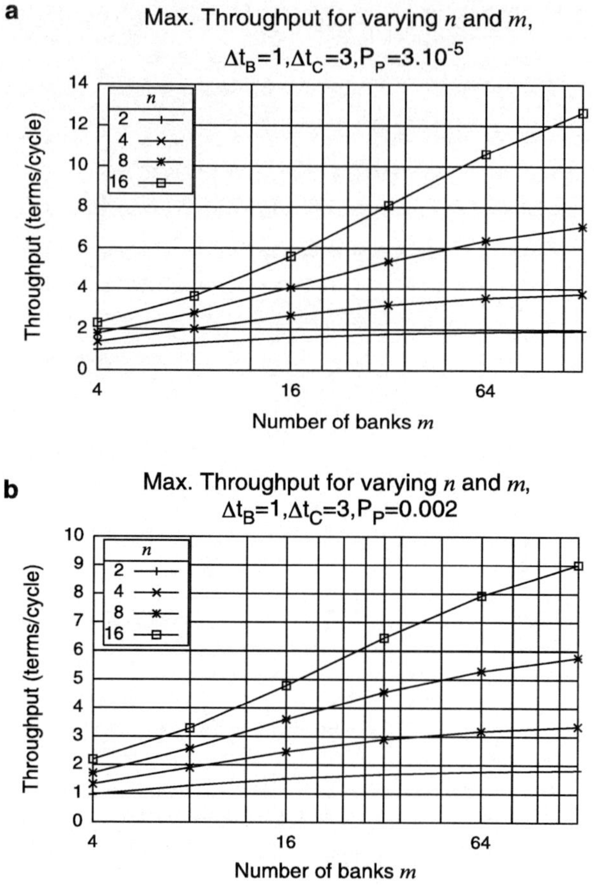

Fig. 8 Average access time for a Bloom filter with m banks and n access ports, $\Delta t_S = 20$. (**a**) $P_{\mathcal{P}} = 3.10^{-5}$; (**b**) $P_{\mathcal{P}} = 0.002$

low hit rates, and still only around 25 % for a relatively high hit rate of 1/512. This demonstrates that the assumptions underlying our design are justified.

3.5.4 Impact of Bloom Filter Access Time

A further illustration of the impact of Δt_B is given in Fig. 9, which plots the throughput as a function of Δt_B on a log/log scale. This figure illustrates clearly how a reduction in throughput as result of slower Bloom filter access can be compensated for by increasing the number of access streams. Still, with $\Delta t_B = 4$, we would need 32 parallel streams per input stream, or we would need a very large number ($\gg$128) Bloom filter banks. On the one hand, the upper limit is 512 (the number of M9K blocks on the Stratix-IV 530 FPGA); on the other hand, the size of the demultiplexers and arbiters would become prohibitive as it grows as $m \cdot n$.

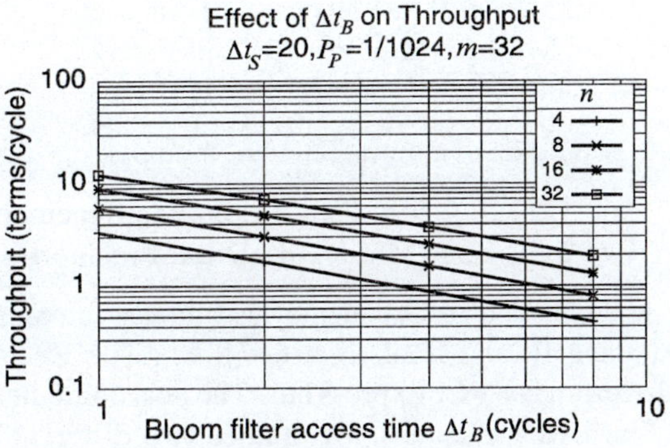

Fig. 9 Impact of Bloom filter access time on throughput

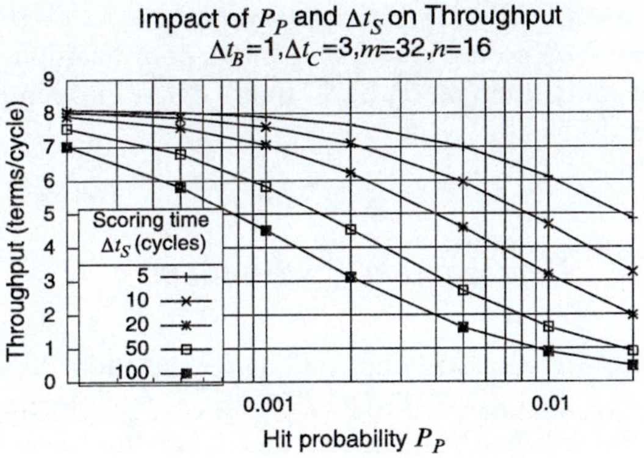

Fig. 10 Impact on throughput of hit probability and external memory access time

3.5.5 Impact of Profile Hit Probability and External Memory Access Time

The final figure (Fig. 10) is probably the most interesting one. It confirms that for very selective profiles (i.e. profiles resulting in very low hit rates), the effect of long external memory access times is very small.

4 FPGA Implementation

We implemented our design on the GiDEL PROCStar-IV development board (Fig. 11). This system provides an extensible high-capacity FPGA platform with the GiDEL PROC-API library-based developer kit for interfacing with the FPGA.

4.1 Hardware

Each board contains four Altera Stratix-IV 530 FPGAs running at 150 MHz. Each FPGA supports a five-level memory structure, with three kinds of memory blocks embedded in the FPGA:

- 6,640 MLAB RAM blocks (320 bits each)
- 1,280 M9K RAM blocks (9K bits each)
- 64 M144K blocks (144K bits each)

and two kinds of external DRAM memory:

- One 512 MB DDR2 SDRAM on-board memory (Bank A)
- Two 2 GB SODIMM DDR2 DRAM memories (Bank B and Bank C)

The embedded FPGA memories run at a maximum frequency of 300 MHz, Bank A, Bank B, and Bank C at 667 MHz. The FPGA-board is connected to the host platform via a PCI Express bus. The host computer transfers data to the FPGA using 32-bit DMA channels. We studied two different host systems: a system based on HP BL460 blade servers with quad-core 64-bit Intel Xeon X5570 at 2.93 GHz and 3.5 GB DDR2 DRAM memory, running 32-bit Windows XP, and a second system based on a dual-core 64-bit Intel Atom board at 1 GHz with 4 GB DDR2 DRAM memory, running 64-bit Ubuntu GNU/Linux; in this paper, we primarily focus on the former.

4.2 Development Environment

FPGA-accelerated applications for the PROCStar board are implemented in C++ using the GiDEL PROC-API libraries for interacting with the FPGA. This API defines a hardware abstraction layer that provides control over each hardware element in the system—for example, Memory I/O is implemented using the GiDEL MultiFIFO and MultiPort IPs. To achieve optimal performance, we implemented the FPGA algorithm in VHDL (as opposed to Mitrion-C as used in our previous work). We used the Altera Quartus toolchain to create the bitstream for the Stratix-IV.

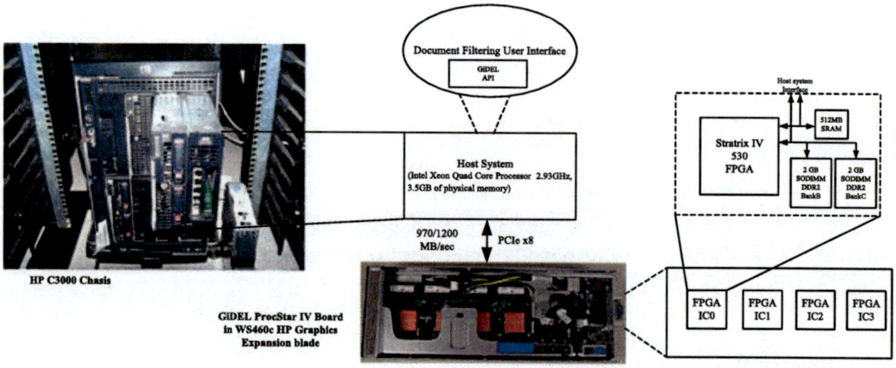

Fig. 11 Block diagram of FPGA platform and photograph of experimental hardware

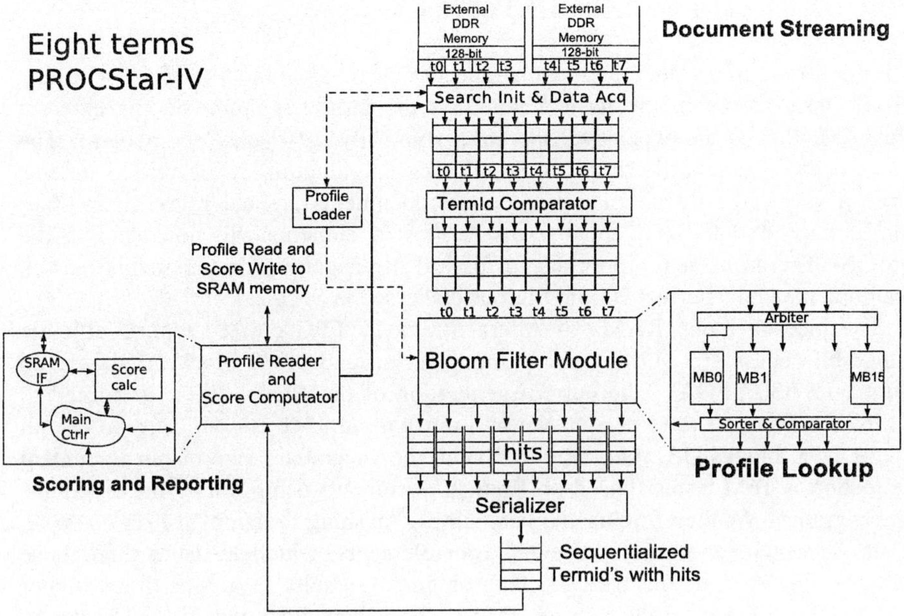

Fig. 12 Overall block diagram of FPGA implementation

4.3 FPGA Implementation Description

Figure 12 presents the overall workflow of our implementation. The input stream of document term pairs is read from the SDRAM via a FIFO. A Bloom filter is used to discard negatives (terms that do not appear in the profile) for multiple terms in parallel. Profile weights are read corresponding to the positives, and the scores are computed for each term in parallel and accumulated to achieve the final score described in (1). Below, we describe the key modules for the implementation: document streaming, profile negative hit filtering, and profile lookup and scoring.

4.3.1 Document Streaming

Using a bag-of-words representation (see Sect. 2.2) for the document, the document stream is a list of *(document id, document term tuple set)* pairs. Physically, the IO width of the FPGA is 64 bits for each SDRAM memory. As the SDRAM is clocked at 667 MHz, the ProcMultiPort FIFO combines two contiguous 64-bit words into a single 128-bit word for each memory bank. The document term tuple $d_i = (t_i, f_i)$ can be encoded in 32 bits: 24 bits for the term id (supporting a vocabulary of 16 million terms) and 8 bits for the term frequency, so the algorithm can process eight terms (256 bits) in parallel. To mark the start and end of a document we insert a marker word (64 bits) followed by the document id (64 bits).

4.3.2 Bloom Filter for Profile Hit Filtering

For every term in the document, the application needs to look up the corresponding profile term to obtain the term weight. As the profile is stored in the external SDRAM, this is an expensive operation (typically 20 cycles per access). The purpose of document filtering is to identify a small amount of relevant documents from a very large document set. As most documents are not relevant, most of the lookups will fail (i.e. most terms in most documents will not occur in the profile). Therefore, it is important to discard the negatives first. For this reason, we implemented a perfect Bloom filter as discussed in Sect. 2.4.

The internal block RAMs of Altera Stratix-IV FPGA that supports efficient single-bit access are M9K memory modules (around 1,280 blocks available on Stratix-IV 530 FPGA). The current generation of the Bloom Filter is limited to 4 Mb; however, in future designs we will move to an 8 Mb Bloom Filter to use all 1,280 M9K blocks [13]. On the other hand, the vocabulary size of our document collection is 16M terms (based on English documents using unigrams, diagrams, and trigrams). We therefore used a very simple "hashing function", $key = elt \gg 2$. Thus we obtain one entry for every four elements, which leads to three false positives out of four on average. This obviously results in a four times higher access rate to the external memory than if the Bloom filter would be 16 Mb. As the number of positives in our application is very low, the effect on performance is limited. The higher internal bandwidth of the MRAMs leads to very fast rejection of negatives. Although the MRAM is fast, concurrent lookups lead to contention. To reduce contention we designed a distributed Bloom filter. The Bloom filter memory is distributed over a large number of banks (16 in the current design) and a cross-bar switch connects the document terms streams to the banks. In this way contention (when multiple tuples access same bank) is significantly reduced.

4.3.3 Profile Lookup and Scoring

Because of the need for lookup, the profile must be implemented as some type of map (dictionary). A hash function is an obvious approach; however, as the size of the profile is not known in advance, it is impossible to construct a perfect hash; imperfect hashes suffer from collisions which deteriorate the performance. When the key space is very large, the contention probability is high.

Our solution is simply to use the term id as the memory address and to implement the weight lookup from the profile as a content-addressable data structure from the on-board SDRAM (Bank A). As the upper limit to the vocabulary size in our case is 16 million terms, we require 128 MB of memory; the PROCStar-IV provides 512 MB of on-board SDRAM. Note that it is possible to increase the vocabulary size to exceed the memory capacity, with a very low performance penalty [14].

As explained in Sect. 2.4, the actual lookup and scoring system is quite straightforward: the input stream is scanned for header and footer words. The header word action is to store the subsequent document ID and to set the corresponding

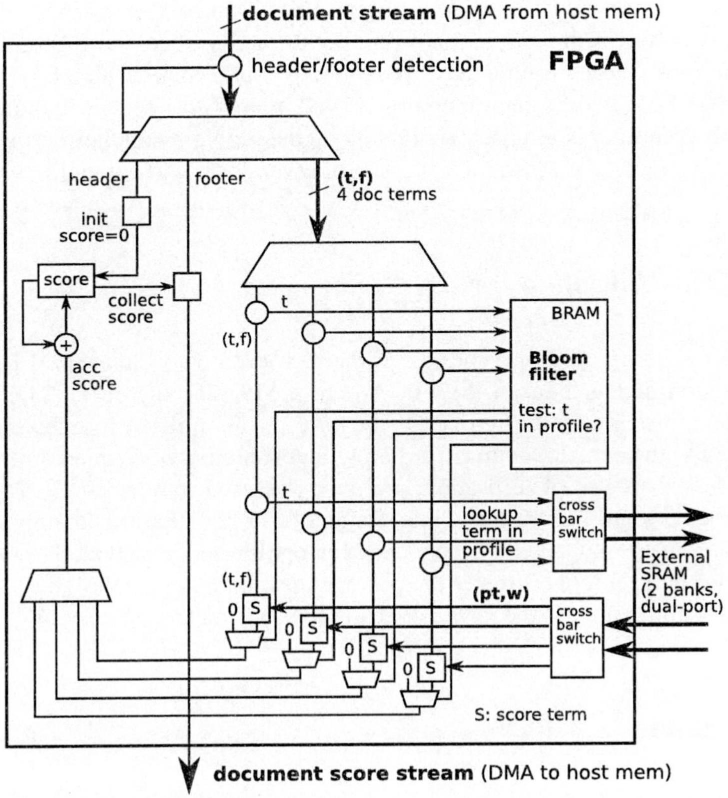

Fig. 13 Implementing profile lookup and scoring

document score to 0; the footer word action is to collect and output the *(document ID, document score)* pair if the score exceeds the threshold. For every two terms in the document, first the Bloom filter is used to discard negatives and then the weights corresponding to positives are read from the SDRAM. The score is computed for each of the terms in parallel and added. The score is accumulated for all terms in the document and finally the score stream is filtered against a limit before being output to the host. Figure 13 summarises the implementation of the profile lookup and scoring.

4.3.4 Discussion

The implementation above leverages the advantages of an FPGA-based design, in particular the memory architecture of the FPGA: on a general-purpose CPU-based system it is not possible to create a very fast, very low-contention Bloom filter to discard negatives. Also, a general-purpose CPU-based system only has a single, shared memory. Consequently, reading the document stream will contend for memory access with reading the profile terms, and as there is no Bloom filter, we have to look up each profile term. We could of course implement a Bloom filter but as it will be stored in main memory as well, there is no benefit: looking up a bit in the Bloom filter is as costly as looking up the term directly. Furthermore, the FPGA design allows for lookup and scoring of several terms in parallel.

4.4 FPGA Utilisation Details

Our implementation used only 17,652 of the 424,960 Logic Elements (LEs) or a 4 % utilisation of the logic in the FPGA, and 4,579,824 out of 21,233,664 bits for a 22 % utilisation of the RAM. Of the 17,652 LEs utilised by whole design on the FPGA, the actual Document Filtering algorithm only occupied 4,561 LEs, which is less than 1 % of utilisation, and rest was used by the GiDEL Memory IPs. The memory utilised for the whole design (4,579,824 bits) was mainly for the Bloom Filter that is mapped on Embedded Memory blocks (i.e. M9k). The Quartus PowerPlay Analyzer tool estimates the power consumption of the design to be 6 W. The largest contribution to the power consumption is from the memory I/O.

5 Evaluation

In this section we discuss our evaluation results. We present our experimental methodology and the data summarising the performance of our FPGA evaluation and comparison with non-FPGA-accelerated baselines, and we conclude with the learnings from our experiments.

5.1 Creating Synthetic Data Sets

To accurately assess the performance of our FPGA implementation, we need to exercise the system on real-world input data; however, it is hard to get access to such real-world data: large collections such as patents are not freely available and governed by licenses that restrict their use. For example, although the researchers at Glasgow University have access to the TREC Aquaint collection and a large patent corpus, they are not allowed to share these with a third party. In this paper, therefore, we use synthetic document collections statistically matched to real-world collections. Our approach is to leverage summary information about representative data sets to create corresponding language models for the distribution of terms and the lengths of documents; we then use these language models to create synthetic data sets that are statistically identical to the original data sets. In addition to

Table 1 Summary statistics from representative real-word collections that we used as templates for our synthetic data sets

Collection	# docs	Avg. doc. len.	Avg. uniq. terms
Aquaint	1,033,461	437	169
USPTO	1,406,200	1,718	353
EPO	989,507	3,863	705

addressing IP issues, synthetic document collections have the advantages of being fast to generate and easy to experiment with, and not taking up large amounts of disk space.

5.1.1 Real-World Document Collections

We analysed the characteristics of several document collections—a newspaper collection (TREC Aquaint) and two collections of patents from the US patent office (USPTO) and the European patent office (EPO). These collections provide good coverage on the impact of different document lengths and sizes of documents on filtering time. We used the Lemur[3] Information Retrieval toolkit to determine the rank frequency distribution for all the terms in the collection. Table 1 shows the summary data from the collections we studied as templates.

5.1.2 Term Distribution

It is well known (see, e.g., [15]) that the rank-frequency distribution for natural language documents is approximately Zipfian:

$$f(k;s;N) = \frac{1/k^s}{\sum_{n=1}^{N} 1/n^s}$$

where f is frequency of term with rank k in randomly chosen text of natural language, N is the number of terms in the collection, and s is an empirical constant. If $s > 1$, the series becomes a value of a Riemann ζ-function and will therefore converge. This type of distribution approximates a straight line on a log–log scale. Consequently, it is easy to match this distribution to real-world data with linear regression.

Special purpose texts (scientific articles, technical instructions, etc.) follow variants of this distribution. Montemurro [16] has proposed an extension to Zipf's law which better captures the linguistic properties of such collections. His proposal is based on observation that in general after some pivot point p, the probability of finding a word of rank r in the text starts to decay much faster than in the beginning. In other words, in log–log scale the low-frequency part of the distribution has a

[3] www.lemurproject.org.

steeper slope than the high-frequency part. Consequently the distribution can be divided into two regions each obeying the power law, but with different slopes:

$$F(r) = \begin{cases} a_1 r + b_1 & r < p \\ a_2 r + b_2 & \text{otherwise} \end{cases}$$

We determine the coefficients a_1, a_2, b_1, b_2 from curve-fitting on the summary statistics from the real-world data collections. Specifically, we use the sum of absolute errors as the merit function combined with a binary search to obtain the pivot. We then use a least-squares linear regression, with χ^2 statistics as a measure of quality (taken from [17]). A final normalisation step is added to ensure that the piecewise linear approximation is a proper probability density function.

5.1.3 Document Length

Document lengths are sampled from a truncated Gaussian. The hypothesis that the document lengths in our template collections have a normal distribution was verified using a χ^2 test with 95 % confidence. The sampled values are truncated at the observed minimum and maximum lengths in the template collection.

Once the models for the distribution of terms and document lengths are determined, we use these models to create synthetic documents of varying lengths. Within each document, we create terms that follow the fitted rank-frequency distribution. Finally, we convert the documents into the standard bag-of-words representation, i.e. a set of unordered *(term, frequency)* pairs.

5.2 Experimental Parameters

Statistically, the synthetic collection will have the same rank-frequency distribution for the terms as the original data sets. Consequently, the probability that a term in the collection matches a term in the profile will be the same in the synthetic collection and the original collection. The performance of the algorithm on the system now depends on

- The size of the collection.
- The size of the profile.
- The "hit probability" of the Bloom filter, i.e. the probability that the profile corresponding to a term has a non-zero weight.

To evaluate these effects, we studied a number of different configurations—with different document sizes, different profile lengths, and different profile constructions. Specifically, we studied profile sizes of 4K, 16K, and 64K terms, the first two are of the same order of magnitude as the profile sizes for TREC Aquaint and EPO as used in our previous work [14] and the third, larger profile was added to investigate the impact of the profile size. We studied two different document collections: 128K documents of 2,048 terms, which is representative for the patent collections, and 512K documents of 512 terms, similar to the Aquaint collection. Note that the total size of the collection is not important for the performance evaluation: for both the CPU and FPGA implementation, the time taken to filter a collection is proportional to its size.

We evaluated four ways of creating profiles. The first way ("Random") is by selecting a number of random documents from the collection until the desired profile size is reached. These documents were then used to construct a relevance model. The relevance model defined the profiles which each document in the collection was matched against (as if it were being streamed from the network). The second type of profiles ("Selected") was obtained by selecting terms that occur in very few documents (less than ten in a million). This is most representative of real-world usage, and we hence focus on these types of profiles in our results. For our performance evaluation purpose, the main difference between these profiles is the hit probability, which was 10^{-5} for the "Random" profiles and 5.10^{-4} for the "Selected" profiles. For reference, we also compared the performance against an "Empty" profile (one that results in no hits).

5.3 FPGA Performance Results

5.3.1 Access Time Measurements

The performance of the FPGA was measured using a cycle counter. The latency between starting the FPGA and the first term score is 22 cycles. For the subsequent terms, the delay depends on a number of factors. We considered four different cases:

- "Best Case": no contention on the Bloom filter access and no external memory access
- "Worst Case": contention on the Bloom filter access and external memory access for every term
- "Full Bloom Filter Contention": contention on the Bloom filter access for every term but no external memory access
- "External Access": no contention on the Bloom filter access, external memory access for every term

These cases were obtained by creating documents with contending/not contending term pairs and by setting all Bloom filter bits to 0 (no external access, which corresponds to an empty profile) or 1 (which correspond to a profile that would contain all terms in the vocabulary).

The results are shown in Table 2. As we read eight terms in parallel, the Best Case (i.e. the case of no contention and no hits) demonstrates that the FPGA implementation does indeed work at I/O rates, i.e. $\Delta t_B = 1$.

Table 2 FPGA cycle counts for different cases

Case	#cycles/term
Best case	0.125
Worst case	27
Full Bloom filter contention	1.2
External access	18

Table 3 Throughput of document filtering application (M terms/s) for (**a**) 256K documents of 4,096 terms and (**b**) 1M documents of 1,024 terms

Profile	System1	System2	FPGA board
(a)			
Random, 4K	269	416	3,090
Random, 16K	245	324	3,090
Random, 64K	223	379	3,090
Selected, 4K	118	232	3,088
Selected, 16K	107	164	3,088
Selected, 64K	82	136	3,088
Empty, 4K	710	1,564	3,090
Empty, 16K	711	1,664	3,090
Empty, 64K	710	1,338	3,090
(b)			
Random, 4K	292	1,118	3,090
Random, 16K	288	1,014	3,090
Random, 64K	253	945	3,090
Selected, 4K	120	309	3,088
Selected, 16K	94	350	3,088
Selected, 64K	72	183	3,088
Empty, 4K	911	2,005	3,090
Empty, 16K	844	1,976	3,090
Empty, 64K	877	1,952	3,090

The most important result in Table 2 is the "Bloom Filter Contention", which shows that in our design $\Delta t_C = 1.2$. The case of "External Access", which means no contention on the Bloom filter and lookup of both terms in the external memory, shows that $\Delta t_S = 18$.

As explained in Sect. 3, the Bloom filter contention depends on the number of Bloom filter banks (8 parallel terms, 16 banks in the current design). The probability for external access depends on the actual document collection and profile, but as the purpose of a document filter is to retrieve a small set of highly relevant documents, this probability is typically very small (<0.00001), as demonstrated by the experiments discussed in the next section. Consequently, the typical performance is determined by the cycle counts for Best Case and Bloom Filter Contention. At a clock speed of 150 MHz this results in a throughput of 772.5 million terms per second (772 MT/s) per FPGA, as seen in Table 3. The model presented in Sect. 3 gives 772.54 Mterms/s as the maximum achievable throughput for a design with $m = 16, n = 8$ with $\Delta t_B = 0.125$, $\Delta t_C = 1.2$, $\Delta t_S = 18$ running at 150 MHz. This proves the validity of the model, as well as showing that the effect of the external memory access is indeed very small for a realistic profile.

5.3.2 Comparison with CPU Reference Systems

Table 3 presents performance results for our FPGA implementation for various workload types. Focusing on a *selected* profile of 16K terms for 128K documents, our measured performance (shown in column 4) is 3,090 million terms/second for the FPGA system (772.5 million terms/second *per FPGA*). Table 3 also shows the sensitivity to various other parameters. The performance of the FPGA design is comparable for different profile sizes and document sizes. However, as expected, the performance varies based on different hit probabilities for different profiles.

To compare the FPGA performance against a conventional CPU, we carried out the experiments discussed in Sect. 5.1 on an optimised multi-threaded reference implementation, written in C++, compiled with g++ with optimisation -O3, and run on two different platforms: *System1* has an Intel Core 2 Duo Mobile E8435, 3.06 GHz and 8 GB RAM, 1,067 MHz bus; *System2* has a quad-core Intel Core i7-2600, 3.4 GHz, with 16 GB RAM, 1,333 MHz bus. A large amount of memory is required for the datasets. We keep the entire data set in memory because the memory I/O is much higher than the disk I/O. While an in-memory approach might not be practical on a CPU-based system, on the FPGA-based system, this is entirely practical as the PROCStar-IV board has a memory capacity of 32 GB. For example, the Novo-G FPGA supercomputer, which hosts 24 PROCStar-IV boards, can support a collection of 768 GB. Note also that the format in which the documents are stored on the disk is a very efficient bag-of-words representation, which is much smaller than the actual textual representation of the document.

5.3.3 Throughput

The throughput results are summarised in Table 3. For example, focusing on one example case, for the selected profile with 64K terms and 1M documents, compared to the 3,090 million terms/second performance achieved by our design, *System2* system achieves 136 million terms/second and *System1* system achieves 82 million terms/second. This translates to a 38× speedup for the FPGA-based design relative to *System1* system and a 23× speedup relative to *System2* system.

Additionally, examining the results for various workload configurations, the FPGA's performance is relatively constant across different workload inputs apart from the "Full" profile. This bears out the rationale for our design: because in general hits are rare, the FPGA works at the speed determined by I/O and Bloom Filter performance. Unlike the FPGA-based design, the *System2* system sees more variation in performance with profile size (degraded performance with increased profile size) and document size (degraded performance with larger documents) and a bigger drop-off in performance between various profile types compared to the FPGA-based design.

Table 4 Power consumption of document filtering application (W) for 1M documents of 1,024 terms, profile: selected, 64K

#threads	System1	System2	FPGA system
0 (idle)	40	67	35
1	67	93	61.5
2	67	107	68
4	67	135	74.5
8	67	141	81

5.3.4 Performance-per-Watt

We also measured the power consumption of our systems using the WattsUp Pro power meter. The measured power consumption for the maximum number of threads (Table 4) is 67 W for *System1*, 141 W for *System2*. In contrast, the FPGA-based design consumes 81 W (35 W of which is consumed by the host system). Clearly, the FPGA-based design achieves improved energy efficiency compared to the baselines. The FPGA-based design achieves energy-efficiency improvements of 31× and 40× for the *System1* and *System2*, respectively. Though the FPGA speedup relative to *System1* was higher than that relative to *System2*, the overall energy-efficiency improvements are very similar. Our results illustrate the potential improvements in performance and energy efficiency relative to traditional baseline implementations.

5.3.5 Performance Versus Cost

We use the cost model presented in [18], which computes the equivalent monthly cost of running a large data center.

The total cost of ownership according to [18] is

$$C_{\text{tot}} = C_{\text{SPC}} + C_{\text{O}}$$
$$C_{\text{SPC}} = C_{\text{S}} + C_{\text{P}} + C_{\text{C}}$$

where C_{SPC} is the cost for space, power, and cooling ($C_{\text{S}}, C_{\text{P}}, C_{\text{C}}$) and C_{O} is the operations cost. We assume that C_{O} is the same for the data centre with and without FPGA acceleration, and leave this factor out of the equation.

The cost for space, power, and cooling C_{SPC} is calculated as

$$C_{\text{SPC}} = \text{uc}_{\text{S}} \cdot \text{nu}_{\text{S}} + (1 + K_1 + L_1 + K_2 L_1) \cdot \text{uc}_{\text{P}} \cdot \text{nu}_{\text{P}} \tag{29}$$

with uc the unit costs (i.e. \$/sqft/month for space, \$/W/month for power/cooling) and nu the number of units (sqft resp. W). The factors K_1, K_2, L_1 represent power and cooling burdening and load, as explained in the paper. Typical values are \$1,000/month for space per rack and \$0.072/W/month for power.

The cost IT_{dep} for the servers is calculated assuming \$100,000 per rack with a 3-year estimated lifetime. The monthly cost for software, licensing and personnel on a per-rack basis is estimated at \$10,000.

Table 5 Performance versus cost

Cost breakdown	CPU	CPU+FPGA
Space	21M$/year	
Power & cooling	52M$/year	29M$/year
IT infrastructure	59M$/year	248M$/year
Total	132M$/year	299M$/year
Performance (single system)	136 Mops/s	3,090 Mops/s
Performance/cost	32 Mops/$	330 Mops/$

To compare the cost of a data center with and without FPGA acceleration, we used the measures results from the previous section, actual prices for the servers and some assumptions similar to Shah and Patel's work:

- The server cost is $2,500 (including the cost of the rack based on 40 servers per rack, as in Shah and Patel's work).
- The measured power consumption for the reference system is 141 W, based on the server used in this work power consumption; the measured power consumption for the FPGA system is 81 W. We assume 10 % overhead for the rack.
- We model the cost of the FPGA system at $8,000. This is not based on current prices for high-end FPGA cards, because these prices tend to be very high because of the low volume involved. A single 10 MW data centre would require about 50,000 cards, which is much more than the total amount of high-end FPGA cards sold today. Instead, we use the cost of an NVIDIA Tesla M2050 as a reference, this is typically $2,500. We believe this is representative as the Tesla board contains a GPU made with a similar process and of comparable gate count as the FPGA, and a similar amount of SDRAM. As the GiDEL PROCStar-IV board combines four FPGAs, it would be slightly cheaper than four separate boards. Our baseline costs comparisons hence use $8,000 for the FPGA system, but we also present results showing the sensitivity to this parameter.
- The license cost for the FPGA tools and the additional cost of developing the application are one-off cost that do not scale with the number of servers, and are therefore negligible for a large data center.
- Finally, based on our measurements, the FPGA-accelerated solution provides an average speed-up of 10× compared to the CPU-only solution. We assume that this speed-up will be exploited to create smaller data centres with the same computational power.

We calculate the increase in costs as a result of adding an FPGA card to each server, and based on these costs we calculate the performance/cost figure using the performance results from the previous section (averaged over all "Random" and "Selected" runs from Table 3). The results are shown in Table 5.

As we can see from the table, the main increase in cost is due to our high estimate for the price of the FPGA cards: the IT infrastructure cost is about 4× higher because of this. The cost for power and cooling decreases with almost a factor of 2 (from 52M$/year to 29M$/year); space costs are not affected. As the FPGA card

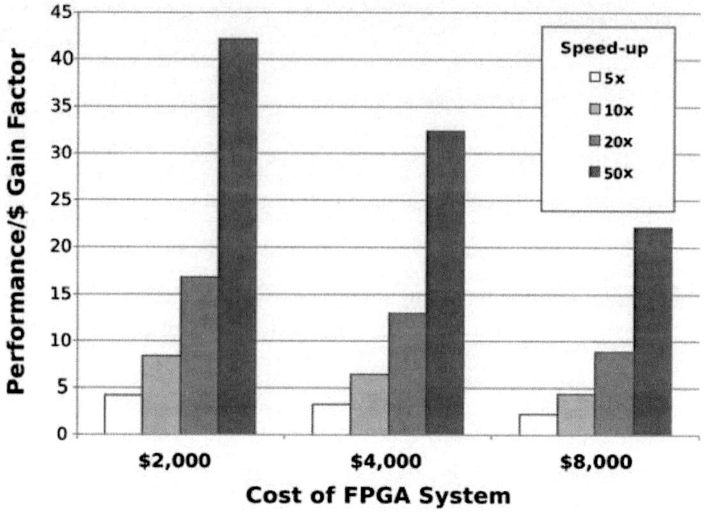

Fig. 14 Performance/cost versus FPGA system cost and performance gains

results in a large increase in performance, the final Performance/Cost figures show that the FPGA-accelerated solution is almost ten times more cost efficient than the traditional one, for the typical case.

To explore the effect of the FPGA system's cost and performance, we varied both parameters as shown in Fig. 14. The graph shows the speedup required for a given FPGA system cost to achieve a given factor improvement in performance/$. The most important points are that even in the worst case the FPGA-based systems still outperform the CPU-only solution, and that the potential for cost reduction is very great.

6 Discussion

Analytical model. In the above sections we have used a preliminary implementation of our proposed design to validate the analytical model. The design does indeed behave in line with the model, for the case of eight parallel terms and a 16-bank Bloom filter. The performance is 772 M terms/s. This design is not optimal; as is clear from the model, for 8 parallel terms, a 16-bank implementation does result in a throughput of 66 % of the I/O rate of 1,200 M terms/s. Using 64 banks would achieve 90 %. Our aim was not so much to achieve optimal performance as to implement and evaluate our novel design and compare it to the analytical model. We therefore decided to limit the number of banks to 16 to reduce the complexity of the design, as the implementation was undertaken as a summer project.

In terms of the analytical model itself, there is some scope for further refinement, in particular for the external access: we currently use a single access time for one and more hits. Just like for the Bloom filter, we can include a fixed cost for concurrent accesses on the external memory. We also want to refine the model to include the effect of grouping terms, i.e. the n parallel terms are usually grouped per two or four depending on the I/O width. This affects the waiting time on contention, as all terms in a single group need to wait before a new group can be fetched. Currently, the model assumes all terms are independent. For the case of two terms this assumption is correct, for more terms there is a slight underestimation of the access time in the case of contention. The counting problem for this case is complicated as it requires enumerating all the possible groupings and working out the effect if one or more accesses per group is in contention.

Improving benefits from FPGA implementation. While our results clearly illustrate the potential benefits from FPGA-based acceleration, they can potentially be improved further. The current design uses a 16-bank Bloom filter, which is not optimal for scoring eight parallel terms. Extending the design to 64 banks would increase the throughput by almost 40 % (from 772 to 1,053 MT/s). Furthermore, the current Bloom filter combines four terms per bit. We can double the Bloom filter size (i.e. 8 Mb), leading to two terms per bit, which will reduce the rate of false positives accordingly. For the given application of scoring a known collection of documents, we could also reorder the terms in each document to reduce contention. Combined, these improvements can potentially result in a throughput very close to the I/O limit of 1,200 MT/s.

Comparison of FPGA versus other alternatives.

ASIC Bloom Filter: As mentioned earlier, the main performance improvement from our approach over a general-purpose CPU is that we can use bit-accessible Bloom filters to discard negatives. If we could create an ASIC for this purpose it could potentially have an order-of-magnitude better performance. It is important to note that the FPGA runs at a very low clock speed (150 MHz), which is the main reason why it is a low power technology. Consequently, the FPGA implementation can only win by having more parallelism or by having a better memory architecture. The Bloom filter is just that: a better memory-optimised architecture compared to the CPU cache. In our particular implementation, we have 8× parallelism (per FPGA) for accessing the document collection, and 16× for accessing the Bloom filter. The former is dictated by the IO width and DRAM clock speed; the latter can still be improved as explained above.

GPGPU Implementation: The same observations as for the CPU also apply to GPGPUs, with different parameters: the GPU clock speed is usually only about 2× lower than the CPU (instead of 20× for the FPGA). Memory I/O is comparable. Again, the crucial difference, and the reason why the FPGA can still outperform the GPU for this application, is in the memory architecture. The GPGPU has a scratch pad per streaming multiprocessor. However, the size of this memory is typically

16 KB, up to 128 KB for high-end GPUs. This is much too small to store a Bloom filter of the size that we require (4 Mb). So we can't implement a high-bandwidth Bloom filter on the GPU local memory to discard negatives. Furthermore, although the GPU has a large number of data parallel threads inside each multiprocessor, they contend for the global memory access, so this becomes the bottleneck. That means that there is little benefit in the large number of parallel cores provided by the GPU for this application.

Avenues for future work. Considering the wide potential application domain of information filtering, and the need for power- and cost-effective system architectures, our future work will explore a number of different avenues: novel algorithms, low-power system architectures and high-level FPGA programming. We see the latter as crucial: the key concerns which hamper the adoption of FPGAs are the programming complexity and the lack of standardised APIs. The former can be addressed by the use of high-level languages such as the C-based Impulse-C, Catapult-C and others [19, 20] or the MORA framework [21], which offers a C++ API for high-level FPGA application programming. To address the latter, it is important to put forward a standard. OpenCL [22] is the emerging industry standard for programming of multicore and manycore devices, in particular GPUs. It provides a flexible API for host–device communication and a C-like language for device kernel programming. Adding FPGAs to the set of platforms supported by OpenCL is therefore very attractive.

7 Conclusion

In this chapter we have presented our work on FPGA-based high-performance real-time Information Filtering applications. Our novel design uses a low-latency perfect Bloom filter to eliminate unnecessary accesses to external memory. We have also presented analytical model for the throughput of the application. This combinatorial model takes into account the access times to the Bloom filter and the external memory, the access probability and the probability and cost of contention on the Bloom filter. We show an excellent agreement between the model and the actual measurements. The approach followed and the intermediate expressions are applicable to a large class of resource sharing problems.

The analysis of the system performance clearly demonstrates the potential of the design for delivering high-performance real-time search: we have shown that the system can in principle achieve the I/O-limited throughput of the design. Our current, sub-optimal implementation works at 66 % of its I/O rate and this already results in speed-ups of up to a factor of 20 at 125 MHz compared to a CPU reference implementation on a 3.4 GHz Intel Core i7 processor. Our analysis indicates how the system should be dimensioned to achieve I/O-limited operation for different I/O widths and memory access times.

Finally, our work demonstrates of the potential of a system consisting of a high-performance FPGA board with a low-power CPU host for use in large data centres for processing information filtering or similar tasks: our results show that this technology can lower the total cost of ownership of such a data centre with an order of magnitude.

Acknowledgements The authors acknowledge the support from HP, who hosted the FPGA board and provided funding for a summer internship. In particular, we'd like to thank Mitch Wright for technical support and Partha Ranganathan for managing the project.

We'd like to acknowledge Anton Frolov who implemented the synthetic document model.

Wim Vanderbauwhede wants to thank Dr Catherine Brys for fruitful discussions on probability theory and counting problems.

References

1. C.L. Belady, In the data center, power and cooling costs more than the it equipment it supports. Electronics Cooling **13**(1), 24 (2007)
2. J.M. Kaplan, W. Forrest, N. Kindler, Revolutionizing data center efficiency, in *Uptime Institute Symposium* (2008)
3. W. Vanderbauwhede, L. Azzopardi, M. Moadeli, in *19th IEEE International Conference on Field Programmable Logic and Applications (FPL09)* (IEEE, New York, 2009), pp. 417–422
4. L. Azzopardi, W. Vanderbauwhede, M. Moadeli, in *Proceedings of the 32nd International ACM SIGIR Conference on Research and Development in Information Retrieval (SIGIR09)* (ACM, New York, 2009), pp. 664–665
5. Google n-gram project (2010), http://ngrams.googlelabs.com/ngrams/info
6. Lemur, The Lemur toolkit for language modeling and information retrieval (2005), http://www.lemurproject.org/. Accessed 25th April 2012
7. V. Lavrenko, W.B. Croft, Relevance based language models, in *Proceedings of the 24th Annual International ACM SIGIR Conference on Research and Development in Information Retrieval* (ACM, New York, 2001), pp. 120–127
8. V. Kindratenko, R. Wilhelmson, R. Brunner, T.J. Martinez, W. Hwu, High-performance computing with accelerators. Comput. Sci. Eng. **12**(4), 12–16 (2010)
9. GiDEL Ltd, PROCStar III, Data Book (2009)
10. B.H. Bloom, Commun. ACM **13**(7), 422 (1970). doi:http://doi.acm.org/10.1145/362686.362692
11. G. Andrews, K. Eriksson, *Integer Partitions* (Cambridge University Press, Cambridge, 2004)
12. C. Chen, K. Koh, *Principles and Techniques in Combinatorics* (World Scientific, Singapore, 1992)
13. Stratix iv handbook, http://www.altera.com/literature/hb/stratix-iv/stratix4_handbook.pdf. Accessed 25th April 2012
14. W. Vanderbauwhede, L. Azzopardi, M. Moadeli, in *International Conference on Field Programmable Logic and Applications, 2009 (FPL 2009)* (IEEE, New York, 2009), pp. 417–422
15. R. Losee, J. Am. Soc. Inf. Sci. Technol. **52**(12), 1019 (2001)
16. M. Montemurro, Phys. A Stat. Mech. Appl. **300**(3–4), 567 (2001)
17. W. Press, *Numerical Recipes: The Art of Scientific Computing* (Cambridge University Press, Cambridge, 2007)
18. C. Patel, A. Shah, Hewlett-Packard Laboratories Technical Report (2005)

19. J. Xu, N. Subramanian, A. Alessio, S. Hauck, in *2010 18th IEEE Annual International Symposium on Field-Programmable Custom Computing Machines* (IEEE, New York, 2010), pp. 171–174
20. B. Holland, M. Vacas, V. Aggarwal, R. DeVille, I. Troxel, A.D. George, Survey of C-based application mapping tools for reconfigurable computing, in *Proceedings of the 8th International Conference on Military and Aerospace Programmable Logic Devices (MAPLD'05)* (NASA, Washington, 2005)
21. W. Vanderbauwhede, M. Margala, S.R. Chalamalasetti, S. Purohit, A C++-embedded domain-specific Language for programming the MORA soft processor array, in 2010 *21st IEEE International Conference on Application-specific Systems Architectures and Processors (ASAP)* (IEEE, Washington, 2010), pp. 141–148
22. J. Stone, D. Gohara, G. Shi, Comput. Sci. Eng. **12**(3), 66 (2010). doi:10.1109/MCSE.2010.69

High-Performance Data Processing Over *N*-ary Trees

Valery Sklyarov and Iouliia Skliarova

Abstract An *N*-ary tree ($N \geq 2$) is a connected graph that does not contain cycles and has up to N children for any node. It can be used efficiently to represent data in well-structured hierarchical clusters and to process the data through the parent–child relationships. Several branches of a tree can be handled concurrently, the data hierarchy is described explicitly, and recursion can easily be applied. Thus this model is very appropriate for parallel high-performance computations in areas such as data processing (e.g. sort and search), priority queue management, combinatorial searches and so forth. *N*-ary trees have been profoundly studied (primarily for $N = 2$) and are supported by software libraries. FPGAs have large embedded dual-port memories with programmable data width for different ports, advanced logic capabilities, and a large potential for parallelism and these features enable *N*-ary trees with data operations associated with their nodes to be represented more compactly and processed more efficiently in FPGAs than in software. A number of recent research efforts are dedicated to high-performance computations in electronic circuits and systems without the direct use of processing elements, which undoubtedly introduce many constraints (e.g. pre-defined operand sizes, fixed instruction sets, limited concurrency and parallelism). This chapter presents recent advances in this area and is composed of four basic parts: (1) an overview of *N*-ary trees, their applications, and potential varieties; (2) a discussion of common techniques for implementing and processing *N*-ary trees in hardware, including their representation in memory, models of computations and algorithms; (3) a description of hierarchical finite-state machines (HFSMs) with extended capabilities (with datapath, in particular) that enable *N*-ary trees to be processed in hardware and provide support for parallelism, hierarchy and recursion; (4) examples, practical applications, experiments and

V. Sklyarov (✉) • I. Skliarova
Department of Electronics, Telecommunications and Informatics, University of Aveiro/IEETA/HIPEAC, Aveiro, Portugal
e-mail: skl@ua.pt; iouliia@ua.pt

W. Vanderbauwhede and K. Benkrid (eds.), *High-Performance Computing Using FPGAs*, DOI 10.1007/978-1-4614-1791-0_8,

comparisons of HFSMs. The last part shows that the circuits that have been implemented are faster than the alternatives, and this conclusion is confirmed by examples and experiments in several application areas.

1 Introduction

Tree-like structures (TLSs) have a wide variety of applications. They are frequently used to support search algorithms [33], to sort data [3, 6], to manage priorities in queues [6], to solve problems in combinatorial optimization [7, 57], to represent Boolean [2] and elementary [26] functions, to simplify image filtering and segmentation [28], etc. Processing TLSs in software has been studied profoundly and is well supported by standard libraries [3]. Advances in microelectronics, embedded systems and application-specific hardware accelerators allow functions and data structures currently in software to be mapped to hardware. Although there is some progress in this area, to our knowledge alternatives to standard software libraries have not been developed in hardware up to now. Design effort has been focused mainly on accelerating particular applications, such as data sorting [25], and Boolean constraints propagation in combinatorial search [7]. Using and taking advantage of application-specific circuits in general and FPGA-based accelerators in particular for such purposes has a long tradition. A number of research works are targeted towards the potential of advanced hardware architectures and recently developed systems such as graphical processing units (GPUs) and design techniques such as systems and networks on chip. For example, the system [11] solves sorting problem over multiple hardware shading units achieving parallelization through SIMD operations on GPUs. The potential use of FPGAs for data processing has been studied within projects [12, 14, 24] that implement traditional CPU tasks on programmable hardware. The advantages of customized hardware as a database co-processor are explored in several publications (e.g. [25]).

The most requested and most frequently used TLS is a binary tree [2,3,6,33,57]. However, N-ary ($N > 2$) trees and incomplete trees can be more advantageous for a number of practical applications. This chapter reviews the features of TLSs and suggests a novel technique for processing TLSs in hardware. The basic topics that the chapter covers can be summarized as follows:

- The optimization of algorithms through advanced processing of TLSs and parallelization.
- The processing of N-ary ($N > 2$) and incomplete trees.
- A new way of using previously developed methods and models, such as the tree-walk tables proposed in [7] and combining a TLS with sorting networks.
- A computational model based on HFSMs.
- The implementation of software methods such as recursive calls in hardware not directly supported by common hardware description languages.
- A demonstration of the effectiveness of the proposed techniques based on prototyping in FPGA, numerous experiments and comparisons.

The remainder of this chapter is in six sections. Section 2 describes how trees can be used to solve different computational problems. Section 3 suggests various representations for trees in memory. Section 4 discusses computations over trees. Section 5 is dedicated to synthesizable HFSMs with datapath that is considered to be the core of computations. Section 6 describes implementations, experiments and comparisons and discusses the results. The conclusion is given in Sect. 7.

2 Tree-Like Structures and Their Use for Solving Computational Problems

2.1 Basic Definitions

A tree is a connected graph that does not contain cycles [3, 6, 33]. Figure 1a gives an example of a binary tree that can be seen as a structure that represents some operations (e.g. A, B, C, D, E) associated with tree nodes (e.g. a, b, c, d, e) and relationships between the operations shown by tree edges (e.g. α, β, χ, δ). The tree with associated operations is called a TLS. Note that in general the considered trees are N-ary ($N \geq 2$).

Operations associated with the nodes can be combinational (executing during one clock cycle) or sequential (executing during more than one clock cycle). Besides, such operations can be compositional, i.e. composed of other operations. Thus, we can associate with a node either a statement in software language or a procedure (subroutine). Branches of a tree can be processed in parallel (see Fig. 1b).

For many practical problems we need to traverse TLS (to visit each node) or to search over TLS (to find out a node that satisfies some optimization criteria). For a binary tree (BT) this (either traversal or search) can be done recursively beginning with the root node and using the following general sequence of operations:

TREE-WALK SEGMENT (**TWS**)
if (node of BT exists) then

-- if required execute operations associated with the node
-- recursively execute TWS for one sub-tree (left sub-tree).
-- if such sub-tree does not exist, then skip this step;
-- if required execute operations associated with the node

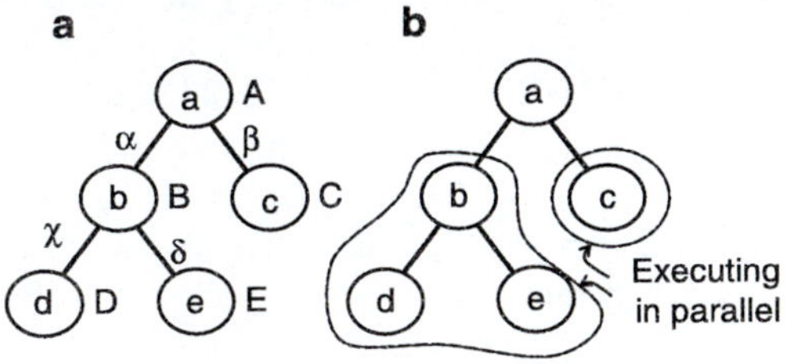

Fig. 1 Tree-like structures: (**a**) operations; (**b**) parallel execution

Fig. 2 Sorting (**a**) and resorting (**b**) integers. Reapplying sorting criteria (**c**)

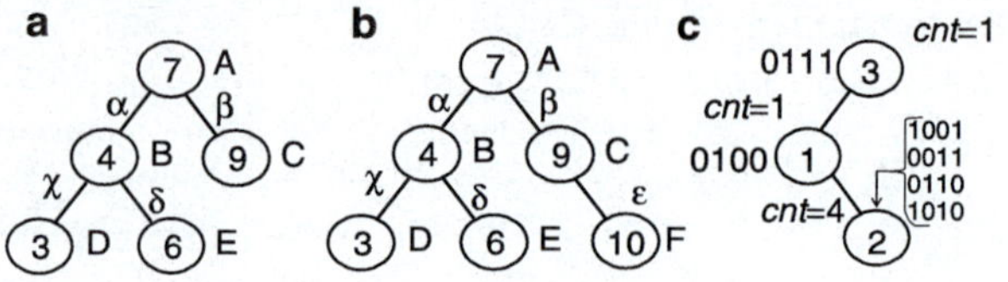

-- recursively execute TWS for the other sub-tree (right sub-tree).
-- if such sub-tree does not exist, then skip this step;
-- if required execute operations associated with the node

end if;

Since the target of the chapter is the implementation of computations in hardware circuits, we will use descriptions (such as the TWS above) in VHDL notation replacing statements and conditions with comments. Note that recursive calls are not directly supported in VHDL and they will be realized using HFSMs described in Sect. 5. We use recursive algorithms because of their clarity and compactness [42]. Alternatively, iterative algorithms can also be implemented relying on the proposed technique.

2.2 *Data Sort (Binary Trees)*

Suppose that the nodes of the tree contain three fields: a pointer to the left sub-tree, a pointer to the right sub-tree, and a value (e.g. an integer or a pointer to a string). The nodes are maintained so that at any node, the left sub-tree only contains values that are less than the value at the node, and the right sub-tree contains only values that are greater. Such *binary tree* can easily be built and traversed permitting data to be sorted and resorted applying and reapplying various criteria [3].

Let us consider examples. Figure 2a depicts a tree that was built for the following sequence of integers: 7, 9, 4, 6, 3 applying criterion "sorting by value". Operations A, B, C, D, E compare a newly arriving value and allocate a new child node either on the left or on the right-hand side of the analyzed node. If a newly arrived value is equal to the value in the analyzed node, then an associated counter (*cnt*) is incremented. Suppose a new integer 10 is arriving. Resorting is executed very fast because just a new node is allocated on the tree (see Fig. 2b). The result of ascending sort is produced by traversing the tree from left to right: 3, 4, 6, 7, 9, 10 [3]. Similarly we can get the result of descending sort traversing the tree from right to left.

Let us now apply a new criterion of sorting, such as that the number of ones in binary codes of the integers looking at each level of the tree from left to right (i.e. 7, 4, 9, 3, 6, 10). The new tree is shown in Fig. 2c. Now the binary codes (7-0111, 4-0100, 9-1001, 3-0011, 6-0110, 10-1010) are considered instead of the integers. Sorting is implemented by the number of ones and such counting operation is frequently required for solving combinatorial problems [57].

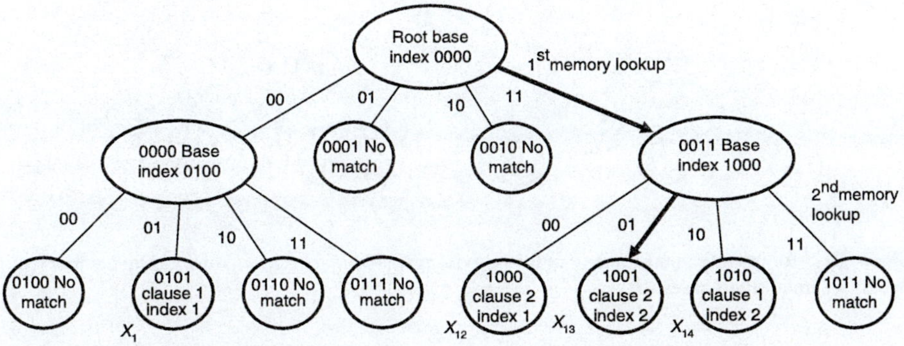

Fig. 3 Clause index tree walk from [7]

An important advantage of trees compared to alternative models is an opportunity of rapid adaptation to potential modifications, such as that considered in Fig. 2.

2.3 *Combinatorial Problems (N-Ary Trees)*

Many practical applications require combinatorial algorithms [7, 13, 35, 57] to be executed. Let us consider the satisfiability problem (SAT) and a search tree that was used in [57]. The root of the tree corresponds to the initial situation in solving the regarded task. Edges of the tree lead to child nodes representing simplified Boolean formulae. Every pair of child nodes permits to remove one variable from the formula assigning it 0 for one child and 1 for another child.

In other applications, such as [7], the tree permits to find out clauses containing a given variable. Figure 3 demonstrates an example [7]. Clauses in a given Boolean formula are partitioned into groups that are processed by multiple inference engines in parallel (one group per engine). Each engine is pipelined and has two stages. The first stage is called clause index walk, which takes a variable assignment from a given queue and searches for a clause with the variables using an N-ary tree ($N \geq 2$) (all necessary details can be found in [7]). Suppose the variable index has a width of M (so, 2^M variables can be handled), and every non-leaf tree node has $N = 2^m$ children (i.e. the tree will be M/m deep). Given a non-leaf node, the address of its leftmost child in the tree-walk table is called the base index [7] of the node. The rest of the children are ordered sequentially, following the leftmost child. Therefore, to locate the ith child, the index can be calculated by adding i to the base index. If a child is not associated with any clause, a *no match* (-1 in [7] and 0 in our case) tag is stored in the entry. If for a node, all of its $N = 2^m$ children have *no match*, then the tree is not expanded and a *no match* tag is stored in the node itself.

An example from [7] is shown in Fig. 3 for $M = 4$, $m = 2$ and $N = 2^m = 4$. There are two clauses, $(X_1 \vee X_{14})$ and $(X_{12} \vee X_{13})$, with variable indexes 0001, 1100, 1101

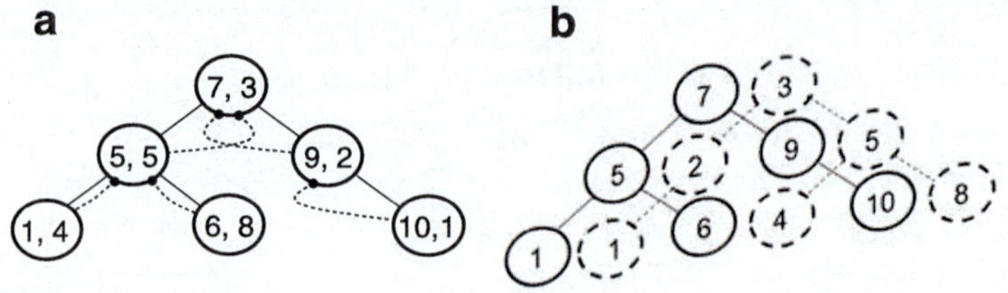

Fig. 4 TLS for priority management: multi-purpose merged binary tree (**a**); unwinding the merged tree to separate binary trees (**b**)

and 1110. Suppose the input variable is 1101. The base index of the root node is 0000 and the first two bits of the input are 11. The table index is the sum of two: $0000 + 11 = 0011$. Using this table index, the first memory lookup is conducted by checking the 0011 entry of the table. This entry shows that the next lookup is an internal sub-tree root with the base index 1000. Following this base index, adding it to the next two bits of the input 01, we reach the leaf node $1000 + 01 = 1001$. This leaf node stores the variable association information; in this case, the variable is associated with X_{13} of the clause two. Note that we process now *N-ary trees* and $N > 2$.

2.4 Priority Management (Merged Binary Trees)

Let us consider a system whose functionality is controlled by a sequential flow of external instructions. The number of instructions is not known in advance and the input instruction transfer rate is not the same as the instruction processing speed in the system. Thus, it is necessary to use input buffering. For some practical applications the instructions have to be processed non-sequentially. Each instruction is provided with additional field(s) indicating priority or some other parameters required for the proper selection of instructions. A priority buffer is a device that stores an incoming (sequential) flow of instructions (or other data) and allows outputs to be selectively extracted from the buffer for processing.

Buffers of such type are needed for numerous practical applications [9, 18, 19, 21, 50, 52]. For example, in [21] a priority buffer (PB) stores pulse height analyzer events. Real-time embedded systems [9] employ priority preemptive scheduling in which each process is given a particular priority (a small integer) when the system is designed. At any time, the system executes the highest-priority process. One of the proposals of [50] is to create a smart agent scanning and selecting data according to their priority. We believe that priority management might be also very useful for such applications that are described in [32].

One potential technique for hardware implementation of a PB is proposed in [44] and it defines the PB as a multi-purpose TLS that is a merged binary tree. Let us consider the graph in Fig. 4 that permits to find out instructions by their priorities. Depending on situation either the first (p_1) or the second (p_2) priority criterion (p_1, p_2) is taken into account. The graph in Fig. 4a is built for the following sequence of priorities: (7,3), (9,2), (5,5), (6,8), (1,4), (10,1). Solid edges correspond to binary tree for the first priority criterion and dashed edges—for the second priority criterion. Thus, the tree in Fig. 4a is a merger of two trees shown in Fig. 4b for the following two sets of priorities 7, 9, 5, 6, 1, 10 and 3, 2, 5, 8, 4, 1. The TLS permits to manage priorities easier [44] than other known methods.

2.5 *Multi-Level Data Sort (K-Trees)*

Suppose we need to sort M-bit data items. Let us apply the method described in Sect. 2.2 to $(M-K)$ most significant bits and sort the remaining K bits using another method. In this case up to 2^{M-K} data items will be sorted by a binary tree and up to 2^K data items associated with the nodes of the tree will be sorted differently applying, for example, sorting networks [6, 15] not requiring control flow of instructions or branches. Besides, sorting networks are parallel in nature and in many cases they can be implemented by combinational circuits. However, sorting networks are suitable in hardware for relatively short sequences of data whose length is known a priori [25]. We suggest combining TLSs with sorting networks (or with some other blocks for fast processing of the attached to the nodes data items) based on two potential techniques that are: (1) using K-trees; and (2) tree-walk that is similar to [7]. Both techniques are explained below on examples.

K-tree is a tree with up to 2^K (data) items associated with each node. Suppose $M = 6$ and we need to sort the following integers: 49, 7, 58, 48, 5, 51, 50, 59, 54, 57, 55 (arriving sequentially from left to right). Figure 5a depicts the binary tree for sorting described in Sect. 2.2. Let us replace the integers with their binary codes and let us sort just data items corresponding to $M-K$ most significant bits. For $K = 2$ $(M-K) = 4$-bit values have to be sorted and they are shown in italic font: *1100*01, *0001*11, *1110*10, *1100*00, *0001*01, *1100*11, *1100*10, *1110*11, *1101*10, *1110*01, *1101*11. Figure 5b presents the resulting binary tree. There are up to $2^K = 4$ integers associated with each node of the tree. For example, there are four integers associated with the root *1100*. They correspond to codes *1100*01, *1100*00, *1100*11, *1100*10 in the sequence given above and they are 49, 48, 51, 50 with the less significant bits 01, 00, 11, 10 respectively. There are four different integers for $M-K = 4$ most significant bits: (1) *1100* (12); (2) *0001* (1); (3) *1110* (14); and (4) *1101* (13) and they are sorted using the tree (see Sect. 2.2). Up to 2^K integers associated with each node of the tree in Fig. 5b are handled using sorting networks, which will be discussed later in Sect. 4.

The tree-walk technique [7] can be used to provide faster traversal of N-ary trees ($N > 2$) instead of binary trees. Let us consider four integers from the example above: 1100 (12), 0001 (1), 1110 (14), 1101 (13). As you can see, the values 12, 1, 14, 13 are the same as indexes of variables in the example of Sect. 2.3 and in [7]. This is done to demonstrate common features of different problems that might be solved

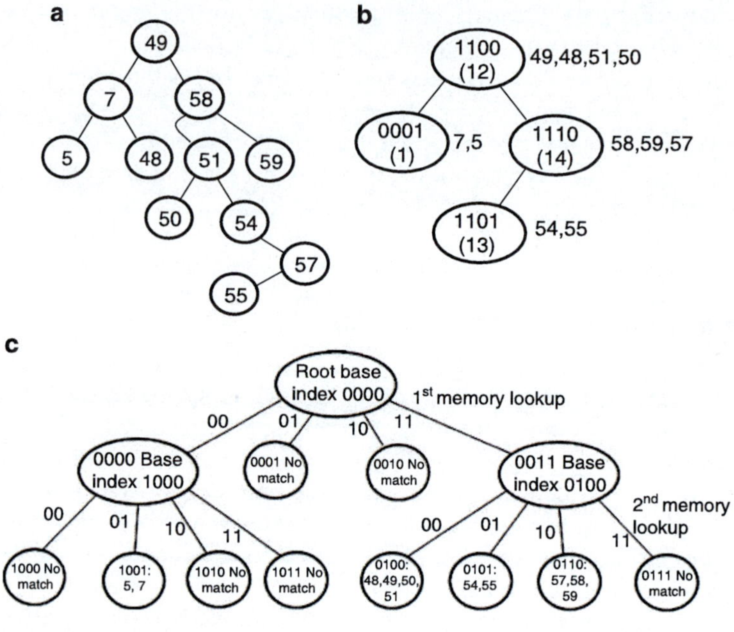

Fig. 5 TLSs for sorting: binary tree (**a**), *K*-tree (**b**), *N*-ary tree (**c**)

over TLS. Suppose, $M - K = 4$, $m = 2$. Figure 5c depicts *N*-ary tree ($N = 2^m = 4$) for our example. Visually the tree in Fig. 5c is more complicated than the trees in Fig. 5a, b. However, Fig. 5c permits to represent *N*-ary trees in memory in a very compact form and to minimize the number of memory accesses.

2.6 TLS and Address-Based Data Sort

The main idea of the address-based data sort [47] is rather simple. As soon as a new data item is received, its value V is considered to be an address of memory to record a flag (1). We assume that memory is zero filled at the beginning. Since data in memory are placed in the sorted order with some "holes" between the values "1" (indicating the availability of data), sorting is done by reading data in increasing or decreasing addresses. This method is obviously simple and effective, but there are some problems listed below.

First, the size of memory is large. Suppose, we need M-bit data to be sorted and $M = 32\ldots64$. Thus, the number of one-bit words becomes $2^{32}\ldots2^{64}$. Second, if we sort M-bit data, many often, the number of input data items Q is significantly less than 2^M ($Q \ll 2^M$) especially for large values of M. Thus, we can expect a huge number of empty positions in memory space without data (i.e. "holes" with zeros).

This situation is somehow similar to the SAT problem where we want to consider a formula with Q clauses and M variables and $Q \ll 2^M$. Thus, we can apply some ideas inherited from the SAT such as the tree-walk tables proposed in [7]. Sorting using N-ary trees ($N > 2$) that is based on tree-walk tables is described in [47].

3 Representation of Tree-Like Structures in Memory

In hardware implementation TLSs have to be kept in memory. Potential representations of trees in software (namely array and pointer-based representations) are discussed in [3]. We briefly describe array-based technique and its implementation in hardware in Sect. 3.1. Then in Sects. 3.2–3.4 we propose alternative representations and additions allowing N-ary and K-trees to be stored.

3.1 Array-Based Representation

Figure 6 demonstrates two ways for array-based representation of the tree from Fig. 1a.

Each tree node in Fig. 6a is placed in one memory word containing a data item (such as a, b, c, d, e) and two addresses: *LA*—address of the left sub-tree and *RA*—address of the right sub-tree. Since address 0 cannot appear in the fields *LA* and *RA*, this value (i.e. 0) is used to indicate absence of sub-trees. Figure 6b lists *LA* and *RA* in memory words at addresses following the relevant data item.

3.2 Coding Nodes in K-Trees

From Fig. 5b, c we can see that more than one item can be associated with each node. For example, in Fig. 5b four items 49, 48, 51, 50 are associated with the node 1100 (12). To represent such items we will use positional encoding with 2^K bits for

a

address	node	LA	RA
0	a	α=1	β=2
1	b	χ=3	δ=4
2	c	0	0
3	d	0	0
4	e	0	0

b

address	0	1	2	3	4	5	6	7	8	9	10	...
node, LA, RA	a	3	6	b	9	12	c	0	0	d	....	

Fig. 6 Array-based representation of tree-like structures in two possible formats (**a**) and (**b**)

Fig. 7 Tree-walk table for Fig. 5c

Data/{LA,RA}	1000	-1(0)	-1(0)	0100	1111	0011	0111	-1(0)	-1(0)	0101	-1(0)	-1(0)
$Address_2$	0000	0001	0010	0011	0100	0101	0110	0111	1000	1001	1010	1011
$Address_{10}$	0	1	2	3	4	5	6	7	8	9	10	11

the relevant K-tree. In such code $Bit_i = 1$ if and only if there exists an item associated with the considered node in which K less significant bits are i_2 (i.e. binary code of i), else $Bit_i = 0$. For example, there are two items 7 and 5 associated with the node 0001(1) in Fig. 5b. Binary code of 7 is *0001*11 and $K = 2$ less significant bits are $11_2 = 3$. Thus, bit 3 in 2^K code is 1. Analogously, bit 1 is equal to 1 and other bits are equal to 0: 0101 (the bits are numbered beginning with 0 from left to right).

3.3 *Tree-Walk Tables*

Let us represent in memory an N-ary tree that is $(M - K)/m$ deep and $N = 2^m$. The method of coding will be explained on an example in Fig. 7 for the tree in Fig. 5c.

We remind that TLS in Fig. 5c was built for the following integers: 49, 7, 58, 48, 5, 51, 50, 59, 54, 57, 55. The first data item is $49_{10} = \mathit{1100}01_2$, and we will sort $M - K = 4$ most significant bits shown in italic. The first $2^m = 4$ words of tree-walk table (with the addresses 0000, 0001, 0010, 0011) have to be used for pointing out child sub-trees from the root of N-ary tree and $N = 2^m = 4$ (see also Sects. 2.3, 2.5). Thus, for the first data item ($49_{10} = \mathit{1100}01_2$) the first free address (0100) can be used as a base index for the child sub-tree 11. The base index for the child sub-tree 11 is 0011 (i.e. *base address of the root* +11 [7]). So, at the address 3 (0011) we will write the value 0100 and at the address $0100 + 00 = 0100$ we have to code the value 49, i.e. to write 1 in the Bit_1: 0100 (the details of coding were given in Sect. 3.2). The second data item is $7_{10} = \mathit{0001}11_2$. Using similar steps we write the first free address 1000 (the addresses 0100, 0101, 0110, 0111 have already been reserved for the sub-tree 0011) at the address 0000 (i.e. *base address of the root +00* [7]). At the address 1001 (base address$=$ $1000 + 01$) the value 0001 has to be written. The third data item is $58_{10} = \mathit{1110}10_2$. Thus, at the address $0100 + 10 = 0110$ the value 0010 has to be recorded. The fourth data item is $5_{10} = \mathit{0001}01_2$. Thus, at the address $1000 + 01 = 1001$ the value 0101 has to be written (a new bit need to be changed in the previous value 0001).

The steps described permit to fill in the tree-walk table in Fig. 7. The size of the resulting table is 12 words * 4 = 48 bits; the size of arrays for Fig. 5a, b is 154 and 32 bits, respectively.

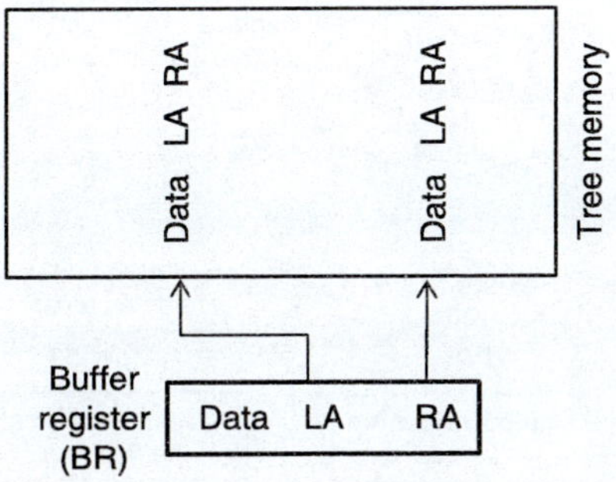

Fig. 8 Using dual-port memory

3.4 Dual-Port Memories

An additional optimization can be achieved for binary trees if we use dual-port memories permitting two words to be accessed simultaneously through *LA* and *RA* of a buffer register (BR) [23]. Thus, we can store in each word *data*, *LA* and *RA* for the left and for the right nodes (see Fig. 8).

As a result, we are capable to make two tests at the same time, namely check if a left/right sub-tree exists for an item in the BR and if there is a left/right sub-tree for any children of the node in BR. This permits to reduce time required for traversal or search [23].

4 Computations Over Tree-Like Structures

The following three types of computations are discussed:

- Sequential computations.
- The use of accelerators executing computations associated with nodes of K-trees.
- Parallel computations.

4.1 Sequential Computations Over Binary Trees

We base the proposed sequential computations on two general procedures that are: (1) building a TLS and (2) traversing/search over the TLS.

In case of simple sorting (see Fig. 5a) the tree is built using the following pseudo code:

-- Receive a new data item i
BUILD TREE SEGMENT (**BTS**)
if (j is less than the value in the node) then
    -- create a new left child or call BTS for the left child
else (j is greater than the value in the node) then

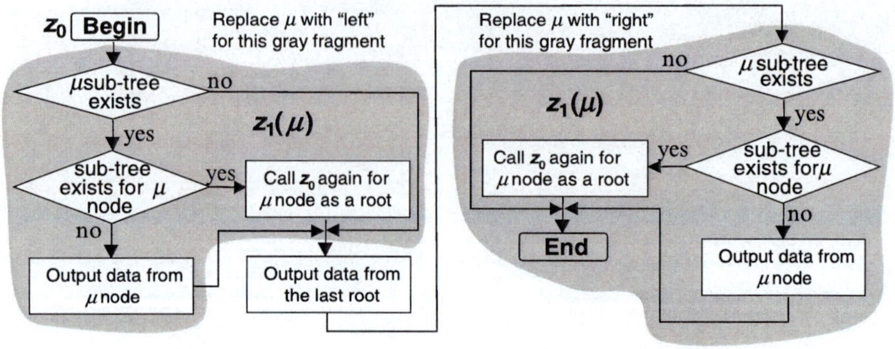

Fig. 9 Traversing algorithm

-- create a new right child or call BTS for the right child
else -- increment a counter associated with the node
end if;

The tree is traversed with the aid of TREE-WALK SEGMENT from Sect. 2.1. Using dual-port memories permits traversal of the tree to be performed faster [23]. There are two basic fragments in the proposed algorithm z_0 (see Fig. 9) designated $z_1(\mu)$ and shown in grey. Initially, the BR (see Sect. 3.4) holds the root of the sorting tree. The top module $z_1(\mu = \text{left})$ examines left sub-trees. If a left sub-tree (node) exists, then it is checked again to determine whether the left sub-tree also has either left or right sub-trees (nodes). This is possible for dual-port memories. If there is no sub-tree from the left node, then the value of the left node is the leftmost data value and can be output as the smallest. In the last case the node in the BR holds the second smallest value and the relevant data value is sent to the output. The right module $z_1(\mu = \text{right})$ performs similar operations for right nodes.

If you look at Fig. 9 you can see that the module $z_1(\mu)$ on the left is exactly the same as the module $z_1(\mu)$ on the right. Indeed, just the argument μ is different: in the first case μ = left and in the second case μ = right. Thus, we can benefit from potential hierarchy and use the same module $z_1(\mu)$ (and, consequently, the same circuit) with different arguments. Acceleration comparing with TREE-WALK SEGMENT is achieved because two subsequent tree nodes can be verified instead of one.

In case of search over TLS the tree is built using similar technique.

4.2 *Sequential Computations Over N-Ary Trees*

One potential way is the design of an application-specific engine, such as that is used for tree-walk tables in [7]. Let us consider the TLS in Fig. 5c. N-ary tree ($N = 4$ for our example) can be built as follows:

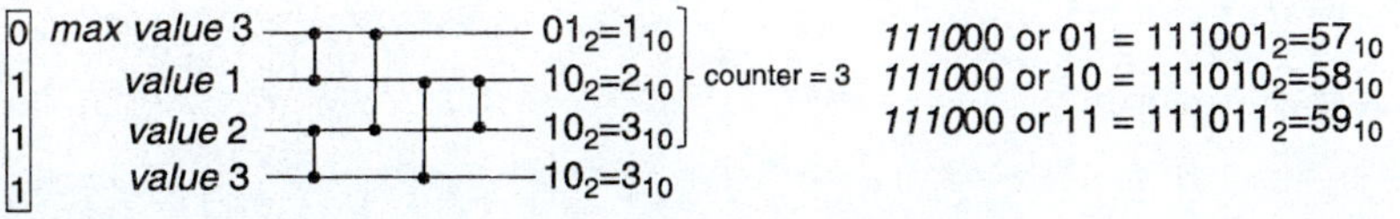

Fig. 10 Implementation details based on sorting networks

1. Allocate the root and N empty children.
2. Take data item.
3. *if (a leaf is not reached) then*

create child or find out and jump to the existing child;
else modify code for the leaf

4. Repeat point 3 for all data items.

Traversing N-ary trees will be considered in Sect. 5.3.

4.3 Processing Incomplete Tree-Like Structures

An example of incomplete TLSs is a K-tree, *i.e.* a tree with up to 2^K (data) items associated with each node (see Sect. 2.5). The idea of incomplete solutions has already been used in some problems of combinatorial search [35, 36, 49], where an initial incomplete solution of a problem was found in software and the produced intermediate results were further processed in fast hardware accelerators giving the final solution. Such decomposition was done because faster customized hardware does not permit to handle data structures above some pre-defined complexity. We will use incomplete TLSs in a similar manner. At the beginning traversal/search procedures over TLSs will be executed in such a way that leads to some constrained intermediate solutions that further can be processed more efficiently using models and methods that differ from TLSs. This technique will be considered separately for traversal and search.

Suppose we build a K-tree (see examples in Fig. 5b, c) and we need to traverse the tree to produce an ordered set of data items. Each node of K-tree holds a set of items. These sets will be ordered using a better way such as sorting networks [6, 15] or some other application-specific combinational circuits (see, for example, [47]). One example is given in Fig. 10. Comparators needed for the networks are shown through the known Knuth notation [15]. Any set of items associated with nodes is represented by a Boolean vector v. For example, a set holded by the node 0110 in Fig. 5c is represented by $v = 0111$ (see Fig. 7). Vector v is used to assign K-bit values associated with a node to inputs of a pre-defined sorting network. It is done through the following expression:

for all i within range $0,\ldots,2^K$-1 ***do*** **if** $v(i) = 0$ **then** 2^K-1 **else** i

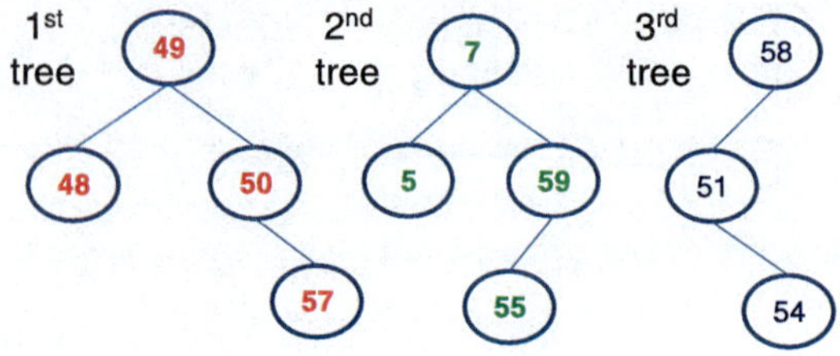

Fig. 11 Sorting data based on $P = 3$ trees that are built and processed in parallel

Thus, if $v(i)$ is 0, the relevant input of sorting network is assigned the maximum value, else the actual value i. The number of values that have to be taken from the network is determined by a counter indicating the amount of associated values for each group. The counter is a combinational circuit that takes input v and outputs the number of ones in v. Figure 10 shows the vector $v = 0111$ for our example (for which the counter is equal to 3) and the network that outputs the sorted values 1, 2 and 3. Since $M - K$ bits of the code for the node 0110 in Fig. 5c are *1110*, then the final sorted values are $1110 01_2 = 57_{10}$, $1110 10_2 = 58_{10}$ and $1110 11_2 = 59_{10}$. $M - K$ bits of the code equal to 1110 are produced when we traverse the tree.

When we use TLSs for search problems we construct the tree incrementally. As soon as a new node of the tree corresponds to an intermediate solution that is under pre-defined constraints, the further steps are executed differently activating an engine for fast processing.

4.4 Parallel Computations

For parallel computations over TLSs we suggest a parallel algorithm (PA), which permits to construct more than one tree in parallel and executes traversal/search procedure over multiple trees.

Let us consider data sort over binary trees discussed in Sect. 2.2. The parallel algorithm permits P trees ($P > 1$) to be created. The implemented method [22] will be demonstrated on an example for $P = 3$. In this case the 1st, the 4th ($P + 1$), the 7th ($2P + 1$), etc. incoming data items are included in the 1st tree. Consequently, the 2nd, the 5th $(P + 1) + 1$, the 8th $(2P + 1) + 1$, etc. data items are included in the 2nd tree, and the 3rd, the 6th $(P + 1) + 2$, the 9th $(2P + 1) + 2$, etc. data items are included in the 3rd tree. Figure 11 shows $P = 3$ trees that are built for the same sequence of input data that was used in Sect. 2.5, i.e.: 49, 7, 58, 48, 5, 51, 50, 59, 54, 57, 55 (arriving sequentially from left to right).

From Figs. 5a and 11 we can see that the maximum depth of each individual tree in the parallel algorithm is less than the depth of a simple binary tree.

Suppose a set of trees (such as that are shown in Fig. 11) is built and each tree is stored in the relevant memory. Since $P = 3$, there are totally $P = 3$ blocks of memory. To output the sorted data the following method is proposed:

1. The trees in $P = 3$ memories are traversed in parallel and there are also $P = 3$ dual-port output memories. Data items from the tree p $(1 \leq p \leq P)$ are saved in the output memory p applying the considered above method for sequential computations and using the first port.
2. This second point is executed in parallel with point 1 above but the second port of the output memories is used. At the beginning, all the addresses point to the first cell of the corresponding output memories. When all P addresses contain data items the smallest one (or the greatest one) is extracted and the address of the appropriate memory (from which the data item has been extracted) is incremented.
3. Points 1 and 2 above are repeated until all data items are sorted.

The considered algorithm builds shallower trees and requires smaller number of steps than for a single tree. Thus, we can reduce time for both the construction of the TLS and the output of data from the TLS [22].

4.5 Hardware Architecture

Handling TLSs is provided by a processing engine (PE) interacting with memory that is used to store initial, intermediate and final results.

The memory keeps TLS. PE implements the required functionality by activating modules that execute operations needed for each node and selects/creates the next node. Since nodes can be processed recursively (see Sect. 2.1), we need a technique allowing recursive activation of the modules.

Figure 12a shows the basic functions of modules. After finishing operations needed for the selected node modules execute either forward or backward steps for jumping to the next node (see Fig. 12a). A forward step jumps to the next node or creates a new node (see Fig. 12). A backward step returns back to the parent node in order to make some changes and to continue forward/backward propagation.

PE is based on HFSM [45] linked with necessary datapath components and allowing hierarchical and recursive execution of modules. Thus, similar to FSMs with datapath [5] we can talk about HFSMs with datapath.

5 Hierarchical Finite State Machines

HFSMs with datapath [5, 41, 45] provide support for hierarchy and recursion and permit the modules described in Sect. 4.5 to be implemented in hardware. PE of other types [10, 16, 20, 29, 51, 53] reviewed in [39] can also be used.

Hierarchy in an HFSM [41, 45] is supported by a stack memory. There are two types of HFSMs [45]: HFSMs with explicit modules and HFSMs with implicit modules. HFSMs with explicit modules includes two stacks (a stack of states—the

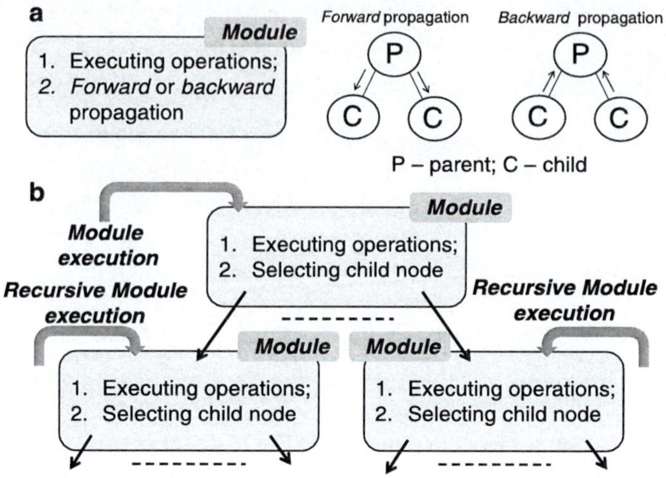

Fig. 12 Modules of PE and their basic functions

FSM_stack and a stack of modules—the *M_stack*) and a combinational circuit (CC) with datapath, which are responsible for state transitions and operations within any *active module* selected by the *stack of modules* (*M_stack*). An HFSM with implicit modules includes just one stack that keeps track of returns from the currently active module. Synthesis of HFSMs with explicit and implicit modules is considered in detail in [45]. So, we focus here on the use of HFSMs for implementing PE.

5.1 Specification of Modules and Synthesis of Hardware Circuits

Let us describe the modules' functionality using hierarchical graph-schemes (HGS) [41], which are constrained flow-charts (all necessary details can be found in [41, 45]). Figure 9 can be seen as an example of HGS.

Synthesis of HFSM with explicit modules from HGS includes the following steps:

1. Defining HGSs for modules $z_0,\dots,z_{Q-1}$ and coding the modules $z_0,\dots,z_{Q-1}$ by codes $K(z_0),\dots,K(z_{Q-1})$, where Q is the number of required modules.
2. Marking the HGS for each module with labels $a_0,\dots,a_{H-1}$ [45] that will be further considered as HFSM states (H is the maximum number of labels that are used for any individual HGS).
3. Describing HFSM stacks in hardware description language (we will use VHDL).
4. Customizing VHDL templates [45] for implementing state transitions, output signals that control memory and functions of datapath.

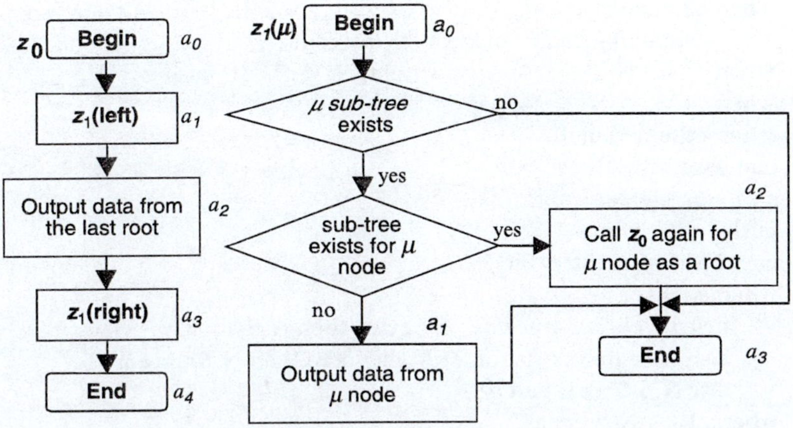

Fig. 13 Specification of modules for synthesis of HFSM

5. Synthesis of hardware circuits from customized VHDL templates.
6. Implementation of the circuits in hardware (we will use FPGAs for prototyping purposes).

Examples of specification, synthesis and implementation in hardware will be discussed in subsequent sections for traversing TLSs and search over TLSs.

5.2 *HFSM for Traversing Tree-Like Structures (Binary Trees)*

In this section we demonstrate all steps needed to design hardware circuits implementing the algorithm in Fig. 9. At the beginning the algorithm is converted to hierarchical specification in form of HGS (see Fig. 13).

The first two hierarchical calls z_1(left) and z_1(right) require argument (either *left* or *right*) for z_1. When we call $z_1(\mu)$ we configure a *mux* that takes data from left sub-tree for z_1(left) and from right sub-tree for z_1(right). These data are supplied for processing in $z_1(\mu)$ (see Fig. 13).

Stack memories in HFSM are described by reusable VHDL code given in [45]. The template [45] in point 4 (see Sect. 5.1) can be customized as follows:

```
process (clk, rst)
--...
case M_stack (StackPtr)
when z0 =>
    push<='0''; pop<='0';
    case FSM_stack(StackPtr) is
    when a0 =>N_S<= a1;
    when a1 =>N_S<= a2; Arg<=left; N_M<= z1; push<= '1';
```

```
        when a2 =>N_S<= a3;
            -- outputting data item from the last root
        when a3 =>N_S<= a4; Arg<=right; N_M<= z1; push<= '1';
        when a4 =>N_S<= a4; pop<='1';
        when others=>null;
        end case;
    when z1 =>
        push<='0'; pop<='0';
        case FSM_stack(StackPtr) is
        when a0 =>
            if(left/right sub-tree does not exist) then N_S<= a3;
            elsif (sub-tree exists for left/right node) then N_S<= a2;
            else N_S<= a1; end if;
        when a1 =>N_S<= a3;
            -- outputting a data item from left/right node
        when a2=>N_S<=a3; N_M<=z0; push<='1';
        when a3=>N_S<=a3; pop<='1';
        when others=>null;
        end case;
    when others => null;
    end case; -- .....
    end process;
```

Here StackPtr is a stack pointer common to both *M_stack* and *FSM_stack*, push/pop—signals that increment/decrement the stack pointer, N_S—signal for the next state, N_M—signal for the next module, Arg—signal that controls the *mux*, selecting data for either left or right sub-trees. Encoding of modules and states is done in VHDL through specifying the proper types:

```
type MODULE_TYPE is (z0, z1);
signal N_M: MODULE_TYPE;
type STATE_TYPE_HFSM is (a0, a1, a2, a3, a4);
signal N_S: STATE_TYPE_HFSM;
```

VHDL code above is synthesizable. Synthesis and implementation of the circuits can be done in commercially available CAD systems, such as Xilinx ISE for implementation on the basis of FPGAs.

5.3 HFSM for Traversing Tree-Like Structures (N-Ary Trees, $N > 2$)

Recursive traversal of N-ary trees for $N = 2$ and $N > 2$ is very similar. Let us consider Fig. 5c assuming that sorting has to be done using N-ary tree ($N = 4$) and networks described in Sect. 4.3 are not used. To satisfy such requirement the tree in

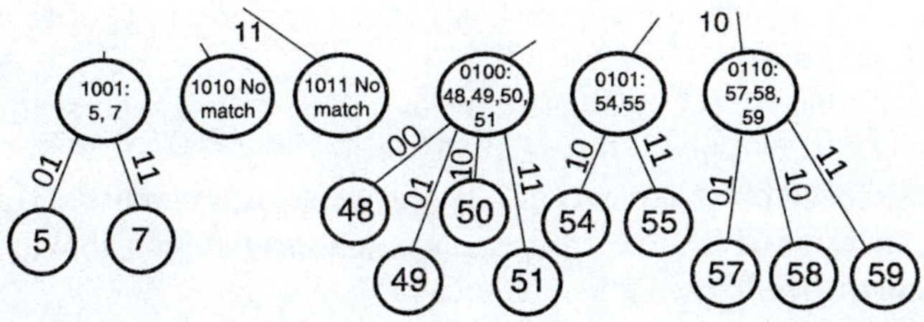

Fig. 14 Extending the tree in Fig. 5c

Fig. 5c has to be slightly modified in a way shown in Fig. 14 where the nodes 1001, 0100, 0101 and 0110 of the tree in Fig. 5c are extended. Non-shown edges lead to *no match* nodes.

Recursive traversal of the updated tree can be done with the aid of the following C function [47]:

```
void sort (tree_node* node)                    // state a0
{ static int level = 0;
  if (node != 0)                               {
    if (level == depth)
      { if (node->l != 0)                      // output data – state a1
        if (node->lm != 0)                     // output data – state a2
        if (node->rm != 0)                     // output data – state a3
        if (node->r != 0)                      // output data – state a4
      }
      else
      { level++;
        if (node->l != 0)                      sort(node->l);      // - state a5
        if (node->lm != 0)                     sort(node->lm);     // - state a6
        if (node->rm != 0)                     sort(node->rm);     // - state a7
        if (node->r != 0)                      sort(node->r);      // - state a8
        level--;                                                   /* - state a9 */}
      }
}                                                                  // state a10
```

where l, lm, rm, r are pointers to child nodes from left to right that are declared in the following structure:

```
struct tree_node
{       int value;
        struct tree_node *l, *lm, *rm, *r;   };
```

The number of steps from the root to leaves (the depth of the tree) is fixed. Note that the function sort is recursive, but iterative technique can also be applied with very similar results in performance and resource consumption. The complete design flow includes the following steps [45]:

- Describing the methods in high-level language (similarly to C code above) and modeling in software
- Selecting the constraints (such as K) for the given M and representing trees in memory
- Associating executable statements in C language (such as output data or recursive calls) with states of HFSM (see comments in the sort function above)
- Optimization of HFSM using the methods [45]
- Customizing VHDL templates [45], i.e. describing transitions between the states, operations in the states and recursive calls/returns using the methods [41, 45]
- Linking with the previously designed circuits for generating input data and presenting the results
- Synthesis and implementation of circuits using commercial tools (we used Xilinx ISE 13.2)
- Uploading the generated bit-stream and testing in FPGA.

All the listed above steps are described in [45]. Optimization technique [45] enables the number of states to be reduced. Recursive calls/returns are implemented using stacks as HFSM memory [41].

5.4 HFSM for Search Over Tree-Like Structures

Figure 15 sketches algorithm described in HGS for solving combinatorial problems. The module z_0 is considered to be the top-level recursive specification. The module z_1 contains problem-specific reduction operations that can activate other lower level modules. Reduction operations permit the initial problem to be simplified. The module z_2 implements problem-specific selection operations. Selection operations make possible to split the problem into sub-problems and to examine them in turn.

At the beginning z_0 is executed for the root of the tree and then it is activated recursively. It should be noted that some (shown in Fig. 15) operations are not needed for certain problems. For example, if we consider the SAT problem, then in the sentence of Fig. 15 "Can we conclude that there is no solution or that the previous result cannot be improved?" the last part, i.e. "or that the previous result cannot be improved?", is not important at all because for the SAT problem any solution is considered to be the final solution. However, in other problems, such as Boolean matrix covering we terminate forward propagation steps that might lead to a solution that is worse than any solution that has already been produced. Note that the goal of this chapter is not an optimization of algorithms over TLSs. Actually we would like just to show that different algorithms over TLSs can be described recursively and can be efficiently implemented using HFSM model.

The HGS in Fig. 15 can be marked with labels (see the labels $a_0,\ldots,a_6$ in Fig. 15) and further steps of synthesis and implementation are similar to those described in Sects. 5.1–5.3.

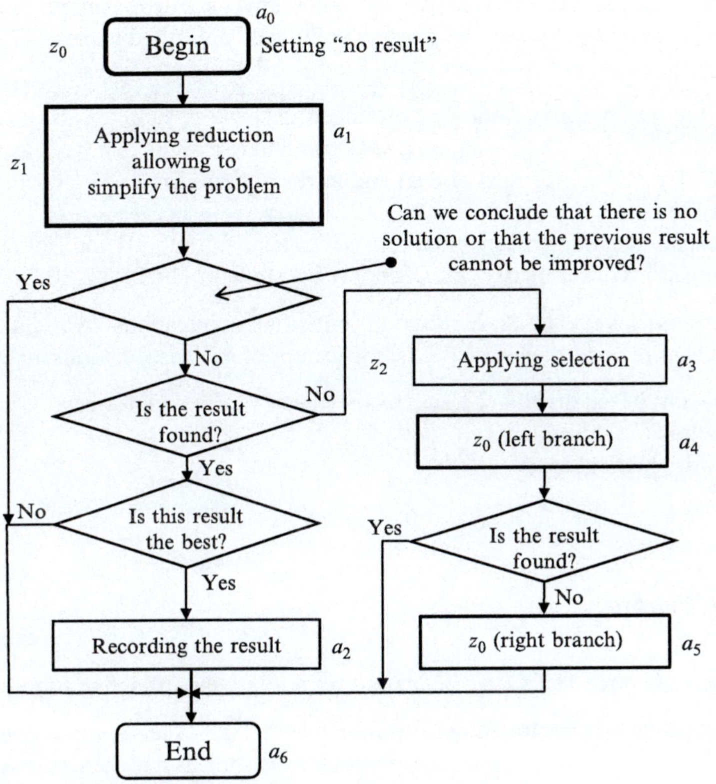

Fig. 15 Top-level description of combinatorial search problems

6 Experiments and Results

Four types of experiments have been performed. Firstly, we verified all the proposed methods in software (in C/C++) running on HP EliteBook 2730p (Intel Core 2 Duo CPU, 1.87 GHz) tablet PC.

Secondly, the synthesis and implementation of the circuits from specification in VHDL were done in Xilinx ISE 13.2 [55] for FPGAs Spartan3E-1200E-FG320 (NEXYS-2 prototyping board of Digilent [8]) and Virtex-4 FX12 (FX12 prototyping board of Nu Horizons). The considered above computations (see Sect. 4) for different TLSs (see Sect. 3) have been implemented and tested using the HFSM model (see Sect. 5).

Thirdly, sequential computations (see Sect. 4) were implemented in PowerPC PPC405 processor embedded to FPGA Virtex-4 FX12 available on the prototyping board FX12. Synthesis and implementations were done using Xilinx ISE and EDK [55].

Fourthly, comparison of recursive and alternative iterative computations was provided for some applications reported in previously published papers [42,43].

Data for the experiments were formed by the following two ways:

- A random-number generator that produced data items supplied in portions to actual (FPGA-based) and simulated (PC computer) circuits that have to sort the previously received portions of data and to resort them for a new portion as fast as possible.
- Data that are taken from benchmark instances available at [34] and from particular applications developed by Ph.D. and M.Sc. students [17,23,27,30,31,37,48].

Since there is a very large number of potential applications over TLSs, the experiments have been chosen just for some groups of such applications, namely:

- For data sort based on binary trees, N-ary trees ($N > 2$) and K-trees
- For combinatorial problems
- For priority management

6.1 Applicability

The considered above TLSs have been used for solving the following problems:

- Dynamic data sort in stream applications where initial data items arrive to the circuits that have to sort data and to resort them as soon as new items are received
- Static data sort where available in memory data have to be sorted
- Combinatorial search problems, namely the SAT and matrix covering
- Priority management in applications [27,48].

For the first two problems input data are either kept in memory common to traditional computers (see Fig. 16a, where the given integers are shown in binary and decimal notations), or represented in form of incoming streams (see Fig. 16b), dynamically generated from different sources, such as distributed sensors in networked embedded systems.

Suppose an FPGA-based hardware implements a sorter that takes input data either from memory (Fig. 16a) or from streams (Fig. 16b) and outputs the sorted sequence that can either replace the original (unsorted) data in memory, be saved in a separate memory, or be transmitted to another device (see Fig. 16).

Physically sorters can be used differently, for example, they can be connected through a system bus of a general-purpose computer and access computer memory (that is a source of data) through allocated windows in memory space; or they can be seen as a standalone accelerator getting external packages of unsorted data and outputting sorted sequences. In some practical applications data have to be resorted dynamically as soon as a new data package/item is received [44]. In all cases we have to make clear what is taken into account when we measure performance and

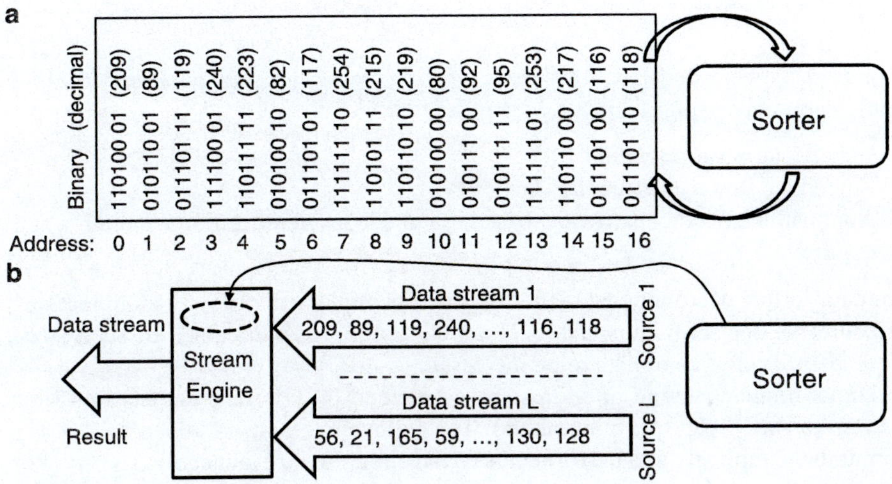

Fig. 16 Input/output data and the role of sorter

resources. For example, if a sorter is operated as it is shown in Fig. 16a we eventually have to measure a time interval beginning from the first access to un-sorted data in external memory until getting the final recorded sorted sequence. However, if input data are preliminary copied to FPGA memory and the result is also kept inside the FPGA, then measurements are similar but since they are done inside the FPGA, the performance is undoubtedly better due to avoiding multiple input/output operations. In some methods, such as address-based data sort [47], getting data can be combined with their processing. Thus, sorting and getting input data can be executed in parallel. Since it is very difficult or even impossible to take into account the performance of input/output operations we assume that all data are available inside FPGA and they are either preliminary copied to built-in FPGA memory (such as block RAM) or produced by built-in random number generator that is faster than the evaluated data sorters.

The third type of problems is even more difficult because implementation of search-specific operations is not a target of this chapter. So, we just compared using TLS with previously developed applications [27, 36, 38] without showing details.

The last problem requires fast resorting for newly arriving data items [44] and this necessity was primarily taken into account.

For all the referred above applications we were able to design, implement in hardware and test HFSM circuits in a single FPGA chip. Thus, we can conclude that the proposed technique is widely applicable. The emphasis was done on cheap FPGA circuits that can be used for different embedded systems.

6.2 Performance

Experiments were done over different types of trees with sequential and parallel implementations using:

- General-purpose software
- Embedded Power PC (just for data sort)
- Application-specific hardware based on the considered HFSM model when required

For data sorting we found that the fastest circuits implement different modifications of address-based technique. Using TLS (N-ary trees, in particular) for such technique permits the size of data to be increased.

Direct implementation of address-based method [47] permits any set of 18-bit data to be sorted (up to 2^{18} numbers). The following results obtained in [23, 47] permit the complexity and performance of the circuit to be evaluated:

- Number of slices (N_s): 326 (3%)
- Number of slice flip-flops (N_{ff}): 57 ($\sim$ 0%)
- Number of 4 input LUTs (N_{LUT}): 578 (3%)
- Number of BRAMs (N_{BRAM}): 16 (57%)
- Maximum clock frequency (F): 155 MHz

The number of clock cycles N_{in} needed to fill in BRAM is equal to Q assuming that each data item can be saved in BRAM during one clock cycle.

The number of clock cycles N_{out} needed to sort data is equal to 2^{14} [23] (one clock cycle is used to read 16-bit word from which up to 16 data items can be extracted during the same clock cycle). 18-bit data are formed through processing data from parallel RAM blocks.

The first implementation of the method based on tree-walk tables for N-ary incomplete trees permits any set of 18-bit data to be sorted (up to 2^{18} numbers). The following results obtained in [23, 47] permit the complexity and performance of the circuit to be evaluated (HFSM with explicit modules was used): $N_s = 562$ (6%); $N_{ff} = 131$ (1%); $N_{LUT} = 1{,}048$ (6%); $N_{BRAM} = 18$ (64%); $F = 76$MHz. The maximum number of clock cycles needed to sort data is 53,556. For different data sets the actual number of clock cycles varies from 7,000 to 53,556.

As you can see, the last implementation needs more hardware resources and operates at lower clock frequency. However, tree-walk tables permit data with bigger value M to be sorted within the same FPGA (the results of experiments are given below). Besides, direct use of address-based data sort requires resetting all one-bit fields to zero when repeating the sort with a new set of data (such resetting is not needed for tree-walk tables).

The second implementation of the method based on tree-walk tables for N-ary incomplete trees enables us to sort sets of 20-bit data when $Q \ll 2^M$. The following results obtained in [23, 47] permit the complexity and performance to be evaluated: $N_s = 586$ (6%); $N_{ff} = 130$ (1%); $N_{LUT} = 1{,}109$ (6%); $N_{BRAM} = 28$ (100%);

$F = 79$ MHz. The actual number of clock cycles in different experiments varied from 8,500 to 86,000.

Note that 20-bit data items cannot be sorted in a single FPGA Spartan3E-1200E-FG320 if we apply direct address-based technique described (i.e. data cannot be sorted without using TLS). The maximum number of sorted data (with tree-walk tables) depends on the distribution of data within the interval from 0 to $2^{20} - 1$. This number is increased if there are many large clusters within the interval that are almost entirely filled in. For example, if all data are within the 20-bits interval 00------------------ and if $N = 4$ then there is just one sub-tree (00) from the root and up to 2^{18} data items can be sorted. If all data are within the two 20-bits intervals 01------------------ and 10------------------ then there are two sub-trees from the root, etc. Some examples from [23, 47] with 18-bit, 19-bit and 20-bit data produced by a random number generator that were implemented and tested in FPGA are given below:

- $Q = 100$, $N_{18} = 7{,}000$–$7{,}700$, $N_{19} = 8{,}500$–$8{,}900$, $N_{20} = 9{,}300$–$10{,}000$, where N_{18}, N_{19} and N_{20} are numbers of clock cycles for 18-bit, 19-bit and 20-bit data accordingly
- $Q = 300$, $N_{18} = 16{,}000$–$16{,}600$, $N_{19} = 19{,}000$–$20{,}000$, $N_{20} = 22{,}000$–$25{,}000$
- $Q = 500$, $N_{18} = 20{,}000$–$24{,}000$, $N_{19} = 27{,}000$–$29{,}000$, $N_{20} = 34{,}000$–$36{,}000$
- $Q = 1{,}000$, $N_{18} = 33{,}000$–$36{,}000$, $N_{19} = 44{,}000$–$47{,}000$, $N_{20} = 55{,}000$–$58{,}000$
- $Q = 2{,}000$, $N_{18} = 45{,}000$–$47{,}000$, $N_{19} = 68{,}000$–$70{,}000$, $N_{20} = 82{,}000$–$83{,}000$

Since the random number generator equally distributes data within the intervals $0,\ldots,2^{19}-1$ and $0,\ldots,2^{20}-1$, the examples above give the worst cases for 19-bit and 20-bit data. There is no worst case for 18-bit data because any set of data can be sorted.

Two strategies shown in Fig. 16 were evaluated. Data were sorted either based on their static representation (Fig. 16a) or sorted and resorted dynamically as soon as new data items are produced by a generator. Let us call the latter case dynamic sort and it is directly applicable to priority buffers [44] requiring fast reordering of new data that can arrive at any time. Dynamic sort can also efficiently be implemented using binary trees (see Sect. 2.2).

We found that the developed circuits are also faster than implementations in general-purpose software and in embedded to FPGA Power PC (the latter two cases were tested in HP EliteBook 2730p and PPC405 embedded to FPGA Virtex-4 FX12). Some results of comparison can be found in [46]. It should be noted that quite complex problems can be processed in relatively simple and cheap FPGAs of Spartan-3 family. Performance is comparable with known results obtained for significantly more advanced FPGAs. Often the known methods were either modeled or just partially tested in available prototyping systems. Frequently, external onboard memories were used. Thus, the exact comparison in hardware is indeed difficult. All the considered here circuits were entirely implemented in a single FPGA and external resources were not used at all.

You can see that there is no data dependency between tree branches. Thus, the algorithm permits individual sub-trees with any desired level of parallelism to be

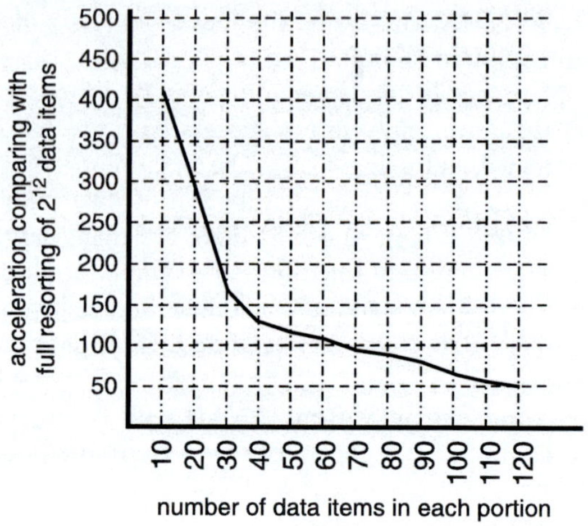

Fig. 17 Experiments with resorting

processed. We found that parallel processing of binary trees is faster than sequential, but hardware resources are also increased. Some experiments that were done with N-ary trees ($N > 2$) have shown that although parallelization permits the number of clock cycles to be reduced, the acceleration is less than for the case of binary trees. Besides, the required hardware resources are increased significantly. On the other hand, using incomplete TLS gives nearly the same results as for binary trees.

Now let us talk about resorting that is important for priority management [44] and for which the proposed processing of TLSs is especially beneficial. Figure 17 permits to compare acceleration of resorting for a new portion (that includes from 10 to 120 data items) with resorting of all data items (that is 2^{12}). We used just binary trees. As you can see from Fig. 17 acceleration is significant.

Experiments with combinatorial search problems (namely the SAT and matrix covering) have shown that using TLS does not permit to increase performance. This is because the relevant algorithms execute a huge number of search-specific operations, which mainly affect the execution time. On the other hand, TLSs enable the algorithm to be described more clearly and to apply more easily methods of software development, such as hierarchy and recursion. Besides, TLS are very helpful for time-consuming parts of the respective algorithms, such as Boolean constraint propagation.

6.3 *Resources*

Resources have been evaluated for:

- HFSMs with explicit modules
- HFSMs with implicit modules

Table 1 HFSMs with explicit vs. HFSM with implicit modules

	$HFSM_e$ (explicit)			$HFSM_i$ (implicit)		
Algorithm	N_s	F (MHz)	N_{BRAM}	N_s	F (MHz)	N_{BRAM}
A_1	2,415	100	13	1,175	107	13
A_2	4,407	100	27	1,965	103	27
A_3	2,609	77	20	1,146	65	20
A_4	2,812	75	20	1,312	83	20
A_5	3,343	73	20	1,609	83	20

Table 1 presents the maximum attainable clock frequency (F) and FPGA resources (the number of slices—N_s and the number of block RAMs—N_{BRAM}) needed for different implementations of HFSMs with explicit ($HFSM_e$) and implicit ($HFSM_i$) modules for different algorithms over TLS designated $A_1,\ldots,A_5$. These algorithms are more complicated than those considered in the previous subsection and that is why the relevant HFSMs need more FPGA slices (N_s).

As you can see $HFSM_i$ [45] are less resource consuming. Another advantage of $HFSM_i$ is that there is an opportunity to apply known optimization methods that have been developed for conventional state machines. The HFSM with explicit modules ($HFSM_e$) [41] is not so well suited for such optimization, mainly because states in different modules can be assigned the same codes. However, modules in $HFSM_i$ become implicit and cannot be updated and refined such easily as for $HFSM_e$. Although the HGSs for $HFSM_e$ and $HFSM_i$ are the same and all features are supported, modularity, hierarchy and recursion become less clear for $HFSM_i$ at the implementation level. The use of HFSMs is profoundly studied and analyzed in [40] with numerous experiments in FPGAs of Spartan-3e and Spartan-6 families of Xilinx and Cyclon-IVe family of Altera. Particularly, it is shown that HFSMs can be deeper optimized and therefore permit to achieve very good results in processing TLSs in hardware.

Particular complexities for some implemented circuits are given in Sect. 6.2 and are summarized in Fig. 18. Please note that for flexibility $HFSM_e$ was used. As you can see, the values of N_s and N_{ff} (the number of slices and flip-flops) are negligible comparing with available FPGA resources. The main limitation, not allowing to increase the size of sorted data, is the number of available block RAMs. Indeed, the maximum size of sorted data (20-bits) for the considered FPGA is set because all block RAMs have already been used (see Fig. 18). Note that the number of embedded block RAMs was essentially increased in new generations of FPGAs (see, for example, [1, 54]). Thus, significantly more complicated problems can be solved in future on a single FPGA microchip.

6.4 *Scalability*

The results of experiments have shown that the main restriction that limits the number of data items is the available embedded block RAMs on the FPGA micro-chip. The algorithms themselves are easily scalable. This conclusion is done

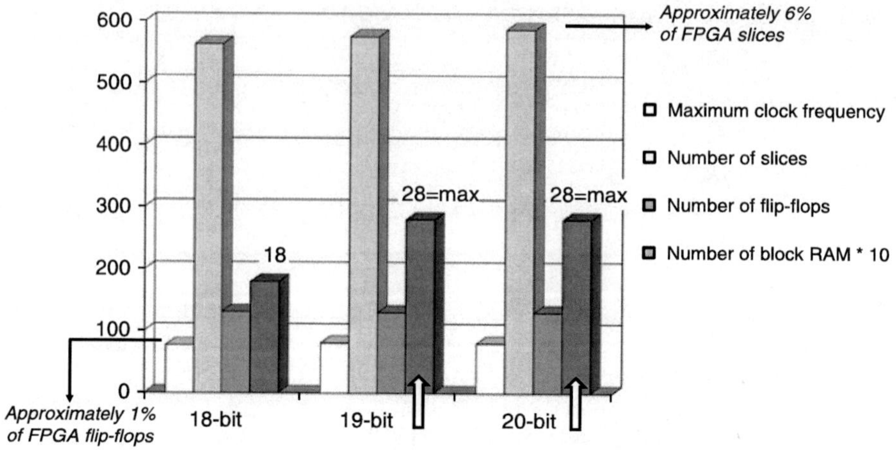

Fig. 18 FPGA resources needed for data sort based on N-ary trees and $HFSM_e$

after experiments with Spartan-3 FPGA and external memory available on the prototyping board NEXYS-2. *N*-ary ($N > 2$) incomplete trees with the relevant accelerators for tree nodes that are associated with multiple items give additional potentialities allowing the number of data items to be increased significantly. We can also replace the cheap Spartan-3E FPGA with a more advanced FPGA and supply external memory. The results of preliminary analysis of using Virtex-6 based ML-605 development system have shown that the number of data items in sorting algorithms can be increased up to hundreds of millions.

6.5 *Recursive Vs. Iterative Algorithms*

Comparison of alternative recursive and iterative algorithms permits to conclude the following:

1. *Resources and performance*:
   - Implementation in hardware of recursive algorithms over TLS with the aid of HFSM model is comparable with iterative implementations based on similar model
   - Recursive implementations in hardware are better than implementations for general-purpose software and embedded general purpose processors
2. *New features*: The proposed technique permits to benefit from the strategy divide and conquer.
3. *Clarity*: In our opinion recursive specifications are better understandable than iterative specifications. Besides, they are more compact.

4. *Reusability*: It is important that hierarchical modular specifications directly support reusability. Indeed, the same module can be reused in different algorithms and even within the same algorithm, which may lead to reduction in the design time and even in hardware resources.
5. *Applicability*: Although emphasis is done on recursive computations (mainly due to their compactness and clearness), the considered modular and hierarchical technique is also well applicable to iterative computations.

6.6 Comparison with Known Results

In [25] the complete median operator (that in general requires less time than data sort) to process 256 MB of data consisting of 4-byte words takes 6.173 s, i.e. 100 ns per word using advanced Virtex-5 FPGA. The best performance achieved in Sect. 6.2 for low-cost Spartan-3 FPGA permits to sort data taking on average 0.41 ns per each item (2^{18} data items require 2^{14} clock cycles using 155 MHz frequency in the simplest case). If we want to increase the size of data with the limited number of embedded RAM blocks, we construct N-ary trees and performance will be decreased. However, the trees enable us to resort data very fast. Indeed, in methods [25] any resorting for a new portion of data requires all data items to be sorted and, thus, for data from Fig. 17 it takes about 4 ms. For our technique that is based on TLS with $N = 16$, $K = 4$ and for examples in Fig. 17, it takes from 1 to 12 microseconds for different number of items in portions. Thus, acceleration is significant. If we use binary trees considered in this chapter, then similar resorting is even faster and requires from 200 to 2,400 ns. The only problem for the latter case is a very limited size of data. We found that the considered Spartan-3 FPGA can sort just up to 2^{14} data items over binary trees. Once again the main problem is the limited number of available block RAMs. Using a more advanced hardware platform would permit more data to be processed. For example, an ML-605 prototyping board contains XC6VLX240T FPGA of Virtex-6 family and includes 512 MB DDR3 SODIMM. If data are stored in this memory, then $10^8$32-bit data items can be sorted using address-based technique. The major problem will be optimizing memory access, i.e. the sorting algorithm will have to be adapted to interact with memory as much as possible in burst mode (for example, access to random individual bits required for filling in memory in address-based data sort is very expensive and has to be optimized).

In [56] the maximum speed of sorting is estimated as 180 million records per second. Thus, resorting of all data in Fig. 17 would require about 23 ms. Once again the achieved results in Fig. 17 are significantly better. Comparison with [4] (that provides very useful data for numerous sorting algorithms) also shows advantages of the proposed technique for stream applications that require fast resorting.

The advantages of TLS for accelerating time-consuming parts of an SAT solver were clearly demonstrated in [7]. It is shown that FPGA co-processor can achieve 3.7–38.6x speedup on Boolean constraint propagation compared to state-of-the-art software SAT solvers. Numerous experiments with combinatorial search algorithms over TLS and HFSMs were done in [31,37] (using benchmarks [34]) demonstrating more clear and compact specifications.

7 Conclusion

This chapter suggests common organization and methods of computations over trees in hardware. The main contributions of the chapter are: (1) study of wide spectrum of trees; (2) suggesting new common ways to represent and to process trees in hardware based on the HFSM model; (3) proof of competence of the proposed technique based on prototyping in FPGA, numerous experiments and comparisons.

Tree-like structures have been successfully used for solving such problems as data sort, combinatorial search and priority management. Different kinds of trees have been explored: binary trees, *N*-ary trees, merged binary trees and *K*-trees. Besides, we proposed to combine different methods of processing allowing performance to be improved. For instance, the chapter shows that tree-like structures can easily be linked with fast sorting networks; different branches of *N*-ary trees can be traversed in parallel; trees can be used with the efficient address-based sorting method, etc. Tree-like structures are processed in hardware using the proposed HFSM model augmented with VHDL templates which are easily customizable to match the respective specification. The suggested templates are fully synthesizable using any appropriate CAD software.

The experiments were done in general-purpose and embedded software and in FPGA-based application-specific hardware. The achieved results prove the applicability of the suggested methods of processing and storing trees in hardware circuits which outperform the respective software implementations. Moreover, the obtained performance is comparable to known results reported for more advanced FPGA-based platforms. Our methods are especially beneficial when fast resorting is required (for example, in priority management).

Two kinds of HFSMs, which provide support for hardware implementation, have been evaluated. The results of experiments demonstrate that the proposed model of HFSM with implicit modules is faster and requires less hardware resources compared to HFSMs with explicit modules. Moreover, the model of HFSM with implicit modules permits the known optimization methods developed for conventional FSMs to be entirely reused. On the other hand, individual modules in HFSMs with implicit modules cannot be updated as easily as in HFSM with explicit modules.

Although the results of experiments clearly demonstrate applicability and high efficiency of the proposed methods, which can easily be implemented in low cost widely available FPGAs, more advanced hardware platforms are required for processing significantly larger sets of data. Currently, up to 2^{18} of 20-bits data can be handled and the main restriction that limits the number of data items is the available embedded block RAMs on the FPGA microchip. The suggested models and algorithms themselves are easily scalable. Preliminary analysis of the

Virtex-6 based ML-605 development system shows that the number of data items can be increased up to hundreds of millions.

Acknowledgements This work was supported by FEDER through the Operational Program Competitiveness Factors—COMPETE and by National Funds through FCT—Foundation for Science and Technology in the context of project FCOMP-01-0124-FEDER-022682 (FCT reference PEst-C/EEI/UI0127/2011). We would like to thank Dmitri Mihhailov from Tallinn University of Technology for numerous experiments that have been done for tree-based data processing.

References

1. Altera product catalog, version 11.0, 2011. Available at: www.altera.com/literature/sg/product-catalog.pdf
2. R.E. Bryant, Graph-based algorithms for boolean function manipulation. IEEE Trans. Comput. 35(8), 677–691 (1986)
3. F.M. Carrano, *Data Abstraction and Problem Solving with C++: Walls and Mirrors* (Addison Wesley, Boston, 2005), 968 pp
4. R.D. Chamberlain, N. Ganesan, Sorting on architecturally diverse computer systems, in *Proc. 3rd Int. Workshop on High-Performance Reconf. Comp. Techn. and App. – HPRCTA'09*, New York, USA, 2009, pp. 39–46
5. P.P. Chu, *FPGA Prototyping by VHDL Examples: Xilinx Spartan-3 Version* (Willey, Hoboken, 2008), 440 pp
6. T.H. Cormen, C.E. Leiserson, R.L. Rivest, C. Stain, *Introduction to Algorithms*, 2nd edn. (MIT Press, Cambridge, 2003), 1180 pp
7. J.D. Davis, Z. Tan, F. Yu, L. Zhang, A practical reconfigurable hardware accelerator for Boolean satisfiability solvers, in *Proc. 45th ACM/IEEE Design Automation Conference – DAC'2008*, Anaheim, California, USA, 2008, pp. 780–785
8. *Digilent Products* (Digilent Inc., Pullman, 2013), Available at: http://www.digilentinc.com
9. S.A. Edwards, Design languages for embedded systems. Computer Science Technical Report CUCS-009-03. Columbia University, 2003
10. A. Girault, B. Lee, E.A. Lee, Hierarchical finite state machines with multiple concurrency models. IEEE Trans. Comput. Aided Des. Integr. Circ. Syst. 18(6), 742–760 (1999)
11. N.K. Govindaraju, J. Gray, R. Kumar, D. Manocha, GPUTeraSort: High performance graphics co-processor sorting for large database management, in *Proc. 2006 ACM SIGMOD Int'l Conf. on Management of Data*, Chicago, IL, USA, 2006, pp. 325–336
12. D.J. Greaves, S. Singh, Kiwi: Synthesis of FPGA circuits from parallel programs, in *Proc. IEEE Symposium on Field-Programmable Custom Computing Machines - FCCM'08*, 2008, pp. 3–12
13. J. Gu, P.W. Purdom, J. Franco, B.W. Wah, Algorithms for the satisfiability (SAT) problem: a survey. DIMACS Ser. Discrete Math. Theor. Comput. Sci. **35**, 19–151 (1997)
14. S.S. Huang, A. Hormati, D.F. Bacon, R. Rabbah, Liquid metal: object-oriented programming across the hardware/software boundary, in *European Conf. on Object-Oriented Programming*, Paphos, Cyprus, 2008, pp. 76–103
15. D.E. Knuth, *The Art of Computer Programming, Volume 3: Sorting and Searching*, 2nd edn. (Addison-Wesley, Reading, 1998), 780 pp
16. S. Lee, S. Yoo, K. Shoi, Reconfigurable SoC design with hierarchical FSM and synchronous dataflow model, in *Proc. 10th Int. Symp. on Hardware/software codesign*, Estes Park, USA, 2002, pp. 199–204
17. J. Lima, Processador com Conjunto de Instruções Variável Remotamente. M.Sc. thesis. University of Aveiro, 2009

18. T. Lin, Mobile Ad-hoc Network Routing Protocols: Methodologies and Applications, Ph.D. thesis, Blacksburg, Virginia, 2004
19. H. Lonn, J. Axelsson, A comparison of fixed-priority and static cyclic scheduling for distributed automotive control application, in *Proc. 11th Euromicro Conference on Real-Time Systems*, York, England, 1999, pp. 142–149
20. T. Maruyama, M. Takagi, T. Hoshino, Hardware implementation techniques for recursive calls and loops, in *Proc. Proc. 9th Int. workshop on Field Programmable Logic and Applications – FPL'99*, Glasgow, UK, 1999, pp. 450–455
21. R.A. Mewaldt, C.M.S. Cohen, W.R. Cook et al., The low-energy telescope (LET) and SEP central electronics for the STEREO mission. Space Sci. Rev. **136**, 285–362 (2008)
22. D. Mihhailov, V. Sklyarov, I. Skliarova, A. Sudnitson, Parallel FPGA-based implementation of recursive sorting algorithms, in *Proc. 2010 Int. Conf. on ReConFigurable Computing and FPGAs - ReConFig 2010*, Cancun, Mexico, 2010, pp. 121–126
23. D. Mihhailov, Hardware Implementation of Recursive Sorting Algorithms Using Tree-like Structures and HFSM Models, Ph.D. thesis, Tallinn University of Technology, 2012
24. A. Mitra, M.R. Vieira, P. Bakalov, V.J. Tsotras, W. Najjar, Boosting XML Filtering through a scalable FPGA-based architecture, in *Proc. 4th Biennial Conference on Innovative Data Systems Research - CIDR*, Asilomar, CA, USA, 2009
25. R. Mueller, J. Teubner, G. Alonso, Data processing on FPGAs, in *Proc. VLDB Endowment*, vol. **2**(1), Lyon, France, 2009, pp. 910–921
26. S. Nagayama, T. Sasao, Complexities of graph-based representations for elementary functions. IEEE Trans. Comput. **58**(1), 106–119 (2009)
27. A. Neves, Interacção Remota com Circuitos Implementados em FPGA, M.Sc. thesis, University of Aveiro, 2009
28. N. Ngan, E. Dokladalova, M. Akil, F. Contou-Carrère, Fast and efficient FPGA implementation of connected operators. J. Syst. Architect. **57**, 778–789 (2011)
29. S. Ninos, A. Dollas, Modeling recursion data structures for FPGA-based implementation, in *Proc. Int. Conf. on Field Programmable Logic and Applications - FPL'08*, Heidelberg, Germany, 2008, pp. 11–16
30. R. Oliveira, Análise e comparação de métodos soft-hard em sistemas reconfiguráveis, M.Sc. thesis, University of Aveiro, 2010
31. B. Pimentel, Synthesis of FPGA-based accelerators implementing recursive algorithms, Ph.D. thesis, University of Aveiro, 2009
32. S. Rajasekaran, S. Sen, Optimal and practical algorithms for sorting on the PDM. IEEE Trans. Comput. **57**(4), 547–561 (2008)
33. K.H. Rosen, J.G. Michaels, J.L. Gross, D.S. Shier, *Handbook of Discrete and Combinatorial Mathematics* (CRC Press, Boca Raton, 2000), 1232 pp
34. H.H. Hoos, SATLIB Benchmark Problems (University of British Columbia, Canada, 2011), Available at: http://www.cs.ubc.ca/~hoos/SATLIB/benchm.html
35. I. Skliarova, A.B. Ferrari, Reconfigurable hardware SAT solvers: a survey of systems. IEEE Trans. Comput. **53**(11), 1449–1461 (2004)
36. I. Skliarova, A.B. Ferrari, A software/reconfigurable hardware SAT solver. IEEE Trans. VLSI Syst. **12**(4), 408–419 (2004)
37. I. Skliarova, Arquitecturas reconfiguráveis para problemas de optimização combinatória, Ph.D. thesis, University of Aveiro, 2004
38. I. Skliarova, V. Sklyarov, Design methods for FPGA-based implementation of combinatorial search algorithms, in *Proc Int. Workshop on SoC and MCSoC Design - IWSOC'2006*, Yogyakarta, Indonesia, 2006, pp. 359–368
39. I. Skliarova, V. Sklyarov, Recursion in reconfigurable computing: a survey of implementation approaches, in *Proc. 19th Int. Conf. on Field Programmable Logic and Applications - FPL'09*, Prague, Czech Republic, 2009, pp. 224–229
40. I. Skliarova, V. Sklyarov, A. Sudnitson, *Design of FPGA-based Circuits Using Hierarchical Finite State Machines* (TUT Press, Tallinn, Estonia, 2012), 240 pp

41. V. Sklyarov, Hierarchical finite-state machines and their use for digital control. IEEE Trans. VLSI Syst. **7**(2), 222–228 (1999)
42. V. Sklyarov, FPGA-based implementation of recursive algorithms. Microprocessors and microsystems. Spec. Issue FPGAs Appl. Des. **28**(5–6), 197–211 (2004)
43. V. Sklyarov, I. Skliarova, B. Pimentel, FPGA-based implementation and comparison of recursive and iterative algorithms, in *Proc. 15th Int. Conf. on Field Programmable Logic and Applications - FPL'05*, Finland, 2005, pp. 235–240
44. V. Sklyarov, I. Skliarova, Modeling, design, and implementation of a priority buffer for embedded systems, in *Proc. 7th Asian Control Conference – ASCC'2009*, Hong Kong, 2009, pp. 9–14
45. V. Sklyarov, Synthesis of circuits and systems from hierarchical and parallel specifications, in *Proc. 12th Biennial Baltic Electronics Conference, Invited paper*, Tallinn, Estonia, 2010, pp. 37–48
46. V. Sklyarov, I. Skliarova, R. Oliveira, D. Mihhailov, A. Sudnitson, Processing tree-like data structures in different computing platforms, in *Proc. Int. Conf. on Informatics and Computer Applications - ICICA' 2011*, Dubai, UAE, 2011, pp. 112–116
47. V. Sklyarov, I. Skliarova, D. Mihhailov, A. Sudnitson, Implementation in FPGA of address-based data sorting, in *Proc. 21st Int. Conf. on Field Programmable Logic and Applications - FPL'2011*, Creete, Greece, 2011, pp. 405–410
48. S. Soldado, FPGA Urban Traffic Control Simulation and Evaluation Platform, M.Sc. thesis, University of Aveiro, 2009
49. J. de Sousa, J.P. Marques-Silva, M. Abramovici, A configware/software approach to SAT solving, in *Proc. 9th IEEE Symposium on Field-Programmable Custom Computing Machines - FCCM'01*, California, USA, 2001, pp. 239–248
50. H.T. Sun, First failure data capture in embedded system, in *Proc. IEEE IIT*, Chicago, USA, May 17–20, 2007, pp. 183–187
51. S. Uchitel, J. Kramer, J. Magee, Synthesis of behavorial models from scenarios. IEEE Trans. Soft. Eng. **29**(2), 99–115 (2003)
52. Proceedings of the 2nd UK Embedded Forum, Newcastle, Leicester, Southampton, 2005, Available at: http://www.staff.ncl.ac.uk/albert.koelmans/books/secondukembforum.pdf, 303 pp
53. J. Whittle, P.K. Jayaraman, Generating hierarchical state machines from use case charts, in *Proc. 14th IEEE Int. Requirements Eng. Conf.*, Minneapolis, USA, 2006, pp. 16–25
54. Xilinx 7 series FPGAs. Product brief (2011). Available at: www.xilinx.com/publications/prod_mktg/7-Series-Product-Brief.pdf
55. Xilinx Products (Xilinx Inc., San Jose, 2013), Available at http://www.xilinx.com
56. X. Ye, D. Fan, W. Lin, N. Yuan, P. Ienne, GPU-Warpsort: a fast comparison-based sorting algorithm on GPUs, in *IEEE Int. Parallel & Distributed Processing Symposium - IPDPS 2010*, Atlanta, USA, 2010
57. A. Zakrevskij, Y. Pottosin, L. Cheremisinova, *Combinatorial Algorithms of Discrete Mathematics* (TUT Press, Tallinn, Estonia, 2008), 193 pp

FPGA-Based Systolic Computational-Memory Array for Scalable Stencil Computations

Kentaro Sano

Abstract Stencil computation is one of the typical kernels of numerical simulations, which requires acceleration for high-performance computing (HPC). However, the low operational-intensity of stencil computation makes it difficult to fully exploit the peak performance of recent multi-core CPUs and accelerators such as GPUs. Building custom-computing machines using programmable-logic devices, such as FPGAs, has recently been considered as a way to efficiently accelerate numerical simulations. Given of the many logic elements and embedded coarse-grained modules, state-of-the-art FPGAs are nowadays expected to efficiently perform floating-point operations with sustained performance comparable to or higher than that given by CPUs and GPUs. This chapter describes a case study of an FPGA-based custom computing machine (CCM) for high-performance stencil computations: *a systolic computational-memory array (SCM array) implemented on multiple FPGAs*.

1 Introduction

Numerical simulation is now an important and indispensable tool in science and engineering, which analyzes complex phenomena that are difficult to subject to experimentation. For a large-scale simulation with sufficient resolution, high-performance computing (HPC) is required. Stencil computation [4] is one of the typical kernels of high-performance numerical simulations, which include computational fluid dynamics (CFD), electromagnetic simulation based on the finite-difference time-domain (FDTD) method, and iterative solvers of a linear equation system.

K. Sano (✉)
Tohoku University, 6-6-01 Aramaki Aza Aoba, Sendai, Miyagi 980-8579, Japan
e-mail: kentah@caero.mech.tohoku.ac.jp

W. Vanderbauwhede and K. Benkrid (eds.), *High-Performance Computing Using FPGAs*, DOI 10.1007/978-1-4614-1791-0_9,

Since most of their computing time is occupied by the computing kernels, acceleration techniques for stencil computation have been required. However, the low operational-intensity of stencil computation makes it difficult to fully exploit the peak performance of recent accelerators such as GPUs. Operational intensity is the number of floating-point operations per data size read from an external DRAM for cache misses [30]. In general, stencil computation requires relatively many data accesses per unit operation, and therefore its operational-intensity is low. Although the recent CPUs and GPUs have been getting higher peak-performance for arithmetic operations by integrating more cores on a chip, their off-chip bandwidth has not been sufficiently increased in comparison with the peak-performance. As a result, stencil computation can enjoy only a fraction of the peak performance because cores are idle for many cycles where required data are not ready. To make matters worse, a large-scale parallel system with many CPUs and/or GPUs usually has the problem of parallel processing overhead, which limits speedup especially for computations with a low operational-intensity due to inefficient bandwidth and latency of the interconnection network. These inefficiencies of the processor-level and system-level execution cause the performance per power of supercomputers to decline. We need to address this efficiency problem because the performance per power is also a big issue in high-performance computation.

Custom computing machines (CCMs) constructed with programmable-logic devices, such as FPGAs, are expected to be another way to efficiently accelerate stencil computation because of their flexibility in building data-paths and memory systems dedicated to each individual algorithm. Especially, state-of-the-art FPGAs have become very attractive for HPC with floating-point operations due to their advancement with a lot of logic elements and embedded modules, such as DSP blocks, block RAMs, DDR memory-controllers, PCI-Express interfaces, and high-speed transceivers/receivers. High-end FPGAs are now capable of performing floating-point operations with sustained performance comparable to or higher than that achieved by CPUs and GPUs.

This chapter presents *a systolic computational-memory array* (SCM array) [20, 22, 23] to be implemented on multiple FPGAs, which is a programmable custom-computing machine for high-performance stencil computations. The SCM array is based on the SCM architecture that combines the systolic array [11, 12] and the computational memory approach [7, 17, 18, 29] to scale both computing performance and aggregate memory-bandwidth with the array size. Processing elements of the array perform 3×3 star-stencil computations in parallel with their local memories. Since this architecture is especially designed to achieve performance scalability on a multi-FPGA system, two techniques are introduced for the bandwidth and synchronization problems of inter-FPGA data-transfer: *a peak-bandwidth reduction mechanism* (BRM) and *a local-and-global stall mechanism* (LGSM).

We describe a target computation, an architecture and a design for the SCM array to be implemented on a multi-FPGA system. For three benchmark problems, we evaluate resource consumption, performance and scalability of prototype implementation with three FPGAs. We also discuss feasibility and performance for implementation with a 2D array of high-end FPGAs. We show that a single state-of-the-art FPGA can achieve a sustained performance higher than 400 GFlop/s and a multi-FPGA system scales the performance in proportion to the number of FPGAs.

2 Related Work

So far, various trials have been conducted to design FPGA-based CCMs for stencil computations in the numerical simulations based on finite difference methods, which include an initial investigation of an FPGA-based flow solver[10], an overview of an FPGA-based accelerator for CFD applications[26], a design of FPGA-based arithmetic pipelines with a memory hierarchy customized for a part of CFD subroutines[16], and proposals for FPGA-based acceleration of FDTD method [3, 6, 25]. Most of them rely on pipelining data-flow graphs on an FPGA. However, their design and discussion lack in scalability of performance and memory bandwidth which is necessary to achieve HPC. Furthermore, discussion and evaluation of the system scalability, particularly for multiple-FPGA implementation, are indispensable for HPC with a large-scale system.

Recently, several systems with a lot of tightly coupled FPGAs have been developed: BEE3, Maxwell, Cube, Novo-G, and SSA. BEE3, the Berkeley Emulation Engine 3, is designed for faster, larger and higher fidelity computer architecture research [5]. It is composed of modules with four tightly coupled Virtex-5 FPGAs connected by ring interconnection. The modules can be further connected to each other to construct a large FPGA computer. Maxwell is a high-performance computer developed by the FPGA high performance computing alliance (FHPCA), which has a total of 64 FPGAs on 32 blade servers [2]. Each blade has an Intel Xeon CPU and two Xilinx Virtex-4 FPGAs. While the CPUs are connected by an interconnection network, the FPGAs are also connected directly by their dedicated 2D torus network. The Cube is a massively parallel FPGA cluster consisting of 512 Xilinx Spartan 3 FPGAs on 64 boards [15]. The FPGAs are connected in a chain, so that they are suited to pipeline and systolic architecture.

The Novo-G is an experimental research testbed built by the NSF CHREC Center, for various research projects on scalable reconfigurable computing [8]. Novo-G is composed of 24 compute nodes, each of which is a Linux server with an Intel quad-core Xeon processor and boards of ALTERA Stratix IV FPGAs. Data transfer can be made between adjacent FPGAs, through a wide and bidirectional bus. The SSA, the scalable streaming-array, is a linear array of ALTERA Stratix III FPGAs for scalable stencil computation with a constant memory bandwidth [24]. The FPGAs are connected by a 1D ring network to flow data through a lot of computing stages on the FPGAs. By deeply pipelining iterative stencil computations with the stages, the SSA achieves scalable performance according to the size of the system.

These systems provide not only a peak computing performance scalable to the system size but also low-latency and wide-bandwidth communication among

FPGAs. Particularly the inter-FPGA communication achieved by direct connection of FPGAs is very attractive and promising for parallel computing with less overhead. Present typical accelerators, such as GPUs, suffer from the latency and bandwidth limitation of their connection via host nodes and a system interconnection network.

However, it is also challenging to efficiently use reconfigurable resources with flexibility of inter-FPGA connection for large-scale custom computing. We have to find an answer to the question of how we should use multiple FPGAs, or what architecture and designs are suitable and feasible for scalable and efficient stencil-computation. This chapter presents *the SCM array* as an answer to this question. The SCM array is a programmable custom-computing machine for high-performance stencil computations. The SCM array is based on the SCM architecture that combines the systolic array [11, 12] and the computational memory approach [7, 17, 18, 29] to scale both computing performance and aggregate memory-bandwidth with the array size. The SCM array behaves as a computing memory, which does not only store data but also perform computations with them. The processing elements of the array perform 3×3 star-stencil computations in parallel with their local memories. To achieve both flexibility and dedication, we give the SCM array a hardware layer and a software layer to execute programs with various stencil computation. The hardware layer has a simple instruction-set for only a minimum of programmability.

Since this architecture is especially designed for scalable computation on a multi-FPGA system with different clock-domains, two techniques are introduced for the bandwidth and synchronization problems of inter-FPGA data-transfer: the peak-bandwidth reduction mechanism (BRM) and the LGSM. To evaluate resource consumption, performance, and scalability for three benchmark problems, we demonstrate that the SCM array prototyped with three FPGAs achieves performance scalable to the number of devices.

The FPGA-based programmable active memory (PAM) [29] is an approach that is similar to our SCM array in terms of *the PAM* concept. PAM is a 2D array of FPGAs with an external local-memory, which behaves as a memory for a host machine while processing the stored data. Extensibility is also given by allowing PAM to be connected with I/O modules or other PAMs. On the other hand, our SCM array and its concept differ in the following ways from PAM. First, PAM is not specialized for floating-point computation. Second, the constructive unit of our SCM array is different from that of PAM. The constructive unit of PAM is PAM itself, which is an FPGA array where custom circuits are configured over multiple FPGAs. In our SCM array, each FPGA is a basic unit and has the same hardware design as a module. The array of FPGAs forms a scalable SCM array, and therefore we can easily extend the system by adding FPGAs, such as is implemented on "stackable mini-FPGA-boards."

Although the peak performance cannot be exploited due to the low operational-intensity of stencil computation, efforts have also been made to optimize the computation on GPUs [19]. Williams et al. reported the performance of 3D stencil computation with a GPU [4]. A single NVidia GTX280 GPU achieves 3D stencil computation of 36 GFlop/s, which is a much higher performance than that of a multi-core processor due to the GPU's high memory bandwidth of 142 GByte/s. However, the efficiency is not high, being 46% of the double-precision peak performance of 78 GFlop/s. It is important to note here that the performance is measured without the data-transfer to/from a host PC. The inefficiency results in a low performance per power, 0.15 GFlop/sW, even though they measured the power consumption of only the GPU board.

Phillips et al. reported parallel stencil-computation on a GPU cluster [19]. By applying optimization techniques, they achieved 51.2 GFlop/s with a single NVidia Tesla C1060 similar to GTX280, which corresponds to 66% of the peak performance. However, in the case where 16 GPUs were utilized, only a $\times 10.3$ speedup was achieved for a $256^2 \times 512$ grid because of the communication/synchronization overhead among GPUs. As a result, the efficiency is reduced to 42% from 66%. These results show that while each GPU has a high peak-performance, it is difficult to obtain high sustained-performance for stencil computation, especially with multiple GPUs. Scalability is significantly limited in a large parallel system with many accelerators, and most of the entire performance is easily spoiled. We need a solution for efficiently scaling performance according to the size of the system.

3 Target Computation and Architecture

3.1 General Form of Stencil Computations

The SCM array targets iterative stencil-computation, which is described by the pseudo-code of Fig. 1. In scientific simulations, such as CFD, the partial differential equations (PDE) that govern the physical phenomena to be simulated are often numerically solved by the finite difference method. The computational kernels for the method are of stencil computations, which are given by discretizing values defined at grid points and numerically approximating the derivatives. Here we explain a 2D case for simplicity. Let $v(n,i,j)$ be values defined on a 2D computational grid, where n, i and j denote a time step, x- and y-positions, respectively. The nested inner-loops iterate the stencil computation for each grid-point (i,j) over the grid. At each grid-point, $v(n,i,j)$ is updated from time-step n to $(n+1)$ by computing $F()$ with only the values in the neighboring grid-points. We refer to the region

```
for(n=0; n<N ; n++) {   // for iterations

  for(j=0; j<y; j++)     // for grid traverse
    for(i=0; i<x ; i++) {
      // Update grid value from n to n+1
      v(n+1,i,j) := F( v(n,i,j) in S(i,j) )
    }
}
```

Fig. 1 Pseudo code of iterative stencil computation

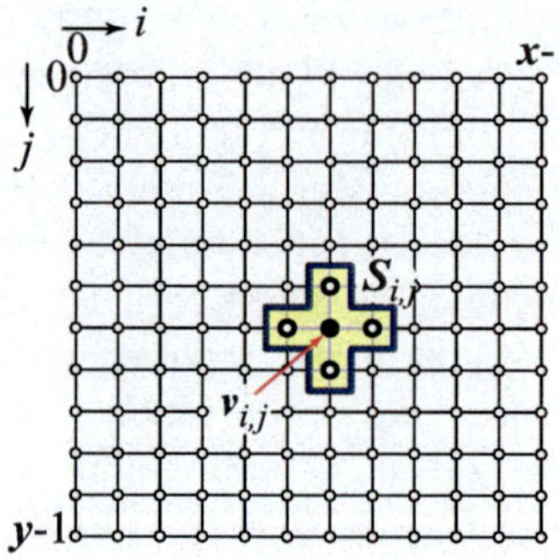

Fig. 2 Simple 3 × 3 star stencil

of these neighboring grid-points as *a stencil* and denote it with $S(i, j)$. Figure 2 illustrates a 2D simple example of a 3 × 3 star-stencil containing the five local grid-points. The outer loop of the pseudo code repeats the grid updates for time-marching computation along successive time-steps.

As reported in [20–22], $F()$ of typical stencil computation is simply a weighted sum of the neighboring values, which is written as:

$$v_{i,j}^{\text{new}} = c_0 v_{i,j} + c_1 v_{i-1,j} + c_2 v_{i+1,j} + c_3 v_{i,j-1} + c_4 v_{i,j+1} \tag{1}$$

for a 3 × 3 star-stencil, where c_0–c_4 are constants. We refer to (1) as *a 2D neighboring accumulation*. Similarly, a 3D neighboring accumulation requires two more terms for a 3 × 3 × 3 star-stencil in 3D. Although wide stencils are used for higher-order differential schemes, they can be decomposed into 2D or 3D multiple neighboring-accumulations. Therefore we consider that neighboring accumulation for 3 × 3 or 3 × 3 × 3 star-stencils is a general form of 2D or 3D stencil computation, respectively, which should be accelerated. The stencil computation has parallelism, which allows neighboring accumulations of grid-points to be independently performed. Furthermore, the computation of each grid-point has locality because the necessary values are localized around the point. We designed the SCM array to accelerate the neighboring accumulations by exploiting their parallelism and locality.

3.2 SCM Architecture

The SCM array is based on the SCM architecture [20,22,23], which is a combination of the programmable systolic array [11,12] and the computational memory approach [7, 17, 18, 29]. The systolic array is a regular arrangement of many processing elements (PEs) in an array, where data are processed and synchronously transmitted between neighbors across the array. Such an array provides scalable performance according to the array size by pipelining and spatially parallel processing with input data passing through the array. However, the external memory access can also be a

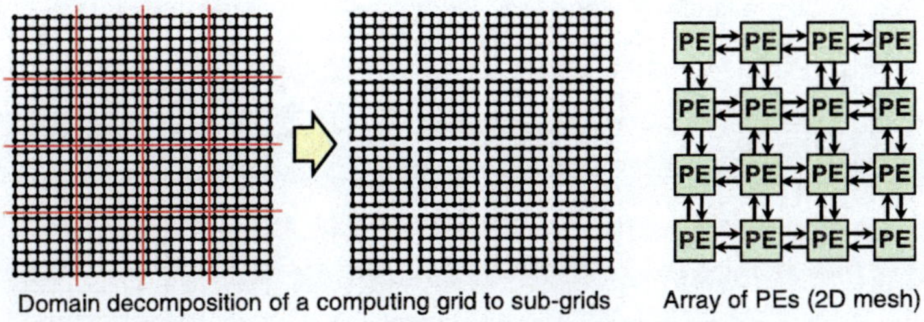

Fig. 3 Grid decomposition into sub-grids and their assignment to a PE array

bottleneck of the performance improvement if the memory bandwidth is insufficient in comparison with the bandwidth required by the array.

To avoid this limitation of the external memory bandwidth, we adopted the computational memory approach. This approach is based on the "processing in memory" concept, where computing logic and memory are arranged very close to each other [7, 17, 18, 29]. In the SCM architecture, the entire array behaves as *a computing memory*, which not only stores data but also performs computations with them. The memory of the array is formed by the local memories distributed to the PEs. Since PEs can concurrently access their local memories and perform computations, the SCM architecture has a computing performance scalable to the array size without any bottlenecks related to external memories.

The SCM architecture is actually a 2D systolic array of PEs connected by a mesh network. To exploit the parallelism and locality of stencil computation, we decompose the entire grid into sub-grids, and assign them to the PEs, as shown in Fig. 3. Since all the grid data are loaded onto the local memories, the PEs can perform stencil computations of their assigned sub-grids in parallel, exchanging the boundary data between adjacent PEs.

4 Design of SCM Array

4.1 2D Array of Processing Elements

Figure 4 shows an overview of the designed SCM array, which is a 2D array of PEs with local memories. The PEs are connected by a 2D mesh network via communication FIFOs. Each PE has a floating-point multiply-and-accumulate (FMAC) unit for the neighboring accumulation with data read from the local memory or the FIFOs. The result can be written to the memory and/or sent to the FIFOs of the adjacent PEs. The four communication FIFOs, $\{S,N,E,W\}$*-FIFOs*, hold the data transferred from the four adjacent PEs until they are read.

Figure 5 shows the data-path of a PE, which is pipelined with the following eight stages: *the instruction sequencing* (IS), *the memory read* (MR), *the five execution*

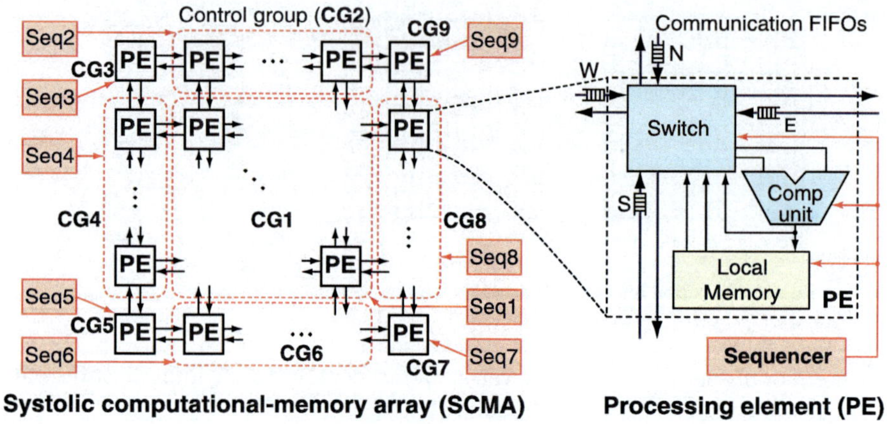

Fig. 4 SCM array of PEs and sequencers for control groups

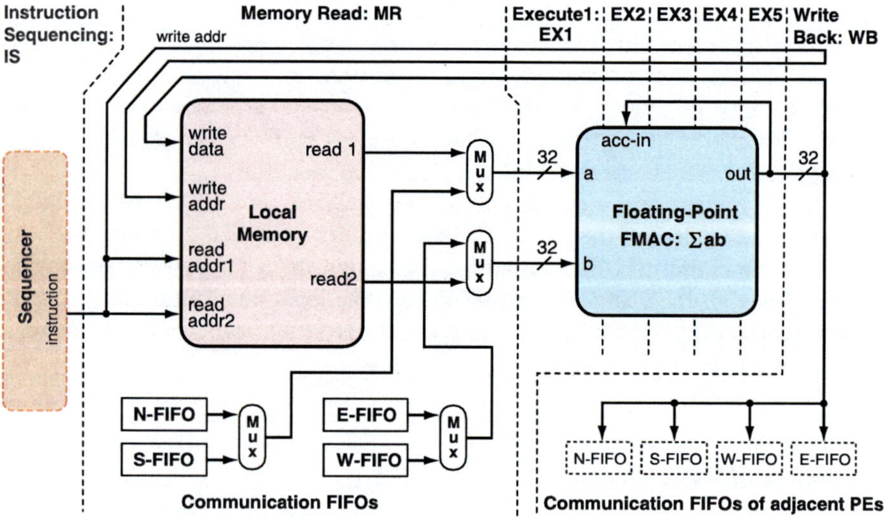

Fig. 5 Pipelined data-path of a PE

stages of the FMAC unit (EX1 to EX5), and *the memory write-back* (WB) [22]. In the MR stage, two values are selected from the read data of the local memory, and the data outputted by the communication FIFOs. The selected values are inputted to the FMAC unit. The FMAC unit can sequentially accumulate an arbitrary number of the products of inputs, which are IEEE754 single-precision floating-point numbers. Since the FMAC unit has the forwarding path from the EX5 stage to the EX2 stage for accumulation, it can sum up the products of inputs fed every three cycles. This means that three sets of (1) are required to fully utilize the FMAC unit.

4.2 Sequencers and Instruction Set

As shown in Fig. 4, the SCM array has sequencers, which give programmability to the array. In scientific simulation with a computational grid, the PEs taking charge of the inner grid-points regularly perform the same computation while the PEs for the boundary grid-points have to execute different computations. In order to achieve such diversity of control, we partition the PEs into several control groups (CGs), and assign a sequencer to each of them. Figure 4 shows a typical example of CGs partitioning. Here the array has the nine CGs: CG1, CG2, ..., CG9, which are given the PEs of the top, left, bottom, and right sides; the four corners; and the inner grid-points. The sequencer of each CG reads instructions from its own sequence memory and sends control signals to all the PEs of the CG based on the instructions. That is, the PEs of each CG are controlled in single instruction-stream and multiple data-stream (SIMD) fashion.

Table 1 shows an instruction set of the SCM array. Instructions are classified into two major groups: *control instructions* and *execution instructions*. Each control instruction supports the nested loop-control, no operation (nop), and halt. Nested-loops are executed with the lset and bne operations. The lset initiates the next-level loop by setting the number of iterations and the starting address of the loop body. After the lset is executed, the loop counter is decremented and the branch to the starting address is taken if it is not zero when the bne is executed. Each execution instruction has an opcode to select operation of the FMAC, two operands to specify the sources of the FMAC inputs, and two operands to specify the destinations of the FMAC output for the local memory and the communication FIFOs. Please note that an execution instruction can be merged with a bne instruction.

The code of Fig. 6 is an example of a sequence for three sets of (1) for grid-points (0,0), (1,0), and (2,0), where all the constants are 0.25. The grid is decomposed into 3×2 sub-grids as shown in Fig. 7. The PE computing the grid points communicates with the east, west, and north PEs through the E-, W-, and N-FIFOs, respectively. Because of the three-cycle forwarding of the FMAC unit, we concurrently perform the three sets of accumulations every three instructions. The lset instruction is used to repeat the computation for 1,600 times. Note that the branch is actually performed after the next instruction to the bne is executed. The code written in the assembly language is assembled and converted to the sequence binary for sequencers.

4.3 Techniques for Multiple-FPGA Implementation

For performance scalability, we design the SCM array to be scalably implemented over multiple FPGAs, by partitioning the array into sub-arrays. We can build a larger SCM array with more FPGAs. Figure 8 shows sub-arrays implemented over FPGAs A, B, and C.

Table 1 Instruction set of the SCM array (in an assembly language of the SCM array)

Type	opcode	dst1,	dst2,	src1,	src2	Description
Control instructions	nop					No operation
	halt					Halt
	lset	ITERS,	ADDR			Loop-counter$_i$:= ITERS Bne-reg$_i$:= ADDR (for the i-th nested loop)
	bne					Branch to Bne-reg$_i$ if Loop-counter$_i$ is not eq. to zero.
Execution instructions	mulp	-,	L1,	L2,	EFIFO	FMACout = M[L2] × E-FIFO, M[L1] := FMAC-out
	accp	SE,	L1,	L2,	L3	FMAC-out = FMAC-out + M[L2] × M[L3], {S,E}-FIFOs := FMAC-out, M[L1] := FMAC-out
Merged	accpbne	-,	L1,	L2,	L3	accp & bne

```
QUAD  =         0.25             // LABEL definition
      lset      1600,        LOOP
LOOP: mulp          ,            ,        v[0;0],      QUAD
      mulp          ,            ,        v[1;0],      QUAD
      mulp          ,            ,        v[2;0],      QUAD
      accp          ,            ,        v[1;0],      QUAD
      accp          ,            ,        v[2;0],      QUAD
      accp          ,            ,         EFIFO,      QUAD
      accp          ,            ,         WFIFO,      QUAD
      accp          ,            ,        v[0;0],      QUAD
      accp          ,            ,        v[1;0],      QUAD
      accp          ,            ,         NFIFO,      QUAD
      accp          ,            ,         NFIFO,      QUAD
      accp          ,            ,         NFIFO,      QUAD
      accp        SW,   vnew[0;0],        v[0;1],      QUAD
      accpbne      S,   vnew[1;0],        v[1;1],      QUAD
      accp        SE,   vnew[2;0],        v[2;1],      QUAD
```

Fig. 6 Example code of (1) for (0,0), (1,0) and (2,0) on the sub-gird of Fig. 7. QUAD and "v[0;0]" are labels for local-memory addresses

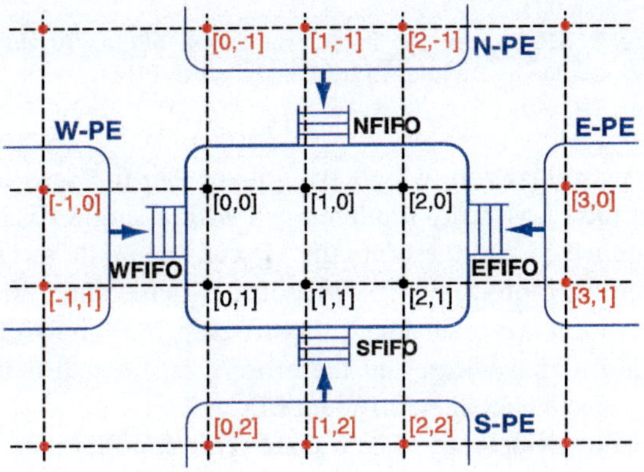

Fig. 7 Assignment of 3×2 sub-grids for Jacobi computation with PEs

In multi-FPGA implementation, the more devices the system has, the more difficult it is to uniformly distribute a single clock source to them. To avoid distributing a single clock to all devices in a large system, we introduce *a globally asynchronous and locally synchronous (GALS) design* [13], where each FPGA is given an independent clock-domain. We transfer data between different clock-domains without meta-stability by using dual-clock FIFOs (DcFIFOs), which have different clock sources for input and output. Multiple clock-domains allow us to easily build an extensible and scalable system with many FPGAs.

4.3.1 Peak-Bandwidth Reduction Mechanism (BRM)

From a programming point of view, we should allow the sub-arrays over multiple FPGAs to logically operate as a single array. However, multiple-FPGA implementation presents several problems related to the off-chip bandwidth and

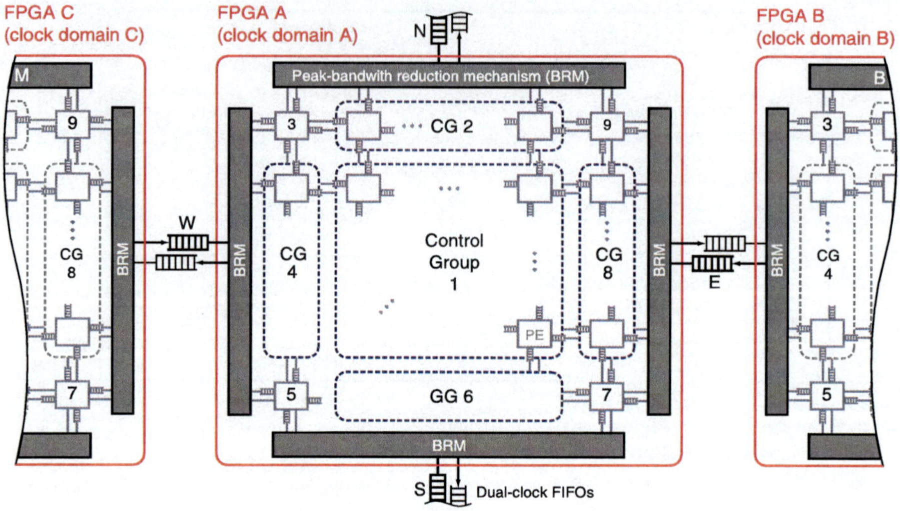

Fig. 8 SCM sub-arrays implemented over FPGAs A and B with different clock-domains. The clock-domains are bridged by the dual-clock FIFOs

synchronization, which are not seen in the case of single-FPGA implementation. In the SCM array implemented with a single FPGA, the adjacent PEs are directly connected and therefore the PE can send data to the adjacent PEs every cycle. The abundant on-chip wiring-resources make such direct connection available among PEs. On the other hand, the off-chip I/O bandwidth is much limited compared to the internal wires, and therefore it is more difficult for all the PEs to be directly connected between different FPGAs.

Fortunately, the data transfer between PEs is less frequent than read/write of the local memory, and we can utilize this characteristic to solve the off-chip bandwidth problem [22]. When a PE sends the result of computing (1), the data-transfer occurs only after the cycles necessary for FMAC unit to accumulate the terms. For example, the code of Fig. 6 has only the three instructions to send data in the fifteen instructions of the loop body. In addition, only the border grid-points of a sub-grid cause data transfer between adjacent PEs. Thus, actual stencil computation requires much less net-bandwidth than the peak bandwidth for sending data every cycle. Accordingly, the inter-FPGA bandwidth does not have to be as high as the aggregate memory-bandwidth of the border PEs in the SCM array.

However we still need a technique to handle the successive data-transfers conducted every cycle, which locally request the peak bandwidth, even if the inter-FPGA bandwidth is more than the average data-transfer rate. We designed a module with a peak-bandwidth reduction mechanism (BRM) to buffer the successive data-transfer requests, which is fully described in [22]. BRM is based on time-division multiplexing (TDM) for multi-cycle data-transfer with a buffer FIFO. For explanation, assume that the inter-FPGA bandwidth has the one n-th of the aggregate memory bandwidth of the border PEs. In this case, a set of words sent by the border PEs at a cycle is buffered by BRM, and takes n cycles to arrive at the adjacent FPGA. We refer to such BRM as "n : 1 BRM," which reduces the peak bandwidth to the one n-th with an n-times longer delay. If BRM has a buffer with a sufficient size for successive data-transfer, it can average out the net bandwidth. Thus BRM has a trade-off between the peak-bandwidth reduction and the delay increase. Therefore, we have to additionally take care of the increased delay in scheduling instructions to send data and use the received data.

4.3.2 Local-and-Global Stall Mechanism

In addition to the off-chip bandwidth problem, multi-FPGA implementation causes a synchronization problem of execution and data-transfer among devices because of the slight but inevitable difference in frequency among different clock domains even if their sources have the same configuration of frequency. Assume that we have two clock oscillators for 100 MHz. They can be different from each other, for example, 100.001 and 100.002 MHz. Since different frequencies cause the PEs to execute at different speeds, we need to synchronize the instructions executed by the PEs to send and receive data. The inter-PE communication is performed by explicitly executing instructions for communication, which send data to the adjacent PEs and read data from the communication FIFOs. If all the PEs are synchronized to a single clock, we can statically schedule instructions for adequate communication. However, if we use different clocks, PEs operating at a higher frequency could read an empty FIFO before the corresponding datum is written to the FIFO, and/or write a datum to a full FIFO before the FIFO is read by another PE. To avoid these data-transfer problems, we need some hardware mechanism to suspend the PEs in the case of a read-empty (RE) or write-full (WF).

We can easily think of the simplest design of the SCM array where all the sequencers and PEs simultaneously stall just after RE or WF is detected. However, this design is impractical. Although each sequencer can stall immediately when it locally detects RE or WF, the stall signal takes more than one cycle to be distributed from the sequencer to the others and make them stall. This is because the stall-signal distribution requires at least one OR operation for RE and WF detection, long wires with large fan-out, and another OR operation to generate the global-stall signal. If we implement the stall-signal distribution within one cycle, it reduces the operating frequency.

To solve this problem, we introduce *a LGSM* to the CGs, which is based on the very simple concept that sequencers stall immediately when they detect RE or WF, and the others stall several cycles later. Figure 9 shows an overview of the LGSM. Sequencer 2 of CG 2 observes $w._{\mathrm{N}}$ and $f._{\mathrm{S}}$ to detect RE and WF, which are the write signal of N(north)-FIFOs of its own PEs and the almost-full signal of S(south)-FIFOs of the PEs in the different clock-domain, B, respectively. With $w._{\mathrm{N}}$, the sequencer counts the number of remaining data in the N-FIFOs for the issued read-operation at present. When the sequencer issues a read operation for an

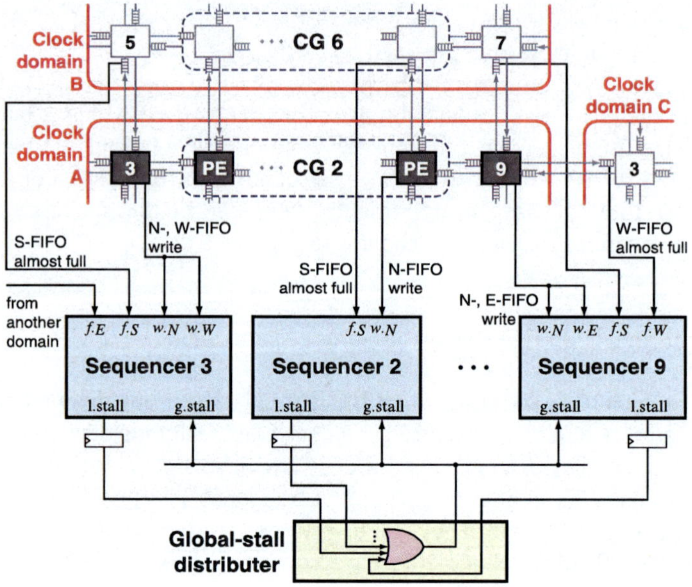

Fig. 9 Local-and-global stall mechanism (LGSM) to guarantee data-transfer synchronization in the GALS design

empty FIFO or a write operation for an almost-full FIFO, the sequencer immediately becomes the stall state. Then it sends a local-stall (l.stall) signal to *the global-stall distributor* (GSD).

The l.stall signal is latched at the output of each sequencer and reaches GSD at the next cycle. GSD has the inputs of the l.stall signals from all the sequencers. In GSD, the OR operation with all the l.stall signals generates a global-stall (g.stall) signal, which is distributed to all the sequencers. If necessary, we can insert additional latches into the distribution tree of the g.stall signal to prevent the operating frequency from decreasing. Sequencers in the execution state stall immediately when they receive the g.stall signal. Note that the sequencers with delayed stall issue one more instruction than Sequencer 2 because it locally stalls prior to the global stall. Therefore, Sequencer 2 issues the instruction before the other sequencers resume execution. This function is provided by a stall-control unit in each sequencer. The details of LGSM are described in [13].

5 Implementation and Evaluation

5.1 Implementation

With implementation of SCM arrays on multiple-FPGAs, we demonstrate that SCM arrays have scalability according to the number of FPGAs. We also evaluate the

Fig. 10 1×3 FPGA-array of 3 DE3 board

overhead caused by LGSM for its frequency degradation and resource consumption by comparing 3×3, 4×4, and 8×8 SCM arrays with and without LGSM. We use Terasic DE3 boards [28] to construct the 1×3 FPGA-array of Fig. 10, while our final goal is to build a 2D FPGA-array as shown in Fig. 11.

Figure 12 shows the block diagram of the DE3 board. The DE3 board has an ALTERA Stratix III EP3SL150 FPGA [1] and four HSTC connectors, which can be used to connect DE3 boards to each other with a single-end or LVDS signaling. Each HSTC connector has 130 I/O pins including clock lines. Each pair of the three DE3 boards is connected by the two HSTC connectors for bi-directional data-transfer. By using 64 pins of the 130 pins at 100 MHz in single-end, each HSTC connector provides a uni-directional data-transfer of 0.8 GB/s between FPGAs. The 4:1 BRM allows at most 8×8 PEs to send or receive data between FPGAs with this connection. The Stratix III EP3S150 FPGA has 113,600 ALUTs (adaptive LUTs), which is equivalent to 142,000 LEs, block RAMs with a total of 6,390 Kbits, and 96 36-bit DSP blocks. The block RAMs consist of 355 M9K blocks and 16 M144K blocks. The size of each M9K block is 9 Kbits while each M144K has 144 Kbits.

Figure 13 shows a block diagram of a system implemented on each FPGA. We wrote verilog-HDL codes of 3×3, 4×4, and 8×8 SCM arrays with or without LGSM. We implement the 4:1 BRMs and the distributors to reduce the peak-bandwidth requirement to one-fourth for inter-FPGA connection. We use DcFIFOs to connect the logics in different clock-domains. We also implement ALTERA's system on programmable chip (SOPC) with an NIOS II processor for the USB interface. The host PC can read and write data and commands to the SCM array's

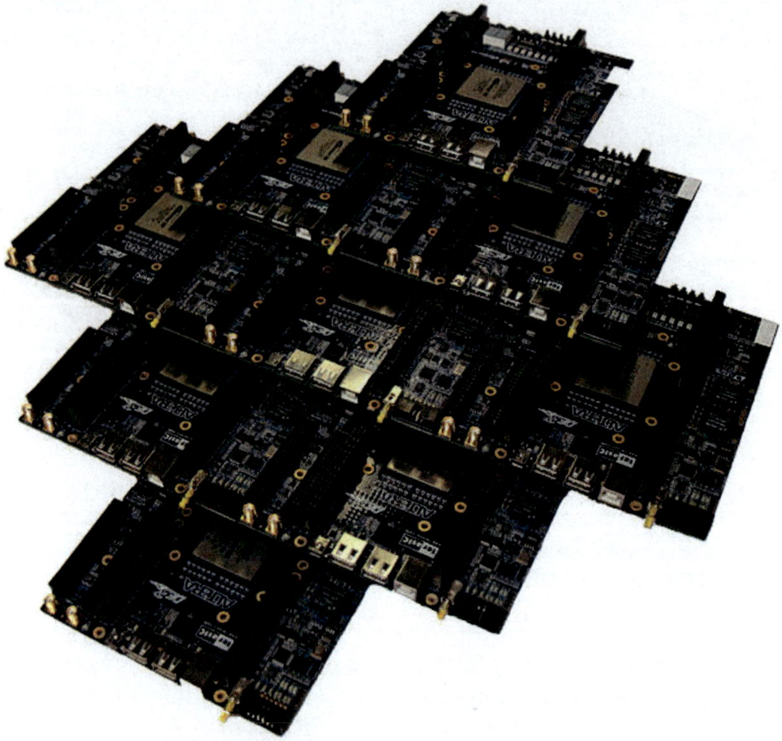

Fig. 11 3×3 FPGA-array of 9 DE3 boards, which is planed to be implemented

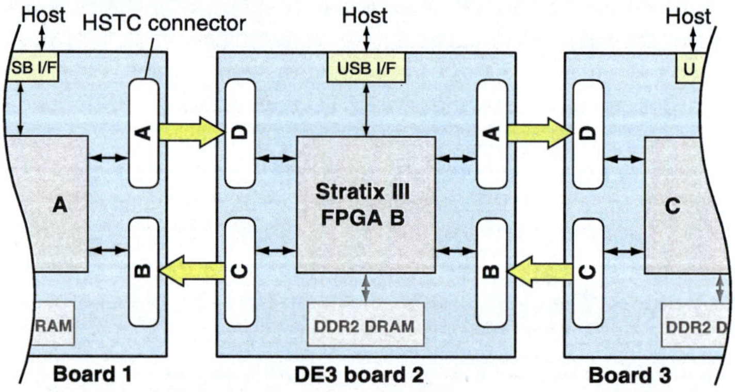

Fig. 12 Block diagram of DE3 board and a 1×3 FPGA array

memories via USB. We compiled the system by using ALTERA Quartus II compiler version 9.1 with the options of "area," "standard fit," and "incremental-compilation off."

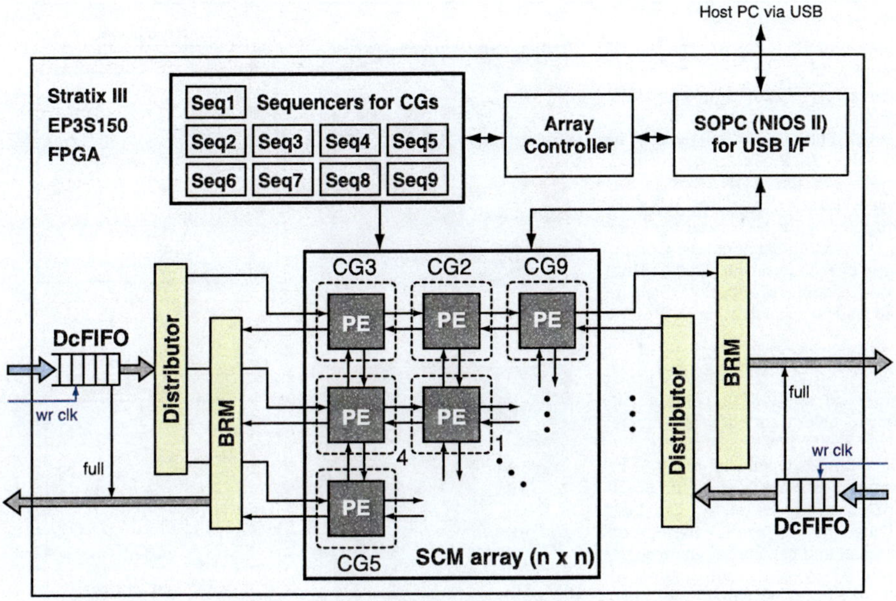

Fig. 13 Block diagram of a system including a sub-SCM-array implemented on an FPGA

5.2 Benchmark Computations

PEs perform floating-point operations for (1), and therefore the prototyped system can actually compute real applications based on the finite difference method. We program the PEs of the SCM array in the dedicated assembler language [22]. For benchmarks, we use the applications summarized in Fig. 14. The red-black successive over-relaxation method [9], *RB-SOR*, is one of the parallelized iterative solvers for Poisson's equation or Laplace's equation. In the RB-SOR method, the grid points are treated as a checkerboard with red and black points, and each iteration is split into a *red*-step and a *black*-step. The red- and black-steps compute the red points and the black points, respectively. We solved the heat-conduction problem on a 2D square plate with RB-SOR. Each PE computes with an 8×24 sub-grid so that the 9×3 SCM array on the three FPGAs computes a 72×72 grid for 2.0×10^6 iterations.

The fractional-method [27], *FRAC*, is a typical and widely used numerical method for computing incompressible viscous flows by numerically solving the Navier–Stokes equations. We simulated the 2D square driven cavity flow with the FRAC, giving the result shown in Fig. 14. The left, right, and lower walls of the square cavity are stable, and only the upper surface is moving to the right with a velocity of $u = 1.0$. Each PE takes charge of a 5×15 sub-grid. For the 9×3 SCM array on the three FPGAs, we compute 5,000 time-steps with a 45×45 grid while the Jacobi computation is performed for 1,000 iterations at each time-step.

Red-black SOR (RB-SOR)

Numerical solver of Laplace's equation: $\nabla^2\phi = 0$. The 2D time-independent heat-conduction is computed with a 2D grid by using the red-black SOR method. The red-black SOR is a parallelized version of the SOR (Successive Over-Relaxation) method, which is computed with a checkerboard-like grid of red and black points. Computation of each iteration is divided into a red step and a black step, which are independent to each other.

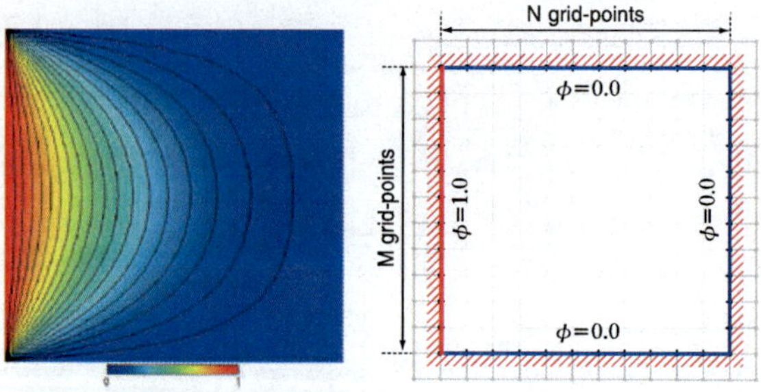

Fractional-step method (FRAC)

Numerical method to compute incompressible viscous flow. 2D square driven cavity flow is computed giving a time-independent result. The fractional-step method is composed of calculating tentative velocities, solving the Poisson's equation of pressure, and calculating the true velocities. We use Jacobi method to solve the Poisson's equation of pressure.

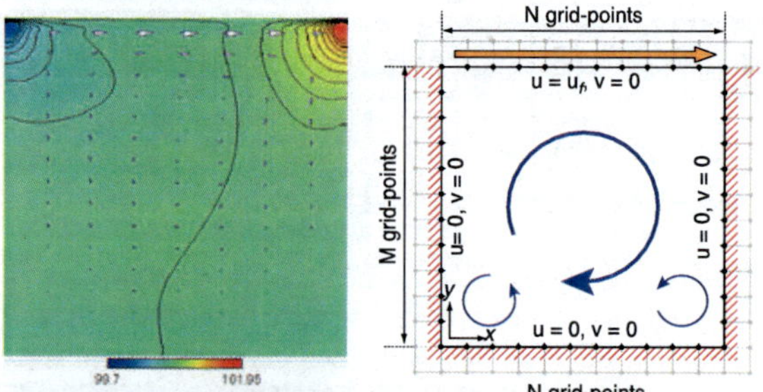

FDTD method (FDTD)

Numerical method to solve the Maxwell's equations for electromagnetic problems. The 2D propagation of electromagnetic waves is computed with a 2D grid. The square-wave source was placed at the left-bottom corner of the grid. We use the Mur's first-order absorbing boundary condition for the border of the grid.

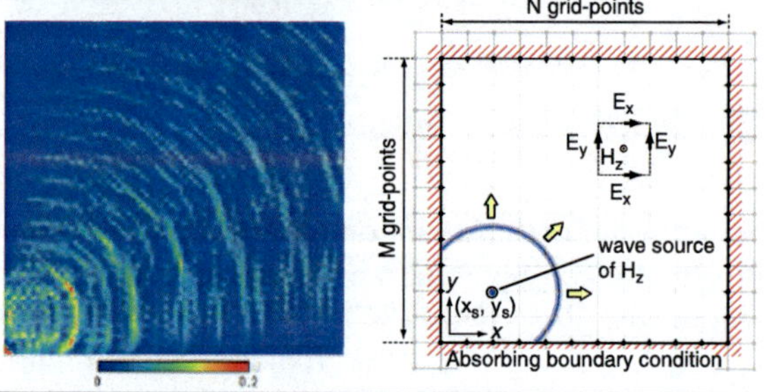

Fig. 14 Benchmark computations

The FDTD method [31], is a powerful and widely used tool to solve a wide variety of electro-magnetic problems, which provides a direct time-domain solution for Maxwell's Equations discretized by difference schemes on a uniform grid and at time intervals. Since the FDTD method is very flexible and gives accurate results for many non-specific problems, it is widely used for solving a wide variety of electro-magnetic problems. We compute the FDTD method to simulate 2D propagation of electro-magnetic waves with a square-wave source. At the left-bottom corner, we put the square-wave source with an amplitude of 1 and a period of 80 time-steps. On the border, Mur's first-order absorbing boundary condition is applied. Each PE takes charge of a 4×6 sub-grid. For the 9×3 SCM array on the three FPGAs, we compute 1.6×10^6 time-steps with a 36×18 grid.

Table 2 Synthesis results for Stratix III FPGA

		Without LGSM			With LGSM		
Array size		3 × 3	4 × 4	8 × 8	3 × 3	4 × 4	8 × 8
f_{max} [MHz]		128	123	123	128	125	118
ALUTs	SCMA	12,248 (10.8%)	20,984 (18.5%)	78,189 (68.8%)	12,420 (10.9%)	21,179 (18.6%)	78,341 (69.0%)
	BRM	881 (0.776%)	1,123 (0.989%)	2,274 (2.00%)	878 (0.773%)	1,128 (0.993%)	2,260 (1.99%)
	Others	10,041 (8.84%)	7,743 (6.82%)	6,133 (5.40%)	9,080 (7.99%)	10,784 (9.49%)	6,133 (5.40%)
36-bit DSP blocks		9 (9.38%)	16 (16.7%)	64 (66.7%)	9 (9.38%)	16 (16.7%)	64 (66.7%)
Total memory bits		1,842,752 (32.7%)	2,072,128 (36.8%)	3,645,592 (64.7%)	1,842,752 (32.7%)	2,072,128 (36.8%)	3,645,592 (64.7%)

5.3 *Synthesis Results*

Table 2 shows the synthesis results of SCM arrays with and without LGSM for $3 \times 3, 4 \times 4$, and 8×8 PEs. The larger the SCM array, the lower the frequency that is available. This is because of the longer critical paths in the larger array. However, the 8×8 array can still operate at more than 118 MHz, which is sufficiently higher than 100 MHz. The 8×8 array consumes about 75% of the ALUTs, 67% of the 36-bit DSP blocks, and 65% of the total memory bits. Based on these data, we estimate that we can implement up to 100 PEs on this FPGA to form a 10×10 array, giving 1.5 times higher performance than that of the 8×8 array.

LGSM slightly decreases the maximum operating frequency, f_{max}, by only a few MHz. The difference in frequency of the 8×8 array between SCM arrays with and without LGSM is only 5 MHz. This means that LGSM does not have a major impact on the scalability to the array size. There is almost no performance degradation caused by introducing LGSM. LGSM slightly increases resource consumption by 0.17% at most for the ALUTs of the SCM array and do not increase the total memory bits. Note that the ratio of resources increased by LGSM is almost constant regardless of the array size. This is because the number of sequencers, 9, is unchanged. These results show that we can use LGSM at very low cost in terms of resources.

5.4 *Performance Results*

Although we set 100 MHz to all the clocks of the FPGAs, they can be slightly different because they are generated by different clock sources. The stall cycles caused by this difference may considerably increase the total cycles of computation

Table 3 Execution and stall cycles, performance and utilizations for 1, 2 and 3 FPGAs

Computation		1 FPGA	2 FPGAs	3 FPGAs
RB-SOR	Exec cycles	1,440,003,040	1,440,003,040	1,440,003,040
	Stall cycles	0 (0.0%)	11,207 (0.00078%)	11,250 (0.00078%)
	GFlop/s	11.4	22.9 (×2.01)	34.4 (×3.02)
	Utilization	89.2%	89.5%	89.7%
FRAC	Exec cycles	977,382,044	977,382,044	977,382,044
	Stall cycles	0 (0.0%)	7,592 (0.00078%)	7,591 (0.00078%)
	GFlop/s	11.2	22.4 (×2.01)	33.7 (×3.02)
	Utilization	87.2%	87.6%	87.7%
FDTD	Exec cycles	950,432,027	950,432,027	950,432,027
	Stall cycles	0 (0.0%)	7,220 (0.00076%)	7,212 (0.00076%)
	GFlop/s	10.2	20.6 (×2.02)	30.9 (×3.04)
	Utilization	79.6%	80.4%	80.6%

and spoil the scalability by using multiple devices. To evaluate the stall cycles of SCM arrays operating on multiple FPGAs, we implement the 8×8 array with LGSM on each FPGA and execute the benchmark programs of RB-SOR, FRAC, and FDTD with SCM arrays on a single FPGA, two FPGAs and three FPGAs connected as a 1D array. For the execution, we gave the constant size of the sub-grid computed by each PE so that the size of the entire grid is proportional to the number of FPGAs. Therefore, the SCM arrays on a different number of FPGAs require the same cycles for computation while larger SCM arrays provide higher performance with larger grids computed by more PEs.

Table 3 shows the numbers of execution cycles and stall cycles for the SCM arrays on the single-, double-, and triple-FPGA arrays. Since the SCM array on a single FPGA operates with a single clock-domain, no stall cycle is observed for all the benchmark computations on the single-FPGA SCM array. On the other hand, the double-FPGA and triple-FPGA SCM arrays have stall cycles due to the difference in frequency among clock domains, which cause a slight increase in the total number of cycles. However, the ratios of the stall cycles to the total execution cycles are very small and ignorable and are about $8 \times 10^{-4}\%$. With these results, we made sure that the clock frequencies of 100 MHz have small but inevitable differences, and that the stall mechanism works well to guarantee the data synchronization with slight loss of performance.

Table 3 also shows the actual floating-point performance for each benchmark computation in GFlop/s. We used the measured cycles including the execution and stall cycles to obtain the performance. We calculated the utilization that is defined as the actual performance divided by the peak performance. The results show that high utilization of 80–90% is achieved for the benchmark computations executed on the SCM arrays irrespective of the number of FPGAs. Note that almost the same utilization is maintained in increasing the number of FPGAs. The utilization is slightly improving rather than declining because the ratio of border grid-points is reduced.

These results show that the SCM array provides complete scalability to the array size and the number of devices, so that m FPGAs achieve m times higher performance to compute an m times larger grid. If we implement a 10×10 sub-array on each FPGA, the three FPGAs are expected to give 53.8 GFlop/s to RB-SOR. Note that the total bandwidth of the local memories is also completely proportional to the size of the SCM array. The single-FPGA, double-FPGA, and triple-FPGA SCM arrays have the internal bandwidth to read and write local memories of 76.8, 153.6, and 230.4 GByte/s at 100 MHz, respectively. If we implement a 10×10 sub-array on each FPGA instead of a 8×8 sub-array, they could provide 120, 240, and 360 GByte/s, respectively.

5.5 *Feasibility and Performance Estimation for State-of-the-Art FPGAs*

Here we discuss the feasibility of implementation with 2D FPGA arrays and estimate their peak performance for high-end FPGA series. We consider the three high-end ALTERA FPGAs: Stratix III EP3SL340 FPGA (65 nm), Stratix IV EP4SGX530 FPGA (40 nm), and Stratix V 5SGSD8 FPGA [1]. The resources of these FPGAs are summarized in Table 4. We obtain the total I/O bandwidth of each FPGA by multiplying the number of transceivers and the bandwidth per transceiver. We assume that implementation optimization allows PEs to operate at 125 MHz on these FPGAs. Since an FMAC requires one single-precision floating-point multiplier to be implemented with one 36-bit or 27-bit DSP block, we assume that the number of PEs available on each FPGA is the same as the number of 36-bit or 27-bit DSP blocks. Each PE can perform both multiplication and addition at 125 MHz. Therefore, the peak performance of each FPGA is obtained by $0.25 \times$ (the number of PEs) [GFlop/s].

As shown in Table 4, while Stratix III and IV FPGAs have a moderate peak-performance, Stratix V FPGA has a peak performance of 491 GFlop/s per chip. This is due to integration of many DSP blocks on state-of-the-art FPGA. Since an SCM array performs stencil computation with a utilization of about 85%, the Strativ V FPGA is expected to achieve a sustained performance of $491 \times 0.85 = 417$ GFlop/s per chip. This performance is higher than the sustained performance of stencil computation on a single GPU. Furthermore, due to the complete scalability of an SCM array, we can efficiently scale the performance by using multiple FPGAs. For example, we can obtain a sustained performance of 41.7 TFlop/s with 100 Stratix V FPGAs. The cluster of 100 FPGAs can be cheaper than a larger GPU cluster to provide comparable performance with much less utilization.

For feasibility, we should discuss whether the I/O bandwidth is sufficient to connect the FPGAs with a 2D mesh network because a 2D FPGA array is suitable to scale a 2D SCM array. We also assume an $N \times N$ square array of PEs where $N = \sqrt{(\text{the number of PEs})}$ on each FPGA, the required bandwidth for four links

Table 4 Estimation of array size and available bandwidth for a 2D array of high-end FPGAs

		Stratix III L EP3SL340	Stratix IV GX EP4SGX530	Stratix V 5SGSD8
Technology		64 nm	40 nm	28 nm
Equivalent LEs		337,500	531,200	695,000
Memory [KBytes]		2,034	2,592	6,417
36/27-bit DSPs		144	256	1,963
Transceivers (Gbps/ch)		132 (0.156)	48 (8.5)	48 (1.41)
Total I/O bandwidth [GB/s]		**20.6**	**51**	**84.6**
Assumed PE freq. [MHz]		125	125	125
Estimated # of PEs		144	256	1,963
Peak GFlop/s per FPGA		36	64	491
Required uni-directional BW for 4 links [GB/s]		**24.0**	**32**	**88.6**
Reduced BW by $n:1$ BRM [GB/s]	$n=2$	**12.0**	**16.0**	**44.3**
	$n=4$	**6.0**	**8.0**	**22.2**
	$n=8$	**3.0**	**4.0**	**11.1**

(Peak GFlop/s) = (# of PEs) $\times$ 2 operations $\times$ 0.125 GHz)
(Req. unidir BW for 4 links) = $4 \times \sqrt{(\# \text{ of PEs})} \times 4$ words $\times$ 0.125 GHz

of a 2D mesh network is calculated by $4 \times N \times 4 \times 0.125$ GByte/s. In Table 4, the required bandwidth for four links is slightly higher than the available I/O bandwidth for Stratix III and V FPGAs. However, BRM reduces the required bandwidth less than the I/O bandwidth. Especially, much less bandwidth is required if we use 4:1 BRM. These estimations show that implementation of an SCM array with multiple FPGAs is feasible and we can build an array of a large number of FPGAs without inter-FPGA communication bottleneck.

6 Summary

This chapter presents the SCM array for programmable stencil computations. The target computation of an SCM array is the neighboring accumulation for 3×3 star-stencil computations. The SCM array is based on the SCM architecture, which combines the systolic array and the computational memory approach to scale both computing performance and aggregate memory-bandwidth in accordance with the array size. After the structure and behavior of processing elements and their sequencers are described, we show GALS implementation for multiple FPGAs with different clock-domains. We also present two techniques, the peak-bandwidth reduction mechanism (BRM) and the LGSM, which are necessary to solve the bandwidth and synchronization problems of inter-FPGA data-transfer.

We implement a prototype SCMA with three Stratix III FPGAs. We demonstrate that the prototyped SCMAs compute the three benchmark problems: RB-SOR, FRAC, and FDTD. By implementing 3×3, 4×4, and 8×8 SCMAs, we evaluate their impact on the synthesis results including operating frequency and resource consumption. We also evaluate the overhead of LGSM in terms of operating frequency and resource consumption. We show that the size of the array and LGSM have a slight influence on the operating frequency, but only frequency degradation is limited. Thus our design of an SCMA with LGSM is scalable for available resources on an FPGA.

To evaluate performance scalability for multiple FPGAs, we compare single-FPGA, double-FPGA, and triple-FPGA SCMA, where each sub-SCMA has an 8×8 array. The number of FPGAs gives complete scalability of sustained performance maintaining a utilization of 80–90% for each benchmark. As a result, the three FPGAs operating at 100 MHz achieve 31–34 GFlop/s for single-precision floating-point computations. We expect that a 10×10 sub-SCMA can be implemented on each FPGA to provide 1.5 times higher performance than that of the 8×8 sub-SCMA. We ensure the LGSM provides the necessary stalls for the differences in frequency of the 100 MHz clocks; however, the number of stall cycles is very small, just 8×10^{-4}% of the total cycles. This means that sub-SCMAs in different clock-domains are synchronized by LGSM, but the overhead is ignorable for computing performance.

A feasibility study of implementation with the high-end FPGA series showed that the Stratix V FPGA is expected to achieve a peak performance of 419 GFlop/s and a sustained performance of about 417 GFlop/s. We showed that BRM allows a large SCM array to be implemented with many FPGAs without a bottleneck in inter-FPGA communication.

In future work, we will construct a 2D FPGA array and implement a large SCMA on it for larger 3D computations. We will also develop a compiler for SCM arrays based on the prototype version of a compiler for a domain-specific language [14].

Acknowledgements This research and development were supported by Grant-in-Aid for Young Scientists(B) No. 20700040, Grant-in-Aid for Scientific Research (B) No. 23300012, and Grant-in-Aid for Challenging Exploratory Research No. 23650021 from the Ministry of Education, Culture, Sports, Science and Technology, Japan.

References

1. Altera Corporation (2012), http://www.altera.com/literature/
2. R. Baxter, S. Booth, M. Bull, G. Cawood, J. Perry, M. Parsons, A. Simpson, A. Trew, A. McCormick, G. Smart, R. Smart, A. Cantle, R. Chamberlain, G. Genest, Maxwell a 64 FPGA supercomputer, in *Proceedings AHS2007 Conference Secound NASA/ESA Conference on Adaptive Hardware and Systems* (2007), pp. 287–294, http://ieeexplore.ieee.org/xpls/abs_all.jsp?arnumber=4291933
3. W. Chen, P. Kosmas, M. Leeser, C. Rappaport, An fpga implementation of the two-dimensional finite-difference time-domain (FDTD) algorithm, in *Proceedings of the 2004 ACM/SIGDA*

12th International Symposium on Field Programmable Gate Arrays (FPGA2004) (2004), pp. 213–222, http://dl.acm.org/citation.cfm?id=968311
4. K. Datta, M. Murphy, V. Volkov, S. Williams, J. Carter, L. Oliker, D. Patterson, J. Shalf, K. Yelick, Stencil computation optimization and auto-tuning on state-of-the-art multicore architectures, in *Proceedings of the 2008 ACM/IEEE Conference on Supercomputing* (2008), pp. 1–12, http://dl.acm.org/citation.cfm?id=1413375
5. J.D. Davis, C.P. Thacker, C. Chang, BEE3: revitalizing computer architecture research. MSR-TR-2009-45 (Microsoft Research Redmond, WA, 2009)
6. J.P. Durbano, F.E. Ortiz, J.R. Humphrey, P.F. Curt, D.W. Prather, FPGA-based acceleration of the 3D finite-difference time-domain method, in *Proceedings of the 12th Annual IEEE Symposium on Field-Programmable Custom Computing Machines* (2004), pp. 156–163, http://ieeexplore.ieee.org/xpls/abs_all.jsp?arnumber=1364626
7. D.G. Elliott, M. Stumm, W. Snelgrove, C. Cojocaru, R. Mckenzie, Computational RAM: implementing processors in memory. Des. Test Comput. **16**(1), 32–41 (1999)
8. A. George, H. Lam, G. Stitt, Novo-G: at the forefront of scalable reconfigurable supercomputing. Comput. Sci. Eng. **13**(1), 82–86 (2011)
9. L.A. Hageman, D.M. Young, *Applied Iterative Methods* (Academic, New York, 1981)
10. T. Hauser, A flow solver for a reconfigurable FPGA-based hypercomputer. AIAA Aerosp. Sci. Meet. Exhib. **AIAA-2005-1382** (2005)
11. K.T. Johnson, A. Hurson, B. Shirazi, General-purpose systolic arrays. Computer **26**(11), 20–31 (1993)
12. H.T. Kung, Why systolic architecture? Computer **15**(1), 37–46 (1982)
13. W. Luzhou, K. Sano, S. Yamamoto, Local-and-global stall mechanism for systolic computational-memory array on extensible multi-FPGA system, in *Proceedings of the International Conference on Field-Programmable Technology (FPT2010)* (2010), pp. 102–109, http://ieeexplore.ieee.org/xpls/abs_all.jsp?arnumber=5681763
14. W. Luzhow, K. Sano, S. Yamamoto, Domain-specific language and compiler for stencil computation on FPGA-based systolic computational-memory array, in *Proceedings of the International Symposium on Applied Reconfigurable Computing (ARC2012)* Springer, (2012), pp. 26–39, http://link.springer.com/chapter/10.1007%2F978-3-642-28365-9_3?LI=true
15. O. Mencer, K.H. Tsoi, S. Craimer, T. Todman, W. Luk, M.Y. Wong, P.H.W. Leong, Cube: a 512-FPGA cluster, in *Proceedings of the IEEE Southern Programable Logic Conference 2009* (2009), pp. 51–57, http://ieeexplore.ieee.org/xpls/abs_all.jsp?arnumber=4914907
16. H. Morishita, Y. Osana, N. Fujita, H. Amano, Exploiting memory hierarchy for a computational fluid dynamics accelerator on FPGAs, in *Proceedings of the International Conference on Field-Programmable Technology (FPT2008)* (2008), pp. 193–200, http://ieeexplore.ieee.org/xpls/abs_all.jsp?arnumber=4762383
17. D. Patterson, T. Anderson, N. Cardwell, R. Fromm, K. Keeton, C. Kozyrakis, R. Thomas, K. Yelick, A case for intelligent RAM: IRAM. IEEE Micro **17**(2), 34–44 (1997)
18. D. Patterson, K. Asanovic, A. Brown, R. Fromm, J. Golbus, B. Gribstad, K. Keeton, C. Kozyrakis, D. Martin, S. Perissakis, R. Thomas, N. Treuhaft, K. Yelick, Intelligent RAM(IRAM): the industrial setting, applications, and architectures, in *Proceedings of the International Conference on Computer Design* (1997), pp. 2–9, http://ieeexplore.ieee.org/xpls/abs_all.jsp?arnumber=628842
19. E.H. Phillips, M. Fatica, Implementing the himeno benchmark with CUDA on GPU clusters, in *Proceedings of International Symposium on Parallel and Distributed Processing (IPDPS)* (2010), pp. 1–10, http://ieeexplore.ieee.org/xpls/abs_all.jsp?arnumber=5470394
20. K. Sano, T. Iizuka, S. Yamamoto, Systolic architecture for computational fluid dynamics on FPGAs, in *Proceedings of the 15th Annual IEEE Symposium on Field-Programmable Custom Computing Machines (FCCM)* (2007), pp. 107–116, http://ieeexplore.ieee.org/xpls/abs_all.jsp?arnumber=4297248
21. K. Sano, W. Luzhou, Y. Hatsuda, S. Yamamoto, Scalable FPGA-array for high-performance and power-efficient computation based on difference schemes, in *Proceedings of the International Workshop on High-Performance Reconfigurable Computing Technology and Applications (HPRCTA)* (2008), http://ieeexplore.ieee.org/xpls/abs_all.jsp?arnumber=4745679

22. K. Sano, W. Luzhou, Y. Hatsuda, T. Iizuka, S. Yamamoto, FPGA-array with bandwidth-reduction mechanism for scalable and power-efficient numerical simulations based on finite difference methods. ACM Trans. Reconfigurable Technol. Syst. (TRETS) **3**(4), Article No. 21 (2010)
23. K. Sano, W. Luzhou, S. Yamamoto, Prototype implementation of array-processor extensible over multiple FPGAs for scalable stencil computation. ACM SIGARCH Computer Architecture News (HEART special issue), **38**(4), 80–86 (2010)
24. K. Sano, Y. Hatsuda, S. Yamamoto, Scalable streaming-array of simple soft-processors for stencil computations with constant memory-bandwidth, in *Proceedings of the 19th Annual IEEE Symposium on Field-Programmable Custom Computing Machines (FCCM)* (2011), pp. 234–241, http://ieeexplore.ieee.org/xpls/abs_all.jsp?arnumber=5771279
25. R.N. Schneider, L.E. Turner, M.M. Okoniewski, Application of fpga technology to accelerate the finite-difference time-domain (FDTD) method, in *Proceedings of the 2002 ACM/SIGDA 10th International Symposium on Field Programmable Gate Arrays (FPGA2002)* (2002), pp. 97–105, http://dl.acm.org/citation.cfm?id=503063
26. W.D. Smith, A.R. Schnore, Towards an RCC-based accelerator for computational fluid dynamics applications. J. Supercomput. **30**(3), 239–261 (2003)
27. J.C. Strikwerda, Y.S. Lee, The accuracy of the fractional step method. SIAM J. Numer. Anal. **37**(1), 37–47 (1999)
28. TERASIC Corp. (2012), Accessed 30th January 2013, http://www.terasic.com.tw
29. J.E. Vuillemin, P. Bertin, D. Roncin, M. Shand, H.H. Touati, P. Boucard, Programmable active memories: reconfigurable systems come of age. IEEE Trans. Very Large Scale Integr. (VLSI) Syst. **4**(1), 56–69 (1996)
30. S. Williams, A. Waterman, D. Patterson, Roofline: an insightful visual performance model for multicore architectures. Comm. ACM **52**(4), 65–76 (2009)
31. K.S. Yee, Numerical solution of inital boundary value problems involving maxwell's equations in isotropic media. IEEE Trans. Antennas Propag. **14**, 302–307 (1966)

High Performance Implementation of RTM Seismic Modeling on FPGAs: Architecture, Arithmetic and Power Issues

Victor Medeiros, Abner Barros, Abel Silva-Filho, and Manoel E. de Lima

Abstract This work presents a case study in the oil and gas industry, namely the FPGA implementation of the 2D reverse timing migration (RTM) seismic modeling algorithm. These devices have been largely used as accelerators in scientific computing applications that require massive data processing, large parallel machines, huge memory bandwidth and power. The RTM algorithm enables you to directly solve the acoustic and elastic waves problems with precision in complex geological structures, demanding a high computational power. To face such challenges we suggest strategies such as reduced arithmetic precision, based on fixed-point numbers, and a highly parallel architecture are suggested. The effects of such reduced precision for storage/processing data are analyzed in this chapter through signal-noise ratio (SRN) and universal image quality index (UIQI) metrics. The results show that SRN higher than 50 dB can be considered acceptable for a migrated image with 15 bits word size. A special stream-processing architecture aiming to implement the best possible data reuse for the algorithm is also presented. It was implemented by an FIFO-based cache in the internal memory of the FPGA. A temporal pipeline structure has also been developed, allowing that multiple time steps to be performed at the same time. The main advantage of this approach is the ability to keep the same memory bandwidth needs of processing just one time step. The number of time steps processed at the same time is limited by the amount of FPGA internal memory and logic blocks. The algorithm was implemented on an Altera Stratix 260E, with 16 processing elements (PEs). The FPGA was 29 times faster than the CPU and only 13% slower than the GPGPU. In terms of power consumption, the CPU+FPGA was 1.7 times more efficient than the GPGPU system.

V. Medeiros (✉) • A. Barros • A. Silva-Filho • M.E. de Lima
Center for Informatics, UFPE - Brazil
e-mail: vwcm@cin.ufpe.br; acb@cin.ufpe.br; agsf@cin.ufpe.br; mel@cin.ufpe.br

W. Vanderbauwhede and K. Benkrid (eds.), *High-Performance Computing Using FPGAs*, DOI 10.1007/978-1-4614-1791-0_10, © Springer Science+Business Media, LLC 2013

1 Introduction

Nowadays, accelerators like GPGPU and FPGAs also have emerged as strong streaming computation candidates [6, 8, 9, 17] to increase performance in a specific hardware platform. Thus, compute-intensive algorithms applied to scientific computing modeling such as bioinformatics, image processing, geophysics and seismic applications [8, 12] can be mapped directly into FPGAs [2] or GPGPUs hardware [4, 17] in order to exploit its massive parallel processing power.

Particularly, these reconfigurable platforms have been widely used in applications of digital signal processing in fixed-point arithmetic, achieving gains in performance when compared with general purpose CPUs [2,3] and GPGPUs [4,17].

In fact, to evaluate which platform is optimal for a given application has become a difficult task. For instance, applications such as seismic imaging in petroleum industry, the main focus of this work, involve a lot of data collected based on floating point operations from petroleum fields in order to locate underground oil and gas reservoirs with basis on an impulsive seismic waves source, geophones and a storage system. This process is performed through the acquisition of travel time and the speed of seismic waves which can be reconstructed by applying methods such as Kirchhoff [20] and reverse time migration (RTM) [5, 16], resulting in data that are used to verify the disposition of the rocks on the ground.

However, no matter of the method, not only the processing time but also the time required to transfer huge amounts of data stored in memory for processing phase is still quite costly. In a sense, optimization strategies that aim to reduce access latency and search for data in memory, as well as increased bandwidth become necessary.

FPGAs have, therefore, shown excellent potential as hardware accelerators, allowing to explore issues such as stream processing, considering approaches based on pipeline, parallelism on data processing, as well as more efficient reuse of data when compared with other platforms such as GPGPUs. Also, with the FPGAs, it is possible to customize the formatting of data, enabling data reduction in precision whenever possible, without jeopardizing the quality of results for certain problems like the infeasibility in CPUs and GPGPUs. And, in addition to these, there is the possibility of data compression and further enhancement of the advantages mentioned previously, especially an increase in bandwidth access to data in system memory. These devices can also provide an order of magnitude more performance per joule [17] than other platforms like GPGPUs and massively parallel processors array (MPPAs), and over 250 times on general purpose CPUs.

This work explores the use of FPGAs to perform the acceleration of the seismic modeling problem in 2D, the first computational step of the RTM method. Thus, the contributions of this work are as follows:

(a) Analysis and development of an architecture to accelerate the RTM method of a 2D seismic modeling for three different platforms: CPU, GPGPU and FPGA.
(b) Hardware implementation on the PROCe III Gidel platform.

(c) Arithmetic issues analysis, especially the effect of using a custom arithmetic operator.
(d) Evaluation of power consumption in different platforms.

Through the obtained results this study aims to show that the FPGA can be used not only to accelerate the RTM method, but in many other applications in high performance computing as well.

For the rest of this chapter, it is organized in the following sections. In the next two sections, we present an overview of the RTM algorithm and discuss some recent related work. Section 4 presents the system architecture for the algorithm implementation based on FPGA, an analysis of the effects of using a customized arithmetic operator and a power consumption evaluation for some of the platforms used during the work. Then, in Sect. 5, the results targeting the seismic application with GPGPU and CPU comparisons are presented. Finally, a few conclusions are presented in Sect. 6.

2 The RTM Algorithm

Seismic exploration of oil can be divided into three areas: acquisition, processing and interpretation. The acquisition is responsible for obtaining seismic traces, through the injection of an excitation source (soil or water) along with the capture by receivers (geophones or hydrophones). The processing includes various algorithms such as Kirkchoff and RTM capable of transforming the confusing information obtained in the sets of seismic traces (seismogram) into something more understandable. However, a given implemented algorithm varies significantly according to the strengths and weaknesses of each architecture. This step essentially comprises the seismic modeling and migration stages. In the interpretation stage, information about several geological layers are extracted from a seismic section processed, such as a fault location and oleifera formation, which are finally analyzed by experts. The development of this work is directed at the processing step, particularly at the modeling method.

Some seismic modeling and migration methods have been applied in many different geological areas. The method of Kirchhoff for common offset depth has been the most used algorithm in the petroleum industry, due to its low computational cost relation when compared with other existing methods. However, this method is not suitable for underground areas that have very complex structures, with lateral variations in speeds, or high dips of the layers. Furthermore, the RTM method that allows to solve directly the wave equation acoustic/elastic, produces, in general, remarkably accurate results. However, this method has a quite high computational cost while making use of strategies for parallel processing. In order to face such challenge, this work provides the development of an FPGA-based platform for exploring the power of parallelism of the RTM algorithm aiming 2D modeling.

From the perspective of the wave propagation theory, the modeling and migration processes are subjected to numerical solution of wave equation. The method of solution by finite difference, used in RTM, for the acoustic wave equation, is the most widespread approach and allows the division of the problem into space sub-domains. This aspect contributes to exploit the parallelism of the architecture. However, it is necessary to ensure constant communication with each simulation step due to the dependence of the operations between steps, which is an aspect that reduces the performance in processing.

Thus, the discretized finite difference equation is solved with the RTM algorithm as follows. Initially, the modeling stage is performed. The forward propagation of a modeled source wave-field to the receiver is performed for each shot location through a known gridded velocity model. A wave-field is saved for later application of the "imaging condition" at each time step. A shot corresponds to a seismic source. In the field, this source is a real explosion in a given region of interest. In the forward modeling this seismic source is simulated by a mathematical function. Second, the received wave-field for each shot (as recorded in the field) is backward propagated in time through the same velocity model. At each time step the corresponding source and receiver wave-fields are correlated by applying the imaging condition. Thus, the final wave-field in the source propagation scheme is correlated with the initial wave-field in the receiver propagation scheme, and so on backwards through the receiver propagation. The results are added to form a partial image volume for each shot, and the image volumes for consecutive shot gathers are spatially added to produce the final prestack depth image. This work is intended to implement the first step of the RTM processing responsible for the architecture modeling stage.

3 Related Work

In order to improve the system performance over applications of scientific computing, different parallel platforms have been evaluated for different algorithms. There has been significant debate on which platform produces the most efficient solution. However, only through a careful optimization for each platform, with the engagement of hardware, computer-science and algorithmic scientists, we can come up with a reasonable assessment of the alternatives available today.

Recently, multicore clusters are usually composed by large memories and mechanical storage, thus wastefully burning a lot of power. An energy-efficient solution is to design systems that are more specific for the applications they're running. It is possible to customize a system with dedicated accelerators to better match the requirements of the application, thus resulting in better efficiency.

Solutions based on FPGAs have focused on solving some of the toughest digital problems of industry. Over the past decade, very large FPGAs have been implemented in applications such as radar, cryptography, WiMax/Long Term Evolution (LTE), and software-defined radio (SDR). Factors including cost pressure and stringent size, weight, and power (SWaP) requirements have created constrained environments [13] that require high-performance and high-efficiency architectures. FPGA-based systems can satisfy these requirements. In this sense, hybrid solutions such as CPUs coupled to FPGA-based accelerators can be associated with the best computing configuration available, even when we consider the host CPU power consumption.

For petroleum industry explorations, research has been intense once seismic data processing uses very complex floating point mathematical algorithms, indicated by geophysical specialists to reconstruct the earth subsurface in order to discover where petroleum can be found. These data are detected by hydrophones and recorded in large repositories from an initial seismic pulse, generating a geological model of the earth (imaging method) that can be interpreted by experts to help them to decide where to build petroleum recovery infrastructure.

Therefore, to evaluate different algorithms for different platforms of intensive computation has been the target of several industries and research centers. Every new year, new architectures of intensive computing are proposed in order to efficiently exploit the parallelism of these applications through strategies such as increasing the amount of processing units, memory access, I/Os, operating frequency, as well as other aspects restricted to manufacturing technology. All these features, associated with the most advanced technology, would not have been so successful if each designer's creativity in providing different stream computation solution for a given problem was not available.

Chuan He et al. [9] propose a reconfigurable platform for seismic data processing based on FPGA to speed up the pre-stack Kirchhoff time migration (PSTM) algorithm. The algorithm is expensive for the practical 3D applications because it is computationally intensive and requires large amount of input data. The kernel part of the algorithm, which consumes more than 90% of the CPU time, is tuned to maximize its execution speed in the proposed platform. Simulation results show that the proposed platform operating at a frequency of 50 MHz can calculate the Kirchhoff summations in 50 million points per second and up to 15.6 times faster than a 2.4 GHz P4 workstation considering the particular algorithm. The experimental environment used in this case was composed of the Xilinx Virtex II Pro series FPGA chip.

The same algorithm (PSTM) was re-implemented in a GPGPU GeForce8800GT from NVidia in Shi et al. [14] work in order to improve the efficiency of the CPU code. Prototypes were evaluated and results showed that they were more than 7.2 times faster than their CPU versions on Intel's P4 3.0G. In order to make comparisons with previously cited work, that is based on an FPGA approach, it is important to mention that there are newer GPGPUs than the 8,800 s.

In [6] Haohuan Fu et al. developed a tool for automatic exploration in order to customize number representations on FPGAs and provide acceptable precision for seismic applications. This tuned number format was used to improve the performance of the FK step in downward continued-based migration and the acoustic 3D convolution kernel in the RTM approach. This strategy achieved speedups ranging from 5 to 7 including the transfer time to and from the processors.

An examination of random number generation (RNG) on four different architectures (multi-core CPUs, GPGPUs, FPGAs and MPPAs) was made in Thomas et al. [17] work. Three RNG algorithms (Uniform, Gaussian and Exponential distribution) were evaluated for each platform. The most appropriated algorithm for generating each type of number was determined for each platform. Then the peak generation rate and estimated power efficiency for each device in GSample/s and MSample/joule was calculated. The results are more conclusive when efficiency is compared, as the FPGA provides an order of magnitude with more performance per joule than any other platform, and 250 times more than that of the CPU.

One of the greatest difficulties in comparing platforms such as CPU-based, FPGA-based and GPGPU-based ones is that while a naive, unoptimized algorithm can be implemented and directly compared on each platform, an optimized version of the same algorithm may vary significantly due to the strengths and weaknesses of each architecture. Clapp et al. [5] evaluate RTM algorithm for GPGPU and FPGA accelerators. Their work discusses some algorithmic and optimization decisions made to implement RTM on each platform. In addition, they show how these choices are directly tied to the characteristics of the underlying architecture.

This work will explore the use of FPGAs to perform the acceleration of the problem in 2D seismic modeling, which is the first computational step of the RTM method. A real implementation showed that the proposed architecture is up to 29 times faster than a CPU and only 13% slower than a Nvidia Tesla GPGPU. We also analyze the power efficiency of these platforms. In this context the FPGA is 1.7 times more efficient than the GPGPU. We also discuss that a top FPGA platform, using the same techniques implemented in this work, could have an even better performance.

4 System Architecture

The RTM algorithm is implemented in an FPGA co-processor based architecture, on a Gidel PROCe III board [7] connected to a host machine, which supports an Intel Core 2 Quad powered PC, through a PCIe $\times 4$ bus. The host part is composed of: the application core; the application driver, an API to access the co-processor board resources provided by Gidel; and the board device driver. The co-processor board is composed of: a PCIe bridge core, implemented in an additional FPGA; the main FPGA that supports all wrap cores provided by Gidel and the stream processing core that implements the RTM seismic modeling algorithm; and the board DDR2 memories. These elements are described in details in the following sections. Figure 1 illustrates the system architecture.

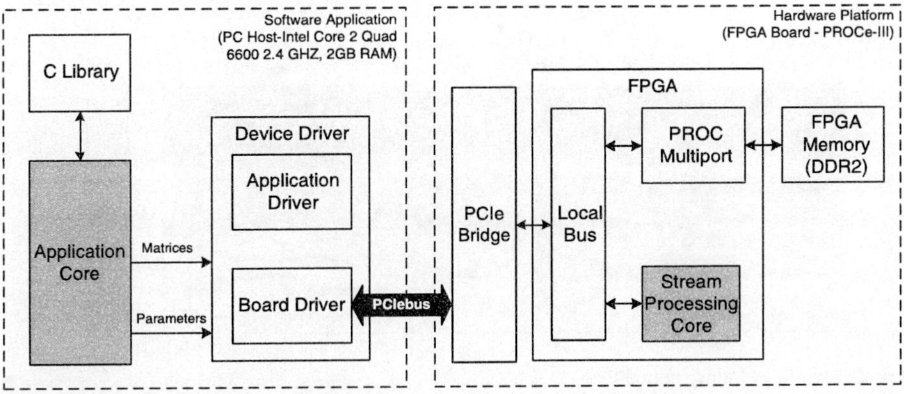

Fig. 1 System architecture overview

4.1 System Overview

The following sections give us an overview of the software and hardware parts of the whole system.

4.1.1 Software Application

The software application is directed towards supporting the RTM modeling algorithm execution. The main module of this application is composed of some basic functions. The first of these functions is meant to organize the data, from host disk, to the FPGA inside memories. This RTM algorithm database is basically composed of three matrices: the velocity model matrix (VEL); the current pressure field matrix (CPF); and the previous pressure field matrix (PPF).

The VEL matrix stores the velocity model, proposed by a geologist and represents the different wave propagation velocities on the ground layers. For example, Fig. 2 illustrates the Marmousi velocity model, a 2D synthetic data set which is used to verify if an algorithm produces correct images. As the implemented algorithm is a second-order temporal finite difference, two matrices are involved in the processing: the CPF matrix that stores the pressure field values on the current time step; and the PPF matrix that stores the pressure field values on the previous time step. These three matrices are fitted into the FPGA memory blocks in a way to allow an optimized and efficient data access. This data organization is explained in more detail in the Sect. 4.2.1.

The second function of the application core is to configure some algorithm parameters like the size of matrices, seismic pulse data, seismic pulse position, and the number of time steps.

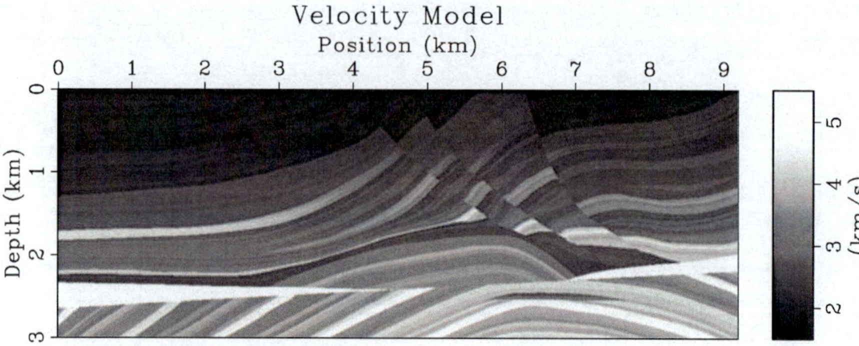

Fig. 2 Marmousi velocity model

In the current version, the system only deals with a single data partition because only one board with one FPGA has been used. A multi-FPGA board version in a multi-board cluster is under development. In this last approach, a third basic function to split and allocate the data partition among these co-processor resources should also be developed. A fourth function to visualize the processed data is also under way.

PROCWizard

The Gidel PROCWizard IDE has been developed to provide a simple interface to integrate hardware and software components of a design. PROCWizard works closely with the Gidel boards and it is able to generate both, hardware and software codes, in HDL and C++ codes, respectively, and a communication channel through a PCIe bus. This IDE also automatically integrates all IP cores provided by Gidel for the design, as the PROCMultiport (see Sect. 4.1.2), provides a device driver, used by the application core to access the board resources and a debug environment that allows, in runtime, the visualization of the board memories and CPU registers.

Device Driver Platform

The device driver generated by the PROCWizard IDE allows the access to the board resources through a C++ class. The user can instantiate it in their applications and call the class methods that implement the device driver API. The device driver is also responsible for the hardware initialization (FPGA configuration) and the board clock frequency setup.

4.1.2 Hardware Platform

The hardware platform includes the communication elements that enable the data transfer between the host and the FPGA board; the Multiport ip core provided by Gidel that enables an easy interface for memory access; and the customized stream processing core that implements the RTM algorithm.

Communication and Memory Structures

The Gidel PROCe III board provides an FPGA meant for the PCIe bridge core. Thus, all the logic and memory resources in the main FPGA remain available to the user's design.

The PCIe bridge, the PROCMultiport and the stream processing core are all connected through an internal FPGA local bus. This bus allows the data transfer between the host and the board memory. It also supports data transfer between the stream processing core and the board memory through the PROCMultiport core.

On this FPGA-board, there are three memory banks available: a 512 MB DDR2 SDRAM; and two DDR2 SODIMMs with up to 8 GB each. All three memories have a 64-bits interface and are all accessible through the Gidel PROCMultiport core.

PROCMultiport

The PROCMultiport module is also created by the PROCWizard IDE. It acts as an abstraction layer between the memory modules and the user core, providing an efficient and simple way to read and write data. This module can provide a virtual multiple port access to a memory that has only one port. It is possible to define up to 16 ports for each memory module, a specific clock domain and data width for each port.

The data in each port can be accessed on a sequential or random scheme. In this work, the sequential access has been adopted because of the special data organization scheme for the stream processing. In this case, each port is accessed as an FIFO data structure. Thus, it is not necessary to generate addresses for the memory. So, the interface becomes quite simple. The sequential access also improves the system performance by taking advantage of memory data bursts.

When the PROCMultiport is joint with the PCIe bridge core, the DMA channels and the host device driver also allow DMA data transfers between the host and FPGA memory releasing the CPU to run other important tasks during the data transfer.

In this design, six memory ports have been configured. Four of them are reading ports for the CPF, PPF, and VEL matrices and the seismic pulse vector. The fifth and sixth ports are writing ports to the subsequent CPF and PPF pressure values.

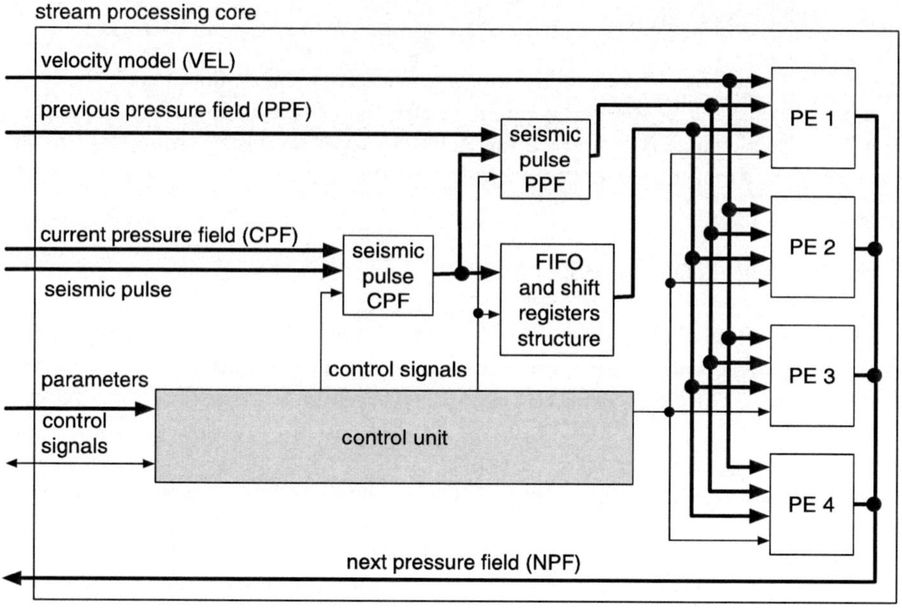

Fig. 3 Stream processing core components

These ports are directed to the same memory space of the last processed CPF and PPF matrices, since this data is not useful in the subsequent time step anymore. This strategy optimizes memory usage.

Stream Processing Core

The streaming processing core is composed of the processing elements and support structures to provide data efficiently for them. These processing elements implement the finite difference equation. This equation has a high degree of parallelism due to the data independence on each element calculation. This characteristic enables a great performance improvement through the instantiation of multiple processing elements in the stream processing core. In this work four processing elements have been instantiated spatially. That means that four elements can be processed at the same time, in each clock cycle. Actually, a sixteen processing elements solution was implemented based on the time domain parallelism exploration. This approach is explained in detail in Sect. 4.2.2. The number of processing elements in the architecture is basically limited by the memory bandwidth and the amount of resources in the FPGA. The internal components of the stream processing core can be visualized in Fig. 3.

The developed architecture aims to maintain a constant data flow from the memory to the processing elements and back to memory after the processing.

The data flow starts with the data transfer from the host memory to the FPGA memory. After that, the data starts to be read from the FPGA board memory to the PROCMultiport CPF, PPF, and VEL ports. So, these data are read by the stream processing core and through the above-mentioned supporting structures sent to the processing elements. After this data processing, they are also sent back to memory through a port of the PROCMultiport core. The set of elements that provide data for the processing elements is named data flow architecture.

The processing elements are detailed in Sect. 4.3. The data flow architecture is detailed in Sect. 4.2.

4.2 Data Flow Architecture Implementation

As mentioned previously the internal structure of the processing core is composed of the PEs that implement the finite difference equation, and data structures that provide data for these processing elements. These data flow structures includes: the FIFO and shift registers structure responsible for the data flow and data reuse implementation; the seismic pulse CPF and seismic pulse PPF modules responsible for inserting the seismic pulse; and the control unit, as can be visualized in Fig. 3. In this section all of these elements are detailed. This section also explains the memory organization and temporal pipeline strategies to improve the system performance.

4.2.1 Memory Organization

In order to improve performance, the data from the seismic pulse vector and the matrices CPF, PPF, and VEL are especially organized in the FPGA memory to reduce their access latency. The main idea is to organize the data in a way to support a sequential access, allowing the memory controller to work in burst mode and, consequently, decrease memory latency. The data organization has also an essential role in the data reuse as it is explained in Sect. 4.2.3.

The VEL is developed by a geologist and represents the different wave propagation velocities on the subsurface layers. As the model is a second-order temporal finite difference algorithm, the CPF matrix stores the pressure field values on the current time step and the PPF matrix stores the pressure field values on the previous time step.

These three matrices are then stored into the three DDR2 memory banks connected to the FPGA. Figure 4 shows the data distribution inside the FPGA memories. The first memory bank stores the seismic pulse vector and then the VEL matrix. The second stores the CPF matrix and the third one stores the PPF matrix.

The seismic pulse vector is a representation of an excitation source applied to the seismic model and is disposed sequentially. The matrices are divided into slices and each one corresponds to four columns in the original matrix. That means that the first slice has all the lines of the first four matrix columns. One can consider

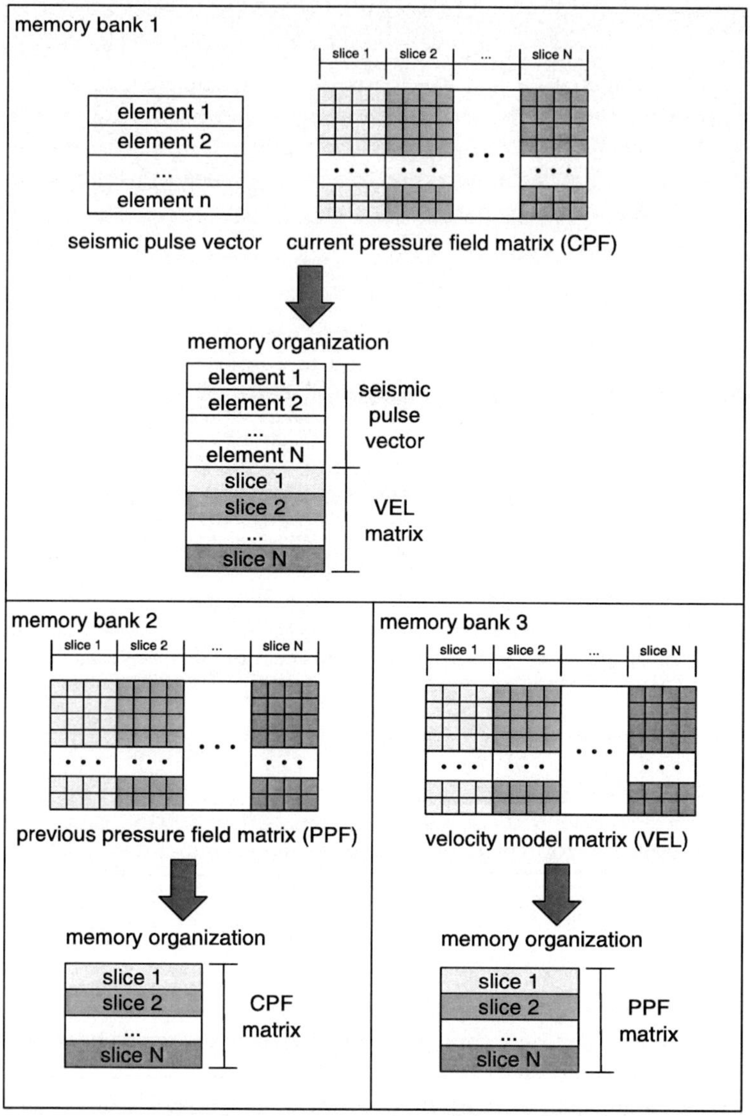

Fig. 4 Memory data organization

that the memory is a sequence of 128-bits wide words, as the PROCMultiport provides a 128-bits wide interface for the CPF, PPF and VEL ports. Each word read corresponds to four single precision floating points (32-bits wide) that represent one slice row.

All the three matrices have the same size and the seismic pulse vector has the size that equals the number of time steps that are being processed. The next pressure field

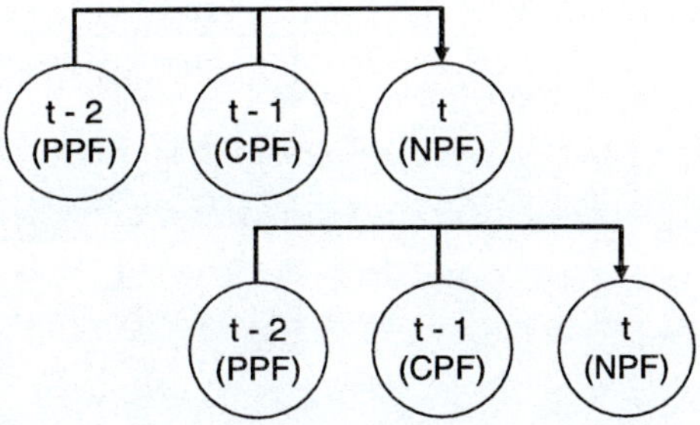

Fig. 5 Normal process (one only time step at once)

matrix (NPF) is written as the CPF matrix for the next four time steps processing. The CPF matrix is written as the next PPF matrix. These writings are made in the original memory space as the current data is not necessary in the next four subsequent time steps due to the four temporal pipeline depth stages. The temporal pipeline is explained in detail in Sect. 4.2.2.

4.2.2 Temporal Pipeline

In the presented architecture each PE demands one point to be fetched from the memory to produce one result point (when the pipeline is full). This fact demonstrates how the optimized data path of the architecture is, concerning data reuse. But at the same time it shows that the quantity of PEs is bounded mainly by the memory bandwidth. This happens because all PEs are at the same time step. Nevertheless it is possible to process more than one step concurrently (the time domain parallelism) to improve performance without demanding more memory bandwidth.

Figure 5 shows the general scheme of the process step to produce one time step at once. The matrices PPF at time -2 and CPF at time -1 are used to generate matrix NPF at time 0. After that, the CPF matrix becomes the new PPF and NPF becomes CPF. As one can see, two matrices are read from the memory in order to compute one result matrix. The disadvantage of this method is the need of two matrices to be written into memory instead of one. But, due to the increase in the number of temporal pipeline stages, this overload per time step is significantly decreased.

Otherwise, in Fig. 6a, as soon as the first elements of the matrix at time 0 are computed they are carried out to the next step to proceed the computation of the matrix at time 1. Then, two time steps are processed at once with the same bandwidth used to produce one only matrix. In the new iteration the PPF matrix is the matrix at time 0. In addition, in Fig. 6b, the number of temporal pipeline stages is increased to three. As expected only two matrices are read from memory and others two are written into memory. As can be seen the matrix at time 0 never reaches external memory of the FPGA.

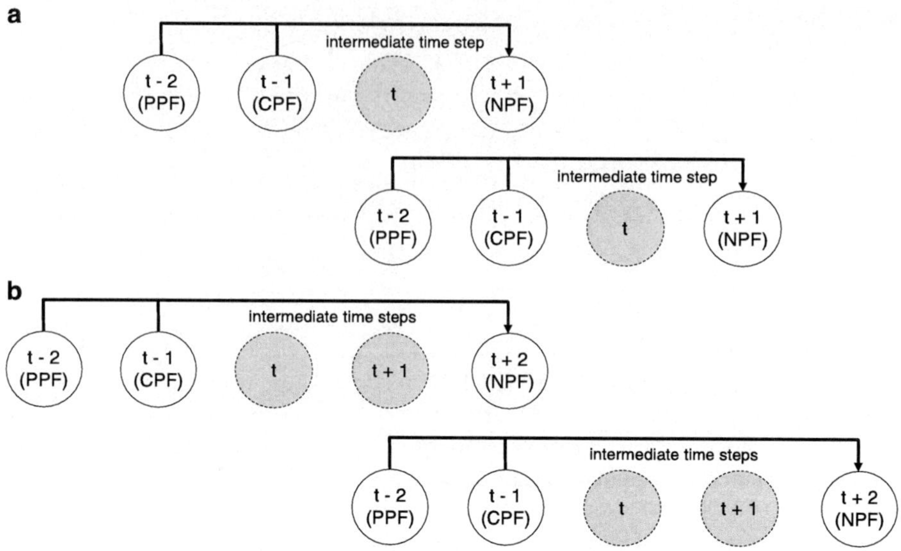

Fig. 6 Multiple time steps processing. (**a**) Two time steps at once (**b**) three time steps at once

The multiple time steps approach is only possible because of the temporal data reuse. By this strategy, all the modules used in one time step are replicated internally and the output of each step feeds the input of the next one.

The current FPGA implementation has a four stage temporal pipeline. That means that this architecture processes four time steps at the same time as soon as the pipeline is full. In other words, this solution has a total of 16 PEs, each four process a different time step. The maximum number of PEs with this approach is now bounded by the logic density of the FPGA. As soon as a larger FPGA device is available, more time steps can be added to increase the performance. In such scenario, doubling the time steps would double the performance of this implementation.

4.2.3 Data Flow Architecture Components

FIFOs and Shift Register Structure

The FIFOs and shift register structure have two basic functions in the system. The first is to provide data correctly and in an efficient way for the processing elements. The second is to implement data reuse. The main idea is to process four adjacent elements (a slice row) at the same time. Each of them in a specific PE. Using this strategy, the stencil becomes something like what is pictured in Fig. 8. A register structure that implements this four processing element stencil is created.

$$C_{i,j} = 2 * B_{i,j} - A_{i,j} - \{Vel^2_{i,j} * fat * [16 * (B_{i,j+1} + B_{i,j-1} + B_{i+1,j} + B_{i-1,j}) \\ - 1 * (B_{i,j+2} + B_{i,j-2} + B_{i+2,j} + B_{i-2,j}) - 60 * B_{i,j}]\}$$

$A \rightarrow$ previous pressure field matrix (PPF)
$B \rightarrow$ current pressure field matrix (CPF)
$C \rightarrow$ next pressure field matrix (NPF)
$Vel \rightarrow$ velocity model matrix (VEL)

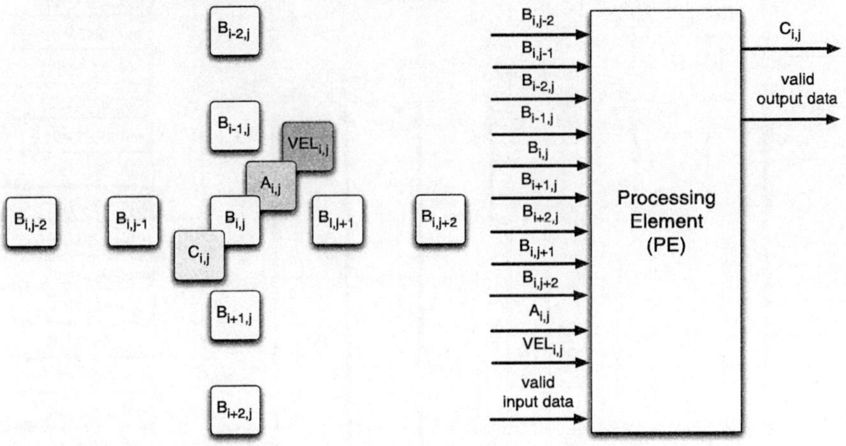

Fig. 7 Wave propagation equation, stencil and PE interface

The great advantage here is the reuse of some data as adjacent elements are being processed at the same time. This is possible to be observed in the PE interface in Fig. 7, and in the outputs of the FIFO and shift register structure in Fig. 8.

Each PE has nine inputs from the CPF matrix. Thus, for the four processing elements, in principle, 36 pieces of output data from the FIFOs and shift registers' structure to the processing elements should be generated. However, as some data are common to the adjacent processing elements, only 24 pieces of output data are needed.

Moreover, the FIFOs structure has the capacity of storing an entirely data slice, enabling this slice to be used in two processing slices. At the same time that data is used on the right side of the shift registers they are inserted into the FIFOs and are used in the next processing slice. For this approach to work properly, an FIFO initialization at the beginning of data processing is necessary.

In the initialization process, all FIFOs in Fig. 8 must be filled. Each FIFO has a size equal to the number of rows in the matrix, in other words, stores a whole matrix column. At the beginning of the process, the FIFOs 0, 1, 2, and 3 will store the first four columns of our matrix, the first slice. When the initialization process is finalized, the normal processing flow starts. Then the subsequent slice starts to

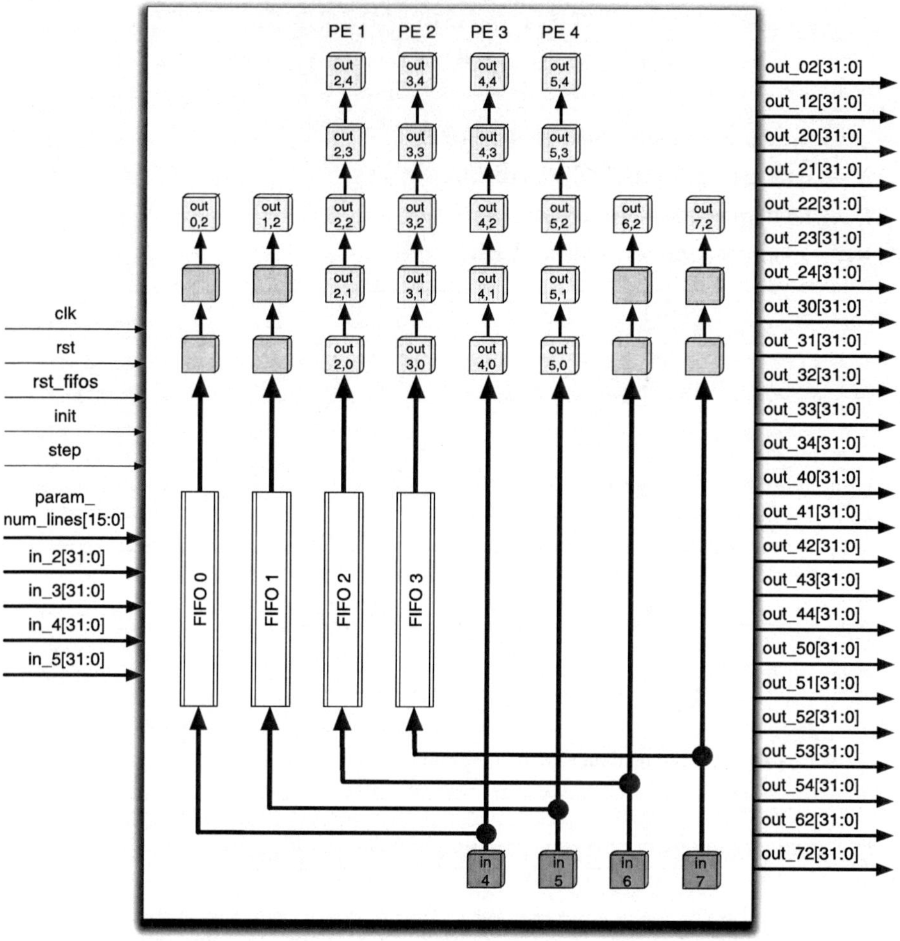

Fig. 8 FIFO and shift registers component architecture

be read. At the beginning of each slice it is necessary to run four data flow steps without processing, in order to align the data correctly in the respective registers in the control structure.

A data flow step consists in taking a slice row, four single precision floating point elements, from the CPF matrix port in the PROCMultiport core. These values are illustrated in Fig. 8 by the in_4, in_5, in_6, and in_7 inputs. Then, these elements are enqueued in FIFOs 0, 1, 2, and 3, respectively. At the same time, all FIFOs are dequeued and their outputs shifted into their respective shift registers located above them. The inputs in_4, in_5, in_6, and in_7 are also shifted into the four last shift registers. All the elements in the shift registers structures are then shifted up. All these steps happen in each clock cycle of the normal processing state. This strategy ensures that data taken from the memory will be used until it is no more necessary for processing. It also ensures that this architecture implements a 100% data reuse. The 100% data reuse maintains the data generation equal to consumption keeping the data stream without halt.

Seismic Pulse CPF and Seismic Pulse PPF

The seismic pulse vector is used to simulate a real explosion that occurs in the field survey. This simulated energy source allows the correct ground modeling. The seismic pulse vector is generated by the application (in software) and its size equals the number of time steps, which means that there is a seismic pulse value for each time step. The application also provides parameters that define the seismic pulse insertion position.

The seismic pulse CPF and seismic pulse PPF components are responsible for reading the seismic pulse vector values, from the memory and provide them for the processing elements. At each time step, one value from the seismic pulse vector must be read and added to one element in the CPF matrix, at one specific position, defined by the parameters of the application.

The control unit signalizes the correct moment to add these values to the seismic pulse CPF component. Inside it, there is one single precision floating point adder and some multiplexers that, controlled by the control unit signals, generates the correct data flow to the FIFO and shift registers structure and then to the processing elements.

At each time step, when the seismic pulse CPF generates one seismic pulse value at the specified matrix position, this value is stored by the seismic pulse PPF in an internal register. This stored value will be used as the element in the same previously defined seismic pulse position in the PPF matrix in the next time step.

Control Unit

The control unit component is responsible for controlling all the data flow inside the core. This component has been built in such a way that the algorithm parameters can be configured by the software application. The main algorithm parameters are the number of rows and columns, the number of time steps and the row and the column where the seismic pulse must be inserted. The control unit stores all these parameters in internal registers.

All the component behavior is defined by some parallel finite-state machines (FSM). The main function of the first FSM is to start the reading of the software application parameters and the input matrices CPF, PPF, and VEL through the PROCMultiport core.

The second FSM in the core is responsible for writing the outputs from the processing elements back to the memory through the PROCMultiport core. It works reading the valid output flag in the PE interface when it becomes active, meaning that the output from all the processing elements can be written back in the memory through the PROCMultiport.

The third and fourth FSMs are related to the seismic pulse data insertion. One machine inserts the seismic pulse value in the CPF matrix and the other in the PPF matrix, based on the seismic pulse position, defined by the software parameters.

4.3 Arithmetic Issues

Historically, seismic data processing has been performed using floating-point arithmetic standard, following the pattern IEEE 754/854. This pattern has the advantage of virtually increasing the power of representation of real numbers in computer systems. However, due to the high area cost of floating point IP cores and due to high cost of hardware implementation, in this project we choose to work with customized number representation, using fixed point standard, which allows us to obtain significant improvements in performance, power consumption, and data throughput. However, all change in numerical representation that involves substituting floating point by fixed point, results in reduction of precision. In seismic data processing, the effects of reducing precision may be verified by the reduction in signal-to-noise ratio (SNR) and the degree of similarity, between these images and those generated by standard floating point representation [1, 6].

In order to simplify the process of choosing the configuration of fixed point to be adopted, a fully parameterizable model for processing elements was implemented. This model allows an easy exploration of the effects in changing the number representation, from floating point to fixed point, on the data processed by the PE, as well as the effects of reducing the precision on fixed point configuration. In the IEEE 754 floating point single precision, the word has 32 bits. The idea of the model is to explore words with fewer than 32 bits in order to improve the data throughput in memory. The idea is to represent with fewer bits than the IEEE single precision standard, composed of 32 bits. This may improve data memory throughput, reduce PE implementation cost and logic area into FPGAs. The processing element model was described using the high-level hardware description language, SystemVerilog. The PE equation described in Fig. 9 below has been divided into a set of functions in order to facilitate the development of the model. Each one of the model functions was transformed into a PE internal module.

As the PE model is completely parametrized, all fixed point configurations of fixed point format used in the input and output arguments in each function are defined automatically in accordance with the setting of three parameters: NUMBITS, indicating the size of the binary word used to store the data value in the pressure field in fixed point format; NUMBITS_PINT, indicating the number of bits of the integer part of the adopted format; and NUMBITS_PFRAC, indicating the number of bits of the fractional part. All other model parameters are adjusted as a function of these three main parameters. These parameters are adjusted to ensure

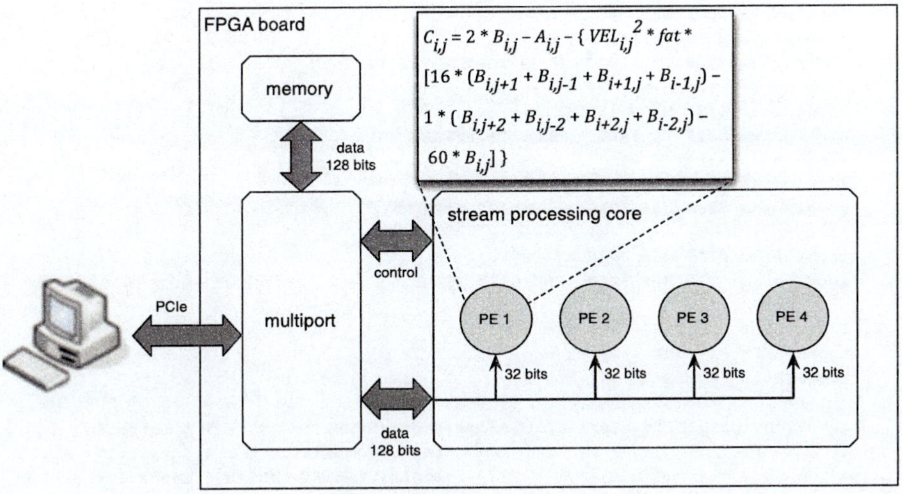

Fig. 9 Processing element on system overview

that the range and precision of generated values in each internal operation do not exceed the capacity of the fixed point used, ensuring an acceptable precision but without using registers larger than necessary. Figure 10 illustrates one part of the model parameters reported in the signature of some shown functions.

Figure 11 presents the main part of PE of RTM algorithm, described in SystemVerilog. The functions called in this algorithm perform the operations of the equation shown in Fig. 9.

The parameter NUMBITS_PINT defines the range of values that may be stored and processed in the pressure field. As this range may vary from model to model, it is important to prevent that higher values, which may occur in the pressure field, do not saturate, causing information loss and degradation of the final result of all process.

Once the range of values is adjusted, they will never be changed. The next step is to define the number of bits of the fractional part of the fixed point representation. The fractional part determines the precision of the PE and directly influences the quality of the results.

The main objective of this PE model is to allow exploring the effects of precision reduction, looking for the smaller fixed point configuration that may be used. Each bit reduced in the data width represents improvements in hardware resources, data access throughput, and power consumption reduction.

After the definition of the fixed point precision, the last part of the process consists in verifying the results quality. This is performed by measuring the quality of the images generated by the system. These are extremely important, because a misinterpretation caused by a low quality image of then may result in losses of millions of dollars. Thus, it is essential that the data quality obtained through this methodology can be evaluated within reliable technical criteria.

```
localparam NUMBITS = 32;
localparam NUMBITS_PINT = 8;
localparam NUMBITS_PFRAC= NUMBITS-NUMBITS_PINT-1;
//-------------------------------------------------
localparam BITS_M1 = (NUMBITS+2);
localparam BITS_M1_FRAC = NUMBITS_PFRAC;
//-------------------------------------------------
localparam BITS_M16 = (NUMBITS+2)+4;
localparam BITS_M16_FRAC = NUMBITS_PFRAC;
//-------------------------------------------------
localparam BITS_M60 = (NUMBITS)+6;
localparam BITS_M60_FRAC = NUMBITS_PFRAC;

logic[BITS_P16_SOMA-1:0] M16_soma_r;
logic[BITS_P1_SOMA-1:0] M1_soma_r;
logic[BITS_P60_SOMA-1:0] M60_r;

function logic[BITS_M16-1:0] M16Soma(logic[NUMBITS-1:0] b_i_j_menos_1,
                                     logic[NUMBITS-1:0] b_i_j_mais_1,
                                     logic[NUMBITS-1:0] b_i_menos_1_j,
                                     logic[NUMBITS-1:0] b_i_mais_1_j
                                    );
function logic[BITS_M1-1:0]  M1Soma(logic[NUMBITS-1:0] b_i_j_menos_2,
                                    logic[NUMBITS-1:0] b_i_j_mais_2,
                                    logic[NUMBITS-1:0] b_i_menos_2_j,
                                    logic[NUMBITS-1:0] b_i_mais_2_j
                                   );
function logic[BITS_P60-1:0] p60Soma(logic[NUMBITS-1:0] b_i_j);
```

Fig. 10 Model parameters and functions signatures

As was stated before, the change in the number format will introduce some of imprecision in the results because of the arithmetic operations and data precision representation.

From the viewpoint of the observer, these inaccuracies may represent changes from light, unnoticeable, to most severe, that can make the resultant images unusable. In order to measure these changes two different metrics have been used: the SNR, which explicitly measures the percentage of error introduced in the results; and the proposed by Zhou Wang in [18] universal image quality index (UIQI), which measures how this changes affects the human perception of results quality. The UIQI seeks to measure different aspects that may be relevant to human observer, such as contrast, brightness, and strange artifacts introduced in the image. The two metrics use (1) and (2), respectively.

In these equation x = original data set, processed and stored in float point; y = data set processed and/or stored in fixed point; Vrms (x) = energy of the original signal, Vrms (xy) = energy of the noise introduced, and $\sigma_x\sigma_y$ = respectively, the variance of the original signal and the variance of the less precise signal; σ_{xy} = covariance matrix between the original signal and the less accurate signal.

```
// Loop em T
for (t=0;t<cont_snap+2;++t) begin
...
// Loop em X
for (x=2;x<(nnoib-2);++x) begin
// Loop em Z
    for (z=2;z<(nnojb-2);++z) begin
        vv_fat = vel2Fat(VEL[i][j],fat_fixo);
        M16_r = M16Soma(B[i][j-1],B[i][j+1],B[i-1][j],B[i+1][j]);
        M60_r = M60(B[i][j]);
        M1_r =  M1Soma(B[i][j-2],B[i][j+2],B[i-2][j],B[i+2][j]);
        M1_16_60_r = MSoma1_16_60(M16_r, M1_r, M60_r);
        MVel2_fatS1_16_60_r = velFatSoma(vv_fat,M1_16_60_r);
        MmenosAij_r = MmenosAij(vel_fat_soma,A[i][j]);
        absixabsj_r = Mabsixabsj(absi[i],absj[j]);
        MmultBorda1_r = MmultBorda1(absixabsj,MmenosAij_r);
        Msoma2Bij_r = Msoma2Bij_r(B[i][j],MmultBorda1_r);
        C[i][j] = MmultBorda2(absixabsj,Msoma2Bij_r);
        end
    end
end
```

Fig. 11 Processing core

$$\mathrm{UIQI} = \frac{4\sigma_{xy}\overline{x}\overline{y}}{(\sigma_x^2+\sigma_y^2)[(\overline{x})^2+(\overline{y})^2]} \tag{1}$$

$$\mathrm{SNR} = 20 * \log\left(\frac{Vrms(x)}{Vrms(x-y)}\right) \tag{2}$$

Both techniques were used to verify the results obtained in several system configurations. The graphs in the Figs. 12 and 13 illustrate the obtained results.

As may be verified by graphics in Fig. 12, in the first 6 configurations, between 10- and 15-bit precision, a significant improvement in SNR was obtained as the result for each bit precision was added. However, after this, all the measurements are stable and show no noticeable improvements when compared with 15-bit precision configuration to the configuration with 15-bit precision.

Analyzing the graph of Fig. 13, it is possible to verify that just as the SNR, the degree of similarity has a rapid growth between the configurations with 10 and 15 bits of precision. From this point, this improvement gets smaller for each added bit, approximately 0.0056 per bit.

To verify this strange behavior presented in the results, one experiment was conducted aiming to measure explicitly the error level introduced by changing the number representation from floating point to fixed point. One new PE was implemented, this time using C++, which processes using standard floating point, but that convert the results obtained to fixed point format. The results of this experiment are shown in Fig. 14.

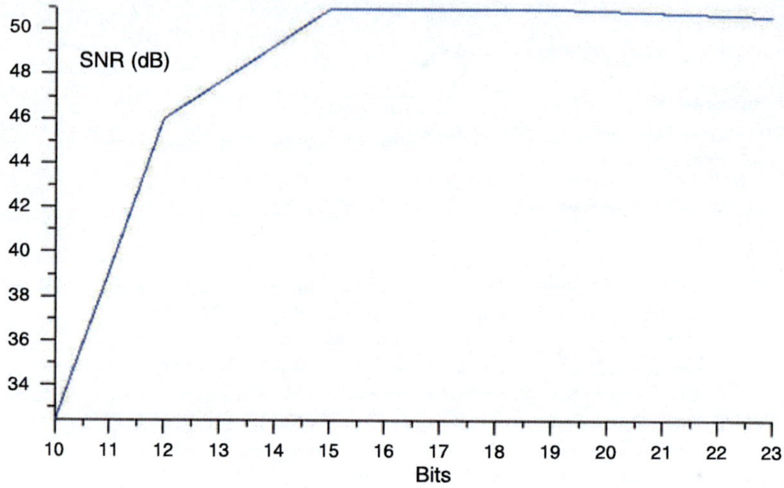

Fig. 12 Results obtained with the SNR metric

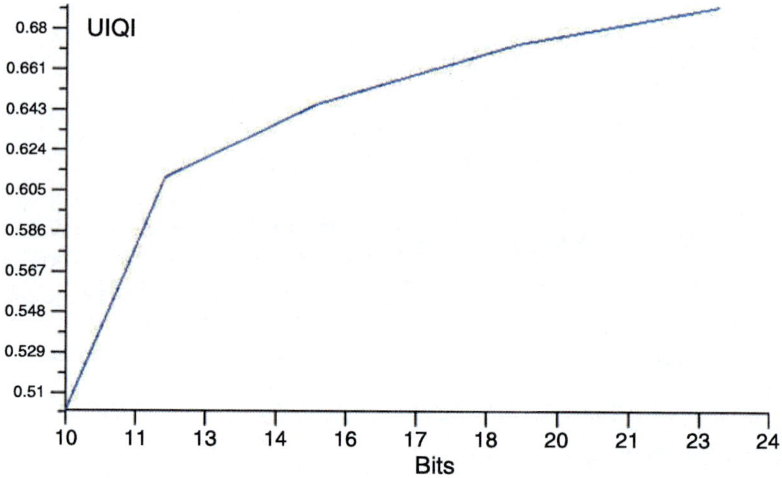

Fig. 13 Results obtained with the UIQI metric

As may be verified by graphics in Fig. 12, in the first 6 configurations, between 10- and 15-bit precision, a significant improvement in SNR was obtained as the result for each bit precision was added. However, after this, all the measurements are stable and show no noticeable improvements when compared with 15-bit precision configuration.

Another important result in this experiment was that the metric UIQI did not show significant improvement from the configuration with 15-bit precision. This result, although requiring expertise in seismic data processing, points that this is the lowest precision required for representing seismograms.

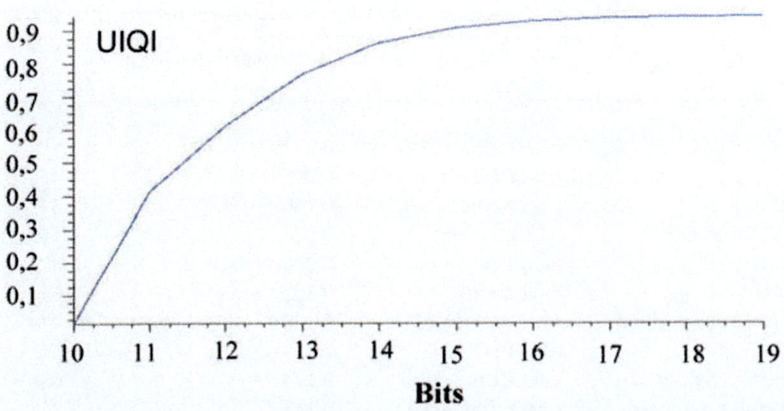

Fig. 14 Results obtained with the processing using floating point and the storage using fixed point

5 Results

This section presents performance and power results. It also discusses some image quality issues related to the use of floating point and fixed point number representations.

5.1 Performance Results

In order to evaluate the performance of the developed system, we implemented the same RTM modeling algorithm in a set of configurations such as: simple code without any special configuration (sequential C code); multi-threaded code (12 threads); OpenCL code; SIMD intrinsics (both SSE and AltiVec); threads and SIMD combined; and the HDL version for the FPGA. These configurations ran in the following platforms: CPU Intel Xeon E5405; CPU AMD Athlon 64 X2 6000+; CPU IBM PPC970MP; GPGPU Nvidia Tesla C1060; and FPGA Altera Stratix III 260E. Such platforms and configurations are summarized in Table 1.

In our experiments, subsets from the Marmoursi velocity model were used. The most important parameters on the algorithm execution are the size of matrices and the number of time steps that will be processed.

Originally, the Marmousi velocity model has 751 rows and 2,301 columns. As the number of necessary time steps is only a function of the model number of lines (the model depth), it was also fixed and equalized to 14,960. The number of rows was fixed in 748 and the number of columns were changed in the following sequence: 600, 800, 1,000, 1,200, 1,400, 1,600, 1,800, 2,000, 2,200, and 2,300.

The average performance of the CPU, GPGPU, and FPGA is presented in Fig. 15 and in Table 1. The Nvidia Tesla C1060 implementation, using OpenCL, had the

Table 1 Platforms and configurations used in experiments and their performance results

Configuration	Platform	Average performance (Gsample/s)	Speed-up
Sequential C code	Intel Xeon	0.065	1
	AMD Athlon X2	0.0719	1.108
	IBM PPC 5	0.095	1.458
Multi-thread (12 threads)	Intel Xeon	0.150	2.313
	AMD Athlon X2	0.074	1.153
	IBM PPC 5	0.226	3.479
SIMD (SSE and AltiVec)	Intel Xeon	0.109	1.676
	AMD Athlon X2	0.128	1.974
	IBM PPC 5	0.233	3.582
Multi-thread and SIMD combined	Intel Xeon	0.174	2.686
	AMD Athlon X2	0.143	2.211
	IBM PPC 5	0.674	10.381
OpenCL	AMD Athlon X2	0.102	1.568
	Nvidia Tesla C1060	2.143	33.001
HDL code	Altera Stratix III	1.878	28.912

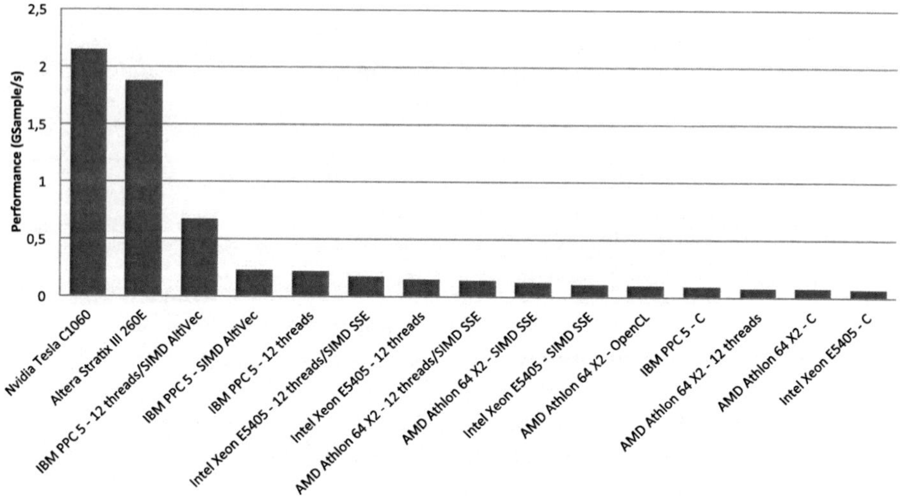

Fig. 15 CPU, GPGPU, and FPGA performance comparison

best results in our experiments. Table 1 also presents the speeding-up of all solutions related to the slowest one, which is the sequential C code running on the Intel Xeon E5405.

A Stratix III FPGA implementation is 29 times faster than the slowest CPU solution and the GPGPU solution (Nvidia Tesla C1060) is only 14% times faster than the FPGA one. The FPGA and the GPGPU presents almost the same performance however, the FPGA runs on an around ten times slower frequency than the GPGPU.

That is possible due to FPGA's intrinsic parallelism and pipeline structures. FPGAs can perform processing on streams, in which various data are processed by various operations in different stages of different pipelines (multiple data and multiple instructions), unlike GPGPUs which, in processing streams, multiple identical threads work on different data (single instruction and multiple data). The FPGA platform also has ten times slower memory bandwidth than the GPGPU [7, 11].

Through these results it is possible to conclude that a more powerful FPGA platform that provides better memory bandwidth and more available logic can beat the GPGPU performance by improving the time and space domain parallelism when both apply the same techniques mentioned previously.

Another FPGA advantage is obtained when energy efficiency is considered. In Sect. 5.3 a power analysis is made.

5.2 *Arithmetic Analysis*

The results obtained in this project confirm the feasibility of using standard numerical representation of lower precision in the storage and operation of seismic data.

When comparing the images obtained by using fixed point, Fig. 16b–f, with the reference image obtained by using floating point (Fig. 16a), is not noticeable that only the seismogram images with 12 and 10 bits of precision have substantial differences. The image with 15 bits of precision shows subtle differences that require further evaluation. The images produced with 19 and 23 bits of precision can be regarded as identical to the reference image.

The synthesis results listed in Table 2 shows that the strategy of adopting fixed point format allowed the obtainability of both a reduction in the hardware area and an increase in the PE operating frequency in all configurations.

Specifically about the hardware area improvement, the results indicated that the area improvement would be of more than three times as much. However, in this project, constraints in memory throughput of the used platform, restricted it to 50%.

The results demonstrate the benefits and risks that evolved changing the standard floating point representation to fixed point in applications of seismic data processing.

The metrics used were effective to verify the quality of the results, confirming the validity of the methodology used.

The synthesis results pointed to the possibility of gains by approximately three times as much in the number of processing elements and more than 27% in the system operating frequency, with good final results quality when a configuration of fixed point with 21 bits of precision was adopted.

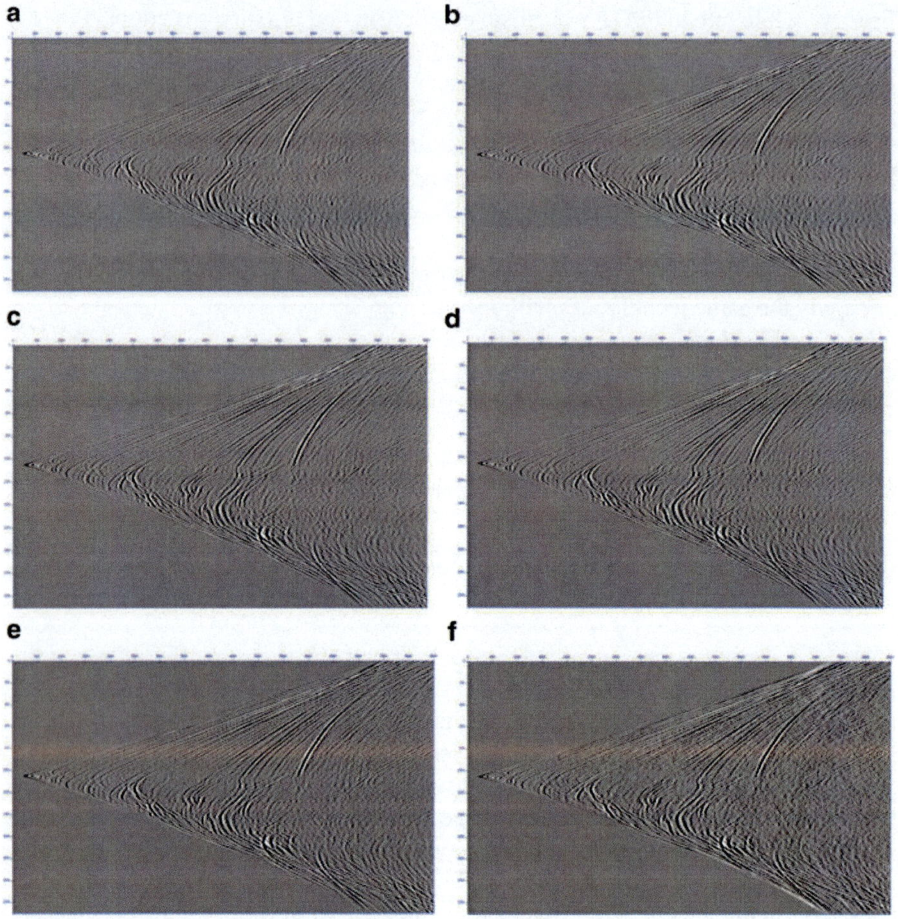

Fig. 16 Seismogram images for different precisions. (**a**) Standard single precision floating point; (**b**) fixed point 23-bits precision; (**c**) fixed point 19-bits precision; (**d**) fixed point 15-bits precision; (**e**) fixed point 12-bits precision; (**f**) fixed point 10-bits precision

5.3 Power Analysis

Recently [19], new metrics for evaluating efficiency in applications that consider energy as a crucial factor have been proposed. The GFLOP/Joule metric indicating the compute density per Joule considers that the higher this ratio, the better the efficiency of the proposed solution. Therefore, approaches can be evaluated based on a metric that takes into account the performance of the application including the expenditure of energy.

Table 2 Synthesis results for some PE configurations targeting the FPGA

PE number representation	ALUTS	Registers	Frequency (MHz)
Floating point	6,491 (3%)	6491 (2%)	178
Fixed-point (32 bits)	1,888 (<1%)	1,888 (<1%)	226
Fixed-point (25 bits)	1,464 (<1%)	1,435 (<1%)	226
Fixed-point (21 bits)	1,393 (<1%)	1,224 (<1%)	227

Table 3 Summary of the characteristics of the platforms compared

Technology	Platform	Frequency	Power (W)
CPU	Athon 64 × 2 6,000	3.0 GHz	125
GPGPU	Tesla T10 c1060	1.3 GHz	160
FPGA	Stratix III 260e	125 MHz	18
	Gidel ProcStar III		

Table 4 Comparison of performance and efficiency of platforms

Technology	Processing time (s)	Performance (Gsample/s)	Efficiency (MSample/J)
CPU	354.980	0.072	0.6
CPU+GPGPU	11.923	2.143	7.5
CPU+FPGA	13.610	1.878	12.8

In this section, a comparison between platforms CPU, GPU, and FPGA is made the aspect of performance (GSample/s), as well as with respect to efficiency in energy consumption (MSample/Joule). For this purpose, we consider the algorithm RTM was considered for the processing of 14,960 time steps of the Marmousi model with 748 rows and 2,300 columns. And in order to make fair comparisons, the full system: CPU, CPU+GPGPU, and CPU+FPGA, was considered, once it is not possible to use the FPGA or the GPGPU on their own.

Table 3 shows the characteristics of each platform through its description, operation frequency operation (GHz) and average power (Watts), indicated in each column.

For the calculation of the efficiency of these platforms, the average power consumption of each device is obtained in [10, 11], and [15]. The performance results are shown in Table 4.

The results presented allow the conclusion that the CPU+FPGA has a superior performance in both architectures. When compared to the CPU, the CPU+FPGA has achieved more than 26 times better level of performance. When compared to the CPU+GPGPU, the performance came to be about 14% lower. The results are more striking when comparing the efficiency in terms of MSample/Joule. The total amount of work is the same for all platforms, i.e. 25.6 GSamples. Considering MSample/Joule metric to evaluate the results, the figures in Joules for each platform CPU, CPU+GPGPU, and CPU+FPGA are, respectively, 125*355, 285*12, and 143*14 J. In MSamples per Joules these figures are 0.6, 7.5, and 12.8 for CPU,

CPU+GPGPU and CPU+FPGA, respectively, as shown in last column in the previous table. In this case, the figure of the CPU+FPGA was more than 21 times higher than the one of the CPU. Compared to the CPU+GPGPU, CPU+FPGA efficiency was about 1.7 times better. This high efficiency is mainly due to low operating frequency when compared to other FPGA architectures and its built-in feature that allows the exploitation of parallelism.

6 Conclusions

In this work, an FPGA implementation of the modeling part of the RTM seismic algorithm was presented. A GPGPU and a CPU version of the same algorithm to evaluate the FPGA performance also was developed. It was observed that the FPGA is faster than the CPU but slower than the GPGPU. However, the FPGA has around ten times less memory bandwidth than the GPGPU and runs on an around ten times slower frequency that enables a lower power consumption. Despite the fact that the implementation of the spatial domain parallelism on a GPGPU is straightforward when using OpenCL, optimizations like the time domain parallelism and change in the floating point precision are too costly or impossible to implement. Thus, an FPGA implementation offers much more potential for more sophisticated optimizations.

Results for the FPGA approach are, on average, only 13% slower than our best GPGPU implementation. Considering the memory bandwidth limitations on the Gidel board, the FPGA can be regarded as a very promising solution. A more powerful FPGA platform that provides better memory bandwidth and more available logic can beat the GPGPU performance by improving the time and space domains parallelism.

Our research team is already exploring other optimization possibilities like data compression techniques and different numerical representations usage in order to further improve the application performance without jeopardizing the quality of results. Through these optimizations it is possible to free more logic elements enabling to explore a deeper temporal pipeline and consequently improve performance. Some experimental results show that using low precision numerical representation may significantly increase the performance without affecting considerably the resultant images. These studies were associated with impact evaluation in the quality of the results of the generated final image through SNR and UIQI metrics. Results indicated a speedup increase of 6.75% when compared to a software approach. This is a great advantage of the FPGA over other architectures like CPUs and GPGPUs that do not feature this flexibility in customizing the floating point precision.

Analysis in terms of MSample/Joule also was performed considering CPU, CPU+GPGPU, and CPU+FPGA technologies, and results indicated CPU+FPGA as an energy efficient approach, being approximately 1.7 times better when compared with CPU+GPGPU running the same algorithm.

Acknowledgments The authors would like to thank the Petrobras Research Center (CENPES) for technical support in seismic concepts, FINEP/CNPq, RPCMod network coordination and FACEPE. Additionally, the authors gratefully acknowledge to continuous support by Gidel, including the availability of prototype boards for performance measurements.

References

1. A. Barros, B. Dutra, V. Brito, M. Lima, A. Silva-Filho, R. Gandra, R. Braganca, Performance evaluation model based on precision reduction and FPGAs applied to seismic modeling, in *Simpasio em Sistemas Computacionais (WSCAD-SSC), 2011* (Vitória, Brazil, 2011), p. 2. doi:10.1109/WSCAD-SSC.2011.24
2. S. Brown, Performance comparison of finite-difference modeling on cell, FPGA, and multicore computers. SEG Tech. Program Expanded Abstr. **26**(1), 2110–2114 (2007). doi:10.1190/1.2792905, http://link.aip.org/link/?SGA/26/2110/1
3. C. Chang, J. Wawrzynek, R. Brodersen, Bee2: a high-end reconfigurable computing system. IEEE Des. Test Comput. **22**(2), 114–125 (2005). doi:10.1109/MDT.2005.30
4. S. Che, J. Li, J.W. Sheaffer, K. Skadron, J. Lach, Application specific processors, 2008. SASP 2008. Symposium on Accelerating Compute-Intensive Applications with GPUs and FPGAs, pp. 101–107, June 2008. doi:10.1109/SASP.2008.4570793
5. R.G. Clapp, H. Fu, O. Lindtjorn, Selecting the right hardware for reverse time migration. Lead. Edge **29**(1), 48–58 (2010). doi:10.1190/1.3284053, http://tle.geoscienceworld.org/cgi/content/abstract/29/1/48
6. H. Fu, W. Osborne, R.G. Clapp, O. Mencer, W. Luk, Accelerating seismic computations using customized number representations on FPGAs. EURASIP J. Embed. Syst. **2009**, 1–13 (2009). doi:http://dx.doi.org/10.1155/2009/382983
7. Gidel: PROCe III board specifications (2012), http://www.gidel.com/PROCe%20III.htm. Accessed 29th January 2013
8. C. He, M. Lu, C. Sun, Accelerating seismic migration using FPGA-based coprocessor platform, in *FCCM '04: Proceedings of the 12th Annual IEEE Symposium on Field-Programmable Custom Computing Machines* (IEEE Computer Society, Washington, 2004), pp. 207–216
9. C. He, G. Qin, M. Lu, W. Zhao, Optimized high-order finite difference wave equations modeling on reconfigurable computing platform. Microprocess. Microsyst. **31**(2), 103–115 (2007). doi:http://dx.doi.org/10.1016/j.micpro.2006.02.010
10. N. Kirsch, AMD Athlon 64 X2 6000+ Processor Review (2012), http://www.legitreviews.com/article/463/2/. Accessed 29th January 2013
11. Nvidia: Tesla C1060 specifications (2012), http://www.nvidia.com.br/docs/IO/56050/Tesla_C1060_boardSpec_v03.pdf. Accessed 29th January 2013
12. C. Petrie, C. Cump, M. Devlin, K. Regester, High performance embedded computing using field programmable gate arrays, in *Proceedings of the 8th Annual Workshop on High-Performance Embedded Computing* (MIT Lincoln Laboratory, Lexington, United States, 2004), pp. 124–150
13. T. Scofield, J. Delmerico, V. Chaudhary, G. Valente, Xtremedata dbx: an FPGA-based data warehouse appliance. Comput. Sci. Eng. **12**(4), 66–73 (2010). doi:10.1109/MCSE.2010.93
14. X. Shi, X. Wang, C. Zhao, H. Yang, Practical pre-stack kirchhoff time migration of seismic processing on general purpose gpu, in *CSIE '09: Proceedings of the 2009 WRI World Congress on Computer Science and Information Engineering* (IEEE Computer Society, Washington, 2009), pp. 461–465. doi:http://dx.doi.org/10.1109/CSIE.2009.78
15. G. Stitt, FPGA-based Scientific Computing: A Bright Future? (2012), http://cas.ee.ic.ac.uk/people/gac1/DATE2011/Stitt.pdf. Accessed 29th January 2013

16. W.W. Symes, Reverse time migration with optimal checkpointing. Geophysics **72**(5), SM213–SM221 (2007). doi:10.1190/1.2742686, http://link.aip.org/link/?GPY/72/SM213/1
17. D.B. Thomas, L. Howes, W. Luk, A comparison of CPUs, GPUs, FPGAs, and massively parallel processor arrays for random number generation, in *FPGA '09: Proceeding of the ACM/SIGDA International Symposium on Field Programmable Gate Arrays* (ACM, New York, 2009), pp. 63–72. doi:http://doi.acm.org/10.1145/1508128.1508139
18. Z. Wang, A. Bovik, A universal image quality index. IEEE Signal Process. Lett. **9**(3), 81–84 (2002). doi:10.1109/97.995823
19. J. Williams, C. Massie, A.D. George, J. Richardson, K. Gosrani, H. Lam, Characterization of fixed and reconfigurable multi-core devices for application acceleration. ACM Trans. Reconfigurable Technol. Syst. **3**(4), 19:1–19:29 (2010). doi:10.1145/1862648.1862649, http://doi.acm.org/10.1145/1862648.1862649
20. Ö. Yilmaz, in *Seismic Data Analysis*. Society of Exploration Geophysicists (Tulsa, OK, 2001). doi:DOI:10.1190/1.9781560801580, http://link.aip.org/link/doi/10.1190/1.9781560801580

High-Performance Cryptanalysis on RIVYERA and COPACOBANA Computing Systems

Tim Güneysu, Timo Kasper, Martin Novotný, Christof Paar, Lars Wienbrandt, and Ralf Zimmermann

Abstract Special-purpose computing platforms based on reconfigurable hardware have shown to typically exhibit a much better performance-cost ratio than off-the-shelf computers populated with general-purpose processors. In this chapter we introduce two different FPGA-based cluster architectures, called COPACOBANA and RIVYERA. These high-performance computing clusters are populated with up to 256 Xilinx Spartan or Virtex FPGAs per system and can be interconnected to form an even larger system with 2,560 FPGA per rack. In this chapter, we present a wide range of applications from the fields of cryptanalysis that have been successfully implemented on both architectures.

1 The Evolution: COPACOBANA and RIVYERA

Massively parallel special-purpose hardware systems are often the best choice to tackle computationally complex algorithms and applications. However, for many such applications the cost-performance ratio is a crucial criterion. In other words, an optimally designed hardware platform for a specific application does not provide any features or peripherals that are not required, usually at the cost of limited flexibility and reusability of the system for any other applications. In this chapter,

T. Güneysu (✉) • T. Kasper • C. Paar • R. Zimmermann
Horst Görtz Institute for IT-Security, Ruhr-University Bochum, Germany
e-mail: tim.gueneysu@rub.de; timo.kasper@rub.de; christof.paar@rub.de; ralf.zimmermann@rub.de

M. Novotný
Faculty of Information Technology, Czech Technical University in Prague, Czech Republic
e-mail: martin.novotny@fit.cvut.cz

L. Wienbrandt
Department of Computer Science, Christian-Albrechts-University of Kiel, Germany
e-mail: lwi@informatik.uni-kiel.de

W. Vanderbauwhede and K. Benkrid (eds.), *High-Performance Computing Using FPGAs*, DOI 10.1007/978-1-4614-1791-0_11, © Springer Science+Business Media, LLC 2013

we present two hardware architectures that aim to combine both goals—to provide an optimal cost-performance ratio for a wide range of applications from the field of cryptanalysis.

1.1 The COPABOBANA Computing System

The Cost-Optimized Parallel Code Breaker (COPACOBANA) was designed to provide a significant amount of computing resources to applications with only a minor demand on memory and communications. The majority of other FPGA-based computing clusters or supercomputers, however, focus on data-oriented applications requiring large amounts of memory and widely dimensioned bandwidth. Examples for such universal supercomputing systems are Cray's XD1 system [30] as well as the SGI RASC technology [31] that also include reconfigurable devices in their design. Unfortunately, such platforms are inappropriate for most tasks in cryptanalysis due to their high costs and the related non-optimal cost-performance ratio. Here, to the best of our knowledge, COPACOBANA is the only low-cost alternative to commercial supercomputers offering no nameable amount of memory but a significant amount of computing resources for less than € 10,000 hardware production costs [40].

The hardware architecture of COPACOBANA was developed according to the following design criteria [39]: First, we assume that computationally costly operations can be parallelized. Second, all concurrent instances have only a very limited requirements to communicate with each other. Third, the demand for data transfers between host and nodes is low due to the fact that computations heavily dominate communication requirements. Ideally, (low-speed) communication between the hardware and a host computer is only required for initialization and the transfer of results. Hence, a single conventional (low-cost) PC should be sufficient to transfer required data packets to and from the hardware, e.g., connected by a standardized interface. Fourth, all presented algorithms and their corresponding hardware nodes demand very little local memory which can be provided by the on-chip block RAM modules of an FPGA.

Since our initial goal was to satisfy an application's demand for plenty of computing power, we installed a total of 120 FPGA devices on the COPACOBANA cluster.[1] Building a system of comparable dimension with commercially available FPGA boards is certainly feasible but rather expensive. By stripping down functionality to the bare minimum and producing the hardware ourselves, we are able to achieve with COPACOBANA an optimal cost-performance for the set of applications shown in Sect. 2.

[1]This number was determined by the size of the FPGA modules (DIMM form factor) and the maximum bus length.

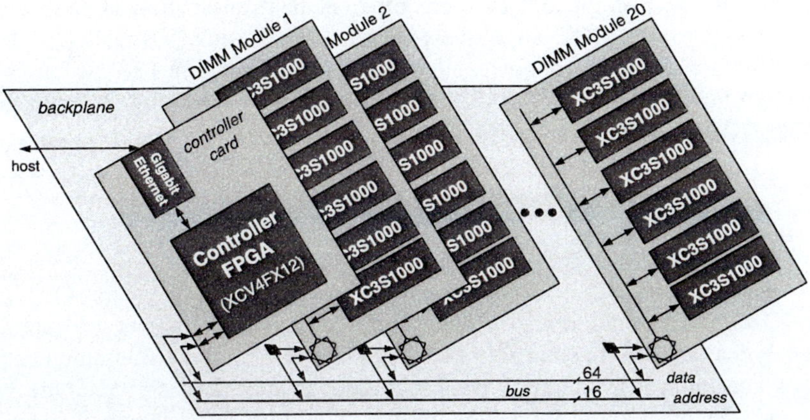

Fig. 1 Schematic architecture of COPACOBANA

For a modular and maintainable architecture, we designed small FPGA modules which can be dynamically plugged into a backplane. Each of these modules in DIMM form factor hosts 6 low-cost Xilinx Spartan3-XC3S1000 FPGAs which are directly connected to a common 64-bit data bus on board. The data bus of the module is interfaced to the global data bus on a backplane. While disconnected from the global bus, the FPGAs on the same module can communicate via the local 64-bit data bus. Additionally, control signals are run over a separate 16-bit address bus. Figure 1 gives an overview of the architecture of COPACOBANA. For simplicity, a single master bus was selected to avoid interrupt handling or bus arbitration. Hence, if the communication scheduling of an application is unknown in advance, the bus master need to poll the FPGAs.

The top level entity of COPACOBANA is a host-PC which is used to initialize and control the FPGAs, as well as for the accumulation of results. Programming can be done simultaneously for all or a specific subset of FPGAs. Data transfer between FPGAs and a host-PC is accomplished by a dedicated control interface. The controller has also been designed as a slot-in module so that COPACOBANA can be connected to a computer either via a USB or Ethernet controller card. A simple API software library on the host-PC provides low-level functions that allow for addressing individual FPGAs, storing and reading FPGA-specific application data. With this approach, we can easily attach more than one COPACOBANA device to a single host-PC.

For some applications in the field of cryptanalysis, arithmetic intensive operations need to be performed using *Digital Signal Processors* (DSP) blocks that have become available with many latest FPGA devices (cf. Sect. 2.4). In 2007, we therefore designed new slot-in modules for use with COPACOBANA that host 8 Xilinx Virtex-4 XC4VSX35 FPGAs, each providing 192 DSP slices. Due to the larger size of the FPGAs (FF668 package with dimension of 27×27 mm) we enlarged the modules what includes also modifications of the corresponding connectors on the backplane. For more efficient heat dissipation at high clock frequencies up to 400 MHz, an actively ventilated heat sink is attached to each FPGA. With a more powerful power supply providing 1.5 kW at 12 V, we are able to run a total of 128 Virtex4-SX35 FPGAs distributed over 16 plug-in modules in the COPACOBANA V4.

1.2 The RIVYERA Computing System

COPACOBANA was tailored to provide pure computational power for (simple) cryptanalytical applications but was known to be very limited on communication facilities and on-system memory. Hence, to support advanced cryptanalytic processes and even other application domains (e.g., bioinformatics), we finally decided to come up with an enhanced and more powerful architecture. For a wide range of advanced applications beyond simple exhaustive key search attacks, it turned out that requirements for on-system memory and fast communication systems are crucial. This insight led to a major redesign of the original COPACOBANA architecture. The hardware platform RIVYERA was introduced in 2008 [51] and includes a completely reworked communication and memory infrastructure. The RIVYERA platform is developed and distributed by SciEngines GmbH [55] and consists of two elements. First, a cluster of FPGAs spread over multiple plug-in cards and, second, a standard server-grade mainboard, running a Linux operating system on an Intel Core i7-930 processor with 12 GB of RAM. The FPGA cluster is equipped with up to 128 user configurable Xilinx Spartan3-5000 or Spartan6-LX150 FPGAs respectively, distributed over 16 plug-in cards, each containing 8 user FPGAs. Additionally, a DRAM module with a capacity of 32 MB is attached to each user FPGA in the RIVYERA S3-5000. RIVYERA S6-LX150 provides 512 MB DRAM per default to each user FPGA, but can be extended with a memory extension module providing an additional amount of 2 GB DRAM per FPGA.

RIVYERA offers a high-performance bus system. The interconnection between all FPGAs is organized as a one-dimensional array or systolic chain. The general idea of a systolic chain is to provide fast point-to-point connections between every two neighbors. On the one hand, a systolic-like system typically results in shorter wires achieving higher frequencies and therefore higher data throughput. On the other hand, a typical problem becomes the latency of large chains and its usability. However, the latency has been reduced significantly by adding shortcuts, i.e., each FPGA on an FPGA card is connected with two neighbors forming a ring including a communication controller. Additionally, the FPGA card slots are connected in a ring as well via the communication controllers on each FPGA card.

To link the host software to the FPGA application, at least one communication interface of the FPGA cards is connected via PCIe to the host mainboard. Furthermore, communication bandwidth can be increased by adding more communication

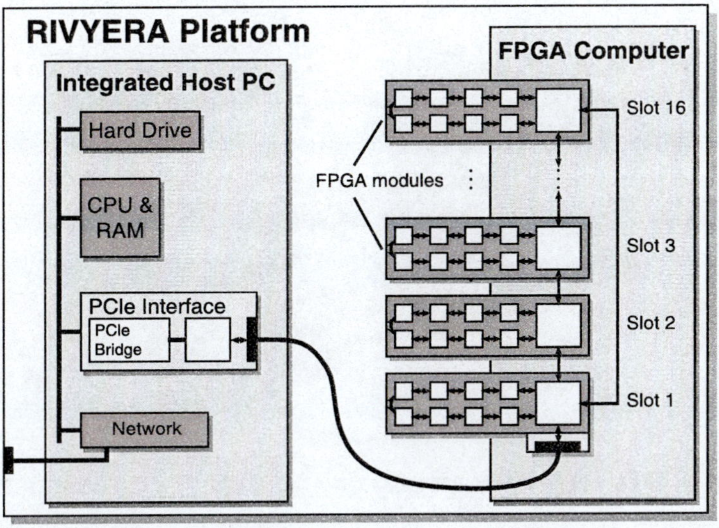

Fig. 2 Schematic architecture of RIVYERA

controllers to the host, connected to other available FPGA cards. Even an uplink to other RIVYERA machines can be established, allowing even more computational resources for particular applications. An overview of the architectural concept is illustrated in Fig. 2.

The provided API, as already mentioned, supports the development of software and hardware design. In particular, the API provides the communication interface for software and hardware, making any potential host-FPGA or FPGA–FPGA connection transparent to the developer by an automatic routing of data packets. The API includes broadcast facilities, methods for configuring the user FPGAs and a communication interface for the FPGA-attached DRAM.

A picture of the RIVYERA S3-5000 is shown in Fig. 3. Due to its small form factor, RIVYERA can be installed in 3U of a standard rack and powered by only two redundant 650 W supplies.

Besides cryptanalytic applications, algorithms from bioinformatics were implemented on RIVYERA, e.g., the Smith–Waterman alignment [54, 61], BLAST database search [62, 63] and short-read alignment with BWA. Descriptions of these applications can be found in Chap. 3. Research is also in progress on Genome-Wide Association Studies (GWAS) and de-novo assembly.

Another recently explored field of applications is the accelerated stock market analysis methods based on FPGA-based optimizations of investment strategies. This can lead up to speedups of more than 17,000 compared to a standard PC platform [58, 59].

Fig. 3 View on RIVYERA S3-5000 computing system

2 Cryptanalysis on COPACOBANA and RIVYERA

The security of symmetric and asymmetric ciphers is usually determined by the size of their security parameters, in particular the key-length. Hence, when designing a cryptosystem, these parameters need to be chosen according to the assumed computational capabilities of an attacker. Depending on the chosen security margin, many cryptosystems are potentially vulnerable to attacks when the attacker's computational power increases unexpectedly. In real life, the limiting factor of an attacker is often the financial resources. Thus, it is quite crucial from a cryptographic point of view to not only investigate the complexity of an attack but also to study possibilities to lower the cost-performance ratio of attack hardware. For instance, a cost-performance improvement of an attack machine by a factor of 1,000 effectively reduces the key lengths of a symmetric cipher by roughly 10 bit (since $1{,}000 \approx 2^{10}$). In this section we discuss cryptanalytical implementations on COPACOBANA and RIVYERA which can offer, depending on the application, a cost-performance that is several orders of magnitude better than that of current PCs.

2.1 *Exhaustive Key-Search Attacks*

Brute-force attacks aim at extracting secret keys by means of exhaustive key-search: Given a pair of plain- and ciphertext, all possible keys are tested until the correct one, i.e., generating the correct ciphertext for the given plaintext or vice versa, is identified. The attacks usually rely on optimized implementations of the investigated cryptographic primitives but do not involve any special cryptanalytic methods. The performance of an exhaustive key-search for a given cipher—even when not revealing the secret key in a practical time—is an important estimator for thereal-world security of cryptographic primitives. In the following we illustrate practical implementations of brute-force attacks targeting DES, Hitag2, KeeLoq, electronic passports and PRESENT on COPACOBANA.

2.1.1 Brute-Force Attack on DES

The data encryption standard (DES) with a 56-bit key size was chosen as the first commercial cryptographic standard by NIST in 1977 [47]. This blockcipher remained the standard for symmetric cryptography for more than 20 years and is still (in 2012) employed in quite a few (recent) products. In the time of specification, a key size of 56 bits was considered to be a good choice considering the huge development costs for computing power in the late 1970s, that made a search over all the possible 2^{56} keys appear impractical. There have been a lot of feasibility studies on the possible use of parallel hardware and distributed computing for breaking DES. The first estimates were proposed by Diffie and Hellman in 1977 [13] for a brute-force machine that could find the key within a day at a cost of US$ 20 million. In 1998, the electronic frontier foundation (EFF) built a DES hardware cracker called *Deep Crack* which found the key of DES Challenge II-2 in 56 h [17]. Their DES cracker consisted of 1,536 custom designed ASIC chips at a cost of material of around US$ 250,000 and could search 88 billion keys per second. In 2006 we reimplemented the DES cracking application on the COPACOBANA computing system. Four fully pipelined DES cores can be placed on each of the 120 Spartan-3 XC3S1000 FPGAs that are capable to test one key per cycle. Each core can test 2^{42} keys in $2^{40} \times 7.35$ ns per FPGA, which is approximately 135 min. The entire machine can instantiate $4 \times 120 = 480$ cores on all FPGAs testing 480 keys every 7.35 ns at 136 MHz, i.e., 65.28 billion keys per second. To find the correct key, COPACOBANA has to search through an average of 2^{55} different keys. Thus, it can find the right key in approximately $T = 6.4$ days on average. Of course, more than one COPACOBANA can be attached to a single host and the key space shared, so that the search time is reduced to $\frac{T}{n}$, where n denotes the number of machines.

We like to briefly compare our solution to a software-based approach to undermine the power of dedicated (FPGA) hardware for this application. With an expense of € 10,000 required for the material of a COPACOBANA system, we can afford around 50 low-cost PCs (Celeron@3 GHz including necessary peripherals) for € 200 each in equal measure. A standard software implementation of DES can compute roughly five million DES encryptions per second on such a PC. Hence, with the fixed investment of € 10,000, we here yield a throughput of 250 million DES keys per second with the PC cluster. Compared to the 65.28 billion DES keys searched by a single COPACOBANA per second, we can outperform the PC cluster in this case by a factor of more than 260. Regarding the power consideration, we measured a fully equipped COPACOBANA running a DES key search to consume less than 600 W. Related to this, we assume a single Pentium4-based computer to

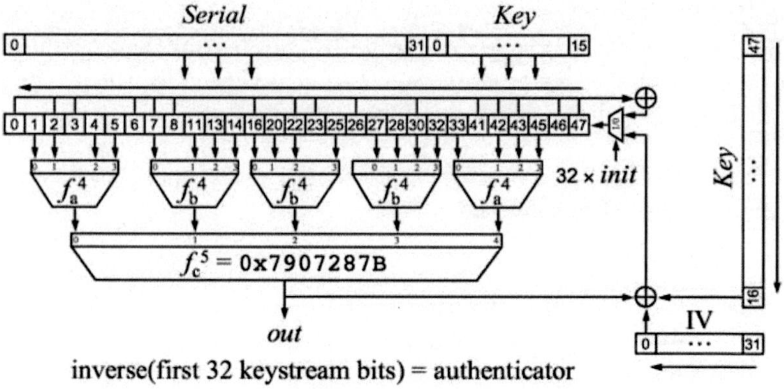

$f_a^4 = \texttt{0x2C79} = abc+ac+ad+bc+a+b+d+1$

$f_b^4 = \texttt{0x6671} = abd+acd+bcd+ab+ac+bc+a+b+d+1$

Fig. 4 Internal structure of Hitag2 initialization and encryption, adopted from [9]

require 150 W on average. Hence, comparing the power consumption of the entire key search on a COPACOBANA and the PC cluster, a worst-case key search on COPACOBANA will take 184 kWh whereas the PC cluster consumes the immense amount of approximately 1.5 GWh during runtime.

2.1.2 Brute-Force Attack on Hitag2

Hitag2 is a stream cipher primarily used in radio frequency identification (RFID) applications, such as car immobilizers. It has been developed and introduced in late 1990s by Philips Semiconductors (currently NXP). Hitag2 is reported to be used in models produced by many leading car manufacturers. It is not clear whether the Hitag2 is still used in newly produced cars, however, due to its relatively recent introduction it is sure that many cars with Hitag2 are still in daily use.

The internal structure of Hitag2 is (see Fig. 4) very similar to its predecessor, i.e., the Crypto1 cipher [10, 64] used in Mifare-Classic cards. Hitag2 uses a 48-bit key and its internal state also has the corresponding length of 48 bits. Due to the relatively short key-length of the cipher, brute-force attacks on Hitag2 are feasible and practical, as described below. However, due to its internal structure and operation, Hitag2 is also vulnerable to algebraic attacks: the authors of [11] are able to extract the secret 48-bit key within 45 h, on the basis of four sniffed transactions, by running MiniSat 2.0 on a PC.

Figure 5 depicts the protocol between the reader (RW device) embedded in the car, and the transponder (tag) embedded in the immobilizer. The transponder and the reader share a common 48-bit secret key. To prevent replay attacks, a unique initialization vector (IV) is generated for every transaction between the reader and

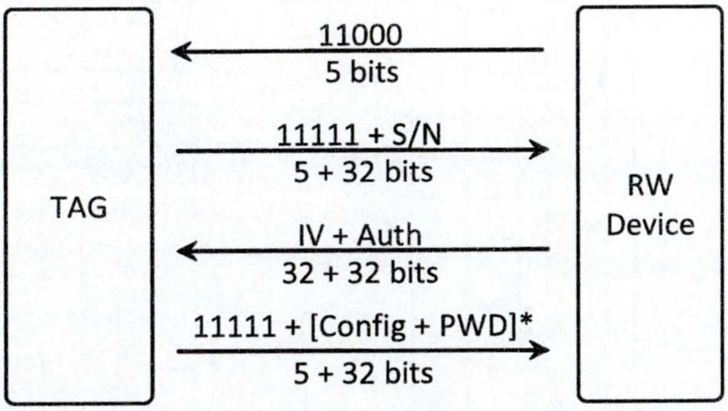

Fig. 5 Hitag2 protocol in *Crypto mode*

the transponder. The secret key, the initialization vector, and the serial number of the tag are used for the initialization of a cipher. After initialization, the Hitag2 cipher produces a keystream. The first 32 bits of the keystream are used as an *authenticator* and the remaining bits are used for encryption like in any standard stream cipher.

For a brute-force attack the serial number, the initialization vector, and the authenticator have to be sniffed from one transaction. From this data, 32 bits of keystream can be recovered, while the secret key has 48 bits. Hence, at least two transactions between tag and reader have to be monitored as a prerequisite for a successful recovery of the key. A corresponding brute-force attack on Hitag2 has been efficiently implemented on COPACOBANA [60]. To parallelize the attack, the search space is divided into 512 subspaces and distributed to the FPGAs by the host computer. If the search in a particular subspace finishes without success, another subspace is assigned to the FPGA. Each FPGA internally generates all 2^{39} keys of an assigned subspace. The block-level structure of the Hitag2 breaker as implemented in each FPGA is illustrated in Fig. 6. The breaker consists of the *control module* and 256 *Hitag2 executional cores*, denoted as *H2 Core*. Therefore, each FPGA verifies 256 keys concurrently against the data obtained from the first transaction. If any H2 Core produces a correct authenticator, the found key candidate is passed to the *H2 Core—final* stage (see bottom of Fig. 6) for verification against data from second transaction between the transponder and the reader.

The brute-force attack outperforms all previous implementations by several orders of magnitude. Just two monitored communications between a Hitag2 transponder and a reader, instead of four sniffed transactions required in other published attacks, are sufficient to reveal the secret key: each FPGA of COPACOBANA verifies 378 million keys per second, therefore, one fully equipped COPACOBANA with 120 FPGAs is able to determine the correct key in less than 2 h (103.5 min) in the worst case. The proposed design is almost linearly scalable, which allows further reduction of the attack time by employing more COPACOBANA machines. The attack is also energy efficient. Considering COPACOBANA to have a consumption of less than four standard PCs, the attack needs the same amount of energy as one PC running for less than 8 h, while the algebraic attack reported in [11] needs one PC running for 45 h.

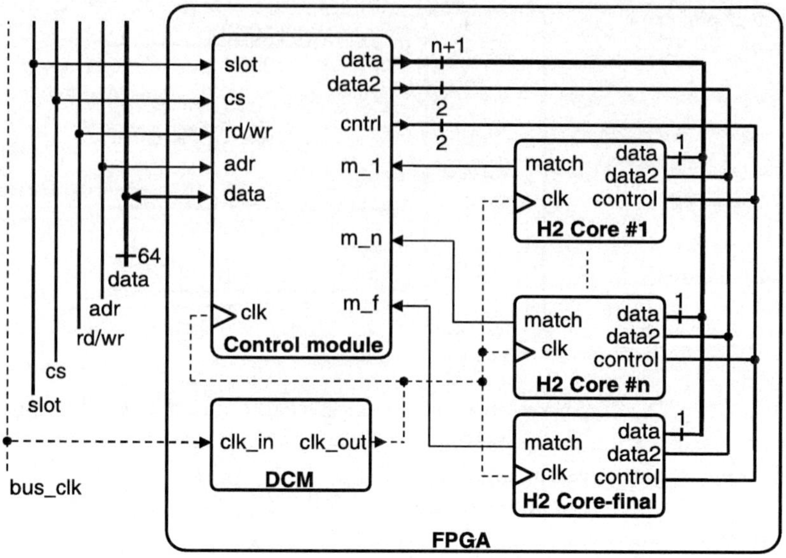

Fig. 6 Block-level structure of the Hitag2 breaker in one FPGA

Fig. 7 Generating a hopping code with KEELOQ

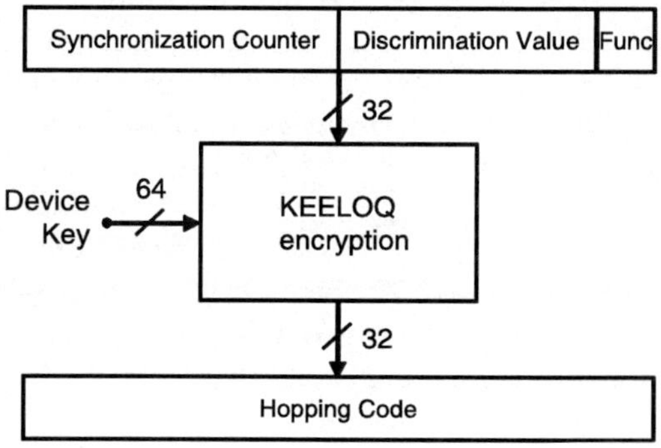

2.1.3 Brute-Force Attack on KeeLoq

Many real-world car door systems and garage openers are based on the KEELOQ cipher. These electronic systems consist of remote controls, which replace traditional keys, and receivers which control the door. On having its button pressed a remote sends a so-called hopping code to the receiver to open or close the door. A hopping code is generated by a KEELOQ encryption with a device key, incorporating a 16-bit counter value, a 12-bit discrimination value, and a 4-bit function value, as shown in Fig. 7. While the counter is incremented in the remote each time a new hopping code is generated, the discrimination and function values remain constant.

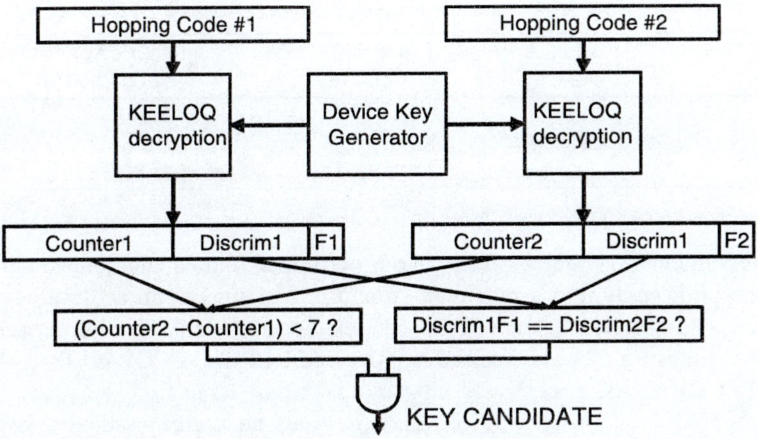

Fig. 8 KEELOQ breaker

To obtain the unique device key of a remote control on the side of the receiver, the serial number of the remote is typically decrypted by means of KeeLoq using a manufacturer key. Alternatively, for the key derivation a randomly generated seed value may by combined with the serial number in three different scenarios. Altogether we get four different possible inputs for the generation of device keys:

1. 64 bits of the serial number (N), 0xNNNNNNNNNNNNNNNN
2. 28 bits of the serial number (N) combined with 32 bits of the random seed (S) according to 0x0NNNNNNNSSSSSSSS
3. 12 bits of the serial number combined with 48 bits of the seed according to 0x0NNNSSSSSSSSSSSS, or
4. 60 bits of the seed following the pattern 0x0SSSSSSSSSSSSSSS

Since the KEELOQ cipher has been extensively studied [6, 16, 32], several different types of attacks have been proposed. The attack described in [16] reveals the manufacturer key by means of power analysis. In the following, breaking the system is straightforward for those commercial products deriving the device keys according to Scenario 1: the secret device key of the remote control can be instantly derived after monitoring its serial number. However, the attacks are not applicable if a seed is used, even when the manufacturer key is known. The goal of our implementation [49] is hence to find the correct device key for the respective Scenarios 2–4, in which a random seed is incorporated. For the key-search we need to intercept two hopping codes of the same remote control, generated with the same device key that have a small difference between their counter values.

We implemented a brute-force attack of KEELOQ on COPACOBANA. The diagram of the circuit implemented in each FPGA is shown in Fig. 8. A candidate for the device key is found by means of exhaustive key-search, if the decryptions of two intercepted hopping codes reveal identical discrimination values and

Table 1 Worst case times for the brute-force attack on KEELOQ

SEED length (bits)	1 FPGA (<80 $)	1 COPACOBANA (<10,000 $)	100 COPACOBANAs (<1,000,000 $)
32	39 s	0.33 s	3.3 ms
48	29.6 days	5.9 h	213 s
60	332 years	1,011 days	10.1 days

moderately increased counter values. In our optimized implementation we unrolled both decryption units into a pipelined structure. The maximum achievable clock frequency is 110 MHz, i.e., each FPGA can test up to 110 million keys per second. Worst-case times for all possible seed lengths, and 1 FPGA, 1 COPACOBANA and 100 COPACOBANAs, respectively, are summarized in Table 1.

We conclude that using a 32-bit seed provides no security, since a key can be found in real time. While a seed with 48 bits can be broken in less than 6 h by one COPACOBANA, employing a 60-bit seed can provide reasonable security despite the flawed cipher and the usage of a manufacturer key.

2.1.4 Brute-Force Attack on Electronic Passports

The ePass (electronic passport) is deployed in many countries all over the world. A contactless chip embedded in the passport, compliant to the ISO 14443 [33] standard, holds private data of the holder that is secured by cryptographic primitives. The security of the first generation of passports is questionable, as detailed in this section: our key-search attacks, as presented in more detail in [43], tackle the basic access control (BAC) security mechanism in the first generation of passports issued in Germany since November 2005. Our implementation on COPACOBANA is also applicable to passports of various other countries. After publicizing our findings a new version of the German ePass was released in November 2007 with an improved variant of the here attacked key derivation scheme.

The security and privacy threats of the ePass have been widely discussed (e.g., [28, 34, 35, 45]) and have provoked public debates. A recent attack [8] implies that tracking the movements of a particular passport—without breaking the passport's cryptographic keys—is feasible from a maximal range of some centimeters. In contrast, the brute-force attacks presented in this section reveal the secret encryption keys (and enable tracking of individuals) from a distance of several meters, by means of eavesdropping.

The cryptographic keys k_{ENC} and k_{MAC} for the encryption and generation of a Message Authentication Code (MAC), respectively, are derived from a machine readable zone (MRZ) that is printed on the paper document. Then, a mutual three-pass authentication according to the BAC protocol [29] is executed via the wireless interface. With our key-search implementation, the MRZ—and thus the secret keys—can be recovered from a distance.

In a typical attack scenario a suitable receiver and an antenna have to be mounted nearby an e-passport inspection system to monitor the bits transmitted via the air channel. An attacker then has two options for gaining the plain- and ciphertext needed for the exhaustive key search: The *first option* targets k_{ENC}. The plaintext for this case is a random number $\text{RND}_{\text{Epass}}$ transmitted by the passport at the beginning of the BAC. The corresponding ciphertext $\text{ENC}_{k_{\text{ENC}}}(\text{RND}_{\text{Epass}})$ is sent some steps later during the BAC. The encryption function $\text{ENC}_{k_{\text{ENC}}}(\cdot)$ is Triple-DES in CBC mode, with the initialization vector being publicly known [29]. Our implementation decrypts the most significant 8 bytes of the ciphertext with varying keys and compares the results with the plaintext $\text{RND}_{\text{Epass}}$—in case of a match, the passport's MRZ and thereby k_{ENC} is found. The data signal of the reader is much stronger than that transmitted by the passport [20]. For our *second approach*, monitoring the signal of the reader is sufficient, hence greater eavesdropping ranges can be achieved. This time, k_{MAC}, is targeted by intercepting a message of the reader and its respective MAC.

The keys k_{MAC} and k_{ENC} are derived from the MRZ according to

$$k = msb_{16}\left(\text{SHA-1}\left(msb_{16}\left(\text{SHA-1}\left(\text{MRZ}\right)\right) \| C\right)\right).$$

Since two subsequent rounds of SHA-1 [46] have to be executed, each FPGA is provided with two pipelined SHA-1 units. The first option for the attack requires a Triple-DES to be executed after hashing the key. This is achieved by implementing one DES round and executing it for 48 times (16 times for each single DES), instead of a fully parallel DES. Despite this slow approach for the DES, $80 - 48 = 32$ clock cycles of idle time remain until the next output of the SHA-1 is determined. For the second proposed attack targeting the MAC, four more DES rounds have to be executed in addition to the Triple-DES. The corresponding $7 \times 16 = 112$ clock cycles when implementing one DES engine would imply $112 - 80 = 32$ clock cycles of idle time each time a new hash value is delivered. Thus, a second DES engine is placed on each FPGA, after some optimizations of the control logic. Finally, the post-processing after the SHA-1 takes 56 clock cycles and is thus again $80 - 56 = 24$ clock cycles faster than the SHA-1.

Since 120 FPGAs are available on COPACOBANA, the key space is split into 120 appropriate subspaces: each FPGA contains a fixed portion of the MRZ, while an MRZ generator produces all remaining combinations of the MRZ and supplies them to four engines on each FPGA that process the plaintext in parallel, as sketched above. The delivered ciphertexts are compared to the correct one and, in case of a match, the sought-after MRZ is returned to the data bus. The MRZ generator is crucial for the run-time of the attack, as it minimizes the communication via the data buses. It further enables to adapt the key search to different scenarios: the entropy of the MRZ can be considerably reduced [43, 53] with an increasing knowledge about the passport holder and the issuing system of the e-Pass.

Due to the memory limitations of COPACOBANA the SHA-1 is the slowest part of our implementation, requiring 80 clock cycles. Its critical path also sets the upper limit for the clock frequency, i.e., $f_{\text{clk}} = 40\,\text{MHz}$. The 120 FPGAs run in parallel

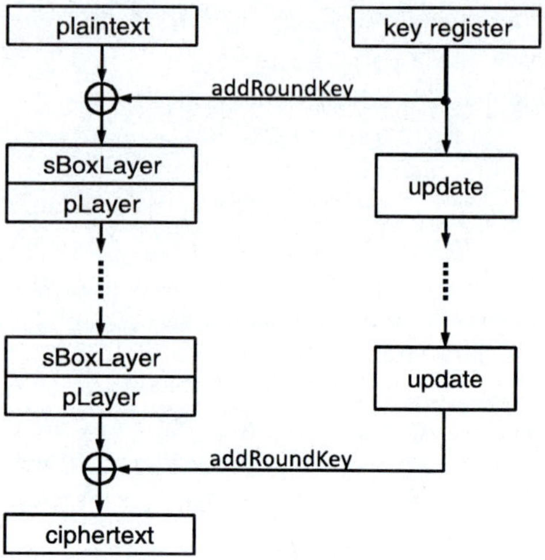

Fig. 9 Structure of PRESENT

and each possess four encryption engines, hence $4 \times 120 = 480$ key candidates are tested every 2 μs, resulting in a throughput of $2^{27.84} \approx 240$ million keys per second. In a typical attack scenario, only 2^{33} keys need to be tested on average (see [43] for details about the entropy)—COPACOBANA reveals the correct secret in $\frac{2^{32}}{2^{27.84}} \approx 18$ s.

2.1.5 Brute-Force Attack on PRESENT

PRESENT [7] is a symmetric block cipher with a block size of 64 bits and a key length of 80 (or 128) bits. PRESENT was designed as a replacement for lightweight ciphers such as Crypto1, KeeLoq, or Hitag2, which can all be broken either by means of brute-force, algebraic attacks or side-channel analysis [11, 16, 32, 38, 60].

The cipher is illustrated in Fig. 9. It is based on a substitution-permutation network which consists of 31 rounds. Each round consists of three parts: a byte-wise XOR of the input data and the round key (*addRoundKey*), a nonlinear S-box modification (*sBoxLayer*), and finally a permutation (*pLayer*). The round keys are computed by using the same S-box.

All 16 S-boxes in the substitution layer are identical, which allows down-scaling the chip area at the cost of an increased number of clock cycles. As a result, circuits implementing PRESENT may occupy a significantly smaller chip area compared to other ciphers, e.g., DES, which makes PRESENT an ideal solution for large-scale applications, such as tickets for public transport, RFID tags identifying goods in shops, car immobilizers, or door openers. On the other hand, PRESENT can be implemented as a pipelined circuit with a high throughput (and a large area), hence trading area for speed is possible in a very wide range. Performing a brute-force attack against a cipher with an 80-(or 128-bit) key is evidently infeasible with common resources. However, our implementation aims at evaluating the resources that are practically required to break PRESENT with such an attack.

For our implementation we opted for the "weaker" 80-bit variant of PRESENT, which has an architecture similar to that of DES. Both ciphers work on 64-bit blocks, but PRESENT requires twice as many rounds compared to DES. Thus, just two pipelined cores fit into one FPGA, implying a throughput of one key per clock cycle. The maximum achieved frequency is 100 MHz, i.e., one FPGA verifies 200 million keys per second and the whole COPACOBANA with 120 FPGAs verifies 24 billion keys per second. Searching the whole key space of 2^{80} keys takes $\frac{2^{80}}{24 \cdot 10^9 \cdot 86{,}400 \cdot 365.25} \approx 1.596 \cdot 10^6$ years with one COPACOBANA. A brute-force attack on an 80-bit PRESENT hence takes almost 800,000 years on average, which makes PRESENT a good solution for lightweight cryptography, unless other cryptanalytical attacks were found.

2.2 *Guess-and-Determine Attack on A5/1*

In contrast to block ciphers, where the key used for an encryption is directly targeted, cryptanalysis of stream ciphers often focuses on recovering the internal state of the cipher [5, 24], i.e., the content of internal registers at a certain time of execution. The internal state is derived on the basis of a known keystream—once it is revealed the cipher may be run forward to decrypt the remainder of the ciphertext. In some cases the cipher may be also run backwards to obtain a previous state of the secret key, e.g., in case of Crypto1 or A5/1.

A5/1 is a synchronous stream cipher used for voice encryption in the GSM mobile communication system. The communication between the mobile phone and the base transceiver station (BTS) is divided into frames with a length of 114 bits. For each frame, a new keystream is produced. All frames of one phone call share the same 64-bit session key K. The 22-bit initialization vector IV is unique for each frame of the phone call; however, it is equal to the 22-bit frame number FN which is publicly known.

The A5/1 cipher consists of three linear feedback shift registers (LFSRs) $R1$, $R2$, and $R3$ with lengths of 19, 22, and 23 bits, as depicted in Fig. 10. The most significant bits of all three registers are added modulo 2 to produce one bit of a keystream per clock cycle. The registers are clocked irregularly, while the bits $R1[8]$, $R2[10]$ and $R3[10]$ are used as clocking bits: in each clock cycle, the majority of all three clocking bits is calculated. The register Ri is clocked only if its associated clocking bit is equal to the majority of all three clocking bits, thus either two or three registers are clocked per clock cycle.

The security of A5/1 has been extensively analyzed. Pioneering work in this field was done by Anderson [1], Golic [24], and Babbage [3]. Anderson's basic idea was to guess the complete content of the registers $R1$ and $R2$ and bits $R3[10] \dots R3[0]$ of register $R3$. In this way the clocking of all three registers is determined; the upper

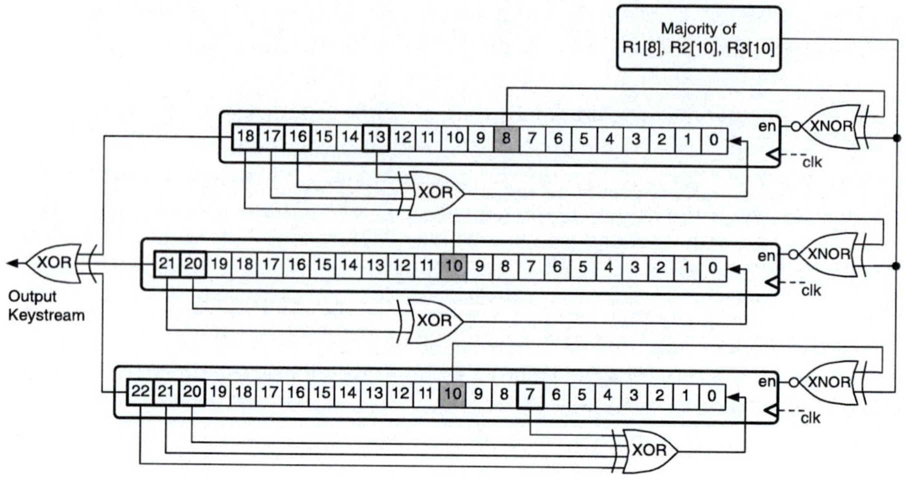

Fig. 10 A5/1 cipher

bits $R3[22]\ldots R3[11]$ can be derived from known keystream bits and guessed bits can be verified against the remainder of the keystream. As the internal state has 64 bits, at least 64 bits of keystream need to be known in advance. In the worst-case, each of the 2^{52} derived state candidates need to be verified.

The hardware-assisted attack by Keller and Seitz [36] is based on Anderson's idea. However, they propose a method to exclude a significant fraction of possible candidates at a very early stage of the verification process. Similar to Anderson, they guess shorter registers $R1$ and $R2$ and they determine upper bits of $R3$ from know keystream. In contrast to Anderson, the lower bits of $R3$ are successively guessed and, as they serve as clocking bits, wrong guesses are eliminated by recognizing contradictions in clocking. Unfortunately, the approach given in [36] does not only immediately discard wrong candidates but also a priori *restricts* the search for candidates to a certain subspace, which reduces the success probability of their attack to only 18 %.

We have improved the attack of Keller and Seitz eliminating the drawback of their attack and thus increasing the success probability to 100 %. The attack [23] was implemented on COPACOBANA. In our attack around 2^{50} guesses need to be checked for consistency, as shown in Table 2. The verification of each guess takes $17\frac{2}{3}$ clock cycles on average. One FPGA can host up to 36 guessing engines.

The guessing engine is passing through a binary decision tree. By storing intermediate states at some level of the decision tree we reduce the time complexity of the attack to 10.08 clock cycles per guess on average. Implementation results for standard and optimized guessing engines are summarized in Table 3. This table also demonstrates that "less is sometimes more": implementing only 32 standard engines instead of 36 enables to increase the maximum clock frequency, which in turn reduces the time of the attack. Nevertheless, we obtained the best results

Table 2 Complexity of attacks on A5/1—number of guesses that need to be checked for consistency

Attack	Complexity
Plain brute-force attack	2^{64}
Plain guess-and-determine attack (Anderson)	$2^{41} \cdot 2^{11} = 2^{52}$
Attack by Keller–Seitz (success rate 18 %)	$2^{41} \cdot (2 - \frac{1}{2})^{11} \approx 2^{47.43}$
Smart guess-and-determine attack (our attack)	$2^{41} \cdot (2 - \frac{1}{4})^{11} \approx 2^{49.88}$

Table 3 Implementation results for standard and optimized A5/1 guessing engines

# Guessing engines	Clock cycles		f_{max}	f_{test}	Worst-case time	
per FPGA	per guess	Slices	[MHz]	[MHz]	Estimated	Measured
36 standard	17.67	6,953 (91 %)	81.85	72.00	16.31 h	–
32 standard	17.67	6,614 (86 %)	102.42	92.00	14.36 h	13.58 h
23 optimized	10.08	7,494 (98 %)	104.65	92.00	11.40 h	11.78 h

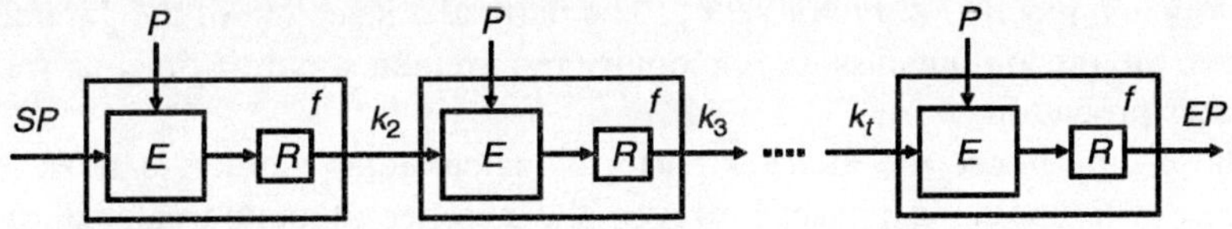

Fig. 11 Chain generation according to Hellman's TMTO

for 23 optimized engines per one FPGA. The worst-case time of our guess-and-determine attack on A5/1 is below 12 h with one COPACOBANA, i.e., it takes less than 6 h on average.

2.3 *Time-Memory Trade-Off Attacks*

In cryptanalysis, a time-memory trade-off (TMTO) aims to find a compromise between two well-known extreme approaches, i.e., performing exhaustive searches (c.f. Sect. 2.1) and precomputing extremely large tables, in order to efficiently compute cryptographic schemes. A TMTO method introduced by Hellman in 1980 [27] offers a way to reasonably reduce the actual search complexity, while keeping the amount of precomputed data reasonably low.

A TMTO attack is divided into two phases: the precomputation phase and the online phase.

In the *precomputation phase* a large amount of calculations is performed. The time complexity of this phase is comparable to that of a brute-force attack, however, in case of TMTO the calculations are executed only once and their results stored. The calculations are organized in chains, as shown in Fig. 11. The entry of the chain is denoted as a starting point SP and the final result is denoted as an end point EP. In the ith step of the computations, the encryption function E encrypts a plaintext P with key k_i, producing the ciphertext $C_i = E_{k_i}(P)$. C_i is then modified by means of a re-randomization function R producing the key for the next step, $k_{i+1} = R(C_i)$. The combination of the encryption function E and the re-randomization function R is called the *step function* f. The chain is terminated after t steps and a new chain is generated from another starting point. For each chain only the pair (SP, EP) is stored in a table, which significantly reduces the memory requirements compared to a full precomputation table. The pairs are sorted by the end points.

Hellman calculated that (due to the birthday paradox) the chains of one table cover only $N^{\frac{2}{3}}$ distinct keys (where N is the number of all possible keys). To cope with this problem he suggested to generate multiple tables, each associated with a different re-randomization function. If we denote t to be the length of the chain, m the number of chains in one table and r the number of tables, then the selection of the parameters is typically $t \approx m \approx r \approx N^{\frac{1}{3}}$. The particular selection of the parameters influences the precomputation time, memory requirements, online time, as well as the success probability.

In the *online phase* the attacker performs calculations similar to those ones from the precomputation phase, however, the number of online calculations is significantly smaller. The results of these calculations are compared to end points stored in the table: in case of a match the key may be retrieved by reconstructing the appropriate chain from its start point. In practice, the time required to complete the online phase of Hellman's TMTO is dominated by the high number of table accesses. Random accesses to the storage, e.g., a harddisk, can be several orders of magnitude slower than all necessary calculations. Employing the so-called *distinguished points* (DPs), as recommended by Rivest [12] in 1982, significantly reduces the amount of required table accesses. A DP is a point with the property that it can be easily verified, e.g., a point with the ten most significant bits set to zero. Following the DP method, chains do not have a uniform length, as they are generated until a DP is produced. Another variant of TMTO method, called *rainbow tables* [50], was proposed by Oechslin in 2003. He suggested not to use the same re-randomization function R when generating a chain for a single table but a (fixed) sequence $R_1, R_2, R_3, \ldots, R_{t-1}, R_t$ of different re-randomization functions. As the effect of birthday paradox problem is suppressed by this strategy, more chains can be generated in one table and the number of tables may be significantly lower. In practice, the number of chains generated in one table is $m \approx N^{\frac{2}{3}}$, i.e., there is only one table ($r = 1$) or the number of tables is in order of units. The length of one chain is typically again $t \approx N^{\frac{1}{3}}$. The number of table accesses is comparable to the distinguished-points method while performing twice as fast in the online phase. Currently, rainbow tables are considered to be the best TMTO method for single data.

Table 4 Expected runtimes and memory requirements for TMTO on DES

Method	SR	DU	PT (COPA)	OT
Hellman	0.80	1,897 GB	24 days	$2^{40.2}$ TA + $2^{40.2}$ C
Distinguished points	0.80	1,690 GB	95 days	2^{21} TA + $2^{39.7}$ C
Rainbow tables	0.80	1,820 GB	23 days	$2^{21.8}$ TA + $2^{40.3}$ C

2.3.1 TMTO Attacks on DES and PRESENT

The performance results of our DES breaker (see Sect. 2.1.1) serve as the basis for the following initial extrapolation of the complexity of various TMTO methods applied to DES. Table 4 presents our worst-case estimations concerning the success rate (SR), disk usage (DU), and the duration of the precomputation phase (PT) on COPACOBANA, as well as the number of table accesses (TA) and calculations (C) during the online phase (OT). Note that these figures are based on estimations given in [27, 50, 57] that are applied to our implementations on COPACOBANA.

Based on the results presented in Table 4 we have chosen the rainbow tables method for implementing a TMTO attack on DES. For this implementation, we have developed another DES core which operates with 16 pipeline stages [25]. Using four parallel DES units with 16 stages each, we can run 64 chain computations in parallel per FPGA. Figure 12 illustrates our architectures for generating rainbow tables in further detail.

On the given Spartan-3 devices, our entire implementation including I/O and control logic consumes 7,571 out of 7,680 (98 %) available slices of each FPGA and runs at a maximum clock frequency of 96 MHz. A single COPACOBANA is then able to compute more than 46 billion iterations of the step function f per second. The actual duration of the precomputation phase for generating the rainbow tables lasts slightly less than 32 days. Then, the calculations in the online phase may be accomplished in just around 30 s. The limiting factor of the attack is the number of $2^{21.8}$ table accesses.

For mounting a further TMTO attack targeting PRESENT we also have selected the rainbow tables method. We have chosen the "weaker" variant of PRESENT working with an 80-bit key. The precomputation engine is similar to that one build for DES, however, one FPGA contains two TMTO units having 32 pipeline stages each in this case. Again, we can run 64 chain computations in parallel per FPGA and the maximum clock frequency is 100 MHz. If we set the chain length to $t = 2^{27}$ steps and calculate just one rainbow table containing $m = 2^{53}$ chains, then the precomputation of the table would take the same time as the brute-force attack, i.e., PT $\approx 1.6 \cdot 10^6$ years and all pairs of start points and end points (SP, EP) would occupy roughly DU = 128 PB of disk space.

The online phase of the attack can then be accomplished in less than nine days with one COPACOBANA, which looks promising; however, there are two limiting factors: first, the resources necessary for the precomputation phase are too high.

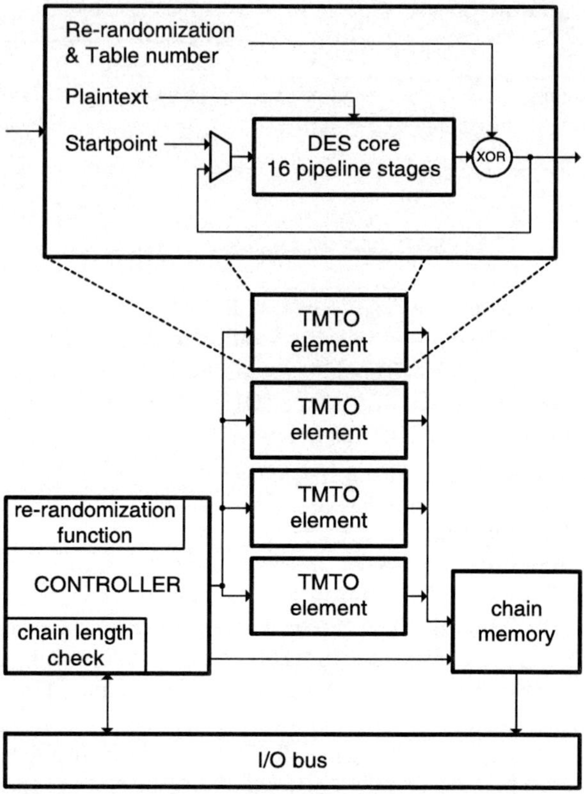

Fig. 12 Implementation for generating DES rainbow tables

Second, even if we neglect the precomputation phase (we suppose that we are given the table from a very powerful third party), then nine days with one COPACOBANA can still be regarded as too expensive, with respect to the application area of the cipher, such as tickets for public transport and RFID tags.

2.3.2 Time-Memory-Data Trade-Off Attack on A5/1

As stated above, in case of stream ciphers we aim at finding the n-bit internal state that generated corresponding w consecutive bits of known keystream. For an attack on a stream cipher often $w > n$ holds, e.g., in case of the A5/1 one frame has $w =$ 114 bits, while the internal state has $n = 64$ bits only. In this situation it is possible to derive $D = w - n + 1$ set of data points from the stream bits $(b_1, \ldots, b_w)$, namely $y_1 = (b_1, b_2, \ldots, b_n)$, $y_2 = (b_2, b_3, \ldots, b_{n+1})$ and so on. For breaking the cipher it is sufficient to find the internal state for any of the D data sets $y_1, \ldots, y_D$ [3, 24], i.e., we have D times more chances for mounting a successful attack.

The common approach to exploit the existence of multiple data sets in case of time-memory-data trade-off (TMDTO) is to reduce the coverage of the tables by a factor of D. When applied to the original Hellman method [5] or to the distinguished point method, one simply generates D times less tables. As a result, both the precomputation time and the memory requirements are reduced by a factor of D, while the online complexity remains unchanged.

Unfortunately, rainbow tables cannot gain from multiple data, as shown in [4]. In the same paper a new variant of the rainbow method, called *thin-rainbow method*, was sketched providing a better TMDTO. In this variant one does not use a different re-randomization function in each step of the chain computation but applies a sequence of S different re-randomization functions ℓ-times periodically in order to generate a chain of length $\ell S + 1$. More precisely, the corresponding step-functions $f_1, \ldots, f_S$ are applied in the following order:

$$f_1 f_2 \cdots f_S f_1 f_2 \cdots f_S \cdots f_1 f_2 \cdots f_S$$

To reduce the number of disk accesses in the online phase one can combine the thin-rainbow scheme with the distinguished point method. This is done by looking for a distinguished point after each application of the f_S-function. During the precomputation we only keep a chain if it exhibits its first DP after $\ell_{\min} \le \ell \le \ell_{\max}$ applications of f_S, for certain parameters $\ell_{\min}$ and $\ell_{\max}$, and if this DP is furthermore different from the end points of the chains that are already stored.

For our implementation of a TMDTO attack on A5/1 [25] we have selected the sketched thin-Rainbow DP method. In the case of multiple data set this approach allows for a simple and efficient hardware implementation, while exhibiting a low number of disk accesses during the online phase and an efficient trade-off curve. The TMDTO engine used for precomputation phase is depicted in Fig. 13. A modified engine has been designed for the online phase [48]. Calculations are executed in 234 TMTO elements which are connected in series. Each TMTO element is calculating one chain. The result after each rainbow sequence $f_1 f_2 \cdots f_S$ is checked for the DP-criterion by means of a DP checker.

An A5/1 TMDTO engine runs at a maximum frequency of 156 MHz. Computing a step-function f_i takes 64 clock cycles. Since one FPGA contains 234 TMTO elements (each consisting of two A5/1 cores), the whole COPACOBANA can perform approximately 2^{36} step-functions per second. Selecting the TMDTO parameters requires special attention, since this highly influences the precomputation time (PT), the disk usage (DU), as well as the time needed in the online phase for the chain computations (OT), the number of table accesses (TA), and the success rate (SR). Table 5 summarizes the results for different sets of parameter choices under the assumption that $D = 64$. Furthermore, we assume that COPACOBANA is used both for the precomputation and during the online phase. Thus it is worth to trade a higher online complexity, e.g., for a lower demand of disk space (compare with rows 4 and 5). For our implementation, we have selected the set of parameters presented in the third row, since it produces a reasonable precomputation time and a reasonable size of the tables, as well as a relatively small number of table accesses. The success rate of 63 % may seem to be small, but it increases significantly if more data samples are available: for instance, if four frames of known keystream are available, then $D = 4 \cdot 51 = 204$ and thus the success rate increases to 96 %.

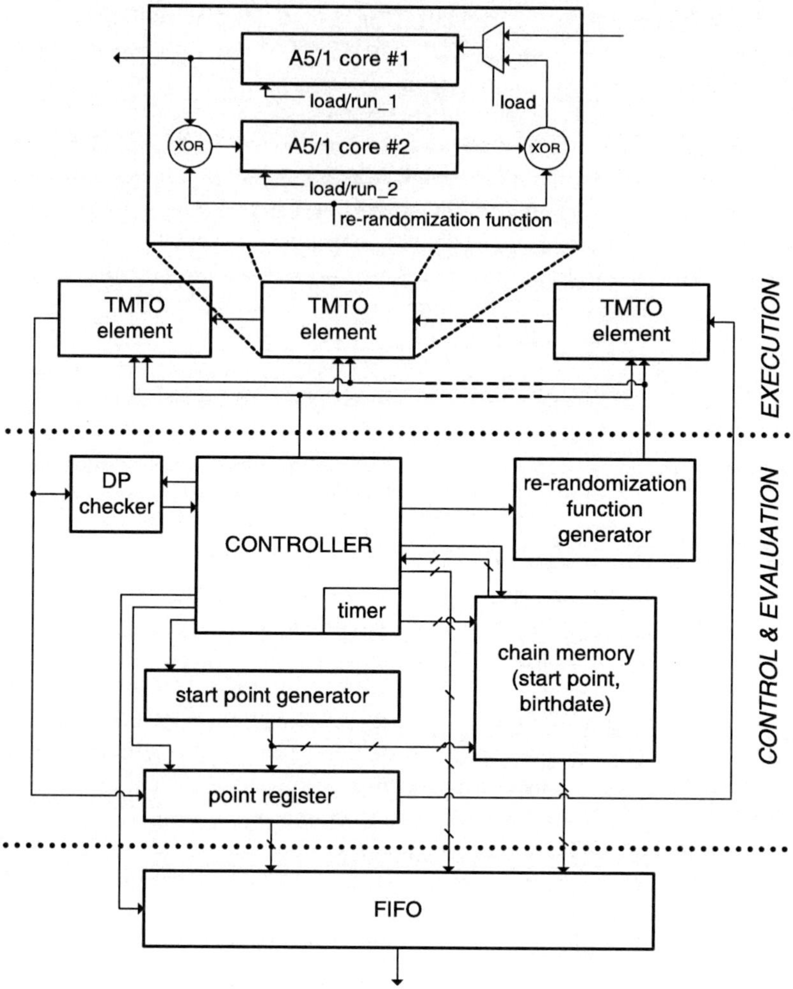

Fig. 13 Diagram of an A5/1 TMDTO engine implemented in one FPGA

2.4 Factoring Large Numbers with Elliptic Curves

The factorization of a large composite integer n is a well-known mathematical problem which has attracted special attention since the invention of public-key cryptography. RSA is known as the most popular asymmetric cryptosystem and was originally developed by Ronald Rivest, Adi Shamir, and Leonard Adleman in 1977 [52]. Since the security of RSA relies on the attacker's inability to factor large numbers, the development of a fast factorization method could enable a key recovery from RSA messages and signatures. Recently, the best known method for factoring large RSA integers is the General Number-Field Sieve (GNFS). An important step

Table 5 A5/1 TMDTO: expected runtimes and memory requirements

#	m	S	d	I_l	PT [days]	M [TB]	T [s]	TA	P_{total}
1	2^{41}	2^{15}	5	$[2^3,2^6]$	337.5	7.49	70.5	2^{21}	0.86
2	2^{39}	2^{15}	5	$[2^3,2^7]$	95.4	3.25	92.0	2^{21}	0.67
3	2^{40}	2^{14}	5	$[2^4,2^7]$	95.4	4.85	27.6	2^{20}	0.63
4	2^{40}	2^{14}	5	$[2^3,2^6]$	84.4	7.04	17.7	2^{20}	0.60
5	2^{39}	2^{15}	5	$[2^3,2^6]$	84.4	3.48	70.5	2^{21}	0.60
6	2^{40}	2^{14}	5	$[2^4,2^6]$	84.4	5.06	21.5	2^{20}	0.55
7	2^{37}	2^{15}	6	$[2^4,2^8]$	47.7	0.79	186.3	2^{21}	0.42
8	2^{36}	2^{16}	6	$[2^4,2^8]$	47.7	0.39	745.3	2^{22}	0.42

in the GNFS algorithm is the factorization of mid-sized numbers for smoothness testing. For this purpose, the elliptic curve method (ECM) has been proposed by Lenstra [42]. The ECM algorithm proved to be suitable for parallel hardware architectures in [22, 44, 56], particularly on FPGAs.

The ECM algorithm performs a very high number of operations on a very small set of input data and is not demanding in terms of high communication bandwidth. Furthermore, the implementation of the first stage of ECM requires only little memory since it is based on point multiplication on an elliptic curve. The operands required for supporting GNFS are well beyond the width of current computer buses, arithmetic units, and registers, so that special-purpose hardware can provide a much better solution.

In [44], de Meulenaer et al. show that the utilization of DSP slices in Virtex-4 FPGAs for implementing a Montgomery multiplication can significantly improve the ECM performance. In that work, the authors used a fully parallel multiplier implementation, which provides the best known performance figures for ECM phase 1 so far. However they did not provide details how to realize ECM phase 2.

To accelerate integer arithmetic using a similar strategy, we employed the new slot-in module for use with the second release of COPACOBANA. These slot-in modules host 8 Xilinx Virtex-4 XC4VSX35 FPGAs, each providing 192 DSP slices. Based on the DSP cores in the Virtex-4, we created a multi-core ECM design per FPGA (in contrast to [44], which uses a single core only).

Figure 14 describes the layout of our ECM System, which contains multiple engines, a data scheduler and a buffered system bus. As the right side of the figure shows, one ECM engine contains two different clock regions. The faster clock region is reserved for the elliptic curve processor, which uses DSP cores to compute point operations (addition and doubling) on the curve and contains instructions to compute phase 1 and phase 2 of the algorithm. The slower clock region is used by the control unit to execute the complete algorithm. Please note that only point operations on the elliptic curve are performed on FPGAs. This means that the setup of the Montgomery curve needs to be done on the host-PC and then transferred to the FPGAs.

In addition to the comparison shown in Table 6, our implementation adjusts to a block size parameter to natively support different integer target sizes. As an

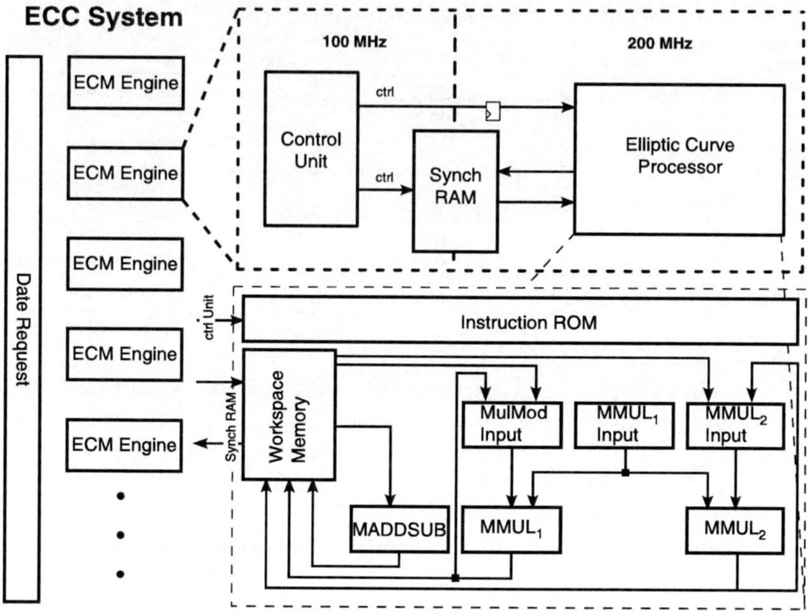

Fig. 14 FPGA-layout of our ECM system, composed of multiple ECM engines, as well as the structure of the engine and its elliptic curve processor

Table 6 Comparing our results using $b = 13$ ($L = 202$ bit) with Gaj et al. for Virtex-4 FPGAs

Aspect	Gaj et al. [22]	This work	Factor
FPGA device	V4LX200-11	V4SX35-10	
Supported bits	198	202	1.02
Max. ECM cores	24	24	1.00
Max. frequency	104 MHz	200 MHz	1.92
Cycles for addition	41	29	0.71
Cycles for multiplication	216	201	0.93
Cycles for phase 1	1,666,500	1,473,596	0.88
Time for phase 1	16 ms	7.37 ms	0.46
(# Phase 1)/s	1,448	3,240	2.24
(# Phase 1+2)/s	696	1,560	2.24

example, with a block size $b = 10$, the design factors integers up to 151 bit in 4.73 ms (phase 1) and 5.12 ms (phase 2). This corresponds to 5,064 computations per second for the first phase and 2,424 computations per second for phase 1 and 2.

A main issue of the ECM is memory. Although phase 1 has only very moderate memory constraints, phase 2 involves a significant amount of precomputations, as well as storage for prime numbers. Since memory on FPGA devices is rather limited (192×18 kbit BRAM elements per device), a fast accessible, external memory could help to improve the impact of phase 2 by storing even larger tables than presented above.

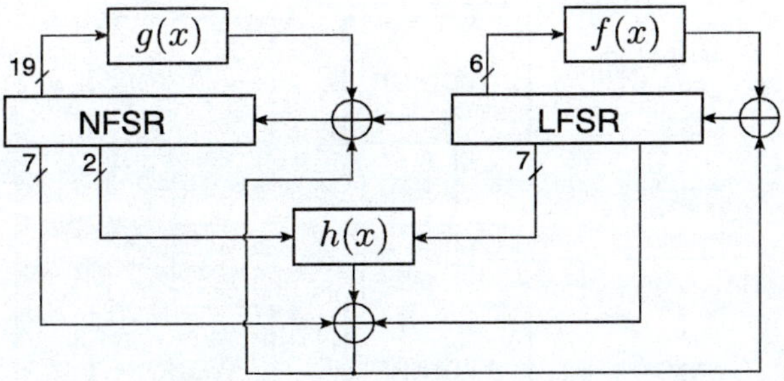

Fig. 15 Overview of the Grain-128 initialization based on the presentation in [26]

2.5 *Cryptanalysis Using Cube Attacks*

In order to analyze a cryptographic scheme, we can split the output into its separated output bits and describe each by a multivariate polynomial over the Galois field GF(2) on secret variables, e.g., the key bits, and public variables, e.g., bits of the plaintext or initialization vector. Even though the full expression is much too large to write down, evaluate, and benefit from, we can still draw conclusions by running the target algorithm as a black box with some restrictions to the assigned input data by the cryptanalyst.

In the specific case we discuss here, we will use Cube Testers [2], a type of distinguishers [18, 19, 21, 37], which are related to higher-order differential attacks [41]. We implement such Cube Testers on RIVYERA with its Spartan-3 5000 FPGAs to evaluate an algebraic attack called Dynamic Cube Attack [14, 15]. This attack can—in theory—be used for a key-recovery, but due to the high complexity the choice of its parameters and the verification needs to be verified by experiments: a full computation is not possible with the current technology.

The exemplary target of our implementation is the stream cipher Grain-128. Figure 15 shows its construction: Basically, this cipher consists of two 128-bit feedback shift registers—one linear (LFSR) and one nonlinear (NFSR)—with the primitive polynomials of degree 128, defined as $f(x) = 1 + x^{32} + x^{47} + x^{58} + x^{90} + x^{121} + x^{128}$ and $g(x) = 1 + x^{32} + x^{37} + x^{72} + x^{102} + x^{128} + x^{44}x^{60} + x^{61}x^{125} + x^{63}x^{67} + x^{69}x^{101} + x^{80}x^{88} + x^{110}x^{111} + x^{115}x^{117}$.

In addition, a nonlinear function $h(x)$ depending on the state of both registers generates the output bit sequence. The cipher uses a 96-bit initialization vector (IV) and a 128-bit secret key to initialize the keystream. During initialization, the output is fed back into the shift registers and clocked 256 times. As this construction effectively uses only bit-level logic the cipher fits very well into hardware (considering area and speed), but will be considerably slower when implemented on a standard CPU (even though techniques like bit-slicing can be used to increase the speed of the software implementation).

As we start out by evaluating an attack on the Grain-128 of theoretical nature, our first goal is to find the best suitable parameters and to verify the overall effectiveness based on these. One parameter with a high impact on the complexity is the cube

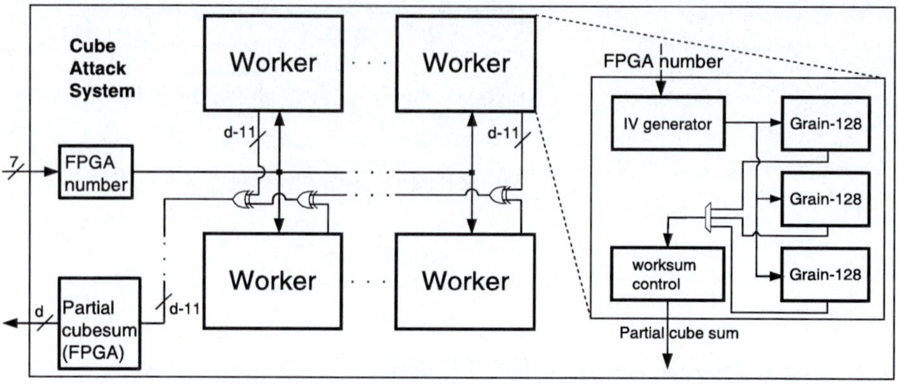

Fig. 16 Overview of the cube attack system on the Xilinx Spartan-3 5000 FPGA including the structure of the worker components

dimension d: We need to compute the first bit after the initialization of the Grain-128 cipher 2^d times for each key for which we evaluate the parameter set and the success probability of the attack.

The previous attack (with more restrictions) used $d = 46$ and was evaluated by a software cluster. To remove the restrictions and design the new attack, it was necessary to raise the dimension to $d = 50$. As a consequence, the number of Grain-128 initializations increased from 2^{46} to 2^{50}, where each initialization needs (depending on the implementation) at most 2^8 computations. To evaluate the parameters and estimate the effectiveness of the attack, 2^7 to 2^{10} random keys should be tested. While this is not feasible with the software cluster anymore, we can try to use the FPGA cluster RIVYERA to test and evaluate random keys.

Figure 16 shows the layout of the FPGA implementation. Each FPGA works on a subset of the Cube and computes 2^{d-11} Grain initializations (we use 128 FPGAs and 16 Workers per FPGA). To relax the routing, every Worker keeps its own IV generator close to the three Grain instances necessary to process the internal pipeline. The result is first updated per Worker and summarized per FPGA to keep the communication and post-processing on the host PC at a minimum.

Table 7 shows the results of our FPGA implementation. We implemented the evaluation process of the online phase for the RIVYERA cluster using 128 FPGAs. In order to evaluate multiple parameter sets and evolve the attack, we created a flexible design and started with the small cube dimensions of 46 up to a dimension of 50. While with the smallest dimension, computing all 2^{46} initializations takes less than 18 min, computing 2^{50} initializations takes about 4 h on the full cluster.

The row *Configurations Built* already indicates that the problem is more complex than a simple brute-force attack: as several parameters of the attack, i.e., the polynomials are chosen specifically for each key and the dynamic cube attack requires a lot of flexibility on the FPGA, implementing a generic design, e.g., for all possible polynomials over a 96-bit vector, would result in huge multiplexer cascades, leading to an impossible scenario.

Table 7 Results of the bitstream generation process for cubes of dimension 46, 47, and 50

Cube dimension d	46			47	50	
Clock frequency (MHz)	100	110	120	120	110	120
Configurations built	1	7	8	6	60	93
Percentage	6.25	43.75	50	100	39.2	60.8
Online phase duration	17.2 min	15.6 min	14.3 min	28.6 min	4 h 10 min	3 h 49 min

The duration is the time required for the RIVYERA cluster to complete the online phase, computing 2^d Grain initializations. The Percentage row gives the percentage of configurations built with the given clock frequency out of the total number of configurations built with cubes of the same dimension. This indicates the impact of the key-dependent parameters, i.e., the polynomials, on the complete design

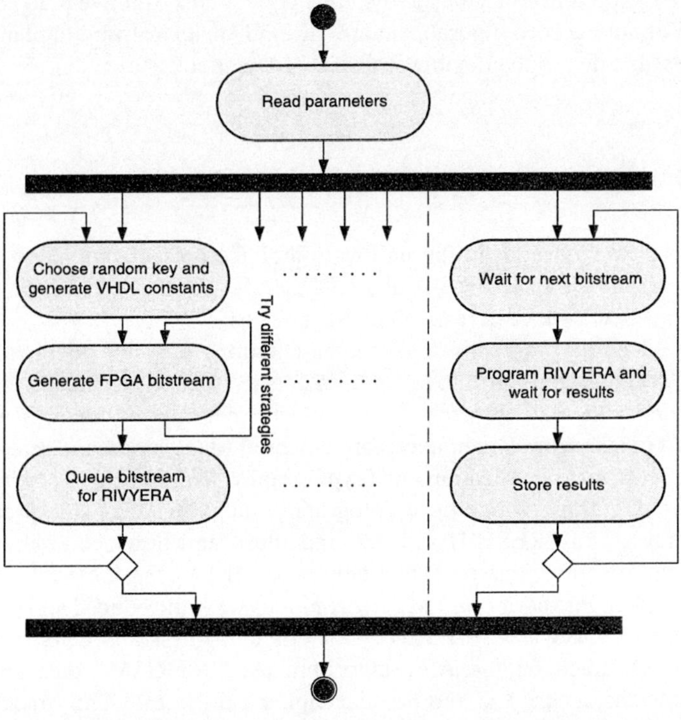

Fig. 17 Cube attack implementation on the RIVYERA FPGA cluster utilizing the cores of the i7 host CPU to generate VHDL code and place and route key-dependent designs while the 128 FPGAs compute the result in parallel to estimate the effectiveness of the attack parameters and the feasibility of the attack

To circumvent this problem, we created a hardware–software solution. Figure 17 shows how we can utilize this co-design using the integrated i7 CPU of the RIVYERA and its FPGAs: the software handles the generation of key-specific FPGA bitstreams, while the FPGAs work on the bruteforce-like attack on these previously created configurations. At the start of each iteration, we begin with generic code. The software part chooses a random key for each CPU core to evaluate the current parameters. Afterwards, it derives the key-dependent polynomials and creates the dependent VHDL code. This allows us to reduce the complexity of the hardware requirements to achieve a valid, usable configuration.

But the automatic generation of key-dependent FPGA configurations has its price: it is not possible to fully optimize the hardware implementation, as the parameters modified by the (randomly chosen) key have a major impact on the design. This is the reason for the difference in the achieved clock frequencies in Table 7. Whenever the verification of the design constraints fail, a different strategy re-implements the design and continuously decreases the target frequency.

To conclude this section, with this setup it is possible to obtain an estimation of the complexity and feasibility of the dynamic cube attack and tweak its attack parameters. Without the reconfigurable nature of FPGAs, a hardware implementation seems impossible due to the flexible structure of the attack.

3 Summary

In this chapter, we introduced the hardware architectures of the special-purpose computing clusters COPACOBANA and RIVYERA. We then detailed on a variety of applications that have been realized on the devices.

In the field of cryptography, our implementations are able to break several widespread cryptographic primitives and enable to extrapolate attacks on highly secure ciphers with realistic security parameters in terms of financial cost and attack time. Thus, our implementations are essential to (re)assess the security level of different real-world applications of cryptography. We presented several brute-force attacks targeting symmetric cryptography, including the widespread DES, KeeLoq, Hitag2, and PRESENT ciphers, and illustrated how secret keys can be extracted from electronic passports that employ the SHA-1 hash algorithm and the Triple-DES cipher. Furthermore, TMTO techniques were introduced and practically applied to attack DES and PRESENT. An efficient guess-and-determine attack and a TMDTO attack on the A5/1 algorithm used for GSM voice encryption illustrate, how the security of the popular cipher can be efficiently tackled with different approaches. For cryptanalysing asymmetric cryptosystems, such as RSA, we developed further efficient implementations. They are in particular relevant for estimating the longevity of the used key lengths. In this context we presented an implementation for the factorization of large integers based on the ECM. Finally, an evaluation of a key-recovery attack using cube testers targeting the stream cipher GRAIN-128 is illustrated.

In conclusion, the hardware architectures and implementations introduced in this chapter cover a broad range of applications in cryptanalysis.It turned out that more complex applications such as TMDTO and biological sequence alignment cannot be run on the plain COPACOBANA system due to its limitations with respect to data throughput and reconfiguration performance. Nevertheless, with the evolution to the

RIVYERA cluster, hardware–software co-designs can simplify the requirements of very complex tasks such as the verification of a complex algebraic attack, when using the full reconfigurable potential of the device. For other attacks (e.g., the exhaustive key search on DES) the cost-performance ratio of the COPACOBANA platform can be considered optimal. With the recent development of the RIVYERA S6-LX150, providing at least 2.5× more resources than RIVYERA S3-5000 at a roughly twofold increase of frequency, existing implementations may be easily ported with an expected speedup of roughly a factor 4 and 16, respectively. For the same reasons, applications originally addressing COPACOBANA expect a speedup of more than 12 to 16 on this latest machine.

References

1. R. Anderson, A5 (was: Hacking digital phones) (17 June 1994), http://yarchive.net/phone/gsmcipher.html, Sci.crypt
2. J.P. Aumasson, I. Dinur, W. Meier, A. Shamir, Cube testers and key recovery attacks on reduced-round md6 and trivium, in *Fast Software Encryption*, ed. by O. Dunkelman (2009), pp. 1–22
3. S. Babbage, A space/time tradeoff in exhaustive search attacks on stream ciphers, in *European Convention on Security and Detection*, vol. 408 (IEEE Conference Publication, Los Alamitos, 1995)
4. E. Barkan, E. Biham, A. Shamir, Rigorous bounds on cryptanalytic time/memory tradeoffs, in *Proceedings of CRYPTO'06*. Lecture Notes in Computer Science, vol. 4117 (Springer, Berlin Heidelberg, 2006), pp. 1–21
5. A. Biryukov, A. Shamir, Cryptanalytic time/memory/data tradeoffs for stream ciphers, in *Proceedings of the 6th International Conference on the Theory and Application of Cryptology and Information Security: Advances in Cryptology* (2000), pp. 1–13
6. A. Bogdanov, Attacks on the KeeLoq block cipher and authentication systems, in *3rd Conference on RFID Security 2007 (RFIDSec 2007)* (2007). http://rfidsec07.etsit.uma.es/slides/papers/paper-22.pdf.
7. A. Bogdanov, G. Leander, L.R. Knudsen, C. Paar, A. Poschmann, M.J. Robshaw, Y. Seurin, C. Vikkelsoe, PRESENT - An ultra-lightweight block cipher, in *Proceedings of CHES 2007*. Lecture Notes in Computer Science, vol. 4727 (Springer, Berlin Heidelberg, 2007), pp. 450–466
8. T. Chothia, V. Smirnov, A traceability attack against e-passports, in *Financial Cryptography and Data Security*. Lecture Notes in Computer Science, vol. 6052 (Springer, Berlin Heidelberg, 2010), pp. 20–34
9. N.T. Courtois, S. O'Neil, HITAG 2 Stream Cipher – C Implementation and Graphical Description (2006–2007). http://cryptolib.com/ciphers/hitag2/
10. N.T. Courtois, K. Nohl, S. O'Neil, Algebraic Attacks on the Crypto-1 Stream Cipher in MiFare Classic and Oyster Cards (2008). Cryptology ePrint Archive, Report 2008/166, http://eprint.iacr.org/2008/166
11. N.T. Courtois, S. O'Neil, J.J. Quisquater, Practical algebraic attacks on the Hitag2 stream cipher, in *ISC '09: Proceedings of the 12th International Conference on Information Security*. Lecture Notes in Computer Science, vol. 5735 (Springer, New York, 2009), pp. 167–176
12. D. Denning, *Cryptography and Data Security* (Addison-Wesley, Reading, 1982)
13. W. Diffie, M.E. Hellman, Exhaustive cryptanalysis of the NBS data encryption standard. Computer **10**(6), 74–84 (1977)

14. I. Dinur, A. Shamir, Breaking Grain-128 with dynamic cube attacks, in *Fast Software Encryption*, ed. by A. Joux. Lecture Notes in Computer Science, vol. 6733 (Springer, New York, 2011), pp. 167–187
15. I. Dinur, T. Güneysu, C. Paar, A. Shamir, R. Zimmermann, An experimentally verified attack on full Grain-128 using dedicated reconfigurable hardware, in *ASIACRYPT*, ed. by D.H. Lee, X. Wang. Lecture Notes in Computer Science, vol. 7073 (Springer, Berlin Heidelberg, 2011), pp. 327–343
16. T. Eisenbarth, T. Kasper, A. Moradi, C. Paar, M. Salmasizadeh, M.T.M. Shalmani, On the power of power analysis in the real world: a complete break of the KeeLoq code hopping scheme, in *Advances in Cryptology - CRYPTO 2008* (2008), pp. 203–220
17. Electronic Frontier Foundation, *Cracking DES: Secrets of Encryption Research, Wiretap Politics & Chip Design* (O'Reilly & Associates Inc., Springer, Berlin Heidelberg, 1998)
18. H. Englund, T. Johansson, M.S. Turan, A framework for chosen iv statistical analysis of stream ciphers, in *INDOCRYPT*, ed. by K. Srinathan, C.P. Rangan, M. Yung. Lecture Notes in Computer Science, vol. 4859 (Springer, Berlin Heidelberg, 2007), pp. 268–281
19. E. Filiol, A new statistical testing for symmetric ciphers and hash functions, in *ICICS*, ed. by R.H. Deng, S. Qing, F. Bao, J. Zhou. Lecture Notes in Computer Science, vol. 2513 (Springer, Berlin Heidelberg, 2002), pp. 342–353
20. K. Finkenzeller, *RFID-Handbook* (Wiley, New York, 2003)
21. S. Fischer, S. Khazaei, W. Meier, Chosen iv statistical analysis for key recovery attacks on stream ciphers, in *AFRICACRYPT*, ed. by S. Vaudenay. Lecture Notes in Computer Science, vol. 5023 (Springer, Berlin Heidelberg, 2008), pp. 236–245
22. K. Gaj, S. Kwon, P. Baier, P. Kohlbrenner, H. Le, M. Khaleeluddin, R. Bachimanchi, Implementing the elliptic curve method of factoring in reconfigurable hardware, in *Proceedings of CHES'06*. Lecture Notes in Computer Science, vol. 4249 (Springer, Berlin Heidelberg, 2006), pp. 119–133
23. T. Gendrullis, M. Novotný, A. Rupp, A real-world attack breaking A5/1 within hours, in *Proceedings of the 10th Workshop on Cryptographic Hardware and Embedded Systems (CHES 2008)* (Springer, New York, 2008), pp. 266–282
24. J. Golic, Cryptanalysis of alleged A5 stream cipher. in *Proceedings of Eurocrypt'97*. Lecture Notes in Computer Science, vol. 1233 (Springer, Berlin Heidelberg, 1997), pp. 239–255
25. T. Güneysu, T. Kasper, M. Novotný, C. Paar, A. Rupp, Cryptanalysis with COPACOBANA. IEEE Trans. Comput. **57**(11), 1498–1513 (2008)
26. M. Hell, T. Johansson, A. Maximov, W. Meier, A stream cipher proposal: Grain-128, in *2006 IEEE International Symposium on Information Theory* (IEEE, New York, 2006), pp. 1614–1618. doi:10.1109/ISIT.2006.261549
27. M.E. Hellman, A cryptanalytic time-memory trade-off, in *IEEE Transactions on Information Theory*, vol. 26 (IEEE, New York, 1980), pp. 401–406
28. J.H. Hoepman, E. Hubbers, B. Jacobs, M. Oostdijk, R.W. Schreur, Crossing borders: security and privacy issues of the European e-passport, in *Proceedings of IWSEC'06*. Lecture Notes in Computer Science, vol. 4266 (Springer, Berlin Heidelberg, 2006), pp. 152–167
29. ICAO: Machine Readable Travel Documents, PKI for Machine Readable Travel Documents offering ICC Read-Only Access, Technical Report (2004). http://www.mrtd.icao.int
30. Inc., C.: Cray XD1 Supercomputer (2008), Available at http://www.cray.com/downloads/FPGADatasheet.pdf. Accessed April 2012
31. S.G. Incorporated, SGI RASC Technology (2008). http://www.sgi.com/products/rasc/. Accessed April 2012
32. S. Indesteege, N. Keller, O. Dunkelman, E. Biham, B. Preneel, A practical attack on KeeLoq, in *Proceedings of the Theory and Applications of Cryptographic Techniques 27th Annual International Conference on Advances in Cryptology, EUROCRYPT'08* (Springer, Berlin Heidelberg, 2008), pp. 1–18. http://portal.acm.org/citation.cfm?id=1788414.1788415
33. ISO/IEC 14443: Identification Cards - Contactless Integrated Circuit(s) Cards - Proximity Cards - Part 1–4 (2001), www.iso.ch

34. A. Juels, D. Molnar, D. Wagner, Security and privacy issues in E-passports, in *Proceedings of SecureComm'05* (IEEE Computer Society, Los Alamitos, 2005), pp. 74–88
35. G. Kc, P. Karger, *Security and Privacy Issues in Machine Readable Travel Documents (MRTDs)*. (IBM T.J. Watson Research Labs, 2005)
36. J. Keller, B. Seitz, A Hardware-Based Attack on the A5/1 Stream Cipher (2001), URL http://pv.fernuni-hagen.de/docs/apc2001-final.pdf. Accessed April 2012
37. S. Khazaei, W. Meier, New directions in cryptanalysis of self-synchronizing stream ciphers, in *INDOCRYPT*, ed. by D.R. Chowdhury, V. Rijmen, A. Das. Lecture Notes in Computer Science, vol. 5365 (Springer, Berlin Heidelberg, 2008), pp. 15–26
38. G. de Koning Gans, J.H. Hoepman, F. Garcia, A practical attack on the MIFARE classic, in *Smart Card Research and Advanced Applications*, ed. by G. Grimaud, F.X. Standaert. Lecture Notes in Computer Science, vol. 5189 (Radboud University Nijmegen Institute for Computing and Information Sciences/Springer, The Netherlands/Berlin Heidelberg, 2008), pp. 267–282
39. S. Kumar, C. Paar, J. Pelzl, G. Pfeiffer, M. Schimmler, Breaking ciphers with COPACOBANA - A cost-optimized parallel code breaker, in *Proceedings of the Workshop on Cryptograpic Hardware and Embedded Systems (CHES 2006)*, ed. by L. Goubin, M. Matsui. Lecture Notes in Computer Science, vol. 4249 (Springer, Berlin Heidelberg, 2006), pp. 101–118
40. S. Kumar, C. Paar, J. Pelzl, G. Pfeiffer, A. Rupp, M. Schimmler, How to break DES for €8,980, in *SHARCS2006* (Cologne, Germany, 2006)
41. X. Lai, Higher order derivatives and differential cryptanalysis, in *Symposium on Communication, Coding and Cryptography, in honor of James L. Massey on the occasion of his 60th birthday* (1994), pp. 227–233
42. H. Lenstra, Factoring integers with elliptic curves. Ann. Math. **126**, 649–673 (1987)
43. Y. Liu, T. Kasper, K. Lemke-Rust, C. Paar, E-passport: cracking basic access control keys, in *On the Move (OTM) 2007*. Lecture Notes in Computer Science, vol. 4804 (Springer, Berlin Heidelberg, 2007), pp. 1531–1547
44. G. de Meulenaer, F. Gosset, M.M. de Dormale, J.J. Quisqater, Integer factorization based on elliptic curve method: towards better exploitation of reconfigurable hardware, in *Proceedings of FCCM'07* (IEEE Computer Society, Los Alamitos, 2007), pp. 197–206
45. J. Monnerat, S. Vaudenay, M. Vuagnoux, About Machine-Readable Travel Documents, in *Proceedings of RFIDSec'07* (2007), pp. 15–28
46. National Institute of Standards and Technology: FIPS 180-3 Secure Hash Standard (Draft), http://www.csrc.nist.gov/publications/PubsFIPS.html. Accessed April 2012
47. NIST FIPS PUB 46-3, Data encryption standard, in *Federal Information Processing Standards* (National Bureau of Standards, 1977)
48. M. Novotný, *Time-Area Efficient HW Architectures for Cryptography and Cryptanalysis*, 1st edn. (Europäischer Universitätsverlag, Bochum, 2010)
49. M. Novotný, T. Kasper, Cryptanalysis of KeeLoq with COPACOBANA, in *Workshop on Special Purpose Hardware for Attacking Cryptographic Systems (SHARCS 2009)* (2009), pp. 159–164
50. P. Oechslin, Making a faster cryptanalytic time-memory trade-off, in *Proceedings of CRYPTO'03*. Lecture Notes in Computer Science, vol. 2729 (Springer, New York, 2003), pp. 617–630
51. G. Pfeiffer, S. Baumgart, J. Schrder, M. Schimmler, A massively parallel architecture for bioinformatics, in *ICCS2009*. Lecture Notes in Computer Science, vol. 5544 (Springer, Berlin Heidelberg, 2009), pp. 994–1003
52. R.L. Rivest, A. Shamir, L. Adleman, A method for obtaining digital signatures and public-key cryptosystems. Comm. ACM **21**(2), 120–126 (1978)
53. H. Robroch, ePassport Privacy Attack, Presentation at Cards Asia Singapore, http://www.riscure.com. Accessed 26 April 2006
54. M. Schimmler, L. Wienbrandt, T. Gneysu, J. Bissel, COPACOBANA: a massively parallel FPGA-based computer architecture, in *Bioinformatics – High Performance Parallel Computer Architectures*, ed. by B. Schmidt (CRC Press, Boca Raton, 2010), pp. 223–262
55. SciEngines GmbH, http://www.sciengines.com. Accessed April 2012

56. M. Šimka, J. Pelzl, T. Kleinjung, J. Franke, C. Priplata, C. Stahlke, M. Drutarovský, V. Fischer, C. Paar, Hardware factorization based on elliptic curve method, in *Proceedings of FCCM'05* (IEEE Computer Society, Los Alamitos, 2005), pp. 107–116
57. F. Standaert, G. Rouvroy, J. Quisquater, J. Legat, A time-memory tradeoff using distinguished points: new analysis & FPGA results, in *Proceedings of CHES'02*. Lecture Notes in Computer Science, vol. 2523 (Springer, Berlin Heidelberg, 2002), pp. 596–611
58. C. Starke, V. Grossmann, L. Wienbrandt, M. Schimmler, An FPGA implementation of an investment strategy processor, in *ICCS2012*. Procedia Computer Science, Elsevier, vol. 9 (2012), pp. 1880–1889
59. C. Starke, V. Grossmann, L. Wienbrandt, S. Koschnicke, J. Carstens, M. Schimmler, Optimizing investment strategies with the reconfigurable hardware platform RIVYERA. Int. J. Reconfigurable Comput. **2012**, 10 (2012). doi:10.1155/2012/646984
60. P. Štembera, M. Novotný, Breaking Hitag2 with reconfigurable hardware, in *Proceedings of the 14th Euromicro Conference on Digital System Design* (IEEE Computer Society Press, Los Alamitos, 2011), pp. 558–563
61. L. Wienbrandt, S. Baumgart, J. Bissel, C.M.Y. Yeo, M. Schimmler, Using the reconfigurable massively parallel architecture COPACOBANA 5000 for applications in bioinformatics, in *ICCS2010*. Procedia Computer Science, Elsevier, vol. 1 (2010), pp. 1027–1034
62. L. Wienbrandt, S. Baumgart, J. Bissel, F. Schatz, M. Schimmler, Massively parallel FPGA-based implementation of BLASTp with the two-hit method, in *ICCS2011*. Procedia Computer Science, Elsevier, vol. 1 (2011), pp. 1967–1976
63. L. Wienbrandt, D. Siebert, M. Schimmler, Improvement of BLASTp on the FPGA-based high-performance computer RIVYERA, in *ISBRA2012*. Lecture Notes in Bioinformatics, vol. 7292, (Springer, Berlin Heidelberg, 2012), pp. 275–286
64. I. Wiener, Crypto1 specification, reference implementation and test vectors (2007–2008), http://cryptolib.com/ciphers/crypto1/. Accessed April 2012

FPGA-Based HPRC Systems for Scientific Applications

Tsuyoshi Hamada and Yuichiro Shibata

Abstract Modern FPGAs are promising candidates for energy-efficient and high-performance computing platforms also in the field of floating point scientific applications. In this chapter, we show two case studies of applying FPGA-based systems for this application domain. First, implementation of ocean model simulation is discussed, especially focusing on how use of a high-level design tool and data transfer optimization techniques affect the execution performance. Then, FPGA implementation of astrophysical *N*-body simulation is presented and compared to ASIC, GPUs and general purpose processors (GPPs) using a variety of criteria. In this comparative study, the advantages of FPGAs in terms of performance per Watt are emphasized.

1 Introduction

The flexibility of FPGAs with which custom arithmetic datapaths can be built and tailored to the needs of each application has been recognized to offer promising prospects also in scientific applications [1]. While there are endless varieties of scientific applications, in this chapter, we focus on floating point performance of reconfigurable computing systems which is also emphasized in the field of supercomputing [2].

In an early stage of reconfigurable computing machines, executing a floating point arithmetic operation on FPGAs itself was somewhat challenging [3, 4]. Most successful experiences of the first generation of FPGA-based custom computing machines were made in application domains of integer or bit-level arithmetic,

T. Hamada (✉) • Y. Shibata
Nagasaki University, Japan
e-mail: hamada@nacc.nagasaki-u.ac.jp; shibata@cis.nagasaki-u.ac.jp

W. Vanderbauwhede and K. Benkrid (eds.), *High-Performance Computing Using FPGAs*,
DOI 10.1007/978-1-4614-1791-0_12, © Springer Science+Business Media, LLC 2013

such as pattern matching [5]. However, rapid and continuous progress of FPGA technologies has removed this restriction mainly in two ways. First, the advance in transistor integration density simply made FPGAs larger and faster. Second, modern FPGA architectures started providing hardware macros which can be utilized as efficient building blocks for floating point arithmetic logic. As of 2012, high-end FPGAs are able to execute fully pipelined IEEE 754 double precision floating point arithmetic operators with a frequency of around 400 MHz [6].

As a naive comparison in terms of the frequency shows, performance of a floating point arithmetic unit configured on an FPGA is rather suppressed compared to those of general purpose processors (GPPs) or GPUs. At the same time, however, FPGAs offer unique architectural features that can lead to efficient high-performance reconfigurable computing (HPRC) systems. Especially, the following two advantageous features of FPGAs are important.

First, very deeply pipelined custom datapaths with multiple arithmetic units are able to be configured with FPGAs for each individual application program. Such datapaths are infeasible on any other platforms where general versatility needs to be provided. By making the best use of the application-specific deeply pipelined datapaths, HPRC systems enable high throughput execution of applications with a relatively low clock frequency, achieving a good performance per Watt value especially when rich thread level parallelism is found.

The second advantage of FPGAs is customizability in an arithmetic unit level. Rather than standard single or double precision floating point arithmetic, any degree of precision is available by designing custom arithmetic units with arbitrary bits of significant and exponent. When a dynamic range or accuracy required for a given arithmetic operation is known to be limited than the standard ones, it can be optimized to be smaller and faster. Even when an application needs a higher degree of precision than the standard arithmetic, FPGA-based custom arithmetic units are often more efficient than other computing platforms where multiple standard arithmetic must be sequentially combined.

Therefore, in order to bring out the full potential of the aforementioned advantages of FPGAs and to achieve energy-efficient high-performance computing, application developers need to make a number of right implementation decisions at many levels from arithmetic to algorithms. While a variety of software environments for HPRC systems have been implemented to improve the productivity of application development [7, 8], performance of HPRC systems still seems to strongly depend on programming skills with intimate understanding of architectural issues [9].

In this chapter, we demonstrate two case studies of the two use of HPRC systems in the floating point scientific computing application domain. In the first example, we discuss how the execution performance is affected by implementation and optimization techniques with a high-level design tool, especially focusing on DMA transfer in ocean model simulation. Next, we present detailed comparative study in N-body simulation among FPGAs, ASIC, GPUs and GPPs, revealing that FPGAs are competitive in terms of performance per Watt figures.

2 Ocean Model Simulation: Optimization of DMA Transfer

Coming into the 2000s, many commercial high-end reconfigurable systems equipped with a GPP as well as FPGAs became available. As a case study, this section presents the implementation of ocean model simulation kernels with a high-level software design environment on an SRC-6 reconfigurable computer [10], especially focusing on an optimization method of DMA transfer which often becomes a performance bottleneck.

2.1 Computing Framework

Figure 1 shows the structure of our target framework, the SRC-6 computer. It consists of Xeon 2.8 GHz microprocessors and reconfigurable processors called MAP, which are connected each other with a crossbar switching fabric called Hi-Bar Switch. The MAP is composed of three Xilinx Virtex II XC2V6000 FPGAs; programmers can use two of them (FPGA1 and FPGA2) to configure their application circuits while the other one is dedicated to a system controller. The MAP also has six banks of 4-MB On-Board Memory (OBM), where data processed by the

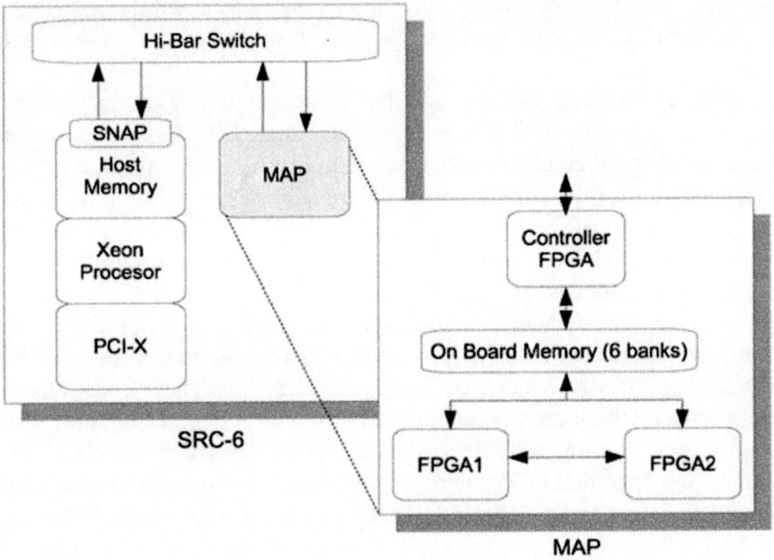

Fig. 1 Structure of the SRC-6 and its reconfigurable processor (MAP). The MAP consists of three FPGAs, but one of them is dedicated for a system controller. A total of six banks of on-board memory (OBM) are shared by the controller and the two user FPGAs. In this case study, a system consisting of a single host processor and single MAP module was utilized

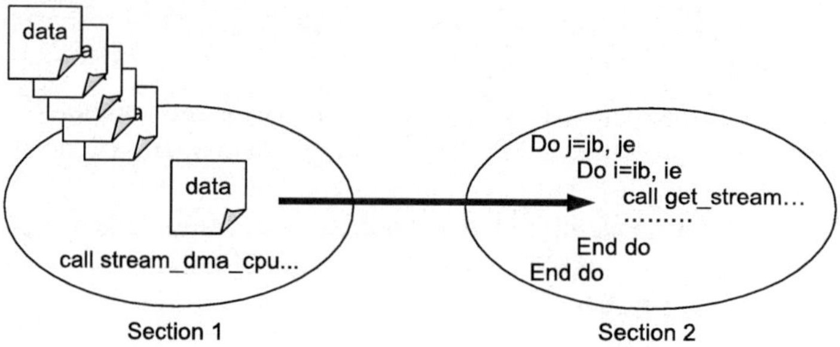

Fig. 2 Single streaming DMA. The two sections work in parallel in individual control on the FPGA. Thus, DMA transfer in Sect. 1 and data access in Sect. 2 can be efficiently overlapped

FPGAs are stored. The FPGAs can access multiple OBM banks at the same time, since they are connected via individual interactive ports. This means proper data allocation is important to prevent access conflicts on the OBM banks. Arbitration of access to the OBM can be explicitly controlled by programmers. The OBM banks are also shared by a DMA controller and are used to exchange data among the FPGAs and the host memory. The two FPGAs also have another interconnection which can be used for direct exchange of data.

The SRC-6 has a unique compiler which generates a relocatable object file from C and FORTRAN source files. The compiler translates the loops specified by users into hardware circuits to be configured on FPGAs. The compiler automatically analyzes structure of the loops and generates hardware that processes the repetition in a pipelined manner.

DMA transfer between FPGAs and the host memory is a key to efficient execution of applications on this type of architecture. SRC-6 supports two styles of DMA transfer: regular DMA and streaming DMA. In the regular DMA, the FPGAs have to wait for the completion of the transfer before starting the operation. With the streaming DMA, on the other hand, the transfer and the operation can be parallelized. In addition, SRC-6 provides two modes of the streaming DMA; single streaming DMA and dual streaming DMA.

Figure 2 shows the concept of the inbound single streaming DMA. In the programming environment for SRC-6, a pair of system functions are available for streaming DMA; `stream_dma_cpu` and `get_stream`. These functions create separated *sections*, which are operated in parallel with independent control on the FPGA. In the first section, a series of data are continuously transferred from the host memory to a specified OBM bank. In the second section, the transferred data is individually fetched and processed typically in a loop. That is, the OBM bank is used as a buffer that absorbs a speed gap between the DMA transfer and the FPGA operation.

In the dual streaming DMA transfer, two banks of the OBM are used as different buffers as shown in Fig. 3. A contiguous series of data on the host memory are alternately transferred to the specified two OBM banks, so that the FPGA can fetch

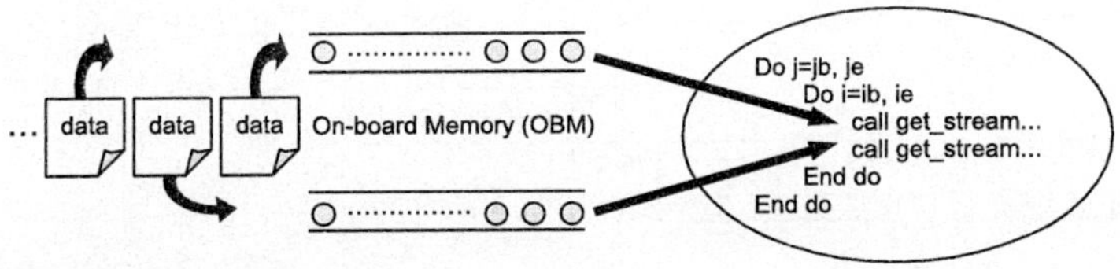

Fig. 3 Dual streaming DMA. Two banks of OBM are used for buffering transferred data, doubling the access bandwidth in Sect. 2

two elements of the transferred data at the same time. This automatically interleaved transfer facility effectively enables the FPGA to extract more data parallelism in applications.

2.2 DMA Optimization Strategies

Figure 4 conceptually shows how execution time of a kernel function implemented on an FPGA is changed by DMA transfer methods. The X axis of the figure indicates the elapsed time. Here, we assume the function needs N inbound data strings and one outbound data string to be transferred. For the sake of simplicity, we will only focus on a DMA method for the inbound data transfer in the following discussion.

The most basic and simplest strategy is to utilize only the regular DMA as shown in Fig. 4a. The execution of the operation on the FPGA starts after N strings of inbound data are transferred. Then, the outbound data are transferred after the execution is completed. By using the single streaming DMA for the last inbound data string, the total time can be reduced as shown in Fig. 4b. Since the time required to transfer the last data string is overlapped with the FPGA execution, the FPGA can start execution after $(N-1)$ data strings are transferred. In a precise sense, the streaming DMA and FPGA execution are not able to start at the same time. This is because the FPGA needs at least the first element of the last string to kick off the operation, but this time difference is almost negligible [11]. Note that the SRC-6 DMA controller does not allow to initiate multiple DMA transfers at the same time, since only one channel is provided between the host processor and the FPGA. That is, a streaming DMA transfer cannot be overlapped with any other DMA transfers.

Use of the dual streaming DMA further improves the FPGA execution time as shown in Fig. 4c. Since the DMA controller interleaves the last data string between two OBM banks, the FPGA can boost the operation by simultaneously fetching two data elements. The time the FPGA must wait before launching the operation remains to be the transfer time for $(N-1)$ data strings. This method is obviously effective in terms of the performance, but the circuit size on the FPGA is needed to be increased so as to extract data parallelism. As a result, this method is often difficult to adopt especially for functions that require a relatively large amount of hardware.

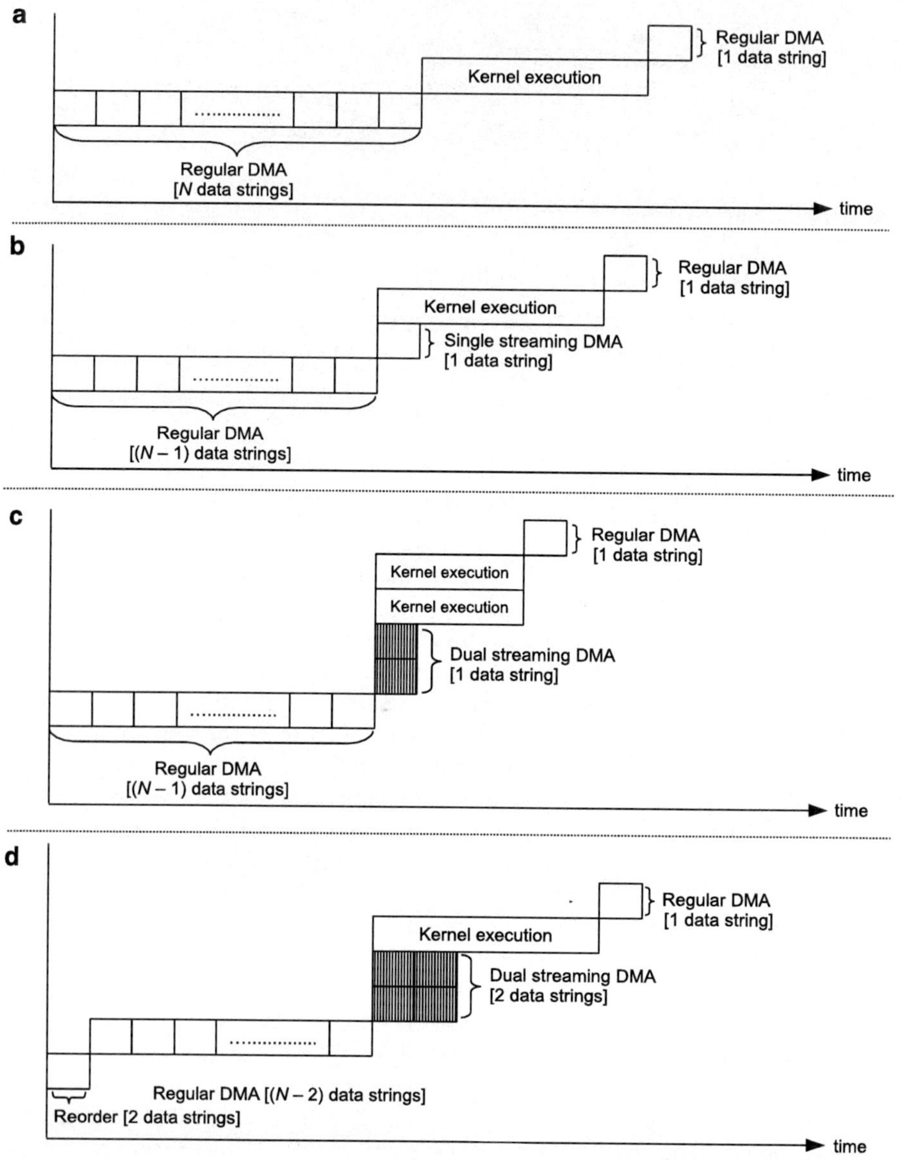

Fig. 4 DMA optimization strategies for N input data strings and single output data string. (**a**) No optimization is made. (**b**) The single streaming DMA is applied, but multiple DMA transfers can not be overlapped due to channel limitations. (**c**) The dual streaming DMA can increase the degree of parallelism of the FPGA execution, but often this is difficult due to limited FPGA resources. (**d**) The twisted streaming DMA transfers two individual data strings in the dual streaming manner. The host processor needs to reorder the data strings in advance

```
Initial values AX(:,:,bid) = 0.0_8
do j=jb, je
  do i=ib, ie
    AX(i,j,bid) = A0 (i  ,j  ,bid) * X(i  ,j  ,bid) + &
                  AN (i  ,j  ,bid) * X(i  ,j+1,bid) + &
                  AN (i  ,j-1,bid) * X(i  ,j-1,bid) + &
                  AE (i  ,j  ,bid) * X(i+1,j  ,bid) + &
                  AE (i-1,j  ,bid) * X(i-1,j  ,bid) + &
                  ANE(i  ,j  ,bid) * X(i+1,j+1,bid) + &
                  ANE(i  ,j-1,bid) * X(i+1,j-1,bid) + &
                  ANE(i-1,j  ,bid) * X(i-1,j+1,bid) + &
                  ANE(i-1,j-1,bid) * X(i-1,j-1,bid)
  end do
end do
```

Fig. 5 Computational flow of the barotropic operator function used in the parallel ocean program (POP), which is a kind of a nine-point stencil operation

Our proposed method shown in Fig. 4d is able to reduce the DMA transfer time even under a sever FPGA resource constraint [12]. The basic idea is to transfer two individual data string to two OBM banks, respectively, using dual streaming DMA instead of interleaving a single data string into the two banks. Thus, the circuit required for the FPGA operation is almost the same as that for Fig. 4b. Moreover, the FPGA can start the operation after $(N-2)$ data strings are transferred, since the last two inbound strings are transferred overlapped with the FPGA operation. However, unlike ordinary dual streaming DMA, the two data strings must be interleaved on the main memory before the transfer. Therefore, the host processor has to reorder the two strings in advance and this could be a considerable overhead especially when the string size is large. We call this method *twisted streaming DMA* in the followings.

Here, we explain the methods, taking the barotropic operator function in an ocean circulation model called the parallel ocean program (POP) as an example [12]. The arithmetic flow of the function is shown in Fig. 5. This function requires a total of five inbound arrays ($A0$, AN, AE, ANE, and X) and one outbound array (AX). Our previous work has revealed that the function requires more than 60% of the FPGA slices and thus use of simple interleaved streaming DMA is infeasible due to the resource constraint. One optimization strategy for inbound DMA transfer is to use the single streaming DMA for the array $A0$. Since the array $A0$ is not reused, any registers to store the streamed values are not required. As mentioned above, streaming DMA cannot transfer multiple data strings concurrently, the other arrays have to be transferred by the regular DMA. The other strategy is to use the twisted streaming DMA for the arrays $A0$ and AE. Before transferring them, the host processor needs to reorder the two arrays alternatively on the main memory to form a double sized single array. Then the double sized array is transferred with the dual streaming DMA, pulling the arrays $A0$ and AE apart again on individual two OBM banks. Since use of streaming DMA hardly affect the FPGA operation time, an essential trade-off will arise between the following two strategies; (1) use of the regular DMA and the single streaming DMA and (2) use of the twisted streaming DMA with reordering process on the host processor.

2.3 Evaluation

First, we analyzed the trade-off point where the proposed DMA optimization method becomes useful. Then, the effects of the method were empirically evaluated through implementation of two kernel functions used in the POP.

2.3.1 Trade-Off Analysis

As previously indicated, our preliminary evaluation results had revealed that the streaming DMA effectively made the data transfer overlapped with the FPGA operations in practical applications. Hence, we compared the regular DMA transfer time for one data string with the reordering time of two data strings, changing the data string size in order to analyze the trade-off point of the twisted streaming DMA method.

Figure 6 plots the execution time of the regular DMA and the reordering process on SRC-6. The time required for the regular DMA was almost proportional to the amount of data transferred. On the other hand, rapid increase in the reordering time appeared when the string size exceeded around 60 KB. As a result, the streaming DMA with reordering was shown to be superior to the regular DMA when the string size was smaller than 150 KB. In many scientific applications including POP, users can choose desired computation granularity of function calls, that is, how much data is processed by a single call of the kernel function. Choosing the string size of 150 KB or smaller is expected to bring about a positive effect with the twisted streaming DMA on SRC-6.

2.3.2 Performance Impact on Kernel Functions

Next, the proposed DMA transfer optimization technique was applied for two kernel functions in the POP; the grad function which calculates the gradient of an energy field at each simulation time step, and the barotropic operator function which calculates a coefficient matrix for a barotropic equation. The kernel functions were described in Fortran and compiled into FPGA configurations with the SRC compiler. We also implemented a code translator which finds optimized data allocation for the OBM banks using an integer programming approach, extending the method proposed in [13]. This translator automatically generates additional code required for the optimized DMA transfer, alleviating the productivity costs of applying the optimization technique.

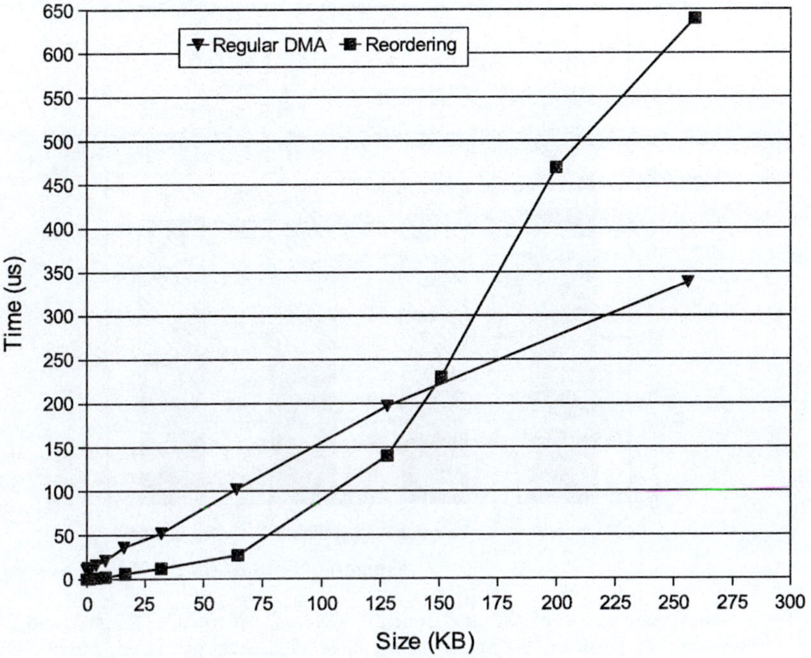

Fig. 6 Measured times for DMA transfer and reordering. The reordering process was compiled with Intel Fortran Compiler 8.1 and executed on the 2.8 GHz Xeon processor, which is the host processor in the SRC-6 system

Figures 7 and 8 summarize the results of performance evaluation of the generated code for the grad function and the barotropic operator function, respectively. In order to highlight how the optimization technique improved the execution performance, the evaluation results when the DMA optimization was partially or completely disabled are also shown. For the grad function, full use of the twisted DMA streaming achieved 1.46 times speedup compared to the software execution on the Xeon processor. The data reordering process on the host processor accounted only for approximately 2% in the total execution time. The barotropic operator function required more FPGA resources and was implemented using the two FPGAs provided in the MAP module. Again, the proposed optimization technique showed good results; 1.37 times speedup compared to the software execution. For this function, the optimized code was 16% faster even than the implementation in which the twisted streaming DMA was applied only for the inbound transfers, suggesting the importance of this kind of optimization process on HPRC systems.

Our first case study described in this section highlighted the current situation of a high-level design environment for commercial HPRC systems as well as the challenges. The SRC compiler we used can take FORTRAN source files in their

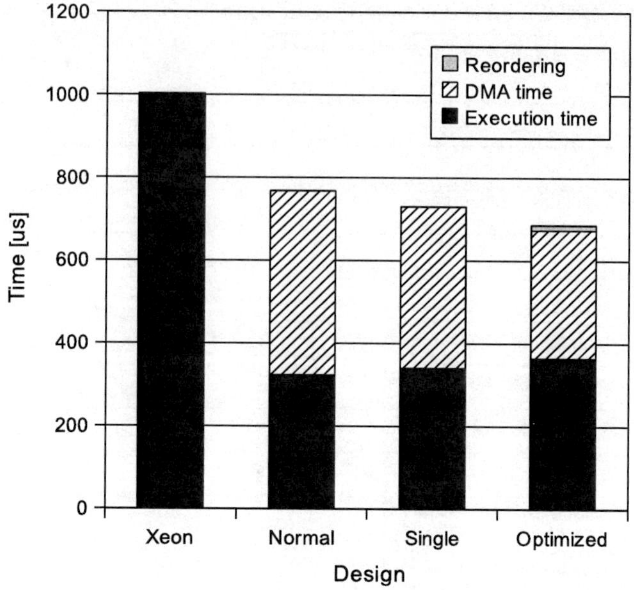

Fig. 7 Performance comparison of the grad function. "Xeon," "Normal," "Single," and "Optimized" corresponds to software execution on the host processor, execution without DMA optimization, execution with the single streaming DMA, and execution fully optimized with the twisted streaming DMA, respectively

almost original form and generate pipelined hardware to be mapped on FPGAs. This effectively improves productivity of applications for HPRC systems compared to conventional register transfer level (RTL) design methodology. However, the naive implementation with original source code for software is not enough to harness the full potential of HPRC systems especially in terms of DMA transfer and memory bank usage. The experiments with our code translator suggest the possibility that the unproductive optimization process can be automated to some extent and offers one future direction of high-level design environments for HPRC systems.

3 *N*-Body Simulation: A Comparative Study

Our next case study is gravitational force calculation for *N*-body simulations in the context of Astrophysics. First, we describe how the algorithm can be efficiently implemented on GPPs, GPUs and FPGAs, and then these implementations are compared in terms of a number of criteria including speed performance, power efficiency, and cost of development.

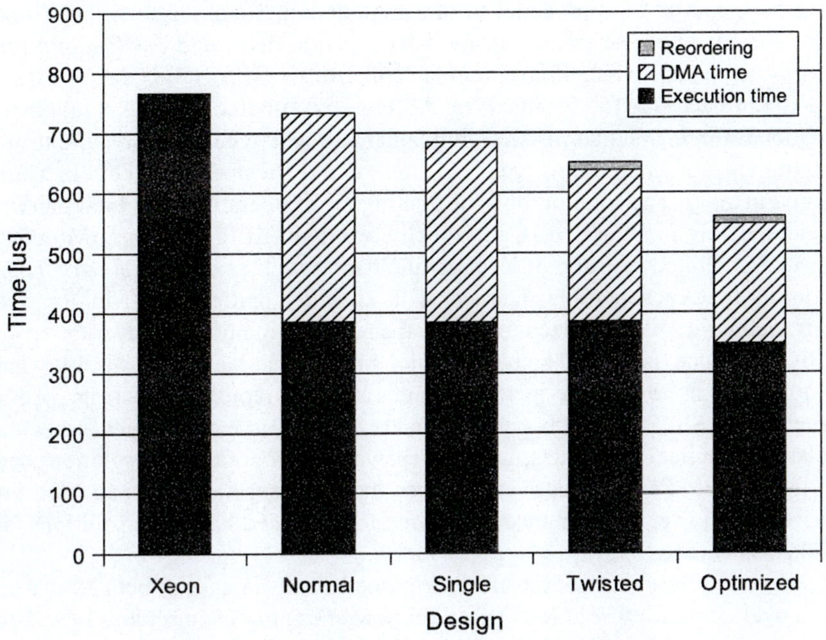

Fig. 8 Performance comparison of the barotropic operator function. While only inbound transfers were optimized in "Single" and "Twisted," both inbound and outbound transfers were fully optimized with the twisted streaming DMA in "Optimized"

3.1 HPRC Solutions to Astrophysical N-Body Simulation

N-body simulation technique is one of the most powerful tools to address a wide range of scientific and engineering applications including the problem of investigating the formation and evolution of astronomical systems [14, 15], molecular dynamics simulations [16–19], fluid mechanics[20–24], acoustics and electromagnetic simulations [25]. The main part of *N*-body simulations consists of the calculation of interaction forces between all bodies or particles in a given system. In the case of astrophysical simulations, planets are modeled as particles for instance. Interactions between the particles are numerically evaluated and then the particles are advanced according to Newton's equation of motion. Since the gravity is a long-range interaction, the computational complexity of this calculation for each time step is $\mathcal{O}(N^2)$, where N is the number of particles in the system. Although some approximation algorithms such as the Barnes–Hut tree-code [26] have been proposed to reduce this complexity to $\mathcal{O}(N\log N)$ at the cost of a large scaling coefficient, the size of the *N*-body simulation (i.e. *N*) has been always limited by the available computational resources. Thus, ever more high performance and efficient computing platforms for *N*-body simulations are desired to meet the increasing need for larger system simulations [19, 27–35].

A wide spectrum of approaches to offering computational platforms to N-body simulations have been examined so far, varying from dedicated ASIC-based platforms [14, 15] to off-the-shelf programmable platforms [31]. In the Grape ("GRAvity piPE") project [14, 15] for instance, ASIC-based supercomputing solutions for gravitational force calculations were implemented, where calculations of pairwise gravitational force interactions were executed on a dedicated ASIC chip in a fully pipelined manner. The total number of floating point operations for each pairwise gravitational force interaction is 20, which are pipelined in hardware. Moreover, the system exploited instruction-level parallelism as well as data-level parallelism, considering that each particle interacts with all other particles. In addition, data cashing techniques were utilized to reduce the required memory bandwidth.

ASIC solutions, however, face two major limitations: lack of flexibility which does not allow us to execute new algorithms and the rapidly increasing costs of building new ASIC chips with state-of-the-art fabrication technologies, especially for relatively moderate volumes. Less expensive and programmable solutions such as off-the-shelf GPPs are attractive. However, their higher power and area consumption as well as lower performance, compared to dedicated hardware solutions, are possible disadvantages.

In order to address some of the aforementioned disadvantages of both ASICs and GPPs, novel computing solutions based on new emerging technologies have been proposed for N-body simulations [30, 32, 36]. Among these, we focus on FPGA-based HPRC solutions [31, 37, 38] and Graphics Processing Unit (GPU)-based platforms [39–41]. In the following, we present comparative implementation of gravitational force calculations on FPGAs, GPUs, ASICs, and GPPs, and compare the implementation results, revealing the advantages and disadvantages of these computing platforms in the context of N-body simulations.

3.2 Design Overview

As mentioned above, the basic operation in N-body simulations is calculation of the interaction forces between pairs of particles in a given system. The position, velocity, and acceleration of each particle in the next time-step are dictated by the accumulation of the forces acting on the particle by all other particles in the system. In a general way, the accumulated force in N-body simulations is described as:

$$f_i = \sum_j G(r_i, r_j), \tag{1}$$

where G is a user-defined function. In the case of gravitational forces, this function is given by:

$$G_i = \frac{m_j r_{ij}}{(r_{ij}^2 + \varepsilon^2)^{3/2}}, \tag{2}$$

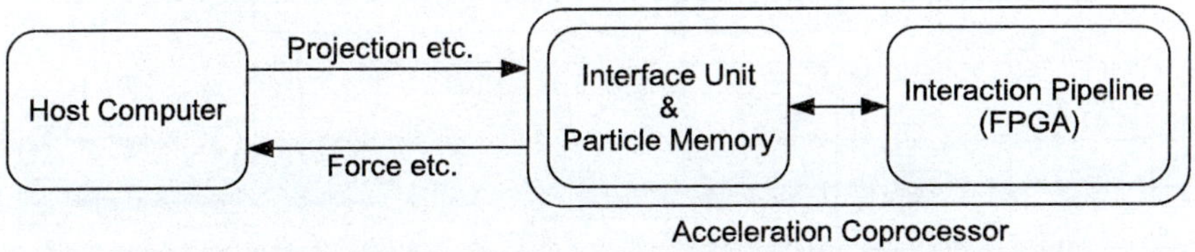

Fig. 9 Basic structure of a hardware-accelerated *N*-body simulation system

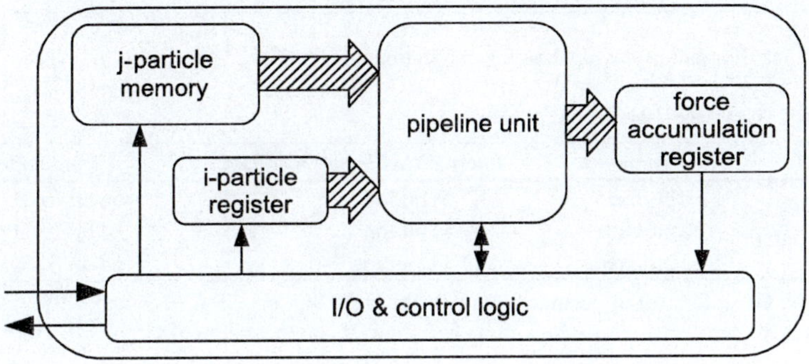

Fig. 10 Basic structure of the hardware accelerator

where r_i and m_i are the position and mass of particle i. $r_{ij} = r_j - r_i$, and ε is a softening parameter. Thus, the accumulated gravitational force is expressed as:

$$f_i = m_i \sum_j \frac{m_j r_{ij}}{(r_{ij}^2 + \varepsilon^2)^{3/2}}. \tag{3}$$

Another benefit of HPRC solutions is that we can utilize any function G through FPGA reconfiguration unlike the ASIC approaches, hence resulting in a more generic implementation. FPGAs also allow us to use and experiment various arithmetic types and precision levels on the same hardware platform.

As illustrated in Fig. 9, a basic structure of a hardware accelerated *N*-body simulation system consists of a host computer and an acceleration coprocessor for force calculation. While the interaction pipelines are implemented on the coprocessor which could consist of a number of FPGA chips, all other calculations such as updates of particle position and velocity are performed on the host computer. Communication between the acceleration coprocessor and the host computer is managed by an interface unit. A particle memory, which is locally accessed by the coprocessor, is also provided to cache information of all the particles involved in a particular round of interaction force calculation with respect to a particular particle i. The information stored in the particle memory includes particle position, velocity, and mass in the case of gravitational force calculation. A block architecture of a hardware accelerator instance, e.g., implemented on a single FPGA chip is shown in Fig. 10.

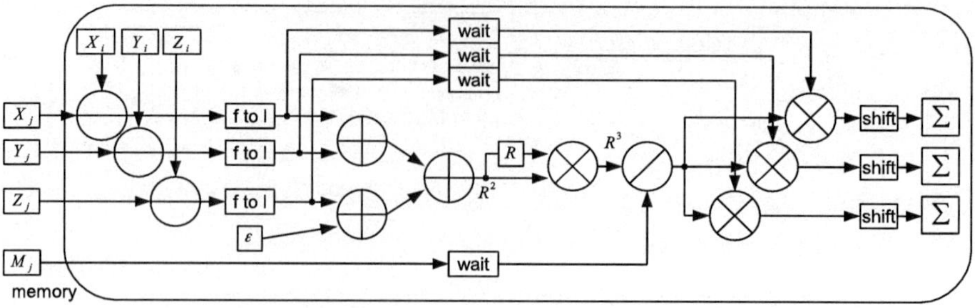

Fig. 11 Block diagram of the pipeline for gravitational force

Table 1 Evaluated accuracy models

Model	Position	Internal (exponent, mantissa)	Accumulation
G3	20 bit fixed	14(7,5) bit log	56 bit fixed
G5	32 bit fixed	17(7,8) bit log	64 bit fixed
G5+	32 bit fixed	20(7,11) bit log	64 bit fixed

G3 and G5 refer to the accuracy used in the ASIC-based GRAPE-3 and GRAPE-5 systems, respectively, while G5+ refers to a more accurate version of G5

Figure 11 depicts an architecture of a fully pipelined hardware for gravitational force calculation [28, 31, 42]. In this design, various data formats are utilized to reduce the required hardware resource while maintaining the accuracy of calculation; the position data for both particle i and particle j are handled in fixed-point format, while m_j is dealt in logarithmic format. The position vectors r_i and r_j are expressed in their three Cartesian components x_i, y_i, z_i, respectively. After subtraction, $x_j - x_i$, the results are converted to logarithmic format, and all subsequent calculations are also performed in logarithmic format.

3.3 FPGA Implementation Results

The fully pipelined hardware for the force calculation depicted in Fig. 11 has been implemented on three different FPGA platforms: the Virtex2Pro-based for Bioler-3 system [33–35, 43], the Virtex-Pro-based Cray XD1 system[43, 44], and the Spartan3-based PROGRAPE-3 system. The hardware architectures of the three platforms are illustrated in Figs. 12, 13, and 14, respectively. For comparison, we also implemented the force calculation pipeline with three different accuracy levels as shown in Table 1.

Table 2 summarizes the implementation results of these pipeline variations with different numbers of pipeline stages on two FPGA chips: Virtex2Pro-5 (XC2VP70-5) and Spartan3-5 (XC3S5000-5). As shown in the results, increase in the number

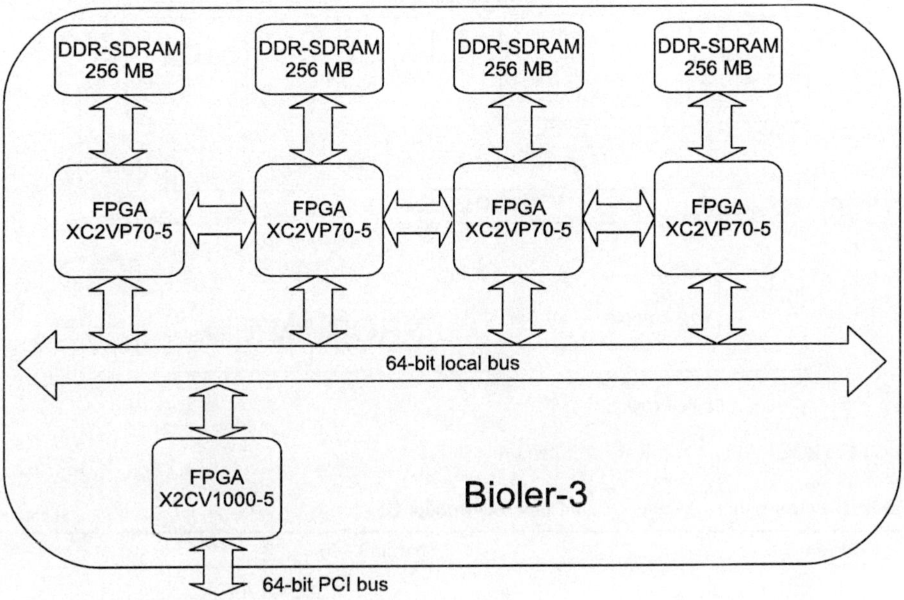

Fig. 12 Bioler-3 hardware architecture

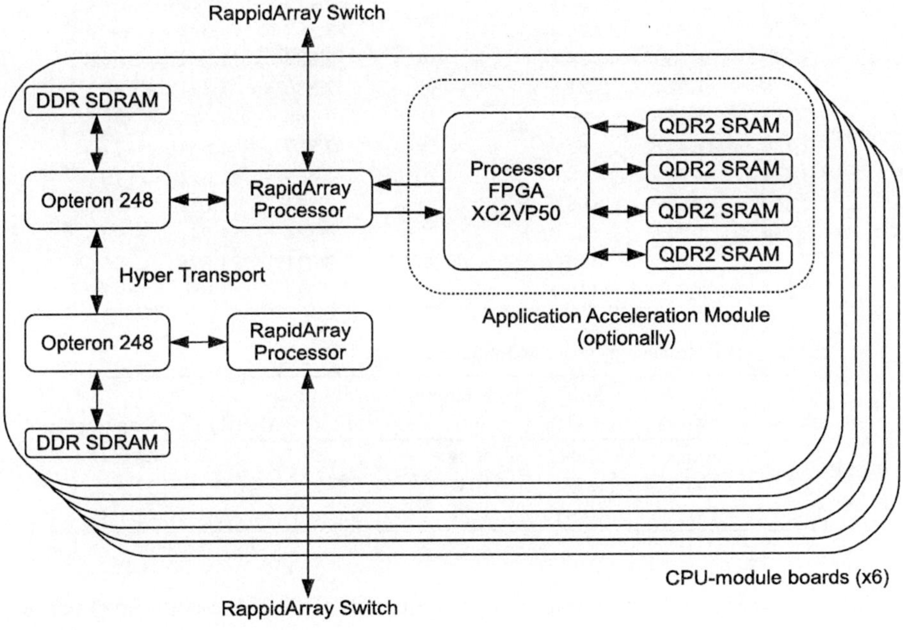

Fig. 13 Cray XD1 hardware architecture

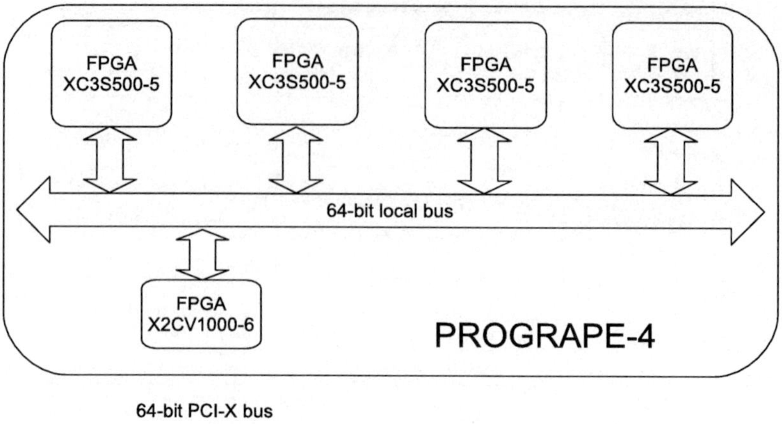

Fig. 14 PROGRAPE-4 hardware architecture

Table 2 Performance of the generated pipeline (model G5)

Virtex2Pro (-5)				Spartan3 (-5)			
	Size				Size		
Stage	LUTs	FFs	f_{max} (MHz)	Stage	LUTs	FFs	f_{max}(MHz)
10	3,338	940	47.1	10	3,366	940	33.1
12	3,402	1,089	57.1	12	3,411	1,089	51.6
13	3,276	1,225	78.4	13	3,302	1,243	67.4
15	3,297	1,360	78.8	14	2,889	1,207	60.0
16	2,878	1,324	81.3	16	2,886	1,321	80.2
17	2,883	1,351	89.7	18	2,888	1,428	79.4
18	2,889	1,477	85.0	19	2,889	1,554	77.5
19	2,871	1,564	88.8	20	2,889	1,572	70.7
21	2,860	1,659	108.4	21	2,861	1,659	73.3
22	2,754	1,747	107.0	22	2,754	1,747	90.8
23	2,860	1,826	110.3	24	3,021	1,875	86.6
24	3,015	1,875	110.7	25	3,033	1,950	92.0

Table 3 Performance of generated pipelines (Virtex2Pro)

Model	f_{max} (MHz)	Size (LUT)	Memory (bit)	Multiplier (18×18 MULT)	Stage
G5	150.5	2,690	108k	3	42
G3	154.6	2,020	108k	0	37
G5+	150.5	3,097	432k	3	42

of pipeline stages in FPGAs comes at little logic (LUT) overheads but the expense of increased usage of flip-flops. However, this does not impose any extra slice costs if the slice LUTs are used for combinational circuits, since FPGAs are rich in flip-flops. This is a stark contrast to ASICs where flip-flops are very costly (Table 3). For comparison, Fig. 3 shows the performance for the pipeline with three different accuracy levels.

3.4 *Discussion on Comparison Results*

We compare our FPGA implementations of the gravitational force pipeline in the G5 accuracy with alternative technologies: ASIC, GPU, and GPP platforms. Table 4 summarizes the comparison results in terms of speed performance, power consumption, performance in Gflops per chip, performance in Gflops per Watt. In addition, other information such as chip technology and year of development is also presented.

Here, an $\mathcal{O}(N^2)$ of leap-flog scheme is used for time integration, since the ratio of computation time for the access to the host or off chip becomes less than 1% with this scheme. With other schemes such as the hierarchical tree algorithm or individual time-step scheme, the results of the performance measurement could strongly depend on the performance of host computer and communication speed or memory bandwidth, and thus the performance analysis would become more difficult to discuss.

The power consumption listed in Table 4 is a difference between measured power values in idle and computing conditions. After measuring the power of a system in the idle condition, we measure the power again in computing. Then, the power consumption is obtained by subtracting these values.

Comparing the different computing platforms based on various CMOS technologies ranging from 500 to 45 nm in a completely fair manner is quite difficult. However, considering the spiraling cost of state-of-the-art ASIC fabrication, it seems to be difficult to justify the ASIC solutions for these types of applications as previously described. Among the programmable and reconfigurable solutions, we first note that the GPU implementations achieve the highest performance in terms of Gflops; the G92 GPU is 11 times faster than the Spartan3-based FPGA implementation. In addition, the Q6600 Core2Quad implementation called the Phantom-GRAPE,[1] which is highly optimized with SSE instructions, is slightly faster than the Spartan3 implementation. Taking into account the difference in chip technologies used (90 nm vs. 65 nm), we can anticipate state-of-the-art FPGA solutions to outperform GPPs, although not by much. In addition, the relatively higher purchases cost of FPGAs and their low level programming model such as VHDL that we used for our implementations are obvious drawbacks of FPGAs to GPPs. However, if we focus on the performance per Watt figure, the Core2Quad Q6600 implementation is 34 times less than the Spartan3 FPGA implementation. Moreover, the G92 GPU implementation is also 15 times less than the Spartan3 implementation in this criterion. This suggests that FPGA-based HPRC could be a viable approach for very high performance and large-scale N-body simulations, on an energy cost basis.

[1] http://progrape.jp/phantom/.

Table 4 Implementation results (GRAPE-5 and G5 model)

	GRAPE-5	Bioler-3	Cray XD1	PROGRAPE-4	ASUS EN8800GTX	MSI N9800GTX+	Core2Quad Q6600	Atom 230
Device Chip	ASIC	**FPGA**	**FPGA**	**FPGA**	GPU	GPU	CPU	CPU
	300k gates	XC2VP70-5	XC2VP50-7	XC3S5000-5	G80	G92	SSE	SSE
Development Year	1997	2004	2004	2006	2007	2008	2007	2008
Chip technology	500 *nm*	130 *nm*	130 *nm*	90 *nm*	90 *nm*	65 *nm*	65 *nm*	45 *nm*
Chips/board	8	4	1	4	1	1	1	1
Pipelines/chip	2	16	10	16	N/A	N/A	N/A	N/A
Frequency (MHz)	80	133.3	120	100	1350	1890	2400	1600
Gflops/chip	24.3	81	45.6	60.8	470.8	687.1	70.3	6.35
Gflops/board	48.6	324.2	45.6	243.2	470.8	687.1	70.3	6.35
ratio of performance (against GRAPE-5)	1.0	6.7	0.9	5.0	9.7	14.1	1.45	0.13
Cost($) per board	N/A	15000	N/A	2400	790	268	200	74
Mflops per Cost	N/A	21.6	N/A	101	596	2663	352	85.8
Power Consumption per board (without host)	80 W	30 W	N/A	5 W	148 W	122 W	49 W	3.1 W
Power Consumption per chip	8 W	7.5 W	N/A	1.3 W	148 W	122 W	49 W	3.1 W
Gflops/Watt	0.61	11	N/A	**49**	3.2	5.6	1.43	2.05

Best performance in Gflops/Watt indicated in bold.

3.5 *Impact on the Performance per Watt*

The use of parallel computing technologies in the form of FPGAs and SIMD architectures can offer very large speed-ups for N-body simulations, since this application is characterized by a high degree of data and instruction parallelism as well as data locality. However, these acceleration technologies vary in their cost, programming abstraction level, and power consumption. Aiming at exploring these trade-offs, we compared various implementation platforms in the context of gravitational force calculations in N-body simulations. The results showed that GPU platforms achieve the highest performance and thus are very competitive in performance and performance per cost figures. In addition, lower cost of GPUs and their higher programming abstraction level are also advantageous. However, the performance per Watt figure favored FPGAs with a factor of 15 to 1 and 34 to 1 compared to GPUs and GPPs, respectively. This suggests large-scale high performance simulations, where power consumption is a bottleneck, are expected to be a niche but promising market for FPGAs. On the other hand, GPUs could well be clocked down to reduce their energy consumption, at a relatively lower performance penalty. Our future work includes such experiments as well as extending our hardware pipelines to molecular dynamics simulations where more complex calculations with high accuracy levels are required.

References

1. T. El-Ghazawi, E. El-Araby, M. Huang, K. Gaj, V. Kindratenko, D. Buell, The promise of high-performance reconfigurable computing. IEEE Comput. **41**(2), 69–76 (2008)
2. J.L. Hennessy, D.A. Patterson, *Computer Architecture: A Quantitative Approach*, 5th edn. (Morgan Kaufmann, Waltham, 2012)
3. N. Shirazi, A. Walters, P. Athanas, Quantitative analysis of floating point arithmetic on FPGA-based custom computing machines, in *Proceedings of IEEE Symposium on FPGAs for Custom Computing Machines* (IEEE, Piscataway, NJ, 1995), pp. 155–162
4. T.A. Cook, L. Louca, W.H. Johnson, Implementation of IEEE single precision floating point addition and multiplication on FPGAs, in *Proceedings of IEEE Symposium on FPGAs for Custom Computing Machines* (IEEE, Piscataway, NJ, 1996), pp. 107–116
5. J. Arnold, D. Buell, E. Davis, SPLASH2, in *Proceedings of ACM Symposium on Parallel Algorithms and Architectures* (ACM, New York, NY, 1992), pp. 316–322
6. Xilinx, Inc., LogiCORE IP Floating-Point Operator v6.0. DS816, Product Specification (2011)
7. T. Hamada, N. Nakasato, PGR: a software package for reconfigurable super-computing, in *Proceedings of International Conference on Field Programmable Logic and Applications* (IEEE, Piscataway, NJ, 2005), pp. 366–373
8. H. Yamada, Y. Ogawa, T. Ooya, T. Ishimori, Y. Osana, M. Yoshimi, Y. Nishikawa, A. Funahashi, N. Hiroi, H. Amano, Y. Shibata, K. Oguri, Automatic pipeline construction focused on similarity of rate law functions for an FPGA-based biochemical simulator. IPSJ Trans. Syst. LSI Des. Methodol. **3**, 244–256 (2010)
9. O. Mencer, ASC: a stream compiler for computing with FPGAs. IEEE Trans. Comput. Aided Des. Integrated Circ. Syst. **25**(9), 1603–1617 (2006)
10. SRC Computers, Inc., MAPstation (2005), http://www.srccomp.com/. Accessed 31 Mar 2013

11. S. Shida, Y. Shibata, K. Oguri, D.A. Buell, Implementation of a barotropic operator for ocean model simulation using a reconfigurable machine, in *Proceedings of International Conference on Field Programmable Logic and Applications* (IEEE, Piscataway, NJ, 2007), pp. 589–592
12. S. Shida, Y. Shibata, K. Oguri, D.A. Buell, An optimization method of DMA transfer for a general purpose reconfigurable machine, in *Proceedings of International Conference on Field Programmable Logic and Applications* (IEEE, Piscataway, NJ, 2008), pp. 647–650
13. M. Weinhardt, W. Luk, Memory access optimization for reconfigurable systems. IEE Proc. Comput. Digit. Tech. **148**, 105–112 (2001)
14. D. Sugimoto, Y. Chikada, J. Makino, T. Ito, T. Ebisuzaki, M. Umemura, A special-purpose computer for gravitational many-body problems. Nature **345**, 33–35 (1990)
15. J. Makino, M. Taiji, *Scientific Simulations with Special-Purpose Computers — The GRAPE Systems* (Wiley, New York, 1998)
16. T. Fukushige, M. Taiji, J. Makino, T. Ebisuzaki, D. Sugimoto, A highly parallelized special-purpose computer for many-body simulations with an arbitrary central force: Md-grape. Astrophys. J. **468**, 51 (1996)
17. R. Susukita, T. Ebisuzaki, B.G. Elmegreen, H. Furusawa, K. Kato, A. Kawai, Y. Kobayashi, T. Koishi, G.D. McNiven, T. Narumi, K. Yasuoka, Hardware accelerator for molecular dynamics: MDGRAPE-2. Comput. Phys. Comm. **155**, 115–131 (2003)
18. T. Narumi, R. Susukita, T. Ebisuzaki, G.D. McNiven, B.G. Elmegreen, Molecular dynamics machine: special-purpose computer for molecular dynamics simulations. Mol. Simul. **21**, 401–415 (1999)
19. M. Taiji, T. Narumi, Y. Ohno, N. Futatsugi, A. Suenaga, N. Takada, A. Konagaya, Protein explorer: a petaflops special-purpose computer system for molecular dynamics simulations, in *Supercomputing, 2003 ACM/IEEE Conference* (IEEE, Piscataway, NJ, 2003), pp. 15–15
20. R.A. Gingold, J.J. Monaghan, Smoothed particle hydrodynamics-theory and application to non-spherical stars. Mon. Not. R. Astron. Soc. **181**, 375–389 (1977)
21. J.J. Monaghan, Smoothed particle hydrodynamics. Annu. Rev. Astron. Astrophys. **30**, 543–574 (1992)
22. M. Steinmetz, GRAPESPH: cosmological smoothed particle hydrodynamics simulations with the special-purpose hardware GRAPE. Mon. Not. R. Astron. Soc. **278**, 1005–1017 (1996)
23. R. Klessen, GRAPESPH with fully periodic boundary conditions: fragmentation of molecular clouds. Mon. Not. R. Astron. Soc. **292**(1), 11–18 (1997)
24. G.R. Liu, M.B. Liu, *Smoothed Paricle Hydrodynamics — A Meshfree Particle Method* (World Scientific, Tuck Link, 2003)
25. C.A. Brebbia, *The Boundary Element Method for Engineers* (Pentech Press, London, 1978)
26. J. Barnes, P. Hut, A hierarchical O(NlogN) force-calculation algorithm. Nature **324**, 446–449 (1986)
27. S.K. Okumura, J. Makino, T. Ebisuzaki, T. Fukushige, T. Ito, D. Sugimoto, E. Hashimoto, K. Tomida, N. Miyakawa, Highly parallelized special-purpose computer, GRAPE-3. Publ. Astron. Soc. Jpn. **45**, 329–338 (1993)
28. A. Kawai, T. Fukushige, J. Makino, $7.0/Mflops astrophysical N-body simulation with treecode on GRAPE-5, in *Supercomputing, ACM/IEEE 1999 Conference* (IEEE, Piscataway, NJ, 1999), pp. 67–67
29. M.S. Warren, J.K. Salmon, D.J. Becker, M.P. Goda, T. Sterling, A 55 TFLOPS simulation of amyloid-forming peptides from Yeast Prion Sup35 with the special-purpose computer system MDGRAPE-3, in *Proceedings of Supercomputing 97, in CD-ROM* (IEEE, Los Alamitos, 1997)
30. T. Hamada, T. Fukushige, A. Kawai, J. Makino, Progrape-1: a programmable special-purpose computer for many-body simulations, in *Numerical Astrophysics* (Springer, 1999), pp. 427–428
31. T. Hamada, T. Fukushige, A. Kawai, J. Makino, PROGRAPE-1: a programmable, multi-purpose computer for many-body simulations. Publ. Astron. Soc. Jpn. **52**, 943–954 (2000)
32. G. Lienhart, A. Kugel, R. Manner, Using floating-point arithmetic on FPGAs to accelerate scientific N-body simulations, in *Proceedings of the 10th Annual IEEE Symposium on Field-Programmable Custom Computing Machines, 2002* (IEEE, 2002), pp. 182–191

33. N. Nakasato, T. Hamada, Astrophysical hydrodynamics simulations on a reconfigurable system, in *13th Annual IEEE Symposium on Field-Programmable Custom Computing Machines, 2005. FCCM 2005* (IEEE, Piscataway, NJ, 2005), pp. 279–280
34. T. Hamada, N. Nakasato, Massively parallel processors generator for reconfigurable system, in *13th Annual IEEE Symposium on Field-Programmable Custom Computing Machines, 2005. FCCM 2005* (IEEE, Piscataway, NJ, 2005), pp. 329–330
35. T. Hamada, N. Nakasato, T. Ebisuzaki, A 236 Gflops astrophysical simulation on a reconfigurable super-computer, in *SuperComputing 2005*, Seattle, 2005
36. T. Hamada, T. Fukushige, A. Kawai, J. Makino, Progrape-1: a programmable special-purpose computer for many-body simulations, in *Numerical Astrophysics* (Springer, New York, 1999), pp. 427–428
37. T.A. Cook, H.-R. Kim, L. Louca, Hardware acceleration of n-body simulations for galactic dynamics, in *Photonics East'95* (International Society for Optics and Photonics, 1995), pp. 115–126
38. K.H. Tsoi, C.H. Ho, H.C. Yeung, P.H.W. Leong, An arithmetic library and its application to the N-body problem, in *12th Annual IEEE Symposium on Field-Programmable Custom Computing Machines, 2004. FCCM 2004* (IEEE, 2004), pp. 68–78
39. L. Nyland, M. Harris, J. Prins, N-body simulations on a GPU, in *Proceedings of the ACM Workshop on General-Purpose Computation on Graphics Processors* (2004), p. 60
40. M. Harris, GPGPU: general-purpose computation on GPUs, in *SIGGRAPH 2005 GPGPU COURSE* (2005), http://www.gpgpu.org/s2005/
41. T. Hamada, T. Iitaka, The chamomile scheme: an optimized algorithm for n-body simulations on programmable graphics processing units. arXiv preprint astro-ph/0703100, 1–26 (2007)
42. T. Ito, J. Makino, T. Fukushige, T. Ebisuzaki, S.K. Okumura, D. Sugimoto, A special-purpose computer for N-body simulations: GRAPE-2A. Publ. Astron. Soc. Jpn. **45**, 339–347 (1993)
43. T. Hamada, N. Nakasato, PGR: a software package for reconfigurable super-computing, in *International Conference on Field Programmable Logic and Applications, 2005* (IEEE, Piscataway, NJ, 2005), pp. 366–373
44. L. Zhuo, V.K. Prasanna, High performance linear algebra operations on reconfigurable systems, in *Supercomputing, 2005. Proceedings of the ACM/IEEE SC 2005 Conference* (IEEE, 2005), pp. 2–2

Accelerating the SPICE Circuit Simulator Using an FPGA: A Case Study

Nachiket Kapre and André DeHon

Abstract Spatial processing of sparse, irregular, double-precision floating-point computation using a single FPGA enables up to an order of magnitude speedup and energy-savings over a conventional microprocessor for the simulation program with integrated circuit emphasis (SPICE) circuit simulator. We develop a parallel, FPGA-based, heterogeneous architecture customized for accelerating the SPICE simulator to deliver this speedup. To properly parallelize the complete simulator, we decompose SPICE into its three constituent phases—Model Evaluation, Sparse Matrix-Solve, and Iteration Control—and customize a spatial architecture for each phase independently. Our heterogeneous FPGA organization mixes very large instruction word (VLIW), Dataflow and Streaming architectures into a cohesive, unified design. We program this parallel architecture with a high-level, domain-specific framework that identifies, exposes and exploits parallelism available in the SPICE circuit simulator using streaming (SCORE framework), data-parallel (Verilog-AMS models) and dataflow (KLU matrix solver) patterns. Our FPGA architecture is able to outperform conventional processors due to a combination of factors including high utilization of statically-scheduled resources, low-overhead dataflow scheduling of fine-grained tasks, and streaming, overlapped processing of the control algorithms. We expect approaches based on exploiting spatial parallelism to become important as frequency scaling continues to slow down and modern processing architectures turn to parallelism (e.g. multi-core, GPUs) due to constraints of power consumption.

N. Kapre (✉)
Nanyang Technological University, 50 Nanyang Avenue, Singapore
e-mail: nachiket@ieee.org

A. DeHon
University of Pennsylvania, Philadelphia, PA 19104, USA
e-mail: andre@acm.org

W. Vanderbauwhede and K. Benkrid (eds.), *High-Performance Computing Using FPGAs*, DOI 10.1007/978-1-4614-1791-0_13,

1 Introduction

SPICE (Simulation Program with Integrated Circuit Emphasis) is an analog circuit simulator used extensively to simulate and verify operation of silicon circuits. It models the analog behavior of semiconductor circuits using a compute-intensive, nonlinear, differential equation solver. This can take days or weeks of runtime on real-world circuits. SPICE is notoriously difficult to parallelize due to its irregular compute structure and a sloppy sequential description [34]. It has been observed that less than 7% of the floating-point operations in SPICE are automatically vectorizable [15].

Spatial parallelism provides a suitable model for constructing accelerators for challenging problems like SPICE. It offers a natural way to express the heterogeneous computational structure in SPICE and exposes the inherent parallelism available in the problem. Furthermore, modern FPGAs can be configured to efficiently support spatial parallelism with multiple floating-point operators coupled to hundreds of distributed, on-chip memories and interconnected by a flexible routing network. In Table 2, we observe that modern FPGAs can match and even surpass the peak floating-point capacity of modern multi-core processors while dissipating far less power. Spatial parallelism allows us to configure the FPGA to deliver a higher fraction of this floating-point peak through a combination of careful static scheduling and low-overhead distributed processing.

As shown in Table 1, a SPICE simulation accepts a netlist description of the circuit to be simulated along with input stimulus. The simulator then returns the response of the circuit in the form of output analog waveforms as shown in Fig. 1. The simulation algorithm discretizes circuit response and repeatedly solves circuit equations at each discrete step to generate output waveforms. We also show an abstract internal representation of the simulation algorithm in Fig. 2. This iterative simulation consists of two key computationally intensive phases per iteration: *Model Evaluation* (② in Fig. 2) followed by *Matrix Solve* (③ in Fig. 2). This organization allows the nonlinear, differential equation solver to be simplified to a system of linear equations $A\mathbf{x} = \mathbf{b}$ which is handled in the *Matrix Solve* phase. The nonlinear, time-varying circuit elements are linearized using a Newton–Raphson loop and discretized using Trapezoidal integration in the *Model-Evaluation phase*. These two loops are managed in the third phase of SPICE which is the *Iteration Controller* (① in Fig. 2). A well-balanced, scalable, parallel architecture must accelerate all three phases of SPICE.

Table 1 Example SPICE netlist

```
* R-C-D Circuit Topology
V1 1 0 PWL(0 1 1e-11 2 2e-11 3)
R1 1 2 1
D1 2 0 DNOM
C1 2 0 10e-11
.MODEL DNOM D (IS=1E-15 Vj=0.02 cjo=1e-9)
* SPICE Analysis options
.TRAN 1e-12 3e-9
.PLOT TRAN v(2)
.END
```

Table 2 Raw floating-point throughput and power (double-precision)

Chip	Tech. (nm)	Clock (GHz)	Peak GFLOPS (Double)	Power (Watts)
Intel Core i7 965	45	3.2	25	130
Xilinx Virtex-6 LX760	40	0.2	26	20–30

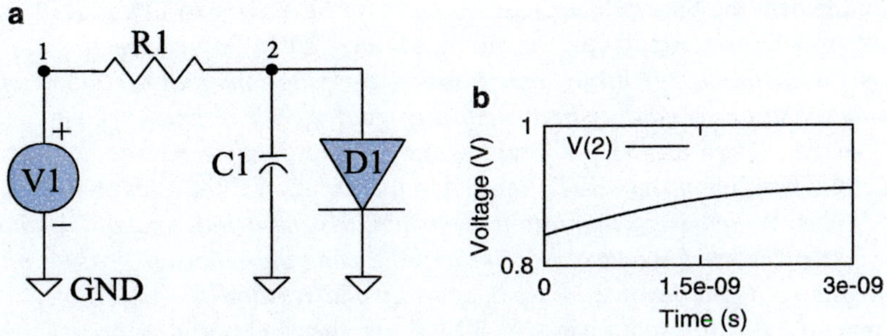

Fig. 1 Example of a SPICE simulation. (**a**) Input circuit; (**b**) output waveform

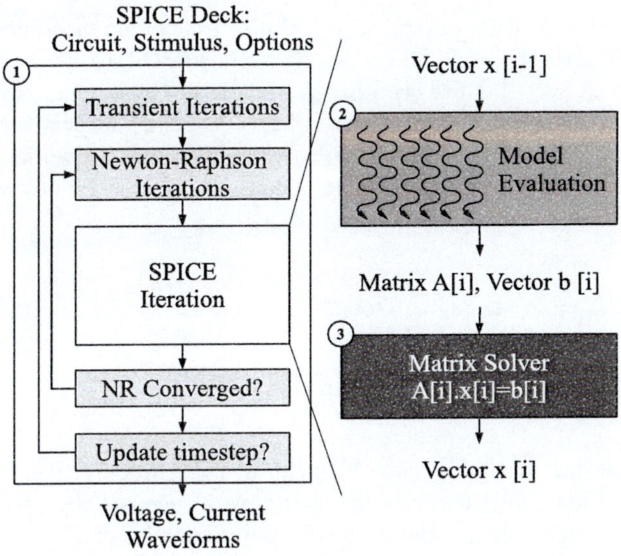

Fig. 2 Flowchart of a SPICE simulator

This chapter reviews our previous research [19–22] that systematically solves different subproblems emerging from the complete SPICE acceleration challenge.

We now list the key themes of this chapter:

- We accelerate the Model-Evaluation phase of SPICE using a custom VLIW organization overlayed on top of the FPGA [19]. We show scalability of our approach across different Verilog-AMS models [21]. *This idea will be broadly applicable to other irregular, data-parallel problems that underutilize multi-core CPUs or GPUs.*
- We implement the Sparse Matrix-Solve phase of SPICE on an FPGA [20] using a Token Dataflow architecture customized for Matrix-Solve. *Token Dataflow designs are suitable for large-scale parallel computations that are either challenging for static scheduling or structurally resolved at runtime.*
- We design a hybrid, VLIW architecture for implementing the apparently sequential fraction of the SPICE simulator, the Iteration Controller phase, on the FPGA [22]. *We believe this solution is particularly important to avoid Amdahl's Law bottleneck once we have accelerated the compute-intensive portion of the application but still desire additional speedup and scalability.*
- We integrate the different phases of SPICE together and outline a programming methodology and execution flow to use the accelerator for different simulations. *Our approach highlights the benefits of avoiding an expensive per-circuit-instance compilation flow while still delivering the benefits of spatial parallelism*

The rest of this chapter is organized as follows. We explain the underlying computational characteristics of SPICE and briefly explain FPGA-based implementation of computation in Sect. 2. Next, we discuss suitable FPGA compute organizations for implementing the three SPICE phases in Sects. 3, 4 and 5 respectively. In Sect. 6 we provide details about the composed FPGA compilation framework and quantify the performance and energy of complete SPICE accelerator. Finally we wrap up with some key insights and lessons in Sect. 8.

2 Background

2.1 Summary of SPICE Algorithms

SPICE simulates the dynamic analog behavior of a circuit described by nonlinear differential equations. SPICE solves the nonlinear differential circuit equations by computing small-signal linear operating-point approximations for the nonlinear and time-varying elements until termination (① in Fig. 2). We show an example R-C-D circuit topology and a corresponding transient simulation in Table 1. The linearized system of equations is represented as a solution of $A\mathbf{x} = \mathbf{b}$ handled in the *Matrix-Solve* phase (③ in Fig. 2), where A is the matrix of circuit conductances, $\mathbf{b}$ is the vector of known currents and voltage quantities and $\mathbf{x}$ is the vector of unknown voltages and branch currents. The simulator calculates entries in A and $\mathbf{b}$ from the device model equations that describe device transconductance (e.g. Ohm's law for resistors, transistor I-V characteristics) in the *Model-Evaluation* phase (② in Fig. 2).

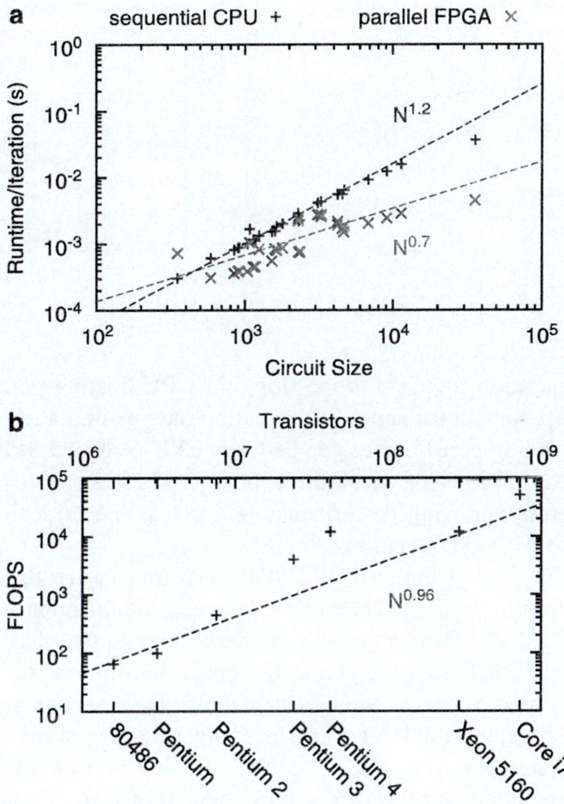

Fig. 3 Scaling trends for FLOPS and runtime (`spice3f5`). (**a**) Sequential runtime scaling of SPICE simulator. (**b**) Peak FLOPS scaling of Intel CPUs

2.2 *SPICE Performance Analysis*

Since the SPICE simulation is an iterative algorithm, we can understand key characteristics of the complete simulation by analyzing a single iteration. In Fig. 3a, we show performance scaling trends for a single iteration of the SPICE solver for two scenarios. First we show data for sequential implementation of the open-source `spice3f5` package on an Intel Core i7 965 across a range of benchmark circuits shown later in Appendix. We also show data for our parallel FPGA implementation across the same benchmarks. We observe that sequential runtime for one iteration scales as $O(N^{1.2})$ as we increase circuit size, N, while parallel runtime scales faster as $O(N^{0.7})$. These trends have been previously reported in [34]. Our experiments re-examine this claim on modern circuits and modern architectures and observe that they still continue to hold true. In Fig. 3b, we show the peak floating-point scaling trends of Intel CPUs obtained from Intel datasheets to contrast against SPICE runtime trends. We observe that the sequential CPU peak (FLOPS) have barely scaled as $O(N)$ while SPICE runtime has scaled super-linearly as $O(N^{1.2})$. While Moore's Law continues to deliver increasing circuit sizes (for both circuit

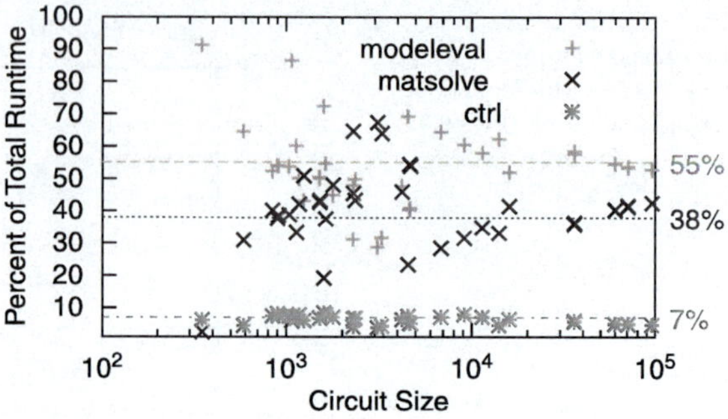

Fig. 4 Sequential runtime distribution of SPICE simulator

simulation and CPU processing), the CPU floating-point peaks have been unable to keep up with the super-linear scaling rate of simulation times. This means there is a widening performance gap between CPU peak and SPICE runtime. In contrast, the FPGA processing capabilities shown in Table 2 can be organized entirely in parallel thereby allowing performance to scale as the critical latency of the computation $O(N^{0.7})$ as shown in Fig. 3a.

To further understand SPICE performance trends, we break down the contribution to total SPICE runtime between the different phases in Fig. 4. We observe that Model-Evaluation and Sparse Matrix-Solve phases account for over 90% of total SPICE runtime across the entire benchmark set. For circuits dominated by nonlinear devices, Model-Evaluation phase accounts for as much as 90% (55% average) of total runtime since the runtime of this phase scales linearly with the number of nonlinear devices in the circuit. Simulations of circuits with a large number of resistors and capacitors (i.e. linear elements) generate large matrices and consequently the Sparse Matrix-Solve phase accounts for as much as 70% of runtime (38% average). This phase empirically scales as $O(N^{1.2})$ which explains the super-linear scaling of overall SPICE runtime. Finally, the Iteration Controller phase of SPICE comprises a small but nontrivial fraction ($\approx$7%) of total runtime. While this represents a small fraction of total runtime, once we accelerate Model-Evaluation and Sparse Matrix-Solve phases, it can become an Amdahl's Law bottleneck limiting overall application speedup. Thus, our parallel FPGA architecture must parallelize all three phases of SPICE.

2.3 *SPICE Model-Evaluation*

In the Model-Evaluation phase, the simulator computes conductances and currents through different elements of the circuit and updates corresponding entries in the matrix with those values. For resistors this needs to be done only once at the start of the simulation. For nonlinear elements, the simulator must search for an operating-point using Newton–Raphson iterations that requires repeated evaluation of the

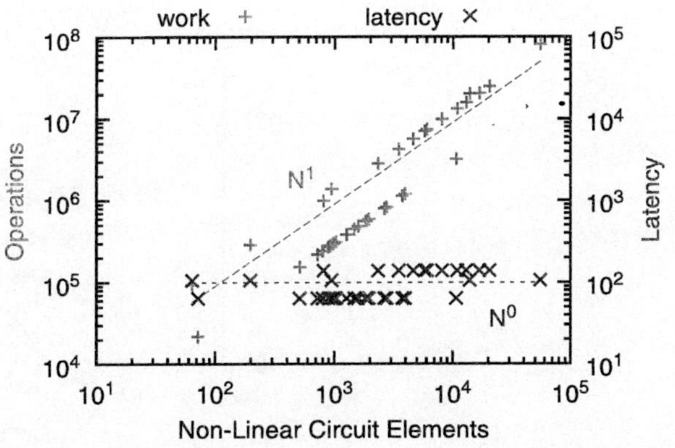

Fig. 5 Work-vs-latency of model-evaluation phase

model equations and a linear solve multiple times per time-step as shown by the innermost loop in step ① of Fig. 2. For time-varying components, the simulator must recalculate their contributions at each timestep based on voltages at several previous timesteps in the outer loop in step ① of Fig. 2.

In Fig. 5, we plot the number of floating-point operations and the latency of evaluation (floating-point operations along critical path from input to output) as a function of the number of nonlinear elements in the circuit. Since each device contributes a fixed number of floating-point operations per instance, we see a linear growth in the number of operations. However, the latency of evaluation stays constant since each evaluation is completely independent and can be processed simultaneously. This highly data-parallel computation is suitable for implementation on FPGAs, GPUs, as well as multi-cores.

2.4 *SPICE Matrix Solve* ($\mathbf{Ax} = \mathbf{b}$)

Modern SPICE simulators use modified nodal analysis (MNA) [5] to assemble circuit equations into the matrix A. This generates highly sparse, asymmetric matrices which are processed using sparse, direct LU factorization techniques to deliver robust simulation results. Our approach uses the state-of-the-art KLU matrix solver [35] optimized for SPICE circuit simulation and avoids per-iteration changes to the matrix structures. The static nonzero pattern enables reuse of the matrix factorization graph across all SPICE iterations and allows us to perform a one-time distribution of computation across a parallel architecture. The solver reorders the matrix A to minimize fillin using block triangular factorization (BTF) and column approximate minimum degree (COLAMD) techniques. It then uses the left-looking Gilbert–Peierls [13] algorithm to compute the LU factors of the matrix column-by-column such that $A = LU$. Finally, it calculates the unknown $\mathbf{x}$ using Front-Solve $L\mathbf{y} = \mathbf{b}$ and Back-Solve $U\mathbf{x} = \mathbf{y}$ operations.

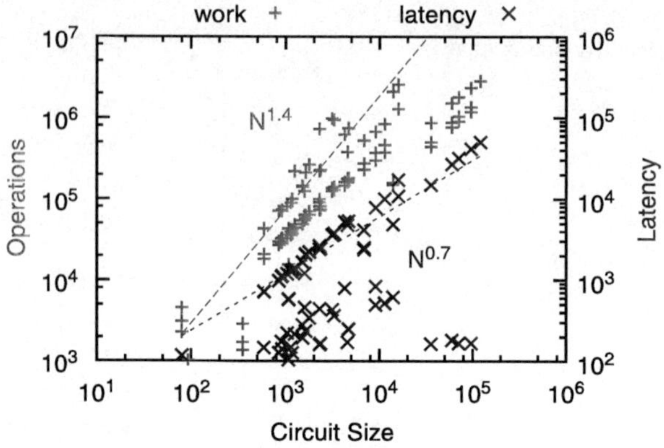

Fig. 6 Work-vs-latency of sparse matrix-solve phase

In Fig. 6, we plot the number of floating-point operations in the factorization and latency of evaluation as a function of the size of the circuit. We observe that the number of floating-point operations in the Matrix-Solve computation scale as $O(N^{1.4})$ while the latency of the critical path through the compute graph scales as $O(N^{0.7})$. This suggests a parallel potential of $O(N^{0.7})$ which can be realized by distributing the dataflow graph across ideal parallel hardware (e.g. no communication delays, perfect distribution, unlimited internal processing bandwidth).

2.5 SPICE Iteration Controller

The SPICE iteration controller shown in Fig. 2 is responsible for two kinds of iterative loops: (1) *inner loop*: Newton–Raphson linearization iterations for nonlinear devices and (2) *outer loop*: adaptive time-stepping for time-varying devices. The Newton–Raphson algorithm is responsible for computing the linear operating-point for the nonlinear devices like diodes and transistors. Additionally, an adaptive time-stepping algorithm based on truncation error calculation (Trapezoidal approximation, Gear approximation) is used for handling the time-varying devices like capacitors and inductors. The controller implements customized convergence conditions and local truncation error estimations that determine how the transient analysis state machines are advanced at runtime in a data-dependent manner. The state-machine and breakpoint-processing logic are highly data-dependent and determine the total number of SPICE iterations required for the complete simulation.

As we saw earlier in Fig. 4, the Iteration Control phase only accounts for $\approx 7\%$ of total sequential runtime. However, our parallel SPICE implementation takes care to efficiently implement this portion to avoid an Amdahl's Law bottleneck. We show the danger of ignoring this phase for parallelization in Fig. 7 which shows the runtime breakdown for the `r4k` netlist in different implementation scenarios. We observe that we can get a speedup of $\approx 6\times$ when parallelizing the Model-Evaluation

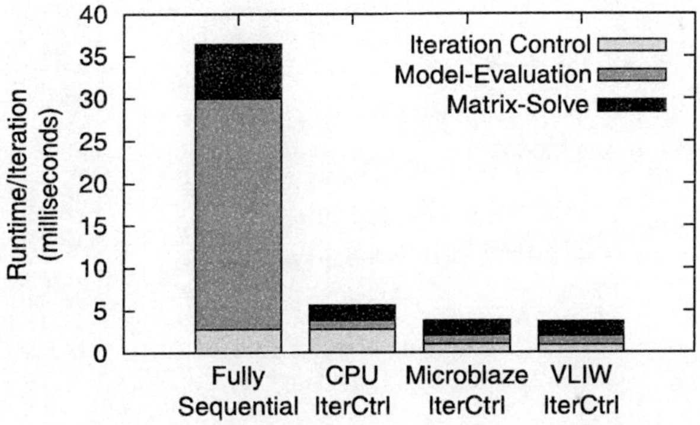

Fig. 7 Parallel potential for iteration control (`r4k` netlist)

and Sparse Matrix-Solve phase of SPICE (parallel FPGA runtimes obtained from Sect. 7). If we parallelize the Iteration Control phase, we can improve overall speedup to ≈9×. The Iteration Control phase of SPICE is dominated by *data-parallel* operations in convergence detection and truncation error-estimation which can be described effectively in a *streaming* fashion. The loop management logic for the Newton–Raphson and Timestepping iterations is control-intensive and highly irregular. We can capture both these computational structures effectively using a streaming framework.

2.6 Promise of FPGAs

We briefly review the FPGA architecture and highlight some key characteristics of an FPGA that make it well suited to accelerate SPICE. A Field-Programmable Gate Array (FPGA) is a massively parallel architecture that implements computation using hundreds of thousands of tiny programmable computing elements called k-LUTs (k-input lookup tables that can implement any boolean function of k inputs, typically $k = 4$–6) connected to each other using a programmable bit-level communication fabric. An FPGA allows us to configure the computation in space rather than time and evaluate multiple operations concurrently in a fine-grained fashion. In Fig. 8, we show a simple calculation and its conceptual implementation on a CPU and an FPGA. For a CPU implementation, we process the instructions stored in an instruction memory temporally on an ALU while storing the intermediate results (i.e. variables) in a data memory. Thus, a single evaluation of the graph takes several CPU cycles. On an FPGA, we can implement the operations as pipelined spatial circuits while implementing the dependencies between the operations physically using pipelined wires instead of variables stored in memory to get high performance. This allows the FPGA mapping to start a new evaluation in each cycle delivering higher throughput than the CPU. Modern FPGAs also include

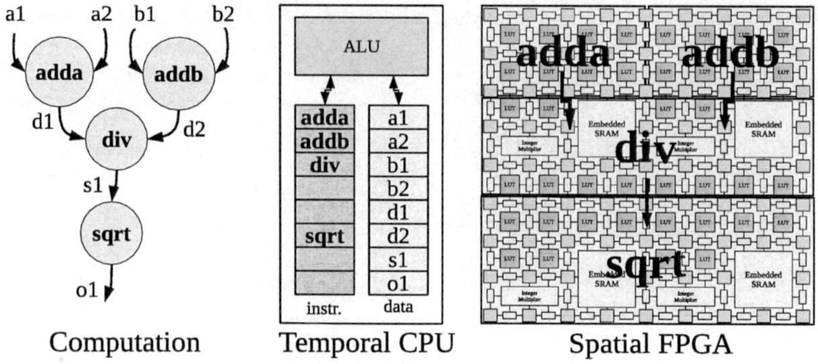

Fig. 8 Implementing computation

hundreds of embedded memories distributed across the fabric that deliver 10–100× higher on-chip bandwidth compared to a processor [10]. The spatial FPGA fabric can be configured to implement hundreds of specialized datapaths connected to high-bandwidth on-chip memories thereby providing a higher overall throughput compared to modern multi-core processors. This potential provides a foundation for customizing a specialized SPICE accelerator using the FPGA fabric.

2.7 Historical Review

We now review the various studies and research projects in the past three-and-a-half decades that have attempted to build parallel SPICE systems. Some of these studies accelerate SPICE by devoting expensive hardware resources to squeeze additional performance while others reorganize the computation to use lower-precision evaluation that is easier to parallelize. Our approach expands on certain ideas from the past while delivering a cheaper, SPICE-accurate accelerator.

We can refine the classification of parallel SPICE approaches by considering underlying trends and characteristics of the different systems as follows:

1. *Compute Organization*: We see parallel SPICE solvers using a range of different compute organizations including conventional multi-processing, multi-core, VLIW, SIMD, and Vector.
2. *Precision*: Under certain conditions, SPICE simulations can efficiently model circuits at lower precisions.
3. *Compiled Code*: In many cases, it is possible to generate efficient instance-specific simulations by specializing the simulator for a particular circuit.
4. *Numerical Algorithms*: Different classes of circuits perform better with a suitable choice of a matrix factorization algorithm. Our FPGA design may benefit from new ideas for factoring the circuit matrix.

One of the early parallel SPICE designs Awsim-3 [27, 28] uses a compiled code approach and a special-purpose system with lower-precision, table-lookup Model-Evaluation (*Compiled Code*, *Precision*) to provide a speedup of 560× over a Sun 3/60. However, a bulk of these speedups are due to dedicated hardware floating-point unit since the Sun 3/60 implements floating-point in software (tens of cycles/operation). Additionally, table-lookup approximations result in a simulation with accuracy trade-offs. A message-passing, parallel SPICE implementation [16] on an expensive, 40-node SGI Origin 2000 supercomputer (MIPS R10K processors) was able to speed up SPICE for certain specialized benchmarks by 24× (*Compute Organization*). More recently, in [25], a multi-threaded version of SPICE is developed using PThreads. It achieves a speedup of 5× using 8 SMP (Symmetric Multi-Processors) on a small benchmark set which is amenable to parallel matrix factorization (*Compute Organization*). GPUs have been used to speed up the data-parallel Model-Evaluation phase of SPICE by 50×[1] (double-precision on ATI GPU) or 32×[14] (lower-accuracy, single-precision on NVIDIA GPU) but can only accelerate the SPICE simulator in tandem with the CPU by 3× for the 2-chip GPU-CPU processing system (*Compute Organization*). Recent approaches [37] have used coarse-grained domain-decomposition techniques to parallelize SPICE by 31×–870×(mean 119×) across a 32 processor grid at SPICE-level accuracy (*Numerical Algorithms*).

FPGAs have traditionally enjoyed limited use for accelerating SPICE due to limited logic capacity of older FPGA families and lack of tools and methodology for attacking a problem of this magnitude. A compiled code, partial evaluation approach for timing simulation (lower precision than SPICE) using FPGAs was demonstrated in [42] where the processing architecture was customized for each SPICE circuit using fixed-point computation (*Compiled Code*, *Precision*). Our FPGA-based approach accelerates the SPICE computation while retaining the accuracy of `spice3f5` and developing an economical single-FPGA system for accelerating SPICE. We reuse the idea of compiled-code methodology pioneered by many previous approaches. We can compose our technique with KLU-based domain-decomposition approaches [37] to scale to even large problems and system sizes e.g. multi-FPGA systems. Additionally, we can integrate lower-precision techniques (e.g. table lookup) into our mapping flow to get cumulative benefits.

3 Model Evaluation

In this section, we show how to compile the nonlinear differential equations describing SPICE device models using a high-level, domain-specific framework based on Verilog-AMS. This approach is broadly applicable to other HPC workloads with a dataflow compute kernel e.g. mathematical expressions which can be evaluated in parallel in dataflow fashion. We sketch a hypothetical *fully spatial* design that distributes the complete Model-Evaluation computation across the chip as a configured *circuit* to achieve the highest throughput. We then develop a

Table 3 Diode Verilog-AMS equations (*left*), dataflow graph (*right*)

```
module diode (a, c);
{
   parameter real is=10f;
   parameter real vj=0.3;

   inout a,c;
   electrical a,c;
   branch (a, c) ac;

   I (ac) <+ is*(exp (V (ac)/vj) - 1);
}
```

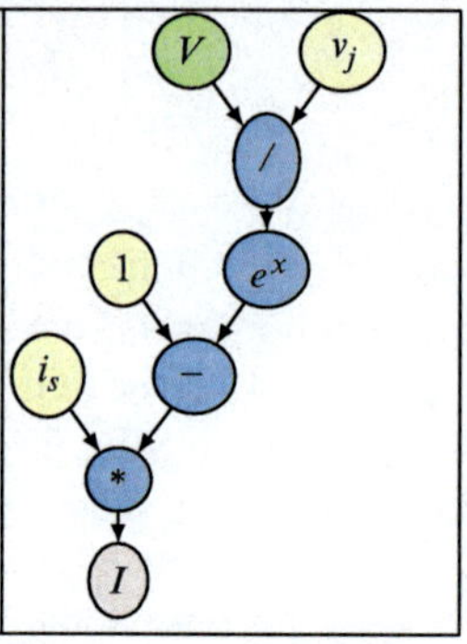

realistic spatial organizations that can be realized on a single FPGA using statically scheduled time-multiplexing of FPGA resources. This allows us to use less area than the fully spatial design while still achieving high performance. Our automated compilation and tuning approach can scale the implementation to larger system sizes when they become available.

3.1 Structure

As discussed earlier, the Model-Evaluation phase has high *data parallelism* consisting of thousands of independent device evaluations each requiring hundreds of floating-point operations. Additionally, we make other structural observations that will help simplify and enhance our FPGA mapping. We note that there is a limited diversity in the number of nonlinear device types in a simulation (e.g. typically only `diode` and `transistors` models). There is high pipeline parallelism within each device evaluation as operations can be represented as an acyclic feed-forward dataflow graph (DAG) with nodes representing operations and edges representing dependencies between the operations. These DAGs are static graphs that are known entirely in advance and do not change during the simulation enabling efficient offline scheduling of instructions. Individual device instances are predominantly characterized by constant parameters (e.g. V_{th}, Temperature, T_{ox}) that are determined by the CMOS process leaving only a handful of parameters that vary from device to device (e.g. W, L of device). This specialization potential in the form of constant-folding, identity simplification and other compiler optimizations can eliminate 70–80% of repeated, unnecessary work.

We compile the device equations from a high-level domain-specific language called Verilog-AMS [26] which is more amenable to parallelization and optimization than existing C description in `spice3f5`. We show a simple code example for the diode in Table 3. In contrast to Verilog-AMS, the `spice3f5` C descriptions make extensive use of pointers into shared data-structures that are harder to analyze

Table 4 Device model instruction counts

	Instruction distribution (optimized)						
Model	Add	Mult.	Divide	Sqrt.	Exp.	Log.	Rest
`bjt`	22	30	17	0	2	0	8
`diode`	7	5	4	0	1	2	9
`jfet`	13	31	2	0	2	0	8
`mos1`	24	36	7	1	0	0	21
`vbic`	36	43	18	1	10	4	9
`mos3`	46	82	20	4	3	0	38
`hbt`	112	57	51	0	23	18	60
`bsim4`	222	286	85	16	24	9	137
`bsim3`	281	629	120	9	8	1	117
`mextram`	675	1,626	397	22	52	37	238
`psp`	1,345	2,319	247	30	19	10	263

(Rest includes mux, bool and integer operations)

and do not provide a clean way to separate variables from constants. The Verilog-AMS compilation also allows us to capture the device equations in an intermediate form suitable for performance optimizations and parallel mapping to several target architectures. We use open-source Verilog-AMS nonlinear models from Simucad ranging from the small, simple `diode` model to the large, complex `bsim3`, `psp` models.

Our Verilog-AMS compiler generates a generic feed-forward dataflow graph (see diode example in Table 3 of the computation that is processed by the backend tools). The compiler currently performs simple dead-code elimination, mux-conversion, constant-folding, identity simplification and common-subexpression elimination optimizations. We tabulate the optimized instruction counts for the different device models in Table 4.

3.2 *Fully Spatial Architecture*

A spatial circuit implementation of computation is a straightforward embodiment of a dataflow graph on an FPGA. Such a circuit contains physical operators for every instruction in the dataflow graph and uses physical wires to implement dependencies between the instructions. These operators can evaluate in parallel and communicate results directly using the programmable FPGA interconnect. Furthermore, if the computation is data parallel, we can exploit pipeline parallelism by adding a suitable number of registers along the wires to balance dataflow. This will then permit us to start a new evaluation of the dataflow graph in each cycle and deliver results of the computation after the pipeline latency of the graph. This pipelined, spatial circuit implementation of data-parallel computation will deliver the highest possible performance for our Model-Evaluation computation. In contrast, a conventional von-Neumann architecture (e.g. Intel CPUs) will implement this computation by

Table 5 Estimated speedup (vs. Intel Core i7 965) and FPGA costs (Virtex6 LX760) of multi-FPGA designs

Device models	Total speedup	FPGAs required	Speedup per FPGA
`bjt`	14	1	14
`diode`	34	1	34
`jfet`	17	1	17
`mos1`	14	1	14
`vbic`	17	1	17
`mos3`	12	1	12
`hbt`	62	3	20
`mextram`	204	18	11
`bsim3`	47	6	8
`bsim4`	69	4	17
`psp`	155	21	7

fetching a binary representation of computation stored in memory. The binary implicitly encodes the dataflow structure using a sequence of instructions that communicate results using registers (i.e. memory). The dataflow parallelism hidden in this implicit encoding must be rediscovered by the von-Neumann architecture in hardware, often limiting the amount of parallelism that can be exploited from the dataflow graph.

Ideal Mapping We can imagine implementing the data-parallel operations in Model-Evaluation as a pipelined dataflow circuit on the FPGA. If cost is not a concern, this approach provides up to two orders of magnitude speedup over an implementation using Intel Core i7 965 microprocessor when using a Xilinx Virtex-6 LX760 FPGA (see Table 5). We compute a lower-bound on the number of FPGAs required to implement the dataflow graph based on total operator area (ignoring FPGA external IO limitations and pipelining area costs). This model provides a lower-bound on cost and an upper-bound on the speedup possible with the spatial approach. For the designs that fit in a single FPGA, this model only needs to be refined with pipelining costs and can avoid the complexities of the multi-FPGA distribution. A single-FPGA, fully spatial implementation of all devices will be eventually possible with the increasing FPGA densities made possible by Moore's Law. From Table 5, the `bsim3` model currently requires only 6 Virtex-6 LX 760 FPGAs to fit. This means an FPGA that is $4\times$ denser will fit the complete device evaluation graph. This FPGA will become possible two technology nodes into the future at 22 nm (Virtex-6 is manufactured at the 40 nm technology node).

3.3 *Custom VLIW Architecture*

In the previous section, we saw how fully spatial implementations (circuit-style implementation of dataflow graphs) are too large to fit on current FPGAs. Hence, computation must be *time-shared* over limited FPGA resources. These graphs

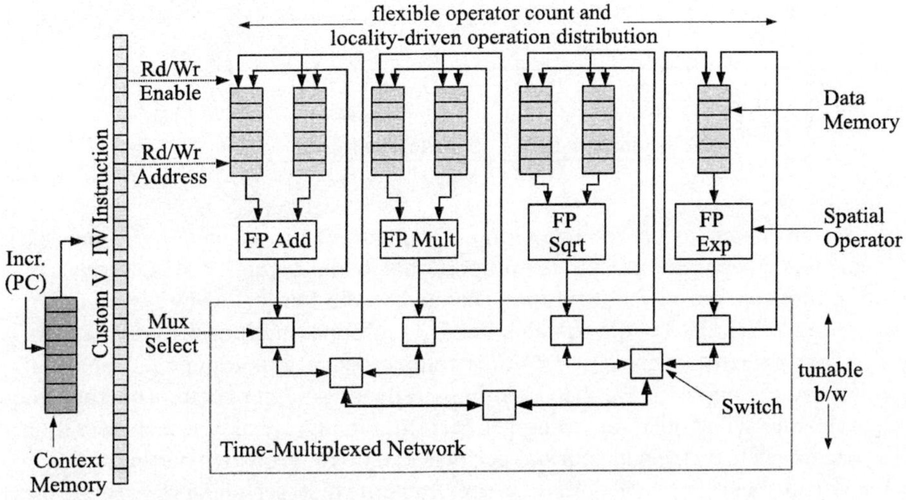

Fig. 9 Custom VLIW organization

contain a diverse set of floating-point operators such as adds, multiplies, divides, square-roots, exponentials, and logarithms. We map these graphs to custom VLIW *processing tiles* with spatial implementation of the floating-point operators.

Pipelined, spatial FPGA implementations of elementary functions like `exp` and `log` operate at a high throughput of one evaluation/cycle (250 MHz) while the processor implementations require 100s of instructions (10–20 cycles at 3 GHz) [17]. Additionally, we support these spatial operators by coupling them to local, distributed, high-bandwidth memories, as shown in Fig. 9, which is not possible with fixed-function CPUs or GPUs. We statically schedule these resources offline in VLIW [12] fashion and perform loop-unrolling, tiling and software pipelining optimizations to improve performance.

Each *tile* in the time-shared architecture consists of a heterogeneous set of floating-point operators coupled to local, high-bandwidth memories and interconnected to other operators through a communication network. We use a time-multiplexed fat tree [23] to connect these operators which allows us to tune the interconnect bandwidth to match communication requirements between the operators. A time-multiplexed switch in this architecture consists of a multiplexer for each IO port with a small context memory for storing the routing instruction i.e. select bits for the multiplexers. Thus, a VLIW instruction for the complete *tile* is a combination of read/write addresses for local on-chip memories, address control bits, datapath multiplexer controls and switch route decisions for each statically scheduled cycle of operation. We develop a Verilog-AMS compiler for the nonlinear device models that is capable of recognizing a suitable subset of the language specification while performing useful optimizations such as constant-folding,

Table 6 Parallel software environments

Arch.	Compiler	Libraries	Timing
Intel CPUs	`gcc-4.4.3`(-O3)	OpenMP 3.0 [7]	PAPI 4.0.0 [33]
Xilinx FPGA	Synplify Pro 9.6.1,	Coregen [43]	–
	Xilinx ISE 10.1	Flopoco [8]	–

if-mux conversion, and dead-code elimination. Our VLIW scheduling framework first chooses an operator mix per *tile* proportional to the frequency of occurrence of floating-point operations in the graph generated by the Verilog-AMS compiler. For example, a `bsim3` VLIW *tile* contains 1 add, 3 multiplies and 1 each of divide, sqrt, log and exp operators while a `bsim4` *tile* contains 2 adds, 2 multiplies and 1 each of divide, sqrt, log and exp. We then partition the floating-point operations across the heterogeneous set of operators using MLPart [3]. Finally, we assign each operation to a specific cycle on the datapath and perform a static route on the time-multiplexed network using a greedy LPT (longest processing time first) scheduler to generate the VLIW instruction context. For a nonlinear device model, we configure the FPGA with multiple *tiles* in SIMD-like (Single Instruction Multiple Data) manner where each VLIW instruction is SIMD broadcast to all *tiles*.

3.4 Experimental Setup

In our experimental flow, we compare the performance of the Intel Core i7 965 quad-core CPU with a Xilinx V6 LX760 FPGA. We measure runtime averaged across thousands of device evaluations. We map the data-parallel model equations to the CPU and FPGA with the help of software frameworks listed in Table 6. To target these architectures we use a combination of *automated code-generation* and *auto-tuning* to generate optimized implementations across these different systems. Our code-generator writes out multiple configurations of parallel code for the CPU and the FPGA-VLIW architecture based on architecture-specific templates. For the CPU, we perform loop-unrolling and generate vector instructions for certain operations to optimize performance. For the FPGA, we can choose the number of datapaths (PEs) and the richness of the BFT interconnect between these datapaths (PEs) to tune performance. Our auto-tuner exhaustively explores several implementation parameters e.g. loop unroll factor for the different architectures as shown in Table 7. We also show the range of possible values taken by these parameters as well as the increment step for the exploration. Such an exhaustive exploration is possible in our case since the Model Evaluation graphs are completely known in advance and the design space is small. This framework is also capable of targeting GPUs and other multi-core devices [21].

Table 7 Auto-tuning parameters

Architecture	Parameter	Range	Increment
Intel	Loop-unroll factor	1–5	+1
	Threads	1–8	+1
	MKL vector	True/false	
FPGA	Loop-unroll factor	1–15	+1
	Operators per PE	8–64	×2
	BFT rent parameter	0.0–1.0	+0.1

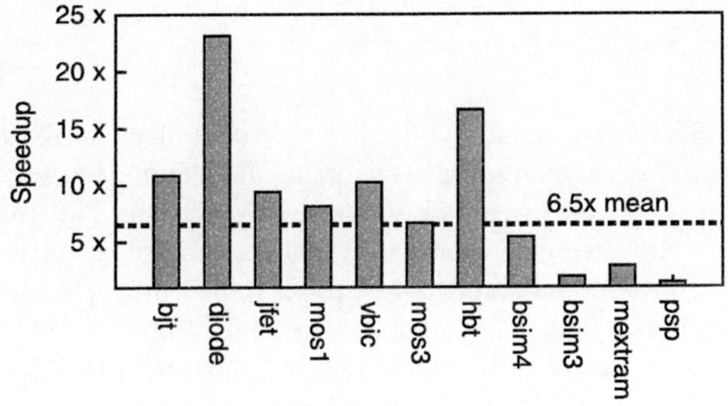

Fig. 10 Speedups for model-evaluation

3.5 *Results*

In Fig. 10, we compare the performance achieved by Model-Evaluation implementations between a quad-core Intel Core i7 965 (loop-unrolled and multi-threaded) and a Xilinx Virtex-6 LX760 FPGA (loop-unrolled, tiled and statically scheduled). We observe speedups between 1.4×–23× (geomean 6.5×) across our nonlinear device model benchmarks. We deliver these speedups due to higher utilization of statically-scheduled floating-point resources (up to 70%), explicit routing of graph dependencies over physical interconnect and spatial implementation of elementary floating-point functions. The FPGA is able to achieve higher speedups for smaller, simpler nonlinear devices e.g. `diode`, `bjt` since they require smaller interconnect switch programming context and a lower memory footprint to store the intermediate values in the evaluation.

3.6 *Future Work*

We now identify additional opportunities for improving the performance of the parallel FPGA design of the Model-Evaluation phase.

1. Double-precision floating-point operators consume a large amount of area on FPGAs. Custom floating-point or fixed-point operators that operate at just enough precision might provide an opportunity for improving the compute density on FPGAs. We can redesign the Model-Evaluation datapaths with lower precision by adapting existing techniques [29] to obtain additional speedup. We demonstrate some preliminary results using this technique for small devices [30].
2. Additionally, we can improve the performance of the Model-Evaluation phase with extra loop-unrolling and the use of off-chip memory capacity.

4 Sparse Matrix Solve

In Sect. 2, we identified the Matrix-Solve phase of the SPICE circuit simulator as the most challenging phase for parallelization. Large-scale, sparse matrix factorization is also a commonly-used HPC kernel. The computation is characterized by sparse, irregular operations that are too fine-grained to be effectively exploited on conventional architectures (e.g. multi-cores). In this section, we show how to parallelize Sparse Matrix Solve using a combination of the KLU algorithm (*better software*) and an efficient dataflow FPGA architecture (*better hardware*). We start by introducing the KLU algorithm that extracts the exact compute graph of matrix factorization and then describing a dataflow architecture for efficiently mapping the compute graph to an FPGA.

4.1 Structure

The SPICE simulator `spice3f5` assembles the sparse circuit left-hand side (LHS) matrix and the right-hand side (RHS) vectors in $A\mathbf{x} = \mathbf{b}$ using the MNA approach. Since circuit elements tend to be connected to only a few other elements, the MNA circuit matrix is highly sparse (except high-fanout nets like power lines, etc). The underlying nonzero structure of the matrix is defined by the topology of the circuit and consequently remains unchanged throughout the duration of the simulation. As discussed earlier, the KLU matrix solver performs a one-time partial pivoting at the start of the simulation to deliver a static compute graph that can be efficiently distributed and evaluated in parallel. This static pattern is reused across all SPICE iterations and enables us to generate a specialized static dataflow architecture that processes the graph in parallel. The KLU Gilbert–Peierls algorithm has irregular, *fine-grained task parallelism* during *LU* factorization. We will now look at the pseudocode for the computation to understand the nature of parallelism available in the algorithm.

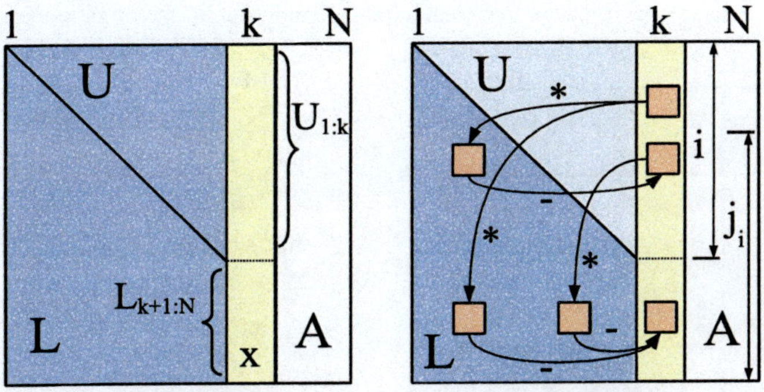

(Gilbert-Peierls) (Sparse L-Solve)

```
%--------------------------------
% input: sparse matrix A
% output: factored L and U
%--------------------------------
L=I; % I=identity matrix
for k=1:N
    b $=$ A(:,k);           % kth column of A
    x $=$ L \ b;            % \ is Lx=b solve
    U(1:k) $=$ x(1:k);
    L(k+1:N) $=$ x(k+1:N) / U(k,k);
end;
```

Listing .1 Gilbert-Peierls Algorithm (A=LU)

```
%--------------------------------
% input: matrix L (1:k-1)
% output: kth column of L
%--------------------------------
x=b;
% symbolic analysis predicts non-zeros
for i = 1:k-1 where x(i)$!$=0
    for j = i+1:N where L(j,i)$!$=0
        x(j) $=$ x(j) - L(j,i)*x(i);
    end;
end;
% returns x as result
```

Listing .2 Sparse L-Solve (Lx=b, x=unknown)

In Listing .1, we illustrate the key steps of the factorization algorithm. It is the Gilbert–Peierls [13] left-looking algorithm that factors the matrix column-by-column from left to right (shown in the figure accompanying Listing .1 by the sliding column *k*). For each column *k*, we must perform a sparse lower-triangular matrix solve shown in Listing .2. The algorithm exploits knowledge of nonzero positions of the factors when performing this sparse lower-triangular solve (the

Table 8 Sparse circuit matrix (*left*) and dataflow graph for *LU* factorization (*right*)

(1,1)			(1,4)
	(2,2)		(2,4)
(3,1)	(3,2)	(3,3)	(3,4)
		(4,3)	(4,4)

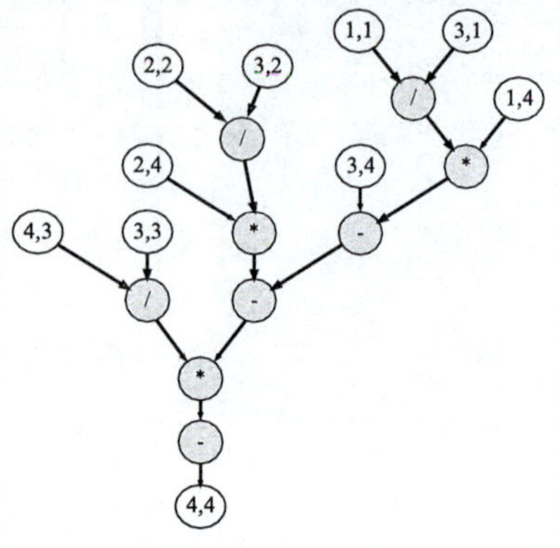

$x(i) \neq 0$ checks in Listing .2). This feature of the algorithm reduces runtime by only processing nonzeros and is made possible by the early symbolic analysis phase. It stores the result of this lower-triangular solve step in x (Line 8 of Listing .1). The kth column of the L and U factors is computed from x after a normalization step on the elements of L_k. Once all columns have been processed, L and U factors for that iteration are ready. From the pseudo-code in Listings .1 and .2 it may appear that the matrix solve computation is inherently sequential. However, we can visualize this computation as a dataflow graph by unrolling the loops from code listings. We show the dataflow graph corresponding to a small example matrix in Table 8. From the dataflow graph, we observe that there are two forms of parallel structure in the Matrix-Solve factorization computation that we can exploit in our parallel design: (1) factorization of independent columns organized into parallel subtrees and (2) fine-grained dataflow parallelism within the column. We now describe our parallel architecture capable of exploiting this parallelism.

4.2 *Token-Dataflow Architecture*

The Sparse Matrix-Solve computation can be represented as a sparse, irregular dataflow graph that is fixed at the beginning of the simulation. We recognize that static online scheduling of this parallel structure may be infeasible due to the prohibitively large size of these sparse matrix factorization graphs (millions of nodes and edges, where nodes are floating-point operations and edges are dependencies). Hence, we organize our architecture as a dynamically scheduled Token Dataflow [36] machine. This organization is capable of exploiting parallelism across a sparse, irregular graph with fully decentralized, distributed control. We automatically generate the dataflow graphs for LU factorization as well as Front/Back solve steps from symbolic analysis and evaluation of the sparse factorization

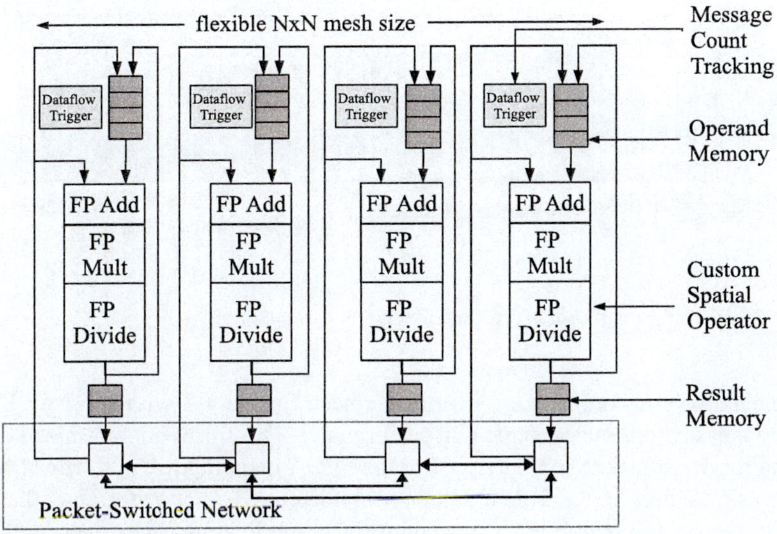

Fig. 11 Custom dataflow organization

in the KLU solver as shown in Fig. 18. Our parallel FPGA architecture, shown in Fig. 11, consists of multiple interconnected *Processing Elements* (PEs) each holding hundreds to thousands of graph nodes. We partition our graph across the PEs thereby assigning several nodes to each PE in our parallel architecture. We place nodes on the PEs using MLPart [3] to exploit locality. Each PE can fire a node dynamically based on a fine-grained dataflow triggering rule. This allows parallel evaluation of multiple graph nodes which have received all their inputs as computation proceeds down the graph. The *Dataflow Trigger* in the PE keeps track of ready nodes and issues operations when the nodes have received all inputs. Tokens of data representing dataflow dependencies are routed between the PEs over a packet-switched network. Each switch in the network is assembled using simple *split* and *merge* blocks as described in [23]. The switches implement dimension-ordered routing (DOR [11]) on a Bidirectional Mesh topology. The *Send Logic* in the PE injects messages into the network for nodes that have already been processed. For very large graphs, we partition the graph and perform static prefetch of the subgraphs from external DRAM. This is possible since the graph is completely feed forward. We show the performance possible with this architecture in Sect. 7.

4.3 Experimental Framework

In our Matrix-Solve experimental flow, we compare the performance of the optimized CPU implementation with the FPGA dataflow architecture. We first use `spice3f5` simulator with its Sparse 1.3 [24] solver to obtain a reference functional

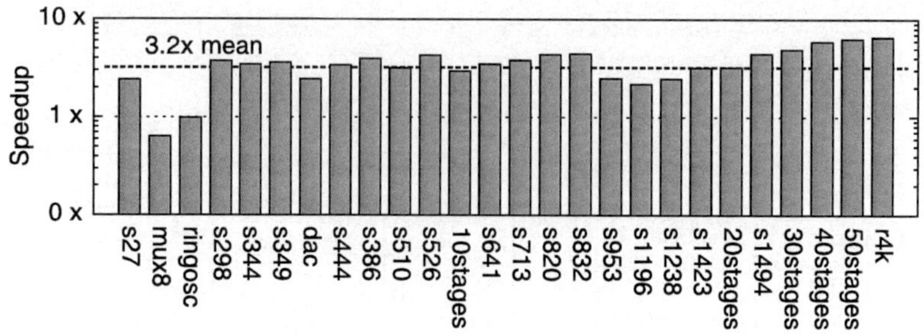

Fig. 12 Speedups for matrix-solve (vs. Core i7 965)

implementation for comparison. We then replace Sparse 1.3 with the new KLU solver to measure optimized sequential performance. This forms our optimized CPU baseline for performance comparison. For the FPGA mapping, we perform a cycle-accurate simulation of the Token dataflow architecture. For large graphs that do not fit in the onchip memories, we account for graph loading times over a DDR3 memory interface. We report cycle counts from the simulation to compute speedups. We use a rich and diverse set of benchmark circuit-simulation matrices detailed later in Appendix.

4.4 Evaluation

When we integrate the KLU matrix solver in `spice3f5` instead of the default Sparse 1.3 solver, we are able to speed up the software implementation by ≈35% across our benchmark circuits. We achieve higher improvements for larger benchmarks since the symbolic analysis overheads can be amortized easily for large matrices. We use this as our software baseline for comparing with the FPGA implementation. In Fig. 12, we compare the performance of our FPGA architecture implemented on a Virtex-6 LX760 with an Intel Core i7 965. We observe speedups of 0.6–6.5× (geomean 3.2×) for the 25-PE FPGA mapping that devotes all FPGA resources for Matrix-Solve acceleration over a range of benchmark matrices. For the complete SPICE system, we can only fit a 9-PE system for Matrix-Solve as discussed in Sect. 6. Our FPGA implementation allows efficient processing of the fine-grained factorization operations which can be synchronized at the granularity of individual floating-point operations. To better understand the speedups we plot the distribution of parallel runtime across the different steps of the Matrix-Solve implementation in Fig. 13. We observe that performance is dominated by the cost of loading the large dataflow graph from offchip memory. We may be able to reduce this overhead with better DRAM memory interfaces and higher on-chip capacity.

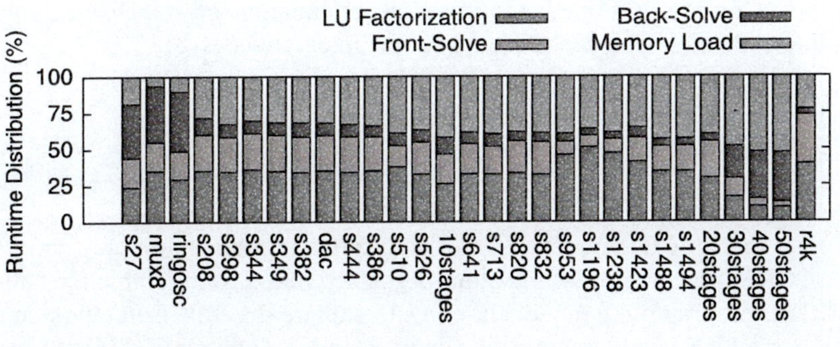

Fig. 13 Parallel runtime distribution for matrix-solve

4.5 Future Work

We now identify some ideas for achieved higher performance for the parallel Matrix Solve design.

1. We need to explore newer domain decomposition [37] and associative reformulation [18] strategies for improved scalability of the bottleneck Sparse Matrix-Solve phase of SPICE. With domain-decomposition, we can break up the large matrix into multiple submatrices that can be solved independently and possibly even distributed across multiple FPGAs.
2. Sparse matrix solve operations on large matrices can generate large dataflow graphs with millions of nodes and edges. These large graphs are challenging to distribute across multiple PEs. We can accelerate the placement algorithm itself using parallelism to minimize the one-time setup cost of the parallel simulation.
3. Apart from these approaches, it may be useful to consider completely different algorithms (iterative matrix-free fixed-point simulation [6] or constant-Jacobian [47]) for SPICE simulations that completely eliminate the need for performing per-iteration matrix factorization.

5 Iteration Control

In Sects. 3 and 4, we discussed the two computationally intensive phases of the SPICE simulator. In this section, we explain how to implement the sequential, control-intensive SPICE state-machines. We caution against mapping this sequential Iteration Control computation to a lightweight embedded microcontroller (e.g. Xilinx Microblaze, Altera NIOS) as it creates a performance bottleneck and decreases overall speedups. This is broadly true for any parallel application (including HPC problems), with a minimal degree of sequential control that could become a performance bottleneck. We discuss a streaming approach that will permit a high-level expression of the Iteration Control computation using the SCORE [4] framework. Our FPGA organization uses a combination of static and dynamic scheduling to deliver balanced speedups for the integrated design.

5.1 SCORE Framework

We express the SPICE Iteration Control algorithms in a stream-based framework called SCORE [4] (Stream Computation Organized for Reconfigurable Execution). The SCORE programming model allows us to capture the SPICE iteration control algorithm at a high-level of abstraction and permits exploration of different implementation configurations for the parallel SPICE solver. The streaming abstraction naturally matches the processing structure of the control algorithms and the overall composition of the solver. However, the SCORE compute model was originally designed for rapidly reconfigurable, time-multiplexed FPGAs. Modern FPGAs offer poor dynamic reconfiguration support and are unsuitable for the coarse-grained, dynamically reconfigurable implementation of SCORE. Consequently, we develop a new implementation model for SCORE based on resource-sharing and static scheduling. We adapt the backend flow from our Model-Evaluation infrastructure described in Sect. 3 to support dataflow graphs generated from the SCORE description of the Iteration Control computation.

SCORE allows description of streaming applications using dynamic dataflow. A SCORE program consists of a graph of operators (compute) and segments (memory) linked to each other via streams (interconnect). Computation within an operator is described as a finite-state machine (FSM). The operations within a state can be described as a dataflow graph, while the state machine transitions are captured using a state transition graph. This suits the control-intensive nature of the SPICE iteration control algorithm.

We show the high-level SCORE representation of the SPICE Iteration Controller in Fig. 14. We describe the control algorithms as SCORE operators and state-machines interconnected by streams. The stream connection allows pipelined, parallel evaluation of the different operators when possible. The white nodes in Fig. 14 represent the state-machine and breakpoint logic. For calculating convergence and local truncation error (LTE), we stream voltages, currents and charges through the operation graph for the respective equations. The gray nodes are the data-parallel stateless nodes that calculate LTE and compute convergence as a function of voltage $\mathbf{x}$, current $\mathbf{b}$ and charge $\mathbf{Q}$ vectors. We represent the Model-Evaluation and Sparse Matrix-Solve phases of SPICE as black boxes. Internally these are implemented differently using FPGA organizations described earlier.

In Table 9 we show the number of floating-point instructions and their types in the different SCORE operators. These statistics are obtained from the optimized operation graphs generated by `tdfc`, the SCORE compiler. As expected, we observe that the *If-Mux*, *Comparison*, and *Boolean* instructions constitute the bulk

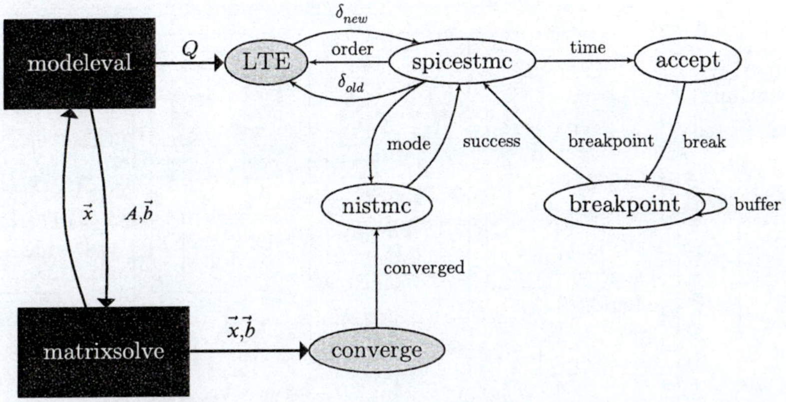

Fig. 14 High-level SCORE operator graph for `spice3f5`

Table 9 SCORE compiler optimized instruction counts for iteration control

Operator	Add	Mult.	Divide	Sqrt.	If-Mux	Cmp.	Bool	Rest	Total
`converge`	7	1	0	0	6	5	1	0	20
`LTE`	16	8	9	1	21	20	0	0	75
`breakpoint`	95	2	1	0	110	76	35	11	330
`nistmc`	2	0	0	0	8	7	5	2	24
`spicestmc`	29	15	6	0	79	42	24	17	212
Total	149	26	16	1	224	150	65	32	513

Column Rest includes floor, ceiling, and other special functions

Table 10 SCORE operator activation frequency for a simple resistor–capacitor-diode circuit

Operator	Total activations/iteration	Percent of total
`converge`	1,088,465	64.394
`LTE`	601,076	35.560
`accept`	299	0.017
`breakpoint`	48	0.002
`nistmc`	152	0.009
`spicestmc`	262	0.015

of the control-intensive computation in this phase of SPICE. We also note that we need only one *SQRT* floating-point operation and no other expensive elementary floating-point functions. In Table 10, we show the dynamic activation counts for the different SCORE operators in the Iteration Control phase of SPICE. An activation is when a state within that SCORE operator gets fired. We observe that the `LTE` and `Convergence` calculation dominate the dynamic activation counts.

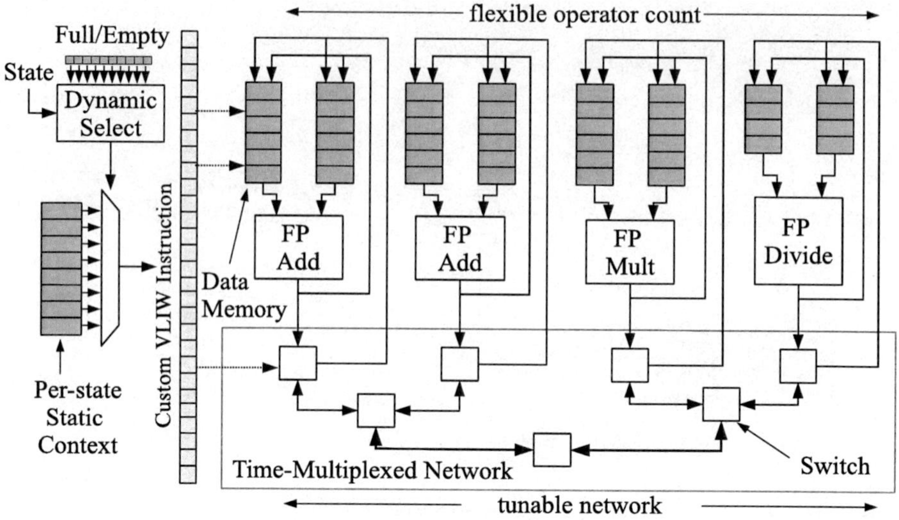

Fig. 15 Hybrid VLIW organization for iteration control

5.2 *Hybrid VLIW Architecture for Iteration Control*

Traditionally, FPGA designs offload the sequential control portion of a spatial design either to host CPUs or embedded Microblaze [45] controllers. Such techniques are unsuitable for stand-alone accelerator systems (no host CPU) or double-precision floating-point computation (poor support on Microblaze). Hence, we consider spatial designs that can implement this computation in the FPGA fabric directly. As discussed earlier, we observe that the computation is a combination of (1) data-parallel convergence detection and truncation error calculation and (2) sparsely activated, control-intensive SPICE analysis state-machine logic. We express this parallel structure using the streaming SCORE [4] framework and compile this parallelism to a hybrid VLIW architecture. The underlying FPGA architecture is organized as *tiles* (one *tile* is shown in Fig. 15) interconnected through streams. Each *tile* is a collection of floating-point operators (limited to add, multiply, divide and square-root) that are internally connected with a time-multiplexed network. Each operator is managed by a hybrid controller that dynamically selects between statically-scheduled configurations. The spatial mapping flow combines loop-unrolled, software-pipelined scheduling for data-parallel components like truncation error calculation and convergence detection logic along with dataflow scheduling for sparsely activated state-machine logic. The hybrid VLIW architecture is mostly similar to the Model-Evaluation design and we reuse its backend scheduling framework. The difference in this architecture is the support for limited dynamic processing. The data-parallel convergence detection, truncation error estimation operations, and sparsely activated individual state computations in the SPICE

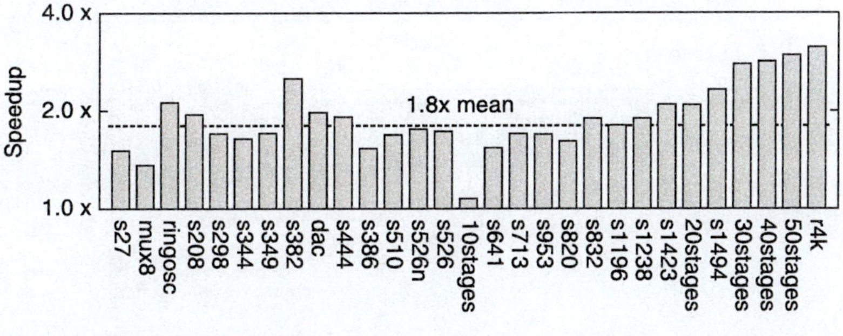

Fig. 16 Speedups for iteration-control (vs. Core i7 965)

analysis state-machines are statically scheduled in VLIW fashion to exploit dataflow parallelism. In contrast, the loop control state machine transition operations are evaluated dynamically using spatial implementation of state transition making this a hybrid VLIW design that combines static and dynamic scheduling.

5.3 Experimental Framework

We compare the performance of different partitioning strategies for implementing the Iteration Controller. We evaluate (1) CPU-FPGA (PCIe), (2) Microblaze-FPGA logic, and (3) our proposed hybrid VLIW-FPGA logic partitionings. For the CPU backend, we generate multi-threaded C++ code from the SCORE compiler [4, 9]. This also allows us to perform a functional software verification with `spice3f5`. We use PAPI to measure the CPU runtime of the Iteration Control phase. For the Microblaze backend, we develop a SCORE runtime customized for the Microblaze soft processor that enables stream operations. This is done through automated code-generation in a flavor of C suitable for use with an embedded operating system running on the Microblaze (Xilkernel [46]). We use a hardware counter to measure the Microblaze clock cycles. For the FPGA-VLIW mapping, we develop a code-generation backend for SCORE that uses the scheduler developed for the Model-Evaluation phase described previously in Sect. 3. We report cycle counts from the FPGA-VLIW scheduler.

5.4 Results

In Fig. 16, we show a 1.8× speedup for the Iteration Control phase of SPICE when comparing a spatial implementation with a host CPU-offload implementation. This suggests that a stand-alone FPGA accelerator execution can deliver better performance. We consider the impact of parallelizing the Iteration Control phase on

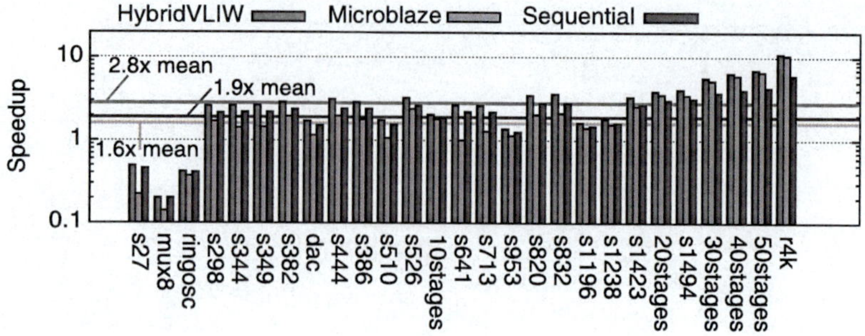

Fig. 17 Speedup for the overall SPICE simulator for different iteration control implementations

the overall speedups of the FPGA accelerator. In Fig. 17, we show the overall SPICE speedups under three implementation scenarios (1) offload to sequential host CPU over PCI (2) offload to Microblaze soft-processor, and (3) spatial implementation over hybrid VLIW design. We observe that the spatial implementation can deliver modest improvements of 2.8×(geomean) over the sequential CPU implementation. We can show this benefit by localizing all communication within the FPGA system and exploiting data parallelism in the convergence detection and truncation error calculation steps. However the amount of overall improvement is not very high since the Iteration Control phase accounts for merely ≈7% of sequential SPICE runtime. Other FPGA studies [38] prefer to implement such sequential fraction of the application on embedded soft-processors like the Xilinx Microblaze. We see the limits of using the Microblaze (1.6× geomean speedup) to implement this sequential computation as it can be worse than even offloading the processing to the host CPU over PCI (1.9× geomean speedup). The Microblaze soft-processor offers poor double-precision floating-point support and schedules computation sequentially over the ALU thereby limiting potential performance. In contrast, the spatial VLIW design exploits the available data parallelism and implements the state-machine processing with lightweight decision-making hardware thus delivering better performance.

5.5 *Future Work*

We now sketch a few ideas for improving the performance of the Iteration controller FPGA design.

1. We can overlap the Model-Evaluation phase with the Sparse Matrix-Solve phase of SPICE. The overlap scheduler needs to statically compute a suitable ordering of the device evaluation in Model-Evaluation to match the dataflow ordering in Matrix-Solve.

2. Both Xilinx and Altera have announced FPGA platforms that are closely coupled with fast sequential cores e.g. Xilinx-ARM Zynq platform and the Altera-Intel Atom platform. We need to reexamine the hardware–software partitioning problem for these platforms to determine if our hybrid VLIW architecture continues to offer better scalability.

6 FPGA Implementation Methodology

We now explain the complete methodology and framework for mapping and running SPICE simulations on an FPGA. As shown in Fig. 18, at a high-level, the SPICE user provides a SPICE netlist for simulation acceleration. Our automated FPGA mapping flow first selects a logic bitstream (marked ⓐ in Fig. 18) based on the type of the nonlinear device model being used e.g. `bsim3` or `bsim4` and the kind of SPICE analysis requested (e.g. DC and Transient analysis). We pre-compile a handful of

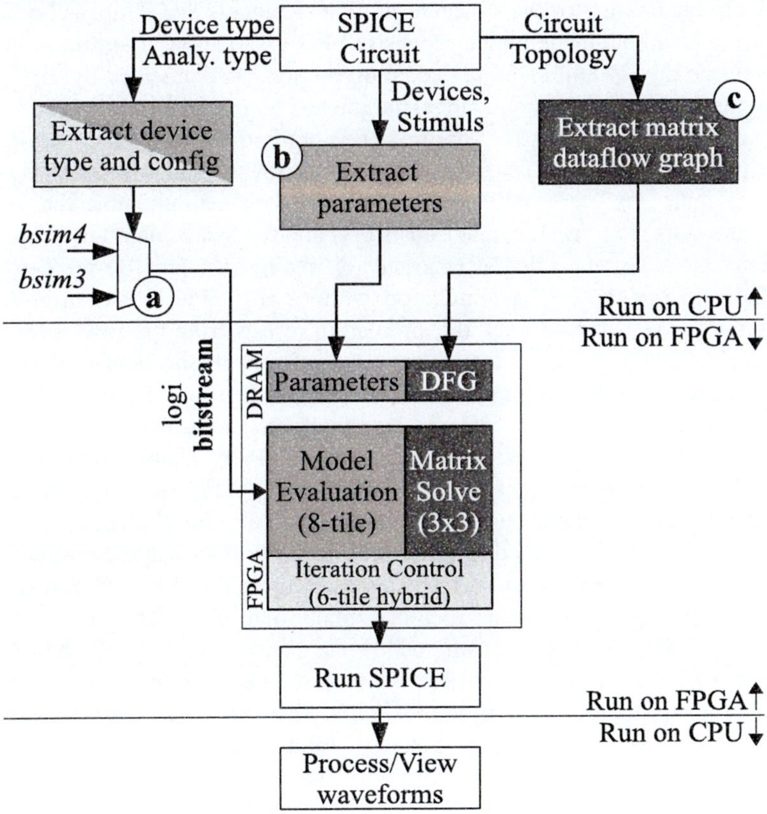

Fig. 18 High-level FPGA SPICE usage flow

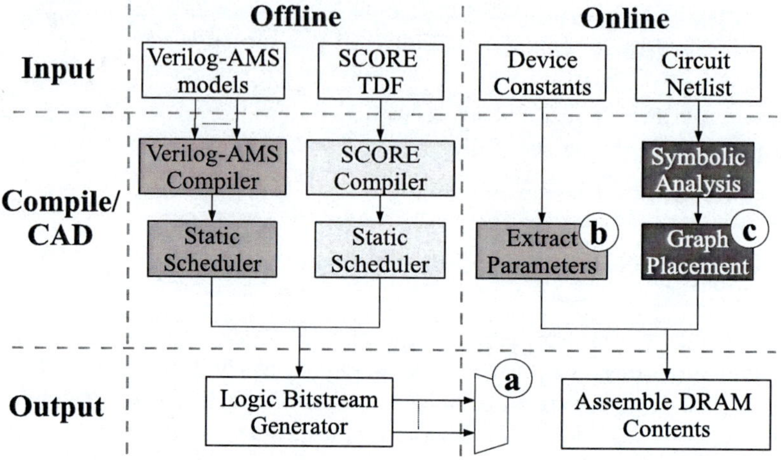

Fig. 19 FPGA SPICE mapping toolflow

FPGA logic bitstreams for the different nonlinear models and simply choose the right bitstream without invoking the expensive FPGA CAD flow at runtime. We pick an appropriate mix of nonlinear device configurations as demanded by the circuit with constant-folding of common model parameters (e.g. parameters specific to the CMOS process like Tox, Vth0) to optimize this configuration. We then exploit the circuit structure to build the sparse matrix and extract the dataflow graph through a one-time static analysis (marked ⓑ in Fig. 18). We then assemble the device-specific constants (e.g. W, L of the transistors) and the sparse dataflow graph into a memory image for the DRAM (marked ⓒ in Fig. 18). Finally, we configure the FPGA and run the FPGA simulation without any CPU intervention during the simulation. We can read back the simulation results from the FPGA for post-processing and analysis. At present, we are required to fit all phases of SPICE on the FPGA because dynamic reconfiguration is too slow to be useful for our benchmark circuits (reconfiguration itself takes 1–2 ms compared to a few milliseconds of FPGA SPICE iteration time). We show the complete FPGA mapping flow in greater detail in Fig. 19 and cross-label the key steps from Fig. 18. The mapping flow is organized into different paths customized for the specific SPICE phase.

In Fig. 20 we show how the different portions of the FPGA fabric co-operate to execute a SPICE simulation. In Step (①), we configure the FPGA with a suitable logic bitstream and download the memory image onto the DRAM to set up the simulation. In Step (②) we stream the device parameters through the Model-Evaluation VLIW circuit to process each device and generate the current (RHS $\mathbf{x}$) and conductance (matrix A) contributions while also checking for convergence in the Iteration-Control. These contributions are inputs to the Matrix-Solve phase in Step (③). Next, in Step (④), the dataflow graphs are streamed through the token dataflow architecture from the off-chip memory to solve for unknown (vector $\mathbf{x}$) along with a convergence check in the Iteration Control. The voltage vector ($\mathbf{x}$) is

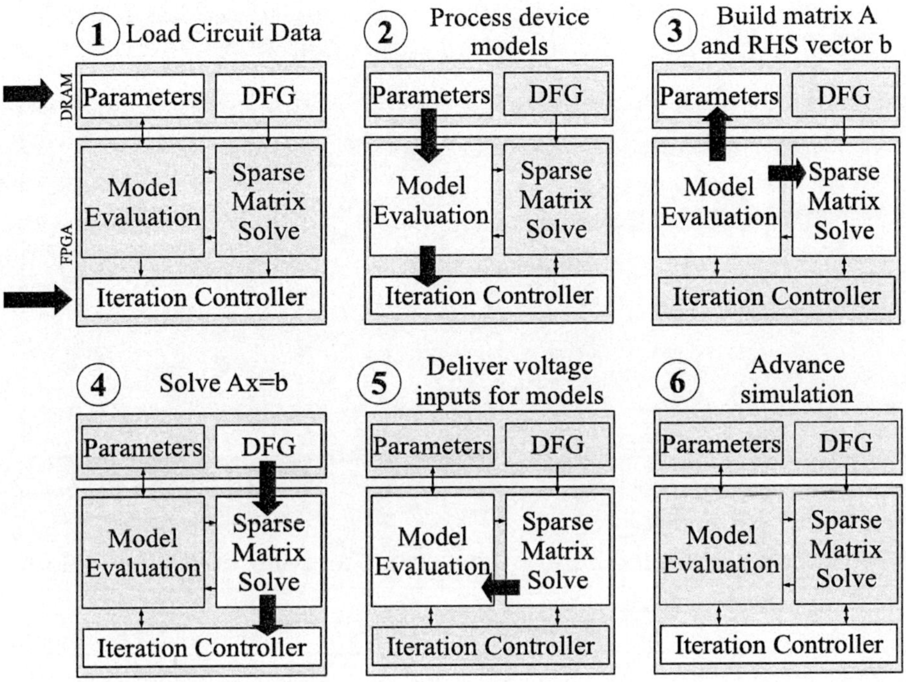

Fig. 20 FPGA SPICE execution flow

input for the next iteration of Model-Evaluation as shown in Step (⑤). The Iteration Controller runs the analysis state-machines that advance or terminate the simulation as appropriate.

6.1 *Offline Logic Configuration*

We generate the logic for implementing the VLIW, Dataflow and Streaming architectures by choosing an appropriate balance of area and memory resources through an area-time trade-off analysis. In Fig. 21, we show Area-Time trade-offs for the different phases of SPICE and pick a feasible configuration by rapidly evaluating all possible configurations in the small space of possible configurations. We mark the feasible configurations in Fig. 21 and show exact resource utilization of these feasible points in Table 11. For the composite design, we are restricted to an 8-tile VLIW engine for Model-Evaluation, a 6-tile VLIW engine for Iteration Control and a 3×3 tile architecture for Matrix-Solve. The FPGA logic configuration includes the VLIW programming for the PEs and switches of the Model-Evaluation and Iteration Control blocks.

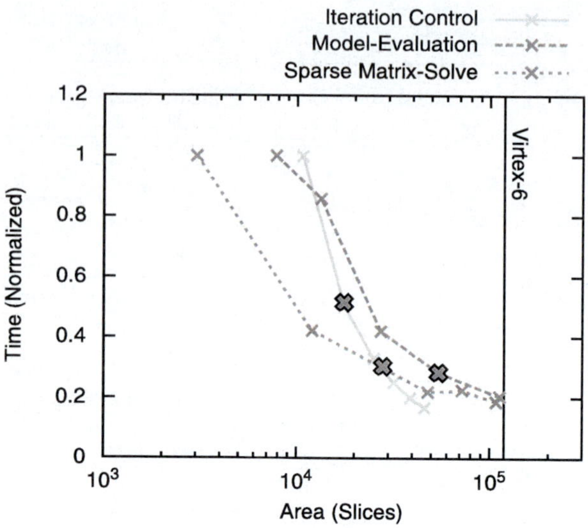

Fig. 21 FPGA SPICE area-time trade-offs

Table 11 FPGA resource distribution for complete SPICE solver (Virtex-6 LX760, bsim4 model)

	Area		Memory		DSPs	
SPICE phase	Slices	%	BRAMs	%	48E1	%
Model-evaluation	62,512	53	448	62	176	20
Sparse matrix-solve	27,090	23	180	25	99	11
Iteration control	17,848	15	32	5	77	9
Total	107,450	91	660	92	352	40

6.2 *Runtime Memory Configuration*

For each circuit, we must program memory resources to store the circuit-specific variables and data structures relevant for the simulation. This is primarily necessary to support the circuit-specific matrix factorization graph required for the Sparse Matrix Solve phase. For the nonlinear devices and independent sources, we store the device-specific constant parameters from the circuit netlist in FPGA on-chip memory or off-chip DRAM memory if necessary. We load a few simulation control parameters (e.g. `abstol`, `reltol`, `final_time`) to help the Iteration Control phase declare convergence and termination of the simulation. We also need to generate a static dataflow graph for the Matrix-Solve phase at the start of the simulation through symbolic analysis. We distribute the sparse dataflow graph across the Matrix-Solve processing elements (shown by the "Graph Placement" block in Fig. 19) and store the graph in off-chip DRAM memory when it does not fit on-chip capacity. We compute a static ordering of loads from the off-chip memory to appropriately stream the graph structure on-chip. Once we have the dataflow graphs, we assign nodes to PEs of our parallel architecture using placement for locality with MLPart [3].

Table 12 Area and latency model for SPICE hardware (Virtex-6 LX760), multiply block uses 11 DSP48 units

Block	Area (slices)	Latency (clocks)	Speed (MHz)	Ref.
Double-precision floating-point operators				
Add	334	8	344	[43]
Multiply	131	10	294	[43]
Divide	1,606	57	277	[43]
Square root	822	57	282	[43,44]
Exponential	1,022	30	200	[8]
Logarithm	1,561	30	200	[8]
Network elements				
TM BFT T-Switch	48	2	300	[23,31]
TM BFT Pi-Switch	64	2	300	[23,31]
PS Mesh Switch	642	4	312	–
Switch-Switch	32	2	300	–
Processing elements and miscellaneous				
VLIW Tile Ctrl.	82	–	300	–
Dataflow PE Ctrl.	297	–	270	–
Microblaze Ctrl.	1,504	–	100	–
DDR2 Ctrl.	1,892	–	250	[32]

6.3 Hardware Library and Cost Model

We tabulate the resource requirements and performance characteristics of the compositional hardware elements in Table 12. We use spatial implementations of individual floating-point *add*, *multiply*, *divide*, and *square-root* operators from the Xilinx Floating-Point library in CoreGen [44]. For the *exponential* and *logarithm* operators we use FPLibrary from Arénaire [8] group. For the Model-Evaluation and Iteration Control architectures, we interconnect the operators using a time-multiplexed butterfly-fat-tree (BFT) network that routes 64-bit doubles (or 32-bit floats when considering Single-Precision implementation) using time-multiplexed switches. For the Matrix-Solve architecture, we interconnect the floating-point operators using a bidirectional mesh packet-switched network that routes 84-bit, 1-flit packets (64-bit double and 20-bit node address) using DOR. We use a hardware generation framework to automatically generate structural VHDL code for the system based on selected implementation parameters such as system size, network topology, and network bandwidth. The software infrastructure to support time-multiplexed scheduling and packet-switched simulation is extended to provide this hardware generation functionality. We store the static schedules as read-only constants in local on-chip distributed memories.

Fig. 22 Measuring FPGA cycle count

$$Cycles = max(T_{modeleval} + T_{matsolve}, T_{iterctrl}(dp)) + T_{iterctrl}(stmc)$$
$$T_{modeleval} = \text{VLIW Model-Evaluation cycles}$$
$$T_{matsolve} = \text{Dataflow Matrix-Solve cycles}$$
$$T_{iterctrl}(dp) = \text{Data-Parallel Iteration-Control cycles}$$
$$T_{iterctrl}(stmc) = \text{State-Machine Iteration-Control cycles}$$

6.4 FPGA Cycle Measurement

We express the total number of cycles required by our FPGA implementation as shown in Fig. 22. This model assumes we must fit all three phases of the SPICE iteration on the FPGA simultaneously. The model also assumes an overlapping of a part of the Iteration Control phase with the other two phases of SPICE. In our model, we only consider the execution times of SPICE iteration and exclude the initial simulation setup time (e.g. circuit parsing, matrix construction, matrix static analysis). This setup cost is small (usually proportional to 1–2 SPICE iteration times [20]) and is common to both sequential CPU and parallel FPGA implementations. This cost is easily amortized over sufficiently large number of iterations where the FPGA-SPICE accelerator is expected to be used.

We report cycle counts from the time-multiplexed schedule (Model-Evaluation and Iteration Controller) and a cycle-accurate simulation (Matrix-Solve). We estimate memory load time for large matrices using streaming loads over the external DDR2-500 MHz memory interface using lowerbound bandwidth calculations. For our speedup calculation, we consider graph loading times as well as vector and device constant loading times from external memory.

7 Evaluation

We report the achieved performance and energy requirements of our parallel SPICE implementation. In Fig. 23a, we compare SPICE runtime on an Intel Core i7 965 with a Virtex-6 LX760 FPGA across benchmark circuits of increasing sizes. We observe a geomean speedup of 2.8× across our benchmark set with a peak speedup of 11× for the largest benchmark. We also show the ratio of energy consumption between the two architectures in Fig. 23b. We estimate power consumption of the FPGA using the Xilinx XPower tool assuming 20% activity on the flip-flops, on-chip memory ports and external IO ports. When using this model, the FPGA consumes up to 40.9×(geomean 8.9×) lower energy than the microprocessor implementation. To the first order, observed speedup and energy benefits are proportional to the size of the benchmark. Larger benchmarks admit greater parallelism across multiple independent devices for the Model-Evaluation phase. Regular circuits with low fanout and high locality also generate good parallelism for the Sparse Matrix-Solve phase. Thus, the variations in acceleration

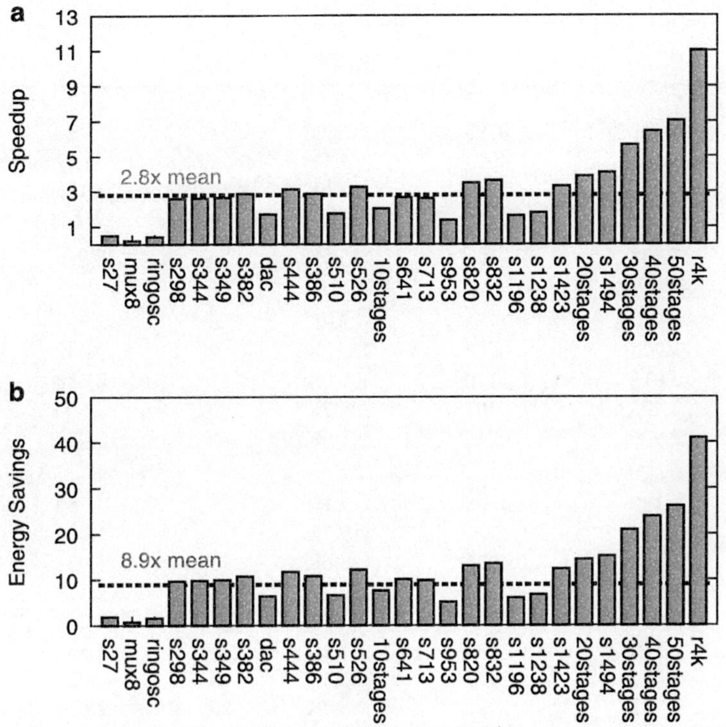

Fig. 23 Comparing Xilinx Virtex-6 LX760 FPGA (40 nm) and Intel Core i7 965 (45 nm) implementations. (**a**) Total per-chip speedup; (**b**) energy ratio

can be explained in terms of the size and differences in circuit structure across our benchmark set. We expect this accelerator to be particularly useful for speeding up SPICE simulations of large circuits (millions to billions of transistors) where the sequential implementations can take days or weeks or runtime. Our performance scaling trends in Fig. 23a do suggest a favorable increase in speedup with increasing circuit size.

8 Conclusions

We show how to use FPGAs to accelerate the SPICE circuit simulator up to an order of magnitude while also delivering an order of magnitude energy reduction when comparing a Xilinx Virtex-6 LX760 with an Intel Core i7 965. We were able to deliver these speedups by exposing available parallelism in all phases of SPICE using a high-level, domain-specific framework and customizing FPGA hardware to match the nature of parallelism in each phase. We were able to compose the overall

Table 13 Circuit simulation benchmark matrices

Bmarks.	Matrix size	Sparsity (%)	Total ops.	Fanout	Fanin	Latency (cycles)
Simucad [39]						
mux8	42	15.0793	626	8	20	1.9K
ringosc	104	6.4903	1.6K	4	92	3.7K
dac	654	1.5849	23.6K	10	1,136	7.7K
Clocktrees [40]						
r4k1	39,948	0.0131	515.5K	6	29,910	127.8K
Wave-pipelined Interconnect [41]						
10stages	3,920	0.1753	72.7K	8	2,384	18.6K
20stages	11,225	0.0618	219.2K	9	9,442	46.2K
30stages	16,815	0.0410	306.0K	11	4,688	88.6K
40stages	22,405	0.0307	395.7K	9	600	134.2K
50stages	27,995	0.0245	493.9K	10	484	169.7K
ISCAS89 Netlists [2]						
s27	189	3.4405	2.1K	6	50	3.6K
s208	1,296	0.5277	19.7K	11	1,414	11.3K
s298	1,801	0.4026	32.6K	13	1,938	13.1K
s344	1,992	0.3522	32.3K	12	2,178	14.7K
s349	2,017	0.3512	33.9K	14	2,218	14.7K
s382	2,219	0.3184	37.2K	16	2,358	16.1K
s444	2,409	0.2952	41.4K	16	2,526	16.6K
s386	2,487	0.2927	46.4K	20	2,626	15.7K
s510	2,621	0.3124	105.3K	54	2,722	21.4K
s526n	3,154	0.2362	66.1K	25	3,280	21.9K
s526	3,159	0.2376	68.1K	26	3,294	20.7K
s641	3,740	0.2000	100.2K	39	4,066	26.5K
s713	4,040	0.1890	126.4K	47	4,380	30.3K
s820	4,625	0.1655	103.2K	29	4,766	26.1K
s832	4,715	0.1629	105.7K	29	4,846	26.6K
s953	4,872	0.1876	353.9K	85	5,212	37.9K
s1196	6,604	0.1399	475.3K	83	7,146	46.4K
s1238	6,899	0.1325	457.9K	78	7,454	46.6K
s1423	9,304	0.0820	296.0K	64	10,384	64.5K
s1488	9,849	0.0827	354.7K	49	10,606	54.8K
s1494	9,919	0.0817	352.4K	50	10,646	54.6K

heterogeneous design that mixes VLIW, Dataflow and Streaming organizations into a unified implementation with the assistance of suitable SCORE composition framework. The tools and techniques we develop for mapping SPICE to FPGAs are general and applicable to a broader range of designs. We believe the ideas explored in this research are relevant across an important class of problems where computation is characterized by static, data-parallel processing and where the algorithm operates on sparse, irregular data structures. Such high-level approaches based on exploiting spatial parallelism will become important for improving performance and energy-efficiency of general-purpose computation.

Appendix

We show the matrix characteristics of the circuit benchmarks used in our experiments in Table 13. We use RAM netlists (Simucad [39]), clocktrees (University of Michigan [40]), wave-pipelined circuits (UBC [41]) and the ISCAS 1989 benchmark set (IBM [2]).

References

1. A.M. Bayoumi, Y.Y. Hanafy, Massive parallelization of SPICE device model evaluation on GPU-based SIMD architectures, in *Proceedings of the 1st International Forum on Next-Generation Multicore/Manycore Technologies*, Cairo, Egypt (ACM, New York, 2008), pp. 1–5
2. F. Brglez, D. Bryan, K. Kozminski, Combinational profiles of sequential benchmark circuits. IEEE Int. Symp. Circ. Syst. **3**, 1929–1934 (1989)
3. A. Caldwell, A. Kahng, I. Markov, Improved algorithms for hypergraph bipartitioning, in *Proceedings of the 2000 Asia and South Pacific Design Automation Conference* (2000), pp. 661–666
4. E. Caspi, Design Automation for Streaming Systems. Ph.D., University of California, Berkeley, 2005
5. Chung-Wen Ho, A. Ruehli, P. Brennan, The modified nodal approach to network analysis. IEEE Trans. Circ. Syst. **22**(6), 504–509 (1975)
6. B. Conn, XPICE Circuit Simulation Software. (unpublished) (2008)
7. L. Dagum, R. Menon, OpenMP: an industry standard API for shared-memory programming. IEEE Comput. Sci. Eng. **5**(1), 46–55 (1998)
8. F. de Dinechin, J. Detrey, O. Cret, R. Tudoran, When FPGAs are better at floating-point than microprocessors, in *Proceedings of the International ACM/SIGDA Symposium on Field-Programmable Gate Arrays* (ACM, New York, NY, USA, 2008), p. 260
9. A. Dehon, Y. Markovsky, E. Caspi, M. Chu, R. Huang, S. Perissakis, L. Pozzi, J. Yeh, J. Wawrzynek, Stream computations organized for reconfigurable execution. Microprocess. Microsyst. **30**(6), 334–354 (2006)
10. M. DeLorimier, N. Kapre, N. Mehta, D. Rizzo, I. Eslick, R. Rubin, T.E. Uribe, T.F.J. Knight, A. DeHon, GraphStep: a system architecture for sparse-graph algorithms, in *IEEE Symposium on Field-Programmable Custom Computing Machines* (IEEE, Piscataway, NJ, USA, 2006), pp. 143–151
11. J. Duato, S. Yalamanchili, N. Lionel, *Interconnection Networks: An Engineering Approach* (Morgan Kaufmann, Los Altos, 2002)
12. J.A. Fisher, The VLIW machine: a multiprocessor for compiling scientific code. IEEE Comput. **17**(7), 45–53 (1984)
13. J. Gilbert, T. Peierls, Sparse partial pivoting in time proportional to arithmetic operations. SIAM J. Sci. Stat. Comput. **9**(5), 862–874 (1988)

14. K. Gulati, J.F. Croix, S.P. Khatri, R. Shastry, Fast circuit simulation on graphics processing units, in *Proceedings of the Asia and South Pacific Design Automation Conference* (IEEE, Piscataway, NJ, USA, 2009), pp. 403–408
15. J. Hennesey, D. Patterson, *Computer Architecture A Quantitative Approach*, 2nd edn. (Morgan Kauffman, Los Altos, 1996)
16. S. Hutchinson, E. Keiter, R. Hoekstra, H. Watts, A. Waters, R. Schells, S. Wix, The Xyce parallel electronic simulator - An overview, in *IEEE International Symposium on Circuits and Systems* (IEEE, Piscataway, NJ, USA, 2000)
17. Intel, Intel Math Kernel Library 10.2.5.035 (Intel, USA, 2005)
18. N. Kapre, A. DeHon, Optimistic parallelization of floating-point accumulation, in *IEEE Symposium on Computer Arithmetic* (IEEE Computer Society, Washington DC, USA, 2007), pp. 205–216
19. N. Kapre, A. DeHon, Accelerating SPICE model-evaluation using FPGAs, in *IEEE Symposium on Field Programmable Custom Computing Machines* (IEEE, New York, 2009), pp. 37–44
20. N. Kapre, A. DeHon, Parallelizing sparse matrix solve for SPICE circuit simulation using FPGAs, in *International Conference on Field-Programmable Technology* (IEEE, Piscataway, NJ, USA, 2009), pp. 190–198
21. N. Kapre, A. DeHon, Performance comparison of single-precision SPICE model-evaluation on FPGA, GPU, Cell, and multi-core processors, in *International Conference on Field Programmable Logic and Applications* (IEEE, Piscataway, NJ, USA, 2009), pp. 65–72
22. N. Kapre, A. DeHon, VLIW-SCORE: beyond C for sequential control of SPICE FPGA acceleration, in *International Conference on Field-Programmable Technology* (IEEE, Piscataway, NJ, USA, 2011)
23. N. Kapre, N. Mehta, M. DeLorimier, R. Rubin, H. Barnor, M. Wilson, M. Wrighton, A. DeHon, Packet switched vs. time multiplexed FPGA overlay networks, in *IEEE Symposium on Field-Programmable Custom Computing Machines* (IEEE, Piscataway, NJ, USA, 2006), pp. 205–216
24. K.S. Kundert, A. Sangiovanni-Vincentelli, Sparse User's Guide: A Sparse Linear Equation Solver (1988)
25. P. Lee, S. Ito, T. Hashimoto, J. Sato, T. Touma, G. Yokomizo, A parallel and accelerated circuit simulator with precise accuracy, in *Proceedings of the 2002 Asia and South Pacific Design Automation Conference* (IEEE, Piscataway, NJ, USA, 2002), pp. 213–218
26. L. Lemaitre, G. Coram, C. McAndrew, K. Kundert, M. Inc, S. Geneva, Extensions to Verilog-A to support compact device modeling, in *Proceedings of the Behavioral Modeling and Simulation Conference* (IEEE, Piscataway, NJ, USA, 2003), pp. 7–8
27. D. Lewis, A programmable hardware accelerator for compiled electrical simulation, in *Proceedings of the 25th ACM/IEEE Design Automation Conference* (IEEE, Piscataway, NJ, USA, 1988), pp. 172–177
28. D. Lewis, A compiled-code hardware accelerator for circuit simulation, in *IEEE Transactions on Computer-Aided Design of Integrated Circuits and Systems* (IEEE, Piscataway, NJ, USA, 1992), pp. 555–565
29. M. Linderman, M. Ho, D. Dill, T. Meng, G. Nolan, Towards program optimization through automated analysis of numerical precision, in *Proceedings of the IEEE/ ACM International Symposium on Code Generation and Optimization* (ACM, New York, 2010), pp. 230–237
30. H. Martorel, N. Kapre, FX-SCORE: a framework for fixed-point compilation of SPICE device models using Gappa ++, in *IEEE Symposium on Field Programmable Custom Computing Machines* (IEEE, Piscataway, NJ, USA, 2012)
31. N. Mehta, *Time-Multiplexed FPGA Overlay Networks On Chip*. Master's thesis, California Institute of Technology, 2006
32. Microsoft Research, DDR2 DRAM Controller for BEE3 (Microsoft Research, USA, 2008)
33. P. Mucci, S. Browne, C. Deane, G. Ho, PAPI: a portable interface to hardware performance counters, in *Proceedings of the Department of Defense High Performance Computing Modernization Program Users Group Conference* (IEEE Computer Society, Washington DC, USA, 1999), pp. 7–10

34. L.W. Nagel, *SPICE2: A Computer Program to Simulate Semiconductor Circuits*. Ph.D. thesis, University of California Berkeley, 1975
35. E. Natarajan, *KLU A High Performance Sparse Linear Solver for Circuit Simulation Problems*. Master's thesis, University of Florida Gainesville, 2005
36. G. Papadopoulos, D. Culler, Monsoon: an explicit token-store architecture. Proc. Annu. Int. Symp. Comput. Archit. **18**(3a), 82–91 (1990)
37. H. Peng, C.K. Cheng, Parallel transistor level circuit simulation using domain decomposition methods, in *Proceedings of the Asia and South Pacific Design Automation Conference* (IEEE, Piscataway, 2009), pp. 397–402
38. A. Putnam, S. Eggers, D. Bennett, E. Dellinger, J. Mason, H. Styles, P. Sundararajan, R. Wittig, Performance and power of cache-based reconfigurable computing, in *Proceedings of the International Symposium on Computer Architecture*, vol. 37 (ACM, New York, 2009), p. 395
39. Simucad/Silvaco, BSIM3, BSIM4 and PSP benchmarks from Simucad (Simucad (now Silvaco), USA, 2007)
40. C. Sze, P. Restle, G. Nam, C. Alpert, ISPD2009 clock network synthesis contest, in *Proceedings of the 2009 International Symposium on Physical design* (ACM, New York, 2009), p. 149
41. P. Teehan, G. Lemieux, M. Greenstreet, Towards reliable 5Gbps wave-pipelined and 3Gbps surfing interconnect in 65nm FPGAs, in *Proceeding of the ACM/SIGDA International Symposium on Field Programmable Gate Arrays* (ACM, New York, 2009), pp. 43–52
42. Q. Wang, D.M. Lewis, Automated field-programmable compute accelerator design using partial evaluation, in *Proceedings of the 5th Annual IEEE Symposium on FPGAs for Custom Computing Machines*, Napa Valley, 1997, pp. 145–154
43. Xilinx, Xilinx CoreGen Reference Guide, 2100 Logic Drive, SanJose, CA, 95124, USA (2000). www.xilinx.com
44. Xilinx, Floating-Point Operator v5.0, 2100 Logic Drive, SanJose, CA, 95124, USA (2009). www.xilinx.com
45. Xilinx, MicroBlaze Processor Reference Guide, 2100 Logic Drive, SanJose, CA, 95124, USA (2010). www.xilinx.com
46. Xilinx, OS and Libraries Document Collection. Technical report, 2100 Logic Drive San Jose, CA 95124, USA (2010). www.xilinx.com
47. X. Ye, W. Dong, P. Li, S. Nassif, MAPS: multi-algorithm parallel circuit simulation, in *Proceedings of the IEEE/ACM International Conference on Computer-Aided Design* (IEEE, Piscataway, NJ, USA, 2008), pp. 73–78

Part II
Architectures

The second part of the book is concerned with architectural developments in High-Performance Reconfigurable Computing. It starts with a contribution from Klauer of the Helmut Schmidt University, Germany, on Convey computing – one of the most ambitious and pragmatic commercial attempts to bring the power of FPGAs into mainstream computing. This is followed by a contribution from Castillo et al. from Spain which presents an academic effort to build a low cost high performance reconfigurable computer, called SMILE. Next, Baity-Jesi et al. from Spain and Italy present a domain-specific FPGA-based supercomputer used for statistical physics, called Janus.

The following three chapters deal with networking and communications issues in FPGA-based parallel systems. First, Nüssle et al. from the University of Heidelberg, Germany, present an FPGA-based low-latency interconnection network called EXTOLL. Then, Baier et al. from Germany and the USA present an FPGA-based high-speed torus interconnect and its implementation for two parallel machines, namely QPACE and AuroraScience. After that, Fröning et al. from the University of Heidelberg, Germany, and Universitat Politècnica de València, Spain, present an FPGA-based cluster memory architecture which allows for dynamic resource provisioning.

This part ends with a contribution from Minoru Watanabe of Shizuoka University, Japan, which presents a radically novel high performance reconfigurable computing paradigm based on high-speed optical dynamic reconfiguration.

The Convey Hybrid-Core Architecture

Bernd Klauer

Abstract *Hybrid Computing* is a term that has originally been used for computations performed on analog/digital hardware and was popular until the late 1970s. Complex computations done under realtime conditions, such as signal processing, were left in the analog domain as the conversion times of A/D converters, sampling rates, and clock speeds of processors were significantly too low to solve complex equations in reasonable times. Today, processor layouts still contain analog components in the I/O areas as amplifiers, sensors or A/D converters. In the area of logical or arithmetical computations they became absolutely irrelevant.

The renaissance that the term *Hybrid Computing* has experienced in recent years comes from a combination of hardwired multicore microprocessors and configurable integrated circuits (FPGAs[1]). This chapter focusses on Hybrid Computing and Hybrid Core Computing which is a special form of Hybrid Computing introduced by Convey Computer Corporation.[2]

1 Hybrid Computing

Two main features have driven the computer performance and development in the past:

- Clock speed and
- Parallelism

[1]Field Programmable Gate Arrays.

[2]Convey Computer Corporation, 1302 East Collins Boulevard, Richardson, TX 75081, www.conveycomputer.com.

B. Klauer (✉)
Helmut Schmidt University, University of the Federal Armed Forces of Germany,
Holstenhofweg 85, 22043 Hamburg, Germany
e-mail: bernd.klauer@hsu-hh.de

W. Vanderbauwhede and K. Benkrid (eds.), *High-Performance Computing Using FPGAs*, DOI 10.1007/978-1-4614-1791-0_14, © Springer Science+Business Media, LLC 2013

Clock speed was the main performance parameter until ~2000. Since then, the core-based parallelism has taken the lead performance parameter role. As Gordon Moore's law [26] is still applicable to predict the development of the density of transistors in future integrated circuits, theoretically processors with hundreds of cores may appear within the next 10 years.

Question 1. Can future microprocessors hundreds of cores really perform efficiently?

The first counter argument against large core counts is Amdahl's law [1, 16, 28]. Amdahl's law shows that each parallel algorithm has its maximum of processing elements that can efficiently speed up the algorithm. This observation can directly be translated into the fact that total applications that can efficiently take profit from a many core architecture decrease with growing core counts.

Only few applications can take advantage from large processor totals [20]. The second counter argument is that many core architectures are facing a lot of architectural problems by increasing core totals, such as bottlenecks in memory and I/O. Those simple observations lead to the prediction that core-based parallelism will face its borders, as clock speeds have in the last 10 years (2000...2010).

Question 2. How can the fact, that transistor totals will still increase in the near future, be utilised efficiently?

The integration of configurable components in computers and even into CPU architectures will be a key issue in the future competitions of performance. This can be derived from the following facts:

- Off-the-shelf computer architectures, even parallel, multi, and many-core architectures and dedicated graphics cores, need algorithms to be optimized and parallelized to perform efficiently on a parallel architecture. This leads to the fact that algorithms are partly modified to suit their environment. The hybrid computing approach works vice versa: Hybrid computing focuses on the optimization of hardware to optimally support algorithms. So the performance is gained from algorithms running on optimized hardware or algorithms that have directly been mapped into the hardware rather than algorithms optimized for a hardware architecture.
- As configurability also means reconfigurability the hardware features can be changed between different applications.
- Hybrid computing also means that there is a hardwired section that runs a standard operating system to provide a convenient user interface and to provide all standard interfaces from a desktop computer.
- The hardwired section can also be used as a development platform for configurations and programs—and also for applications that do not need support by configurability.

During the last decades we observed that initially computer performance was driven by clock speeds. Facing several physical walls, the enabling factor for performance moved from clock speed to parallelism. Taking a look at Amdahl's law

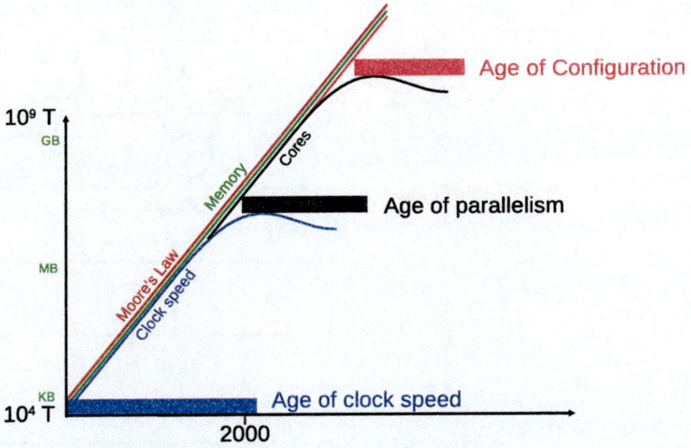

Fig. 1 Gordon Moore's law and its derivatives; T: Transistor total

and the opportunities to parallelize algorithms, we can expect that also parallelism will soon be limited for nonscientific computations.

Thus, computer performance will probably need to seek another driving factor. It is our prediction here that configurability can probably take over the role as enabling factor for computer performance as shown in Fig. 1. Another view on our prediction that future will be hybrid is given in [11, 12].

The reason for this assumption is the fact that a lot of potential parallelism cannot be exploited on hardwired parallel computers. The parallelism of algorithms is indicated by their dataflow graph. To parallelize an algorithm on a hardwired architecture means to map the dataflow graph as well as possible on the given topology of the parallel computer. Typical topologies were vectors, fields, cubes, and hypercubes. All those structures were regular structures suitable to accelerate algorithms like matrix computations perfectly. This leads us to our first observation that—besides some rectangular computation schemes—the dataflow graphs of typical algorithms are usually not as regular as the topology of the parallel computers. The second observation is that the granularity of parallelizable computations is usually not optimal to be mapped on hardwired parallel hardware. For example: Solve the following computation on parallel hardware: $abcd - 42 - wxyz$. We consider the dataflow graph as shown in Fig. 2.

The dataflow graph shows that the multiplications in the first row can perfectly be parallelized as well as the multiplications in the second row. The last two subtractions need to be computed sequentially. Applicable solutions with legacy hardwired architectures would be a VLIW programm with four instructions or a sequential program with eight instructions parallelized by a superscalar issue logic resulting in approximately the same dataflow-based parallelization. A multicore approach to improve performance by exploiting the parallelism of our example would not really be promising.

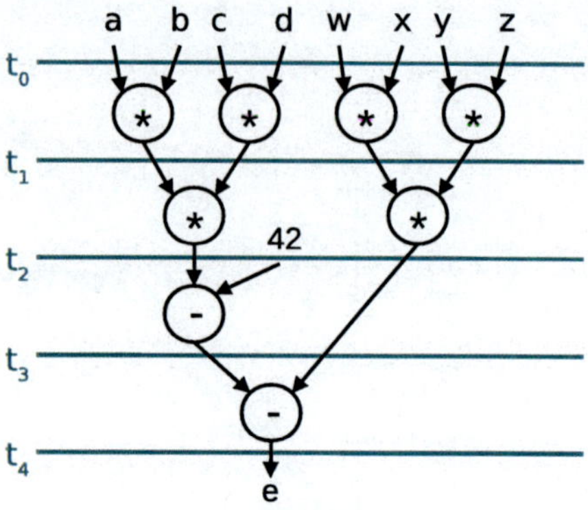

Fig. 2 Dataflow graph of the computation: $abcd - 42 - wxyz$

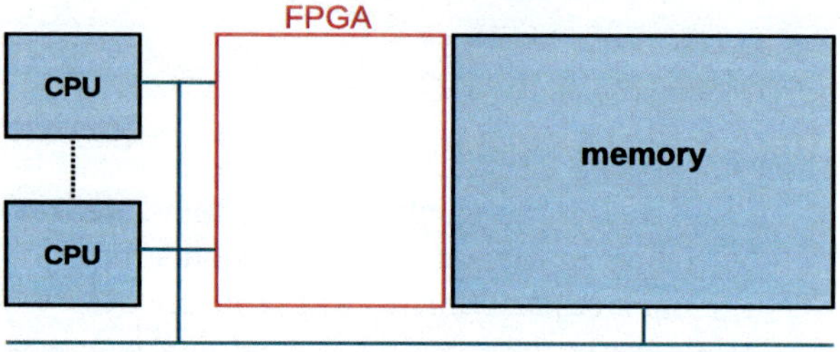

Fig. 3 Class 1. The FPGA is embedded in a legacy CPU environment as a configurable component with some kind of interface to open the FPGA for applications

The same problem solved on a machine with configurable capabilities would result in a digital circuit directly derived from the dataflow graph. Different approaches to connect standard processors to configurable logic (see Sect. 3) can then be used to provide the circuit in the configurable part of the computer with recent values and to collect the results.

2 Classification

We distinguish here three types of hybrid (configurable/hardwired) architectures.

1. Standard Processors which are tightly coupled with configurable circuits, as in Fig. 3
2. FPGAs with embedded hardwired processor kernels, as in Fig. 4
3. FPGAs with embedded processors configured on the FPGA. This class is also shown with Fig. 4. The difference is that the processors are not hardwired but subject to configuration.

We allow the last class in the sense that the processor core is part of the configurable space—but it will not be reconfigured or changed. It is used in the sense of the architecture listed above.

All three classes can host different types of hardware support. The next figures are showing different examples how configurability can be used in a hybrid environment (Figs. 5–8).

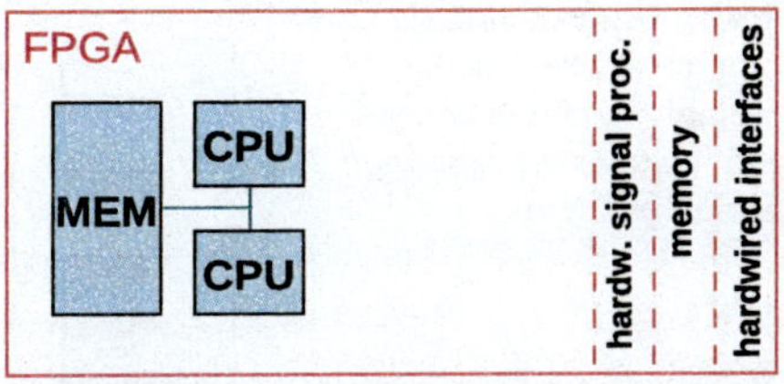

Fig. 4 This figure stands for class 2 and three hybrid architectures. Type one uses hardwired CPUs, type 2 uses soft cores configured on the FPGA

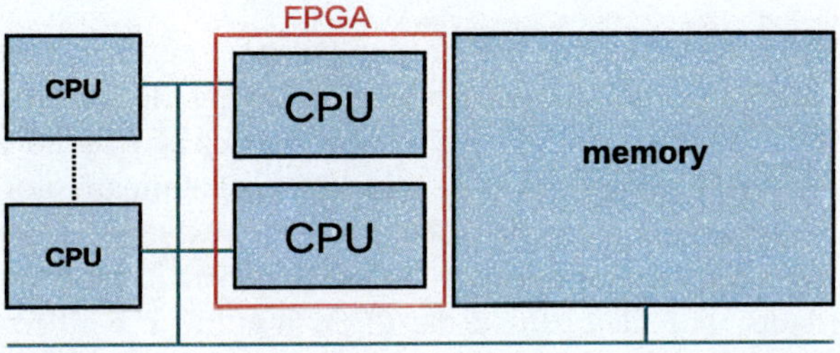

Fig. 5 This example shows two hardwired CPUs extended by two additional CPUs on the FPGA. The extensions can be of the same type as the hardwired CPUs as well as CPUs with a specific instruction set

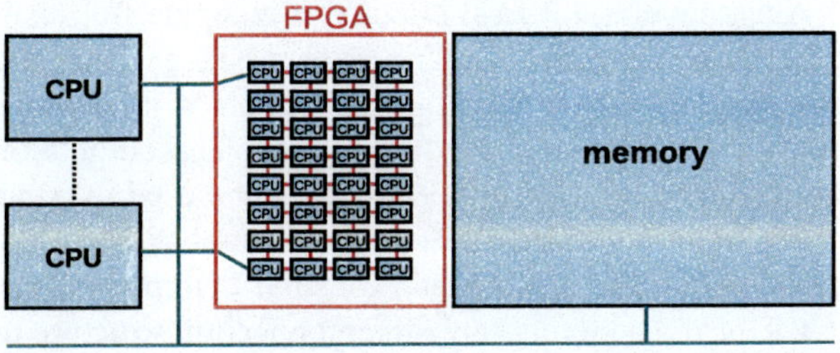

Fig. 6 This example extends the hardwired CPUs by a grid of CPUs on the FPGA. Such extensions have a simple but massively parallel executable instruction set

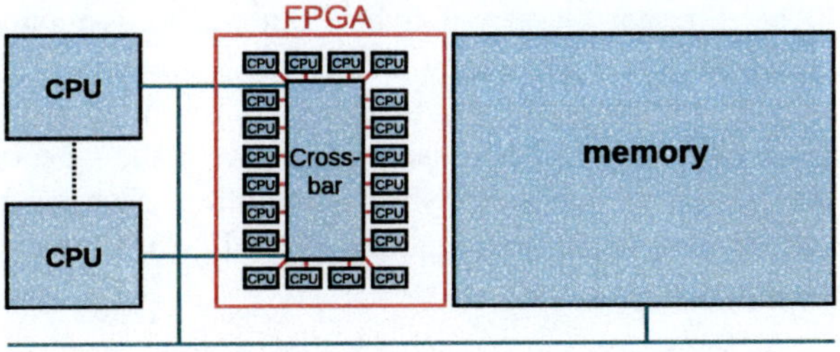

Fig. 7 This example is similar to Fig. 6 with a crossbar to support high communication bandwidth between the simple CPU modules

3 Related Work

The idea of hybrid computing in the sense of mixed hardwired and configurable hardware is not new. The first ideas have been published in the early 1960s [10] when (hardwired) integrated circuits were manufactured with some ten transistors.

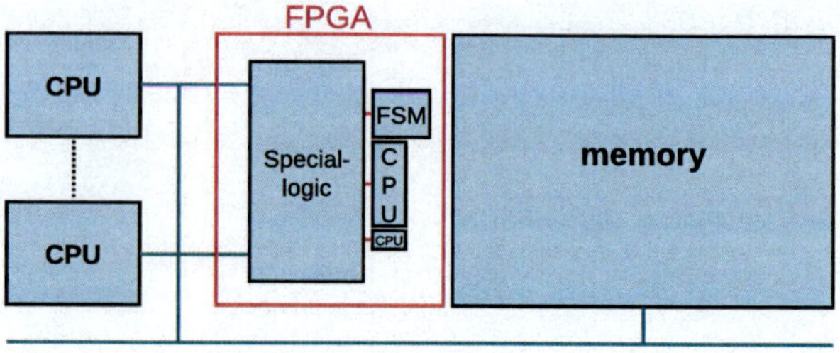

Fig. 8 This example shows a complete heterogeneous assembly of components working together to support a specific problem

Silicon circuits were supposed to be structures that were engraved into a stone-like substrate. In those days the idea of having electronic circuits with an electrically alterable structure and functionality was considered to be lunatic. In the late 1970s a new technology came up with the programmable logic devices (PLDs) such as programmable read only memories (PROMs), programmable array logic (PAL) or programmable logic arrays (PLAs) [19]. The PLDs were the first configurable logic circuits. The idea behind the PLDs was to store all results of a Boolean function (BF) in a memory (PROM). The literals of the BF were fed into the address signals. The memory then answered with the result of the BF.

Alternatively universal circuits to compute full disjunctive normal forms were used. They provided a plane with AND gates, a plane with OR gates and a switching network to feed the AND plane with the input literals and the OR plane with the outputs from the AND plane. The outputs of the AND plane were the results of the BF. The first PLDs could be used to compute only one BF. The first PLDs providing more than one configurable logic block (CLB) raised the need for interconnection schemes. The first commercially available FPGA, providing 64 CLBs and a configurable interconnection structure to connect the CLBs among themselves and to the I/O pins of the circuit was the XC2064 introduced by Xilinx in 1985.

The different classes of hybrid computers as mentioned in Sect. 2 have been implemented by connecting hardwired processors with FPGA technology or by embedding microprocessors into FPGAs. Before focusing on the Convey HC hybrid core computer in the next section we give a short overview on other architectures available in the area of hybrid computing. Surveys of recent architectures for (re)configurable computing are given in [7, 25, 30]. A nice overview on compiling techniques is given in [5].

Intel Stellarton

Intel Stellarton stand for an Intel Atom processor combined with an Altera FPGA in a single package. This arrangement allows the design of hardware optimized embedded systems on a low cost basis as glue logic or interface components can individually be designed into the component—without a specific full custom design [22].

Cray XD1

The XD1 consists of AMD Opteron 64-bit CPUs together with Xilinx Vertex-II Pro FPGAs. The FPGAs can be used to accelerate applications. A chassis with 12 CPUs can be used as building block for supercomputers: Up to 12 chassis can be combined in a rack. Racks can be combined to multirack systems [8, 32].

Axel Cluster

An Axel Cluster consists of so-called heterogeneous computing nodes (HCN) interconnected by fast standard network technologies (GB Ethernet or Infiniband). Each HCN consists of an x86 CPU, a GPU (up to 240 cores) and an FPGA on a PCIe bus [31].

Xilinx Zynq

With the Zync architecture Xilinx provides a variety of ARM Cortex-based designs combining hardwired processors and interface circuits with memory and configurable logic [34].

Garp

The Garp architecture [18] consists of a reconfigurable datapath attached as coprocessor to a MIPS processor. The data cache contains data for the MIPS processor as well as configuration parts for the FPGA section. In [3] it is shown how C compiling techniques can be applied to derive efficient configurations for the Garp architecture.

COPACOBANA

The *Cost-Optimized Parallel Code Breaker* COPACOBANA [24] consists of a backplane with sockets for FPGA modules. Six Xilinx Spartan-3 FPGAs fit into one FPGA module. The backplane provides power supply and communication signals and an FPGA based interface logic to connect the whole system to a PC-based host by USB. Although the architecture is published as special purpose configurable system for code breaking it is (as indicated by its name) a low cost solution for an FPGA-based computing system.

Novo-G

The Novo-G architecture [13, 14] is based on PCIe x8 boards. Each board has four Stratix-III E260 FPGAs. The FPGAs have high bandwidth interconnects and each FPGA is directly connected to two 2 GB DDR2 RAM modules. A Novo-G cluster consists of a network of PCIe backplanes with the FPGA modules.

The Berkeley Emulation Engine 2

The Berkeley Emulation Engine 2 (BEE2) [6] is an architecture with a hierarchical assembly of Xilinx Virtex II FPGAs. Each compute Module has five FPGAs with next neighbour connections. Each FPGA has a private RAM. Each module is connected to its adjacent modules and each module has a 10-GB network connection for non-adjacent communications and a 100-MB network connection to interface the modules with external I/O or memory.

Chimera

The Chimera architecture is a triple hybrid architecture as it encorporates not only a common multicore architecture and FPGAs but also a GPU cluster. The hardwired CPU, the GPU cluster and the FPGA cluster are interconnected by a high-speed backplane [21].

HDL Descriptions of Optimizable Processors for FPGAs

As shown in Sect. 2 hybrid computing can be performed by the configuration of processors on FPGAs. Such processors can be optimized due to application-based information such as a dataflow analysis or known types of parallelism. Coarse grain parallelism can be exploited by placing multiple CPUs on the FPGA. The total CPUs as well as their internal structure are subject to reconfiguration if the application changes [4, 15]. An architecture with self-optimizing properties is shown in [27]. Niyonkuru and Zeidler [27] describe a superscalar processor with an ARM instruction set. A configuration manager (an additional finite state machine) evaluates the load of the arithmetical and logical pipelines. As soon as it detects a need for special functional units it reconfigures the datapath of the ARM-compatible processor accordingly. In case of good reconfigurations the issue logic can issue more instructions concurrently to suitable functional units. This is a heuristical approach with good results in case of strong locality or in case of long sequences of identical instructions. In [17] this approach is enhanced in two ways. With the first enhancement the configuration manager is removed. Therefore the ARM instruction set is extended by operations to control the self-reconfiguration of parts of the ALU. A C-compiler (SUIF [33]) then performs dataflow and instruction flow analysis to decide the best performing configurations and the places in the instruction sequences where reconfiguration should occur.

4 The Convey HC-1 Hybrid Core Computer

4.1 Convey Computer Corporation

Convey Computer Corporation is a venture capital-based firm founded in 2006 by Steve Wallach, Bruce Toal, and Tony Brewer. The company designs and markets high performance servers based on a novel hybrid-core architecture that combines one or more x86 processors with reconfigurable hardware implementing application-specific instructions. The company's first product (the HC-1) was announced in 2008, and a later generation with higher density FPGAs (the HC-1ex) was announced in 2010.

4.2 The Convey HC-1 Architecture

The Convey HC family incorporates an Intel hardwired x86-64 processor and Xilinx FPGAs for reconfigurable hardware. The reconfigurable hardware is implemented as a coprocessor which shares physical memory and virtual addressing with the Xeon host processor. Figure 9 shows its building blocks.

The hardwired host processor components include a Xeon multicore processor, an Intel 5400 series memory controller hub, and Intel-based I/O system. The host processor executes the operating system, executes the sequential parts of applications, and initiates and controls execution of application-specific instructions implemented in the reconfigurable logic of the coprocessor. The FPGA logic designs that implement these instructions are referred to as *personalities* and are loaded dynamically by the operating system as needed by applications. Each personality has a unique ID and defines a particular set of semantics that can range from a single instruction that implements a specific function to an entire instruction set architecture that can be programmed like a conventional processor. Coprocessor instructions are embedded in a programs text region and executed by a transfer of control from a thread executing on an x86 referred to as a dispatch. Host x86 instructions and coprocessor instructions execute in the same process address space and have the same view of memory.

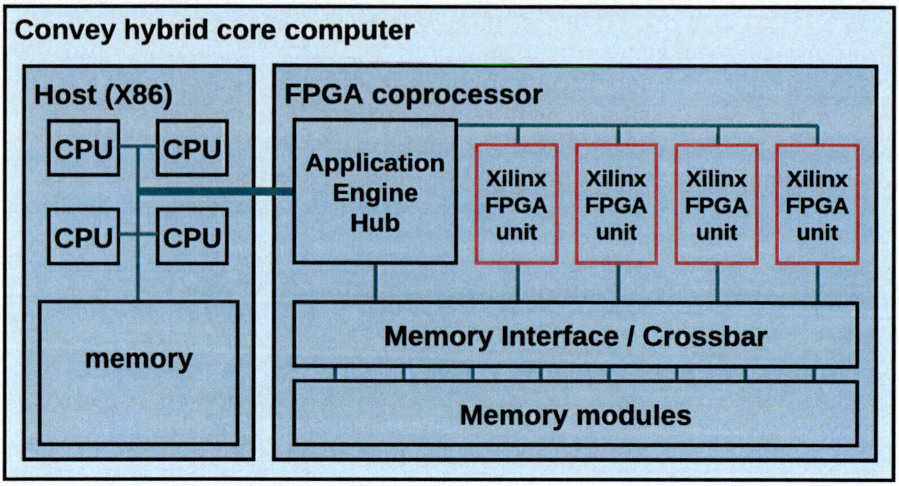

Fig. 9 The convey HC-1 hybrid core computer

Fig. 10 The Application Engine Hub (AEH) consists of multiple FPGAs that collectively implement the interface to the host system, route memory requests between the coprocessor and host, fetch and decode coprocessor instructions, and execute coprocessor scalar instructions. These functions are required for all personalities and are therefore not reconfigurable by users

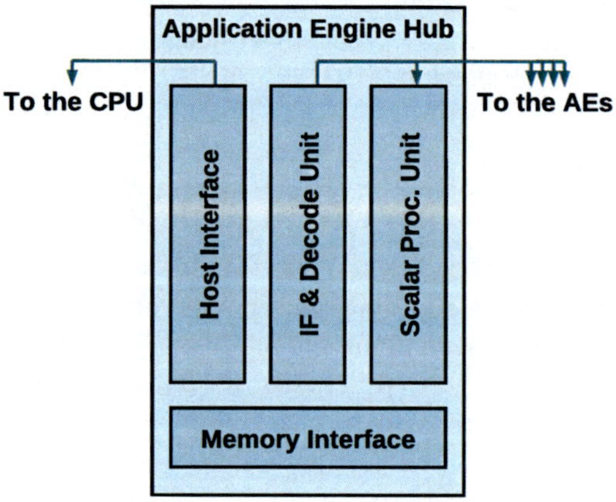

4.2.1 Coprocessor Functional Components

The HC-1 coprocessor (see Fig. 9) has three main building blocks, the Application Engine Hub (Fig. 10), the Application Engines, and the Memory Controllers.

Application Engines (AEs) execute application-specific instructions. There are four of them in the HC-1 and HC-1ex, and they are loaded dynamically by the OS with the configuration files that implement a personality.

Memory Controllers (MCs) are connected to the AEs and to the AEH by point-to-point links. There are eight memory controllers in the HC-1 and HC-1ex, each of which controls two DDR2 memory channels. Memory requests from the AEs consist of virtual addresses which are translated to physical addresses via

Translation Lookaside Buffers in the MC. The native unit of transfer for the memory system is a 64-bit word, and the memory channels support a Convey-designed DIMM type that divides the channel into 8 subchannels, each of which can transfer a word from a different address. This maximizes bandwidth for random and non-unit stride memory accesses. Like the AEH, the MCs implement a fixed set of functions that are common to all personalities and are not reloadable by the user. Moreover, data stored in the coprocessor memory is not affected by reloads of the AEs. This allows the OS to swap personalities without having to reload coprocessor memory.

4.2.2 Coprocessor Execution

Upon receipt of a dispatch, the AEH begins fetching and executing coprocessor instructions. Scalar instructions from the base instruction set are executed by the scalar engine in the AEH, while AE instructions are sent to all four AEs. Scalar instructions are defined by the coprocessor architecture and are part of the coprocessor infrastructure, but the semantics of AE instructions depends on the current personality. Both scalar instructions and AE instructions operate on virtual addresses and are constrained by the same memory protection mechanisms as x86 instructions. Memory requests containing virtual addresses are sent to the MCs, where they are translated to physical addresses. If the requests are to physical addresses in coprocessor memory, they are scheduled on the appropriate DIMM channel. A snoop filter containing tags for all coprocessor memory locations that are encached by the host processor ensures coherency with the host while minimizing snoop traffic. References to physical addresses in host memory are routed through the AEH to the host interface. A cache containing recently accessed host data reduces the latency for accesses to host data.

4.3 Application Development

The Convey architecture supports a heterogeneous model in which part of an application executes on the x86 cores in the host processor and part executes on the coprocessor. The design is intended to preserve the model of a multithreaded Linux process, without constraining the architecture of the algorithm implemented on the coprocessor. A variety of programming tools from Convey and from other vendors can be used to develop both the host-based portion of applications and the personalities that execute on the coprocessor. An overview on the design flow is given in Fig. 11. The HDL sources for the personality, typically VHDL or Verilog are compiled into configuration strings by an HDL compiler, e.g. Xilinx, Cadence, Mentor, and Synopsys. The configuration strings, see 12 are loaded into the application engines as personality. Their function can be envoked by coprocessor instructions executed in the AE hub (see Fig. 9).

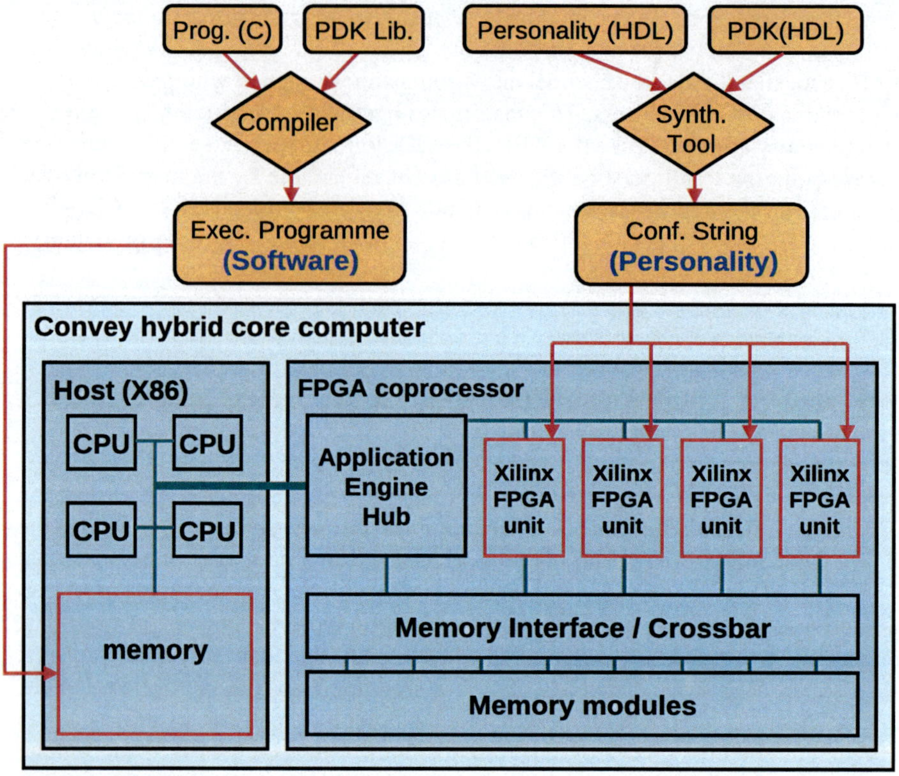

Fig. 11 The design flow for the convey hybrid core computers

Software runs through a standard software development flow with standard compilers. A library that comes with the PDK provides convenient interfacing to the AE hub and the login provided by the HDL description and is linked with the application objects.

4.3.1 Programming Model

The Convey model of execution assumes that execution by the coprocessor occurs on behalf of threads executing on the host and in the same process address space as those threads. A thread may invoke execution on the coprocessor via a "dispatch." A dispatch contains contains an address of coprocessor code to begin execution and parameters to be passed to the coprocessor routine and behaves like an asynchronous procedure call, except the routine being called contains coprocessor instructions instead of x86 instructions. Dispatches are handled one at a time by the coprocessor; multiple dispatches from different threads are queued and handled one by one.

```
> ls /opt/convey/personalities

1.1.1.1.0       32020.1.1.1.0  32201.1.1.2.0  6.1.1.1.0
2.1.1.1.0       32020.1.1.2.0  33001.1.1.1.0  7.1.1.1.0
2.1.1.2.0       32021.1.1.1.0  33001.2.1.1.0  7.1.1.2.0
32001.1.1.1.0   32021.1.1.2.0  4.1.1.1.0      db
32001.1.1.2.0   32030.1.1.1.0  4.1.1.2.0      DefaultBasePersonality
32002.1.1.1.0   32030.1.1.2.0  42.1.1.1.0     sampleDefaultBasePersonality
32002.1.1.2.0   32040.1.1.1.0  44444.1.1.1.0
32004.1.1.1.0   32040.1.1.2.0  5.1.1.1.0
32004.1.1.2.0   32201.1.1.1.0  5.1.1.2.0
```

Fig. 12 A typical directory with personalities

A coprocessor supports multiple instruction sets, but only one may be loaded at a time. The coprocess compares the instruction set required for each dispatch to the personality currently loaded—if they are different the required personality is loaded automatically by the OS.

Coprocessor instructions are divided into two types, scalar instructions and AE instructions. The scalar instruction set is implemented in the scalar engine and is architecturally defined to be common to all personalities. AE instructions are executed by the AEs; their function is defined by the currently loaded personality. AE instructions can range in functionality from simple scalar or SIMD instructions, or may initiate or control execution of a complex state machine that is free running and executes independently of the scalar engine.

There are few restrictions on the semantics of AE instructions, but they are constrained to operate on virtual addresses within the process address space of the controlling process. Since virtual to physical translation is implemented in the coprocessor infrastructure, AEs operate on virtual addresses and can operate on the same datastructures as the host threads, using the same pointers.

The coprocessor implements traps and exceptions that are passed to the operating system as if they were generated by a processor. The coprocessor can therefore generate floating point exceptions, address violations, page faults, and other exceptions just as a processor would. The coprocessor supports a context save mechanism that allows gdb to set breakpoints on coprocessor instructions and examine coprocessor state.

4.3.2 Creating and Using Personalities

Personalities in the Convey architecture are hardware designs that can be dynamically loaded on the Convey coprocessor and accessed via the dispatch mechanism described above. Physically they consist of the configuration files that are loaded into the AE FPGAs along with some additional information required by the Convey system software. Each personality is tagged by a 64-bit ID and stored in a common system directory so it can be loaded as needed by the OS. Figure 12 shows such a directory listing. Convey and other vendors sell prebuilt personalities that accelerate popular applications. These personalities may be installed and used without any hardware design.

Convey also licenses a Personality Development Kit that provides tools for users to create personalities based on their own logic designs. This kit includes interfaces to the Convey infrastructure components such as the memory system and scalar engine, along with useful components such as a memory crossbar. The system is open, and is supported by a variety of high level synthesis tools.

4.3.3 Application Programming

Convey systems run a variant of the Linux operating system, and application development is similar to other Linux systems. Programs may be compiled with the GNU compilers or with an enhanced version of the Open64 compiler suite from Convey. At the lowest level, personalities may be accessed via a procedure call-like mechanism from either gcc or the Convey compilers. This mechanism is most commonly used for personality designs that have just one or two instructions that perform very complex tasks. Threads running on the x86 set up data structures in memory, migrate or copy them to coprocessor memory, then call the coprocessor routine, passing pointers to the input data. The coprocessor routine runs until it has completely processed the data, then either waits on a semaphore in memory for more data, or returns control to the invoking host thread (possibly allowing another dispatch to be processed).

The Convey enhanced Open64 compilers provide additional ways to invoke personalities. Interfaces to coprocessor instructions can be coded as user-written intrinsics, allowing them to be called from optimized code. The compilers also support automatic vectorization and code generation for the Convey extended vector instruction set. This allows customized, complex SIMD instructions to be called from with vector loops, taking advantage of the compiler's dependence analysis and vectorization to manage the distribution of work across multiple function pipes on the AEs.

4.4 Operating System

The Operating System has not been taken under consideration with a focus on configurability. The Convey approach is a runtime library providing functions to load configuration strings into the application engines of the configurable part of the machine. The operating system aspects of configuration are subject to current research. The following services need to be migrated into the OS:

- From the viewpoint of an operating system the configuration space of the FPGAs is a resource to be managed similar to memory. OS functionality for the allocation of config space as well as loading and unloading of configurations needs to be integrated into common operating systems. As different applications may profit from the config space concurrently the reconfigurations need to be done without conflicting running applications.

- As whole processors may be configured into FPGAs it will be possible to revolutionize virtualization. Current virtual machines take profit from emulation with more or less hardware support. Taking into account that FPGAs can host whole processors, virtualization procedures by emulation can be replaced by configuration—leading back to real processors to be provided as configuration as soon as they are requested.

5 Applications and Performance

The personalities that accelerate key algorithms rely on parallelism rather than a high clock rate to achieve performance. High performance personality designs typically through the use pipelining and replication of functional units.

The reconfigurability of the HC-1 coprocessor allows implementation of instructions that are much more complex than the instructions typically found in a hardwired CPU. Coprocessor instructions are often highly pipelined state machines (referred to as function pipes) that implement the same function as dozens or hundreds of x86 instructions, yet are still capable of producing a result each clock cycle. This allows an application-specific design that operates at a relatively slow clock rate to deliver higher performance than a conventional processor executing simpler instructions at a much higher clock rate. The memory system of the coprocessor is also highly parallel and capable of delivering many operands to the AEs and/or storing results on each cycle. To fully saturate the memory system the function pipes are typically replicated to match the available bandwidth. The function pipes do not necessarily operate in lock step and may generate irregular patterns of access. The HC-1 memory system is designed to sustain high bandwidth for random or non-sequential memory access. Figure 13 illustrates this with a relatively simple gather operation.

```
Table2[i] = Table1[Index[i]];
```

The above operation generates three streams of memory requests:

- One sequential load,
- One random load,
- And one sequential store.

On a cache-based system the random load produces a very low effective bandwidth if Table1 is larger than the cache, as the system must transfer an entire cache line to deliver one operand. The HC-1 with its cacheless design delivers a much higher effective bandwidth.

Fig. 13 Memory bandwidth

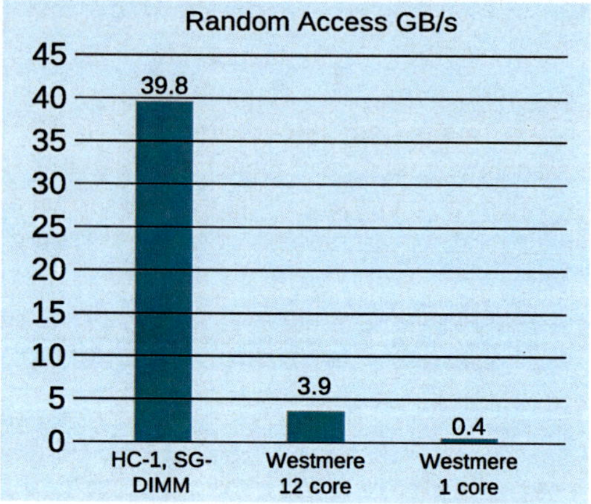

The combination of high functional parallelism and efficient support of random memory accesses is particularly well suited to algorithms which match queries against a large reference, such as those used in genome assembly and alignment. An example of an algorithm that is particularly well suited to acceleration is the Smith–Waterman algorithm [29], which used to compare strings that may have substitutions, insertions, or deletions against a reference database.

Convey's SWSearch Smith–Waterman alignment tool is a protein or nucleotide sequence search program that relies on a hardware personality to perform large numbers of local alignments in parallel. Figure 14 gives an impression of the hardware supported Smith–Waterman implementation. Each of the AEs is divided into a number of tiles that can operate independently to compare different query strings or be combined to process larger strings. On the Convey HC-1ex with Virtex-6 LX760 FPGAS, up to 1,280 characters can be processed on each FPGA on each 150 MHz cycle.

The SWSearch tool preloads the reference database into coprocessor memory for fast access. It then reads query sequences from a query file and assigns them to tiles or sets of tiles according to their length. The tiles then begin loading the references and scoring them. Query sequences too large to be processed by the largest grouping of tiles are aligned using a software implementation of the Smith–Waterman algorithm on the host processors.

A performance test which compared 1,000 randomly chosen proteins against the entire non-redundant (nr) protein database is shown in Fig. 15. The hardware implementation on an HC-1ex to be 13.2 times faster than an SSE2-based implementation (ssearch) running on a fast (2.93 GHz) 12 core x86 system. The newer HC-2ex was 14.5 times faster. This performance is a function of the very large number of cell updates (5,120) that can be performed on each cycle by the hardware.

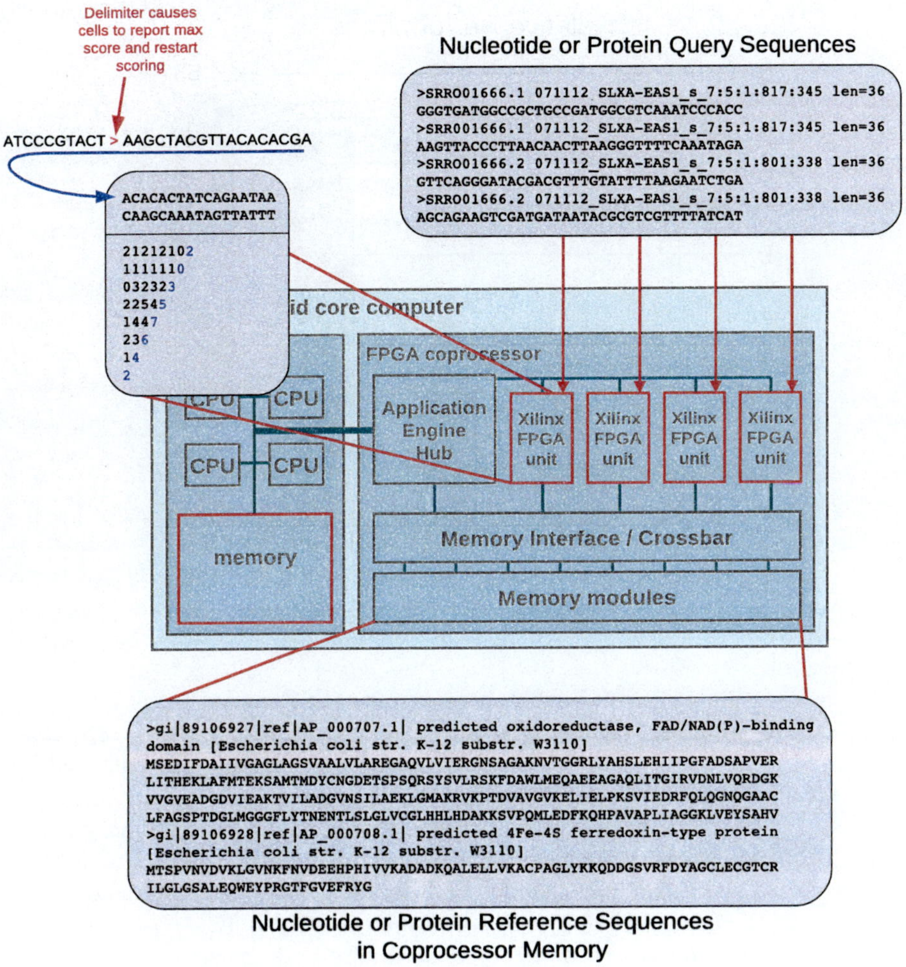

Fig. 14 Convey Smith–Waterman implementation

5.1 *Programming Effort*

In the time when clock speeds were the driving factor in computer performance programming was relatively easy as concurrent programming techniques did not need to be considered. There was nearly no impact from clock speed to programming. Some impacts could be observed in realtime programming by decreasing clock cycle times.

With the upcoming age of parallelism (see Fig. 1) programming was massively affected as algorithms now had to be parallelized. This was a tremendous challenge for programmers as they now needed parallel programming skills as good and efficient automatic parallelization methods are still a matter of research.

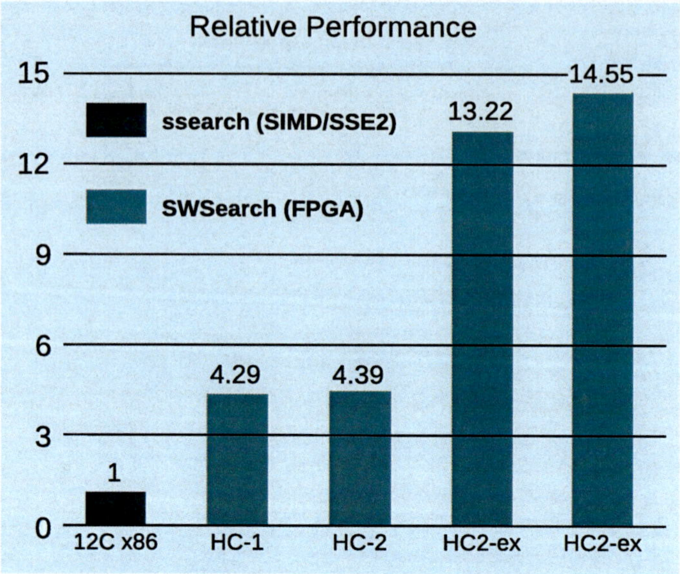

Fig. 15 A relative performance comparison

The programming effort for hybrid computers as the Convey HC machines are can be derived from Fig. 16. It corresponds to Fig. 11. The design flow separates after the programm analysis into a hardware and a software path. The partitioning stage contains the parallelization analysis and decisions which parts of the parallelized algorithms need special parallel hardware support. *Parallel and special* mean in this case parallel hardware that is not provided by standard multicore architectures. This partitioning requires hardware and software design skills. As there are very few excellent designers with expertise in both areas, hardware and software design, good design decisions here require both experts. This doubles approximately the size of design teams. The pure forward design time remains the same as far as testing is not considered and if codesign engineering techniques can be applied [9]. Testing is much more complex as in homogeneous design projects as hardware and software are together the source of errors. Things might work separately but not together. Codesign friendly architectures and test methods still need to be developed [23].

Besides the general propositions above the design complexity of applications for the Convey HC architectures depends strongly on the applications themselves and on the question if a personality needs to be developed or if already designed and tested personalities can be used. In the first case the complete design flow needs to be done by a development team. In the second case the hardware components come with a library that needs to be linked with an application. In this case the whole design process is easy and similar to just software development.

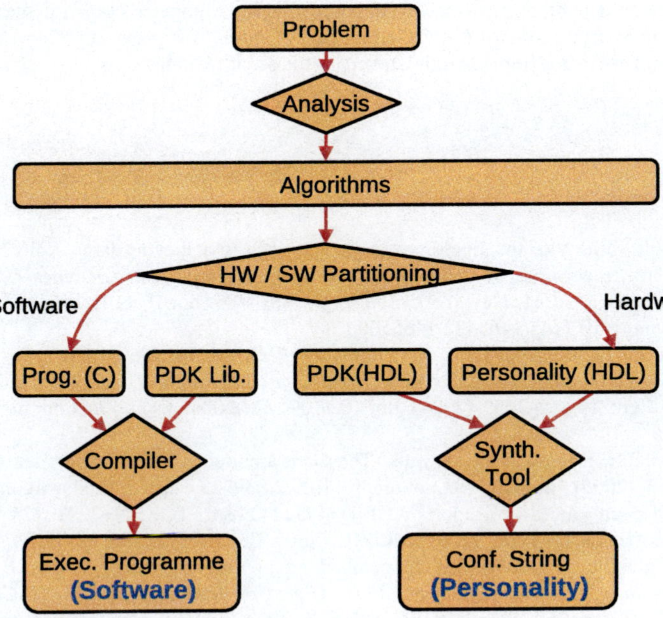

Fig. 16 A design flow for hybrid computing

Hardware and software development in general are well supported as the Convey HC machines come as Intel Linux machines. Design tools and compilers can run on the multicore Linux part of the machines. Synthesis results can then easily be loaded into the FPGA coprocessor. Convey also provides predesigned interfaces to the hardware structures like I/O and (cache coherent) memory interfaces. Also the hardware/software interface is easy to use by the coprocessor concept.

6 Conclusion

As we learn from Fig. 1 hybrid architectures are at least one—probably the only way to exploit performance from the fact that Gordon Moores law is still effective. In Sect. 3 we have observed that academia as well as industrie developed first architectures containing reconfigurable components. Configurable Hardware allows for optimized hardware supporting algorithms instead of declining algorithms to perform on hardwired concurrent computers. We also learned that performance can be derived from the convergence of hardware and software in the term *configware*. This benefit needs to be paid with a significantly increased design effort as far as there is no new generation of codesign tools incorporating the development of hardware, software and *configware* [2].

Acknowledgements I express my sincere thanks to Convey computer corp. for their support of this contribution, especially to Kirby Collins from Convey for providing Convey original documents and to Ian Gregor from Griffith University for proof reading.

References

1. G.M. Amdahl, Validity of the single processor approach to achieving large scale computing capabilities, in *Proceedings of the April 18–20, 1967, Spring Joint Computer Conference, AFIPS '67 (Spring)* (ACM, New York, 1967), pp. 483–485. doi:10.1145/1465482.1465560, http://doi.acm.org/10.1145/1465482.1465560
2. D. Andrews, D. Niehaus, R. Jidin, M. Finley, W. Peck, M. Frisbie, J. Ortiz, E. Komp, P. Ashenden, Programming models for hybrid fpga-cpu computational components: a missing link. IEEE Micro **24**(4), 42–53 (2004). doi:10.1109/MM.2004.36, http://dx.doi.org/10.1109/MM.2004.36
3. T.J. Callahan, J.R. Hauser, J. Wawrzynek, The Garp architecture and C compiler. Computer **33**(4), 62–69 (2000). doi:http://dx.doi.org/10.1109/2.839323, http://portal.acm.org/citation.cfm?id=621455&dl=ACM&coll=portal&CFID=11111111&CFTOKEN=2222222#
4. F. Campi, M. Toma, A. Lodi, A. Cappelli, R. Canegallo, R. Guerrieri, A VLIW processor with reconfigurable instruction set for embedded applications, in *2003 IEEE International Solid-State Circuits Conference, 2003. Digest of Technical Papers. ISSCC* (IEEE, New York, 2003), pp. 250–491. doi:10.1109/ISSCC.2003.1234288, http://dx.doi.org/10.1109/ISSCC.2003.1234288
5. J.M.P. Cardoso, P.C. Diniz, M. Weinhardt, Compiling for reconfigurable computing: a survey. ACM Comput. Surv. **42**(4), 1–65 (2010)
6. C. Chang, J. Wawrzynek, R.W. Brodersen, Bee2: A high-end reconfigurable computing system. IEEE Des. Test Comput. **22**(2), 114–125 (2005), http://doi.ieeecomputersociety.org/10.1109/MDT.2005.30
7. K. Compton, S. Hauck, Reconfigurable computing: a survey of systems and software. ACM Comput. Surv. **34**(2), 171–210 (2002), http://doi.acm.org/10.1145/508352.508353 [An excellent survey paper on reconfigurable computing]
8. Cray Inc., 411 First Avenue S., Suite 600, Seattle, WA 98104-2860 USA: Cray XD1 Datasheet, http://www.hpc.unm.edu/~tlthomas/buildout/Cray_XD1_Datasheet.pdf
9. G. de Micheli, R.K. Gupta, Hardware/software co-design. IEEE Micro **85**, 349–365 (1997)
10. G. Estrin, Organization of computer systems: the fixed plus variable structure computer, in *Papers Presented at the May 3–5, 1960, Western Joint IRE-AIEE-ACM Computer Conference, IRE-AIEE-ACM '60 (Western)* (ACM, New York, 1960), pp. 33–40. doi:10.1145/1460361.1460365, http://doi.acm.org/10.1145/1460361.1460365
11. F. Feinbube, Joint forces: the era of hybrid computer environments, in *HPI Symposium @ SAP* (SAP, Walldorf, 2011)
12. F. Feinbube, The future is hybrid – Developer support for accelerator-based technologies, in *HPI-UCT Workshop*, Capetown, 2011
13. A. George, H. Lam, G. Stitt, Novo-g: at the forefront of scalable reconfigurable supercomputing. Comput. Sci. Eng. **13**, 82–86 (2011). http://doi.ieeecomputersociety.org/10.1109/MCSE.2011.11
14. A.D. George, H. Lam, A. Lawande, C. Pascoe, G. Stitt, Novo-g: a view at the hpc crossroads for scientific computing, in *Proceedings of the 2010 International Conference on Engineering of Reconfigurable Systems & Algorithms, ERSA 2010*, 12–15 July 2010, ed. by T.P. Plaks, D. Andrews, R.F. DeMara, H. Lam, J. Lee, C. Plessl, G. Stitt (CSREA Press, Las Vegas, 2010), pp. 21–30
15. J. Gray, Designing a simple fpga-optimized risc cpu and system-on-a-chip [Online] Available: http://www.dte.eis.uva.es/Docencia/PDF/soc-gr0040-001201.pdf (2000)

16. J.L. Gustafson, Reevaluating Amdahl's law. Comm. ACM **31**, 532–533 (1988)
17. D. Hallmannseder, B. Klauer, Compiler Unterstützung für die dynamische Rekonfiguration eines Mikroprozessors (Compiler support for the dynamic reconfiguration of a microprocessor), in *1. Workshop Innovative Rechnertechnologien, Nanotechnologien für die IT* (Helmut Schmidt Universität, Universität der Bundeswehr Hamburg, Hamburg, 2009), pp. 40–46
18. J.R. Hauser, J. Wawrzynek, Garp: a mips processor with a reconfigurable coprocessor, in *Proceedings of the 5th Annual IEEE Symposium on Field–Programmable Custom Computing Machines* (1997), pp. 12–21
19. Hayden Publishing, Monolithic memories announces: A revolution in logic design. Electron. Des. **26**(6), 148B–148C (1978)
20. M.D. Hill, M.R. Marty, Amdahl's law in the multicore era. Computer **41**(7), 33–38 (2008)
21. R. Inta, D.J. Bowman, S.M. Scott, The "chimera": an off-the-shelf cpu/gpgpu/fpga hybrid computing platform. Int. J. Reconfig. Comput. **2012**, 2:2 (2012). doi:10.1155/2012/241439, http://dx.doi.org/10.1155/2012/241439
22. Intel/Altera Atom Processor E6x6C Series, Santa Clara and San Jose, http://www.altera.com/devices/processor/intel/e6xx/proc-e6x5c.html
23. O. Khan, S. Kundu, Hardware/software codesign architecture for online testing in chip multiprocessors. IEEE Trans. Dependable Secure Comput. **8**, 714–727 (2011). http://doi.ieeecomputersociety.org/10.1109/TDSC.2011.19
24. S. Kumar, C. Paar, J. Pelzl, G. Pfeiffer, M. Schimmler, Breaking ciphers with copacobana – A cost-optimized parallel code breaker, in *Workshop on Cryptographic Hardware and Embedded Systems – CHES 2006, Yokohama* (Springer, Heidelberg, 2006), pp. 101–118
25. I. Kuon, R. Tessier, J. Rose, Fpga architecture: survey and challenges. Found. Trends Electron. Des. Automa. **2**(2), 135–253 (2007). http:/dx.doi.org/10.1561/1000000005 [Modern survey of FPGA Architecture]
26. G.E. Moore, Cramming more components onto integrated circuits. Electronics **38**(8), 114–117 (1965)
27. A. Niyonkuru, H.C. Zeidler, Designing a runtime reconfigurable processor for general purpose applications, in *IPDPS*, Santa Fe, 2004
28. R.R. Schaller, Moore's law: past, present, and future. IEEE Spectr. **34**(6), 52–59 (1997). doi:10.1109/6.591665, http://dx.doi.org/10.1109/6.591665
29. T.F. Smith, M.S. Waterman, The identification of common molecular subsequences. Mol. Biol. **147**, 195–197 (1981)
30. T. Todman, G. Constantinides, S. Wilton, O. Mencer, W. Luk, P. Cheung, Reconfigurable computing: Architectures and design methods. IEE Proc. Comput. Digit. Tech. **152**(2), 193–207 (2005). http://dx.doi.org/10.1049/ip-cdt:20045086 [A recent survey paper on reconfigurable computing platforms and design with a wealth of references]
31. K.H. Tsoi, W. Luk, Axel: a heterogeneous cluster with fpgas and gpus, in *FPGA*, ed. by P.Y.K. Cheung, J. Wawrzynek (ACM, New York, 2010), pp. 115–124. http://doi.acm.org/10.1145/1723112.1723134
32. C. Ulmer, R. Hilles, D. Thompson, Reconfigurable computing aspects of the cray xd1, in *Proceedings of the CUG 2005* (2005). http://www.craigulmer.com/portfolio/unlocked/050516_cug_rc_aspects_of_xd1.pdf
33. R. Wilson, R. French, C. Wilson, S. Amarasinghe, J. Anderson, S. Tjiang, S. Liao, C. Tseng, M. Hall, M. Lam, J. Hennessy, The suif compiler system: a parallelizing and optimizing research compiler. Technical Report, Stanford University, Stanford, CA, 1994
34. Xilinx Zynq Product Brief, San Jose, http://www.xilinx.com/support/documentation/zynq-7000.htm

Low Cost High Performance Reconfigurable Computing

Javier Castillo, Jose Luis Bosque, Cesar Pedraza, Emilio Castillo, Pablo Huerta, and Jose Ignacio Martinez

Abstract High Performance Reconfigurable Computing (HPRC) has emerged as an alternative way to accelerate applications using FPGAs. Although these HPRC systems have a performance comparable to standard supercomputers and at a much lower cost, HPRC systems are still not affordable for many institutions. We present a low-cost HPRC system built on standard FPGA boards with an architecture that can execute many scientific applications faster than in Graphical Processor Units and traditional supercomputers. The system is made up of 32 low-cost FPGA boards and a custom-made high-speed network interface using RocketIO interfaces. We have designed a SystemC methodology and CAD framework that allow the designer to simulate any MPI scientific application before generating the final implementation files. The software runs on the PowerPC processor embedded in the FPGA on a light ad-hoc implementation of MPI, and the hardware is automatically translated from SystemC to Verilog, and connected to the PowerPC. This makes the SMILE HPRC system fully compatible with any existing MPI application. The proof of the concept of the SMILE HPRC has been exhaustively tested with two complex and demanding applications: the Monte Carlo financial simulation and the Boolean Synthesis using Genetic Algorithms. The results show a remarkable performance, reasonable costs, small power consumption, no need of cooling systems, small physical space requirements, system scalability and software portability.

J. Castillo (✉) • P. Huerta • J.I. Martinez
Universidad Rey Juan Carlos, Madrid, Spain
e-mail: javier.castillo@urjc.es; pablo.huerta@urjc.es; joseignacio.martinez@urjc.es

J.L. Bosque • E. Castillo
Universidad de Cantabria, Santander, Spain
e-mail: joseluis.bosque@unican.es; emilio.castillo@unican.es

C. Pedraza
Universidad Santo Tomas, Bogota, Colombia
e-mail: cesar.pedraza@urjc.es

W. Vanderbauwhede and K. Benkrid (eds.), *High-Performance Computing Using FPGAs*, DOI 10.1007/978-1-4614-1791-0_15, © Springer Science+Business Media, LLC 2013

1 Introduction

The use of hardware accelerators to speed computationally intensive applications up has become a major trend in the supercomputing field. As the size of supercomputers increases, different problems arise regarding communication bottlenecks and other common problems such as power consumption, cooling or the physical space needed to install a machine of these characteristics.

Due to these problems, researchers have been proposing architectures based on different types of accelerators for many years. In this context, the arrival of two key technologies: Reconfigurable logic (RC) and graphical processor units (GPU) has been of an enormous importance. In both cases the concept behind is basically the same: finding the best suitable implementation of the algorithm for the specific characteristics of the target architecture. In the context of RC, the target is an FPGA device mainly full of programmable logic resources, whilst for the GPUs the device is made up of a large number of execution cores. These two alternatives are not mutually exclusive: there are hybrid machines that use both technologies in order to make the most of their intrinsic characteristics[1].

In this context, the use of RC and FPGAs has created a new field of research, named high-performance reconfigurable computing (HPRC). HPRC systems are usually classified into two main groups [2]: uniform node non-uniform systems (UNNSs) where the nodes are made up of FPGAs or microprocessors connected through a high-speed network and non-uniform node uniform systems (NNUSs) where only one type of node is used, each one containing a microprocessor with an FPGA tightly coupled. The well-known Cray XD1[3] supercomputer belongs to this category.

The SMILE project (Scientific Parallel Multiprocessing based on Low-Cost Reconfigurable Hardware) presents a new HPRC architecture and a development methodology using commercial off-the-shelf (COTS) FPGA boards[4]. The SMILE architecture is an NNUS architecture where all the nodes are equal, containing a built-in PowerPC processor and the FPGA logic tightly coupled through the system bus. The main features of the SMILE architecture are the following:

- *High performance*: The architecture, where parts of the applications are accelerated using custom hardware, has a performance similar to standard general-purpose microprocessor clusters and to GPU-accelerated systems.
- *Low-power consumption*: Each SMILE node consumes only 5 W, according to our lab measurements, which is an order of magnitude smaller than the power consumption of a personal computer. This is a really important advantage for building a supercomputer because the SMILE architecture does not need any cooling system.
- *Scalability*: For most applications adding more nodes to a SMILE cluster is just as easy as connecting new FPGA boards to the high-speed interconnection network.
- *Portability of the parallel applications*: the system runs an MPI library implementation, therefore any parallel application already programmed can

be ported to the new environments. The only additional effort required is the development of the custom hardware to speed the application up, keeping all the communication scheme unchanged.

- *Low-cost*: By using low-cost commercial FPGA boards the nodes of the system have a reasonable price.
- *Low-space utilization*: The FPGA boards can be placed in a stacked configuration that takes up a very small physical space.

The main contributions of the SMILE project are the design and implementation of a new HPRC architecture based on low-cost FPGA boards. A full SystemC-based methodology was developed to help porting any MPI Application to the SMILE system, as well as a complete CAD framework to use this methodology. This framework enables the simulation and debugging of the whole system before generating and downloading the final implementation to the HPRC system. Two different applications were developed and tested using the methodology: a Monte Carlo financial simulation and a combinational circuit synthesis using Genetic Algorithms. The whole framework, methodology and HPRC SMILE system have been fully tested through an exhaustive experimental evaluation, as well as his results compared to two other high-performance architectures: GPUs and cluster. The today's GPUs importance, as mentioned in the second paragraph of this section, can be appreciated in the TOP500 [5] where a total of 28 systems on the list use GPU technology. The two Chinese systems, Tianhe in the top one and Nebulae in the third position, and the new Japanese Tsubame 2.0 system as number 4 [6] are all using NVIDIA GPUs to accelerate the computations. In all the cases the GPUs are coupled to the processor through an extension bus (PCI-e, HyperTransport, etc).

Section 2 continues with a review of similar research work in the literature. In Sect. 3 the SMILE HPRC architecture is described in detail. Section 4 presents the methodology developed to create any SMILE application using our SystemC framework. In Sect. 5 the two case studies (benchmarks) are described based on the three different architectures. Section 6 presents the results of both benchmarks running on the SMILE HPRC system, GPUs and the ALTAMIRA cluster. Finally, we extract some conclusions and suggest future work.

2 Background

In the HPRC field, different vendors have started to offer machines that include both FPGAs and high-end processors, introducing the concept of HPRCs [7, 8]. These machines combine the hardware customization of the FPGAs and the flexibility of the software running on a general-purpose processor, providing new ways to explore the design space to obtain the best performance.

One of the main HPC vendors, Silicon Graphics International (SGI), provides SGI Reconfigurable Application Specific Computing (SGI-RASC) [9], a technology that can be used with the Altix line of HPC servers. The RC100 model [10] includes

2 Virtex-4 LX200 FPGAs and two NUMAlink interfaces with 12.8 GB/s bandwidth. Another vendor, SRC Computers, offers the H MAP and I MAP reconfigurable processors used in the SRC-7 family of products [11].

Another well-known HPC vendor, Cray Inc., developed the XD1 system as a distributed memory HPC system [3]. The Cray XD1 entry-level supercomputer range uses AMD Opteron 64-bit CPUs and incorporates Xilinx Virtex-II Pro FPGAs for application-specific implementations.

The research and academic community also had several proposals for HPRC systems. Splash 2 was an attached processor system using Xilinx XC4010 FPGAs as its processing elements, developed in the Supercomputing Research Center [12]. Another system that uses FPGAs in a cluster of independent operating systems was developed in the project called Sepia [13], which used the DEC PCI Pamette FPGA board for building a low-cost cluster for image processing applications.

The reconfigurable computing cluster (RCC) project worked on the feasibility of using FPGAs to build cost-effective petascale cluster computers, building a cluster of 64 Xilinx ML-410 Development boards with a Virtex-4 FPGA [14]. Yoshimi et al. [15] designed a 512 FPGA cube made up of eight 64-FPGA boards mainly to be used for physics, financial simulation and massively parallel cryptographic key cracking.

Cathey et al. [16] present a reconfigurable data flow processing architecture that explicitly targets at the same time both fine- and course-grained parallelism. This architecture is based on multiple FPGAs organized in a scalable direct network.

One of the most relevant projects on HPRC is the Research Accelerator for Multiple Processors (RAMP) [17]. In this project there are several universities researching on the next generation tools for computer architecture and computer science. RAMP seeks to take advantage of the high degree of parallelism and density of the FPGAs to emulate new highly parallel computer systems. The project uses a custom platform named Berkeley Emulation Engine 2 (BEE2). BEE2 provides a large amount of FPGA resources, DRAM memory, and high bandwidth I/O channels on one single board. One of the main milestones of the project is to build a proof of concept HPRC named the RAMP Blue cluster.

Although the capacity of current FPGAs has grown enormously in recent years, sometimes the number of functions to be executed in hardware exceeds the FPGA resources limit. To tackle this issue El-Gahzawi et al. [2] propose a technique called Virtualization, i.e. using partial run-time reconfiguration to switch between the different hardware functions programmed in the FPGA, multiplexing the FPGA resources over time.

The main benefit of the HPRC systems: the design of new hardware functions for every application to take full advantage of the FPGA logic and the characteristics of each application is, at the same time, the main drawback when the standard scientific community, with no experience in hardware design tries to use HPRC systems. To facilitate the hardware design the EDA vendors are constantly making efforts to help with high-level synthesis tools based on C-like languages. For example, Mitrion C and Handel-C tools work for the SGI RASC, and Impulse C for the Cray XD1.

Another important trend is the use of hybrid machines combining GPU+FPGA. This provides a designer different targets for the different application tasks so that the designer can find the most appropriate resource for every task in terms of performance or any other considerations like timing, usage, etc. Tsoi and Luk [1] presents a heterogeneous machine using nodes made up of a mix of different accelerators, as well as a map-reduction framework to map these tasks into the different resources. Showerman et al. [18] describes another hybrid system of Virtex-4 FPGAs and Nvidia Quadro GPUs tested in weather forecast, molecular dynamics and cosmology applications, with up to a 48x speedup in some tests.

3 SMILE Architecture

The SMILE project is built as a custom distributed-memory parallel computing machine with low-cost FPGA boards. The programming model is based on the execution of concurrent processes with message-passing communication and synchronization. The communication uses an MPI standard library that provides the portability of applications to different platforms. The concurrent processes are divided into a software part, running on the PowerPC processor of each FPGA, and a custom hardware accelerator designed for the application using the methodology presented in Sect. 4.

The SMILE cluster is made up of up to 32 FPGA nodes and a host computer monitoring the cluster operations and sharing the storage space (hard disks). Currently, each node is a Diligent XUPV2P Board, selected for the low price, low power consumption and high performance. Although all the infrastructure can be migrated to more advanced FPGA families (plug and play), Virtex2P is still use due to the board price. With a moderate cost of $499 per node it is possible to compete with supercomputers in some applications, as can be seen in Sect. 6. The power consumption of the SMILE cluster is not only several orders of magnitude smaller than the power consumption of a traditional supercomputer, but also means no cooling system needed for the SMILE cluster. The physical size needed to accommodate the SMILE cluster is also several orders of magnitude smaller than the space needed for a conventional supercomputer (table size vs. room size).

The board includes a Xilinx V2P30 FPGA with two PowerPC 405 microprocessors and 8 Multi-Gigabit transceivers that are used for high-speed communications between the boards. The dedicated hardware implemented in the FPGA logic is connected to the PowerPC processor through the on-chip peripheral bus (OPB). The board also contains the peripherals needed to develop complex applications such as DDR-SRAM controllers, System ACE controllers (for compact flash memories) and RS232 interfaces.

With all these elements it is possible to run a full version of the Linux kernel in the PowerPC microprocessor, using all the programs and libraries available for this operating system.

The main tool for the SMILE Cluster is an ad-hoc MPI implementation that provides the management of the cluster communication using a standard API. For the initial versions of the SMILE cluster we used a standard MPI implementation, more precisely LAM/MPI, freely available on the web. But the overhead introduced by this MPI version was unacceptable: more than 46 s only for executing the processes in all the nodes. This overhead forced the development of our own MPI implementation called SMPI: a lightweight implementation of the MPI standard. The SMPI library offers the possibility of sending data between nodes through the Ethernet connection or through the Rocket IOs of the board. The library implements just a subset of the MPI standard targeting the best performance with the minimum overhead possible.

The MPI_Init function opens all the sockets needed to communicate the processes, and once the Send and Receive functions are open they act as simple collective communication links between nodes. The data can also be sent through the RocketIO interfaces of the boards because they are now integrated in the system as any other standard Ethernet device of the Linux kernel (use of standard TCP sockets, etc.).

3.1 Network

The network system is a critical issue in parallel architectures. The Ethernet network interface provided by Xilinx is not fast enough to support cluster communications, so this network interface is used only for management tasks. To avoid the communication bottleneck introduced by this low-speed network, a high-performance network has been designed using the three-bidirectional SATA channels included on the board. This network works at 1.5 Gbps offering a performance similar to current parallel system networks.

The management of the RocketIO transceivers has being simplified by the use of the Aurora core provided by Xilinx. This communication channel can now be used in the SMILE Cluster thanks to the new ad-hoc interface developed by our team for the Aurora Core and the Linux operating system. This interface, called SMILE Communication Element (SCE), has two main goals: appears as a conventional network resource in Linux (a call to the SMPI library) and provides the network routing channels between the boards.

The SCE has three different parts (Fig. 1). Since the board has three connectors, the SCE includes: one Aurora core for each connector to manage the data exchange in that link, the Send and Receive FIFOs to store the packets and deal with congestion problems, and some ad-hoc logic to implement the routing algorithm.

The network topology is defined in terms of groups of four nodes named SMILE Block Elements (SBE). Every node in the SBE is connected to its SBE neighbors with a bidirectional channel. To allow the routing between nodes into different SBEs

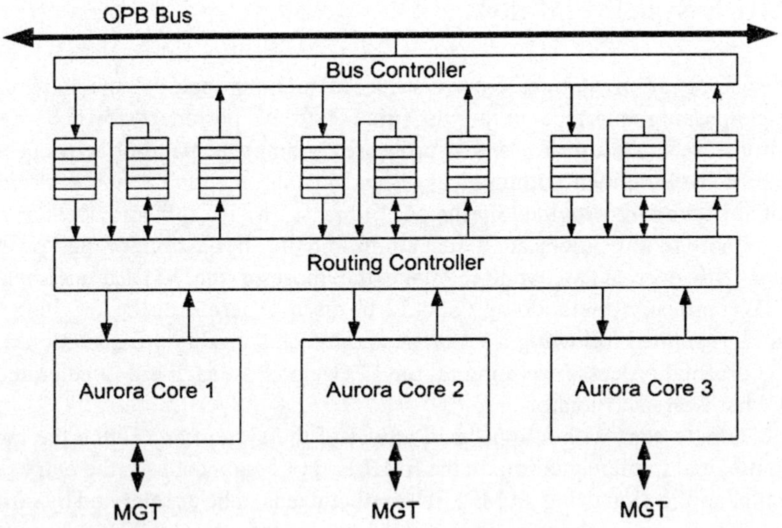

Fig. 1 SMILE communication element

every top node of each SBE is connected to the top nodes of the other SBEs in a ring topology. With this configuration, the SMILE cluster of 32 nodes has a diameter of 10 steps.

The routing algorithm uses a Reservation packet that finds the path between the nodes. Once the path is found, the SCEs of the nodes in the path are marked and reserved, appearing as a shortcut for the following packets. When the frame transmission ends the SCEs are released and free to be used in other request. If a congestion problem appears, two transmissions on the same link, the packets are stored in the SCE FIFOs until the path is free again.

A simple routing algorithm has been designed to find the best path in the SMILE cluster. If the packet destination is in the working node, the SCE delivers the packet to the PowerPC processor in the working node. If not, the packet needs to be routed. In this case the SCE selects which one of the other two interfaces in the board is going to be used. When the destination is in the same SBE, the address can be lower than the working node (the package is sent to the previous node) or higher (the package is sent to the next node). When the destination is in a different SBE, the data goes up to the next node in its SBE and is sent to the SBE destination. Once in the correct SBE destination, the packet goes down the SBE nodes to its final node destination. All the information needed by the routing algorithm (working node address, neighbors, SBE neighbor, etc.) is included in the SCE by the Linux driver during the system start-up.

4 SMILE SystemC Model

Any MPI application running on a distributed memory parallel machine using a message-passing interface library is suitable to be ported to SMILE. These applications are made up of a set of processes running in parallel in processing nodes and a well-defined communication and synchronization scheme based on the standard functions provided by the MPI library. These applications can run on SMILE with no modifications at all, just compiling the application for the PowerPC processor. However, if we want to make the most of the SMILE architecture a HW/SW co-design methodology should be used. Therefore, the original MPI application is refined following a set of steps to reach a final implementation made up of the original processes running on the FPGA boards, each one accelerated by a custom hardware accelerator.

The idea is to start with a high-level model of the system and refine the model down to the final implementation. In the SMILE application context, the entry point is a parallel application using an MPI library that needs to be accelerated by custom hardware. The steps involved in obtaining the final system implementation are:

- Profiling. The first step is to profile the application, finding the time-consuming parts that are appropriate to be implemented into hardware (GNU profiling tools).
- Development of the SystemC model. A SMILE application SystemC model is an MPI application that runs as a SystemC thread. Under this approach, it is possible to run a set of SystemC models in parallel that communicate data through the MPI primitives. This model is fully equivalent to the original application.
- Design of the hardware high-level model. This hardware high-level model is a functional implementation of the final hardware used to speed the application up and is later added to the SystemC system model. The connection between the software and the hardware models is done through untimed sc_fifo channels, as shown in Fig. 2.

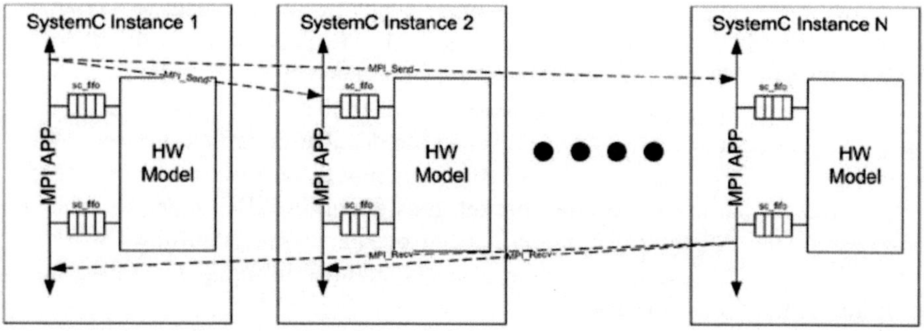

Fig. 2 SystemC untimed model

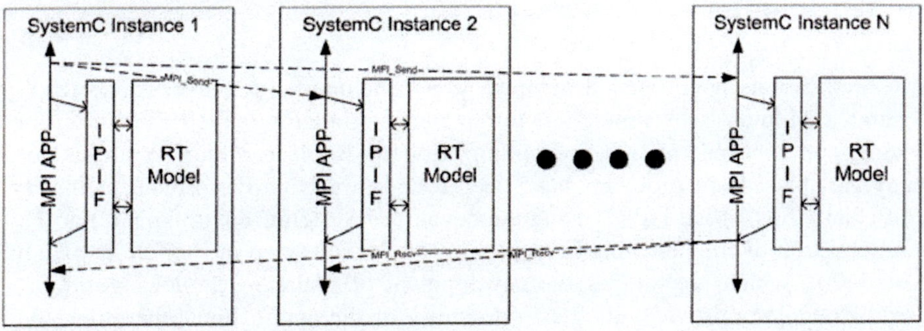

Fig. 3 SystemC model ready to synthesis

- Refinement of the hardware high-level model. There are two different options: using a high-level synthesis tool such as Impulse C or Autopilot to synthesize, or designing an ad-hoc RT-level model and follow a traditional RT-level synthesis methodology. In any case, this step produces a final hardware version ready to be implemented and simulated in the framework.
- Redesign of the communication link between software and hardware. A SystemC model of the Intellectual Property Interface (IPIF) provided by Xilinx is connected to the RT model. The IPIF is a hardware block that communicates the PowerPC processor with the IP-Cores implemented in the FPGA through the OPB and PLB bus. It supports different configurations such as DMA and interrupt-driven or register-based communications. The SystemC IPIF models all these configurations and enables the connection of the RT model to the MPI application.
- Replacement of the functions used by the MPI Application to communicate with the hardware model with the Xilinx functions provided to communicate with the IP-Core through the IPIF (Fig. 3)

These are the steps to follow in order to test the final SMILE system:

1. The MPI Application is compiled for the platform using the Xilinx IPIF libraries.
2. The hardware is synthesized using the appropriate CAD tool. We translate the hardware modules to Verilog using a custom tool, developed by the authors and called sc2v [19].
3. The hardware netlist is connected to the physical IPIF interface in the EDK environment.
4. The system is then synthesized and the bitstream, ready to program the FPGAs, is generated and downloaded.

5 Benchmarks

Two applications have been developed as a benchmark for the SMILE HPRC system: a Monte Carlo simulation for financial problems (called the European Pricing option problem) and the optimization of Boolean circuit synthesis for many variables. Both problems have been tested in a High-Performance Cluster, a GPU and the SMILE HPRC using the developed SystemC environment. For the parallelization of the benchmarks, we followed the four-step methodology of Ian Foster [20], which encourages the development of scalable parallel algorithms. The methodology provides the best portability of the applications; therefore, the processes and communication schemes are identical for both the SMILE HPRC and for the cluster. The portability of the applications is thus assured at the design level. However, it is worth mentioning that each specific implementation has been optimized to obtain the maximum performance for the corresponding architecture, so that we can make a fair comparison between the different architectures.

5.1 Monte Carlo Financial Simulation

The Monte Carlo simulation is widely used in many problems. Its main drawback is that is very demanding from a computational point of view, with a relatively slow convergence rate. As a result, lots of effort have been made to accelerate the Monte Carlo simulation [21–25].

The Monte Carlo simulation is used to solve the European Pricing Option problem, as described in this subsection. In financial terms an option is an agreement: a buyer buys the right, but not the obligation, to buy or sell a value, at a certain price. Scholes proposed a differential equation that is able to calculate a good approximation to the option value based on several assumptions [26]. The Black–Scholes model assumes a perfect market hypothesis where financial markets are *efficient* and prices on traded assets already reflect all the known information. With this hypothesis, the security price changes mathematically with the Markov processes and the value of the asset can be represented as a Brownian motion. Solving this equation, it is obtained:

$$S_T = S_0 \cdot e^{(r-0.5\mu^2)T+(\mu\sqrt{T}N(0,1))} \tag{1}$$

where r is the rate we can expect in a riskless market. S_T is the price of the option depending on the random variations of the market modeled by the normal distribution. It is possible to calculate the expectation of $V_{\text{call}}(S,T)$ by generating a large number of $N(0,1)$ samples and computing the average estimated profit.

$$V_{\text{mean}} = \frac{1}{N}\sum_{i=1}^{N} V_i(S,T) \tag{2}$$

If the return value of the money in a riskless investment is subtracted from the expected profit, we obtain the current value of the option. From this equation, the evaluation of the expected profit is just a question of generating a large number of Gaussian random samples and evaluating the expected profit for each one.

5.1.1 Hardware Implementation on SMILE

The hardware implementation on SMILE follows the SystemC methodology presented in Sect. 4 and the SMILE application runs in a set of nodes with a hardware coprocessor attached to the PowerPC in the FPGA. The development steps begin with the profiling of the MPI application. From this profiling we found out that the biggest time consuming part of the algorithm is the path calculation, therefore a custom hardware coprocessor has to be implemented for this computation. The system operation can be described as follows:

- The application running on the SMILE host sends the simulation parameters and the iteration number to the nodes.
- Each node receives the data and sends the parameters to the coprocessor through the system bus.
- The coprocessor calculates the expected profit and the confidence value for the indicated number of paths.
- The expected profit and confidence is sent back to the host where the final values are calculated and displayed.

Following the SystemC methodology, we first develop a SystemC untimed model of the coprocessor. Once the simulation is working properly we write a SystemC RT model of the coprocessor, changing the communication channels from the untimed models to the modeled IPIF interface. In this benchmark, the SystemC RT Monte Carlo coprocessor uses several external IP-Cores: a VHDL exponentiator, a Mersenne Twister random number generator and some floating point adders and multipliers. Therefore, we design all the SystemC models of the IP-Cores, so that they can be replaced later in the netlist for the Place&Route process. When the SystemC RT simulation of the whole system is running properly, the core is translated into Verilog using the sc2v tool and then synthesized, placed and routed.

The inputs of the coprocessor are: the parameters of the simulation, the output of the Gaussian random number generator, and the expected profit and confidence values from the previous iterations. The generation of normally distributed random samples can be divided into two steps: the generation of uniform-distributed samples and their conversion to Gaussian-distributed samples. This conversion is usually carried out by the Box–Muller equations that generate the Gaussian samples from two uniform samples. Our approach implements a new algorithm that represents the equations in the polar coordinate system instead of the Cartesian system. This representation dramatically reduces the number of operations because it is no longer necessary to implement the sin and cos functions. The general structure of the Gaussian random number generator is shown in Fig. 4. It is important to notice that

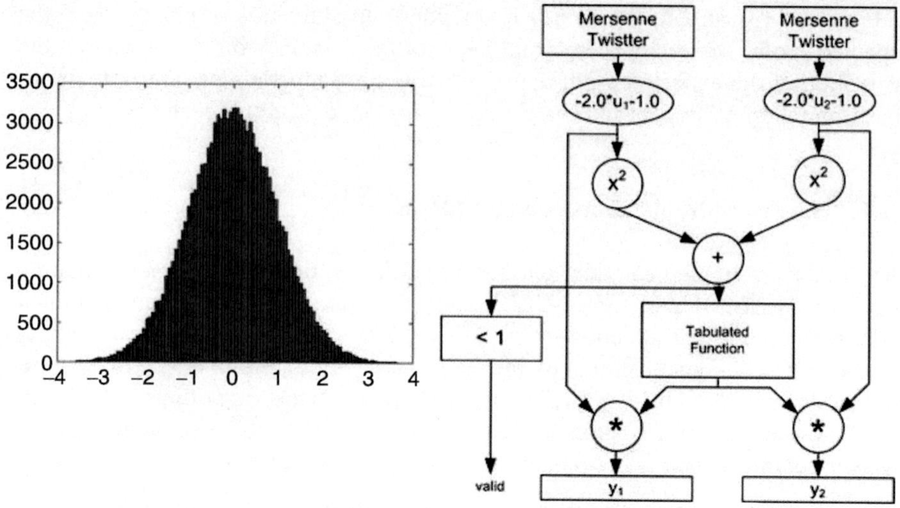

Fig. 4 Gaussian random number generator

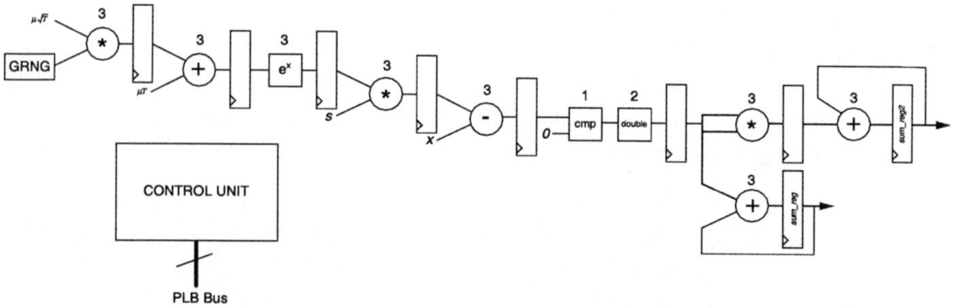

Fig. 5 Pipelined datapath

in order to have the function tabulated and reduce the area of the generator, a fixed-point representation with one sign bit, 23 decimal part bits and 8 integer part bits has been used.

The inputs to the Box–Muller conversion module are two samples of a uniform distribution in the range [0,1) generated by two Mersenne–Twister random number generators. The result goes into the tabulated function that produces two samples per cycle.

Figure 5 shows a pipelined version of the datapath (each element shows its number of cycles). The number of cycles in each stage is limited by the latency of the non-pipelined exponentiation that takes 3 cycles for each calculation. The process is divided into two parts: the first calculates the expected profit and compares the expected profit with zero to check if the option is valid. This part of the algorithm is in single precision IEEE754 floating-point arithmetic. Afterwards, the expected profit of the whole operation and the confidence value are computed in double precision arithmetic. This implies a conversion step from single to double precision. At the end of the process, *sum_reg* and *sum_reg2* contain the expected profit and the confidence value. The coprocessor's control unit, connected to the PowerPC system bus, controls the number of iterations of the coprocessor.

5.1.2 GPU Architecture and Programming Model

For this subsection, we used the Monte Carlo pricing option calculation program provided by NVIDIA inside the NVIDIA CUDA SDK 2.0 in order to compare SMILE HPRC with the optimized version of the GPU algorithm. The process is the same as described in Sect. 5.1.1.

The goal of this implementation is to generate a large number of threads to keep the GPU efficiently busy. The number of options is typically in the hundreds range, but the number of paths per option is in the millions. Therefore, the most appropriate distribution is to use multiple blocks per option to hide the latency of reading the random input values.

Finally, a data distribution is done splitting a bi-dimensional grid into blocks, in terms of the number of options and the number of paths per option. With this approach, each thread computes and sums the payoff for multiple simulation paths of different options.

5.1.3 Parallel Implementation

The best way to carry out the parallelization of the Monte Carlo algorithm is the domain decomposition, because of the independence of the data and the high degree of data parallelism. In this context, we consider a bi-dimensional problem where options are the first dimension and paths for each of the options the second. Given that there are no data dependences, each task will generate a set of pseudo-random numbers (the paths) and compute a subset of the pay-off values for some of the options. This approach has several advantages: selecting the most suitable parallelism degree and balancing the computation and communication times to optimize the performance. Additionally, each random number in the sequence is used for all the options, therefore increasing the locality and reducing the memory requirements.

In this context, all the tasks need the Mersenne–Twister parameters to generate the sequence of pseudo-random numbers; therefore, a general broadcast is compulsory. Once each task has obtained the pay-off value, the average of all of these paths should be calculated. Hence, a reduced parallel operation can be used on a tree communication pattern. Each option has to be reduced but all of them can be done in parallel. Finally, a node has to gather the results of all the options, therefore, a gather operation of the nodes with the final results is needed. However, this approach can increase the communication overhead because of the reductions.

To minimize the impact of the communication overhead in the performance, a single node is in charge of collecting all the partial results and computing the average for each option.

Taking all these considerations in mind, there are two different kinds of processes: a master process that is in charge of broadcasting the Mersenne–Twister parameters to the rest of the processes and gathering the partial results, and a set of slave processes that calculate the values of the pay-off function. The master process will be assigned to the front-end of the cluster and the slave processes will be allocated in the computational nodes.

5.2 Boolean Synthesis with SMILE

The Boolean Synthesis is a design flow process that optimizes and reduces the number of logic gates of a circuit in order to minimize costs, chip area, and increase performance. The use of Evolutionary Algorithms (EA) is a new trend to find original solutions to the problem. In EA, hardware is represented with a chromosome and managed with the Darwinian concept of Natural Selection [27]. The chromosomes mutate and cross with others to create a new population of individuals. As in Nature, when a population of individuals is generated, a fitness function determines which are suitable for accomplishing the target function requirements, and then a selection process excludes some members while the rest mutate and cross again, creating a new population. This process is repeated until a set of individuals that accomplishes the requirements and restrictions of the target function is obtained. For any combinational system problem the fitness functions evaluate the truth table to see if the individual solves the problem and other optimization parameters like number of gates or number of logic levels. It is important to notice that for hardware synthesis it is necessary to use a variation of the simple genetic algorithm (SGA) known as genetic programming (GP) [27] that is able to modify the chromosome length and create new mutating and crossing operators.

Chromosome Representation

The representation in GP is the way a logic circuit is coded using a bit array in order to be managed in the evolution process [28]. This representation must be able to manage all the different solutions of the problem and, moreover, the crossing and muting operators should not generate invalid individuals, and must cover all the solution space so the search is really random. There are different ways of representing combinational hardware for a genetic algorithm [27, 29, 30]. The 2-D tree representation is appropriate for implementing parallel systems because it enables the chromosomes to be split to balance the computational load [31]. Figure 6 shows the selected cell-based structure representation. Each cell has 3 functions f

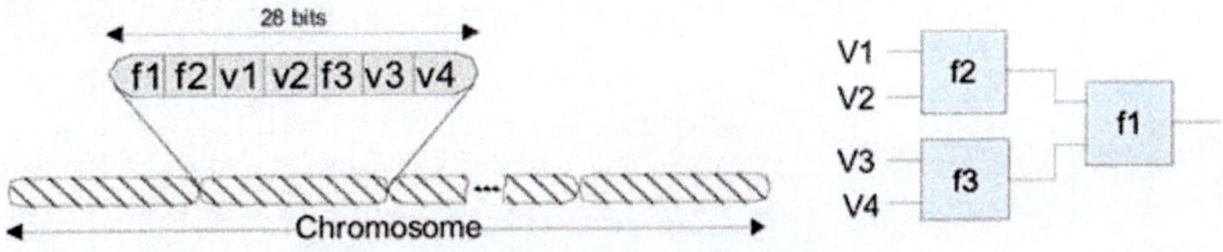

Fig. 6 Cell structure and its representation inside the chromosome

and 4 input variables *v* coded in binary. This representation allows more cells to be added to represent larger circuits (if a more complex solution is necessary) and is suitable to directly translate each cell to the FPGA-LUT architecture.

Fitness Function

Equation (3) shows the fitness function of our GA, the responsible for quantifying the way a chromosome or individual fulfills the requirements. Constants ω_1, ω_2, and ω_3 are used to establish the weights of each of the parameters that will determine the fitness function. The double-summation term calculates the number of coincidences of the individual X for all the possible combinations at the output with the target function Y. The $P(X)$ function calculates the number of logic gates of a chromosome taking into account some of the *introns* or segments of the genotype string that will not have any associated function and that do not contribute to the result of the logic circuit represented. The function $L(X)$ determines the number of levels of the circuit, or in other words, the number of gates in the critical path. The constant m refers to the number of outputs in the circuit and n the number of possible input combinations in the circuit.

$$[H]\text{fitness} = \omega_1.[\sum_{j=1}^{m}\sum_{i=1}^{n} \mathrm{Y(j,i)} - \mathrm{X(j,i)}] + \omega_2.\mathrm{P(x)} + \omega_3.\mathrm{L(x)} \tag{3}$$

Genetic Operators

The *selection* operator is responsible for the identification of the best individuals in the population, taking into account the exploitation and the exploration [31]. The former allows the individuals with better fitness to survive and reproduce more often, and the latter searches in more areas, i.e., finding better results. On the other hand, the *mutation* operator modifies the chromosome randomly in order to increase the search space. It changes: (1) an operator or variable and (2) a segment in the chromosome. Both are executed randomly and with a certain probability. A variable mutating probability during the execution of the algorithm (evolvable mutation) [32] is more effective for Evolvable Systems. Finally, the *crossing* operator combines two selected individuals to obtain two additional individuals to add to the population. A crossing system with one or two randomly selected crossing points has been implemented because it is more efficient for Evolvable Systems [30].

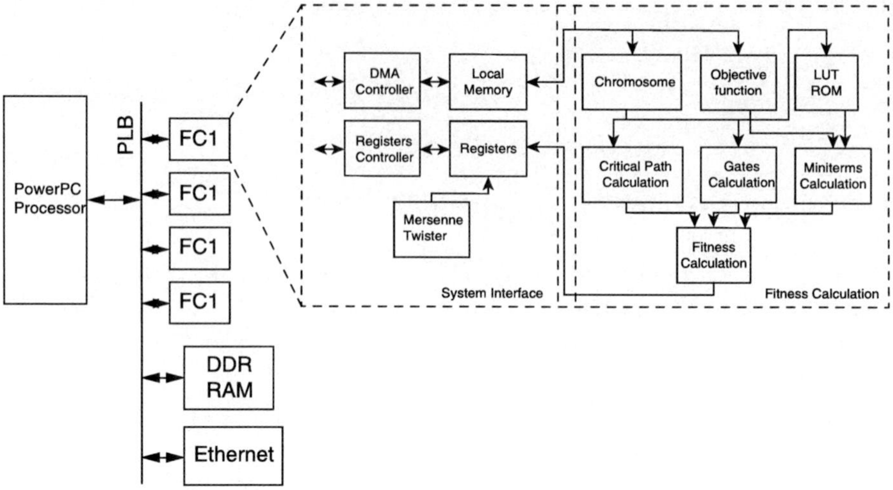

Fig. 7 FCU block diagram

5.2.1 Hardware Implementation on SMILE

Once again we use the SystemC proposed methodology to generate a SMILE HPRC working system. The profiling of the algorithm determined that the biggest time consuming part of the algorithm is the fitness function calculation and the new individual generation (25 and 35% of the execution time, respectively). Therefore, these two have been specifically accelerated with a coprocessor connected to the PowerPC processor.

The *fitness calculation unit* (*FCU*) calculates the three parameters using the objective function, the chromosome and the number of variables as inputs. This coprocessor is connected to the PowerPC 405 processor through the PLB bus using a custom interface. The interface allows register-based and DMA communication to transfer the objective function and the chromosome efficiently. Figure 7 shows the FCUs structure. Once the chromosome has been read from the DDR memory by the memory controller, all the basic cells are converted into their equivalent in Look-Up Table (LUT) through an ROM-based translation. The next block computes the midterm value (number of hits of that individual) using the information from the objective function and a counter as inputs. After computing the number of gates and the logic levels, the fitness calculation block computes the final fitness value that will be sent back to the PowerPC processor. Finally, in order to accelerate the new generation of individuals, crossing and mutation, a Mersenne–Twister-based pseudo random number generator was inserted between the registers of the PLB interface. To further accelerate the fitness evaluation process, 4 coprocessors were implemented in each FPGA of the cluster; therefore, 4 individuals can be evaluated at the same time in each node. All the components were modeled in SystemC and then translated to Verilog, in order to get to the final SMILE implementation.

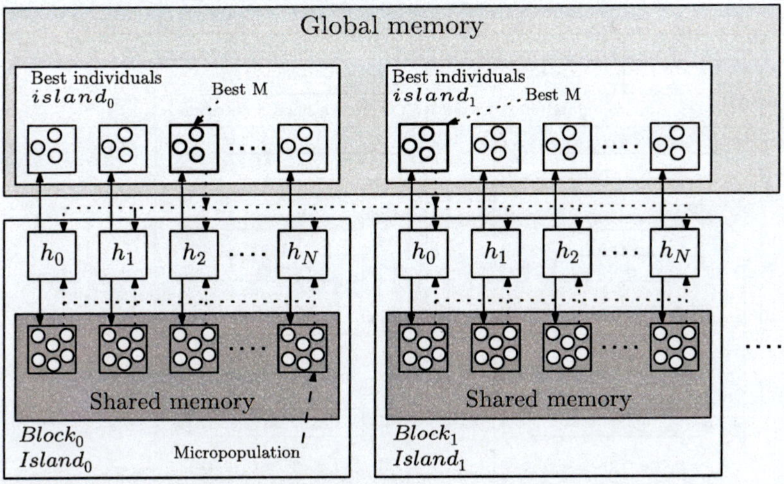

Fig. 8 Block diagram of the CUDA implementation

5.2.2 GPU Architecture and Programming Model

Random Number Generation

During the GP execution, a great amount of random numbers are required to generate an initial population and to mutate and cross individuals. Generating these numbers in the CPU and moving them into the GPU is not feasible because it takes a long time. For this reason, a Mersenne–twister algorithm is executed on the GPU before the $kernel - GP$ to generate a buffer of random numbers in the global memory.

Kernel Structure

Figure 8 shows the way the GP has been implemented in the graphics device. A *kernel* is executed in a thread and is able to generate a μ-population, and perform the select, mutation and crossing operations the number of generations required (Fig. 9). After P generations, M individuals are transferred to the global memory and then to the host device (CPU system). The number P is known as the frequency of migration and the number M is called the migration factor.

It is important to highlight that each thread can cooperate with other threads inside the same block, through the shared memory, sharing the best individuals and improving the efficiency of the GP.

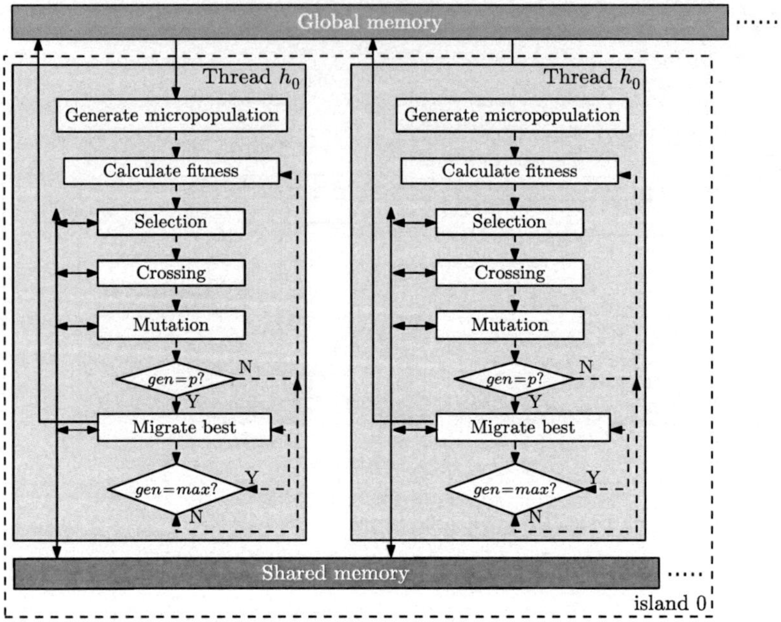

Fig. 9 Thread execution model

5.2.3 Parallel Implementation

As Natural Evolution works with a whole population and not with a single individual (except for selection and reproduction), some operations can be done separately, meaning that almost all operations in a GP are implicitly parallel. Using the island approach, the population is divided into subpopulations that evolve in each processor of the cluster or parallel architecture. When the system starts, each processor creates its subpopulation and starts the evolution process, made up of fitness evaluation, selection, crossing, mutation, and reproduction. These processes are asynchronous because each node starts and ends independently. Once the system reaches a number of generations, a percentage of the individuals are selected to be transferred from one processor to another. A master processor is in charge of collecting the in-transfer individuals and moving them to the rest of the nodes (slaves), increasing the probability of convergence of the algorithm. The ratio of data exchange (the number of the best individuals to be exchanged increases the probability of finding a better solution) and the migration frequency are important parameters in improving the performance of the algorithm.

6 Evaluation

This section presents a set of empirical experiments to evaluate the SMILE HPRC system and compares the results with the GPU and conventional cluster approaches. Again, the main goals are validating the viability of the SMILE HPRC architecture to efficiently solve high-performance computing applications and to verify the performance and scalability of the SMILE HPRC system.

As mentioned in Sect. 5, it is worth mentioning that each specific implementation has been optimized to obtain the maximum performance for the corresponding architecture, so that we can make a fair comparison between the different approaches.

Three different systems have been used for the experiments to compare the difference architectures and implementations.

1. The graphics processing unit, **GPU**, is an NVIDIA GeForce 330M with up to 96 1436 MHz stream processors, connected to the PC host by a PCI Express Bus. It has 1 GB of GDDR3 memory at a 2-GHz clock rate.
2. The cluster set-up, **ALTAMIRA**, is made up of 18 eServer BladeCenters, with 256 JS20 nodes (512 processors) linked together using a 1-Gbps Myrinet network.
3. The **SMILE** configuration is made up of up to 32 FPGA nodes, with the architecture described in Sect. 3.

6.1 Experimental Results for Monte Carlo Simulation

Figure 10 shows the speed-up of the GPUs vs. SMILE HPRC whilst Fig. 11 shows the speed-up of SMILE HPRC vs. the ALTAMIRA cluster, in number of paths per option and for different number of options. In the GPU-CUDA combination, the number of threads is 4,096, and the number of nodes of SMILE and ALTAMIRA is 32 nodes (the largest configuration available in SMILE at that time).

The excellent performance of the GPU compared with the SMILE HPRC has to be highlighted. However, as can be seen in Fig. 10, the speed-up decreases with the workload growth, either due to the number of options or due to the number of paths per option. This can be explained by the limitations of the GPU memory. Managing a large amount of data increases the number of global memory accesses, which forces a much higher latency, heavily decreasing the response time. This leads to a serious problem of scalability. Figure 10 only shows values up to 50 million paths because, above this value, there is a memory overflow and the GPU stops working.

As a result, the GPU not only has a better performance than the SMILE HPRC, but it also has serious scalability problems because of the data size, even getting to a state of system crash.

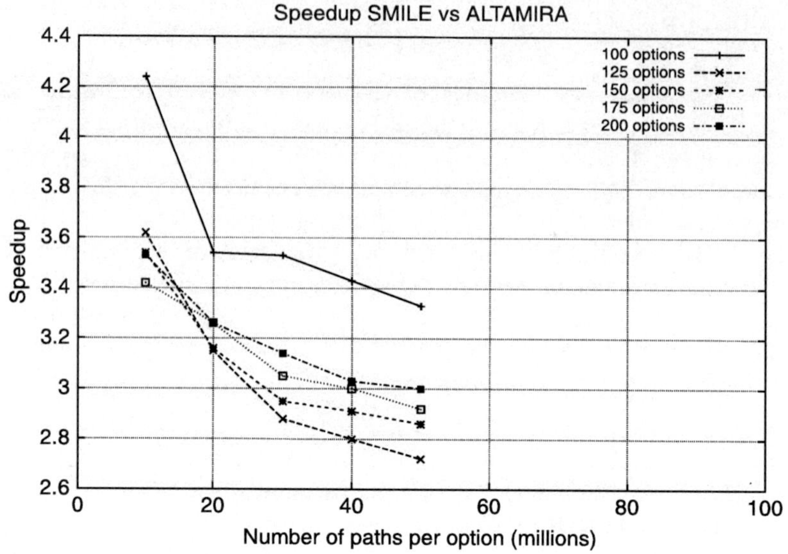

Fig. 10 Speed-up of GPU vs. SMILE

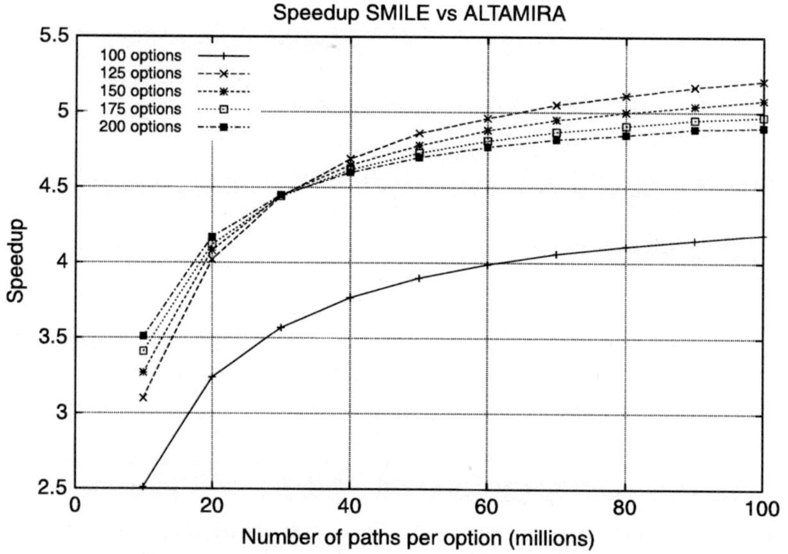

Fig. 11 Speed-up of SMILE vs. Altamira

On the other hand, from Fig. 11, it should be pointed out that there is an excellent performance improvement in SMILE HPRC compared to the ALTAMIRA cluster for the same number of nodes (32 nodes). Additionally, both architectures lack of

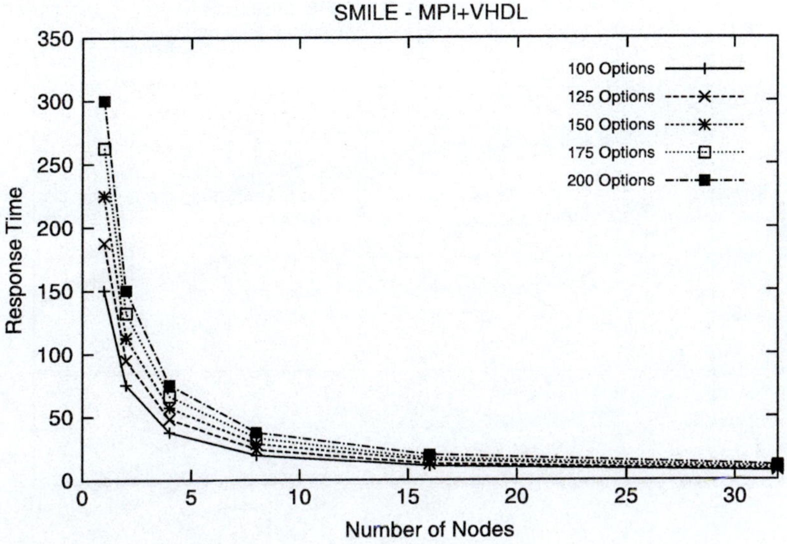

Fig. 12 Elapsed time for different number of nodes (s)

the GPU scalability problems already seen. It is worth mentioning that the speed-up of the SMILE HPRC vs. the ALTAMIRA cluster grows with the workload.

This also means that the SMILE architecture is even more scalable than the ALTAMIRA cluster for this size problem. With these settings, SMILE HPRC is about 5 times faster than the ALTAMIRA cluster. Another key to understanding the results is that even though the ALTAMIRA cluster has CPU nodes running at 2.2 GHz compared with the 100 MHz of the hardware accelerator in the SMILE HPRC, the hardware is able to generate a valid result each clock cycle. In comparison the ALTAMIRA's CPUs spent millions of cycles running the code needed to generate random numbers and generate a result.

Finally, Fig. 12 shows the elapsed time of the SMILE HPRC for different number of nodes. The elapsed time decreases quickly as the number of nodes increases. This behavior is explained because the communication time is practically negligible for this application. Hence the SMILE HPRC presents excellent scalability features. The same behavior is observed for the ALTAMIRA cluster.

6.2 *Experimental Results of Boolean Synthesis*

The experimental results obtained with the implementations of the Boolean synthesis problem described in Sect. 5.2 are presented in this section. In these experiments the number of nodes in the SMILE HPRC and the ALTAMIRA cluster goes from 2 to 16, 16 being the largest configuration of the SMILE architecture, and the population size goes from 512 to 2,048 individuals.

Figures 13–15 show the response time for all the architectures in terms of the number of nodes (SMILE and ALTAMIRA) and the number of threads (GPU).

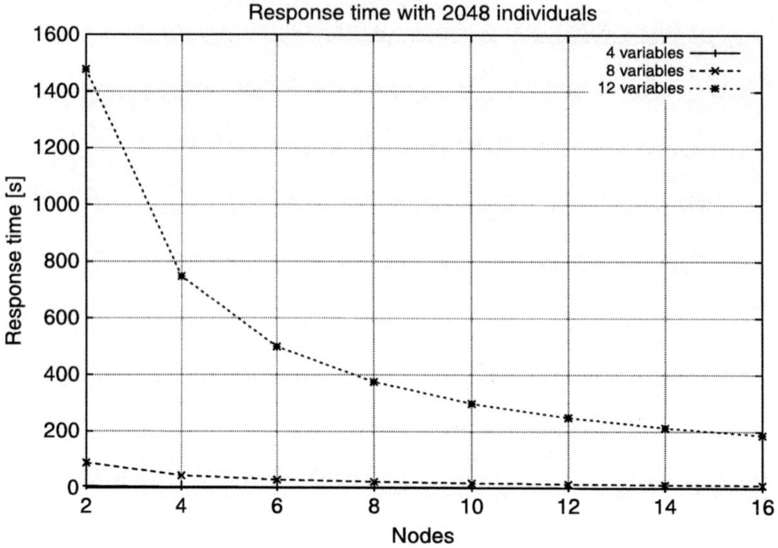

Fig. 13 Altamira response time with 2,048 individuals with different number of variables and 16 nodes

In general, there is a significant decrease in the response times with the increase of nodes (threads) for all the architectures. In the GPU, the response time gets stable at about 200 threads.

The first key aspect that should be noted is the strong impact of the number of variables in the response time in both the ALTAMIRA cluster and the GPU. The increase in the response time when the number of variable goes from 8 to 12 is quite remarkable. However, in the SMILE HPRC this effect is really small or virtually disappears. This can be explained because the number of variables produces an exponential growth in the search space of the genetic algorithm and leads to a great increase in the amount of computation needed to simulate the circuit in the ALTAMIRA cluster and in the GPU. However, in SMILE the circuit is directly tested in hardware and also takes far less time. Thus, the impact on the response time is much smaller.

Talking about scalability, all the three architectures present good features. The GPU does not have the limitations observed in the Monte Carlo Simulation because the size of the data used in the Boolean synthesis is much smaller. This concludes that the GPU architecture is very sensitive to the memory requirements of the application. This situation is also the case when using a single FPGA, but not when using a cluster. If the memory requirements grow, we only need to increase the number of nodes in the cluster to ensure the system scalability.

Finally, Fig. 16 and Table 1 show the speed-up of the SMILE HPRC architecture vs. the ALTAMIRA cluster and GPU, respectively. In both cases, there is an excellent improvement in the performance offered by the SMILE HPRC when

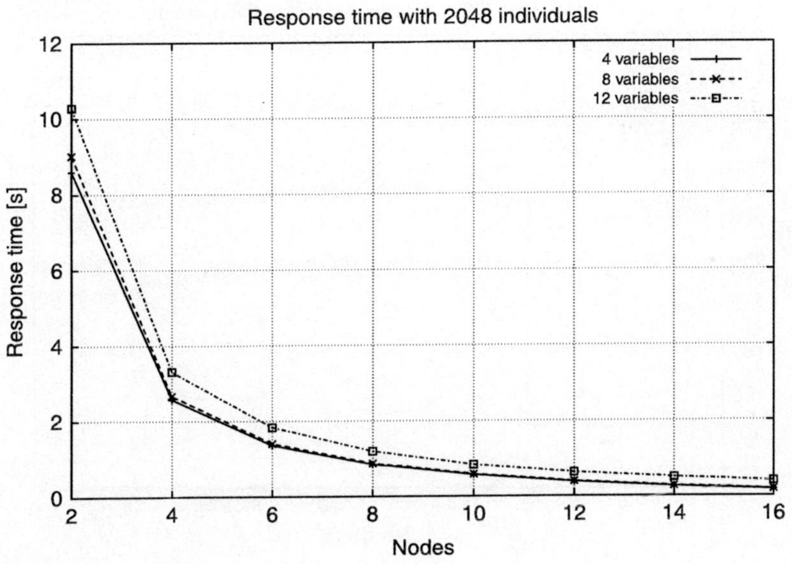

Fig. 14 Altamira response time with 2,048 individuals with different number of variables

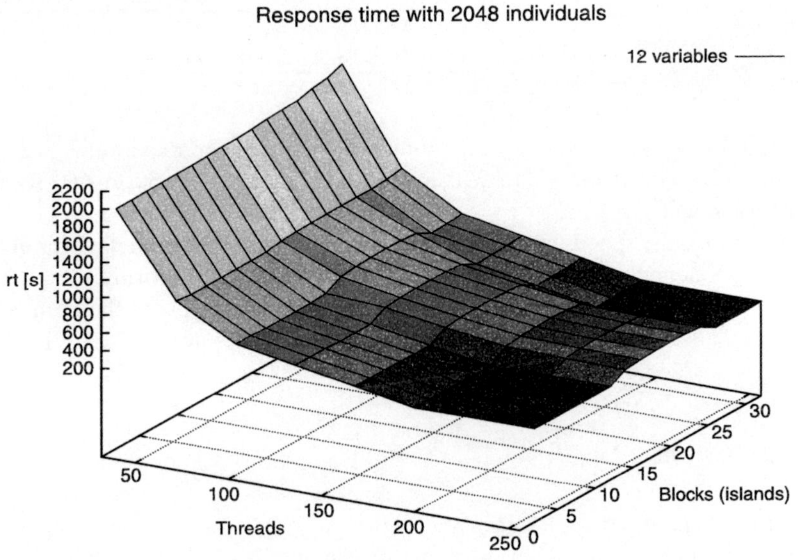

Fig. 15 Response time for the GP on GPU from 32 up to 256 threads

compared to the other two alternatives. Table 1 shows that the speed-up increases when 256 threads are launched instead of 32, no matter the number of islands or CUDA blocks. The reason is because the utilization of the processing elements

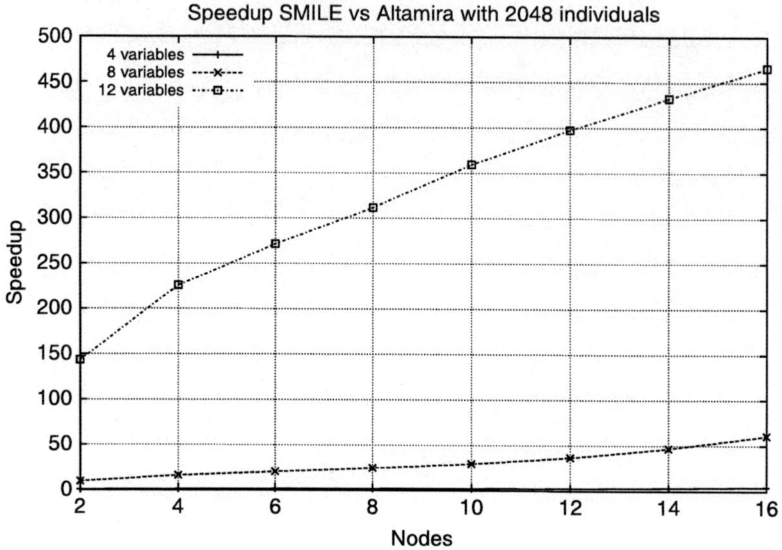

Fig. 16 Speed-up of SMILE vs. ALTAMIRA

Table 1 Speed-up of SMILE vs. NVIDIA 450GTS with 12 variables

	32 threads	256 threads
NVIDIA 450GTS 2 islands	41.6	250
NVIDIA 450GTS 32 islands	41.6	250.6

inside the GPU (192) grows with the number of threads. The maximum speed-up for 12 variables goes from 41 to 250 for the GPU, and from 150 to 450 for the ALTAMIRA cluster.

In the figures, the speed-up increases dramatically with the number of variables, due to the exponential growth in the search space of the genetic algorithm, explained before. Likewise, the speed-up increases, although moderately, with the number of processors. This confirms the excellent scalability properties of the SMILE HPRC.

7 Conclusions

In this chapter the SMILE HPRC, a new HPRC architecture based on a cluster of FPGA boards has been proposed and fully described. The nodes are interconnected through a specific design network with a bandwidth in the Gigabit/s range. The most significant features of the SMILE HPRC are the reasonable costs, the small power consumption, the no need of cooling systems, the small physical space requirements, the high performance offered for specific applications, the system scalability and the software portability. The architecture can execute any MPI parallel application, and also take advantage of the FPGA adaptability, re-configurability and performance. Moreover, a new SystemC methodology has been developed to facilitate the development and debugging of applications for the SMILE HPRC architecture. This methodology and its associated CAD framework enable the simulation of the full system architecture at system level (the parallel program and the communication patterns) as well as at node level (custom hardware developed for an application).

An empirical evaluation has determined both the performance and the scalability of the SMILE HPRC architecture. As benchmarks for these experiments, two well-known applications, the Monte Carlo simulation for financial problems and the Boolean synthesis of digital circuits, have been used and fully detailed. The experiments compared the three different architectures: a GPU programmed with CUDA, a high-performance cluster with a parallel MPI application and the SMILE HPRC with hardware ad-hoc implementations. The experimental results highlight the excellent behavior of the SMILE HPRC in performance and scalability for both applications. For the Boolean Synthesis problem, the SMILE HPRC delivers an outstanding performance compared to the ALTAMIRA cluster and the GPU. However, in the case of Monte Carlo simulation, the GPU overcomes the SMILE HPRC with the current configuration. In terms of scalability, the properties of the SMILE HPRC are much better than the rest of the architectures for all the experiments. Another important fact is the portability that enables any parallel application developed with MPI to be implemented in the SMILE HPRC by replacing the slow software functions by faster custom hardware.

We want to add a discussion about the scalability of SMILE in terms of memory and communications. One of the biggest advantages of SMILE is the distributed memory architecture. Each node can upgrade its memory being able to process more data. The direct connection of the memory with the FPGA its one of the key aspects of SMILE, when more data is needed to process, SMILE gets bigger speedups against GPUs and CPUs.

On the other hand communication is one of the weak points of SMILE as in any parallel architecture. SMILE is suitable for algorithms with low data transfers between nodes and large sets of data to process in the FPGA. The presented examples follow that schema. With other algorithms it is still possible to get big speedups, but depends enormously of the communications patterns. In some cases could be possible to rearrange the high-speed board connections to optimize its behavior for a given application.

As future work, we propose to extend the system to support any available board in the market through the SystemC framework and continue our research in order to add new applications that can take full advantage of the proposed SMILE HPRC architecture.

References

1. K.H. Tsoi, W. Luk, Axel: a heterogeneous cluster with fpgas and gpus, in *FPGA '10: Proceedings of the 18th Annual ACM/SIGDA International Symposium on Field Programmable Gate Arrays* (ACM, New York, 2010), pp. 115–124
2. T.A. El-Ghazawi, E. El-Araby, M. Huang, K. Gaj, V.V. Kindratenko, D.A. Buell, The promise of high-performance reconfigurable computing. IEEE Comput. **41**(2), 69–76 (2008)
3. C. Inc., The supercomputing company cray xd1 supercomputer. IEEE Computer, http://www.hpc.unm.edu/~tlthomas/buildout/Cray_XD1_Datasheet.pdf. Accessed 27 Mar 2013
4. K. Morris, Cots supercomputing (2007), http://www.fpgajournal.com/articles_2007/20070710_cots.htm/
5. TOP500, Top500 list June (2010), http://www.top500.org/list/2010/06/. Accessed 27 Mar 2013
6. S. Matsuoka, The tsubame cluster experience a year later, and onto petascale tsubame 2.0, in *Proceedings of the 14th European PVM/MPI User's Group Meeting on Recent Advances in Parallel Virtual Machine and Message Passing Interface* (Springer, Berlin, 2007), pp. 8–9
7. D.A. Buell, T.A. El-Ghazawi, K. Gaj, V.V. Kindratenko, Guest editors' introduction; High-performance reconfigurable computing. IEEE Comput. **40**(3), 23–27 (2007)
8. M.B. Gokhale, P.S. Graham, *Reconfigurable Computing Reconfigurable Computing, Accelerating Computation with Field-Programmable Gate Arrays* (Springer, Dordrecht, 2005)
9. Renwick Ron: SGI's Approach to Multi-paradigm Computing (2007), http://www.arsc.edu/files/arsc/news/archive/fpga/Tue-1330-Renwick.pdf. Accessed 27 Mar 2013
10. SGI: Sgi RASC RC100 blade (2006), http://www.sgi.com/pdfs/3939.pdf. Accessed 27 Mar 2013
11. S. Comp., Src-7: reconfigurable general purpose computing system, Tech. Rep., SRC Computers Inc (2007), http://www.srccomp.com/techpubs/docs/SRC_MAP_69226-JA.pdf. Accessed 27 Mar 2013
12. J.M. Arnold, D.A. Buell, E.G. Davis, Splash 2, in *SPAA '92: Proceedings of the Fourth Annual ACM Symposium on Parallel Algorithms and Architectures* (ACM, New York, 1992), pp. 316–322
13. L. Moll, M. Shand, A. Heirich, Sepia, Scalable 3d compositing using pci pamette, in *FCCM '99: Proceedings of the Seventh Annual IEEE Symposium on Field-Programmable Custom Computing Machines* (IEEE Computer Society, Washington, DC, 1999), p. 146
14. R. Sass, W.V. Kritikos, A.G. Schmidt, S. Beeravolu, P. Beeraka, Reconfigurable computing cluster (rcc) project: investigating the feasibility of fpga-based petascale computing, in *FCCM '07: Proceedings of the 15th Annual IEEE Symposium on Field-Programmable Custom Computing Machines* (IEEE Computer Society, Washington, DC, 2007), pp. 127–140
15. M. Yoshimi, Y. Nishikawa, M. Miki, T. Hiroyasu, H. Amano, O. Mencer, A performance evaluation of cube: one-dimensional 512 fpga cluster, in *ARC*. Lecture Notes in Computer Science, vol. 5992 (Springer, Berlin, 2010), pp. 372–381
16. C.L. Cathey, J.D. Bakos, D.A. Buell, A reconfigurable distributed computing fabric exploiting multilevel parallelism, in *Proceedings of the 14th Annual IEEE Symposium on Field-Programmable Custom Computing Machines (FCCM'06)* (IEEE Computer Society, Washington, DC, 2006), pp. 121–130
17. J. Wawrzynek, M. Oskin, C. Kozyrakis, D. Chiou, D.A. Patterson, S.-L. Lu, J.C. Hoe, K. Asanovic, Tech. Rep. UCB/EECS-2006-158, EECS Department, University of California, Berkeley (November 2006), http://www.eecs.berkeley.edu/Pubs/TechRpts/2006/EECS-2006-158.html
18. M. Showerman, J. Enos, A. Pant, V. Kindratenko, C. Steffen, R. Pennington, W. Hwu, Qp: a heterogeneous multi-accelerator cluster, in *Proceedings of the 10th LCI International Conference on High-performance Clustered Computing* (Linux Cluster Institute, 2009)
19. J. Castillo, P. Huerta: sc2v, Systemc to Verilog translator (2004), http://opencores.org/project,sc2v(2004). Accessed 27 Mar 2013

20. I. Foster, *Designing and Building Parallel Programs: Concepts and Tools for Parallel Software Engineering* (Addison-Wesley Longman Publishing Co. Inc., Boston, 1995)
21. J.S. Kim, S.J. Byun, A parallel Monte Carlo simulation on cluster systems for financial derivatives pricing, in *Congress on Evolutionary Computation* (IEEE, Edinburgh, 2005), pp. 1040–1044
22. G. Morris, M. Aubury, Design space exploration of the European option benchmark using hyperstreams, in *International Conference on Field Programmable Logic and Applications, FPL 2007*, Amsterdam, 2007, pp. 5–10
23. D.B. Thomas, J.A. Bower, W. Luk, Hardware architectures for Monte-Carlo based financial simulations, in *IEEE International Conference on Field Programmable Technology, FPT 2006*, Bangkok, 2006, pp. 377–380
24. G. Zhang, P. Leong, C. Ho, K. Tsoi, C. Cheung, D.-U. Lee, R. Cheung, W. Luk, Reconfigurable acceleration for Monte Carlo based financial simulation, in *Proceedings of IEEE International Conference on Field-Programmable Technology*, Singapore, 2005, pp. 215–222
25. V. Agarwal, L.-K. Liu, D.A. Bader, Financial modeling on the cell broadband engine, in *2008 IEEE International Symposium on PDPS*, Miami, FL, 2008, pp. 1–12
26. F. Black, M.S. Scholes, The pricing of options and corporate liabilities. J. Polit. Econ. **81**(3), 637–654 (1973)
27. J. Koza, F. Bennett, D. Andre, M. Keane, Genetic programming iii: Darwinian invention and problem solving. Evol. Comput. **7**, 451–453 (1999)
28. F. Rothlauf, *Representations for Genetic and Evolutionay Algorithms* (Springer, Heidelberg, 2006)
29. T. Higuchi, T. Niwa, T. Tanaka, H. Iba, H. deGaris, Evolving hardware with genetic learning: a first step towards building a Darwin machine, in *Proceedings of the Second International Conference on from Animals to Animats*, (MIT Press, Cambridge, MA, USA, 1993), pp. 417–424
30. J. Miller, P. Thomson, Aspects of digital evolution: Evolvability and architecture, in *Proceedings of International Conference Parallel Problem Solving from Nature—PPSN V*, 927–936 (Springer, 1998)
31. Q. Yu, C. Chen, C. Pan, Parallel genetic algorithms on programmable graphics hardware. Lect. Notes Comput. Sci. **3612**, 1051–1059 (2006)
32. R. Krohling, Y. Zhou, A. Tyrrell, Evolving fpga-based robot controllers using an evolutionary algorithm, in *Proceedings of I International Conference on Artificial Immune Systems*, Canterbury, 2002, pp. 41–46

An FPGA-Based Supercomputer for Statistical Physics: The Weird Case of Janus

M. Baity-Jesi, R.A. Baños, A. Cruz, L.A. Fernandez, J.M. Gil-Narvion, A. Gordillo-Guerrero, M. Guidetti, D. Iñiguez, A. Maiorano, F. Mantovani, E. Marinari, V. Martin-Mayor, J. Monforte-Garcia, A. Mũnoz Sudupe, D. Navarro, G. Parisi, M. Pivanti, S. Perez-Gaviro, F. Ricci-Tersenghi, J.J. Ruiz-Lorenzo, S.F. Schifano, B. Seoane, A. Tarancon, P. Tellez, R. Tripiccione, and D. Yllanes

M. Baity-Jesi (✉) • L.A. Fernandez • V. Martin-Mayor • B. Seoane
Departamento de Física Teórica I, Universidad Complutense, 28008 Madrid, Spain

Instituto de Biocomputación y Física de Sistemas Complejos (BIFI), Zaragoza, Spain
e-mail: marcobaityjesi@fis.ucm.es; laf@lattice.fis.ucm.es; victor@lattice.fis.ucm.es; seoane@lattice.fis.ucm.es

R.A. Baños • A. Cruz • J. Monforte-Garcia • A. Tarancon
Departamento de Física Teórica, Universidad de Zaragoza, 50009 Zaragoza, Spain

Instituto de Biocomputación y Física de Sistemas Complejos (BIFI), Zaragoza, Spain
e-mail: raquel.alvarez@unizar.es; andres@unizar.es; jmonforte@bifi.es; tarancon@unizar.es

J.M. Gil-Narvion • S. Perez-Gaviro • M. Guidetti
Instituto de Biocomputación y Física de Sistemas Complejos (BIFI), Zaragoza, Spain
e-mail: jmgil@bifi.es; mguidetti@bifi.es; spgaviro@unizar.es

A. Gordillo-Guerrero
Departamento de Ingeniería Eléctrica, Electrónica y Automática, Universidad de Extremadura, 10071 Cáceres, Spain

Instituto de Biocomputación y Física de Sistemas Complejos (BIFI), Zaragoza, Spain
e-mail: anto@unex.es

D. Iñiguez
Fundación ARAID, Diputación General de Aragón, Zaragoza, Spain

Instituto de Biocomputación y Física de Sistemas Complejos (BIFI), Zaragoza, Spain
e-mail: david.iniguez@bifi.es

A. Maiorano • D. Yllanes
Dipartimento di Fisica, La Sapienza Università di Roma, 00185 Roma, Italy

Instituto de Biocomputación y Física de Sistemas Complejos (BIFI), Zaragoza, Spain
e-mail: andrea.maiorano@roma1.infn.it; yllanesd@roma1.infn.it

F. Mantovani
Dipartimento di Fisica, Università di Ferrara and INFN – Sezione di Ferrara, Ferrara, Italy
e-mail: filimanto@fe.infn.it

E. Marinari
Dipartimento di Fisica, IPCF-CNR and INFN, La Sapienza Università di Roma, 00185 Roma, Italy
e-mail: enzo.marinari@uniroma1.it

W. Vanderbauwhede and K. Benkrid (eds.), *High-Performance Computing Using FPGAs*, DOI 10.1007/978-1-4614-1791-0_16, © Springer Science+Business Media, LLC 2013

Abstract In this chapter we describe the Janus supercomputer, a massively parallel FPGA-based system optimized for the simulation of spin-glasses, theoretical models that describe the behavior of glassy materials.

The custom architecture of Janus has been developed to meet the computational requirements of these models. Spin-glass simulations are performed using Monte Carlo methods that lead to algorithms characterized by (1) intrinsic parallelism allowing us to implement many Monte Carlo update engines within a single FPGA; (2) rather small data base (2 MByte) that can be stored on-chip, significantly boosting bandwidth and reducing latency. (3) need to generate a large number of good-quality long ($\geq$32 bit) random numbers; (4) mostly integer arithmetic and bitwise logic operations.

Careful tailoring of the architecture to the specific features of these algorithms has allowed us to embed up to 1024 special purpose cores within just one FPGA, so that simulations of systems that would take centuries on conventional architectures can be performed in just a few months.

A. Muñoz Sudupe
Departamento de Física Teórica I, Universidad Complutense, 28008 Madrid, Spain
e-mail: sudupe@fis.ucm.es

D. Navarro
Departamento de Ingeniería, Electrónica y Comunicaciones and Instituto de Investigación en Ingeniería de Aragón (I3A), Universidad de Zaragoza, 50018 Zaragoza, Spain
e-mail: denis@unizar.es

G. Parisi • F. Ricci-Tersenghi
Dipartimento di Fisica, IPCF-CNR, UOS Roma Kerberos and INFN, La Sapienza Università di Roma, 00185 Rome, Italy
e-mail: giorgio.parisi@roma1.infn.it; federico.ricci@roma1.infn.it

M. Pivanti
Dipartimento di Fisica, La Sapienza Università di Roma, 00185 Roma, Italy
e-mail: pivanti@fe.infn.it

J.J. Ruiz-Lorenzo
Departamento de Física, Universidad de Extremadura, 06071 Badajoz, Spain

Instituto de Biocomputación y Física de Sistemas Complejos (BIFI), Zaragoza, Spain
e-mail: ruiz@unex.es

S.F. Schifano
Dipartimento di Matematica e Informatica, Università di Ferrara and INFN – Sezione di Ferrara, Ferrara, Italy
e-mail: schifano@fe.infn.it

P. Tellez
Departamento de Física Teórica, Universidad de Zaragoza, 50009 Zaragoza, Spain
e-mail: ptellez@unizar.es

R. Tripiccione
Dipartimento di Fisica and CMCS, Università di Ferrara and INFN – Sezione di Ferrara, Ferrara, Italy
e-mail: tripiccione@fe.infn.it

1 Overview

This chapter describes Janus, an application-driven parallel and reconfigurable computer system, strongly tailored to the computing requirements of spin glass simulations.

A major challenge in condensed-matter physics is the understanding of glassy behavior (see, for instance [1]). Glasses are materials that do not reach thermal equilibrium in human lifetimes; they are conceptually important in physics and they have a strong industrial relevance (aviation, pharmaceuticals, automotive, etc.). Important material properties, such as the compliance modulus or the specific heat, significantly depend on time even if the material is kept for months (or years) at constant experimental conditions [2]. This sluggish dynamics, a major problem for the experimental and theoretical investigation of glassy behavior, places numerical simulations at the center of the stage.

Spin glasses are the prototypical glassy systems most widely studied theoretically [3, 4]. Simulating spin glasses is a computing grand challenge, as their deceivingly simple dynamical equations are at the basis of complex dynamics, whose numerical study requires large computing resources. In a typical spin-glass model, the dynamical variables, one calls them spins, are discrete and sit at the nodes of discrete D-dimensional lattices. In order to make contact with experiments, we need to follow the evolution of a large enough lattice, say a 3D system with 80^3 sites, for time periods of the order of 1 s. One Monte Carlo step (MCS)—the update of all the 80^3 spins in the lattice—roughly corresponds to 10^{-12} s, so we need some 10^{12} such steps, that is $\sim 10^{18}$ spin-updates. One typically wants to collect statistics on several ($\sim 10^2$) copies of the system, adding up to $\sim 10^{20}$ Monte Carlo spin updates. Therefore, performing this simulation program in an acceptable time frame (say, less than 1 year) requires a computer system able to update on average one spin per picosecond or less.

This analysis shows that accurate simulations of spin glasses have been a major computational challenge; the problem has been attacked in different ways, and the development of application-specific computers has been one of the options considered over the years. This chapter describes the Janus project, which has led to the development of the Janus reconfigurable computer, optimized for spin-glass simulations. Janus has played a major role in making the simulations described above possible on a reasonable time scale (order of months); it has provided the Janus collaboration with a major competitive advantage, which has resulted in ground-breaking work in the field of spin glasses as will be described later.

There are several reasons that make traditional computer architectures a poor solution for spin-glass simulations and at the same time suggests that a reconfigurable approach may pay very large dividends:

- The dynamical variables describing these systems only take a small number of discrete values (just two in the simplest case); sequences of bitwise logical operations are appropriate to compute most (not all) quantities involved in the simulation;

- A large amount of parallelism is easily identified; a large number of lattice locations can be processed independently, so they can be handled in parallel.
- The structure of the critical computational kernels is extremely regular, based on ordered loops that perform the same sequence of operations on data values stored at regularly stridden memory locations; the control structure of the program can therefore be easily cast in the form of simple state-machines. The control sequence is the same for all lattice locations, so a single instruction multiple data (SIMD) approach is appropriate and the control structure can be shared by many computational threads.

These points suggest an ideal architecture for a spin-glass engine, based on a very large number of computational cores; cores are extremely slim processors, able to perform only the required mix of logical manipulations and a limited set of arithmetic operations; many cores work concurrently, running the same thread under just one control structure; they process data fetched from memory by just one memory control engine. Seen from a different point of view, one may think of a streaming processor, working on a steady flow of data extracted from and flowing back to memory. As discussed in detail later on, the logical complexity of one such computational core is in the order of just a few thousand logical gates, so one can assemble them by the thousands in just one integrated circuit. This promises significant benefits, provided that the huge amount of data needed to keep all these processors busy can be supplied by the memory system; this is a serious problem that can be handled in this case as the size of the simulation database is small enough to be accommodated on chip.

The requirements described above are slightly at variance with traditional architectures. On the one hand standard CPUs offer features not really exploited by our regular programming paradigm (out of order execution, branch prediction, cache hierarchy); on the other hand they are very limited in the extraction of the available parallelism. Indeed, at the time the Janus project started (early 2006), state-of-the-art simulation programs running on state-of-the-art computer architectures were only able to exploit a tiny fraction of the available parallelism and had an average update time of one spin every $\sim$1 ns, meaning that the simulation campaign outlined above would proceed for centuries.

Curiously enough, the time frame that has seen the development of the Janus project coincides with that in which computer architectures have strongly evolved towards wider and more explicit parallelization: many-core processors with $\mathcal{O}(10)$ cores are now widely available and graphics processing units (GPUs) now have hundreds of what can be regarded as "slim" cores. Today, one might see an ideal spin-glass simulation engine as an *application-specific* GPU, in which (1) data paths are carefully tailored to the specific mix of required logical (as opposed to arithmetic and/or floating-point) operations; (2) the control structure is shared by a much larger number of cores than typical in state-of-the-art GPUs; (3) data allocation goes to on-chip memory structures, and (4) the memory controller is optimized for the access patterns typical of the algorithm.

Architectures available in 2011–2012 have indeed improved performance for spin glass simulations by about one order of magnitude with respect to what was available when the Janus project started (slightly better than one would predict according to Moore's law, see later for a detailed analysis), but standard commercial computers are even today not a satisfactory option for large-scale spin-glass studies.

We already remarked that, over the years, this state of affairs has motivated the development of several generations of application-driven, spin-glass-optimized systems; this approach has been often taken by computational physicists in several areas, such as Lattice QCD [5–7] or the simulation of gravitationally coupled systems [8]; early attempts for spin systems were performed more than 20 years ago [9], and—more recently—an approach based on reconfigurable computing was pioneered [10].

The Janus[1] project has continued along this line, developing a large reconfigurable system, based on field programmable gate-arrays (FPGAs). FPGAs are slow with respect to standard processors. This is more than offset by large speedup factors, allowed by architectural flexibility. A more radically application-driven approach would be to consider an application-specific integrated circuit (ASIC), a custom-built integrated circuit, promising still larger performance gains, at the price of much larger development time and cost, and much less flexibility in the design.

The remainder of this chapter is organized as follows: in Sect. 2 we describe the physics systems that we want to simulate, elaborating on their relevance both in physics and engineering; Sect. 3 provides details on the Monte Carlo simulation approach used in our work; Sect. 4 describes the Janus architecture and its implementation, after which Sect. 5 gives a concrete example. Section 6 summarizes the main physics results obtained after more than 3 years of continuous operation of the machine. Section 7 assesses the performance of Janus on our spin-glass simulations, using several metrics, and compares with more standard solutions. We consider both those technologies that were available when Janus was developed and commissioned and those that have been developed since the beginning of the project ($\approx$2006). We also briefly discuss the performance improvements that may be expected if one re-engineers Janus on the basis of the technology available today. Our conclusions and outlook are in Sect. 8.

2 Spin Glasses

What makes a spin glass (SG) such a complex physical system is frustration and randomness (see Fig. 1). One typical example is a metal in which we replace some of its metallic atoms with magnetic ones. Qualitatively, its dynamical behavior is as follows: the dynamical variables, the spins, represent atomic magnetic moments, interacting via electrons in the conduction band of the metal and inducing an

[1] From the name of the ancient Roman god of doors and *gates*.

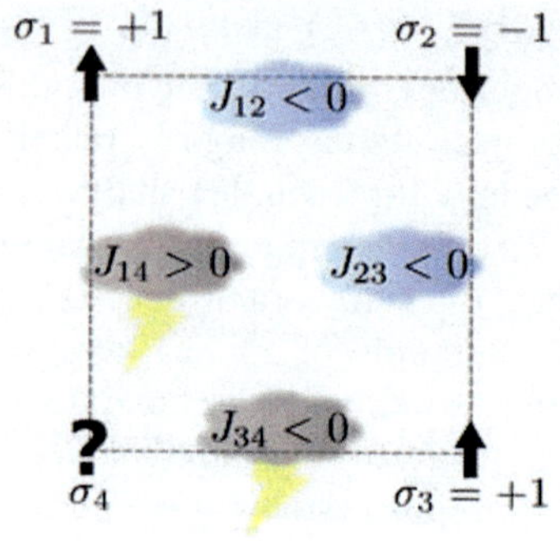

Fig. 1 Four neighboring points of a regular Ising spin lattice (for clarity we show a 2D system). Spins ($\sigma_i = \pm 1$) sit at the edges of the lattice; each link joining two edges has a coupling constant $J_{ij} = \pm 1$. If $J_{ij} > 0$ a link is *satisfied* if $\sigma_i = \sigma_j$, while $J_{ij} < 0$ requires that $\sigma_i \neq \sigma_j$. One can easily check that, for the J_{ij} values in the picture, no assignment of the σ_is exists that satisfies all links (in general, this happens if an odd number of J_{ij} along the circuit has the same sign). This is called *frustration*

effective interaction which changes in sign (the RKKY interaction) depending on the spatial location. In some materials, it is easy for the magnetic moments to lie in only one direction (and not in the original three-dimensional space) so we can consider that they only take values belonging to a finite set. Finally we can assume that spins sit at the nodes of a crystal lattice.[2]

At some sites (i and j) in the lattice, neighbor spins (σ_i and σ_j) may lower their energy if they have the same value: their *coupling constant* J_{ij}, a number assigned to the lattice link between i and j, is positive. However elsewhere in the lattice, with roughly the same probability, two neighboring spins may prefer to have different values (in this case, $J_{ij} < 0$). A lattice link is *satisfied* if the two corresponding spins are in the energetically favored configuration. In spin glasses, positive and negative coupling constants occur with the same frequency, as the spatial distribution of positive or negative J_{ij} is random; this causes *frustration*. Frustration means that it is impossible to find an assignment for the σ_i that satisfies all links (the concept is sketched in Fig. 1 and explained in the caption).

Models that describe this behavior are defined in terms of the following energy function:

$$\mathcal{H} = -\sum_{\langle i,j \rangle} J_{ij} \delta(\sigma_i, \sigma_j); \tag{1}$$

σ_i is the spin at lattice site i; it takes discrete values belonging to a finite set of q elements, δ is the Kronecker delta function and J_{ij} are the coupling constants between the two spins; angle brackets mean that the sum is restricted to pairs of nearest neighbors in the lattice

---

[2] A typical example of an Ising spin glass is $Fe_{0.5}Mn_{0.5}TiO_3$.

Models described by (1) are usually referred to as Potts spin glasses. One typically considers values of q ranging from just two to eight or ten. The simplest case ($q = 2$) has been especially considered since it was first proposed more than 30 years ago; it is known as the Edwards–Anderson spin glass [11]. One usually writes its energy function in a slightly different form:

$$\mathcal{H} = -\sum_{\langle ij \rangle} \sigma_i J_{ij} \sigma_j; \tag{2}$$

Here, the symbols have the same meaning as in (1) but in this case $\sigma = \pm 1$; Eq. (2) goes over to (1)—apart from a constant term that does not affect the dynamics—if one appropriately rescales the values of the J_{ij}.

The coupling constants J_{ij} are fixed and chosen randomly to be ± 1 with 50% probability. A given assignment of the $\{J_{ij}\}$ is called a *sample*. Some of the physical properties (such as internal energy density, magnetic susceptibility, etc.) do not depend on the particular choice for $\{J_{ij}\}$ in the limit of large lattices (self-averaging property). However, in the relatively small systems that one is able to simulate, it is useful to average results over several samples.

Frustration makes it hard to answer even the simplest questions about the model. For instance, finding the spin configuration that minimizes the energy for a given set of $\{J_{ij}\}$ is an NP-hard problem [12]. In fact, our theoretical understanding of spin-glass behavior is still largely restricted to the limit of high spatial dimensions, where a rich picture emerges, with a wealth of surprising connections to very different fields [13].

In three dimensions, we know experimentally [14] and from simulations [15] that a spin-glass reaches an ordered phase below a critical temperature T_c. In the cold phase ($T < T_c$) spins *freeze* in some disordered pattern, related to the configuration of minimal free energy. For temperatures (not necessarily much) smaller than T_c spin dynamics becomes exceedingly slow. In a typical experiment one quickly cools a spin glass below T_c, then waits to observe the system evolution. As time goes on, the size of the domains where the spins coherently order in the (unknown to us) spin-glass pattern, grows.

Domain growth is sluggish, however: in typical spin-glass materials after 8 h at low temperature ($T = 0.73 T_c$), the domain size is only around 40 lattice spacings [16]. The smallness of the spin-glass ordered domains precludes the experimental study of equilibrium properties, as equilibration would require a domain size of the order of $\simeq 10^8$ lattice spacings. However, an opportunity window opens for numerical simulations. In fact, in order to understand experimental systems we only need to simulate lattices sufficiently larger than the typical domain. This crucial requirement has been met for the first time in the simulations made possible by Janus.

3 Monte Carlo Simulations of Spin Glasses

Spin glasses have been heavily studied numerically with Monte Carlo techniques and the Janus architecture has been designed with the main goal of exploiting every performance handle available in this computational area. In this section we provide a simple overview of the relevant algorithms, focusing on those features that will have to be carefully optimized on our reconfigurable hardware. For simplicity, we only treat the Edwards–Anderson model of Eq. (2), defined on a 3D lattice of linear size L; the Monte Carlo algorithms that apply to more general Potts model are similar and—most important in this context—they have essentially the same computational and architectural requirements.

We focus on the Heat-Bath (HB) algorithm ([17]) that ensures that system configurations $\mathcal{C}$ are sampled according to the Boltzmann probability distribution

$$P(\mathcal{C}) \propto \exp\left(-\frac{H}{T}\right), \tag{3}$$

describing the equilibrium distribution of configurations of a system at constant temperature $T = \beta^{-1}$. This is one well-known Monte Carlo method; see, e.g., [18] for a review of other approaches;

Let us focus on a spin at site k of a 3D lattice; its energy is

$$E(\sigma_k) = -\sigma_k \sum_{m(k)} J_{km}\sigma_m \;=\; -\,\sigma_k\phi_k\,, \tag{4}$$

where the sum runs over the six nearest neighbors, $m(k)$, of site k; ϕ_k is usually referred to as the *local field* at site k. In the HB algorithm, one assumes that at any time any spin is in thermal equilibrium with its surrounding environment, meaning that the probability for a spin to take the value $+1$ or -1 depends only on its nearest neighbors. Following (3), the probability for the spin to be $+1$ is

$$P(\sigma_k = +1) = \frac{e^{-E(\sigma_k=+1)/T}}{e^{-E(\sigma_k=+1)/T} + e^{-E(\sigma_k=-1)/T}} = \frac{e^{\phi_k/T}}{e^{\phi_k/T} + e^{-\phi_k/T}}\,, \tag{5}$$

The algorithm then is an iteration of just two steps:

1. Pick one site k at random and compute the local field ϕ_k (4).
2. Assign to σ_k the value $+1$ with probability $P(\sigma_k = +1)$ as in (5). This can be done by generating a random number r, uniformly distributed in $[0,1]$, and setting $\sigma_k = 1$ if $r < P(\sigma_k = 1)$, and $\sigma_k = -1$ otherwise.

A full MCS is the iteration of the above scheme for L^3 times. By iterating many MCS, the system evolves towards statistical equilibrium. Figure 2 shows a snapshot of a large spin-glass lattice at a later stage of the Monte Carlo evolution for a

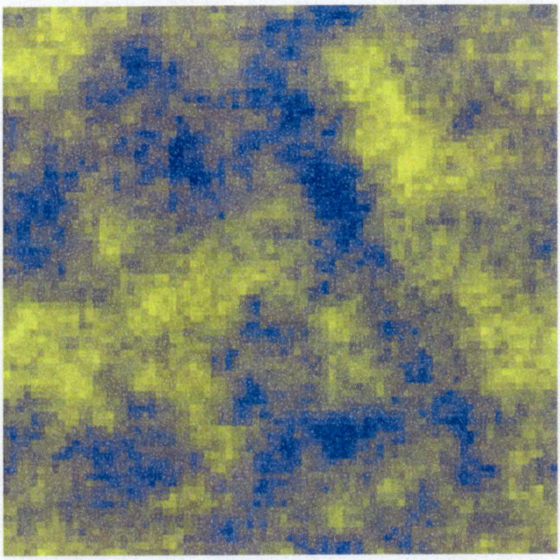

Fig. 2 Domain growth for an Edwards–Anderson spin glass of size $L = 80$ at $T = 0.73T_c$, after 2^{36} Monte Carlo steps, corresponding to a time scale of ≈ 0.1 s

temperature lower than the critical one, showing the build up of an ordered domain structure.

A further critically important tool for Monte Carlo simulations is Parallel Tempering (PT) [19]. Dealing with frustration (as defined above) means handling rough free-energy landscapes and facing problems such as the stall of the system in a metastable state. PT helps to overcome these problems by simulating many copies of the system in parallel (hence the name) at different (inverse) temperatures β_i and allowing copies, whose β (energy) difference is $\Delta\beta$ (ΔE), to exchange their temperatures with probability equal to $\min\{1, \exp(\Delta\beta\Delta E)\}$. Following PT dynamics, configurations wander from the physically interesting low temperatures, where relaxation times can be long, to higher temperatures, where equilibration is fast and barriers are quickly traversed; they explore the complex energy landscape more efficiently, with correct statistical weights. For an introduction to PT, see, for instance, [20].

We now sketch the steps needed to implement a Monte Carlo simulation on a computer. First, one maps physical spin variables onto bits by the following transformation, $\sigma_k \to S_k = (1 - \sigma_k)/2$, allowing to turn most (not all) steps of the algorithm into logic (as opposed to arithmetic) operations. The following points are relevant:

1. The kernel of the program is the computation of the local field ϕ_k, involving just a few logic operations on discrete variable data.
2. The local field ϕ_k takes only the 7 even integer values in the range $[-6, 6]$, so probabilities $P(\sigma_k = +1) = f(\phi_k)$ can be stored in a look-up table.
3. High-quality random numbers are necessary to avoid spurious spatial correlations between lattice sites, as well as temporal correlations in the sequence of spin configurations.

4. Under ergodicity and reversibility assumptions, the simulation retains the desired properties even if each Monte Carlo step visits each lattice site exactly once, in any deterministic order.
5. Several sets of couplings $\{J_{km}\}$ (i.e., *different samples*) are needed. An independent simulation has to be performed for every sample, in order to generate properly averaged results.
6. One usually studies the properties of a spin-glass system by comparing the so-called *overlaps* of two or more statistically independent simulations of the *same* sample, starting from uncorrelated initial spin configurations (copies of a sample are usually referred to as *replicas*).

The last three points above identify the parallelism available in the computation; farming easily takes advantage for 5 and 6, while for 4 we need a more accurate analysis. In fact, if we label all sites of the lattice as *black* or *white* in a checkerboard scheme, all black sites have their neighbors in the white site set, and *vice versa*: in principle, we can perform the steps of the algorithm on all white or black sites in parallel.

We will see in the following that the Janus architecture allows us to exploit to a very large degree the parallelism of point 4 above. If one tries the same approach with a standard processor, mapping independent spins of one sample to the bits of one machine word and applying bitwise logical operations, one quickly meets a bottleneck in the number of required random numbers (this approach is known in the trade as synchronous multi-spin coding, SMSC). An alternate approach (known as asynchronous multi-spin coding, AMSC) maps the same spin of independent samples to the bits of the machine word and uses the same random number to decide on the evolution of all these spins (this introduces a tolerable amount of correlation). This strategy does increase overall throughput but does not decrease the time needed to perform a given number of MCS, which is a critical parameter.

4 Janus: The Architecture

This section describes the Janus architecture, starting from its overall organization, and then going into the details of its hardware structure, of its reconfigurable components, and of its supporting software environment. The idea to develop Janus was born in the early years of this century. After some preliminary analysis helped estimate the level of performance that one could expect, preliminary work really started in late 2005. Early prototypes were available in late 2006, and a large-scale machine was commissioned before the end of 2007. After acceptance tests were completed, Janus become operational for physics in spring 2008. Since then, it has continuously been up and running, and it still provides computer power for Monte Carlo simulations.

4.1 Global Structure

The Janus supercomputer is a modular system composed of several Janus modules. Each module houses 17 FPGA-based subsystems: 16 so-called scientific processors (SPs) and one input/output processor (IOP). Janus modules are driven by a PC (Janus host). For our application, the Janus module is the system partition that exploits the parallelism available in the simulation of one spin glass sample. Several modules are then used to farm out the simulation of many spin glass samples that evolve independently.

We generically refer to SPs and the IOP as *nodes*. The 16 SPs are connected by a 2D nearest-neighbor toroidal communication network, so an application can be mapped onto the whole set of SPs (or on a subset thereof). A further point-to-point network links the IOP to each SP; it is used for initialization and control of the SPs and for data transfer.

The Janus host PC plays a key role of master device: a set of purpose-made C libraries are written using low levels of Linux operating system in order to access the raw Gigabit Ethernet level (excluding protocols and other unhelpful layers adding latencies to communications). Moreover two software environments are available: an interactive shell written in Perl mostly used for testing and debugging or short preliminary runs and a set of C libraries strongly oriented to the physics user, making it relatively easy to set up simulation programs for Janus.

The FPGA panorama was various and the choice of a device for Janus was driven by the simple idea that the only important feature is the availability of memory and logic elements in order to store lattices as large as possible and to house the highest number of update engines. Large on-chip memory size and many logic elements are obviously conflicting requirements; each FPGA family offered different trade-offs at the time of the development phase of Janus.

Our preliminary prototype was developed in 2005 using a PCI development kit housing an Altera Stratix S60 FPGA providing ~57,000 logic elements and ~5 MB of embedded memory. The first two Janus prototype boards developed in 2006 had Xilinx Virtex-4 LX160 FPGAs while the final implementation of the system was based on Xilinx Virtex-4 LX200 FPGAs.

The choice between Altera or Xilinx FPGAs has not been fully trivial. While both families had approximately the same amount of logic elements,[3] the amount of on-chip memory was different: Altera Stratix-II FPGAs offered ~8 Mb organized in three degrees of granularity allowing us to efficiently exploit only ~ 50% of it. Conversely Xilinx Virtex-4 LX200 FPGAs provided ~6 Mb of embedded memories made up of relatively small blocks that we could use very efficiently for our design.

The main clock for Janus is 62.5 MHz; we set a rather conservative clock frequency, trying to minimize time-closure problems when mapping the reconfig-

[3]We consider the largest devices of both FPGA families available when we had to make a final decision: Altera Stratix-II 180 and Xilinx Virtex-4 LX200.

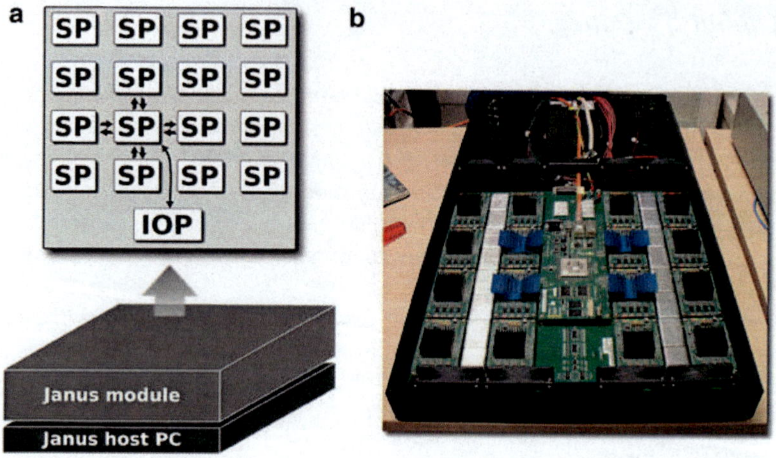

Fig. 3 **(a)** Topology of a Janus board: each SP communicates with its nearest neighbors in the plane of the board. **(b)** Janus board housed in a Janus module

urable portion of an application onto the FPGAs. Selected parts of the system use faster clocks: for instance the Gigabit-ethernet interface within the IOP has a 125-MHz clock (called I/O-clock), as needed by Gigabit protocol constraints. Perfect bandwidth balance is achieved: the Gigabit protocol transfers 1 byte per I/O-clock cycle (i.e., 8 bits every 8 ns) and a so-called *stream router* forwards data to the Janus world with a rate of 2 bytes per system-clock cycle (i.e., 16 bits every 16 ns). Furthermore link connecting the IOP with the SPs is 8 bit wide and runs a double data rate protocol so the bandwidth continues to be balanced.

The 17 FPGA-based nodes are housed on small daughter-cards plugged into a mother-board (see Fig. 3a for a sketchy block diagram and Fig. 3b for a picture of one Janus module). We used daughter-cards for all nodes to make hardware maintenance easier and also to allow an easier technology upgrade. The first large Janus system, deployed in December 2007 at Zaragoza, has 16 modules and 8 Janus host PCs assembled together in a standard 19" rack. More details on the Janus architecture and its implementation are given in [21–24].

4.2 Programming Paradigm

The programming framework developed for Janus is intended to meet the requirements of the prevailing operating modes of Janus, i.e., supporting (re)configuration of SPs, initialization of memories and data structures within the FPGA, monitoring the system during runs, interfacing to memory;

Applications running on Janus can be thought of as split into two sub-applications, one, called *software application* SA, written, for example, in C,

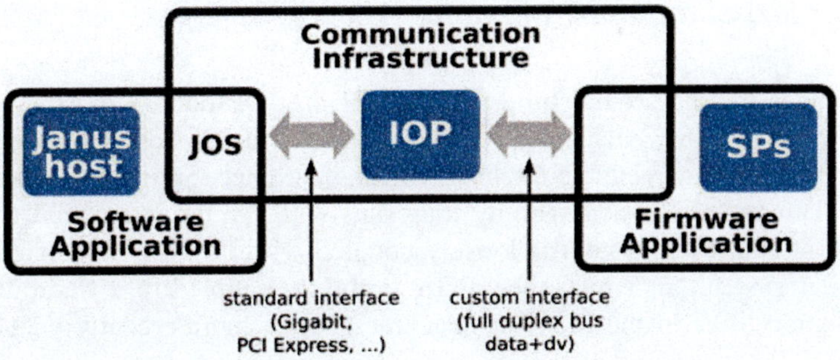

Fig. 4 Framework of an application running on the Janus system

and running on the Janus host. The other, called *firmware application* FA, written, for example, in VHDL, runs on the SP nodes of a Janus board. As shown in Fig. 4, the two entities, SA and FA are connected together by a *communication infrastructure* CI, which is a logical block including physically the IOP and which allows to exchange data and to perform synchronization operations, directly and in a transparent way.

The CI abstracts the low-level communication between SA and FA applications, implemented in hardware by the IOP and its interfaces. It includes, a C communication library linked by the SA, a communication firmware running on the IOP processor, interfacing both the host PC and the SP processor and a VHDL library linked by the FA.

Firmware running on the IOP communicates with the host PC via a dual Gigabit channel, using the standard RAW-Ethernet communication protocol. To guarantee reliability of the communication in the direction IOP to Janus host, we adopt the *Go-back-N* protocol [25] allowing us to reach approximately the 90% of the full Gigabit bandwidth, when transferring messages of the order of 1MB, and using the maximum data payload per frame, 1,500 bytes. This is enough and safe in a context of spin-glass simulations.

Communications from IOP to host PC do not adopt any communication protocol since the IOP interface, barring hardware errors, ensures that no packets are lost. Incoming frames are protected by standard Ethernet CRC code, and errors are flagged by the IOP processor.

Data coming from the SA application packed as burst of frames are routed to one of the devices supported by the IOP firmware. These devices can be internal to the IOP (e.g., memory interface, internal control registers) or external (e.g., SPs).

Developers of Janus-based applications have to provide their SA and FA relaying on the CI communication infrastructure to make the two applications collaborative. A typical SA configures the SPs with the appropriate firmware, using functions provided by the communication library, loads input data to the FA, starts and checks the status of the SP logic and waits for incoming results.

4.3 IOP Structure and Functions

The guidelines for the architectural structure of the IOP come from the original idea of the project that each Janus core is a "large" co-processor of a standard PC running Linux, connected to the host with standard networking interfaces. Spin glass simulations are characterized by long runs with limited interaction with the Janus-host so that each system is loosely coupled with its host. This is different from similar systems in which the FPGA is tightly coupled to a traditional PC and its memory, like in the Maxwell machine [26] or more recently on Maxeler computers [27].

Each simulation starts with the upload of an FA configuring the FPGA of the SPs, followed by the initialization of all the FA data structures (i.e., upload via Gigabit of lattice data, random numbers seeds, and other physical parameters). After this step has completed, the SA starts the simulation and polls the status of each Janus module. The SA detects the end of the FA task and initiates the download of the results (e.g., the final physical configurations) and in some cases runs data analysis. All these operations involve the CI and in particular the firmware of IOP under the control of the Janus-host.

In some cases (e.g., when running the parallel tempering) the SA requires data exchange across different SPs during the run: in this case the IOP performs the additional task of gathering data from all SPs, performing a small set of operations on them and re-scattering data to SPs.

From this simplified operation scheme it is clear that the IOP plays a key role between the Janus operating system (JOS) running on the Janus-host and the FA running on each SP.

The IOP, like the SPs, is based on a Virtex 4 XC4LX200 FPGA but, unlike the SPs, has 8 MB static memory, a PROM programming device for FPGA boot and some I/O interfaces: a dual Gigabit channel, a USB channel and a slow serial link for debug.

The current IOP firmware is not a general purpose programmable processor: its role is to allow data streaming from the Janus-host to the appropriate destination (and back), under complete control of the JOS.

As shown in Fig. 5, the IOP structure is naturally split into two functional areas called *IOlink* and *MultiDev* blocks.

The IOlink block handles the I/O interfaces between IOP and the Janus-host (Gigabit channels, serial and USB ports). It supports the lower layers of the Gigabit Ethernet protocol and performs CRC checks to ensure data integrity.

The MultiDev block contains a logic device associated with each hardware sub-system that may be reached for control and/or data transfer: a memory interface for the staging memory, a programming interface to configure the SPs, an SP interface to handle communication with the SPs (after they are configured), and several service/debug interfaces are present. Each interface receives a data stream, strips header words and forwards the stream to the target hardware component (memory, SPs, etc.) as encoded in the header.

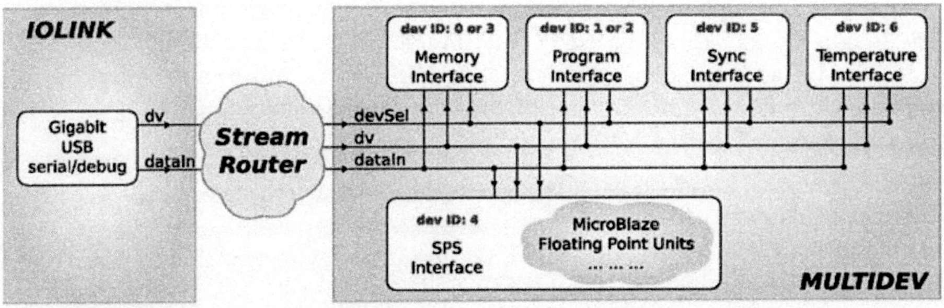

Fig. 5 Block diagram of the IOP architecture

In order to route the input stream coming from IOlink we implemented a module called *Stream Router* that scans the data stream and recognizes the masks associated with each device within the MultiDev.

The current IOP implementation uses ∼4% of the total logic resources and ∼18% of the total memory resources of the FPGA; this means that from the point of view of the Janus developers a future expansion of the IOP functionalities is possible and easy to implement simply adding devices to the MultiDev entity. For instance, one might add floating point units in order to perform floating point arithmetic directly "on module" or include a MicroBlaze™ microprocessor opening the possibility to use the Janus core as a standalone computer.

From the point of view of the Janus user, the IOP, together with a software layer running on the host PC, is meant to be part of the CI and therefore a stable structure in the Janus system. What is not fixed, on the other hand, are the functionalities implemented on the SPs that can perform in principle arbitrary algorithms (of course taking into account the hardware design/limitations).

In the following sections we will describe the details of the software layer interfacing the SA of a user to the IOP and SPs and later an example of just one specific SP firmware used for spin-glass simulations along the lines described in previous sections.

4.4 Software Layer

As part of the CI we developed a software environment running on each Janus host able to cope with any SP-based FA, as long as the latter adheres to a model in which Janus is a memory-based coprocessing engine of the host computer; user programs can therefore use load/store instructions to move their data onto the FA storage devices (e.g., FPGA embedded memories) and activate, stop, and control Janus processes mapped in the FA.

This model is supported by a host-resident run-time environment that we call JOS. It runs on any Linux-based PC and builds on a low-level C library, based on standard Unix raw network sockets. It implements the protocol needed to communicate with the IOP firmware on the Gbit Ethernet link.

For the application user, JOS consists of:

- A multi-user environment for Janus resource abstraction and concurrent jobs management (`josd`);
- A set of libraries with primitives in order to interact with the CI level (e.g., IOP devices), written both in Perl and C (`JOSlib`);
- A set of FA modules for scientific applications and the corresponding C libraries needed to control them via the josd environment (`jlib`).

josd is a background job running on the Janus host, providing hardware abstraction and a stable interface to user applications. It hides all details of the underlying structure of Janus, mapping it to the user as a simple grid of SPs. It interfaces via Unix socket APIs, so high-level applications may be written in virtually any programming language. Whenever a new FA module is developed, new primitives controlling that module are added to JOSlib. User programs, written in high-level languages, use these primitives in order to schedule and control Janus-enabled runs. For debugging and test, an interactive shell (`JOSH`), written in Perl and also based on JOSlib, offers complete (and potentially dangerous) access to all Janus resources for expert users. It provides direct access to the CI, allowing to communicate with the IOP and drive all its internal devices.

5 SP Firmware: An Application Example

SPs are fully configurable devices, so they can be tailored to perform any computational task compatible with the available resources and complying with the communication and control protocols of the CI. In this section we discuss one example, taken from a set of several applications that we developed for spin glass simulations.

Tailoring our FAs for Janus has been a lengthy and complex procedure, justified by the foreseen long lifetime of each application and by an expectation of huge performance gains. Obviously, a high-level programming framework that would (more or less) automatically split an application between standard and reconfigurable processors and generate the corresponding codes would be welcome. Unfortunately the tools available at the time of the development of the machine do not deliver the needed level of optimization and for Janus this work was done manually using a hardware description language (VHDL).

Our implementation of model and algorithm tries to exploit all internal resources in the FPGA in a consistent way (see [21] for a detailed description). Our VHDL code is parametric in several key variables, such as the lattice size and the number of parallel updates. In the following description we consider, for definiteness, a lattice of 80^3 sites corresponding to the typical simulation described in the introduction.

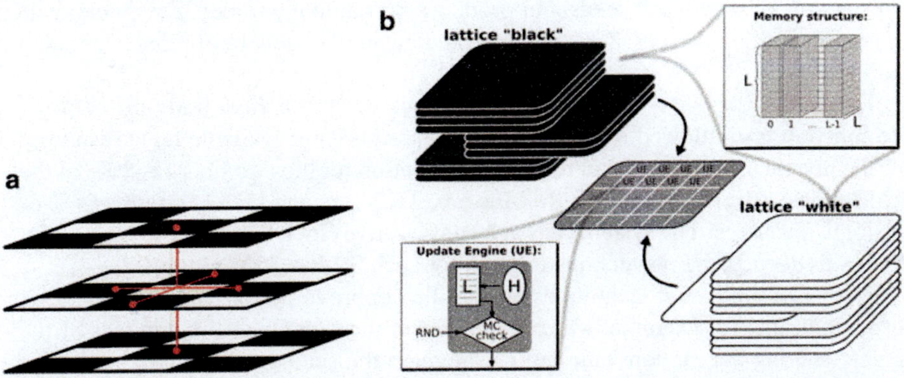

Fig. 6 **(a)** Checkerboard model used for updating spins in parallel. **(b)** Logic diagram of the spin update process performed in each SP

As described in Sect. 3 these algorithms do not allow us to update at the same time spins sitting next to each others in the lattice. On the other hand we can organize the spin update process in two steps such that we process in parallel up to half of the spins at each step. We can in other words split our 3D lattice of spins in a checkerboard scheme and update first all the white spins and then all the black ones (see Fig. 6a).

Virtex 4 LX200 FPGAs come with many small embedded RAM blocks; they can be combined and stacked to naturally reproduce a 3D array of bits representing the 3D spin lattice and making it possible to access data with almost zero latency. We used this memory structure to store the lattice variables, carefully reordered and split in black and white sets. A similar storage strategy applies to the read-only couplings. After initialization, the machinery fetches all neighbor spins of a portion of a plane of spins and feeds them to the update engines.

The flexibility given by the use of FPGAs allows us to implement a number of update engines matching the number of spins processed in parallel. Each update engine returns the processed (updated) spins to be stored at the same given address. For the Edwards–Anderson spin glass, we update from 800 to 1024 spins simultaneously; the update logic is made up of a matching number of identical update engines. Each engine receives the 6 nearest-neighbor spins, 6 couplings and one 32-bit random number; it then computes the local field, which is an address to a probability look-up table; the random number is then compared to the extracted probability value and the updated spin is obtained (see Fig. 6b).

Look-up tables are small 32-bit wide memories instantiated as *distributed* RAM. There is one such table for each update cell. Random number generators are implemented as 32-bit Parisi–Rapuano generators [28], requiring one sum and one bitwise XOR operation for each number. To sustain the update rate we need one fresh random value for each update engine; our basic random number engine has 62 registers of 32 bits for the seeds and produces 80 random numbers per clock cycle by combinatorial cascades; 8–13 of these random number generators are instantiated in the FA.

The number of $800 \cdots 1024$ updates per clock cycle is a good trade-off between the constraints of allowed RAM-block configurations and available logic resources for the update machinery. Total resource occupation for the design is 75–80% of the RAM blocks (the total available RAM is $\simeq$672 KB for each FPGA) and 85–94% of the logic resources. The system runs at a conservative clock frequency of 62.5 MHz. At this frequency, the power consumption for each SP is $\simeq$35 W.

Further optimization techniques are used to improve performances and reduce resource usage: for example, we can simulate at the same time two replicas of the lattice (sharing the random numbers) to increase the statistic, at a small additional cost in terms of memory consumption. Details on this and other tricks are available elsewhere [21].

6 An Overview of Physics Results

In almost 4 years of continuous operation, Janus has provided the Janus collaboration with a major competitive advantage, allowing us to consider lattice sizes and simulation time scales that other researchers could only dream of. This has resulted in ground-breaking work in the field of spin-glasses; major work has been done on the Potts glass model (for several values of q) and on the Edwards–Anderson model with Ising spins described in Sect. 2. We refer the reader to the original papers for a full account; here we provide only a very short and sketchy panorama of our main results.

As for all problems that are not really understood, spin glasses should be studied with a large variety of tools. Indeed, we do not know where the essential clue will come from, so being able to perform different types of simulations efficiently is very important. FPGA reconfigurability is a major asset in this context.

From the experimental point of view, a major limitation is imposed by the slow dynamics, which makes it impossible to study the equilibrium phase diagram. In order to reproduce and understand experimental results, it is very important to perform simulations that match experimental conditions. Experimentalists work with very large samples (containing some $N \sim 10^{23}$ spins) and follow the dynamical evolution for time scales that span several orders of magnitude. They focus their attention on *self-averaging* quantities. These magnitudes are such that, for large enough samples, they take the same value irrespective of the particular configuration of the couplings (technically, their sample variance scales as an inverse power of the number of spins, N). Examples of self-averaging quantities include the internal energy, the magnetic susceptibility (i.e., the derivative of the magnetization density with respect to the applied field), or some correlation functions. Self-averaging is a most important property: it makes it possible to compare data from different experimental teams, working with different spin-glass samples of nominally identical chemical composition.

Hence, if our *dynamic* simulations are to imitate experiments, we need to follow the dynamics of a single system for a time scale spanning several orders of magnitude. The simulated system should be as large as possible, in order to minimize the artifacts introduced by the finite size of the simulated samples. The only good news come from the self-averaging nature of the quantities that one studies: if the simulated systems are large enough, one may average the obtained results over a moderate number of samples (most of the time experimentalists work with just one or two samples!).

The reader may rightly question about how large is "large enough". The point is that, as time proceeds, glassy domains grow in the system, whose size defines a time-dependent "coherence length" $\xi(t_w)$. As long as $\xi(t_w)$ is much smaller than the lattice size, the system behaves as if its size were infinite and reproduces the experimental evolution. However, when the coherence length begins to approach the lattice size, spurious finite-size effects appear. These systematic errors scale as $\exp(-L/\xi(t_w))$; one should make sure that (relative) statistical errors are much larger than this value. So, the precise meaning of "large enough" depends on time, temperature and accuracy. As a rule of thumb, one is on the safe side if the lattice size is at least seven or eight times larger than $\xi(t_w)$ [29]. Since the coherence length grows with the simulation time, a given lattice size may well be large enough for 10^5 MCSs but not for 10^{10} MCSs. Janus has proven an excellent compromise in this respect. It has allowed us to follow the dynamics for some 10^{11} MCSs, on hundreds of samples containing $N = 80^3 \sim 5 \times 10^5$ spins. Since a single lattice sweep corresponds roughly to a picosecond (i.e., 10^{-12} s), this means that we have covered the range from the microscopic time scale to one tenth of a second, which is already long enough to understand what happens in the experimental regime [29,30]

On the other hand, theoretical physicists prefer a different approach. They like to think about a complex phase space, with many local minima, where the system may get trapped for quite a long time. The natural framework for this way of thinking is equilibrium thermodynamics. Hence, we need to reach thermal equilibrium, meaning that the coherence length is as large as the system size. Under these conditions, almost no magnitude is self-averaging. One needs to describe the physics in terms of probability distributions with the disorder. In practice, one needs to reach thermal equilibrium on several thousands of samples, obtain thermal mean values over each of them, and afterwards study the disorder distributions (i.e., quantify how much the *same* quantity can vary, if computed over different samples). An added difficulty is that thermal equilibrium is terribly difficult to reach. Even worse, the larger the system, the harder the equilibration. And, of course, the larger the system, the more significant the reached results.

Fortunately, when one wants to reach equilibrium, it is no longer important that the computational dynamics resemble, in any way, the physical dynamics. The only important mathematical property is *balance* (see, e.g., [17]). This allows an enormous flexibility in the choice of the dynamic rules. In particular, the already discussed parallel tempering dynamics outperforms by orders of magnitude the simple heat-bath algorithm used in the non-equilibrium simulations. Even then, one may need as many as 10^{11} parallel tempering steps (each parallel tempering step is followed by 10 heat-bath full lattice sweeps) in order to reach thermal equilibrium in some samples containing only $N = 32^3$ spins [31].

In our simulations with Janus, we reached equilibrium on thousands of relatively small samples (ranging from $N = 16^3$ to $N = 32^3$, smaller systems were simulated on PCs). This simulation campaign was extremely long: all in all Janus performed 10^{21} spin updates. In the case of the worst samples we estimated that the necessary wall clock time was well over 6 months. For these samples we have accelerated the simulation by increasing the level of parallelism, by running the PT temperature-assignment procedure on the IOP. This has allowed us to distribute the set of temperatures along several FPGAs on a the same module, speeding up the simulation accordingly. These simulations have opened a new window into the nature of the equilibrium spin-glass phase [31–33]

Finally, we combined the results of both the non-equilibrium and the equilibrium simulation to clarify, in a quantitative way, the relation between the dynamical evolution and the equilibrium spin-glass phase. We did this by means of a finite-time scaling formalism, with interesting implications for experimental work [34].

Regarding the Potts glass, we studied its phase transition for $q = 4,5,6$, simulating lattices of up to $N = 16^3$ [35, 36]. We found that, in contrast to the mean-field prediction, this transition remained of the second order for all the considered values of q and that ferromagnetic effects were not relevant.

A further example of the benefits of the FPGA reconfigurability is the possibility of simulating the spin glass under an applied magnetic field. In fact, the fate of the spin-glass phase when an external field is switched on is one of the major open questions in the field. Janus is rather efficient in this context, both for the simpler dynamic simulations, or for the equilibrium simulations that need parallel tempering. Our recent results on this problem have been reported in [37].

7 Janus Performance

In this section we analyze both computing and energy performances of the Janus system for some of the applications described in Sect. 6, and compare with that of systems based on commodity CPUs and GPUs, available both when Janus was developed as well as today. The tables in this section originally appeared in [38] and are reproduced with kind permission of The European Physical Journal (EPJ).

Let us first estimate the effective computing power delivered by a full Janus system, configured to run an Edwards–Anderson simulation. SPs run at clock frequency of 62.5 MHz, and at each clock cycle conservatively update 800 spins. An equivalent C program running on a commodity CPU architecture could require to perform at least the following mathematical operations for each spin-update:

- 1 32-bit integer `sum`
- 2 32-bit `xor`

Table 1 Speed-up factors of one Janus SP with respect to state-of-the-art CPUs available at the time the project was started

Model	Algorithm	Intel Core 2 Duo	Intel i7
3D Ising EA	Metropolis	45×	10×
3D Ising EA	Heat bath	60×	–
$q = 4$ 3D glassy Potts	Metropolis	1250×	–

Table 2 Energy comparison between Janus and commodity PCs

	Janus	AMSC	SMSC
Processor	1 SP	1 CPU	1 CPU
Statistic	1 (16)	1 (128)	1 (4)
Wall-clock time	50 days	770 years	25 years
Energy	2,7 GJ	2,3 TJ	78,8 GJ
Processor	256 SPs	2 CPUs	256 (64) CPUs
Statistic	256	256	256
Wall-clock time	50 days	770 years	25 years
Energy	43 GJ	4,6 TJ	20 (5) TJ

The upper part of the table compares the performance of 1 SP versus a PC. The lower part compares the required time and energy to run the simulation on 256 lattice replicas

- 6 3-bit integer sum
- 6 3-bit xor
- 1 32-bit integer comparison

In the above count we have neglected operations to load data and instructions, and to compute memory address, which are obviously necessary during the run. Counting the 6 short xor and sum operations as one single 32-bit integer operation each, we end up with 6 equivalent operations for each spin update. This translates into a required processing power of $6 \times 800 \times 62.5 \times 10^6$ operations per second, corresponding to a sustained performance of 300.0 Giga-ops for a single SP, and 76.8 Tera-ops for a full Janus system running 256 SPs.

At the time the Janus project started, early 2006, state-of-the-art commodity systems where based on dual-core CPUs. Prior to actually building the system we made an extensive analysis of the performance gain that we could expect from the new machine (see, for instance, [21]). Table 1 contains a short summary of that analysis, listing the relative speed-up for the Ising and the Potts models of one Janus SP with respect to standard processors available in 2006–2007.

Let us now make a comparison of energy efficiency between Janus and commodity computing systems based on PCs at the same point in time as above. Let us consider the case of a simulation campaign of an EA model on a lattice size of 64^3 for 10^{12} MCSs and 256 samples. In comparison, we consider both AMSC and SMCS strategies and estimate the power consumption of one PC at $\simeq$100 W. Table 2 shows the comparison performance of the PC cluster versus the Janus system in terms of energy dissipated and wall-clock time needed to perform the simulation.

Since the deployment of Janus, in spring 2008, significant improvements have been made in the architecture and performance of commodity architectures, and in spite of that, Janus is still a very performing machine.

We have extensively compared [39, 40] Janus with several multi-core systems based on the IBM Cell Broadband Engine, the multi-core Nehalem Intel CPU, and the NVIDIA Tesla C1060 GP-GPU. We have made this exercise for the Ising model (as opposed to the Potts model) as in the former case the relative speed-up is much smaller, so we may expect traditional processors to catch up earlier. We consider these results as state-of-the-art comparisons, assuming that within a factor 2 they are still valid for even more recent multi-core architectures, like the Fermi GPUs. This assumption is indeed verified by an explicit test made on the very recent 8-core Intel Sandy Bridge processor.

As discussed in previous sections, for traditional processor architectures we analyzed both SMSC and AMSC strategies and also considered mixes of the two tricks (e.g., simulating at the same time k spins belonging to k' independent samples), trying to find the best option from the point of view of performance.

A key advantage of Janus is indeed that there is no need to look for these compromises: an SP on Janus is simply an extreme case of SMSC parallelization: if many samples are needed on physics ground, more SPs are used. Equally important, if different samples need different numbers of Monte Carlo sweeps (e.g., to reach thermalization), the length of each simulation can be individually tailored without wasting computing resources on other samples, as would necessarily be the case in an AMSC approach.

Performance results for Janus are simply stated: one SP updates $\approx$1,000 spins at each clock cycle (of period 16 ns), so the spin-update time is 16 ps/spin for any lattice size that fits available memory. In the cases of $L = 96$ and $L = 128$ there is not enough memory in the FPGA to store the lattice and we do not represent the performance in the tables. For standard processors, we collect our main results for the 3D Ising spin-glass in Tables 3 and 4 for SMSC and AMSC, respectively.

We see that performance (weakly) depends also on the size of the simulated lattice: this is an effect of memory allocation issues and of cache performance. All in all, recent many-core processors perform today much better than 5 years ago: the performance advantage of Janus has declined by a factor of approximately 10 for SMSC: today one Janus SP outperforms very latest generation processors by just a factor $5\times \cdots 10\times$. It is interesting to remark that GP-GPUs are not the most efficient engine for the Monte Carlo simulation of the Ising model: this is so, because GP-GPU strongly focus on floating-point performance which is not at all relevant to this specific problem. There is one point where Janus starts to show performance limits, it is associated with the largest system size that the machine is able to simulate: no significant limit applies here for traditional processor.

All in all, for the specific applications we have presented in this chapter, Janus—after 4 years of operation—still has an edge of approximately one order of magnitude, which directly translates on the wall-clock time of a given simulation campaign.

Table 3 SMSC update time (in ns) for a 3D Ising spin-glass (binary) model of lattice size *L*, for Janus and for several state-of-the-art processor architectures

3D Ising spin-glass model, SMSC (ns/spin)						
L	Janus SP	I-NH (8-Cores)	CBE (8-SPE)	CBE (16-SPE)	Tesla C1060	I-SB (16 cores)
16	0.016	0.98	0.83	1.17	–	–
32	0.016	0.26	0.40	0.26	1.24	0.37
48	0.016	0.34	0.48	0.25	1.10	0.23
64	0.016	0.20	0.29	0.15	0.72	0.12
80	0.016	0.34	0.82	1.03	0.88	0.17
96	–	0.20	0.42	0.41	0.86	0.09
128	–	0.20	0.24	0.12	0.64	0.09

I-NH (8-Cores) a dual-socket quad-core Intel Nehalem board, CBE (16-SPE) a dual-socket IBM Cell board, and I-SB a dual-socket eight-core Intel Sandy Bridge board

Table 4 AMSC update time for the 3D Ising spin-glass (binary) model, for the same systems as in the previous table

3D Ising spin-glass model, AMSC (ns/spin)						
L	Janus	I-NH (8-Cores)	CBE (8-SPE)	CBE (16-SPE)	Tesla C1060	I-SB (16 cores)
16	0.001 (16)	0.031 (32)	0.052 (16)	0.073 (16)	–	–
32	0.001 (16)	0.032 (8)	0.050 (8)	0.032 (8)	0.31 (4)	0.048 (8)
48	0.001 (16)	0.021 (16)	0.030 (8)	0.016 (16)	0.27 (4)	0.015 (16)
64	0.001 (16)	0.025 (8)	0.072 (4)	0.037 (4)	0.18 (4)	0.015 (8)
80	0.001 (16)	0.021 (16)	0.051 (16)	0.064 (16)	0.22 (4)	0.011 (16)
96	–	0.025 (8)	0.052 (8)	0.051 (8)	0.21 (4)	0.012 (8)
128	–	0.025 (8)	0.120 (2)	0.060 (2)	0.16 (4)	0.011 (8)

For Janus, we consider one core with 16 SPs. The number of systems simulated in parallel in the multi-spin approach is shown in parentheses

8 Conclusions

This chapter has described in detail the Janus computer architecture and how we have configured the FPGA hardware to simulate spin-glass models on this architecture. We have also briefly reviewed the main physics results that we have obtained in approximately 4 years operating with this machine.

From the point of view of performance, Janus still has an edge on computing systems based on state-of-the-art processors, in spite of the huge architectural developments since the project was started. It is certainly possible to reach very high performances in terms of spin flips per second using multi-spin coding on CPUs or GP-GPUs (or simply by spending money on more computers), thus concurrently updating many samples and achieving very large statistics. In the Janus collaboration, we have instead concentrated on a different performance goal: minimizing the wall-clock for a very long simulation, by concentrating the updating power in a single sample. This has allowed us to bridge the gap between simulations and experiments for the non-equilibrium spin-glass dynamics or to thermalize large systems at low temperatures, thus gaining access to brand new physics. In particular, one single SP of Janus is able to simulate (two replicas of) an $L = 80$ three-dimensional lattice for 10^{11} MCS in about 25 days.

Janus provides one of the few examples of the development of a successful large scale computing application fully running on a reconfigurable computing infrastructure. This success comes at the price of a large investment in mapping and optimizing the application programs onto the reconfigurable hardware. This has been possible in this case as the Janus group has a full understanding of all facets of the algorithms and every performance gain immediately brings very large dividends in terms of a broader physics program. Most potential FPGA-based applications do not have equally favorable boundary conditions, so automatic mapping tools would be most welcome; however, further progress is needed in this area in order to support a widespread use of configurable FPGA-based computing.

Focusing again on the spin glass arena, there is still room for substantial progress. Nowadays, the theoretical analysis of temperature-cycling experiments is still in its infancy. Janus has made possible an in-depth investigation of isothermal aging (i.e., experiments where the working temperature is kept constant). However, isothermal aging reflects only a minor part of the experimental work, where different temperature variation protocols are used as a rich probe of the spin-glass phase.

Janus is not able to support these analyses, as its performance is not enough in this case, and also because memory limits would quickly become a major problem, as the coherence length grows very fast close to the critical temperature. If one wants to work in this direction a new generation Janus system should be developed; this can be done by leveraging on technology progress of FPGAs in the last 5 years and introducing a few limited architectural changes in the memory structure of the SPs and in the interconnection harness with the host system.

If this system is developed, we should be able to reach the same time scales of 10^{11} lattice sweeps, which is roughly equivalent to a tenth of a second, on systems containing some 5×10^7 spins. In other words, we should be able to simulate systems with lattice size up to $L = 400$, large enough to accommodate a coherence length of up to 50 lattice spacings. After 40 years of investigations, a direct comparison between experiments and the Edwards–Anderson model will finally be possible.

Acknowledgements We wish to thank several past members of the Janus Collaboration, F. Belletti, M. Cotallo, G. Poli, D. Sciretti and J.L. Velasco, for their important contributions to the project. Over the years, the Janus project has been supported by the EU (FEDER funds, No. UNZA05-33-003, MEC-DGA, Spain), by the MICINN (Spain) (contracts FIS2006-08533, FIS2009-12648, FIS2007-60977, FIS2010-16587, FPA2004-02602, TEC2010-19207), by CAM(Spain), by the Junta de Extremadura (GR10158), by UCM-Banco Santander (GR32/10-A/910383), by the Universidad de Extremadura (ACCVII-08), and by the Microsoft Prize 2007. We thank ETHlab for their technical help. E.M. was supported by the DREAM SEED project and by the Computational Platform of IIT (Italy); M.B.-J. and B.S. were supported by the FPU program (Ministerio de Educacion, Spain); R.A.B. and J.M.-G. were supported by the FPI program (Diputacion de Aragon, Spain); finally J.M.G.-N. was supported by the FPI program (Ministerio de Ciencia e Innovacion, Spain).

References

1. See, for instance: C.A. Angell, Science **267**, 1924 (1995); P.G. Debenedetti, *Metastable Liquids* (Princeton University Press, Princeton, 1997); P.G. Debenedetti, F.H. Stillinger, Nature **410**, 259 (2001)
2. L.C.E. Struick, *Physical Aging in Amorphous Polymers and Other Materials* (Elsevier, Houston, 1978)
3. J.A. Mydosh, *Spin Glasses: An Experimental Introduction* (Taylor and Francis, London, 1993)
4. A.P. Young (ed.), *Spin Glasses and Random Fields* (World Scientific, Singapore, 1998)
5. P.A. Boyle et al., IBM J. Res. Dev. **49**, 351–365 (2005)
6. F. Belletti et al., Comput. Sci. Eng. **8**, 18–29 (2006)
7. G. Goldrian et al., Comput. Sci. Eng. **10**, 46–54 (2008); H. Baier et al., Comput. Sci. Res. Dev. **25**, 149–154 (2010)
8. J. Makino et al., A 1.349 Tflops simulation of black holes in a galactic center on GRAPE-6, in *Proceedings of the 2000 ACM/IEEE Conference on Supercomputing*, Article n. 43 (2000)
9. A.D. Ogielski, D.A. Huse, Phys. Rev. Lett. **56**, 1298–1301 (1986)
10. J. Pech et al., Comput. Phys. Comm. **106**, 10–20 (1997); A. Cruz et al., Comput. Phys. Comm. **133**, 165–176 (2001)
11. S.F. Edwards, P.W. Anderson, J. Phys. F Met. Phys. **5**, 965–974 (1975); S.F. Edwards, P.W. Anderson, J. Phys. F Met. Phys. **6**, 1927–1937 (1976)
12. J. Barahona, J. Phys. Math. Gen. **15**, 3241–3253 (1982)
13. M. Mézard, G. Parisi, M. Virasoro, *Spin-Glass Theory and Beyond* (World Scientific, Singapore, 1987)
14. K. Gunnarsson et al., Phys. Rev. B **43**, 8199–8203 (1991) See also P. Norblad, P. Svendlidh, *Experiments on Spin-Glasses* in [[4]]
15. H.G. Ballesteros et al., Phys. Rev. B **62**, 14237–14245 (2000)
16. F. Bert et al., Phys. Rev. Lett. **92**, 167203 (2004)
17. See for instance D.J. Amit, V. Martin-Mayor, *Field Theory, the Renormalization Group and Critical Phenomena*, 3rd edn. (World Scientific, Singapore, 2005)
18. M.E.J. Newman, G. Barkema, *Monte Carlo Methods in Statistical Physics* (Oxford University Press, New York, 1999)
19. H. Hukushima, K. Nemoto, J. Phys. Soc. Jpn. **65**, 1604 (1996); E. Marinari, in *Advances in Computer Simulation*, ed. by J. Kerstéz, I. Kondor (Springer, Berlin, 1998)
20. H.G. Katzgraber, *Introduction to Monte Carlo Methods*. Lecture at Modern Computation Science (BIS, Oldenburg, 2009)
21. F. Belletti et al., Comput. Phys. Comm. **178**, 208–216 (2008)
22. F. Belletti et al., IANUS: scientific computing on an FPGA-based architecture, in *Proceedings of ParCo2007, Parallel Computing: Architectures, Algorithms and Applications*. NIC Series, vol. 38 (2007), pp. 553–560
23. F. Belletti et al., Comput. Sci. Eng. **8**, 41–49 (2006)
24. F. Belletti et al., Comput. Sci. Eng. **11**, 48–58 (2009)
25. S. Sumimoto et al., The design and evaluation of high performance communication using a Gigabit Ethernet, in *Proceedings of the 13th International Conference on Supercomputing* (1999), pp. 260–267
26. R. Baxter et al., Maxwell – a 64 FPGA supercomputer, in *Second NASA/ESA Conference on Adaptive Hardware and Systems* (2007), pp. 287–294
27. M. Flynn et al., Finding speedup in parallel processors, in *International Symposium on Parallel and Distributed Computing ISPDC '08* (2008), pp. 3–7
28. V. Parisi, cited in G. Parisi and F. Rapuano, Phys. Lett. B **157**, 301–302 (1985)
29. F. Belletti et al., Phys. Rev. Lett. **101**, 157201 (2008)
30. F. Belletti et al., J. Stat. Phys. **135**, 1121–1158 (2009)
31. R. Alvarez Baños et al., J. Stat. Mech. P06026 (2010)
32. R.A. Baños et al., Phys. Rev. B **84**, 174209 (2011)

33. A. Billoire et al., J. Stat. Mech. P10019 (2011)
34. R. Alvarez Baños et al., Phys. Rev. Lett. **105**, 177202 (2010)
35. A. Cruz et al., Phys. Rev. B **79**, 184408 (2009)
36. R. Alvarez Baños et al., J. Stat. Mech. P05002 (2010)
37. R.A. Baños et al., Proc. Natl. Acad. Sci. USA **109**, 6452–6456 (2012)
38. M. Baity-Jesi et al. (Janus Collaboration), Eur. Phys. J. Spec. Top. **210**, 33–51 (2012)
39. M. Guidetti et al., Spin Glass Monte Carlo simulations on the cell broadband engine, in *Proceedings of PPAM09*. Lecture Notes on Computer Science (LNCS), vol. 6067 (Springer, Heidelberg, 2010), pp. 467–476
40. M. Guidetti et al., *Monte Carlo Simulations of Spin Systems on Multi-Core Processors*, ed. by K. Jonasson. Lecture Notes on Computer Science (LNCS), vol. 7133 (Springer, Heidelberg, 2010), pp. 220–230

Accelerate Communication, not Computation!

Mondrian Nüssle, Holger Fröning, Sven Kapferer, and Ulrich Brüning

Abstract Computer systems are showing a continuously increasing degree of parallelism in all areas. Stagnating single thread performance as well as power constraints prevent a reversal of this trend. On the contrary, current projections show that the trend towards parallelism will accelerate. In cluster computing scalability and therefore the degree of parallelism are limited by the network interconnect and its characteristics like latency, message rate, overlap and bandwidth. While most interconnection networks focus on improving bandwidth, there are many applications that are very sensitive to latency, message rate and overlap, too. We present an interconnection network called EXTOLL, which is specifically designed to improve characteristics like latency, message rate and overlap, rather than focusing solely on improving bandwidth. Key techniques to achieve this are designing EXTOLL as an integral part of the HPC system, providing dedicated support for multi-core environments and designing and optimizing EXTOLL from scratch for the needs of high performance computing. The most important parts of EXTOLL are the network interface and the network switch, which is a crucial resource when scaling the network. EXTOLL's network interface provides dedicated support for small messages for eager communication, and for bulk transfers in the form of rendezvous communication. While support for small messages is optimized mainly for high message rates and low latencies, for bulk transfers the possible amount of overlap between communication and computation is optimized. EXTOLL is completely based on FPGA technology, both for the network interface and the switching. In this work we present a case for accelerated communication, where FPGAs are not used to speed up computational processes, rather we employ FPGAs to speed up communication. We will show that in spite of the inferior performance characteristics of FPGAs compared to ASIC solutions, we can dramatically accelerate communication tasks and thus reduce the overall execution time.

M. Nüssle (✉) • H. Fröning • S. Kapferer • U. Brüning
University of Heidelberg, Germany
e-mail: nuessle@uni-hd.de; holger.froening@ziti.uni-heidelberg.de;
sven.kapferer@ziti.uni-heidelberg.de; ulrich.bruening@ziti.uni-heidelberg.de

W. Vanderbauwhede and K. Benkrid (eds.), *High-Performance Computing Using FPGAs*,
DOI 10.1007/978-1-4614-1791-0_17, © Springer Science+Business Media, LLC 2013

1 Introduction

The need for more powerful high performance computing (HPC) systems continues, and the TOP500 list [1] reveals that most popular installations are clusters, certainly due to their excellent price/performance ratio. Such clusters rely on commodity parts for computing, memory and enclosure, but significant performance improvements can be achieved by replacing commodity Ethernet with specialized interconnection networks like Infiniband [2]. The main reasons for these performance improvements are features that cannot be found in commodity Ethernet, including user-level communication, reliable transmission, off-loading of communication tasks and higher peak bandwidth. These features result in improved performance characteristics like higher sustained bandwidth, lower start-up latency and increased message rate, just to name the most important ones. Besides this, network switching characteristics like congestion management are also improved in these specialized interconnects, while commodity Ethernet still relies on packet dropping in the case of exhausted resources.

Besides clusters, massively parallel processors (MPPs) can also be found in the TOP500 list; however, they are almost completely based on specialized components with a dramatically higher price tag. Their interconnects typically include all features that can be found in the specialized networks for the cluster market, but many details are even further improved. The most prominent examples for such MPP interconnects include Cray SeaStar [3], Cray Gemini [4], Fujitsu TOFU [5] and IBM BlueGene [6]. In particular, they put special attention on scalability by using appropriate routing and switching mechanisms, certain topologies and congestion management techniques.

The goal of EXTOLL is to fill the gap between clusters and MPPs, allowing to rely on cost-effective commodity parts for almost all components, but providing a specialized and optimized interconnection network for HPC demands. In this work we will show that such an approach is feasible. As a research project we rely on FPGA technologies, which allows us to assess our interconnect performance under real-world workloads and to adapt the complete architecture to HPC needs. In spite of the inferior performance characteristics of FPGAs in comparison with ASICs, we can show multiple cases where our architecture outperforms state-of-the-art commercial interconnects, validating the architectural design choices made and constructing a case for *accelerated communication* instead of accelerated computation.

1.1 Vast Increase in Parallelism

The high performance computing landscape is currently driven by the concurrency galore, which tells the reader nothing else but that many small things are way better than few huge ones. Power constraints, limited ILP and bounded signal reach result in stagnating single thread performance, so performance improvements are most

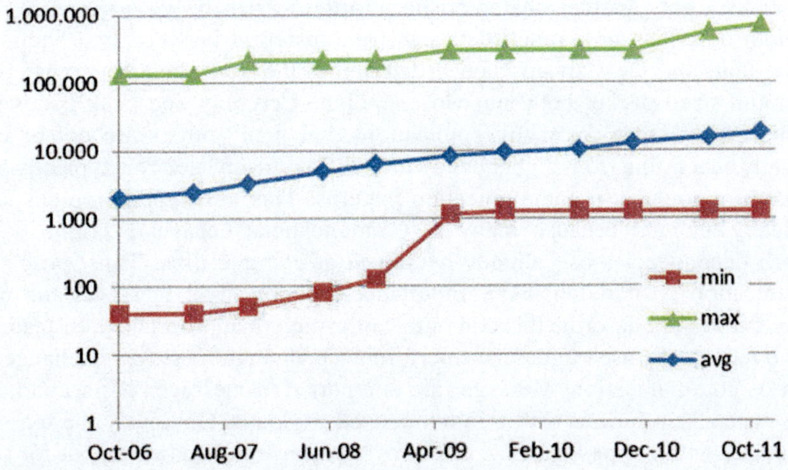

Fig. 1 Number of computational units per system in the TOP500 list over the recent years

easily accomplished by replicating computational units. Also, these facts prevent a reversal of this trend. Recent supercomputers embrace already more than 700,000 cores [1], and the average core count in the TOP500 list grows with a CAGR of more than 50% over the recent 5 years (see Fig. 1). Projections show that this trend will continue in the future. For instance, [7] anticipates that a supercomputer exceeding a performance of 1 EXAFLOP (10^{18} floating point operations per second) will be empowered by about 166M cores. Even today, the interconnection network has to interconnect much more end points than before; additionally the network interface now finds itself between an increasing number of local CPU cores and a vast amount of communication partners. If a network (interface) does not address this issue, communication performance will dramatically suffer.

Another effect of the massive parallelization is an increase of the number of potential communication partners. Although typical problem sizes are continuously increasing, the sheer amount of communication end points results in a shift towards smaller message sizes. In the past, high performance interconnection networks have been almost solely optimized for bandwidth [8], while other characteristics like latency or message rate have not improved significantly. Also, an increase in the number of communication partners typically results in more synchronization overhead, and synchronization primitives typically rely on small messages. With this shift towards higher message counts with smaller payloads, attention must be paid on the performance of small transfers.

1.2 Messaging Characteristics

The most important characteristics for interconnection networks have been peak bandwidth and start-up latency, but the irreversible trend towards higher degrees of parallelism puts another characteristic into the focus: the *message rate*. Peak bandwidth describes how much data can be transferred between two endpoints per time unit, and the start-up latency determines the time spent for transferring a minimum-sized packet between two endpoints. Certainly, the peak bandwidth is of high importance for many applications, but many others are not or only marginally benefitting from higher bandwidths. The start-up latency is typically only important for round-trip communication patterns. This pull type communication is typically avoided because many HPC applications behave deterministically and such dependencies can already be solved at compile time. Thus, push type communication is of much more importance, but for small messages the peak bandwidth does not describe this communication pattern appropriately. Instead, the message rate can be used to describe performance of small messages exchanged in push style communication. Message rate is reported in messages per second, and peak is typically achieved with minimum-sized payloads. Obviously the message rate together with the payload size directly translates into bandwidth, but for such small transfers bandwidth is not important, rather how many messages can be sent out per time unit.

Another important characteristic is overlap, which defines to which extent computation and communication can be performed simultaneously. Related metrics are overhead, defined as "length of time that a processor is engaged in the transmission or reception of each message; during this time, the processor cannot perform other operations" [9], and reported in time units. From the overhead the *application availability* can be derived, which defines the fraction of total transfer time that the application is free to perform non-MPI related work. Obviously it is desirable to maximize application availability, but this is only feasible if the network interface off-loads all communication tasks from the CPU cores and acts as unobtrusively as possible and adds only minimal overhead to the communication.

1.3 The EXTOLL Approach

EXTOLL has been designed from scratch as an interconnection network for HPC demands. Key ideas of the EXTOLL architecture are:

(1) A tight integration into the host system
(2) Inherent support for multi-core environments
(3) A new ratio between on- and off-loading

As a research project, the technology is based on FPGAs due to their reconfigurability and low non-recurrent expenses (NRE). Compared to ASICs, FPGAs are obviously inferior in terms of frequency, data path widths and logic complexity. For this application, this translates in a bandwidth gap of about 50% (FPGA bandwidth/ASIC bandwidth) in comparison with state-of-the-art interconnects. In spite of this, our innovative architecture yields improvements in message rate, overlap and latency that can dramatically speed up applications.

The remainder is organized as follows: In the next section the networking demands of HPC applications are introduced and analyzed using several benchmark and codes. After that the EXTOLL architecture and the implementation of this architecture on an FPGA is described. In the following section the software-stack of EXTOLL is described. Finally, we give a performance evaluation of the FPGA EXTOLL system again using several micro-benchmarks and application codes. The chapter closes with a short overview of related work and a conclusion.

2 Application Demands

A typical HPC application shares many properties with standard programs, including the need for safety and security. However, as contemporary HPC systems rely on the concurrency galore to scale performance up to ultimate levels, HPC applications have to spend a large fraction of their execution time for communication and synchronization.

General computer programs, like web servers and database management systems, often employ a very dynamic task model and allow spawning and terminating threads during run time. Their communication patterns are completely irregular and can change significantly during execution. Often, their concurrency relies on request-level parallelism, which is barely predictable. Such solutions may be appropriate for execution with invariant communication costs, the absence of locality effects or an unpredictable concurrency. Furthermore, many other computer programs are still employed in a sequential way and make no use of concurrency at all.

Completely opposed to this, HPC applications have to make use of the vast amount of parallel resources, which can only be leveraged if the HPC application itself exploits as much concurrency as possible. Strong scaling problems are not subject to scalability constraints by definition, and for weak scaling problems the problem size can typically be increased to ensure their scalability. This huge degree of concurrency is only manageable with simplified task models. So most HPC applications are relying on the single-program-multiple-data (SPMD) paradigm, which enforces that one single program is executed on all resources, and only by assigning process IDs to each instance different behavior can be implemented. Another—even stricter—task model is the bulk synchronous programming (BSP) model, which relies on three phases for computation, communication and barrier synchronization. The static task model also leads to deterministic communication patterns, as the work flow is determined during compile time. This ab-initio knowledge of communication channels helps not only to perform optimizations by overlapping computation and communication but also to pre-post receives, so that unexpected communication by the uncoupled work flows is avoided.

Table 1 MPI Time for various applications

Benchmark	MPI Time (%)	Processes	Source
NPB-EP	0	64	[10]
NPB-MG	9	64	[10]
NPB-BT	10	64	[10]
NAMD	24	64	[10]
HPL	25	64	[10]
NPB-CG	34	64	[10]
NPB-FT	37	64	[10]
NPB-IS	47	64	[10]
WRF	45	64	Own experiments
HPCC MPIFFT	77	64	Own experiments
HPCC RandomAccess	89	64	Own experiments

2.1 Fraction of Communication Time

This extensive use of parallelism together with the execution on partitioned resources with no possibility of data sharing results in huge communication and synchronization overhead. Obviously the time spent in MPI calls is depending on the application, but it is obvious that more parallelism leads to more partitioning. Then, for a fixed problem size more data has to be moved around, so that finally the fraction of time spent for MPI calls is increasing. For an increased problem size this fraction of time is even increasing more. This fraction of time is called MPI time and typically reported in per-cent of the overall execution time. Table 1 gives a brief overview of a variety of benchmarks and their MPI time. Sources of this data are [10] and own experiments (WRF, MPIFFT, RandomAccess). Note that both NAMD and WRF are not only benchmarks but also used in production. Thus, they provide the most insights, and also show that MPI time can vary significantly for different applications.

Note that the MPI time is dependent on the number of processes participating in the execution of the application. The more processes are involved, the more communication is required and thus more time has to be spent in MPI layers. The following Fig. 2 shows the dependency of the MPI time on the process count.

Also note that the MPI time is highly dependent on the used interconnection networks, respectively, software stack. Depending on the capabilities of the network interface, tasks can be off-loaded from the CPUs to the network hardware, reducing the time fraction spent for MPI calls. Obviously the amount of work is the same as before, but now overlap between computation and communication can be leveraged. Thus, above numbers may greatly vary for different platforms. Also, it is desirable to provide as much overlap potential as possible.

MPI time highly depends on how much communication tasks can be off-loaded to the network interface, or in other terms how much overlap between computation and communication can be achieved. As stated in the introduction, overhead [9] is limiting this overlap between computation and communication, and the application

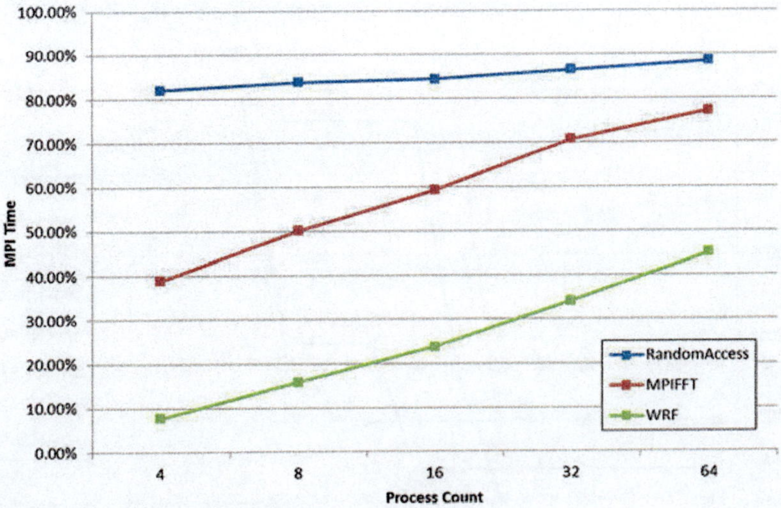

Fig. 2 MPI Time over process count for three selected applications

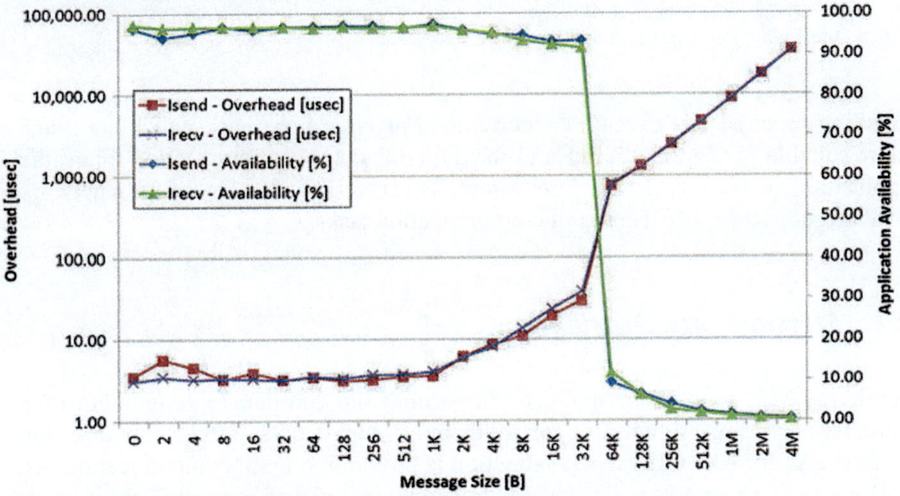

Fig. 3 Application availability and overhead for commodity Gigabit Ethernet

availability indicates how much overlap is possible. In [11] an approach to measure application availability is introduced. Figure 3 shows application availability and overhead for a standard Gigabit Ethernet network.

Because the OS is responsible to handle all communication over Gigabit Ethernet, a system call is required and dramatically increases the time required for this communication call. As a direct result, on multi-core architecture the resulting application availability might be high, because both tasks (calling process and

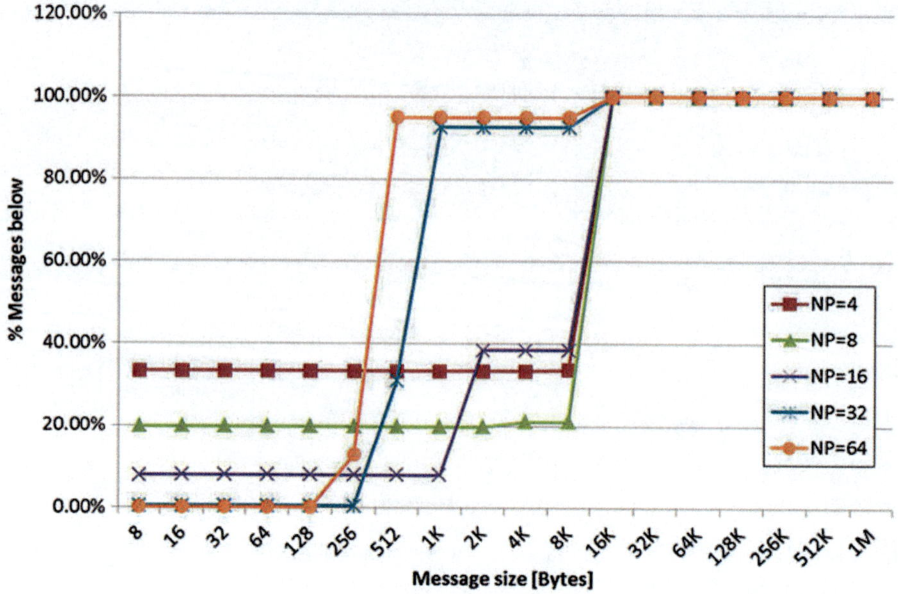

Fig. 4 Message size distribution for RandomAccess

handling process) can execute concurrently. For large messages, which are much more suitable for overlap because of the larger absolute communication times, the application availability drops close to zero. This is a typical behavior for a network with no support for off-loading of communication tasks.

2.2 *Message Size Distribution*

While the MPI time of an application describes the computation/communication ratio, for an optimization it is important to know more details of the communication patterns. As for EXTOLL special attention is put on fine grain communication, we will analyze the message size distribution for a couple of example applications. Other important characteristics of communication patterns are total amount of data moved around, use of collective communication, locality (nearest neighbor, uniform traffic, etc), but this exceeds the scope of this work.

In order to collect message distribution data, we have instrumented our OpenMPI implementation for EXTOLL. Thus, we are able to track message sizes at network level, not at application level. This is important because the MPI layer in between may translate large transfers and collective operations into multiple messages. With this methodology we are able to record the actual transfers on the network. Message sizes are collected in buckets, and message size distribution is reported using cumulative distribution functions (CDF). Figures 4, 5 and 6 show the message size distribution for RandomAccess, MPIFFT and WRF.

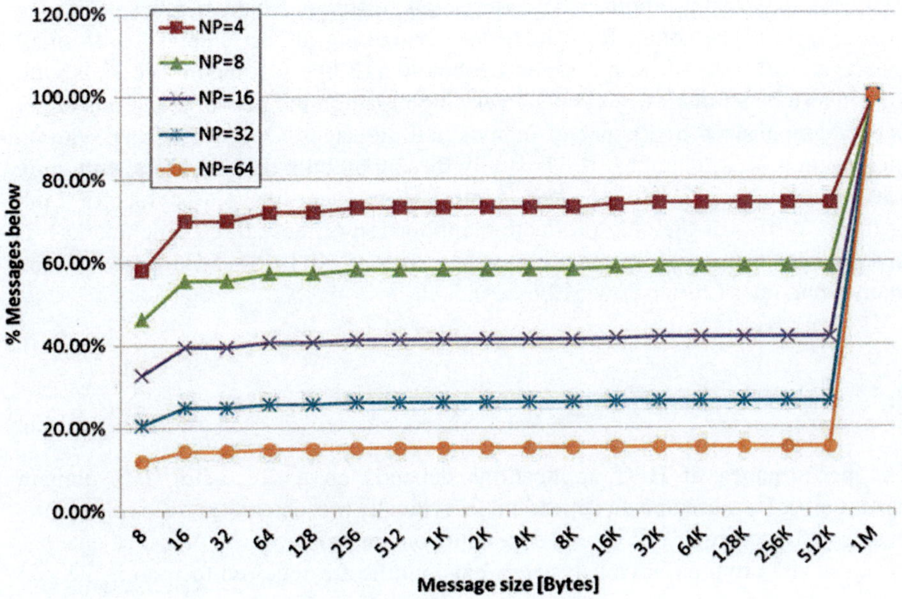

Fig. 5 Message size distribution for MPIFFT

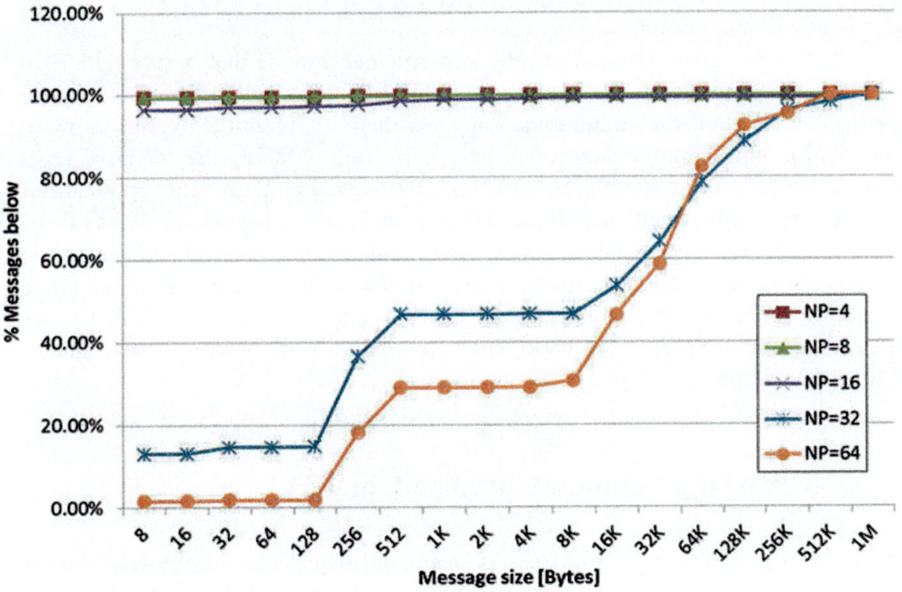

Fig. 6 Message size distribution for WRF

These three applications have completely different characteristics. With an increasing process count, RandomAccess is making almost solely use of small messages, i.e. 95% of all messages are below 512 bytes. RandomAccess is well known as a benchmark which puts a very high load on the network, and this effect is even emphasized by the use of many small messages. Opposed to this, with an increasing process count MPIFFT is only relying on bulk data transfers, with only 15% of all messages below 512 kb. Almost all messages have a payload of 1 mb or more. WRF—as the only production application of these three benchmarks—is making use of many intermediate message sizes (83% below 64 kb), but also on many small ones (29% below 512 bytes).

2.3 *Key Requirements of HPC Applications*

The performance of HPC applications depends on a successful data domain partitioning. Because of data dependencies typically present, this partitioning results in a need for communication; and depending on the problem size, plenty of data has to be moved around. Obviously, high bandwidths are required to reduce the gap between local and remote communication. This is already addressed by commercial solutions, respectively, the available bandwidth is primarily limited by costs: it is easy to increase data path widths to achieve higher throughput, although various practical issues introduce problems. However, opposed to latency and message rate, peak bandwidth is much easier to improve.

What the previous explanations and experiments show is that beside optimizations for peak bandwidth much more attention has to be paid on minimizing the overhead associated with communication. Only then the amount of overlap between computation and communication can be maximized, reducing the effective costs of communication also known as MPI time. Furthermore, many HPC applications make a high use of small messages, and this fine grain communication has to be supported accordingly by the network. Key characteristics here are message rate and latency. Otherwise, the significant fraction of small messages—although much smaller in total volume transferred—will turn into a bottleneck. With EXTOLL, we try to address both issues: maximizing both application availability and performance of fine grain communication.

3 A Networking Solution Optimized for HPC

The EXTOLL hardware architecture is designed to provide solutions to several issues of HPC interconnection networks. In particular, EXTOLL follows these key ideas:

- Native support for multi-core environments
- Single chip solution, integrating both network interface and switching resources

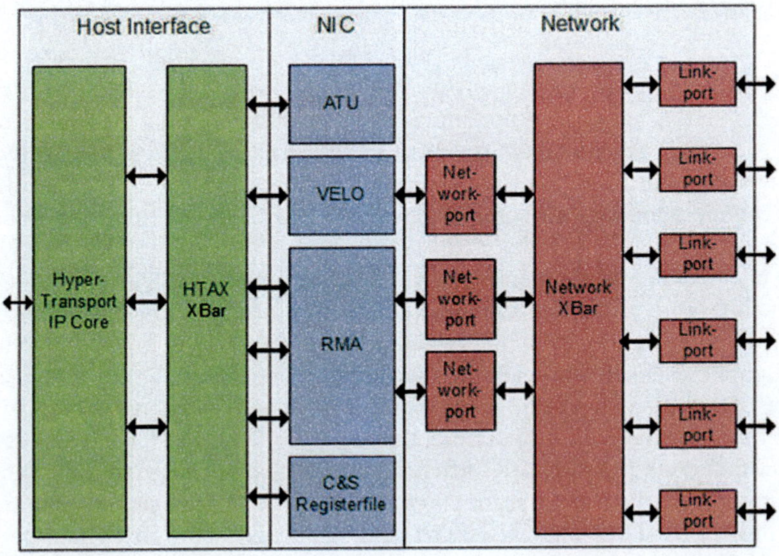

Fig. 7 Top-level architecture of EXTOLL

- Complete hardware-level virtualization, supporting thousands of processes or threads per node
- Very low state overhead, with optimized context sizes resulting in a minimal memory footprint

Figure 7 shows a top-level block diagram of the EXTOLL architecture, integrating the host interface, network interface, and switching. The host interface contains the HyperTransport (HT) Core [12], which directly connects without any intermediate bridges or protocol conversion to one of the node CPUs. The usage of add-in cards is maintained by leveraging the HTX connector, which is the counterpart of the PCIe connector for HT systems. The HTAX on-chip network [13] multiplexes accesses in a non-blocking manner from and to this host interface from multiple units located in the NIC block. This on-chip network protocol is directly derived from HT to minimize protocol conversion overhead. It adds a number of features to overcome some limitations like the limited amount of source tags and improves the addressing scheme. The second large block implements the different modules needed for message handling, thus being responsible for message injection into the network and message reception from the network. We will provide more information on message handling later in this section. The last block implements a complete network switch. It includes a crossbar-based switch, six ports towards the network side and three ports towards the message handling modules on the host side. Specifically, there is one network port for VELO and two network ports for RMA, which handle requests, responses, and completions of Put/Get requests in a distributed fashion.

3.1 Switching

The EXTOLL prototype can run any direct topology with a maximum node degree of six. The routing hardware is not limited to a certain strategy, like dimension order routing, nor to a specific topology. For instance, in the case of smaller networks different topologies from fully interconnected to hypercube and tori are available. Larger configuration will most probably use a 3D torus configuration, though, based on the available number of links.

The integrated switch implements a variant of Virtual Output Queuing (VOQ) on the switch level to reduce Head-of-line (HOL) blocking and employs cut-through switching which enables very low switching latencies. The prototype implements several virtual channels for deadlock avoidance. The switch also supports in-order delivery of packets, which can be leveraged by software components like MPI to simplify protocol design. Furthermore, the architecture allows for the addition of different traffic classes to isolate different traffic streams from each other and to avoid protocol deadlocks. For example, Inter Process Communication (IPC) traffic can be divided from I/O traffic and both could be isolated from system management messages. The whole network acts as a reliable, lossless fabric. Credit-based flow-control is used within the network to keep buffers from overflowing and packets from being dropped. The links employ link-level retransmission and keep the order when retransmitting (replay protocol). On-chip resources are secured by ECC logic. These properties enable a highly efficient, low complexity software layer, in particular with regard to in-order packet delivery and reliable transmission.

3.2 Communication Engines

The two major communication units are the virtualized engine for low overhead (VELO), supporting programmed I/O (PIO) for small transfers, and the remote memory access (RMA) unit that uses DMA to handle large messages. The two supporting units are the address translation unit (ATU) and the control & status register file.

3.2.1 VELO: Support for Small Data Transfers

As the network is designed particularly for ultra-low latencies and high message rates, EXTOLL includes a special hardware unit named VELO that provides optimized transmission for small messages. It offers a highly efficient hardware and software interface to minimize the overhead for sending and receiving such messages. Both optimizations affect the start-up latency as well as the small message rate of the complete system. Using VELO, messages are injected into the network using PIO to reduce the injection latency as much as possible. Details of this scheme, as well as for the general architecture of VELO can also be found in [14]. Here, we concentrate on the features that enable very high small message performance from a system point of view.

Actually, the host-interface must be traversed exactly once, which is crucial to reduce latency as much as possible. For VELO, a single write operation into the memory-mapped I/O (MMIO) space is sufficient to send a message. For additional optimization, some of the message header information is encoded in the address to access VELO. This implementation saves space in the data section of the I/O transaction. The main task of the sending side of VELO is to convert the information from the host-interface into an EXTOLL network packet and to inject it into the network using its associated network port.

The VELO hardware unit is a completely pipelined structure controlled by a number of finite state machines. So no (relatively) slow microcode or even embedded processing resource is involved in sending or receiving data. Another fact that has important consequences on performance is the amount and the location of context or state information for the hardware. VELO is stateless in the sense that each transaction is performed as a whole and no state must be saved for one message to proceed. As a corollary, there is no context information stored in main memory and no caching of such information is necessary. Thus, VELO is able to provide high performance independent of the actual access pattern by different processes, a very important fact in today's multi- or many-core systems. Since the hardware is implemented in such an efficient way, it supports both very low start-up latencies and high message rates. Since the software interface is very lean, the CPU is able to issue messages both with low latency and at a very high rate.

On the receive side, messages are written directly to main memory using a single ring-buffer per receiving process. Each process allocates its own receive buffer, and any source can store messages to this ring-buffer. User-level processes waiting for messages can poll certain memory locations within this ring buffer for new arrivals. This can be done in a coherent way, so polling is done on cache copies. Updates in the ring buffers invalidate the outdated copies, enforcing the subsequent access to fetch the most recent value from main memory.

For the VELO transport, the order is maintained by utilizing the hardware's underlying flow-control principles, i.e. EXTOLL's flow-control in the network and HyperTransport's flow-control for the host side. In extreme cases, this can lead to stalling CPU cores due to missing credits, which are needed to inject messages. To solve this problem a programmable watchdog is provided which prevents system crashes in such a case. On the software side, the problem can be avoided by using a higher-level flow-control. The actual hardware implementation has been improved from the implementation described in [14]. The two most important aspects are that the data path has been widened to 32 bit, and that the clock rate was increased to 156 MHz. Both modifications increase the bandwidth of VELO considerably.

3.2.2 RMA: Support for Bulk Data Transfers

Larger transfers are efficiently supported using the RMA unit. The RMA unit offers Put/Get-based primitives. There is a hardware-based address translation unit (ATU), which is used in conjunction with RMA to enable secure memory access from user-space. Registration and deregistration of pages is very fast and only limited by the time for a system call and the lookup of the virtual to physical translation by the kernel itself [15, 16].

RMA supports a very good CPU off-loading mechanism for large messages. RMA command descriptors are posted to the hardware using PIO, very similar to VELO. Data transmission, however, is executed by integrated DMA engines to read/write payload data from/to main memory.

An interesting feature that has been used for the MPI implementation is the notification framework. RMA operations can trigger a notification at three different points in their lifetime. In each point, a different type of notification is triggered. A notification can be triggered each time a request has left the sending unit, when a response leaves the responding unit and when an operation is finally terminated at the completing unit. Notifications for responses are of course only available to get-style operations since put operations do not generate responses. Each process has exactly one notification queue, in which the hardware stores all incoming notifications for this process regardless of their type. The software process can then check for new notifications in the same manner as it can check for new VELO messages. An individual notification is a 128-bit data structure that encodes information such as the remote process, involved address, operation and type of notification. Details about RMA and a discussion of the RMA engine for MPI-2 use can be found in [16].

4 FPGA-Based Implementation

EXTOLL's overarching goal is to minimize communication overhead, which in more detail translates to low latency and high message rate. In order to achieve this using FPGAs, which are limited in terms of size and clock speed, designers have to face several challenges. In the following sections we will provide an in-depth report of the FPGA-related optimizations.

4.1 FPGA Features for HPC Networking

In Fig. 7, the top-level block diagram of EXTOLL has been shown with its three architectural blocks Host Interface, NIC, and network.

Nowadays, the most common host interface is PCI Express (PCIe). Therefore, almost every FPGA includes a hard IP macro that offers PCIe connectivity in varying speeds to the user. However, PCI Express lacks the one feature that was one of the design goals for EXTOLL: minimum latency. However, AMD processors offer the use of HyperTransport [17] for direct CPU-device communication. Since the specification is openly available from the HyperTransport Consortium, support for HT can be easily added to the FPGA [12]. The main concern for this implementation was the intent to keep the footprint as small as possible because the IP uses valuable device resources and is not included in the FPGA as hard IP.

FPGA vendors offer their devices in many different configurations and sizes, each tailored and suitable for different market areas. Besides the usual distinction between low-cost devices usually targeted at the high volume embedded market and upper-end devices with the highest possible performance, these devices are also available in more than one configuration intended for different applications. These subfamilies differ in integrated IP blocks, number of available logic blocks and high speed connectivity. The most important feature for an HPC network device is the number and the speed of the embedded serializer blocks. These integrated MGTs (multi gigabit transceivers) allow data rates of up to 6.5 GB/s [19] over a differential pair, therefore eliminating the need of many parallel I/Os to transfer large amounts of data between chips. Since the number of I/O pins of an FPGA is limited the use of MGTs allows for the integration of many high speed links in a single FPGA. Because of the large amount of such MGTs a Xilinx Virtex-4 based [20] board with an XC4VFX100 device is used for EXTOLL.

4.2 *Optimizations and Floor Planning*

In general, designing for an FPGA is a straightforward process. After developing the RTL code and verifying it in simulations, the HDL files are loaded in the implementation tool (e.g. ISE for Xilinx FPGAs). Pin locations according to the board layout must be added to the design and timing constraints are set up. Wizards and graphical frontend tools help with both IP integration and I/O planning.

As designs get bigger and bigger, more and more issues come up in the implementation phase:

- Timing closure becomes increasingly difficult
- Increased routing density leads to congestion in certain areas of the FPGA
- Lack of specific device resources like logic slices
- Tool problems like long runtimes

Fortunately, there are several ways to decrease these problems that will be laid out in the following paragraphs. These are solutions that were employed during the design of the EXTOLL interconnect to fit all the required logic inside the FPGA and get timing closure.

Table 2 32 Bit counter implementation results

Resource	DSP implementation	Fabric implementation
FF	1	65
LUT	1	96
Slice	1	75
DSP48	1	0
Target speed	244 MHz	216 MHz

The design goals for EXTOLL were very tough to reach on the targeted Virtex-4 FPGA with an estimated resource utilization of close to 100%. The HyperTransport core must run at 200 MHz internally to be able to process the unidirectional bandwidth of 1.6 GB/s provided through a 16 bit wide HT400 interface. The rest of the logic must run at least at 156 MHz. This is the lowest frequency for the MGT reference clock [19] which is also used to clock the core. Basic principles of good hardware design were already adhered: The logic was heavily pipelined and it was made sure that the pipeline stages did not consist of too many levels of logic. Synthesis results showed that the logic itself was capable of reaching the desired target frequencies; however, after the design was routed timing was not met, even with all optimization switches in the tool set to maximum effort which caused runtimes of several hours for the whole flow. Since the logic was already optimized, other ways to improve the overall design had to be identified.

By analyzing the reports from the implementation tool it was determined that counters were using up a significant number of resources inside the design. The solution to that problem was to implement the counters in DSP slices. The selected device has 160 available DSP slices that are unused because the network data going through the chip is not processed in a computational intensive way.

A single DSP slice described in detail in [18] offers adders, multipliers and logic shift functions and is highly configurable. The adder inside the slice can be used to build a counter which is exactly the function that was previously performed by logic in the fabric. The following Table 2 shows the difference in resource utilization between the two counter implementations.

Although the savings per counter are moderate the overall numbers increase significantly if you keep in mind that the design instantiates not one but dozens of these counters. Since this implementation remaps logic to resources that would otherwise not be utilized it comes for free and as seen in Table 2 without a penalty in performance.

4.2.1 SRL FIFO

The design does not utilize the FIFO IP provided by Xilinx. Although these FIFOs can be operated in an FWFT (First Word Fall Through) configuration they come with a penalty of an additional clock cycle because of the RAM that is used internally. In order to improve the latency of data passing through a new FIFO was

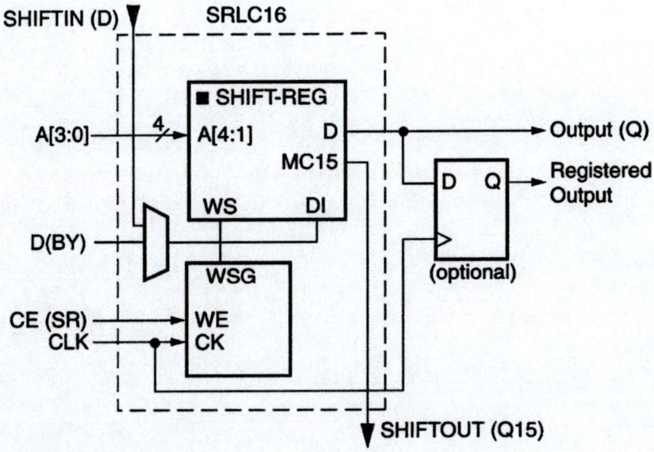

Fig. 8 SRL shift register [21]

developed which possesses a bypass feature. This data bypass is implemented by combining a RAM-based FIFO implementation which constitutes the main storage of the overall FIFO with a register-based FIFO behind the RAM in order to hide the access time of the RAM block. In the case of an empty FIFO the RAM-based FIFO is completely bypassed for the first operation that shifts in data to remove the additional cycle the data would need to appear at the read port of the RAM. Depending on the data width and the number of FIFOs in the overall design this feature can take up lots of registers and therefore valuable resources.

One solution to this dilemma of performance vs. resource allocation is implementing this register based FIFO in shift registers. CLBs (configurable logic blocks) in the Virtex-4 contain SRL primitives like the one shown in Fig. 8. One shift register can also be seen as a register array with 16 entries. By utilizing many SRL instances in parallel an SRL FIFO with 16 entries and variable data width can be constructed. Because every location inside the shift register can be tapped without an access penalty it behaves like a register-based FIFO and can therefore be used to implement the FIFO with bypass feature.

4.2.2 Floorplanning

Designers possess intimate knowledge of their design; they have an idea of the dataflow through the chip and the connectivity between the modules in it, whereas the implementation tool must extract this information by analyzing the design during the tool flow. This is, for once, a time-consuming task in large designs and the algorithms usually do not find an optimal solution. Even after analyzing the relationship of the modules with each other the placer has a huge decision space to decide where single design elements should be placed, especially in a large FPGA.

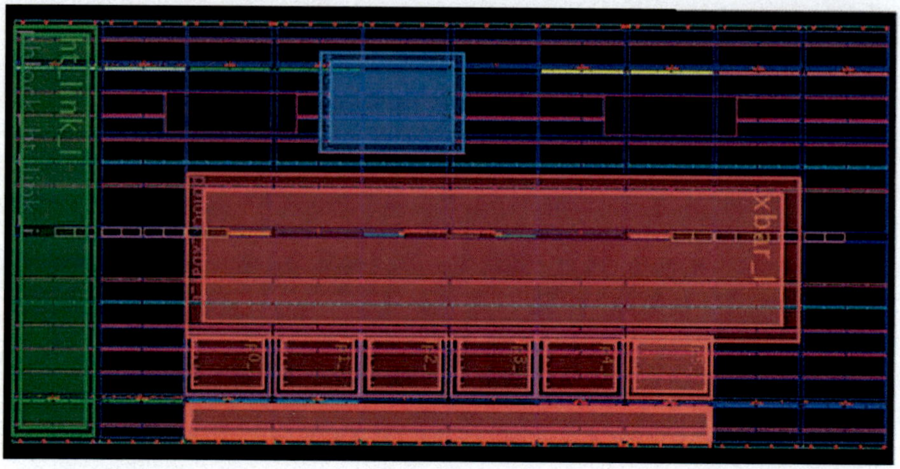

Fig. 9 EXTOLL floorplan

This means that implementation is both a time-consuming process and it is entirely possible that the tool will produce inferior results that might not meet the desired timing requirements.

Therefore, it becomes necessary for projects that push feasibility like the EXTOLL network to guide the tool with a detailed floorplan. The bounding boxes in that floorplan restrict both the placement of modules to a specific area as well as the number of available resources. Giving the tool too much freedom in placing logic turned out to be counterproductive in terms of timing closure. Adjusting size and location of the bounding boxes is usually an iterative process that will require several tool runs to find the best solution. However, as soon as a good floorplan is developed changes in other areas of the design will not have an effect on the timing of the planned modules.

Figure 9 shows the final floorplan of the EXTOLL device. The logic for external communication, the link ports together with the network crossbar (red area) and the host interface (green area), respectively, are constrained to be near the dedicated I/O circuitry used for this purpose. A part of the NIC logic (blue area) was also timing critical and received an additional bounding box in the upper part of the device.

4.3 Implementation Results

By employing HDL optimizations mentioned in the section before which were able to decrease the resource utilization to a more manageable 85% and supporting the tool flow by floorplanning critical section of the design timing was met successfully on the targeted XC4VFX100 device even with a high resource utilization as seen in Table 3. The key points for achieving these results were to take advantage of all the

Table 3 FX100 logic utilization [22]

Resource	Total usage	Utilization (%)
Slices	35,985	85
FlipFlops	28,442	33
LUTs	63,382	75
BRAMs	140	37

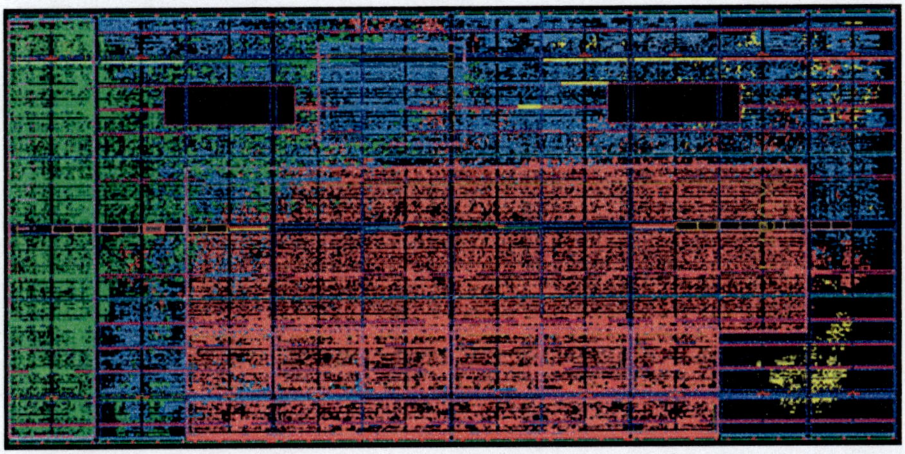

Fig. 10 FX100 utilization map

provided logic blocks like DSP slices and SRL16 blocks that allowed moving logic to previously unused design elements and thus freeing up space for the rest of the design.

The placement map of the final design (Fig. 10) shows that the implementation tool adhered to the guidelines given by the designer and was able to place the modules constrained in the floorplan into the designated area.

All these optimizations allowed for the realization of an interconnection network for High Performance Computing with an FPGA. In Sect. 6 we will show that the resulting performance that was achieved is in many ways competitive to a commercial ASIC implementation.

5 Software Architecture

To make EXTOLL usable for parallel applications, a number of software components had to be developed. Here we will present an overview of the necessary software. The software components to enable MPI applications to use EXTOLL can be divided into three main parts. First, there are operating system kernel level drivers. The second part is formed by low-level application programming interface (API) libraries. Finally, an adaption of a widely used MPI distribution was used to

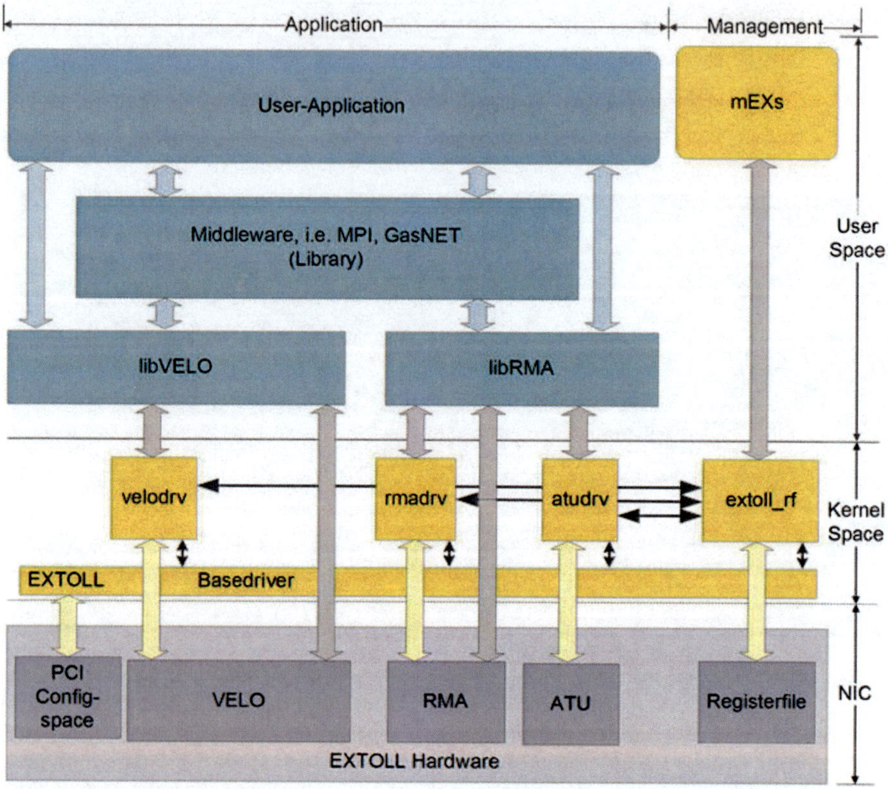

Fig. 11 EXTOLL software stack

enable MPI operations to be mapped to the EXTOLL hardware. Adaptions to other networking APIs are of course also possible (Fig. 11).

Somewhat unrelated to the direct application usage of EXTOLL is the implementation of management and administrative software components. These are important for the administrator of an EXTOLL network as well as in part for developers and will be presented last in this section.

5.1 *Kernel Driver*

As EXTOLL employs kernel-bypass techniques—also known as user-level communication—the kernel level drivers are mainly used to manage the hardware and allocate resources to application processes. Currently the EXTOLL software stack supports modern Linux kernels. The whole driver infrastructure is divided into a number of kernel modules each implementing a specific function.

First, there is the EXTOLL base driver which claims the EXTOLL hardware, remaps necessary I/O memory into kernel space and also contains the interrupt handler logic. The second module is the EXTOLL register file driver. The register file driver consists entirely of code generated from the XML description of the hardware's control and status register file. It offers an interface to the registers to other kernel level modules as well as a Linux *sysfs* interface for user-space software to interact with the registerfile (see also below).

Then, there is one module for each of the major functional units for communication of EXTOLL. Notably there is one module for VELO and one module for RMA. Both modules are similar in their workings, as they provide user-space with the ability to map one of the contexts of the EXTOLL hardware called Virtual Process ID (VPID).

5.2 *Low-Level API Libraries*

There are two low-level API libraries, which manage user-space communication using one of the two main communication units of EXTOLL each: *libvelo* for the VELO functional unit and *librma* for the RMA functional unit.

Libvelo in essence offers two-sided communications using *send()* and *receive()* function calls. Message are sent directly using PIO and then received into a main memory ring-buffer by hardware where they are then retrieved using the receive calls in *libvelo*. The whole implementation is optimized to support very low latency and high message rates (see also Sect. 6).

Librma offers complete control over the RMA communication engine of EXTOLL. There are functions to register and unregister memory for use by RMA operations. These functions perform a system call and the RMA kernel driver then pins the respective pages and updates the EXTOLL ATU translation tables. To actually perform communication a number of functions are made available which post different flavors of put and get operations to the hardware. The hardware then autonomously proceeds with the execution of these operations. As RMA supports the notion of notifications, these are also supported by a group of functions. Each RMA operation can trigger a notification on each endpoint involved with the operation. This feature can then be used to implement higher level protocols and the necessary synchronization.

5.3 *MPI Integration*

For MPI support, components for the popular open-source OpenMPI [23] project have been developed. OpenMPI is a highly modular, component-oriented implementation of the MPI specification. Several possibilities exist to support point-to-point operations in MPI. This can either be accomplished via the *Byte Transfer*

Layer (BTL) where low-level functions to transport unstructured data from point to point have to be implemented. Another option is to implement the *Matching Transfer Layer* (MTL). Here, non-blocking functions for send and receive operations, which also implement the MPI matching semantics, have to be implemented. Most networks (or transports as they are often called in this context) are supported by OpenMPI using BTL. For EXTOLL we chose to implement MTL, which enabled us complete control over the used low-level protocols.

For small messages up to a configurable threshold, MPI messages are sent using an *eager protocol* via VELO. Messages that are longer than one maximum size VELO message are fragmented into multiple VELO messages and reassembled at the receiving side. When a new VELO message is detected, the header information of the message is used to perform the MPI matching operation. If a matching receive has been posted, the receive operation can be completed and the data is copied from the VELO receive queue to the user-specified receive buffer. If no matching receive has been posted, the message becomes an unexpected message and is buffered until a matching receive is posted.

MPI messages longer than the specified threshold are sent using a rendezvous protocol. Here a VELO message carrying the request, i.e. the meta information of the transfer, is sent to the receiver. Again, the VELO message is matched. Once a matching receive has been found, the actual data is transferred using RMA Get operations. Thus, a zero-copy data transfer is implemented for large messages.

A third technique can be employed if the sender and receiver reside on the same physical node of the network. In this case, instead of using VELO and RMA, the data can also be transferred using shared-memory queues. For large intra-node messages, LiMIC [24] is to be employed to reduce the number of necessary *memcpy()* operations. Using this shared memory transport, significant network load can be removed if running on modern multicore nodes, where many MPI processes typically run on the same node.

To support MPI 2.0 one-sided communications, a prototype of an one-sided component (OSC) for OpenMPI was also implemented, showing the potential of high-overlap for one-sided MPI operations when run over the EXTOLL network [16].

In the future, a specialized implementation of the COLL (MPI collectives) component for EXTOLL is planned. The default COLL components implement the MPI collective routines on top of the point-to-point transport. Special hardware features as well as topology-aware optimizations can be leveraged with a specialized COLL component.

5.4 Other Communication Interfaces

EXTOLL with its rich feature set offers also the opportunity to implement interfacing to other common communication libraries besides MPI. One such interface is GasNET [25] which is used for example at the base of the Berkley Universal Parallel C (UPC) runtime. A GasNET prototype implementation for EXTOLL has already

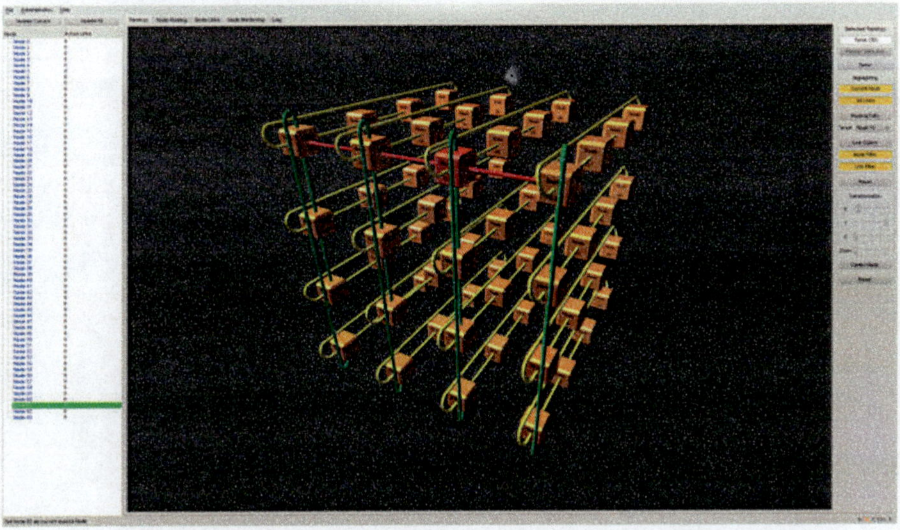

Fig. 12 MEXS GUI showing a 64 node 3-D torus network

been written and shows very promising results. Very similar possibilities exist with most wide-spread or well-known software interfaces for communication in HPC.

5.5 Managing an EXTOLL Network

To manage an EXTOLL network, two interfaces/software packages have been designed. First, there is the possibility to access all control and status registers using the Linux sysfs [26]. Using this feature it is easy to write script code which performs specific management or setup operations.

The second package is more complex and is called MEXS. It consists of a back-end process and a front-end GUI process for the administrator. The back-end process must run on one of the nodes of the EXTOLL network and directly interacts with the networking hardware. MEXS provides functions such as automatic topology enumeration, routing calculation, and distribution of routing information to all nodes of the network. The GUI allows the administrator to monitor the whole network with the aid of a graphical 3-D presentation of the network (see Fig. 12).

6 Evaluation

We developed a custom add-in card [20] combining an FPGA with an HTX connector as host interface and six serial links towards the network side for the implementation of the EXTOLL architecture Fig. 13. We use standard optical

Fig. 13 FPGA-based add-card including network interface and switch

transceivers and fibers to connect these add-in cards using a direct topology, so no additional switching units are required. This allows us to focus on a single FPGA development.

For the following experiments, we have setup an 8 node cluster with these custom add-in cards. A 3D torus is setup for experimentation. Each node includes two AMD Opteron 2380 processors, each one with 4 cores running at 2.5 GHz, and 8 GB of DDR2-800 main memory. Linux version 2.6.31.6 is installed on these machines.

6.1 Limitations

In comparison with ASICs, FPGAs are suffering from lower frequencies, smaller data path widths and often less flexibility due to integrated hard IP blocks. In particular, we are observing the following limitations:

- The host interface is running at HT400, limiting the peak bandwidth to 1.6 GB/s per direction. This bandwidth limitation is even enforced by the link speed of 6.24 GB/s peak, which translates into 624 MB/s peak due to the used 8b/10b encoding.
- The used serializers significantly increase hop latency, resulting in an overall latency of 300 ns per hop.

However, we will see that despite these severe limitations our design shows very good performance results. For several benchmarks we can achieve equal or better performance than recent, commercial ASIC solutions. This in particular demonstrates the architectural advantages of our approach and the possibilities of an FPGA-based networking implementation [27].

6.2 *Micro-Benchmarks*

Here, we will present a couple of micro-benchmarks to characterize the basic properties of EXTOLL, namely bandwidth, latency, message, and overlap. These characteristics are helpful as they indicate which performance can be expected for a given application, depending on this application's communication patterns.

6.2.1 Latency and Bandwidth

Start-up latency (half round-trip) and peak bandwidth are the most often cited network characteristics. The latency reports the time required between issuing the send operation on one node, and completing the corresponding receive call on a different node. The bandwidth reports how much data can be moved between two different nodes per time unit. Historically, latency has been improving much less than bandwidth [8], and this trend is expected to continue in the future. In both cases, latency and bandwidth, there is a huge disparity to in-system performance, i.e. memory access latency and memory bandwidth.

The Intel MPI Benchmarks (IMB) [28] are one of the state-of-the-art tests for basic characteristics like start-up latency and peak bandwidth. While this suite evaluates every important MPI communication function, for the sake of brevity we use only two selected tests here, which are the PingPong test for latency measurement and the SendRecv test for peak bandwidth analysis. We vary the number of hops for the PingPong test, and the number of communication pairs for the SendRecv test, in order to provide more insights.

The results from the Pingpong test are shown in Fig. 14, reporting half round-trip latency over a varying number of hops. Note the logarithmic scale of the x-axis. For up to 32 bytes payload the latency is constant around 1.5 μs, as for these cases we can keep payload and header in a single HyperTransport packet. For 64 bytes and more, we have to use two or more packets, which results in an appropriate latency increase. The three measurements over different hop counts show that independent of the message size each hop adds about 300ns to the overall latency.

A Send–Receive communication pattern is assessed in Fig. 15. Here, a varying number of process pairs are communicating using the SendRecv MPI-function call, and the resulting bandwidth (bi-directional) is reported. For SendRecv, we achieve a peak bandwidth of more than 800 MB/s for a single process pair. For more process pairs, the achievable bandwidth per pair scales as expected, indicating that the network interface overhead due to an increasing number of end points is marginal: for 8 pairs we achieve 99.95 MB/s per pair, which in summary almost matches the bandwidth for a single pair of 837 MB/s.

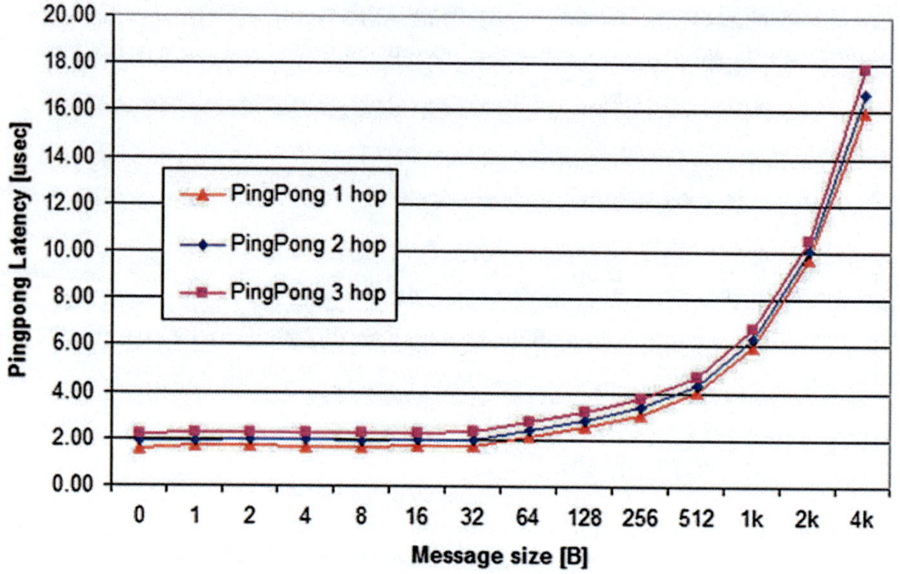

Fig. 14 Half round-trip latency over message size

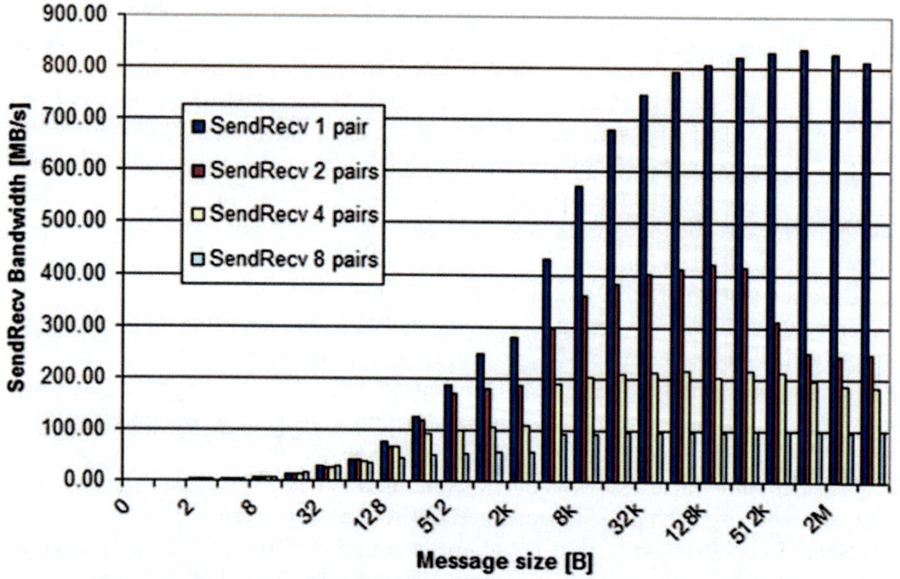

Fig. 15 Sustained bandwidth for a varying number of process pairs

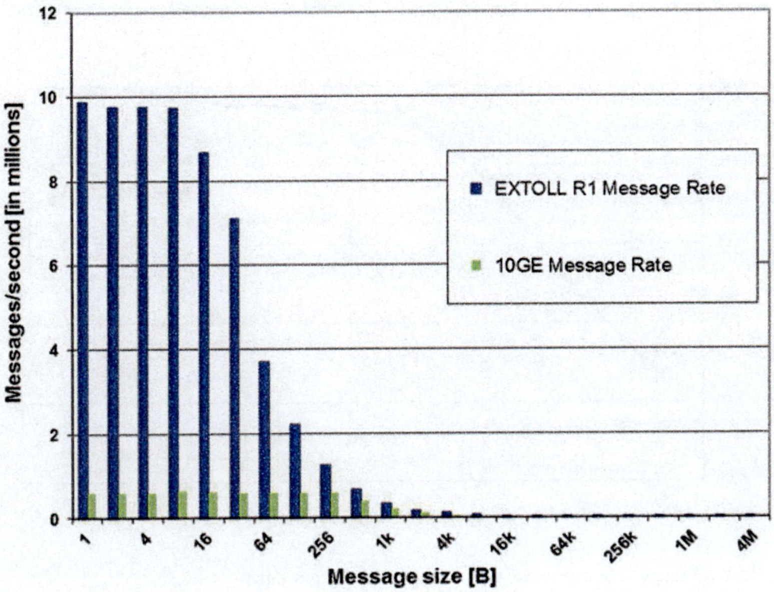

Fig. 16 Message rate for different message sizes

6.3 Message Rate

The message rate is defined as the number of messages that can be injected into a network from a host per second. Thus, it describes the achievable bandwidth for small message sizes. Latency is of paramount importance for round-trip communication patterns; however, for unidirectional transfers its impact is negligible. For such push-style communication patterns, the message rate is much more significant. Taking into account that the increasing degree of parallelism in modern computing systems also leads to an increased number of communication partners, communication pattern characteristics will shift to higher message counts with smaller payloads. Thus, the peak bandwidth is not the only metric that is crucial for the overall performance; instead, an increasing amount of attention must be paid to the performance of smaller messages.

The message rate for EXTOLL is reported in Fig. 16 over different message sizes. Again, note the logarithmic scale of the x-axis. To put the rather unfamiliar message rate into context, we also include results for a 10 GB Ethernet (10GE) adapter (Intel 82598EB), employed in the same system. As one can see, EXTOLL reaches 9.88 million messages per second, while 10GE saturates at 0.62 million messages per second.

Peak message rate is typically not reached with a single process pair, as the MPI overhead requires significant CPU time to send or receive a message. To mask out this CPU overhead, multiple process pairs can be used until the network limits the

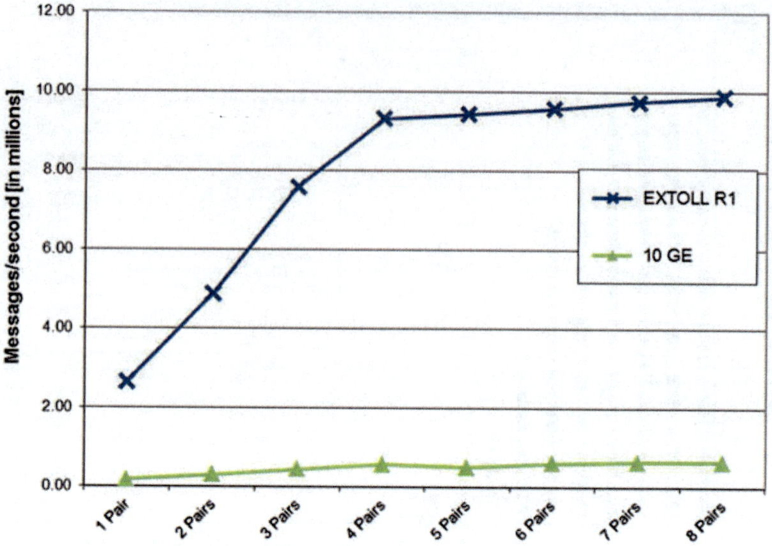

Fig. 17 Peak message rate over a varying number of process pairs

message rate. For the case in Fig. 16, EXTOLL requires 7 or 8 process pairs to reach peak performance, indicating that network-side limitations are only visible for very high loads. For 10GE, the number of process pairs varies in between 2 and 7. For an improved understanding, Fig. 17 reports the maximum message rate for a varying number of process pairs.

6.4 *Overhead and Application Availability*

Latency, bandwidth, and message rate describe the performance of the communication subsystem. However, many HPC applications leverage overlap between communication and computation to minimize their runtime, i.e. while the communication subsystem is busy transferring data, processors can resume computational tasks that have no dependencies on outstanding communication tasks.

Overhead, respectively, application availability characterizes how much time is left during communication for computation. According to [9], "overhead is defined as the length of time that a processor is engaged in the transmission or reception of each message; during this time, the processor cannot perform other operations." Application availability [11] is defined to be the fraction of total transfer time that the application is free to perform work not related to communication. Thus, application availability can be derived as follows:

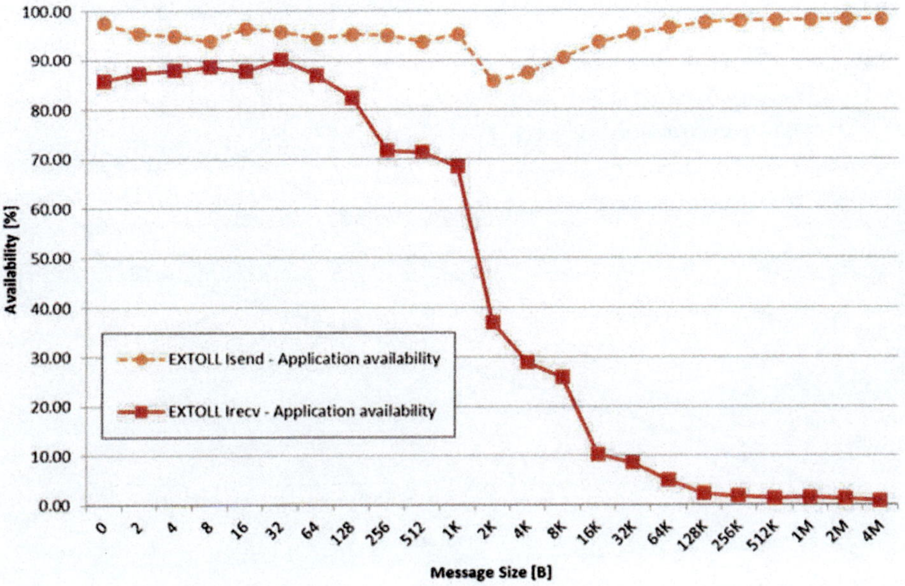

Fig. 18 Application availability for send and receive calls

$$\text{application availability}[\%] = 1 - \frac{\text{over head}[\mu\,\text{sec}]}{\text{transfer time}[\mu\,\text{sec}]} \tag{1}$$

Equation (1): Application availability.

We have used the Sandia MPI Benchmarks (SMB) to determine overhead and the associated application availability for our EXTOLL network. The following Fig. 18 reports in percent how much time is left for non-communication related work.

As one can see, *Isend()* over EXTOLL results in minimal overhead, as communicational work is minimized. On the receive side several tasks have to be performed, like tag matching and copy operations, so that application availability degrades with an increasing message size. While this is quite typical behavior for MPI-based communication, it is worth noting that up to payloads of 1kB more than 70% of the communication time is potentially left for overlapping with computational tasks.

6.5 Complex Benchmarks

The information gathered and presented in the previous section gives a detailed insight into the basic performance characteristics of EXTOLL and therefore in the viability of our architecture. Here, we would like to complete this section by presenting how this performance translates also to the application level. Instead of using the obvious benchmarks like High Performance LinPack (HPL) or NAS

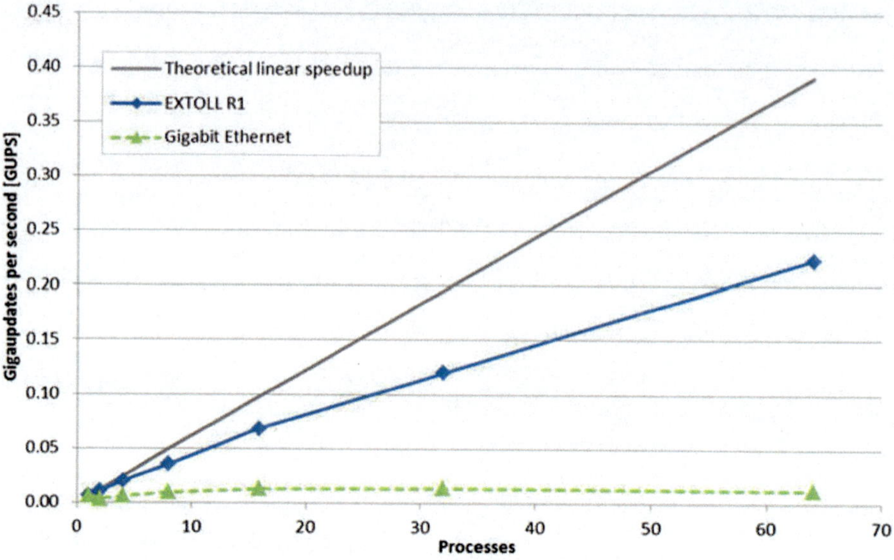

Fig. 19 Performance results for HPCC RandomAccess

Parallel Benchmark (NPB), which provide little insight for modern workloads and are optimized to avoid fine grain communication, we have selected benchmarks which in our opinion are more important and benefit a lot from dedicated support for small messages. Here, these benchmarks are used to assess EXTOLL's application-level performance:

1. HPCC RandomAccess (RA)
2. Weather Research and Forecast (WRF)

6.5.1 HPCC RandomAccess

The RandomAccess benchmark [29] updates a set of memory locations in a randomized fashion and reports the number of updates per second as Giga Updates Per Second (GUPS). It is classified as a benchmark with low locality, both spatial and temporal, and puts the highest pressure on the network with regard to small messages. As can be seen in Sect. 3, this benchmark heavily relies on small transfers; in particular for the highest process count, more than 80% of the transfers have a payload between 256 and 512 bytes.

Figure 19 highlights both performance and scalability. Results are presented for executions using EXTOLL and Gigabit Ethernet as a reference. Also included is the theoretical linear speedup, relative to the performance of 1 process over EXTOLL.

The performance difference between EXTOLL and Gigabit Ethernet is huge, and EXTOLL achieves a speedup close to the theoretical one. Furthermore, while

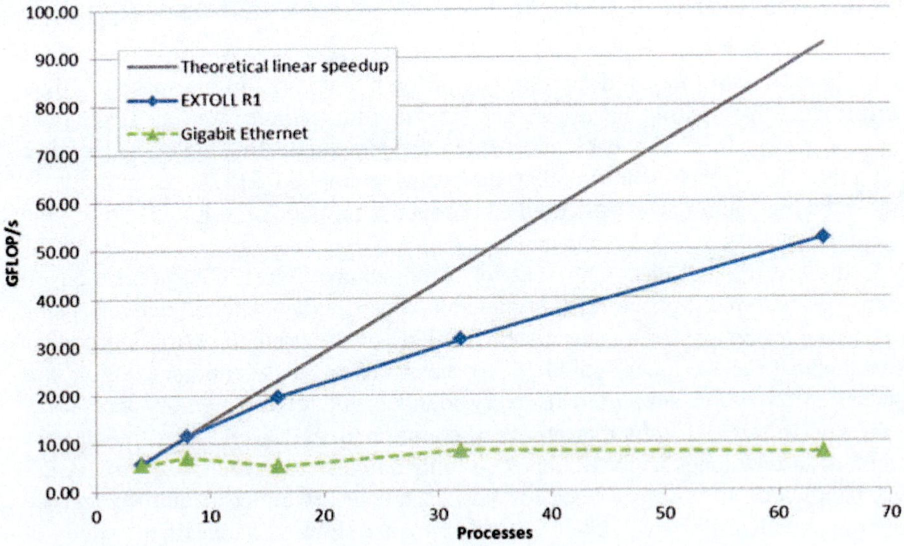

Fig. 20 Performance results for weather research forecast

Gigabit Ethernet degrades in terms of performance already for more than 32 processes, EXTOLL's performance peak is far outside this graph.

6.5.2 Weather Research and Forecast

The Weather Research and Forecasting Model (WRF) [30] is a real-world application rather than a benchmark. It serves for both operational forecasting and atmospheric research. As data input sets are publically available, it can easily be used for benchmarking purposes. We are using version 3.3 for our experiments together with the "CONUS 12 km" data set. Section 3 has unveiled that this benchmark is making use of a broad range of message sizes, although the really large ones (i.e., larger than 512 kB) are not frequently used. This real-world application highlights the importance of dedicated support for small data transfers particularly. The message histogram shows that packet sizes between 128 and 512 bytes (in total approx. 27%) are frequently used, in addition to the packet sizes between 8 and 128 kB (in total approx. 62%). Bulk transfers, however, are almost absent.

Figure 20 shows the performance when executing WRF, results are reported in GFLOP/s. While for 4 processes the performance approximately matches, Gigabit Ethernet is not able to scale performance with process count. Instead, it saturates around 8 GFLOP/s. EXTOLL, however, nicely keeps the pace of an increasing process count. Also shown is the theoretical linear speedup based on EXTOLL's performance for 4 processes. Again, no saturation for EXTOLL is visible, instead it can be expected that adding more computing cores will result in a continuing performance scaling.

7 Related Work

Over the course of the last decades many different networks in the space of High-Performance Computing have emerged. In the context here, we can distinguish between purely HPC networks, most often implemented using ASIC technology, and HPC networking solution that have also employed FPGA technology. In the following section we will briefly introduce a number of solutions from both categories.

In the last few decades, many network architectures for HPC have been developed. For example, Myrinet [31] implements a wormhole switched network using a PCI adapter board and a central switch. Many of the protocol work is offloaded from the host CPU to the on-chip RISC processor of the NIC. Another example was Quadrics [32], which supported also very sophisticated remote memory accesses.

Today, Infiniband networks are very common in HPC, and here mainly the adapters and switches from Mellanox. Infiniband currently reaches up to 56 GB/s link bandwidth and back-to-back low-level latencies of under a microsecond are reported, while MPI level latency measurements show still excellent values of around 1.5 μs.

Even more focused on HPC are the integrated networking solutions from Cray with their SeaStar [3] and Gemini [4] networking chips as well as the TOFU [5, 33] network from Fujitsu which is used within the K-Computer. All three networks are based on direct-topologies, on torus topologies to be more precise. Seastar, the SeaStar+ variant and Gemini are based on 3D topologies, while the TOFU network employs a more complicated 6-D topology. The Cray networks are directly connected to Opteron CPUs using HyperTransport interfaces, very similar to EXTOLL. In [4] some performance numbers of the 90 nm Gemini ASIC are reported. The chip runs with a core frequency of 650 MHz and the router operates at 800 MHz, both significantly faster than what is currently feasible in an FPGA. The TOFU network is also fabricated in 65 nm but runs with a more conservative clock rate of 312.5 MHz. TOFU is directly attached to the I/O interface of SPARC64 CPUs used for the K-Computer.

Besides pure HPC networks, there have been a number of FPGA-based network architectures in the past, many of them in the field of HPC. Two examples are an SCI implementation using FPGAs [34] and the Clint network [35]. In more recent times, a prototype for the DIMMnet-2 network interface controller has been built using a Virtex II Pro 70 FPGA device [36]. The prototype implements only a part of the functionality of the complete DIMMnet-2 design. The maximum packet size is limited for example, but the design runs with nearly 100 MHz and uses 42% of the FPGA device. On the network side, the controller interfaces to an Infiniband fabric. In an evaluation, the NIC showed an excellent latency of about 1 μs, but with an on-chip direct loopback, i.e. without the latency from the actual link and physical layer and without the latency of a switch. In contrast, the 1.16 μs of latency for EXTOLL does include all these latencies, an internal loopback using the EXTOLL switch (thus still including one level of switching) accounts for approximately 800 ns.

In [37] an NIC is described that is able to perform RDMA write operations only. Again it was implemented on a Virtex II device with clock rates of 100 MHz for the core and 78 MHz for the links. The JNIC project [38] implemented a Gigabit-Ethernet NIC on an Altera Stratix EP1S80 FPGA. The specialty here was that the FPGA was directly attached to the Front-Side bus of an Intel Xeon CPU. As the NIC resides in the cache coherent domain it is possible to transfer data using cache accesses which significantly lowers latency and increases bandwidth. While this approach would allow developing a whole new set of architectures and techniques, in [38] the focus is clearly on the software side instead of exploring new hardware ideas.

The QPACE [39] system also employed FPGAs. Here the nearest-neighbor network was implemented using a Virtex IV device. The processors of the system were IBM Cell chips which were directly connected to the FPGAs for high-performance communication. The specialties of the application—Quantum Chromodynamics—defined the communication features implemented in the system.

In fields related to classical HPC, FPGA-based communication acceleration is also used. An example is presented in [40], where an FPGA is used to accelerate the communication in a high-frequency tranding scenario. Actually, several of the underlying IP is shared with the EXTOLL project.

8 Conclusion and Outlook

The need to rely on parallelization to improve performance of parallel computers leads to the problem that the interconnection network becomes the bottleneck. We have presented the EXTOLL interconnection network, which is specifically designed to help alleviating these problems. An analysis of the demand of HPC applications leads to the unique system architecture of EXTOLL.

FGPA technology enables architectures like EXTOLL to be implemented and tested in real systems without the need to tape out an ASIC. Together with the corresponding software architecture, a complete system can be assembled and evaluated. Furthermore, FPGA technology has helped to continuously improve the architecture as well as branch special versions for different research purposes.

The performance reached by EXTOLL—evaluated here on the micro-benchmark level as well as on selected applications—shows that the acceleration of communication is possible with an architecture like EXTOLL. The latency of about 1.5 μs MPI latency and nearly 10 million messages per second is very impressive, especially when compared to commercial ASIC solutions, which generally do not offer such a performance, although they employ higher clock frequencies and ASIC technology. This acceleration of the network also translates into accelerated execution of the parallel application, as shown by the evaluation of WRF. It is thus worthwhile to accelerate communication, especially since the future will bring ever higher parallelism to HPC.

The EXTOLL architecture is continuously improved. While the version described here was implemented on a Xilinx Virtex 4, a second release is being architected to work on Virtex 6 FPGAs. This release will feature an internal clock-rate of 200 MHz and a data-path width of 64 bit. Both measures are possible because of the larger and somewhat faster FPGA and lead to an MPI latency of less than 1.2 μs and an impressing message rate of well over 20 million MPI packets per second. Additionally, new architectural features will improve support for collective operations, scalability and introduce support for direct accelerator to accelerator communication (GPGPU).

References

1. TOP500 list: http://www.top500.org
2. Infiniband Trade Association; *InfiniBand Architecture Specification Volume 1*; Release 1.2.1, 2007
3. R. Brightwell, K.T. Pedretti, K.D. Underwood, H. Trammell, SeaStar interconnect: Balanced bandwidth for scalable performance. IEEE Micro 41–57 (2006)
4. R. Alverson, D. Roweth, L. Kaplan, The gemini system interconnect high performance interconnects (HOTI), in *2010 IEEE 18th Annual Symposium on*, 2010, pp. 83–87
5. Y. Ajima, S. Sumimoto, T. Shimizu, Tofu: A 6D mesh/torus interconnect for exascale computers. Computer 36–40 (2009). IEEE Computer Society
6. The BlueGene/L Team An overview of the BlueGene/L supercomputer, in Proceedings 2002 ACM/IEEE Conf. Supercomputing (SC 02), IEEE CS Press, 2002
7. P. Kogge et al., (eds.), *ExaScale Computing Study: Technology Challenges in Achieving Exascale Systems*. US Department of Energy, Office of Science, Advanced Scientific computing Reasrach, Waschington, DC, (2008). available at http://www.er.doe.gov/ascr
8. D.A. Patterson, Latency lags bandwith. Comm. ACM **47**(10), 71–75 (2004)
9. D. Culler, R. Karp, D. Patterson, A. Sahay, K. Schauser, E. Santos, R. Subramonian, T. von Eicken, LogP: towards a realistic model of parallel computation, in *Fourth ACM SIGPLAN symposium on Principles and Practice of Parallel Programming*, 1993, pp. 262–273
10. S. Sur, M.J. Koop, D.K. Panda, High-performance and scalable MPI over InfiniBand with reduced memory usage: an in-depth performance analysis, in *Proceedings of the 2006 ACM/IEEE conference on Supercomputing (SC '06)*, 2006
11. W. Lawry, C. Wilson, A. Maccabe, R. Brightwell, COMB: a portable benchmark suite for assessing MPI overlap, in *Proceedings of the IEEE International Conference on Cluster Computing (CLUSTER 2002)*, 2002, p. 472
12. D. Slogsnat, A. Giese, M. Nüssle, U. Brüning, An open-source HyperTransport core, in *ACM Transactions on Reconfigurable Technology and Systems (TRETS)*, vol. **1**(3), 2008, p. 1–21
13. H. Litz, H. Fröning, U. Brüning, HTAX: a novel framework for flexible and high performance networks-on-chip, in *Fourth Workshop on Interconnection Network Architectures: On-Chip, Multi-Chip (INA-OCMC), in conjunction with HiPEAC*, 2010
14. H. Litz, H. Fröning, M. Nüssle, U. Brüning, VELO: a novel communication engine for ultra-low latency message transfers, in *37th International Conference on Parallel Processing (ICPP-08)*, 2008
15. M. Nüssle, Acceleration of the Hardware-Software Interface of a Communication Device for Parallel Systems. Ph.D. thesis, University of Mannheim, 2009

16. M. Nüssle, M. Scherer, U. Brüning, A resource optimized remote-memory-access architecture for low-latency communication, in *38th International Conference on Parallel Processing (ICPP-2009)*, 2009
17. Hypertransport Technology Consortium, Hypertransport I/O Link Specification Revision 2.00b, 2005. Document #HTC20031217–0036–0009
18. Xilinx Inc, XtremeDSP for Virtex-4 FPGAs User Guide, UG073 (v2.7), 2008
19. Xilinx Inc, Virtex-4 RocketIO Multi-Gigabit Transceiver User Guide, UG076 (v4.1), 2008
20. H. Fröning, M. Nüssle, D. Slogsnat, H. Litz, U. Brüning, The HTX-board: a rapid prototyping station, in *3rd annual FPGAworld Conference*, 2006
21. Xilinx Inc, Virtex-4 FPGA User Guide, UG070 (v2.6), 2008
22. M. Nüssle, B. Geib, H. Fröning, U. Brüning, An FPGA-based custom high performance interconnection network, in *2009 International Conference on ReConFigurable Computing and FPGAs*, 2009
23. E. Gabriel, G.E. Fagg, G. Bosilca, et al., Open MPI: goals, concept, and design of a next generation MPI implementation, in *Proceedings of the 11th European PVM/MPI Users' Group Meeting (Euro- PVM/MPI04*, 2004
24. H.W. Jin, S. Sur, L. Chai, D.K. Panda, LiMIC: support for high-performance MPI intra-node communication on Linux cluster, in *34th International Conference on Parallel Processing (ICPP-05)*, 2005
25. K. Yelick, D. Bonachea, W.Y. Chen, P. Colella, K. Datta, J. Duell, et al., Productivity and performance using partitioned global address space languages, in *International Conference on Symbolic and Algebraic Computation*, 2007
26. P. Mochel, The sysfs filesystem, in *Proceedings of the Annual Linux Symposium*, 2005
27. H. Fröning, M. Nüssle, H. Litz, U. Brüning, A case for FPGA based accelerated communication, in *9th International Conference on Networks (ICN 2010)*, 2010
28. Intel GmbH, Intel® MPI Benchmarks Users Guide and Methodology Description, 2006
29. V. Aggarwal, Y. Sabharwal, R. Garg, P. Heidelberger, HPCC RandomAccess benchmark for next generation supercomputers, in *Proceedings of the 2009 IEEE International Symposium on Parallel&Distributed Processing (IPDPS)*, IEEE Computer Society, 2009
30. J. Michalakes, J. Dudhia, D. Gill, T. Henderson, J. Klemp, W. Skamarock, W. Wang, The weather research and forecast model: software architecture and performance, in *Proceedings of the 11th ECMWF Workshop on the Use of High Performance Computing In Meteorology*, 2004
31. N.J. Boden, D. Cohen, R.E. Felderman, A.E. Kulawik, C.L. Seitz, J.N. Seizovic, S. Wen-King, Myrinet: a gigabit-per-second local area network. IEEE Micro **15**(1), 29–36 (1995)
32. F. Petrini, et al., The quadrics network: high-performance clustering technology. IEEE Micro **22**(1), 46–57 (2002)
33. Y. Ajima, Y. Takagi, T. Inoue, S. Hiramoto, T. Shimizu, The tofu interconnect. High performance interconnects (HOTI), in *2011 IEEE 19th Annual Symposium*, 2011
34. M. Trams, W. Rehm, SCI transaction management in our FPGA-based PCI-SCI bridge, in *Proceedings of SCI Europe*, 2001
35. N. Fugier, M. Herbert, E. Lemoine, B. Tourancheau, *MPI for the Clint Gb/s Interconnect. Recent Advances in Parallel Virtual Machine and Message Passing Interface*, Vol. 2840/2003, *Lecture Notes in Computer Science* (Springer, Heidelberg, 2003)
36. N. Tanabe, A. Kitamura, et al., Preliminary evaluations of a FPGA based-prototype of DIMMnet-2 network interface, in *IEEE International Workshop on Innovative Architecture for Future Generation High-Performance Processors and Systems*, 2005
37. M. Marazakis, K. Xinidis, V. Papaefstathiou, A. Bilas, Efficient remote block-level I/O over an RDMA-capable NIC, in *Proceedings of the ACM International Conference on Supercomputing (ICS)*, 2006

38. M. Schlansker, N. Chitlur, et al., High-performance ethernet-based communications for future multi-core processors, in *Proceedings of the 2007 ACM/IEEE Conference on Supercomputing (SC '07)*, 2007
39. H. Baier, et al., QPACE - a QCD parallel computer based on Cell processors, in *Proceedings Science (LAT2009)*, 2009
40. C. Leber, B. Geib, H. Litz, High frequency trading acceleration using FPGAs, in *21st International Conference on Field Programmable Logic and Applications (FPL 2011)*, 2011

High-Speed Torus Interconnect Using FPGAs

H. Baier, S. Heybrock, B. Krill, F. Mantovani, T. Maurer, N. Meyer, I. Ouda, M. Pivanti, D. Pleiter, S.F. Schifano, and H. Simma

Abstract In this chapter we describe the architecture of a torus interconnect and its implementation on FPGAs, which so far has been used in two different HPC systems. The network design is optimized for applications which benefit from a tightly coupled network and allows to exchange relatively small messages between nearest neighbours at a high rate. Examples for such applications are lattice quantum chromodynamics (LQCD) simulations and fluid dynamics applications using the Lattice Boltzmann method (LBM). We describe the details of the implementation of our torus network architecture for two massively parallel machines, QCD Parallel

H. Baier • B. Krill • T. Maurer
IBM Deutschland Research & Development GmbH, 71032 Böblingen, Germany
e-mail: hbaier@us.ibm.com; krill@de.ibm.com; tmaurer@de.ibm.com

S. Heybrock • F. Mantovani • N. Meyer
Department of Physics, University of Regensburg, 93040 Regensburg, Germany
e-mail: simon.heybrock@ur.de; filippo.mantovani@ur.de; nils.meyer@ur.de

I. Ouda
IBM Rochester, 3605 HWY 52 N, Rochester, MN 55901-1407, USA
e-mail: ouda@us.ibm.com

M. Pivanti • S.F. Schifano
University and INFN of Ferrara, 44100 Ferrara, Italy
e-mail: marcello.pivanti@fe.infn.it; schifano@fe.infn.it

D. Pleiter (✉)
Forschungszentrum Jülich, 52425 Jülich, Germany

Department of Physics, University of Regensburg, 93040 Regensburg, Germany
e-mail: d.pleiter@fz-juelich.de

H. Simma
Deutsches Elektronen Synchrotron (DESY), 15738 Zeuthen, Germany
e-mail: huber.simma@desy.de

W. Vanderbauwhede and K. Benkrid (eds.), *High-Performance Computing Using FPGAs*,
DOI 10.1007/978-1-4614-1791-0_18, © Springer Science+Business Media, LLC 2013

Computing on Cell (QPACE) and AuroraScience, and present details on the FPGA resource usage. Furthermore, we discuss optimizations which were necessary to fit the design. Finally, we provide an outlook on possible implementation changes when using more recent generations of FPGAs.

1 Introduction

In the past, several projects have developed custom-designed processors optimized for applications from lattice quantum chromodynamics (LQCD). This physics research area investigates the theory of strong interactions, one of the four fundamental forces in nature, by means of numerical simulations. Recent examples of such projects are apeNEXT [1] and QCDOC [2]. These architectures are based on system-on-chip (SoC) designs with a network integrated on chip. However, using modern technologies the development costs of custom processors became too high for an academic project.

The significant increase in floating-point performance in commodity processors and with FPGAs becoming capable of processing large amounts of data opened the path to a different approach to design scalable, application optimized architectures. Using FPGAs to implement a network processor (NWP) which is tightly coupled to a commodity processor, a high performance, custom interconnect can be designed and implemented at a competitive price–performance ratio.

This strategy has been successfully applied in the design of two HPC architectures, QCD parallel computing on cell (QPACE) [3, 4], a supercomputer based on the IBM PowerXCell 8i processor, and AuroraScience [5], a cluster based on Intel Nehalem/Westmere processors. In QPACE a Xilinx Virtex5 FPGA has been directly attached to the processor via a FlexIO interface. In the AuroraScience machine more recent Altera Stratix IV FPGAs have been used which are connected to the southbridge via PCIe. In case of QPACE the interface to the processor turned out to be particularly challenging, mainly because no hard-IP block could be used.

We expect this strategy to be also viable in the near future. During several generations of FPGAs major vendors like Altera or Xilinx have significantly enhanced the capabilities of these devices to receive, process and transmit data. These vendors increased, e.g., both the number and the performance of high-speed transceivers for a given class of FPGAs significantly.

In the next section we will explain in detail the architecture of our torus network. In Sects. 3 and 4 we will discuss details of how this network architecture has been implemented using FPGAs for QPACE and AuroraScience. This is followed by a presentation of performance results in Sect. 5. Before presenting our summary and conclusions in Sect. 7 we give an outlook on options for future implementations in Sect. 6.

2 Torus Network Architecture

2.1 System and Network Processor Architecture

The torus network interconnects computing nodes in a 3-dimensional torus topology. The communication data paths are illustrated in Fig. 1. Each node consists of one or more commodity multi-core CPUs with all necessary peripheral components, like chipset, (local) memory, etc. In addition, the node card hosts an FPGA, which is connected to the CPU (e.g., via a south-bridge) as an IO device and implements the NWP. The main logic blocks of the NWP, as shown in Fig. 2, are the *IO interface* to handle data transfers from CPU to NWP and vice versa, and 6 *link modules* which control the data transmission over the 6 physical links. These interconnect each NWP with the NWPs of the 6 nearest-neighbour nodes on the torus. The physical links are either directly attached to the fast IO pins of the FPGA or through an external PHY component. Each link module has *injection* and *reception* buffers, together with the control logic to implement the torus network link protocol and the interface with the physical link.

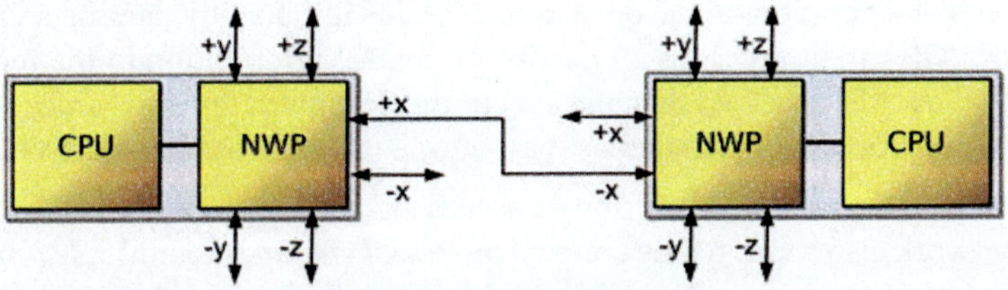

Fig. 1 Main data paths of the torus network. On each node the CPUs are tightly connected to a NWP, which in turn is connected by 6 physical link to the NWPs of the nearest-neighbour nodes in each of the 6 directions

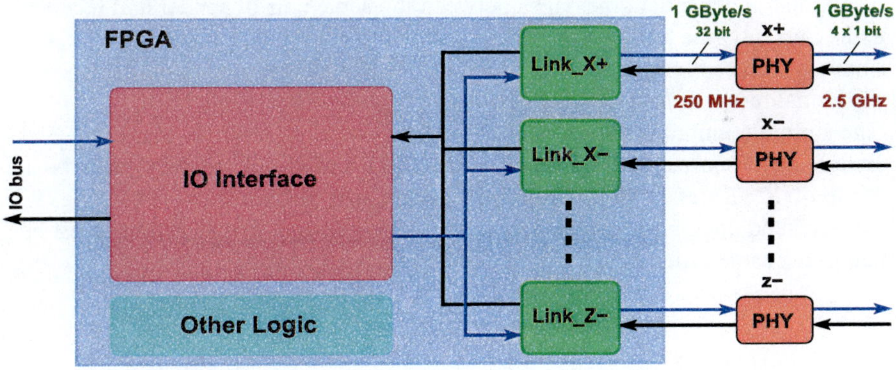

Fig. 2 Block diagram of the network processor

2.2 Communication Model

Data transfer between two nodes proceeds according to a *two-sided* communication model, i.e., explicit operations of both CPUs, sender and receiver, are required to control the data transmission of a message. These operations can be implemented in a blocking or non-blocking way.

Applications access the torus network by (1) moving data into the injection buffer of the NWP of the sending node and (2) enabling data to be moved out of the reception buffer of the NWP of the receiving node. Tracing a CPU-to-CPU data transfer over the torus network the following three transactions occur:

T1: The send operation simply moves the data items of a message into the injection buffer for one of the 6 link modules in the NWP of the sending node. Depending on the architecture and IO interface of the CPU, this operation can be implemented according to different schemes (see below).

T2: As soon as an injection buffer holds data, the NWP breaks it into fixed-size packets and transfers them in a strictly ordered and reliable way over the corresponding link. Reliability implies that cases of data corruption or potential data losses, e.g., due to full reception buffers, are automatically managed at hardware level.

T3: The receive operation on the destination CPU is initiated by passing a *credit* to its NWP. The credit provides all necessary control information to the receiving NWP to move the received data packets to the destination memory location and to *notify* the processor when the last packet of a message has been delivered.

To allow for a tight interconnection of processors with a multi-core architecture, the torus network also supports the concept of *virtual channels* to multiplex multiple data streams over the same physical link. A virtual channel is identified by a tag which is transferred over the link together with each data packet. This is needed to support independent message streams between different pairs of sender and receiver threads (or cores) over the same link. The virtual channels can also be used as a tag to distinguish independent messages between the same pair of sender and receiver threads. Currently the torus network design supports 8 virtual channels per link, but this number can be increased, e.g., to support CPUs with more cores, at the expense of additional resource usage on the FPGA and higher protocol overhead.

The simple communication model of the torus network requires that each send operation has a corresponding receive operation and message sizes must be a multiple of 128 Bytes (which is the fixed packet size of the torus network links). Moreover, send operations which refer to the same link and virtual channel must be issued in the same order as the corresponding receive operations.

2.3 IO Interface

The IO interface of the NWP handles the IO transactions between CPU and NWP and depends on the IO architecture of the CPU. Moreover, for the send operation

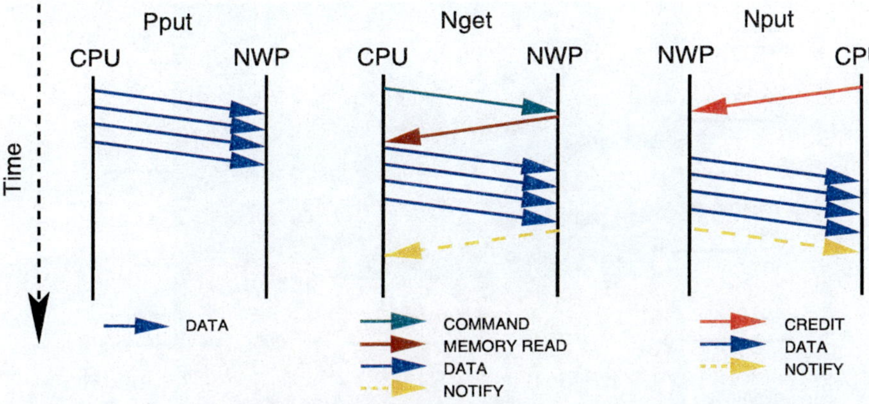

Fig. 3 Schematic view of different schemes for moving data from the CPU to the injection buffers of the NWP and from reception buffers to the CPU

(transaction T1) we have implemented different schemes according to which data is moved from the CPU to the NWP.

In the simplest scheme, which we call Pput in the following (see Fig. 3), the CPU initiates the data transfer of the message to the NWP. It is then convenient to map the injection buffers of the different links and virtual channels directly into disjoint areas of the address space of the CPU. Then, a single IO transaction can be sufficient to move to the NWP both the data and all control information (which can be implicitly encoded into the addresses).

If the CPU has a controller for *direct memory access* (DMA), as in the case of the IBM PowerXCell 8i processor, the transfer can be done by DMA and does not occupy the CPU. In this way, the send operation is non-blocking. It is completed when all data items of the message have been transferred to the NWP. This might be tested (e.g., before an application re-uses the memory locations from where the sent data originated) by querying the status of the DMA engine.

However, if the CPU remains occupied for the entire data transfer, which is the case on x86 CPU architectures when the transfer is done by *Programmed IO* (PIO), then the send operation is always blocking. Therefore, we extended the IO interface of the NWP to support also a second scheme, called Nget in the following. In this case, the CPU only passes the required control information to the NWP. Then the data transfer is controlled by a DMA engine on the NWP, and finally the NWP has to notify the CPU that the data transfer is completed.

Compared to implementing the send operation (T1) through an Nget scheme, a Pput scheme can be more efficient, in particular for short messages, because it requires fewer IO transactions and hence may have a lower latency. However, in the Nget scheme it can be simpler to handle *back-pressure*, which arises when injection buffers are full, while this may require extra transactions in the Pput scheme.

Also for moving data from the reception buffer to the processor (T3) we can choose whether NWP or processor control the operation. We call these schemes Nput and Pget, respectively (see Fig. 3). We have only implemented the case

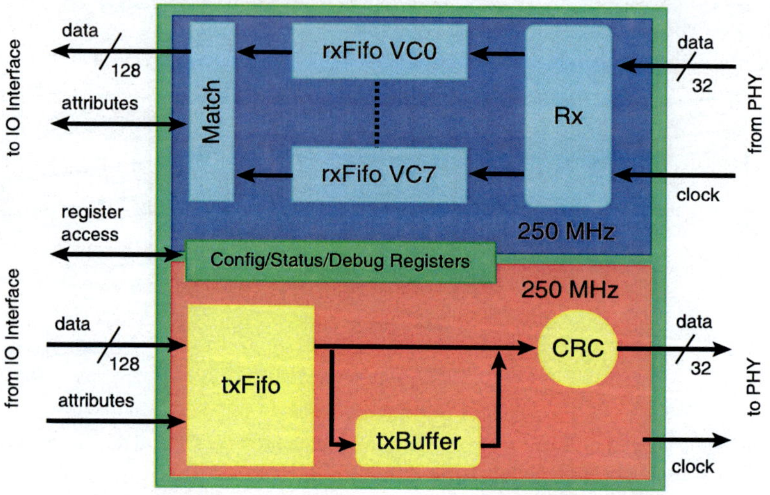

Fig. 4 Block diagram of the link module

Nput where the NWP comprises DMA engines for moving the data to the final memory location. A notification mechanism is required to inform the processor when this final transaction has completed. We have not implemented the Pget scheme, because it would imply larger latencies and does not provide any obvious advantages for the considered processor architectures.

2.4 Link Modules

The link modules, one for each of the six directions of the 3-dimensional torus, control the transfer of data packets over the physical links of the torus network [6]. In the torus network implementations for both, QPACE and AuroraScience, the physical links use an external transceiver PHY (PMC Sierra PM8358a). The corresponding link module is connected to the 10 Gigabit Media Independent Interface (XGMII) port of the PHY through a 32-bit bus. High-speed serial links interconnect the 10 Gigabit Attachment Unit Interface (XAUI) ports of PHYs on neighbouring nodes. These links are routed over backplane and/or cable.

The link module implements a light-weight and robust custom protocol to guarantee data integrity and strict ordering of the packets by making use of the control symbols from 8B/10B coding by the PHY. Each torus network link can simultaneously send and receive data.

The architecture of a link module is shown in Fig. 4 and consists of two basically independent parts for sending and receiving. The data to be sent is stored together with all relevant control information, like address offset and virtual channel index, in the injection buffer (txFifo), which is handled with First-In-First-Out policy. As soon as it holds at least 128 Bytes of data, the transmission logic starts to pop data from txFifo and generates a packet composed of a 32-bit header, 128 Bytes of data payload (32×32 bit), and a 32-bit CRC. The header comprises a 11-bit wide address offset and a 3-bit wide field encoding the virtual channel. The packet is then passed to the external PHY for transmission over the physical link.

On the other side, the RX logic decodes the data packets received from the external PHY, re-computes the CRC and compares it with the CRC of the packet. If they match, the packet is pushed into the reception buffer associated with each virtual channel, and a positive feedback (ACK) is sent back over the link to the link module of the sender. If the reception buffer has no space for further packets (back-pressure) and in the rare cases, when the CRC does not match or other errors occur in receiving the packet, a negative feedback (NACK) is sent back. All further data packets from the PHY are discarded until a RESTART command is received (indicating that the sending link module has started to resend the packets).

The link module temporarily stores each packet, which has been sent, into an internal buffer (txBuffer) until a feedback for that packet is received from the peer-link. If the feedback is positive the corresponding packet is dropped, otherwise the link module recursively enters in *resend* mode, sends a RESTART to the peer, and then begins to re-send all packets from txBuffer until a positive feedback has been received for all of them and further data from the injection buffer can be handled.

The reception buffer is logically organized as separate FIFOs (rxFifo) for each virtual channel. While the received packets are stored in an addressable memory, credits are kept in a separate FIFO for each virtual channel. The destination processor defines the order of the receive operations when it writes the credits. As soon as a credit at the output of the credit FIFO matches a 128-Byte data packet stored in the reception buffer, the receiver logic signals to the processor interface that data is ready and provides the memory address where to store it. The processor interface then transfers the data from the reception buffer into the memory of the CPU. Multiple matches of credits and packets are arbitrated using an arbiter which shifts priorities in a round-robin fashion. The requesting link that has the highest priority will receive a grant to access the link to the processor.

The final destination address of a packet on the receiving node is determined by a local address provided by the receiving processor (encoded in the credit) plus a remote offset. The latter is defined by the sender and included in the packet header.

The status of both the transmitter and receiver part of the link can be monitored via a number of registers. For instance, the following events are counted:

- Checksum mismatch detected by receiver.
- Packet could not be written into the reception buffer (back-pressure).
- Transmit or receive interface of PHY asserts error signal.
- Error symbol inserted by PHY.
- Corrupted feedback commands.

To correct and detect corrupted RESTART or feedback commands, they are transmitted in a redundant way over at least 3 serial lanes. Moreover, the feedback commands (ACK, NACK) carry an 8-bit counter value which allows to detect any loss of a feedback. Since no recovery mechanism for such a rare error event has been implemented the latter error is fatal and triggers abortion of the job execution.

3 Network Implementation for QPACE

3.1 Architecture Overview

QPACE [3, 4] is an application optimized architecture that has been developed by an academic–industrial collaboration. It is based on the IBM PowerXCell 8i multi-core processor which at the time of introduction 2008 provided an exceptionally high peak compute performance of 100 GFlop/s (double precision). Unlike in other architectures based on the PowerXcell 8i (e.g., IBM Roadrunner [7]) the processor is directly connected to the network interface, i.e., the Network Processor (NWP). For QPACE the NWP has been implemented on a Xilinx Virtex5 LX110T with the largest available package option.

The PowerXCell 8i processor is an implementation of the Cell Broadband Engine Architecture [8]. A first implementation has been developed by Sony, Toshiba and IBM with the first major commercial application being Sony's PlayStation 3. The PowerXCell 8i is an enhanced version of that processor with support for high-performance double precision operations, IEEE-compliant rounding, and a DDR2 memory interface. It comprises multiple cores including the power processing element (PPE), which is a standard PowerPC core that can, e.g., be used for running the operating system Linux. From there threads can be started on the 8 synergistic processing elements (SPE). Each SPE comprises a memory flow controller (MFC) and a synergistic processing unit (SPU). The latter has no cache, but provides a private memory, the so-called local store (LS), which has a capacity of 256 kByte and can be directly accessed by the SPU through load and store operations.

The basic building block of the QPACE architecture is the node card. The main components mounted on the node card are the PowerXCell 8i processor, 4 GByte of DDR2 memory, the FPGA and 6 10-GbE PHYs (PMC Sierra PM8358a). 32 node cards are connected to a backplane. Up to 8 backplanes can be mounted in one rack, i.e., the maximum number of node cards per rack is 256.

The node cards mounted on a single backplane can be partitioned in different ways. The largest partition is of size $(x,y,z) = (1,4,8)$. Using the cable connections between different backplanes larger partitions can be obtained, e.g., for an installation with n racks the maximum partition size is $(2n,16,8)$.

3.2 QPACE Network Processor and Network Topology

The FPGA implements an IO fabric comprising the following ports:

- Two 8-bit wide (full-duplex) bi-directional high-speed links connecting the NWP to the PowerXCell 8i processor with a bandwidth of up to 5 GByte/s per direction (on QPACE the link is operated at a data rate of 4 GBytes/s).

- Six (full-duplex) bi-directional links to the 10-Gigabit Medium Independent Interfaces (XGMII) of the torus network PHYs. Per link and clock cycle 32 bits can be sent and received at a clock speed of 250 MHz.
- One common reduced gigabit medium independent interface (RGMII) connecting the NWP to an external Ethernet 1000BASE-T physical transceiver. Running at 250 MHz per direction 4 bits per cycle can be communicated.
- A 4-bit interface (2 differential lines per direction) interface to a global signal tree network.
- Lines connected to the 2 universal asynchronous receiver transmitters (UART) which are interfaces to the serial links.

In Fig. 5 we give an overview of the QPACE NWP (for more details, see [9]). The top area shows the interface towards the processor. For the IO interface of the PowerXCell 8i processor Rambus FlexIO technology has been used. At the physical layer the NWP–processor link connects Rambus FlexIO to Xilinx RocketIO GTP (giga transceiver peripheral) transceivers. The feasibility of bringing up such a link had before been demonstrated on a test platform [10]. Inside the QPACE NWP multiple interfaces have been defined which separate the interface to the processor from the remaining design. The torus and Ethernet network links are connected to the two high-speed interfaces. Transactions via the *master interface* and *slave interface* are controlled by the processor and NWP, respectively. A third interface allows to attach slower ports, like the UART ports, via a simple, shared device control register (DCR) bus (left part of Fig. 5). This bus is also used to access the configuration and status registers of the network interfaces.

To inject packets into the network any of the MFCs initiates a DMA put operation to write the payload into the injection buffer (txFifo) of one of the torus network links. All protocol information, like address offset or virtual channel index, is encoded in the 42-bit DMA destination address.

To receive data the processor has to provide credits to the NWP using a DCR write operation. When the RX logics detects a packet in the reception buffer and a matching credit the data is written using a DMA put operation via the NWP–processor link into any of the Local Stores or the main memory. Once a credit has been consumed the processor is notified by updating a corresponding notification word, which before has been allocated in one of the Local Stores or the main memory, via yet another DMA put operation.

Using the torus network links requires a Linux driver running on the PPE. It implements routines to configure, control and reset the external PHYs and the link logics implemented in the NWP. The driver furthermore takes care of mapping the addresses of the FPGA devices into the address space of the process. In case the main memory is the final destination of data received via the torus network, the

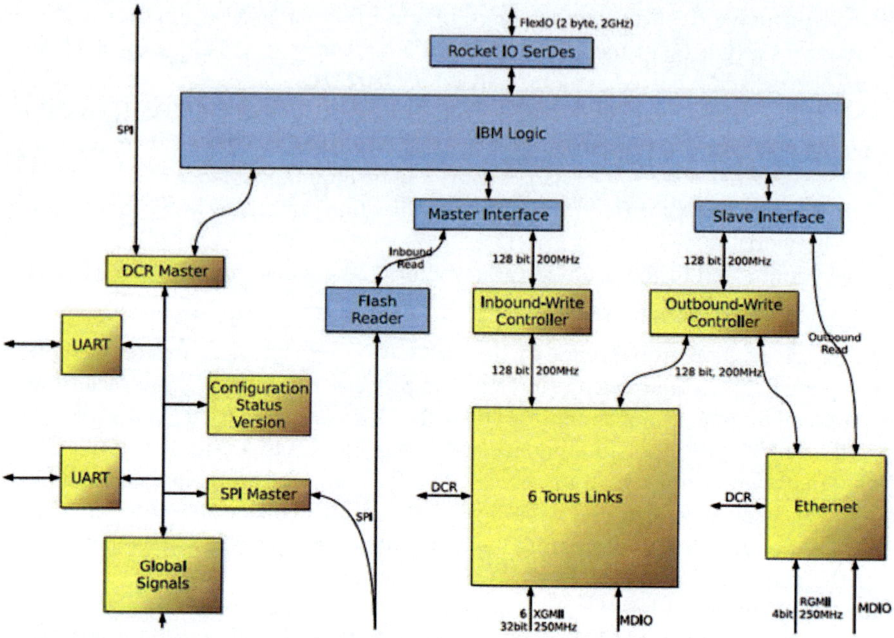

Fig. 5 Overview of the internal architecture of the QPACE network processor. The *blue boxes* refer to entities implemented by IBM, the others have been implemented by the academic partners

driver has to pin the memory page allocated by the user.[1] Using pinned memory pages eliminates the need of letting the kernel allocate memory buffers.

A user space library implements various functions to initialize the network to manage memory and buffers as well as to initiate and control communication operations. Functions needed for data communication have been implemented both on the PPE and SPE.

Within a project where the High-Performance Linpack was ported to QPACE, an OpenMPI Byte Transfer Layer module using the torus network had been implemented [11]. However, for this project a slightly modified version of the network processor design employing the Nget scheme has been used.

To synchronize the nodes both the torus network and the global signal tree network can be used. The latter is a simple network consisting of 2 signals per link and direction. The compute nodes are the leafs of the tree network. Root and branch nodes are implemented on other hardware components in the system. For more details and performance results, see [12].

[1]On QPACE the space where data can be received is limited to a single page. In case huge pages of size 16 MByte are used, only 8 MByte can be addressed from the FPGA.

3.3 FPGA Implementation Details

The main performance parameters for an FPGA being used as an IO fabric including a high performance network are bandwidth and latency. While the latter is mainly limited by functional requirements and clock speed the former is mainly constrained by clock speed and bus width. Additionally, sufficient memory capacity has to be provided to stage data before injecting it into the network or forwarding received data to the processor.

In the interface towards the processor a relatively small number of buffers are needed with a maximum buffer size of 2 kBytes. These have been implemented using distributed RAM. For the torus network the situation is different as a significantly larger amount of memory is needed for each of the links. Here 18- and 36-kBit block RAMs have been instantiated. For instance, the txFifo has been implemented using 36-kBit RAMs which have been configured such that the in- and output ports have a width of 72 bit. Placing 4 of these memory units in parallel allows us to store a 128 bit word of payload plus control information in each clock cycle. By cascading 2 memory units we obtained a FIFO consisting of 32 block RAMs with a total capacity of 32 plus 4 kBytes for data plus control information, respectively.

Most of the user logics is clocked at 200–250 MHz, except for small entities attached to the high-speed transceivers. The design comprises a relatively large number of clock domains including 5 major clock domains which are managed by PLLs. In total there are 21 clock networks, where most of them are managed using Xilinx digital clock managers (DCM). For placement and synthesis of the design it had been mandatory to keep the number of global clock domains as small as possible.

For the QPACE design there had been little options to use hard-IP blocks. Only an Ethernet media access controller (MAC) block could be used for implementing the Gigabit Ethernet interface.

In Table 1 we give an overview of the resources used in the final version of the design. At least 48% of the key resources provided by the Virtex5 LX110T have been used, in a number of cases the resource usage is (almost) 100%. The latter in particular applies to the RocketIO transceivers and general purpose pins. All available high-speed transceivers have been used to implement the NWP–processor link with a bandwidth that roughly balances the aggregate bandwidth of the torus network links. Most of the general purpose pins were required to attach the 10-GbE PHYs with about 80 pins per PHY.

In terms of flip-flops and lookup tables (LUT) about half of the used resources are consumed by the interface to the processor while each torus network link accounts for about 6% and the Ethernet interface for about 2–4%.

Table 1 Fraction of resources available in a Xilinx Virtex5 LX110T which are consumed by the QPACE NWP design

Resource	Usage (%)	Total available
GTP transceivers	100	16
User IO pins	96	680
Occupied slices	95	17,280
PLLs	83	6
LUT–flip-flop pairs	76	69,120
DCMs	58	12
Flip-flops	58	69,120
LUTs	54	37,929
Block RAM	48	148

4 Network Processor with PCIe-Based IO for AuroraScience

The AuroraScience [5] machine is a parallel system using the Aurora hardware built by Eurotech [13]. It is composed of node cards, backplanes and cable connections which are organized in a similar way as in QPACE. The computing nodes are based on the recent generations of commodity multi-core processors developed by Intel: the first AuroraScience boards had Nehalem processors, while more recent versions use Westmere, and Eurotech is currently implementing boards with the latest Sandybridge CPUs.

In the following, we describe the implementation of the NWP on the boards based on Nehalem processors. These boards host two CPU sockets with four-cores, 12 GByte of RAM, an X5520 south-bridge (code-name Tylersburg) and an FPGA. On the AuroraScience nodes the FPGA is an Altera Stratix IV GX-230 and it is connected to the south-bridge by two PCIe x8 Gen2 interfaces. Using one of them provides a peak raw bandwidth of 4 GByte/s, which corresponds to an effective bandwidth of 3.2 GByte/s if we take into account the overhead of the PCIe protocol, as we discuss later.

Apart from porting the VHDL firmware for the link modules (as developed for QPACE) to the Altera FPGA, the NWP of the AuroraScience system required the development of a specific IO interface to handle the transactions between CPU and NWP via the PCIe bus. Moreover, on x86 CPU architectures the processor cannot explicitly instruct the memory-controller to transfer data from memory to IO devices, and for this generation of processors no DMA engines could be used. Thus for moving data from CPU to the NWP device, having the CPU as the initiator must be implemented by using the PIO method. PIO requires that software running on the CPU uses instructions that access IO address space to perform data transfers to or from an IO device. This keeps the CPU busy until all data is transferred to the NWP. Therefore, to allow non-blocking send operations, we have also implemented support for the Nget scheme.

Table 2 Resource usage of the NWP for AuroraScience, including the IO interface for PCIe with reorder buffers, on an Altera Stratix IV GX230

Resource	Usage (%)	Total available
PLL	88	8
IO pins	69	612
PCIe hard-IP block	50	2
GXB transceiver	33	24
Memory bits	17	14,625,792
Logic registers	14	182,400
Combinational ALUTs	9	182,400
Memory ALUTs	1	91,200

The second column shows the used fraction of resources.
The total available resources of the device are reported in the last column

All necessary IO operations between CPU and NWP can be readily implemented through corresponding PCIe transactions. For instance, the basic operations performed by the CPU to implement the Pput scheme, or to read and write configuration and status registers on the NWP correspond to memory read or write PCIe transactions initiated by the CPU. On the other hand, data transfers for send operations according to the Nget scheme and for receive operations correspond to appropriate PCIe transactions initiated by the NWP.

In Table 2 we summarize the FPGA resources used on the Altera Stratix IV GX-230 for the entire NWP firmware of AuroraScience running at a frequency of 250 MHz. The NWP includes 6 link modules and the IO interface with support for the Nget operation and reorder-buffers. It uses about 18% of the logic available on the FPGA, and 2.5 Mbit of memory corresponding to ≈17% of the embedded memory on the FPGA.

4.1 IO Interface

The IO interface of the NWP consists of three major blocks, see Fig. 6. The PCIe IP is a hardware macro embedded in the Stratix IV FPGA implementing the PCIe protocol stack (transaction, data and parts of the physical layer). The PCIe inbound controller (PIC) and PCIe outbound controller (POC) blocks are attached to the *Avalon* interface, a proprietary interface to the PCIe hard-IP block. They handle incoming (down-stream) transaction layer packets (TLP) and generate outgoing (up-stream) TLPs. The other blocks are needed to support specific IO schemes for sending the data.

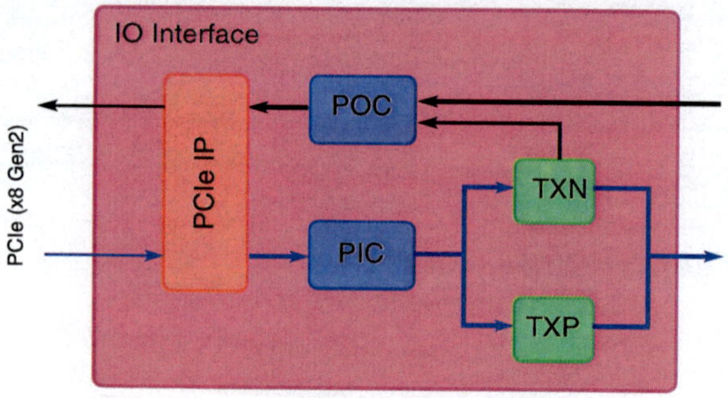

Fig. 6 Block diagram of the IO interface for PCIe

As mentioned above, the Pput scheme for x86 architectures is implemented by PIO method, i.e., the CPU performs store operations to the memory addresses to which the injection buffers have been mapped. The memory controller and south-bridge translate store operations to *memory-write* PCIe transactions with a small payload of only 16 Bytes (the size of an SSE register). To improve performance *write-combining* can be enabled by using *non-temporal* processor instructions. In this case stored data is then written into small temporary *write combining buffers* (WCB) of the CPU. When a WCB is full, it is flushed to the IO bus by a single burst transfer with a payload of 64 Bytes (the size of a cache-line). Since the WCB may be flushed also for other reasons, like thread de-scheduling, message fragments can arrive at the NWP in an interleaved order. Therefore, the NWP implements in the TXP block a *reorder* logic which restores the correct order of data items before they are moved into the injection-buffer of the link module.

During the system boot phase, the processor interface of the NWP requests the CPU to assign a large memory region in the IO address space. All injection buffers are mapped through *Base Address Register* BAR0 into this contiguous memory region. The region of each injection buffer is divided into eight parts corresponding to the eight virtual channels of one link. Data arrives at the NWP together with a PCIe header which includes the destination address. This is then used to decode the injection buffer and the virtual channel index.

To handle the Pput scheme back-pressure, which arises when the injection buffers are full, the NWP writes to a reserved address of the CPU memory (regularly or upon request) the number of packet-items pulled out from each injection buffer. Applications may then determine the space in each injection buffer by subtracting this value from the number of packet-items already written into the buffer.

In order to support non-blocking send operations and to exploit DMA mode, we have implemented extensions of the NWP to support the Nget scheme. In this scheme, the CPU first passes the required control information to the NWP, in particular the memory address of data to be transferred, the size and a logical tag. They are written into an appropriate control register in the TXN block, which implements the necessary control logic and a DMA engine. Then TXN issues the adequate memory read operations through the POC module. After their completion, the NWP notifies the CPU, e.g., by writing to a predefined memory address for the given logical tag. Once the application has detected the data change at this particular memory address it can re-use the memory locations from where the sent data originated.

4.2 Software Layers

To provide convenient and efficient access to the TNW from multi-threaded applications, we have developed a Linux driver and a basic communication library.

For the efficient support of the Pput scheme, the driver marks the address space of the injection buffers (allocated at boot time by the NWP) as write-combining. Moreover it allows to map it into user address space of the application processes through standard mmap function. Applications can thus directly access the injection buffers, saving overheads due to frequent context switches between user- and kernel-mode. Write operations within this memory range are translated by the memory-controller and the south-bridge into an IO write operation consisting of payload data and a header. The header includes the memory address accessed by the thread which is used by the NWP's processor interface to decode the virtual channel tag and the injection FIFO where payload data-item has to be pushed.

To support Nget operations the driver allocates buffers from which the network-processor reads the messages to be sent. These buffers are allocated on contiguous physical memory pages and are marked as un-swappable. For receiving data and notifications from the NWP, the driver allocates additional contiguous memory areas for each virtual channel. Control and status registers of the NWP are mapped on separate addresses through a separate base address register (BAR) and are accessible directly from user-space.

The communication library provides API functions to send and receive messages across the network, and to configure, control and monitor the behaviour of the NWP. The most relevant communication functions are:

- tnwSend is used to send messages over a selected virtual channel. The implementation depends on the scheme used to move data between CPU and network processor. If a Pput scheme is adopted it is implemented as a loop on the length of message. At each iteration data items are read from application buffers and written to the appropriate memory address which corresponds to a particular link and virtual channel. If a Nget scheme is used, the function first copies data from user-buffer to the DMA buffers and then triggers a DMA transfer on the NWP device. A further implementation enables zero-copy transfers by allocating contiguous physical memory regions which can be directly access by the application.
- tnwCredit is used to provide a credit to the network processor. This information is used by the NWP device to move data packets from the reception buffer to the DMA buffer allocated by the kernel.

- tnwPoll polls a specific memory address waiting for the notification which indicates that all data corresponding to a previously issued credit is now available. If DMA buffers are not mapped to user space, the driver has to copy the data to user space.

5 Performance Results

Before discussing the performance of the overall torus network in QPACE and AuroraScience, we first consider only the link performance, i.e., the datapath between two link modules of the network processors. The link module is implemented as a highly pipelined design which runs on the FPGAs, as used in QPACE and AuroraScience, at a frequency of 250 MHz, corresponding to the clock of the parallel (XMGII) ports of the PHY.

The physical link between the PHYs has a raw bandwidth of 2.5 Gbit/s. Taking into account the overhead from 8B/10B coding and from the custom protocol, we have a theoretical bandwidth of 0.941 GByte/s (128 Bytes every 34 clock cycles) if data is transferred only in one direction of a link. When data is sent and received simultaneously over the same link, the theoretical bandwidth is slightly reduced to 0.914 GByte/s per direction (128 Bytes every 35 clock cycles) because data packets and feedback commands from the opposite data streams have to share the same physical link.

From the benchmarks of the CPU-to-CPU transfer rate (see below) we find that this theoretical link bandwidth can be sustained in practice, provided that data is moved sufficiently fast from the CPU into the injection buffer on the NWP (and out of the reception buffer to the CPU).

An important property of the network design is also the latency of the links. We have measured the time starting from the instant when a 128-Byte packet arrives at the injection buffer txFifo until the DMA engine of the receiver NWP can start to move the packet to the processor. We find about 0.5 μs and a more detailed analysis of the breakup of this time is shown in Fig. 7. A large fraction of this latency is due to the logic in the PHYs (where data encoding and decoding is performed). A major delay also arises in the receiving link module. There, all 32 data items of a packet have to be received from the PHY before the CRC can be verified and the entire packet becomes ready for extraction from the reception buffer (if a corresponding credit is available).

5.1 *Network Performance on QPACE*

Let us first consider in QPACE the IO link between the IBM PowerXCell 8i processor and the FPGA. For a fine-grained analysis of the performance of this link we added a FIFO which on request captured control information for all packets

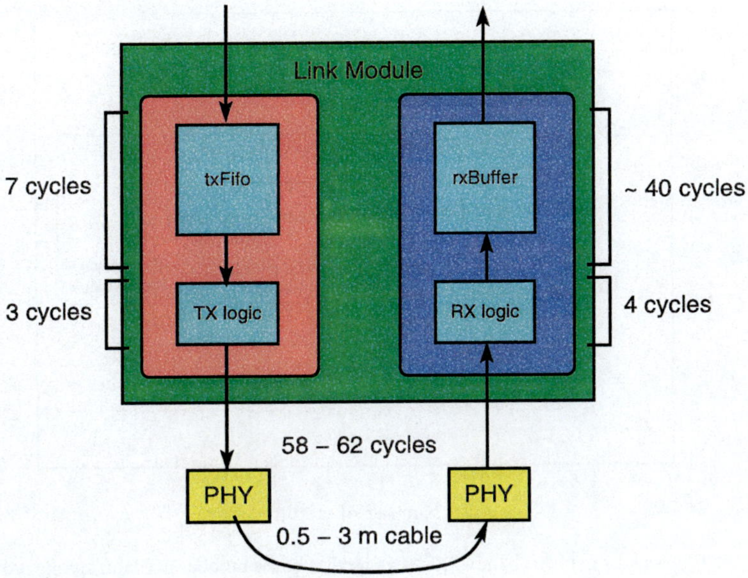

Fig. 7 Breakup of the time to transfer a single packet (128 Bytes) through two link modules and the physical link. One clock cycle corresponds to 4 ns

that are written by the processor via the inbound interface to any of the torus network links plus a time stamp. The flow control of this IO interface is also based on credits (not to be confused with the credits sent from the CPU to the NWP to initiate a receive operation). In Fig. 8 we show the bandwidth as a function of the number of credits provided to the processor. For a small number of credits the bandwidth depends linearly on the number of credits. At some point the bandwidth saturates because all available tags on this link are in use. However, if the data packets are not processed using a sufficiently high clock rate, mis-speculation may occur resulting in a very high performance penalty. Using a fast clock the observed bandwidth depends on whether a packet combining feature is enabled. If this feature is enabled two consecutive packets of length 128 Bytes may be combined into a single packet of length 256 Bytes. In this case tags are freed earlier and thus a larger bandwidth is observed. The effect of this optimization can be clearly observed from the bandwidth measurements with this feature being disabled. If the feature is enabled but the number of concurrent DMA operations NDMA > 1 then combining of packets will often fail and therefore result in a reduced bandwidth. Figure 8 shows that an effective bandwidth of up to 2.8 GByte/s can be reached which is 70% of the nominal peak bandwidth.

To measure the performance of the QPACE torus network we implemented the following micro-benchmarks:

- A ping-pong type of test to estimate the latency. Here a SPU on node *A* sends a message of size 128 Bytes (i.e., 1 packet) from its Local Store to node *B*. Once

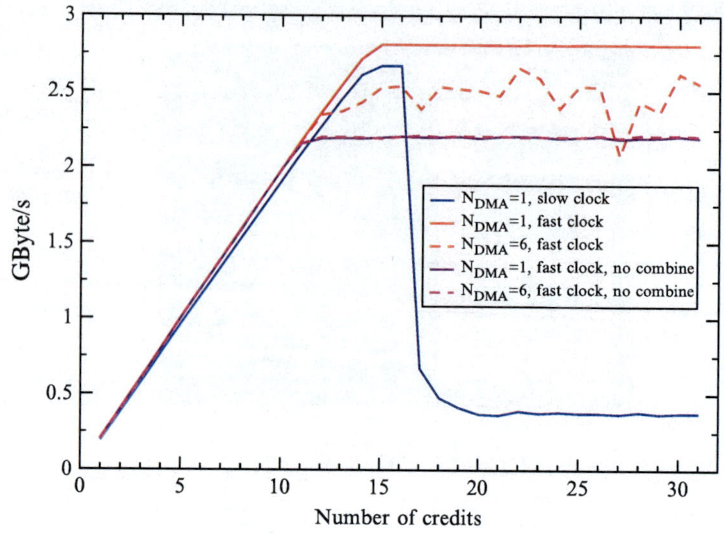

Fig. 8 QPACE processor to NWP bandwidth as a function of the number of (data) credits provided by the NWP to the processor

the data has arrived in the destination LS of node B this node returns the packet back to A. We estimate the latency by measuring the round-trip time on node A and divide this number by 2.

- An exchange type of test to measure the bandwidth. For this benchmark we select $n = 1,2,4$ pairs of SPUs where each pair is distributed over node A and B and connected by the same physical link. Each pair exchanges data, i.e., the SPU on node A sends a message to the corresponding SPU on node B and vice versa.

With the first setup we measured the overall latency (including software overhead) for an LS-to-LS communication to be about 3 μs. It is largely dominated by the time needed to move data from the CPU to the NWP and vice versa.

To understand how well a single torus network link can be saturated, we used the second micro-benchmark where each core of one pair sends and receives to/from the other core. The measured bandwidth is shown in Fig. 9 as a function of the message length. For $n = 4$ pairs of communicating cores we are able to reach the theoretical link bandwidth of 914 MByte/s already for a message length of 2048 Byte.

From these performance results it is clear that the available bandwidth on the NWP–processor link is not sufficient to saturate the bandwidth of all six torus network links. However, this would be a limitation of the design only if the application is able to start communicating with all six neighbouring nodes concurrently.

We may compare these performance results for QPACE with those for other architectures where the PowerXCell 8i processor is used, e.g., the IBM Roadrunner architecture. Here the nodes comprise of 2 QS22 blades with 2 PowerXCell 8i

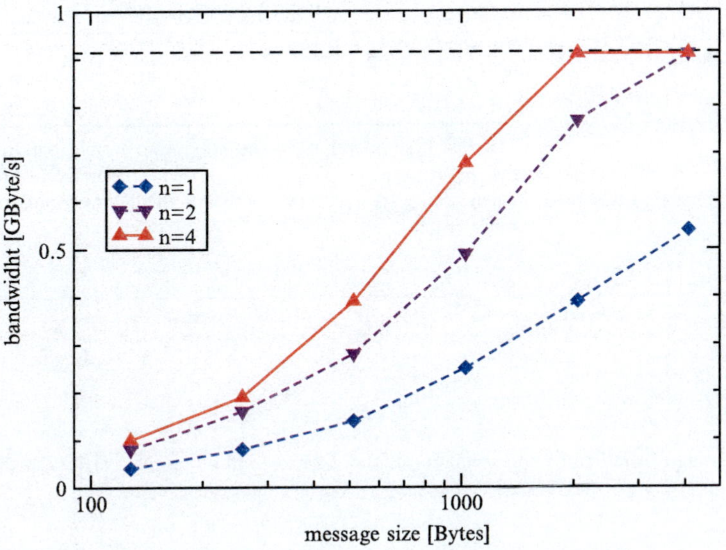

Fig. 9 Bandwidth measured on a single QPACE torus network link as a function of the message size, where n is the number of messages which are in flight concurrently. The *dashed horizontal line* indicates the theoretical maximum bandwidth

processors each plus 1 Opteron blade. Cell processors of different nodes can only communicate via the Opteron processors which leads to a large latency of 8 μs [14]. Using the OpenFabrics ib_rdma_bw benchmark a bandwidth of up to 2.4 GByte/s has been achieved for two QS22 blades interconnected by DDR Infiniband [15].

The performance of a network can be significantly reduced if error rates are high. To estimate the error rate we monitored the error counters of all receivers of a QPACE machine partition consisting of 256 nodes, i.e., 1,536 receivers in total, running a job for 24 h. We repeated this analysis for two different partitions. During both runs a total of 39 (non-fatal) errors had been detected by the receivers (that had been automatically corrected by the network). This corresponds to $1.5 \cdot 10^{-7}$ errors per link, direction and second on average.[2]

5.2 *Network Performance on AuroraScience*

We now discuss the performance of the PCIe-based IO interface used in AuroraScience (for more details, see [16]). In this case, we have a PCIe x8 Gen2 link between CPU and NWP with a nominal bandwidth of 4 GByte/s.

[2]Since we do not monitor the number of packets and because a single error cannot be interpreted as a single bit error, we cannot provide an estimate of the bit or packet error rate.

Table 3 Effective bandwidth measured with different implementations of the Pput scheme

Version	Bandwidth (GByte/s)	Efficiency (%)
No WC	0.133	4.9
WC	0.516	19.4
WC+ROB	2.667	100.0

The last column refers to the maximum usable bandwidth for the Pput scheme, i.e., the maximum theoretical throughput when taking into account overheads of the corresponding PCIe transactions

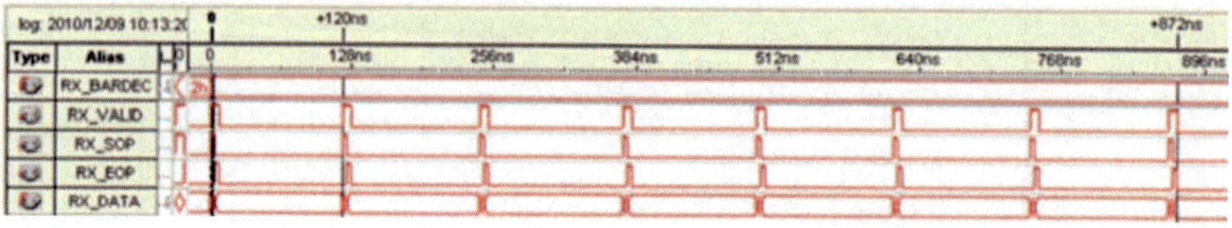

Fig. 10 Timing of the Pput implementation without write-combining. Data arrives at the NWP in portions of 16 Bytes every 120 ns

Let us first consider the Pput scheme for sending data to the network. In this case, a send operation of the processor translates into a sequence of memory write PCI transactions, each with a maximum payload of 64 Bytes. Taking into account the overhead of 16 Bytes for the header and 16 Bytes for padding between the end of a transaction and the start of the next, the maximum usable bandwidth reduces to 2.67 GByte/s. We implemented different versions of the Pput scheme to analyse whether and how this maximum usable bandwidth can be reached. The results are summarized in Table 3.

In the first naive version, denoted by No WC, the use of the *WCB* on the CPU is not enabled. Data is moved to the NWP buffers by issuing individual write operations of 16 Bytes. A timing analysis of the arrival of these write operations at the IO interface of the NWP is shown in Fig. 10. We see that the time separation is at least 120 ns, resulting in a low effective bandwidth of only 0.133 GByte/s.

In the second version, denoted by WC, the use of the WCB is enabled and the effective bandwidth increases to 0.516 GByte/s, i.e., approximately 20% of the usable bandwidth. In this case, the main performance limitation is due to the memory barriers (sfence instructions) that are needed to enforce the correct ordering of data items.

In the third version, denoted by WC+ROB, write combining is enabled on the CPU and we use the reorder-buffer implemented on the NWP. In this setup, the write transactions can be issued in any convenient order, see Fig. 11, and the effective throughput on the PCIe link saturates at the theoretical maximum of 2.67 GByte/s.

We now consider the Nget scheme to move data from the CPU to the NWP for the send operation (and analogous considerations also apply to the Nput scheme which is used to receive data). In this case, the maximum payload of the corresponding PCIe transactions depends on the capabilities of the south-bridge.

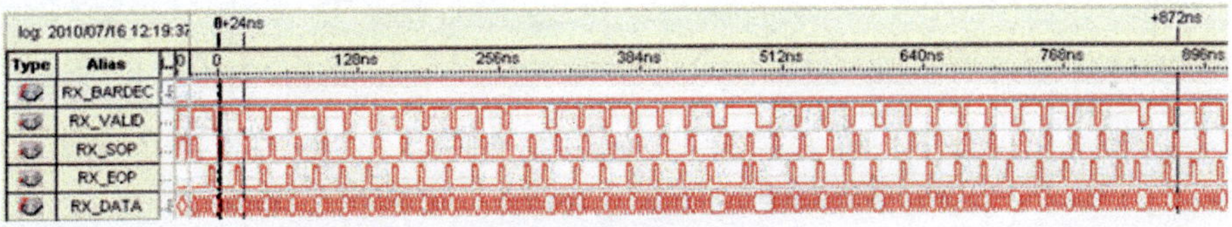

Fig. 11 Timing of the Pput implementation with write-combining and reorder buffer. In most cases the CPU issues data transfers with the maximum size of 64 Bytes, resulting in a full utilization of the PCIe bus. The reorder buffer guarantees the correct order of the individual 16-Byte data items within each 128-Byte packet before it is passed to the injection buffer of the link modules

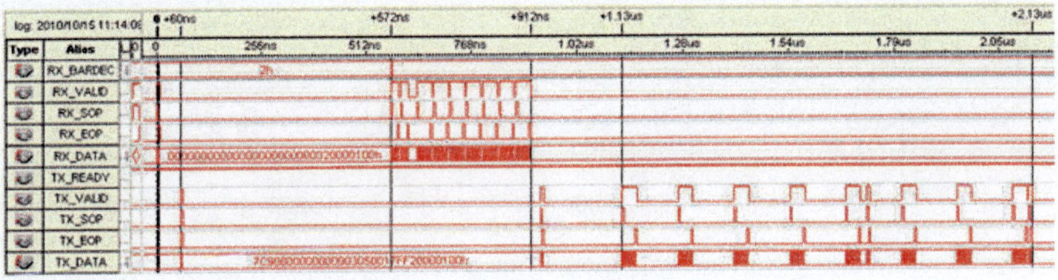

Fig. 12 Timing of the Nget and Nput schemes. The *left* part, between time 0 and 921, shows a data transfer from CPU memory to the NWP device (Nget). The *right* part, starting at time 1130, shows data transfer from NWP to CPU memory (Nput)

For the X5520 chip used on the AuroraScience boards, the maximum payload is 128 Bytes. Taking into account the overheads of the PCIe protocol (header and padding), this yields a maximum usable bandwidth of 3.2 GByte/s.

With our current implementation of the Nget scheme we reach about 3 GByte/s, which is 94% of the maximum usable bandwidth or 80% of the nominal bandwidth. Here, the main performance limitation arises from the overhead to manage the operation. A detailed analysis of the time required to move data from the CPU memory to the injection buffer of the NWP is shown in Fig. 12:

- At time 0 (corresponding to the bold black bar) the Nget command issued by the CPU is received by the NWP device.
- At time $0 + 60$ ns the NWP issues the corresponding memory read request.
- At time $0 + 584$ ns the data from the main memory arrives at the NWP.
- At time $0 + 948$ ns the notify is sent back to the CPU.

From this timing we see that it takes approximately 500 ns from the instant when the NWP issues a memory-read request until the requested data from the memory of the CPU starts to enter in the NWP device. This time includes twice the delay of the PCIe core on the FPGA and the latency of the memory access. Thus, we obtain 250 ns as a rough estimate of the startup latency for transferring any data between the CPU and the NWP device or vice versa.

To determine the overall CPU-to-CPU latency which can be achieved by application programs (including software overhead), we performed a ping-pong

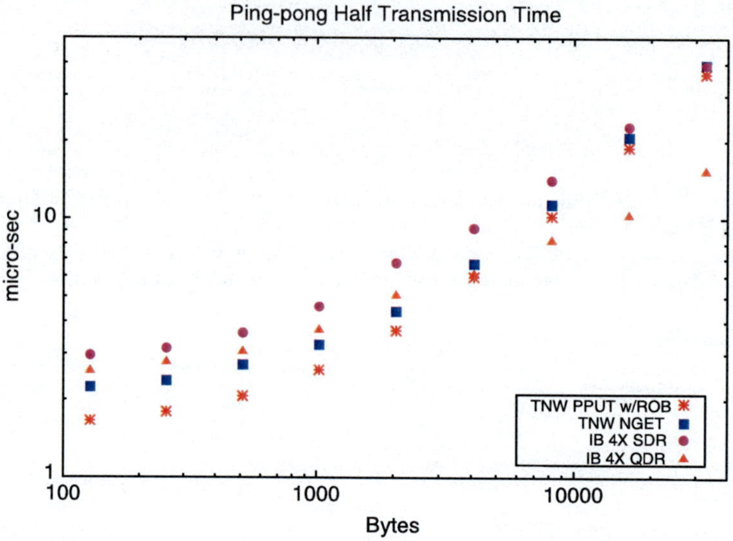

Fig. 13 Half of the ping-pong time measured with an application program as a function of the message size. Results are shown for using the Pput (with write-combining and reorder buffer) or the Nget scheme for the send operations. Additionally we show results using the SDR and QDR Infiniband network

benchmark. Figure 13 shows half of the measured ping-pong time for different message sizes and with the send operation implemented according to both schemes, Pput and Nget. The measured time is well described by the following linear functions of the message size L (in units of kByte):

$$T_{\mathsf{Pput}}(L) = (1.53 + 1.08 \cdot L)\,\mu\mathrm{s}\,,$$
$$T_{\mathsf{Nget}}(L) = (2.06 + 1.17 \cdot L)\,\mu\mathrm{s}\,.$$

From the fitted parameters we estimate the overall latency of a single communication to be approximately 1.5 μs for the Pput scheme, and about 2.1 μs for the Nget scheme.

On AuroraScience also an Infiniband network is available which allows us to perform a direct comparison with the FPGA-based torus network. In Fig. 13 we also show the latency for the ping-pong benchmark using the Infiniband network. Note that now two nodes are connected via a switch. The measured time for SDR and QDR Infiniband can again be parametrized by a linear ansatz

$$T_{\mathsf{SDR}}(L) = (3.75 + 1.13 \cdot L)\,\mu\mathrm{s}\,,$$
$$T_{\mathsf{QDR}}(L) = (3.58 + 0.39 \cdot L)\,\mu\mathrm{s}\,.$$

The latency for SDR and QDR Infiniband is about 3.8 and 3.6 μs, respectively, i.e., about two times larger than for the torus network. In this setup with 2 nodes, the bi-section bandwidth for the SDR Infiniband and the torus network are similar and significantly larger for QDR Infiniband. For a larger number of nodes the bi-section bandwidth will, however, be larger in case of the torus network due to the larger number of links connecting the two parts of the network.

6 Outlook on Future Implementations

With FPGA technologies progressing and more powerful FPGAs becoming available, the design of the torus network presented here may be extended and improved in various aspects.

Both for QPACE and AuroraScience external PHYs have been used to implement the physical layer of the network links. During recent years the number of high-speed transceivers on a single FPGA increased significantly and their performance characteristics improved. For instance, Xilinx Virtex4 FPGAs comprised of up to 24 high-speed transceivers with a maximum signaling rate of 6.5 Gbit/s. A few years later, Xilinx announced Virtex7 FPGAs with up to 96 transceivers with a maximum signaling rate of 13.1 Gbit/s. This makes it feasible to also implement the physical layer on the FPGAs without the need for external PHYs or external signal re-drivers.

Other improvements concern the functional design of the network. In the current implementations the processor has to write control information to the NWP for each communication. Performing these control operations in frequently repeated communication sequences could be avoided by providing hardware support for persistent communication patterns. Various applications where certain communication patterns are repeated many times could potentially benefit from such a hardware feature. Persistent communication patterns could be supported in our design, e.g., by a modification of the credit FIFO. This would extend the usual First-In-First-Out policy and allow reuse of previous credits by implementing programmable read and write pointers to the RAM which hold the credits. These pointers circle continuously through the RAM addresses (possibly only within a specified range) and can be (re-) set individually to a specific start position. This allows to jump back and reuse previously read credits which have not yet been overwritten by new credits.

Another enhancement concerns support of additional deterministic routing features, e.g., for next-to-nearest node communications. The main challenge for such extensions is the risk of introducing deadlocks. Deadlocks can be avoided by making sure that escape paths exists through which packets can be moved towards the final destination where they leave the network. For our torus network architecture it would require an extension of the packet header to encode routing information. Furthermore, forward buffers have to be implemented where received packets are stored that have not reached their final destination. Special care is needed to ensure that these forward buffers do not become full as this would block the link and thus prevent packets to reach the reception buffer. This could be achieved by introducing a flow control mechanism, which would mean a rather significant change of our design. Alternatively, by imposing suitable programming rules the responsibility for keeping the number of packets in flight in any part of the network sufficiently small could be delegated to the application programmer or communication library developer.

7 Summary and Conclusions

In this chapter we have described how FPGAs can be used to implement a network architecture that is optimized for nearest neighbour communications between processors which are interconnected as a 3-dimensional torus. The design has been implemented for two different HPC systems on which it has been extensively used to run scientific applications efficiently on hundreds of processors.

For the FPGA–FPGA interconnect a custom network protocol had been developed which minimizes the overhead. We have demonstrated that it is feasible to reach the theoretical maximum bandwidth in realistic use cases. The network design is robust and on the deployed large systems we observed very low error rates. Correction of data errors is handled at hardware level by resending data between the link modules and does not have a relevant impact on the performance.

Both for QPACE and AuroraScience the implementation and efficient use of the data path between CPU and NWP is more challenging. For QPACE the latency on this link turned out to be relatively high. The maximum measured bandwidth is 70 and 80% of the nominal peak on QPACE and AuroraScience, respectively.

Using FPGAs for implementing such a high-speed network has several advantages. The design remains flexible and the functionality can be adjusted or extended as application requirements evolve. Since costs of ASIC development have become too high for many academic research projects, the use of FPGAs is often the only option unless commodity solutions are an alternative (in terms of both functional and performance requirements and costs). Even if a custom ASIC would be an alternative, the option to fix design errors at late stages of the challenging design and implementation of parallel computer architectures reduces risks and allows to go for a much more aggressive development schedule. Finally, we would like to point out that FPGAs give room for optimizing the interface to the processor which can be interesting for special computing devices, like GPUs and future processor architectures with a large number of cores.

Among the disadvantages, the cost compared to today's commodity network solutions is the critical issue. FPGAs capable of processing very high data rates are typically more expensive than commodity network devices. In case of a torus network this may partially be compensated by avoiding the need of switches. Furthermore, power consumption of a single FPGA tends to be higher compared to, e.g., a commodity Infiniband HCA plus switch port.

The advantages of implementing the communication network with FPGAs and the advances of FPGA technologies towards increased data processing capabilities let us expect that in the near future FPGA-based networks continue to be an interesting option, in terms of both performance and costs.

Acknowledgements We thank all members of the QPACE and AuroraScience teams for their hard and creative work that laid the groundwork for the studies reported on in this paper. In particular, we are grateful to A. Cotta Ramusino, M. Drochner, D. Hierl, A. Nobile, H. Schick, T. Streuer, K.-H. Sulanke, R. Tripiccione, and T. Wettig for their helpful contributions during the design and test phase. The QPACE project was funded by the Deutsche Forschungsgemeinschaft (DFG) in the framework of SFB/TR-55 and by IBM. We furthermore thank the following companies who contributed significantly to the project in financial and/or technical terms: Axe Motors (Italy), Eurotech (Italy), IBM, Knürr (Germany), Xilinx (USA), and Zollner (Germany). This work was supported in part by the European Union (grants 238353/ITN STRONGnet and 227431/HadronPhysics2). The AuroraScience project was funded by *Istituto Nazionale di Fisica Nucleare* (INFN) and by *Fondazione Bruno Kessler* (Trento, Italy).

References

1. F. Belletti et al., Computing for LQCD: apeNEXT. Comput. Sci. Eng. **8**(1), 18–29 (2006)
2. P.A. Boyle et al., Overview of the QCDSP and QCDOC computers. IBM J. Res. Dev. **49**(2), 351–365 (2005)
3. G. Goldrian et al., QPACE: quantum chromodynamics parallel computing on the cell broadband engine. Comput. Sci. Eng. **10**(6), 46–54 (2008)
4. H. Baier et al., QPACE: a QCD parallel computer based on cell processors. PoSLAT **2009**, 001 (2009) [arXiv:0911.2174 [hep-lat]]
5. L. Scorzato, AuroraScience. PoSLAT **2010**, 039 (2010)
6. M. Pivanti, S.F. Schifano, H. Simma, An FPGA-based torus communication network. PoSLAT **2010**, 038 (2010) [arXiv:1102.2346 [hep-lat]]
7. D. Grice et al., Breaking the petaflops barrier. IBM J. Res. Dev. **53**(5), 1:1–1:16 (2009)
8. IBM, Cell Broadband Engine Architecture (2005), Version 1.0, 8 Aug 2005
9. T. Maurer, The QPACE supercomputer, renormalization of dynamical CI fermions, axial charges of excited nucleons, Ph.D. thesis, 2011, http://epub.uni-regensburg.de/21668/. Cited Jan 2012
10. I. Ouda, K. Schleupen, Application note: FPGA to IBM power processor interface setup, IBM Research Report, RC24596 (W0807-021), 2 July 2008
11. H. Boettiger, B. Krill, S. Rinke, QPACE: energy-efficient high performance computing, in *International Conference on Architecture of Computing Systems (ARCS)*, (VDE Verlag, Hannover (Germany), 2010), http://www.vde-verlag.de/books/453222/arcs-10-23th-international-conference-on-architecture-of-computing-systens-2010.html
12. S. Solbrig, Synchronization and error reporting on QPACE (2012), http://www.physik.uni-regensburg.de/strongnet/documents/STRONGnet2010/solbrig.pdf. Cited May 2012
13. http://www.eurotech.com/aurora. Cited Mar 2012
14. K.J. Barker et al., Entering the Petaflop Era: the architecture and performance of roadrunner, in *Proceedings of the 2008 ACM/IEEE Conference on Supercomputing*, (IEEE Press, Austin (Texas, USA), 2008), http://ieeexplore.ieee.org/stamp/stamp.jsp?tp=&arnumber=5217926&isnumber=5213127
15. J.-S. Vogt, R. Land, H. Boettiger, Z. Krnjajic, H. Baier, IBM BladeCenter QS22: design, performance, and utilization in hybrid computing systems. IBM J. Res. Dev. **53**(5), 3:1–3:14 (2009)
16. M. Pivanti, A scalable parallel architecture with fpga-based network processor for scientific computing, Ph.D. thesis, 2011

MEMSCALE: Re-architecting Memory Resources for Clusters

Holger Frönin, Federico Silla, and Hector Montaner

Abstract Using the widespread and cost-effective cluster computing approach, large amounts of different resources can be aggregated in order to solve compute-intensive problems. However, as clusters do not permit any resource sharing because of their shared-nothing approach, resource partitioning is static. While this is typically not a big problem for computing resources as CPUs, the impact of a static partitioning of memory resources is dramatical. As memory resource shortage leads to demand paging with a huge negative impact on performance, this situation is avoided by overprovisioning of resources. However, due to this provisioning for the worst case, for the average case many memory resources are free but not accessible due to the static partitioning. On the other hand, as shared use of memory resources among different nodes in a cluster is not possible, communication tasks are solved by data replication.

MEMSCALE is a new cluster memory architecture that is enabled by a direct low-latency path to remote memory resources. Using MEMSCALE, the static partitioning can be overcome and a dynamic provisioning of resources becomes feasible. Two typical use cases do exist: first, data-intensive applications might benefit from more memory resources, but often not from more computing resources. Such applications can benefit from MEMSCALE by allowing them to borrow memory from remote nodes, rendering the need for overprovisioning obsolete. Therefore, each node can be provisioned for the average case, with significant cost and power savings. The second use case allows applications to rely on shared memory resources for communication and synchronization purposes, avoiding the need for data partitioning and movement. In essence, the shared-memory programming paradigm well known from multi-core computers is expanded to

---

H. Fröning (✉)
University of Heidelberg, Germany
e-mail: froening@uni-hd.de

F. Silla • H. Montaner
Universitat Politècnica de València, Spain
e-mail: fsilla@disca.upv.es; hmontaner@gap.upv.es

W. Vanderbauwhede and K. Benkrid (eds.), *High-Performance Computing Using FPGAs*, DOI 10.1007/978-1-4614-1791-0_19, © Springer Science+Business Media, LLC 2013

clusters. However, as global continuous coherency is a serious concern when scaling shared-memory systems, we revert to a highly relaxed consistency model, which guarantees consistency only for synchronization intrinsics like barriers and locks.

MEMSCALE is implemented using FPGAs in a real cluster prototype. Here, we will use in-memory databases as example workload to demonstrate MEMSCALE's impact for such data-intensive applications.

1 Introduction

Current clusters are based on cost-effective commodity parts and present a huge amount of aggregated computing and memory resources. However, resource partitioning in these clusters is statically performed on a per-node basis and, therefore, no shared use of resources is possible, thus lacking the flexibility required by many uses and hindering the effective deployment of some applications, such as data-intensive ones. However, these applications would dramatically benefit if they could rely on much larger resources, like the overall memory available distributed at cluster level, or, more general, if resources could be dynamically aggregated or disaggregated.

Because of the static resource partitioning, in the event of exhausted memory resources, demand-paging is currently the only viable solution, which comes with dramatic costs, as can be seen in Fig. 1. In this experiment, the available memory is 4GB, with about 500MB associated with the OS kernel. It can be seen that as soon as memory resources are exhausted, performance dramatically drops because of swapping.

In order to reduce the cost of swapping, other second-level storage technologies could be used, like FLASH-based memory. Nevertheless, although they certainly can overcome the capacity limitations of DRAM, they are suffering from a huge performance gap compared to it, as shown in Fig. 2, which provides an overview of the most often used storage technologies and shows the performance gap among them. Notice the logarithmic scale of the vertical axis; it can be seen that even the most-recent FLASH technologies are suffering from a 500 to 1,000 fold performance disparity.

As can be seen, the only option to avoid the huge performance degradation of demand paging is by avoiding it. This leaves no other option but to provision each node in a cluster for the worst case, increasing the amount of memory attached to each cluster node. However, for the average case, this overprovisioning increases overall costs and energy consumption. These costs are further aggravated by an important trend regarding the amount of memory available per core, in particular since the beginning of the multi-core era. Figure 3 shows, for 4-socket servers, the amount of on-board RAM available per core. It can be seen that the memory capacity is not able to keep the pace of the core count increase, even temporally degrading significantly.

In order to provide large amounts of fast memory resources to applications, new technologies are continuously being developed and evaluated for their applicability

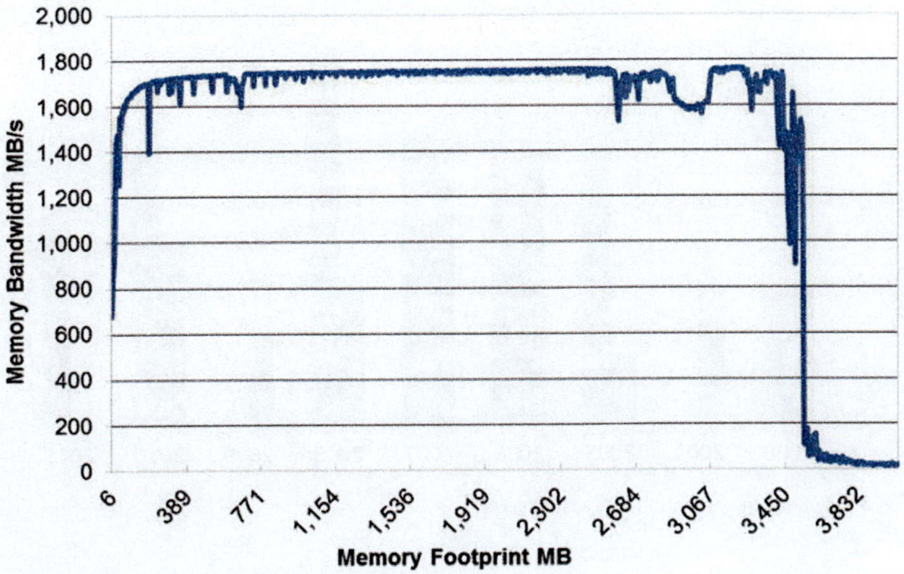

Fig. 1 Performance degradation of memory accesses depending on the memory footprint in relation to the physical memory available

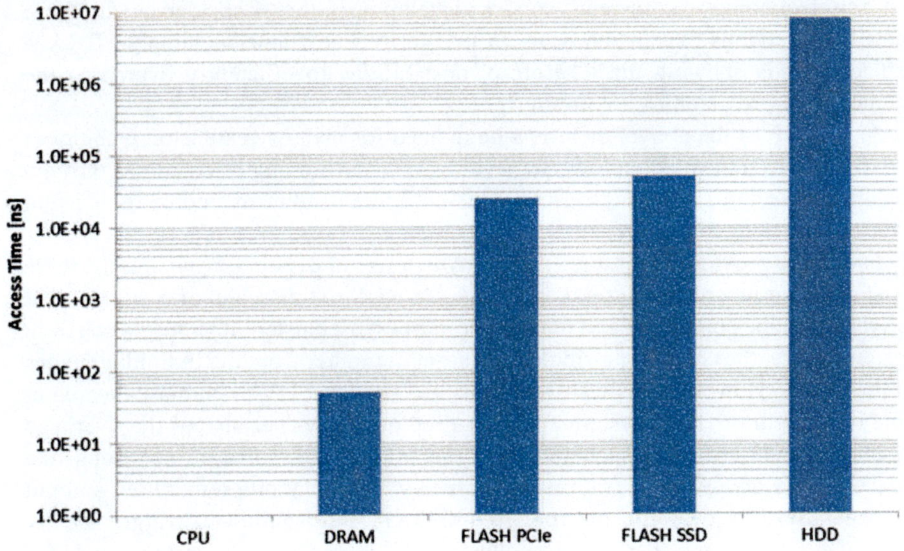

Fig. 2 Performance disparity of different storage technologies

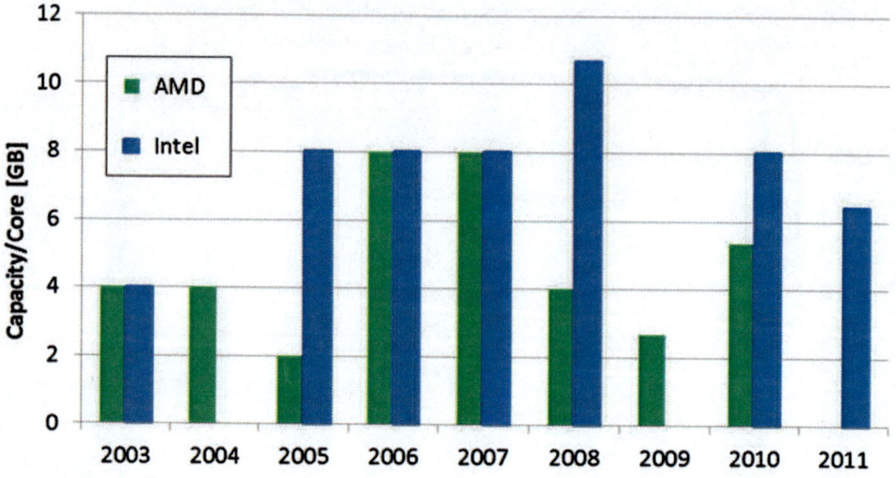

Fig. 3 Memory-to-core ratio for 4-socket servers

to replace DRAM as the fastest storage technology available nowadays. Examples include Phase Change Memories, magneto-resistive random-access memory like spin-transfer torque or the memristor. However, all of them have in common a high uncertainty about their success, as all they currently face significant limitations. Another approach to extend the capacity of the DRAM technology is 3D stacking, but this technology is currently under development too. Considering all these facts, it is obvious that in the near- to mid-term time frame no suitable DRAM replacement is showing up. Therefore, possibilities to efficiently overcome the memory overprovisioning are not in sight.

As a result of the above, there is a huge need for memory extension techniques based on fast DRAM technology that avoid static partitioning of resources through dynamic sharing of scarce resources. This could be used to overcome the current over-provisioning of resources.

In order to provide the required flexibility in terms of dynamic provisioning, there exist some commercial solutions like SGI Altix UV, Numascale, or ScaleMP that aggregate all the computing and memory resources in a cluster into a single coherent system image, thus creating a single and larger computer from the individual nodes of the cluster. However, hardware solutions like the Altix UV or Numascale are not able to scale to the large system sizes required nowadays because of the overhead of cluster-wide coherence schemes. On the other hand, software solutions like ScaleMP rely on a page migration mechanism which, in addition to a high amount of page transfers over the network, requires OS intervention to trigger locally unavailable pages and to maintain coherency. This results in large overheads for fine-grain accesses, as many shared pages and a large amount of synchronization heavily degrade the performance of this software solution. Moreover, both hard- and software approaches share the fact that they are very expensive, which prevents a broad use. Additionally, many applications exist which would rather benefit from more memory resources, but not from additional computing power.

Here we present MEMSCALE, a new shared-memory architecture for clusters that overcomes scalability constraints previously associated with such systems. MEMSCALE allows an application to use remote memory resources independently from remote processing units, thus providing decoupling of resources. In this way, the amount of memory available to a process running in a single node can be increased with the memory from other nodes in the cluster, without having to assign the remote computing units and their associated caches to this process. As in this case no remote caches are involved, there is no need to extend the coherence domain of this process and therefore the scalability problem of cache coherence is avoided. Beside the exclusive use of global memory resources we also support a shared use model by allowing more than one process to use global memory regions. As this use case involves remote caches, coherence domains have to be extended. In order to maintain scalability, we propose a new consistency model. It reverts from continuous coherency to partial coherency, only guaranteeing consistency at certain synchronization points. Synchronization intrinsics like barriers and locks are used as safety net for this relaxed consistency model.

MEMSCALE has been implemented using FPGA technologies, being deployed into a real prototype cluster. This real system serves as a test bed to characterize applications running on MEMSCALE. It not only allows performing a large number of experiments—compared to long-running simulations—but we can also validate the functionality, the performance potential, and the scalability of our approach.

2 Related Work

The problem of insufficient main memory is managed with different approaches, classified into two main categories: exclusive and shared memory. In the first case, an application executed in a single node is granted with exclusive use of memory remotely allocated. In the second case, the memory resources used to extend the address space of an application can be shared among several nodes. In this second case, it is typical that a group of previously independent nodes act like a single shared-memory machine.

2.1 Exclusive Memory

The most traditional approach to extend main memory resources is based on the use of virtual memory and memory page exchanging between main memory and a secondary storage that acts as an additional level in the memory hierarchy. As traditional hard disk is slow due to its mechanical components, during recent years the FLASH memory technology has become popular as a low-latency persistent storage mechanism. Two major companies shipping this product are Fusion-IO and Virident. Fusion-IO last development, ioDrive2 Duo, has a maximum capacity of 2.4TB and 68us read access latency. Regarding Virident, they offer up to 1.4TB with 62us read latency. These high latencies compared to DRAM memory are due to a slower technology but especially to the software overhead required to exchange memory pages.

In the case of using foreign memory, almost every proposal is software-based too. For example, [1–4] leverage idle memory in a peer node in the cluster by exchanging memory pages from local main memory to main memory installed in a peer node. As in the previous case, the extra memory is configured as an additional level apart from regular main memory. This means that processors have no direct access to this memory and, thus, access latency is noticeably high as pages have to be moved by the corresponding software handler every time the requested virtual address is not present in local memory.

An exception to all these software-based approaches can be found in [5], which is focused on a dedicated memory server that makes up for the insufficient main memory at nodes. It consists of a hardware technique for accessing memory in the memory server in a similar way as we do in MEMSCALE. However, they only introduce this technique as an unexplored possibility and end up focusing the study on a software-based alternative, similar to most of the proposed approaches.

Notice, however, that the idea of a dedicated RAM-memory server is not popular nowadays neither in the commercial scene. Therefore, we can find very few solutions. For example, based on FLASH technology, Violin Memories offers a flash-memory rack system connected to the nodes through Fiber Channel. For the new 6000 series, a 22TB memory server presents a read access latency of 140us.

2.2 *Shared Memory*

In this section we present those solutions that increase the amount of available memory by aggregating the resources in a cluster through the use of a shared-memory approach, thus providing a single shared-memory system. The final result can be achieved by software or by hardware.

2.2.1 Software-Based Approaches

There are several academic proposals and commercial solutions for software distributed shared-memory (DSM) systems. We can include in the first group the virtual shared-memory multiprocessor described in [6]. This system is based on a hypervisor that manages the resources from several nodes in order to create a single virtual machine from them where a single operating system is executed.

On the commercial side, ScaleMP offers the currently most popular implementation of software DSM. This system is based on a virtualizing software layer, as in the previous case. The hypervisor is responsible for granting memory accesses by moving pages to the node that requests the corresponding virtual address. This solution, based on Infiniband, achieves a remote memory latency of 25us.

These software techniques present the advantage over hardware-based alternatives that they are easier to install, present shorter time to market, and are typically cheaper. However, despite their cheaper nature, the last version of ScaleMP has an average cost of 800$ per socket in the system. Moreover, in addition to the higher latency due to software managing of remote operations, these approaches usually suffer from congestion when the number of concurrent threads accessing the same memory increases.

Some software DSMs are based on the use of a network feature called Remote Direct Memory Access (RDMA), which requires the appropriate hardware support to provide direct access to the memory installed in a different node. The first open implementation was VIA [7], which set the bases for the RDMA capabilities in InfiniBand. The advantage of this technique is that applications can bypass the operating system and therefore a read or write operation can be done without the intervention of the local or remote operating systems or applications. Notice, however, that even when using this feature processors still do not have direct access to remote memory because the network card does not act as a transparent bridge between processors and remote memory as in the case of MEMSCALE. On the contrary, each data movement has to be programmed, as in the case for traditional DMA operations.

2.2.2 Hardware-Based Approaches

As we have seen, the use of software handlers managing page faults adds latency to remote memory operations. In order to avoid this overhead, the address space of processors can be extended to include remote memory, thus making processors able of directly addressing remote memory and, therefore, no help from software layers is required. To do so, new hardware components have to be developed.

The idea of a hardware-based DSM is not new. The DASH prototype [8] is based on a set of nodes connected by a low-latency network. To extend the addressing capabilities of the processors and to maintain coherence in the inter-node space, each node is equipped with a hardware device called directory controller. As the name suggests, the coherency among nodes is maintained with a directory based on a bit-vector structure. According to the authors, this system scaled up to 64 processors.

The SGI Origin 2000 DSM system [9] by Silicon Graphics is another illustrative example also based on cluster architecture. The difference with DASH is that a similar directory protocol maintains coherence also in the intra-node space. Again, each node is augmented with a hardware device, the Hub chip, in charge of managing remote accesses and the corresponding coherency actions. The maximum theoretical size of this system is 1,024 nodes and 1TB of main memory, although its maximum implementation size has not exceeded 256 processors.

Another directory-based DSM is the FLASH prototype [10]. This system is an evolution of the DASH design. One of its new features is a programmable protocol processor. In this way, this machine has been tested with several coherency protocols, from the previous bit-vector directory to the Scalable Coherent Interface protocol. Scalable Coherent Interface (SCI) [11] is an interconnection proposal for shared-memory multiprocessing, standardized in 1992. It was focused on providing a scalable, low-latency, high bandwidth interconnection with full support for cache coherence. The coherence protocol is based on a directory made up of a set of double linked lists, stored across the cache memories that contain copies of a given memory block. In this way, the size of the directory is always proportional to the size of the memory. However, the time required for traversing the lists increases with system size (number of cache memories) and, thus, the latency of memory operations.

Numascale is a company that provides hardware-based DSMs. This is accomplished with the NumaChip functionality, which implements the SCI protocol in order to extend the coherence domain of isolated nodes to the entire cluster. Numascale leverages commodity hardware to configure a supercomputer with the same memory and computing power as mainframe computers at a lower cost. This is achieved by attaching an add-on card that includes the NumaChip. Actually, nowadays most supercomputers are built from mid-range x86 nodes by using proprietary aggregation technologies. This is also the case of the Bullx Supernode system, which offers a node configuration of four sockets and 1TB of main memory that can be combined with other three to build a supercomputer with 128 cores and 4TB of memory. Another example is the SGI Altix UV supercomputer [12].

The SGI Altix UV is the fifth generation of SGI's scalable global shared-memory architecture, which scales up to 2,560 cores and up to 16 TB of memory. This system is built using the SGI NUMAlink interconnect that provides the high-bandwidth and low-latency required by these global shared-memory systems. This technology is the evolution of the SGI Origin 2000. In this case, the UV-HUB chip is connected to Intel processors through the Quick Path Interconnect links to extend the local coherency domain to the larger cache-coherent NUMAlink environment. In this system, remote memory access latency is 1us.

MEMSCALE fills the gap in the presented categorization: hardware-based exclusive use of remote memory. On the one hand, we achieve much lower latency than those techniques based on page swapping. On the other hand, we achieve higher scalability than both software and hardware DSMs, as we do not maintain the coherence protocol in the inter-node space. Moreover, with the appropriate software layer, MEMSCALE also provides coherence across the entire system. We will later show the benefits of using a software approach for maintaining coherence only when strictly required instead of providing a continuous coherent view of the system.

3 The Memscale Architecture

Although typical x86 server deployments include 64 or 128GB of RAM per motherboard, today's CPUs can support physical addresses with up to 48 bits, which translates into up to 256TB of addressable memory. However, the largest configurations in the market use only a tiny fraction of this, providing up to 2TB. This opens up the opportunity to use the unused huge fraction of the address space to extend the available memory resources by addressing remote memory locations. This is exactly what MEMSCALE does. The main purpose of the MEMSCALE architecture is to provide additional memory resources to processes requiring it by logically assigning them memory that is physically attached to other computers in the cluster. In this way, MEMSCALE allows to extend the address space of a processor in a given node with the memory of other nodes, thus providing the processor with a global address space across the cluster.

MEMSCALE allows two different use models of the memory borrowed from remote nodes. In the first one, known as *exclusive memory*, the memory resources gathered across the cluster by a process are exclusively used by it, thus not sharing them with other processes. Notice that in this configuration, a process is confined to the processors and caches located in the node where it is being executed. In the second possible use, known as *shared memory*, the memory resources borrowed from other nodes can be shared among the threads of an application that spans to several nodes in the cluster. Therefore, in this use model the global address space created can be shared among the processors involved in the execution of a given task.

In the rest of this section we present the key components of our system. To do so, and in order to make the explanation simpler, we will focus first on the exclusive global address space use model. We will later introduce the shared global address space, which builds upon the previous one.

3.1 Overview

MEMSCALE partitions the cluster into memory regions. A memory region is made up of one or more logical portions of main memory that could be located at different nodes of the cluster, and that conform altogether a single coherence domain. A process can freely use the entire memory in the region it belongs to but it has no access to the memory in other regions in the cluster. Similarly, a processor can address any location of its memory region, but cannot address memory locations outside it. Figure 4 shows five nodes of a cluster and five memory regions. Region number 1 is confined to node A and represents the default configuration for a node, that is, processes in that node can access the entire node's memory. On the other hand, region number 4 has been extended to the neighbors of node D, so processes in this node now have direct access to part of the memory located in nodes C and E.

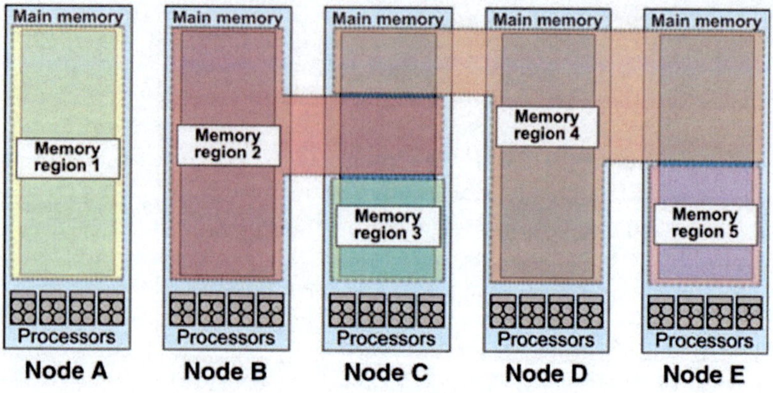

Fig. 4 An example of memory sharing among the nodes of a MEMSCALE cluster

To achieve this, regions 3 and 5 occupy only a portion of the main memory in nodes C and E, respectively. Finally, region 2 has been extended to its neighbor node C, where three memory regions coexist.

In the exclusive use model, MEMSCALE partitions the overall memory resources into as many memory regions as nodes exist in the cluster because all the processors in a given node will always be glued into the same memory region and will additionally be independent from processors in other nodes. In this way, there is one independent operating system at each node. Thus, MEMSCALE does not provide a single system image as the SGI Altix UV or Numascale approaches do, but a collection of independent systems that can borrow and lend memory from each other.

It is important to emphasize that as memory regions are independent coherence domains, all caches in a node will only cache data from one (the same) memory region. This allows a very good scalability of our system, as the size of a memory region has no impact on the performance of the coherence protocol because the number of caches sharing data in that region is limited to the caches in one node. Moreover, because of the independency of the coherence domains, no global coherence protocol across the cluster is required. As can be seen, our system decouples memory from processors, and therefore there is no coherence overhead when aggregating huge amounts of memory.

Nevertheless, this exclusive memory model implies one restriction: the threads of a parallel application have to be executed in one single node, although they may have access to all the memory in the cluster. However, this limitation is not as restrictive as it may seem as in today's mainstream computers it is possible to include up to 8 sockets with up to 10 cores per socket. Therefore, 80 threads is a pretty good level of parallelism. Many applications have even lower parallelization possibilities.

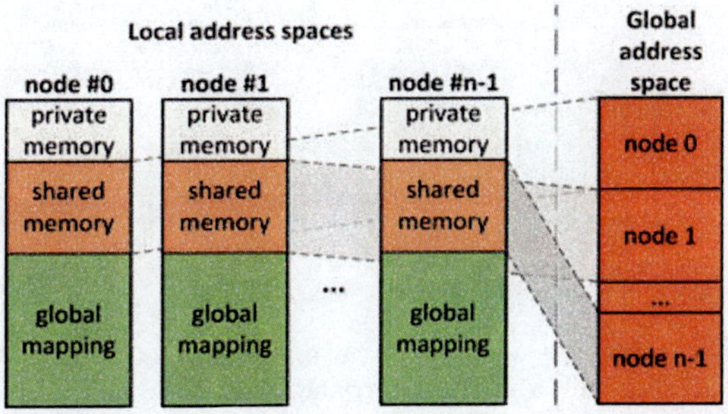

Fig. 5 Global address space layout

3.2 System Architecture

MEMSCALE has been designed with the requirement that applications should not be aware about the fact that the address space they are using is actually achieved by putting together memory from several nodes. In this way, the underlying remote memory system should be completely transparent to them. Moreover, our system does not rely on any kind of run-time or communication library. On the contrary, its core is a small piece of hardware, which allows for a very low remote access time. Thus, accessing remote memory completely relies on hardware and is therefore free of any software overhead. In MEMSCALE, a regular load or store assembler instruction executed by an application will trigger the hardware mechanism to access data from remote memory.

The use of remote memory by a given processor requires that the remote nodes lend part of their memory. Figure 5 shows an example of how the global address space on the right of the figure is built from the memory of all the nodes of a cluster. As shown in the figure, each node's address space is split up into private, shared, and global regions. The private one can be solely used locally, i.e. no remote access to it from other nodes is possible. Typical uses for this memory are to host the OS and similar local processes. The shared-memory region is exported and can be accessed remotely. The aggregation of these shared-memory regions from each of the nodes builds up the global address space, representing the memory resources available within it. The last address region of each node, the global one, allows accessing the global address space. Thus, a source node will issue a load or store operation to an address within its global address region in order to access remote memory locations. As it will be explained later, this (global) memory address encodes a node identifier and a target address within the shared region of the target node.

In order to understand how a processor can determine whether the addresses memory is local or remote (that is, the basic MEMSCALE architecture), it is important to explain that in current mainstream systems, leveraging either Intel or AMD processors, each socket is attached to part of the physical memory by means of its own memory controller, as shown in Fig. 6a. Therefore, as there are several memory controllers in the system to access memory, processors require to know where to forward a given memory request. This is achieved by including at each

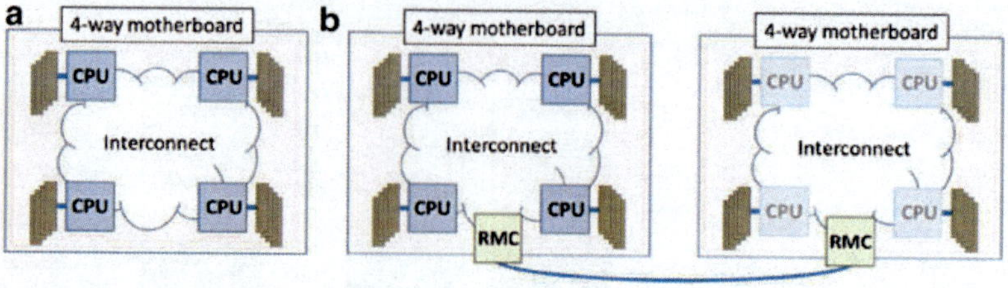

Fig. 6 Memory distribution in a single node and in MEMSCALE. (**a**) Four socket original configuration. (**b**) RMC interconnection

processor a set of base and address registers (BAR) configured at initialization time that reflect the system physical memory distribution. In this way, for each memory operation, the processor compares the requested address with those registers and then forwards the memory request to the corresponding memory controller. Forwarding the memory operation leverages a small network that connects the sockets within the motherboard. This network currently uses either the HyperTransport or the QPI (Quick Path Interconnect) protocols.

This on-board memory distribution and access scheme is the basis upon which we have designed the MEMSCALE technology. We have created a new hardware component, referred to as Remote Memory Controller (RMC), which will be presented to the processors as a new memory controller, as shown in Fig. 6b. However, the RMC will have no memory bank directly connected to it but will rely on the memory banks installed in other nodes in the cluster. In order to enable the RMC functionality, the BAR registers must be reconfigured so that some of the memory accesses are forwarded to the RMC, which will convert those accesses into remote accesses.

The RMC allows accessing remote memory resources across the cluster. Figure 7 shows the shared-memory map seen by each of the nodes in a 16-node example cluster, where four-socket nodes equipped with 16GB of main memory are assumed. Yellow areas labeled as *socket x* represent the 4GB of memory attached to each socket. Yellow areas are the traditional configuration of nodes. However, the addressing capabilities of this cluster have been extended with the global address space created across it. The green area represents the global memory region previously depicted in Fig. 5, made up of the aggregation of the individual main memories of each node. In this example, this global memory area is set between addresses `0x000400000000` and `0x0043ffffffff`, providing 256GB. The RMC is configured to be responsible for this address range. Thus, any memory request related to an address inside that range will be automatically forwarded to the RMC. When the RMC receives a memory request, it needs to obtain the identifier of the target node and the target address to be used inside that remote node. That information is embedded into the requested memory address. The exact procedure to get that information will be described later in the Implementation section.

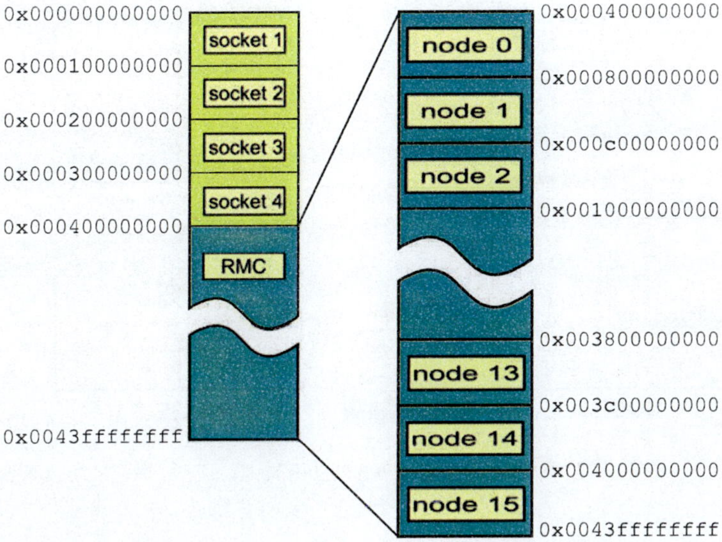

Fig. 7 Example of the memory map of a node in a 16-node MEMSCALE cluster

A very interesting feature of this addressing mechanism is that accessing remote memory does not rely on translation tables. Thus, as no information is stored at the RMCs, the system can scale without limitation. The same applies to the set of base and address registers in the processors: they do not store information about other nodes. Actually, processors are only aware of the local memory distribution described in the left sketch of Fig. 7. Therefore, the number of required registers per node does not depend neither on the number of nodes in the cluster nor on the amount of shared memory.

3.3 Remote Memory Allocation

As with local memory, before using remote memory it has to be reserved. Reserving remote memory is entirely done by software. Although a software approach could reduce performance, notice that the higher latency of a software scheme is not critical because it can be hidden by pre-reserving remote memory. On the contrary, it allows making the RMC design independent from the exact version of the OS kernel used.

Figure 8 shows the diagram of the remote memory allocation process, which is triggered by an application demanding more memory. Typically, applications rely on a library to manage memory allocations, and these libraries end up calling the *mmap* system call to ask the OS kernel for more memory.

We have modified the *mmap* function in order to allocate remote memory. To perform such allocation, the *mmap* function relies on a user-space application, which will be referred to as the client. The communication between these two entities is

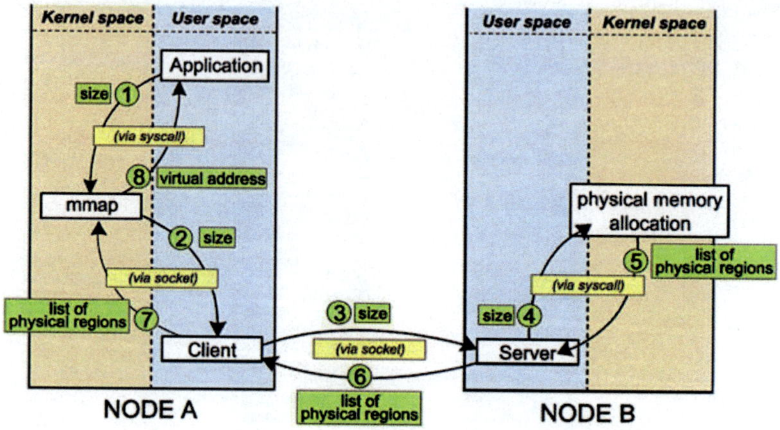

Fig. 8 Diagram of the process of allocating remote memory

based on a socket that facilitates the exchange of information inside a node between kernel and user spaces. The client process receives the desired size of memory and determines which node should be asked for that memory. This decision can follow several policies, like splitting the memory among the nodes in the cluster, or concentrate the allocation in one single node (as in the example in Fig. 8). The decision can also be based on how much available memory exists at the target nodes. This information can be periodically exchanged among nodes in user space.

After selecting the target node, in this case Node B, the client sends a message through the general purpose network present in the cluster to the server application in the remote node (notice that every node executes a client and a server). The server application receives the size of the memory to be allocated and this value is passed to the kernel through a dedicated new system call implemented for this purpose. This system call allocates as much contiguous physical memory as possible. Typically, the system will not be able to allocate contiguous chunks of more than 4MB due to the fragmentation of the memory. Therefore, for larger amounts of memory it would be necessary to allocate several chunks of physical memory. This memory is marked as no swappable because node A needs persistent valid physical addresses. After that, the system call returns the list of physical chunks (physical address and size of the chunk) to the server.

The server then modifies the physical addresses by inserting the identifier of its node (Node B in this case) in the physical address of each segment and sends back that information to the client. After that, the client receives the list of physical segments and adds the start address of the global memory region. The list is passed back, through the socket, to the *mmap* function that was waiting for a response. Then, the different segments are tailored together and mapped into a contiguous portion of virtual memory. Additionally, the appropriate translation between physical and virtual memory pages is also written into the Translation

Look-aside Buffer (TLB) in order to avoid later communication overheads. Finally, the virtual address of this memory is returned to the application and the remote memory allocation process is completed.

3.4 Shared-Memory Across the Cluster

Up to this point, we have seen how a process running in one node is able to use memory from any node within the entire cluster. However, it is possible to go one step further and introduce an evolved system where not only memory is aggregated but also processors are, converting the entire cluster into a powerful distributed shared-memory system featuring a single global address space that can be used by all the nodes. This evolved system corresponds to the shared-memory use model mentioned at the beginning of this section. In this use model a shared-memory parallel application can span to several nodes of the cluster and its threads can share memory locations in the same way as any regular shared-memory parallel application can do within the realm of a single multicore computer.

The starting point for this shared-memory use model is the system presented above. The same RMC functionality presented for the exclusive use model will be used in the shared-memory one. Regarding the remote memory allocation mechanism, it needs to be enriched in order to allow for threads, or processes, running in different nodes to address the same memory locations, so that shared variables can be accessed from different nodes.

However, the most interesting issue related to the shared-memory model of MEMSCALE is that it preserves the feature of having an independent operating system at each cluster node. Remember that this is the characteristic which allows for the very good scalability properties of MEMSCALE. Therefore, although this use model effectively provides real shared memory across the cluster, it has to deal with the intrinsic MEMSCALE nature of not providing coherence across the cluster.

At this point the question is: how would the lack of coherence affect the application programmer's vision of the underlying system? To answer this question, we should take into account that the usual way to program shared-memory applications is by leveraging what is commonly known as a safety net, which basically provides synchronization primitives that avoid race conditions among threads. These synchronization primitives are barriers and locks explicitly inserted within regular parallel code, or implicitly in OpenMP programs. These primitives are programmed by system developers according to the underlying hardware and provide the programmer with an abstraction so that she/he can focus on the application itself. Additionally, it makes no sense that parallel application programmers build their own safety net. Instead, the compiler and synchronization library developers provide the proper implementation for a particular system.

On the other hand, actual requirements from the application programmer's side often do not demand for a continuous maintenance of a system-wide coherent state but it would be enough to provide consistency at synchronization points only.

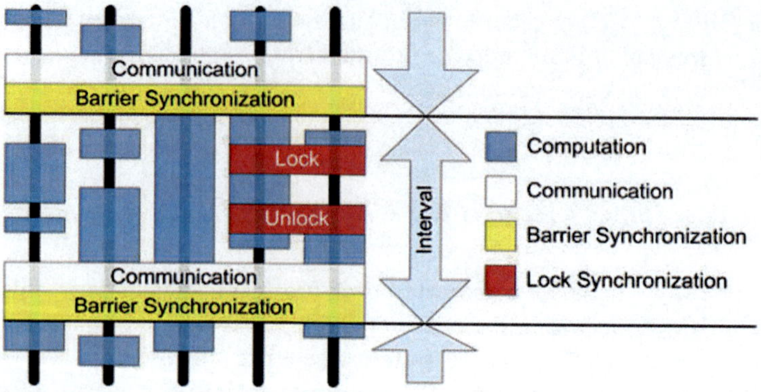

Fig. 9 Task model

Therefore, if safety nets are used by shared-memory application programmers, the lack of coherence will not affect the programmer's vision of the underlying system as long as the safety net is properly tuned to the particular hardware implementation and provides the expected primitives.

In order to provide the safety net that matches the MEMSCALE characteristics, we should take into account that the basic idea in MEMSCALE is enabling efficient distributed shared-memory by avoiding the global coherence across the cluster. Thus, we can base the development of the safety net on the observation that tasks only need the most-recent values of shared data at synchronization points. Therefore, if coherence is guaranteed at such synchronization points by the safety net, then a continuous global coherence is no longer necessary. In other words, providing coherence without synchronization is not required, except for the case that multiple tasks share different elements in a cache line (false sharing), which should be avoided, for instance, by compilers ensuring that different elements within a cache line are not shared between different tasks.

Figure 9 shows a typical execution of a parallel application. Its threads execute uncoupled during computation phases and are also not coupled during communication, which is done in the form of reads and writes to memory. Only at synchronization points, which can be barriers and locks, they cooperate in a coupled fashion. As can be seen, task execution is separated into compute, communication and synchronization phases. These three phases form a so-called interval, which starts after synchronization and includes all computation and communication until synchronization is required again.

With this task model in mind, we can focus now on how to provide the required safety net in this scenario. Let's introduce the concept of *On-Demand Consistency*, which during nonsynchronized operation delays the visibility of writes and allows noncoherent responses for reads, but ensures that at synchronization points both writes are globally visible and reads provide coherent responses. Thus, we are relaxing both coherence and consistency and therefore simplifying the complete system. Notice, however, that in order to ensure coherence and consistency at synchronization points, either an update—or invalidate—policy for cache copies is required. As update-based policies are not used in current commodity processors, in MEMSCALE we leverage invalidations. However, our approach for On-Demand Consistency is not dependent on invalidations; it would also perfectly work with updates.

As mentioned above, the safety net will identify and invalidate the cache line copies used during the last interval. This could be either done by the compiler, which inserts appropriate assembler instructions in the instruction stream, or by the RMC, which dynamically keeps track of copies and invalidates them upon request by the safety net, or by flushing complete caches. In any case, invalidations must take place just before the synchronization point, ensuring that dirty cache lines are written back before any other task might access the associated addresses.

4 Implementation Using FPGAs

The key behind MEMSCALE is to spawn up global address spaces over physically distributed memory and doing so by providing a direct low-latency hardware path to remote memory locations, and by ensuring both global consistency and scalable coherence by reverting from continuous coherenc to a partial one, with guarantees about visibility and consistency only for synchronization intrinsics like barriers and locks.

The use of this relaxed consistency model allows us to reduce the hardware complexity and to focus on minimizing the access costs to remote memory locations. A minimal latency is of upmost importance because any remote access will suffer from a latency disparity compared to local memory. Several facts in the designed hardware contribute to this minimized latency:

1. State-less architecture: no side effects or state-depending actions are required when performing remote memory accesses in a global address space without continuous coherence. Similarly, no process-specific information is necessary, like it is the case for message passing applications. By implementing a state-less design we avoid any time-consuming table lookups or similar techniques required for state-dependent processing. As it will be shown later, only on the target side the Source Tag translation mechanism relies on a table lookup, but without loss of generality this table can be limited in size to facilitate single-cycle access times.
2. Efficient integration into the host system domain: we are using HyperTransport (HT) to directly connect to the host CPUs, without any intermediate bridging or protocol conversion. HT is designed as CPU interconnect and optimized for transactions of cache-line size.
3. Interconnection network among the cluster nodes optimized for low latency: we are relying on the low latency features of the EXTOLL network when forwarding memory accesses through the cluster. In addition to the low-latency

characteristics, EXTOLL provides some mandatory functionality to support global address spaces:

- *In-order packet delivery*: unlike other cluster interconnects, EXTOLL maintains packet ordering. This is important to guarantee that no load can bypass a store and vice versa. While some consistency models allow relaxing some combinations, we have chosen to maintain all of them and not to restrict MEMSCALE to a certain subset of consistency models.
- *Reliability*: EXTOLL guarantees packet delivery by link-level retransmission protocols. Otherwise, an outstanding request might be dropped, leading to stalling CPU cores which expect responses in finite times. Furthermore, it is not feasible to store all outgoing memory operations for potential replay upon packet loss. The resulting memory structures would be too large to maintain the low latency characteristics.
- *Virtual Channels*: separate virtual channels are required to avoid protocol deadlocks by decoupling requests from responses. Otherwise, two nodes with mutual accesses could deadlock each other, as their responses are potentially blocked by requests waiting for previous requests to complete.

Notice that both the interface to the local CPUs and the interconnection network can be chosen quite arbitrarily: it is possible to substitute the HT-Core with a PCIe core, and to replace the EXTOLL network with another one. However, the network has to provide both in-order delivery and reliable transmission, in order to maintain the semantics of global memory accesses.

4.1 Global Architecture

The global architecture of the RMC is shown in Fig. 10. The green units provide connectivity towards the host domain, and the red units on the right to the network domain. In between, the shared memory engine (SME), shown as part of the network interface in blue, directly translates HT packets to EXTOLL network packets and vice versa without any software involvement and minimal conversion overhead. Additionally, it determines target nodes based on the used address and avoids Source Tag collisions in the target domain by re-mapping them to unique values. The SME connects over a custom On-Chip Network (NoC) to the HT Core, which connects directly to one of the host's CPUs. On the network side, it connects over the network port to the switch, whose six links allow setting up direct topologies like meshes and tori, making centralized switching units not required.

An example network configuration is shown in Fig. 11. In this 3×3 2D mesh topology, a source node (Src) is sending a load request to the destination node (Dst), which is hosting the addressed memory region. The path of the request is shown using solid arrows, while the path of the response is shown using dashed arrows. Notice that Dimension Order Routing (DOR) is employed in this example in order to avoid network deadlocks, both for requests and responses.

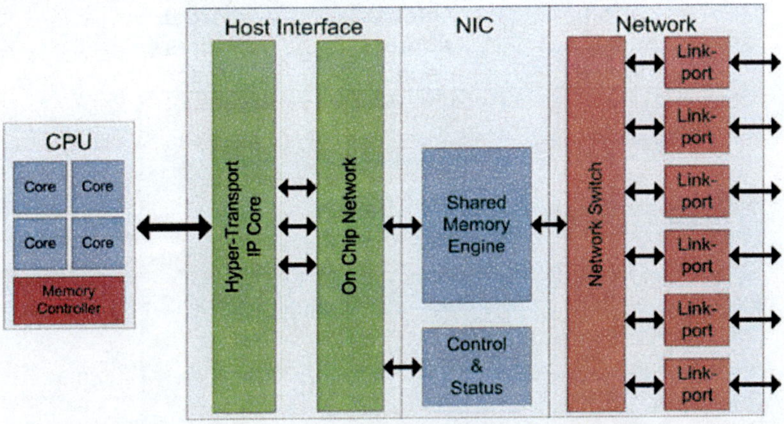

Fig. 10 Shared memory engine as bridging unit between host and network domain

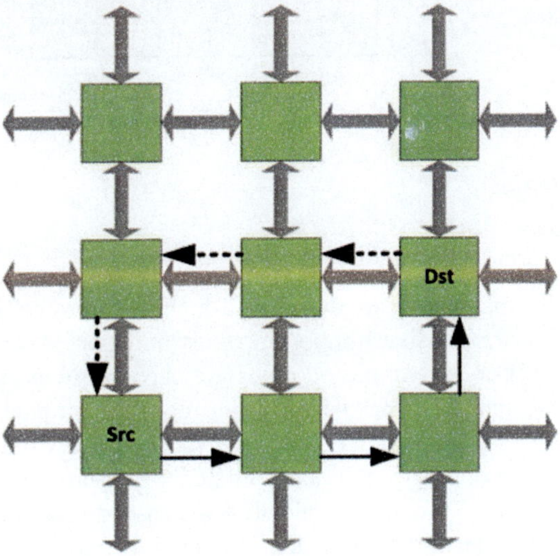

Fig. 11 Example topology, including the path for a remote load request and corresponding response

Finally, note the CPU to the left of Fig. 10. Typical installations employ AMD Opteron CPUs, which integrate computing cores and memory controller in a single die. Thus, CPU modules act both as initiator and completer of memory requests: on the source host side, these cores send out memory accesses to local and remote locations, with the latter being forwarded to the RMC. The HT core forwards them over the NoC to the SME, which translates them into EXTOLL packets which are transferred to the target host. Here, incoming packets are translated back into HT packets, which target a memory controller within one of the remote host's CPU. Then, write requests are completed, or for read requests appropriate responses are generated and sent back to the source host where they complete the appropriate request.

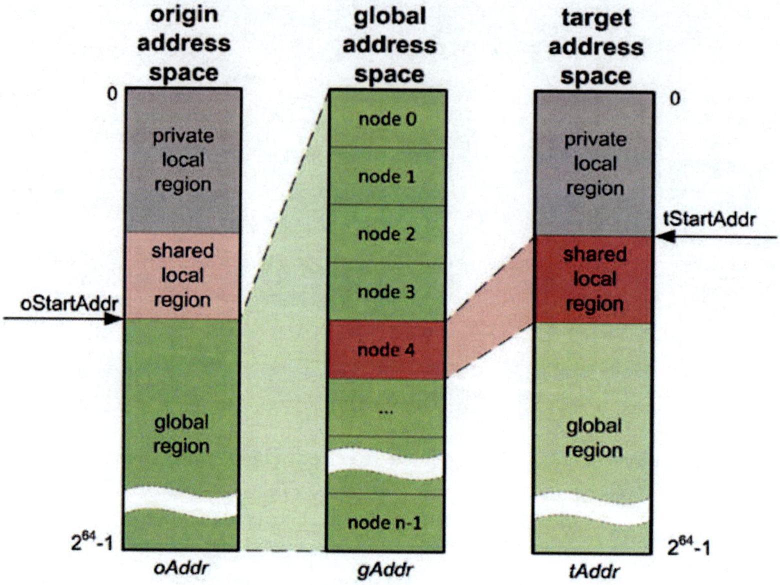

Fig. 12 The three different address spaces: origin, global, and target

4.2 *Implementation*

The major task of the shared-memory communication engine is to forward HT transactions like nonposted reads and posted writes from the origin node to the target node. Nonposted reads are carried out as split phase transactions, where the read request is answered by an independent packet-based response. The response carries a tag which enables the source SME to match the received response to the corresponding sent request. Posted writes are carried out by simply writing data to a specific address. Several sub-tasks can be identified based on this, which are target node determination, address translation, and source tag management.

4.2.1 Target Node Determination and Address Translation

Before describing these sub-tasks in more detail, examining the different address spaces involved is helpful for an improved understanding. In addition to the global address space which was explained in detail in Sect. 3, source- and destination-local address spaces are required for the sub-tasks. Figure 12 shows all three address spaces and how addresses are translated between them.

Three different address spaces exist in this architecture: one global address space and two local address spaces, one for the origin and one for the target side. As a remote memory access is traversing all three address spaces, address translation must take place. To avoid unnecessary latencies for this task the translation scheme should be kept simple but efficient.

Considering Fig. 12, the global address (gAddr) can be determined by subtracting the origin start address of the global partition (oStartAddr) from the origin address (oAddr):

$$\text{gAddr} = \text{oAddr} - \text{oStartAddr} \tag{1}$$

This global address is also used to determine the target node identifier. A bit mask is applied to the address, and a centrifuge operation compacts the masked bits to a value which represents the target node identification (tNodeID):

$$\text{tNodeID} = (\text{gAddr}\&\text{mask}) >> \text{shift\_count}^{1} \tag{2}$$

The operation is very similar to the one described by S. Scott in [13] and adds only minimal latency. Other approaches like table look-ups are also possible, but in favor of less latency a calculation is preferred here. The inverted mask applied to the global address together with the start address of the target local shared partition (tStartAddr) is used to calculate the target address (tAddr), which is the address actually sent to the target node:

$$\text{tAddr} = (\text{gAddr}\& \sim \text{mask}) + \text{tStartAddr} \tag{3}$$

This address translation scheme allows determining the appropriate address to access data on the target node by converting the origin local address to a global address and then to a target address. This calculation is completely implemented on the origin side, i.e. the network packet already contains the target address. The translation only takes place for requests. Responses do not contain an address; they are assigned to their corresponding requests by using source tags.

4.2.2 Source Tag Management

Source tags allow sending back a response to the appropriate source of the request. They are used by the origin CPU in the local node when sending a packet to the RMC, which copies it unmodified into the EXTOLL packet sent to the target node. There, the remote RMC will forward the packet to the right memory controller, which will use the source tag for returning back the response to the RMC in that node. However, as several source nodes can send requests to the same target node, a source tag management at the target RMC is required. In other words, as the existing uniqueness of source tags is limited to each of the individual source domains, if source tags from multiple source domains are mixed in the target node, then the identification is lost.

---

[1]This is a simplified calculation for a contiguous mask. Furthermore, the value of shift_count is dependent on the actual value of the mask.

Because unique source tags are essential for response matching, a source tag translation takes place in the target RMC. Each transaction gets a new source tag assigned, which is unique within the domain of the target node. Additionally, as response matching is handled internally within each target node, there is no need to make source tags unique over multiple target nodes.

The handling of incoming nonposted requests consists of the following steps:

1. Determine free source tag
2. Store origin source tag and origin node identifier in a table indexed by new source tag
3. Send out local nonposted request to access the memory location
4. Upon arrival of response, look-up in table using target source tag to determine origin source tag and origin node
5. Send network packet to origin node, containing nonposted response and origin source tag

For posted requests no source tag translation is necessary because no response is required in this case. Unlike the address translation, the source tag management is implemented completely on the target side. Thus, a packet traversing the network always contains the origin source tag.

4.2.3 Egress and Ingress Modules

From a functional perspective, the SME communication engine can be divided into three parts: Requester, Responder, and Completer. All three units are used for nonposted transactions, while for posted ones only Requester and Completer are required. To simplify the implementation—in particular the number of ports for the NoC, the communication engine is divided into two parts, which are the Egress and Ingress modules. The first acts as Requester and Responder (sending out packets), while the latter acts as Responder and Completer (handling incoming packets).

The Egress module is shown in the upper part of Fig. 13, consisting of a Requester and a Responder module. When acting as a Requester, HT transactions from the local processor to be forwarded to the remote node, are coming in from the left. The address contained in the transaction is translated and the target identifier extracted from the global address, according to the formulas (1), (2), and (3) explained in Sect. 4.2.1, which correspond, respectively, to the three boxes in the Requester module. After the address translation, a network packet containing the transaction is sent out. In the second use case—acting as a Responder for nonposted transactions—the local HT response generated in the remote node is matched against the source tag table (Matching Store) as explained in Sect. 4.2.2: using the origin source tag and origin node id from this table a network packet is generated and sent out to the origin node.

The Ingress module is shown in the lower part of Fig. 13; it acts either as a Responder or as Completer. In both cases it receives incoming transactions from the network. In the case of a nonposted transaction it acts as a Responder, i.e. it

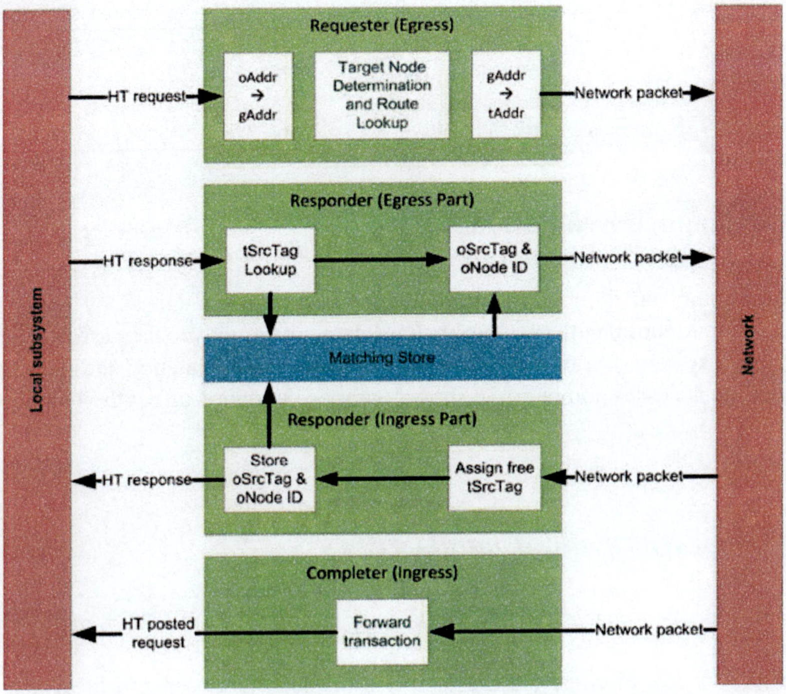

Fig. 13 Internal modules of the SME together with interfaces to host and network

assigns a free target source tag to the transaction, and stores origin source tag and origin node identifier in the appropriate entry of the matching store, as discussed in Sect. 4.2.2. Because the NoC supports 1,024 source tags, the matching store also has 1,024 entries. The transaction with the new (locally unique) source tag is then forwarded over the NoC to the host system. In the case of a posted transaction the Ingress module acts as a Completer (lower part of the figure) and directly forwards the transaction without any source tag translation.

4.2.4 Resource Utilization

The complete RMC design as shown in Fig. 10 has been implemented on a Xilinx Virtex-4 FX100-11. The HT Core is running at 200MHz with a data path width of 64 bits, while the rest of the logic is running at 156MHz with a data path width of 32 bits. Careful pipelining as well as floor planning was necessary to reach timing closure. Table 1 summarizes the FPGA resource utilization.

Table 1 Resource utilization

Resource	Used	Utilization (%)
Occupied Slices	32,696 of 42,176	77
Slice Registers	27,733 of 84,352	32
4 input LUTs	57,057 of 84,352	67
RAM blocks	141 of 376	37

5 Performance Evaluation

After describing our approach to implement global address spaces by leveraging FPGAs in combination with commodity hardware, let us analyze the performance of the proposed system. For doing so, we first characterize the latency and bandwidth of the remote accesses and later we show how these numbers affect the behavior of applications.

5.1 *Basic Performance Characteristics*

The most important characteristics are latency and bandwidth, which are summarized in the Table 2.

As one can see, obviously there is still a disparity between the local and the remote memory access latency (about 90 ns vs. 1,890 ns). However, no other storage technology is able to achieve that low access latency while overcoming main memory capacity limitations. Thus, the remote access latency is still about 25 times faster than a read access to an SSD, and more than 4,000 times faster than spinning disks.

Figure 14 shows the clock cycle distribution for a remote load, based on measurements and simulations. Starting with the latency of a remote load measured using CPU performance counters, several fractions can be identified: the latency fraction originating to FPGA modules is derived from simulations, and using performance counters the latency fraction of the target CPU can be determined. Finally, the remaining latency fraction is associated with the origin CPU.

The previous results have been gathered for an RMC isolated from any application. In the context of an application, Fig. 15 shows the latency and bandwidth numbers obtained from analytical experiments based on the execution of multiple consecutive 8-byte read operations to remote memory.

To study the scalability of our system in terms of cluster size, we have included results for different distances between the node that executes the benchmark and the node that hosts the memory. Case labeled *0 hops* stands for the loopback mode, that is, the target memory controller is in the local node. In this scenario, read requests indirectly arrive to a local memory controller by traversing the RMC.

Figure 15 also gathers different tests to study the potential of our system and predict the performance trend when improving its implementation. These tests

Table 2 Basic performance characteristics

Metric	Value	Comment
Load latency	1.89 usec for a 64-bit load instruction	Memory host is one hop away. Latency increases about 300 ns per direction for each additional hop
Store throughput	300.00 MB/s for 64 byte transactions, using write-combining	Current CPU implementations only allow write-through caching policies, thus no caching effects are visible for this metric

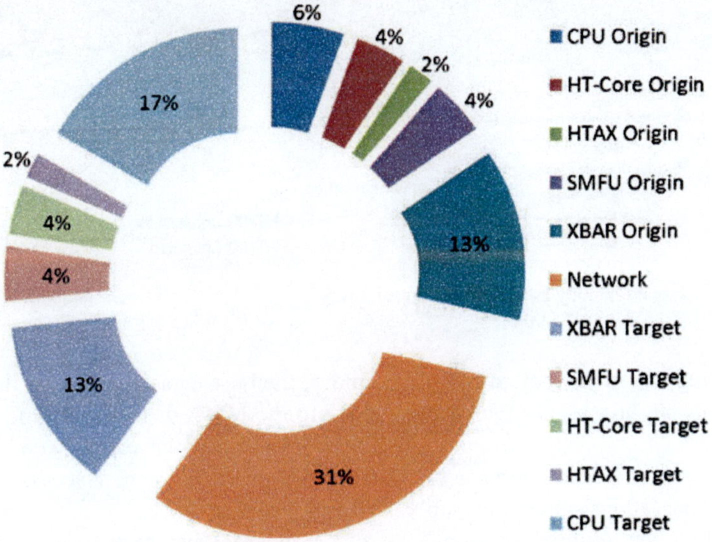

Fig. 14 Latency break-down for a remote load

consist of changing the speed of the HyperTransport interface implemented in the RMC. In this way, we can use a 400MHz interface or we can slow down the FPGA so that it uses a 200MHz interface. In both cases, the core logic of the FPGA is running at 156MHz.

Due to the sequential nature of read accesses, those experiments with caches enabled benefit from locality. In these scenarios, a single read operation retrieves 64 bytes and, thus, the subsequent seven read operations hit the cache achieving a lower average latency.

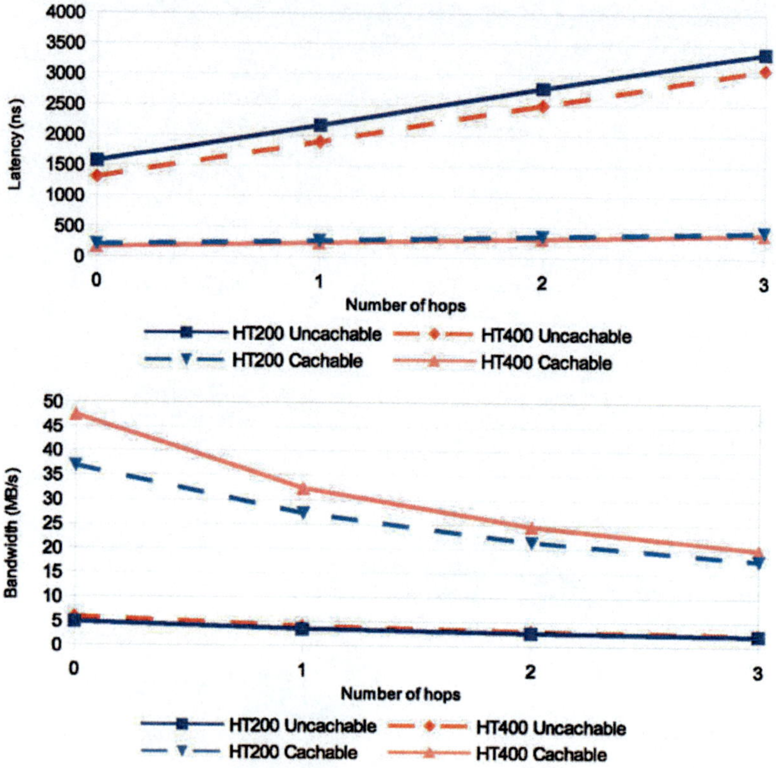

Fig. 15 Analysis on latency and bandwidth scalability

Attending to raw latency to remote memory, the best measurement for the case of 0 hops is slightly higher than 1.3us, showing the effect of the prototype nature of the RMC. An implementation as much optimized as regular memory controllers inside processors would decrease this latency closer to the regular access time to local memory (100ns), thus noticeably improving performance.

Regarding the number of hops, we can conclude that remote load latency increases as the distance between a node and its remote memory increases, as expected. The following equation summarizes this behavior.

$$\text{latency}_{\text{total}} = \text{hops} \bullet \text{latency}_{\text{hop}} + \text{latency}_{\text{loopback}} \tag{4}$$

where hops is the number of nodes between the local node and its remote memory, latency_{hop} is the latency added at each hop (it comprises the propagation time through the fiber optic link and also the routing time at the intermediate FPGAs), and $\text{latency}_{\text{loopback}}$ is a constant time independent of the distance. For example, in the case of HT400 cachable, this constant time is 1.3us and $\text{latency}_{\text{hop}}$ is equal to 600ns. As we can see, distance plays an important role in this system. This is especially true

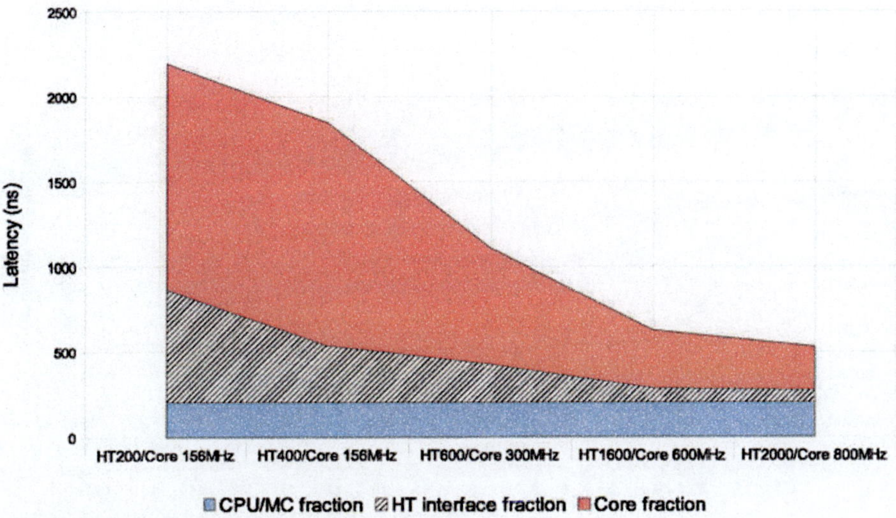

Fig. 16 Latency prediction for remote read operations

when a memory region has been expanded to a large number of nodes. Later in this section we analyze how latency depends on the topology of the network that, in turn, determines the average number of hops.

Attending to bandwidth, for the scenario with only one hop we achieve 32MB/s. Read bandwidth is noticeably increased when caches are enabled due to the higher maximum payload of packets (64 bytes). However, this maximum bandwidth is way lower than other commercial alternatives like InfiniBand or Ethernet. This is because we have focused on latency rather than bandwidth, as shared-memory applications tend to use memory in fine-grained patterns. This limitation shows the difference between our proposal and other interconnection technologies: while InfiniBand or Ethernet is focused on moving data between local and remote memory, our system is focused on moving data between processors and remote memory. The problem is that mainstream processors are designed to face local memory latency. For example, AMD Opteron processors can handle eight outstanding memory requests. Although this number is enough for hiding local memory latency, it is not sufficient to hide remote memory latency. Additionally, due to the current configuration of the RMC, it is located in the IO space where only one outstanding request is allowed. Thus, a new memory request is issued only after the reception of the response to the previous request. For the 1 hop scenario, latency is 1.9us what allows a maximum theoretical bandwidth of 32.12MB/s (with posted write operations, the RMC allows 284MB/s bandwidth).

A second observation according to Fig. 15 is that latency is reduced only 20% when the HT interface frequency is increased from HT200 to HT400. This is due to the fact that the HT interface only constitutes a part of the FPGA, and the RMC core functionality keeps its frequency constant as we said. Figure 16 shows the predicted latency trend when the RMC implementation is improved.

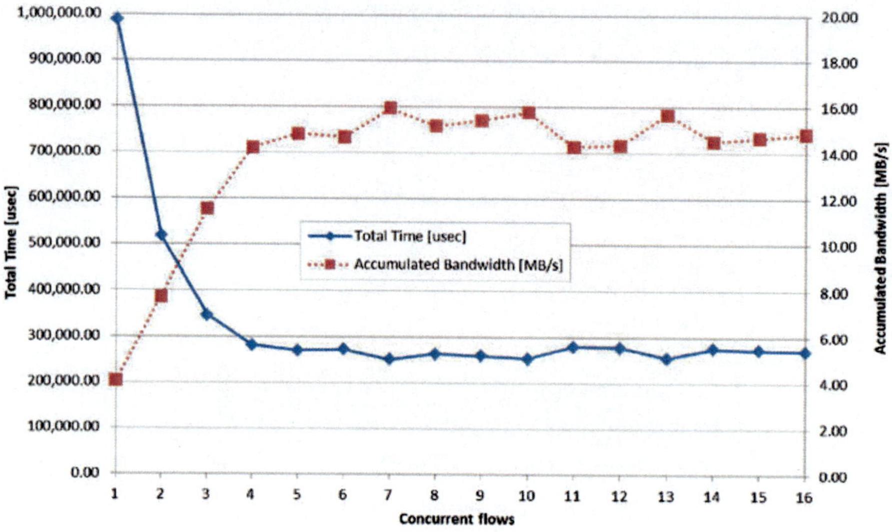

Fig. 17 Load performance for multiple concurrent flows

On the other hand, as current computing systems are inherently parallel and employing an increasing number of computing cores, another important characteristic is the behavior under an increased load due to multiple requesting processes or threads. In this experiment, a multi-threaded application is concurrently issuing load instructions targeting remote memory. Figure 17 shows the load performance for this case.

In this experiment, 512K accesses are evenly divided among a varying number of threads. For each access, a thread issues a load to remote memory. Shown in this figure are the total execution time and the resulting accumulated bandwidth. One can see that performance scales linearly up to 4 concurrent flows. Afterwards, performance saturates and it seems that some system component is preventing a further exploitation of the increased concurrency. Our simulations have shown that the SME is not the source of this behavior. However, the SME is accessible from the CPUs over MMIO space, while memory resides in DRAM space. CPU limits access to MMIO in many regards, including the number of outstanding transactions. We have observed a similar behavior on different CPU generations (including Opteron 41KX HE and Opteron 8354).

Also for this experiment, scattering the load flows to different memory hosts has no impact on performance. On the other hand, if we let multiple clients send several concurrent load flows to one single memory host, performance does not suffer from this increased work load. This rules also the memory controller out as bottleneck for this behavior, indicating that the System Request Queue of the CPU is the bottleneck.

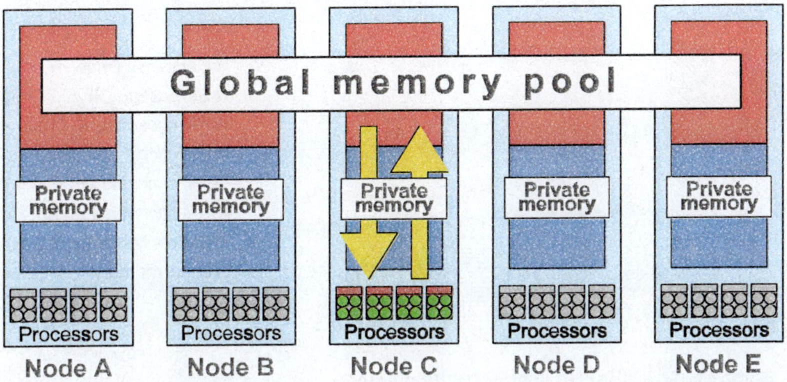

Fig. 18 Example of a global memory pool created from four nodes and accessed from one single node

5.2 Exclusive Memory Performance

After reviewing raw latency to remote memory, let us evaluate how this characteristic affects the behavior of applications. As the objective of our proposal is to allow the use of large amounts of main memory, in the following experiments we focus on the memory-hungry application par excellence: databases[2]. In recent years, main memory has gained importance in databases, not only for caching purposes. Most of the major vendors have developed in-memory solutions for their database servers, like IBM SolidDB or Oracle TimesTen. This shift from secondary storage to main memory allows low latency and high productivity. Moreover, this change is motivated by emerging new uses of information, like social networks, global searchers, and e-mail. This new breed of applications produces an access pattern to data based on a high number of concurrent short queries. This results in a random access to the entire data set. This pattern with no locality shows up the limitations of secondary storage while it perfectly suits the intrinsic characteristics of RAM memory.

This trend towards in-memory databases can greatly benefit from a proposal like ours. To analyze the benefits of a virtually unlimited main memory, we have setup an experiment in our prototype cluster. In this case we chose MySQL server because of its popularity and its open source nature. In a first approach, the configuration of the cluster will be similar to the example depicted in Fig. 18.

In our cluster, each node has 16 GB of main memory. Half of this memory is used to create a global memory pool that, with 16 nodes, sums up 128 GB. The other 8 GBs at each node are left for private usage. As described in Fig. 18, only one node

[2] For a broader coverage of data-intensive applications and other use cases of global address spaces, we'd like to refer to our MEMSCALE-related publications [14–21].

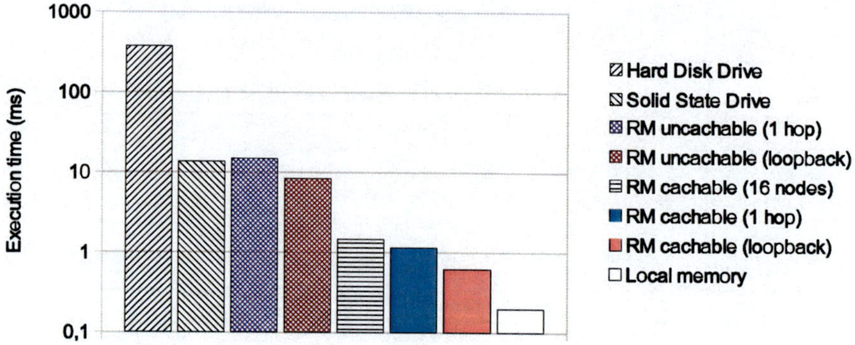

Fig. 19 Comparative analysis of query execution time

will access this memory pool, that is, only one database server is executed and, thus, queries are served from the processors in that single node.

In the following experiment we compare our remote memory approach with local memory and two secondary storage alternatives. For remote memory and local memory we leverage the memory storage engine present in MySQL. To study the lower bound case where the entire database can be loaded into local memory, we use a machine with 128GB main memory. For the other two scenarios we use the MyISAM storage engine. In these two cases the machine is configured with 8GBs of main memory and a Seagate Barracuda SATA 3Gb/s, 32MB cache, 7200RPM and average latency of 4.16ms for the HDD scenario, and two Kingston SNV425-S2 64GB drives configured in RAID 0, each of them with a sequential speed of 200MB/s at reading for the SSD scenario. Regarding workload, we have designed a set of short read-only queries executed against a 100GB database. Figure 19 shows the results.

In this figure we see the average execution time of queries (notice the logarithmic scale of the axis). The first conclusion is that SSD is 28 times faster than HDD, because SSD technology has better random access latency. However, local main memory is 65 times faster than SSD not only due to the fact that RAM memory presents better latency and higher bandwidth, but also because there is no need for accessing secondary storage, so the operating system is not involved in terms of I/O handlers. However, the amount of memory present in a single node has limitations, either economic or technical, and this is the rationale for using our remote memory system for large databases. Regarding the scenario where remote memory is one hop away from processors, we see that our proposal performs ten times better than the SSD configuration, and only seven times worse than the lower bound local memory scenario, with the current FPGA implementation.

Figure 20 shows the throughput of the different scenarios in terms of queries per second. Because main memory is proportionally designed to allow concurrent access to the cores in the motherboard, 16 in our prototype, the local memory scenario scales up to this number. Regarding remote memory scenarios, we see a

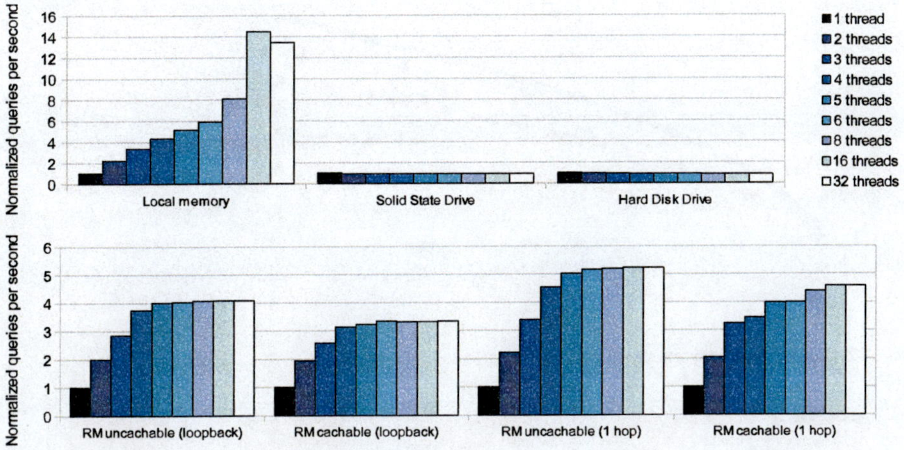

Fig. 20 Maximum throughput in various scenarios

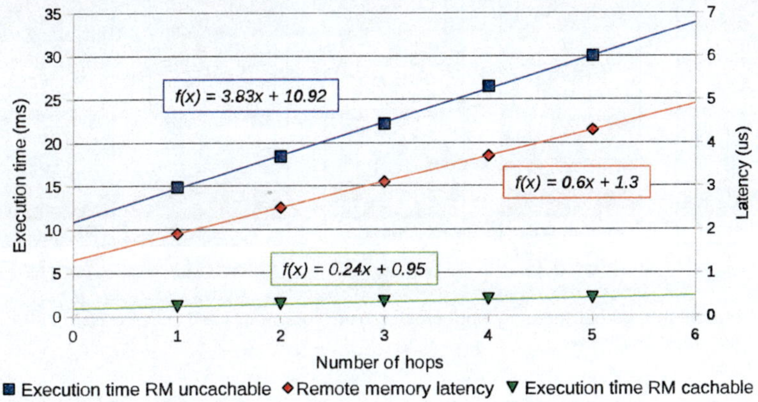

Fig. 21 Relation between remote memory latency and query execution time

similar behavior: they scale up to four concurrent query flows. This is due to a bottleneck in the system that results in congestion with such a number of flows. Thus, an improved implementation would increase latency as well as throughput. Finally, the HDD and SSD scenarios do not scale with more than one query flow because these IO devices act as severe bottlenecks.

To study the scalability of our proposal in large clusters, we have conducted an experiment that analyzes raw latency and query execution time depending on the distance to memory. Results are shown in Fig. 21. From these numbers we can extrapolate the execution time of queries for large systems with different topologies. Results of this study are shown in Fig. 22. For example, with 216 nodes in a 3D torus ($6 \times 6 \times 6$), execution time is only 27% higher than two nodes.

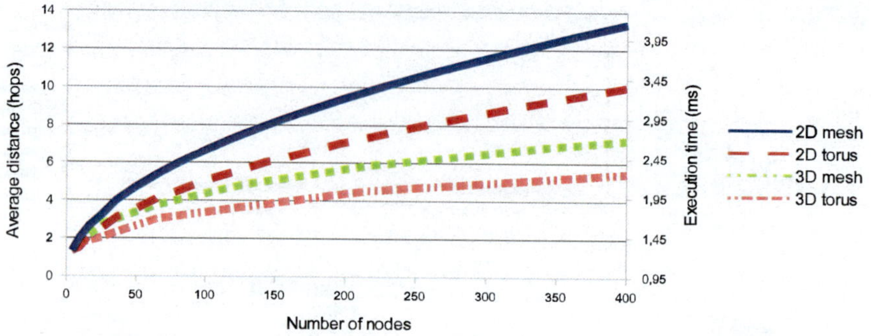

Fig. 22 Predicted query execution time for various topologies

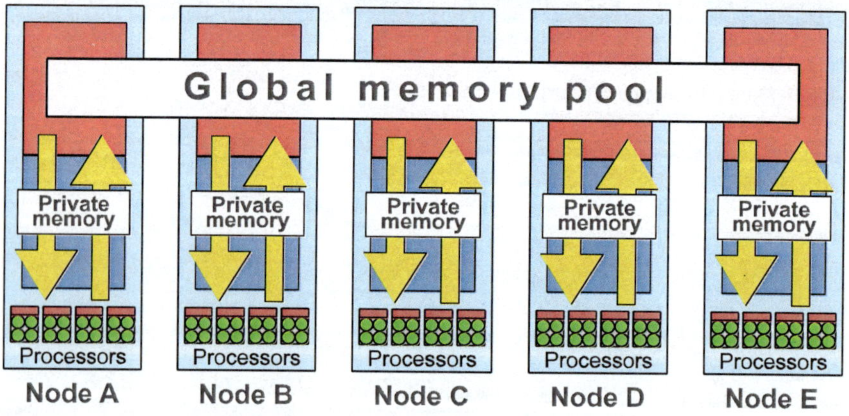

Fig. 23 Example of a global memory pool created from four nodes and accessed from all of them

5.3 *Shared-Memory Performance*

Up to this point data in the global memory pool has been accessed exclusively by one single node. In this section queries are executed in every processor in the cluster, as exemplified in Fig. 23. With this configuration we increase the number of entry points to the database and, thus, the throughput of the server. To avoid problems with the lack of coherence we follow an eventual consistency model, where write operations are buffered and later executed in a cache-disabled period, as illustrated in Fig. 24. However, in these experiments we focus on read-only queries as their completion determines the progress of the client.

Figure 25 shows the throughput of the 16-node cluster in this shared-memory configuration. We see that performance scales up to around 70 concurrent query flows. This is in line with the four flows limitation seen for a single node. In this way, we can progressively add nodes to the cluster to match any throughput and memory need in a horizontal scale fashion.

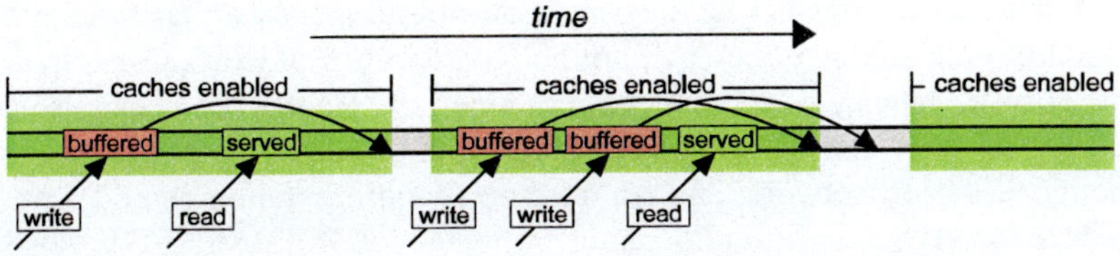

Fig. 24 Functioning of the database under eventual consistency

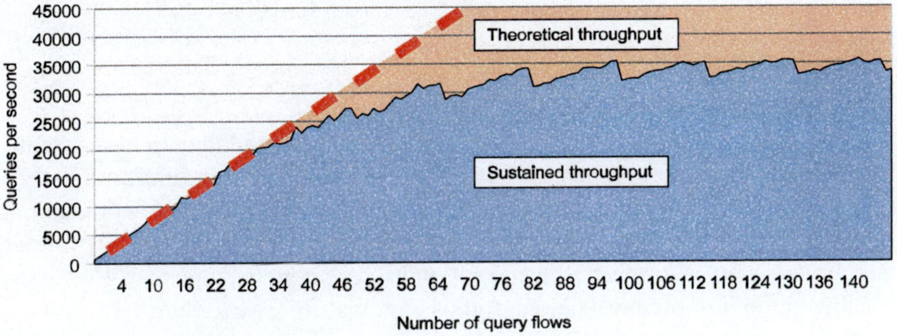

Fig. 25 Results for the cachable multi-node server configuration

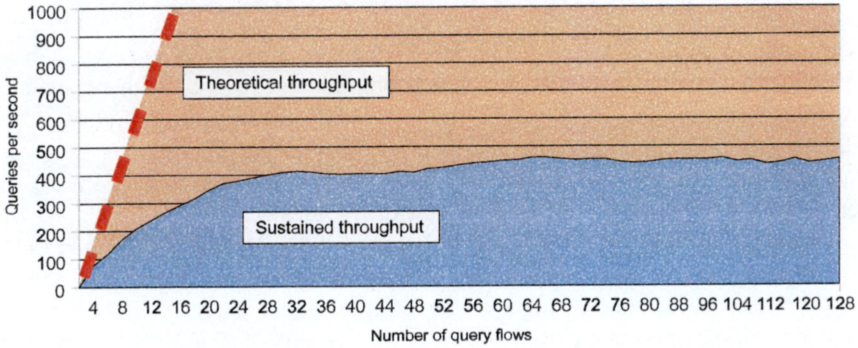

Fig. 26 Results for MySQL cluster

These numbers are compared against MySQL Cluster. MySQL Cluster is a special implementation of the MySQL server intended for clusters. In this way, one instance of MySQL Cluster is executed at each node (16 in this experiment). Data is distributed and stored in main memory, and the different instances collaborate through the cluster interconnection (Gigabit Ethernet). Figure 26 shows the results for this scenario.

The average query latency when only one thread is accessing the MySQL Cluster database is 18ms, slightly higher than SSD. The reason why this RAM-based server has such a high latency is explained in its reference manual: all the records accessed by a transaction should be held in and serviced by the same node, as explained in [22]. But when queries access disperse data in such a way that even a good distribution of data among nodes will not help on locality, MySQL Cluster suffers from the software latency of accessing remote memory. Moreover, we can see that it only scales up to 30 query flows due to congestion problems.

Although a faster network and an optimized configuration would improve performance for MySQL Cluster, it cannot compete with MEMSCALE in this kind of workload.

6 Conclusions

Today, power consumption in clusters is a serious concern and, due to the currently required overprovisioning, many resources in clusters are often unused. Furthermore, memory as a resource is very scarce and an increasing amount of applications is no longer computationally or memory-bandwidth bound. Instead, they are limited by memory capacity. Moreover, the current multi-core trend results in an effectively decreasing amount of memory per core, thus aggravating the problem.

Existing solutions are either both seriously limited in scalability and very costly (hardware distributed shared-memory systems like SGI Altix), or rely too much on locality (software distributed shared-memory systems like ScaleMP). Opposed to this, MEMSCALE allows overcoming the static partitioning found in clusters, thus providing a dynamic use of cluster-level resources. Key is a direct low-latency hardware path to remote resources, allowing to share scarce resources where needed, to avoid overprovisioning of resources and to increase the flexibility of cluster use models in general by enabling a highly dynamic resource partitioning.

We have presented the small hardware unit, which is the core of MEMSCALE, together with a new consistency model that overcomes scalability limitations by reverting to partial coherence. Using a real prototype cluster together with an FPGA-based implementation of the communication infrastructure, we are able to assess MEMSCALE's performance using a variety of applications. Here, we have focused on one of the most important workloads today: in-memory databases. By modifying MySQL, the database can now be stored distributed in the cluster's memory resources, and this aggregation of distributed resources helps to store all data in fast DRAM, without any need to swap out data to slower storage technologies. The performance impact is tremendous. Our experiments show speedups of up to 77x (35,000 queries/second for MEMSCALE vs. 450 queries/second for MySQL Cluster).

In the future, we plan to implement the MEMSCALE design on newer FPGA technologies (e.g., Virtex-6 or Virtex-7), and maybe even try to include it in an ASIC design. Furthermore, we will investigate locality optimizations in detail, in particular for the In-Memory Database. In general, transparency is helpful to keep binary compatibility with legacy software components, but exposing the existence of remote memory might also be beneficial as it facilitates locality optimizations.

References

1. S. Liang, R. Noronha, D.K. Panda, Swapping to remote memory over infiniband: an approach using a high performance network block device, in *IEEE International Conference on Cluster Computing*, 2005, pp. 1–10
2. J. Oleszkiewicz, L. Xiao, Y. Liu, Parallel network RAM: effectively utilizing global cluster memory for large data-intensive parallel programs, in *International Conference on Parallel Processing*, 2004, pp. 353–360
3. M.R. Hines, J. Wang, K. Gopalan, Distributed anemone: transparent low-latency access to remote memory in commodity clusters, in *International Conference on High Performance Computing*, 2006
4. S. Pakin, G. Johnson, Performance analysis of a user-level memory server, in *IEEE International Conference on Cluster Computing*, 2007, 249–258
5. K. Lim, J. Chang, T. Mudge, et al., Disaggregated memory for expansion and sharing in blade servers, in *36th Annual International Symposium on Computer Architecture*, 2009, 267–278
6. M. Chapman, G. Heiser, VNUMA: a virtual shared-memory multiprocessor, in *USENIX Annual Technical Conference*, 2009
7. D. Dunning, G. Regnier, G. McAlpine, et al., The virtual interface architecture. IEEE Micro **18**(2), 66–76 (1998)
8. D. Lenoski, J. Laudon, T. Joe, et al., The DASH prototype: Implementation and performance, in *19th Annual International Symposium on Computer Architecture*, 1992, 92–103
9. J. Laudon, D. Lenoski, The SGI origin: a ccNUMA highly scalable server, in *24th Annual International Symposium on Computer Architecture*, 1997, 241–251
10. J. Kuskin, D. Ofelt, M. Heinrich, et al., The Stanford FLASH multiprocessor, in *21st Annual International Symposium on Computer Architecture*, 1994, 302–313
11. K. Alnaes, E.H. Kristiansen, D.B. Gustavson, D.V. James, Scalable coherent interface, in *IEEE International Conference on Computer Systems and Software Engineering*, 1990, 446–453
12. SGI. Technical Advances in the SGI Altix UV Architecture. White Paper. http://www.sgi.com/products/servers/altix/uv. Accessed April 2012
13. S.L. Scott, Synchronization and communication in the T3E multiprocessor, in *7th International Conference on Architectural Support For Programming Languages and Operating Systems*, 1996
14. H. Fröning, A. Giese, H. Montaner, S. Silla, J. Duato, Highly scalable barriers for future high-performance computing clusters, in *18th annual IEEE International Conference on High Performance Computing*, 2011
15. H. Montaner, F. Silla, H. Fröning, J. Duato, MEMSCALE: in-cluster-memory databases, in *20th ACM Conference on Information and Knowledge Management*, 2011
16. H. Montaner, F. Silla, H. Fröning, J. Duato, Unleash your memory-constrained applications: a 32-node non-coherent distributed-memory prototype cluster, in *13th IEEE International Conference on High Performance Computing and Communications*, 2011
17. H. Montaner, F. Silla, H. Fröning, J. Duato, MEMSCALE: a scalable environment for databases, in *13th IEEE International Conference on High Performance Computing and Communications*, 2011
18. H. Montaner, F. Silla, H. Fröning, J. Duato, *A New Degree of Freedom for Memory Allocation in Clusters. Cluster Computing* (Springer, New York, 2011)
19. H. Montaner, F. Silla, H. Fröning, J. Duato, Getting rid of coherency overhead for memory-hungry applications, in *IEEE International Conference on Cluster Computing 2010*, 2010
20. H. Fröning, H. Litz, Efficient hardware support for the partitioned global address space, in *10th Workshop on Communication Architecture for Clusters (CAC2010), in conjunction with IPDPS 2010*, 2009

21. H. Montaner, F. Silla, J. Duato, A practical way to extend shared memory support beyond a motherboard at low cost, in *19th International ACM Symposium on High-Performance Parallel and Distributed Computing*, 2010
22. Oracle. MySQL Cluster Evaluation Guide - Designing, Evaluating and Benchmarking MySQL Cluster. White Paper. http://www.mysql.com/products/cluster/resources.html. Accessed April 2012

High-Performance Computing Based on High-Speed Dynamic Reconfiguration

Minoru Watanabe

Abstract Currently, demand for implementing all systems including a processor, a peripheral circuit, and a dedicated circuit onto a field programmable gate array (FPGA) is gaining. However, related to the demand, an important issue is that soft-core processors implemented on FPGAs have lower performance than custom processors or FPGA's hard-core processors. Such low performance of soft-core processors on FPGAs is attributable to their look-up table (LUT) and Switching Matrix (SM) architectures. Therefore, under current implementation, such a soft-core processor cannot be used to produce a high-performance system. Instead, a custom processor or an FPGA's hard-core processor must be implemented onto the system along with an FPGA. However, if the FPGA's programmability can be exploited fully, then the performance of soft-core processors and circuits on its programmable gate array can be increased. The key technology is a high-speed dynamic reconfiguration. Therefore, this chapter introduces a new soft-core processor architecture called Mono-Instruction Set Computer (MISC) architecture as high-performance computing based on high-speed dynamic reconfiguration.

As process technologies of Very Large Scale Integration (VLSI) advance, fine-grained field programmable gate arrays (FPGAs) are becoming more widely used in various applications [1, 2]. However, currently, almost no such application systems are purely FPGA-embedded systems in which only an FPGA is implemented, with no other VLSI. Almost all such systems are hybrid systems including an FPGA, a processor, and other custom VLSIs or Application-Specific Integrated Circuits (ASICs). Currently, demand for implementing all systems including a processor, a peripheral circuit, and a dedicated circuit onto an FPGA is gaining.

M. Watanabe (✉)
Electrical and Electronic Engineering, Shizuoka University, 3-5-1 Johoku,
Hamamatsu, Shizuoka 432-8561, Japan
e-mail: tmwatan@ipc.shizuoka.ac.jp

W. Vanderbauwhede and K. Benkrid (eds.), *High-Performance Computing Using FPGAs*, DOI 10.1007/978-1-4614-1791-0_20,

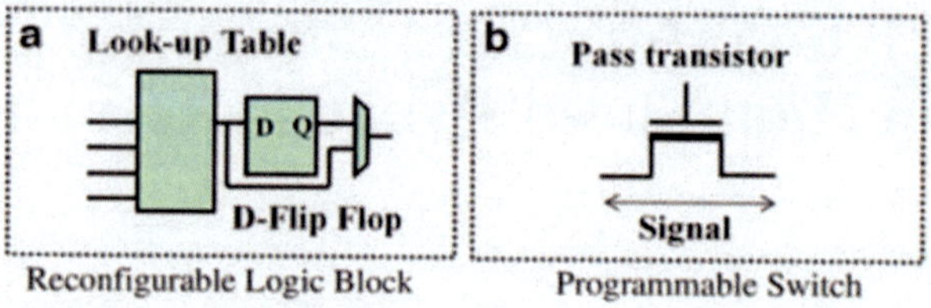

Fig. 1 A reconfigurable logic block and a programmable switch inside a programmable switching matrix

To satisfy that demand, FPGA vendors have come to provide soft-core processors for FPGAs [3–6]. Altera Corp. provides NIOS processors [3, 4], whereas Xilinx Inc. provides Micro Blaze processors [5, 6]. However, the soft-core processors have lower performance than Intel's processors [7–9], ARM processors [10, 11], and other custom processors. Such low performance of soft-core processors on FPGAs results from their look-up table (LUT) and Switching Matrix (SM) architectures, as shown in Fig. 1. Currently available FPGAs always take a fine-grained island style architecture based on LUTs and SMs. The performance of a circuit on LUTs in terms of clock frequency and power consumption is lower than that of a standard cell-based circuit on an ASIC or a custom-designed circuit on a VLSI, given the same function and the same process technology. In addition, the die-size of a circuit on LUTs always becomes larger than that on ASICs or custom VLSIs. Such a large-die circuit always increases the total length of wires so that the parasitic resistance, inductance, and capacitance of the wires are increased [12]. Therefore, such a large-die circuit causes not only a cost increase but also a clock frequency decrease and a power consumption increase. In addition, each pass transistor in a switching matrix to make interconnections between LUTs programmable causes a large propagation delay compared with a simple metal wire connection of ASICs or custom VLSIs [13]. Therefore, the programmable architecture decreases the clock frequency and increases the power consumption and die size or the chip cost of processors and other circuits. Nevertheless, to make a fabricated VLSI programmable, we must adopt the FPGA architecture.

Therefore, when a high-performance system must be produced, such a soft-core processor cannot be used. Instead, a custom processor chip must be implemented onto the system along with an FPGA. In response to those facts, recently, FPGA vendors have begun providing hard-core processors such as a PowerPC processor [14] or an ARM processor [15] inside an FPGA chip. However, in those cases, the advantage of the flexibility of FPGAs is spoiled because of the existence of the fixed hard-core processors. Ironically, vendors have provided proof that the performance of those soft-core processors on FPGAs is inferior to those of the hard-core processors and other custom processors.

Under the use of the FPGA architecture without a hard-core processor, a method remains for us to improve the performance of such soft-core processors: exploitation of their programmability. The performance of an FPGA can be increased if its programmability can be exploited fully. The idea is based on high-speed dynamic reconfiguration. The author calls the new soft-core processor architecture a

Mono-Instruction Set Computer (MISC) architecture. This chapter introduces high-performance computing or the MISC architecture based on high-speed dynamic reconfiguration.

This chapter is organized as follows: First, Sect. 1 introduces a short history of microprocessor development. Based on that history, a hidden hint is clarified: high-speed dynamic reconfiguration can increase the performance of soft-core processors or other operations on a gate array. Section 2 introduces various high-speed dynamically reconfigurable devices to be able to support MISC implementation. Among such high-speed dynamically reconfigurable devices, Sect. 3 introduces Optically Reconfigurable Gate Arrays (ORGAs) as a strong candidate for MISC implementation. Section 4 describes a high-performance implementation scheme of soft-core MISC processors by exploiting high-speed dynamic reconfiguration. A brief discussion and conclusion are presented in Sect. 5.

1 Exploration of the Best Soft-Core Processor Implementation onto FPGAs

To date, microprocessor performance has been progressing dramatically [7–11]. Recently, almost all computer systems use Reduced Instruction Set Computer (RISC) architectures [7–11],[14–17]. Such architectures offer benefits in terms of high clock frequency, low power consumption, and small implementation area or low cost. However, about 35 years ago, Complex Instruction Set Computer (CISC) architectures were widely used for almost all computer systems [18, 19]. In those days, the amount of memory was very small, for example several kilobytes, and memory was very expensive. In addition, high-level programming languages such as C++ and JAVA had not been developed. Invariably, software intended for high-speed computation was designed using an assembler language [20]. Therefore, CISC processors were designed to include various instructions. Each instruction was able to execute a complicated multi-step operation to ease the assembler-level programming and to reduce the necessary amount of memory. Consequently, CISC processor architecture became very complicated. Such complicated architectures prevented an increase in the clock frequency or processing power.

Now, the amount of memory has increased drastically because of progress in semiconductor process technologies. Moreover, memory costs have decreased drastically. In addition, high-level programming languages, such as C++ and JAVA, that can generate various multi-step operations automatically by combining a small number of instructions are available. Consequently, it is possible to choose a simple processor architecture with a small number of single-step instructions, enabling the so-called RISC architecture, which offers benefits in terms of higher clock frequency, smaller implementation area, and lower power consumption than conventional CISC architectures. Their success is based on the fundamental principle that the simplest circuit is the best. The simplest circuit can function with the highest clock frequency, in the smallest implementation area or at the lowest cost, and with

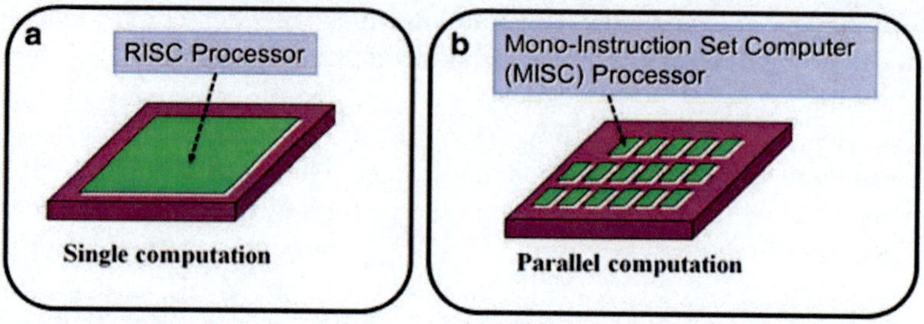

Fig. 2 Concept of a mono-instruction set computer (MISC). (**a**) Conventional static implementation (**b**) New dynamic implementation

the lowest power consumption because the simplest circuit can be constructed with fewer selector passes, less load capacitance of fewer gates, less capacitance of short metal wires, and so on.

Currently, application-specific instruction processors have been reported as one advanced RISC architecture [21–24]. Currently available RISC processors have one or a few arithmetic and logic units (ALUs) with a set of instructions. However, the set of instructions necessary for each target application varies depending on the application. Not all of a processor's instructions are always used for an application. In application-specific instruction processors, unused instructions are removed from the processor architecture, thereby simplifying the architecture. Removal of those instructions improves the clock frequency, the power consumption, and the necessary number of gates. However, although the application-specific instruction processor architecture certainly improves its performance, it is not a dramatic advancement in processor architecture innovation.

A more advanced processor architecture is the MISC architecture [25–27]. Focusing each ALU's operation for a single clock cycle, inside each ALU, only a single instruction is executed during a single clock cycle. The other instructions are never executed simultaneously. Therefore, the non-operation instruction can be regarded as a wasted instruction. Removing waste instructions and further simplifying each ALU of the RISC processor architecture, ultimately the simplest processor architecture or MISC architecture can be achieved. The only condition necessary for the MISC architecture is the use of a high-speed clock-by-clock dynamically reconfigurable device. If a programmable device can be reconfigured clock-by-clock, then any instruction change can be done by reconfiguring the programmable device. Therefore, MISC architecture can be used even for a complicated operation. Such an MISC processor can operate at the highest clock frequency, with the lowest power consumption, and in the smallest implementation area because the complexity of the architecture is the lowest among all processor architectures. Moreover, the extremely small implementation area enables large parallel computation when using the equal implementation area to that occupied by a conventional RISC processor, as shown in Fig. 2. Therefore, the overall performance can be improved drastically. As described earlier, a high-speed dynamic reconfiguration is a very important means to increase the performance of a programmable gate array. The reconfiguration advantage can also be considered as based on the fundamental principle that the simplest circuit can invariably achieve superior performance.

2 High-Speed Dynamically Reconfigurable Devices

2.1 A Variety of Programmable Devices

Currently, various programmable devices are available. For use as small-programmable logic gates, flash-based programmable logic devices (PLDs) and flash-based complex programmable logic devices (CPLDs) are available [1, 2, 28–30]. Flash-based PLDs and CPLDs have a feature enabling them to maintain a configuration context constantly, even without a power supply. However, since the PLDs and CPLDs never support dynamic reconfiguration and since their gate arrays are too small to integrate a large system, including a processor, onto them, the PLD and CPLD are unsuitable for MISC implementation. On the other hand, ACTEL's anti-fuse type FPGAs [31–33] and SRAM-based FPGAs [1, 2] are large gate arrays. However, anti-fuse type FPGAs are one-time programmable FPGAs for which dynamic reconfiguration is impossible, whereas reconfiguration of the SRAM-based FPGAs are too slow or a few hundred milliseconds because of their serial configuration. For them, dynamic reconfiguration is also impossible. Therefore, both FPGAs are unsuitable for MISC implementation.

Of course, recently, SRAM-based FPGAs can also support partial reconfiguration [34, 35]. Their partial reconfiguration time is shorter than the entire reconfiguration time. However, even if the partial reconfiguration period is of microsecond order, the period is insufficient to support a nanosecond-order dynamic reconfiguration necessary for MISC implementation. Moreover, the partial reconfiguration cannot increase the performance of an entire gate array sufficiently. To increase the performance of a programmable gate array or to support MISC implementation, the reconfiguration of an entire gate array is important.

As multi-context devices of another type, high-speed reconfigurable devices have been developed, e.g., DAP/DNA chips, DRP chips, and MuCCRA chips (http://www.ipflex.co.jp) [36, 37]. Such devices package reconfiguration memories and a microprocessor array onto a chip. The internal reconfiguration memory stores reconfiguration contexts of 4–64 banks, which can be changed from one to another during a clock cycle. Thereby, arithmetic logic units of such devices can be reconfigured on every clock cycle of a few nanoseconds. However, since the devices are coarse-grained programmable devices, their flexibility is not good compared with that of fine-grained programmable gate arrays. Therefore, the discussion of such devices is left for another book.

2.2 High-Speed Dynamically Reconfigurable Devices

Multi-context FPGAs have been developed [38–43]. The internal reconfiguration memory stores reconfiguration contexts of 4–16 banks, which can be changed from one to another during a clock cycle. Thereby, the fine-grained programmable gate array can be reconfigured on every clock cycle of a few nanoseconds. Therefore, multi-context FPGAs are the primary candidate for MISC implementation. Nevertheless, an important shortcoming is that their number of configuration contexts is insufficient to execute various reconfigurations continuously. For that reason, MISC-applicable applications and their performance are limited on such multi-context FPGA.

To date, to realize such high-speed dynamic reconfiguration, ORGAs, consisting of a holographic memory, a laser array, and an ORGA VLSI, have been developed [44–58]. A large amount of circuit information or a number of configuration contexts can be stored on a holographic memory and are addressed by a laser array. In ORGAs, configuration contexts can be optically programmed dynamically onto an optically reconfigurable fine-grained gate array. The ORGA architecture can realize high-speed reconfiguration by using an extremely large bandwidth optical bus between a holographic memory and a programmable gate array VLSI. In addition, numerous reconfiguration contexts can be realized since the storage capacity of a three-dimensional holographic memory is greater than that of silicon memories.

The first proposed Optical Programmable Gate Array (OPGA) has demonstrated 50–100 reconfiguration contexts and 16–20 μs reconfiguration time [44–46]. In another demonstration of an optically differential reconfigurable gate array (ODRGA), the reconfiguration frequency was improved to a maximum 72.7 MHz using a differential reconfiguration strategy [47–50]. Furthermore, an ORGA has achieved 144 reconfiguration contexts using microelectromechanical system (MEMS) technology [51, 52]. In addition, a practical 51K gate count Dynamic Optically Reconfigurable Gate Array (DORGA)-VLSI has been reported [53–58]. An ORGA can support high-speed dynamic reconfiguration by exploiting numerous reconfiguration contexts and a high-speed reconfiguration capability. Therefore, ORGAs are well suited for MISC implementation.

3 Optically Reconfigurable Gate Array

3.1 ORGA Architecture

An overview of an ORGA, which comprises a gate-array VLSI (ORGA-VLSI), a holographic memory, and a laser diode array, is portrayed in Fig. 3. The holographic memory stores many configuration contexts. For example, candidates of three-dimensional holographic memories are photopolymer holographic memories and

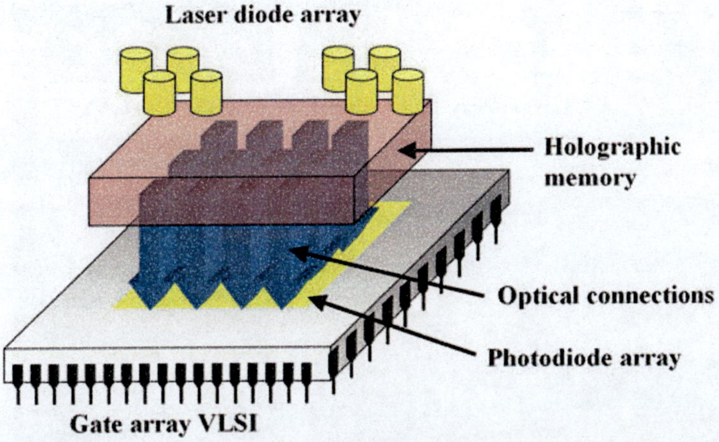

Fig. 3 Overall construction of an optically reconfigurable gate array (ORGA)

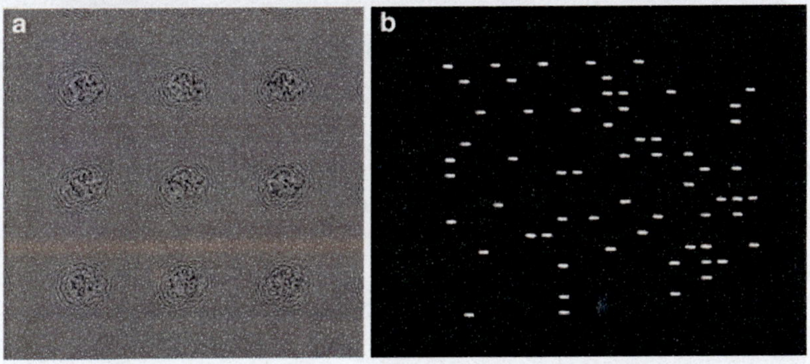

Fig. 4 Example of (**a**) a two-dimensional holographic memory pattern displayed on a spatial light modulator (SLM) and (**b**) its CCD-captured configuration context pattern

photorefractive crystal holographic memories [59–61]. Instead of them, electrically programmable spatial light modulators, liquid crystal spatial light modulators and microelectromechanical system (MEMS) mirror arrays can be used for ORGAs [62–64]. A laser array is mounted on the top of the holographic memory for use in addressing configuration contexts in the holographic memory. One laser corresponds to a configuration context. Turning one laser on, the laser beam propagates into a certain corresponding area on the holographic memory at a certain angle so that the holographic memory generates a certain diffraction pattern, as shown in Fig. 4. A photodiode-array of a programmable gate array on an ORGA-VLSI can receive it as a reconfiguration context. Then, the ORGA-VLSI functions as the circuit of the configuration context. The reconfiguration time of such an ORGA architecture reaches nanosecond-order [47, 50]. Therefore, very-high-speed context switching is possible. In addition, since the storage capacity

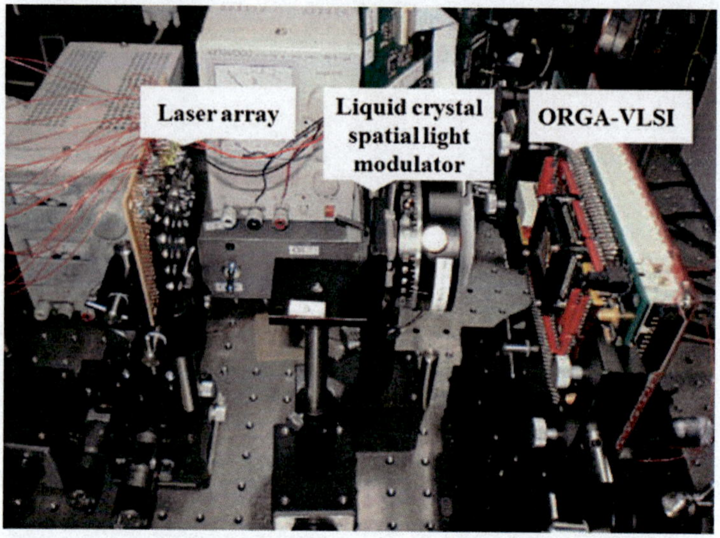

Fig. 5 Photograph of a 16-configuration context ORGA

of a holographic memory is extremely high, numerous configuration contexts can be stored in a holographic memory. The density potential of a three-dimensional holographic memory reaches V/λ^3, where V denotes the recording volume and λ is the wavelength of the recording light source [65]. For example, photorefractive crystal holographic memories $LiNb0_3$ can store 100–155 bits/μm^2 [60, 61]. The area densities are superior to even the latest 32 nm process dynamic random access memory (DRAM) cell with a bit density of 25.6 bits/μm^2 [66]. Therefore, the ORGA architecture enables both the realization of fast reconfiguration and numerous reconfiguration contexts for MISC implementation. A prototype system of a 16-configuration-context ORGA using a liquid crystal spatial light modulator as a holographic memory is shown in Fig. 5 [67].

3.2 *ORGA-VLSI*

This section presents the design of a 51K-gate-count ORGA VLSI [56]. The 51K-gate-count ORGA-VLSI chip was designed using a 0.35-μm standard complementary metal oxide semiconductor (CMOS) process. Figure 6 depicts the gate array structure. Table 1 presents its specifications. The ORGA-VLSI takes an island-style gate array or a fine-grained gate array. The basic functionality of the ORGA-VLSI is fundamentally identical to that of currently available FPGAs. However, the reconfiguration mechanism differs from that of FPGAs. Each programming point of an ORGA's programmable gate array is connected

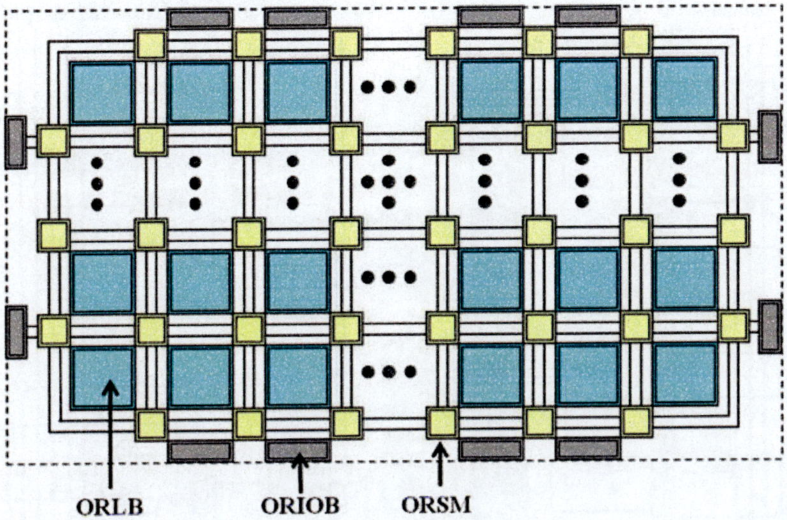

Fig. 6 An island-style gate array consists of optically reconfigurable logic blocks (ORLBs), optically reconfigurable switching matrices (ORSMs), and optically reconfigurable I/O blocks (ORIOBs) (Copyright © 2006 Japanese Journal of Applied Physics [53], Copyright © 2005, International Conference on Solid State Devices and Materials [54], INTECH [55])

Table 1 Specifications of a high-density optically reconfigurable gate array (ORGA) VLSI

Technology	0.35 μm double-poly four-metal CMOS process
Chip size	$14.2 \times 14.2\,\text{mm}^2$
Supply voltage	Core 3.3 V, I/O 3.3 V
Photodiode size	$9.5 \times 8.8\,\mu\text{m}^2$
Horizontal distance between photodiodes	28.5–42 μm
Vertical distance between photodiodes	12–21 μm
Number of photodiodes	170,165
Number of logic blocks	1,508
Number of switching matrices	1,589
Number of I/O bits	272
Gate count	51,272

Copyright © 2006 Japanese Journal of Applied Physics [53], Copyright © 2005, International Conference on Solid State Devices and Materials [54], INTECH [55]

to a photodiode to receive an optically applied configuration context. In common ORGA-VLSIs, photodiodes were constructed between an N-well or N-diffusion and the P-substrate. The photodiode cell was designed as a fully custom design. In the ORGA-VLSI, the acceptance surface size of photodiode is $8.8 \times 9.5\,\mu\text{m}^2$. The photodiode cells were arranged at 28.5–42.0 μm horizontal intervals and at 12.0–21.0 μm vertical intervals: in all, 170,165 photodiodes were used. The fourth

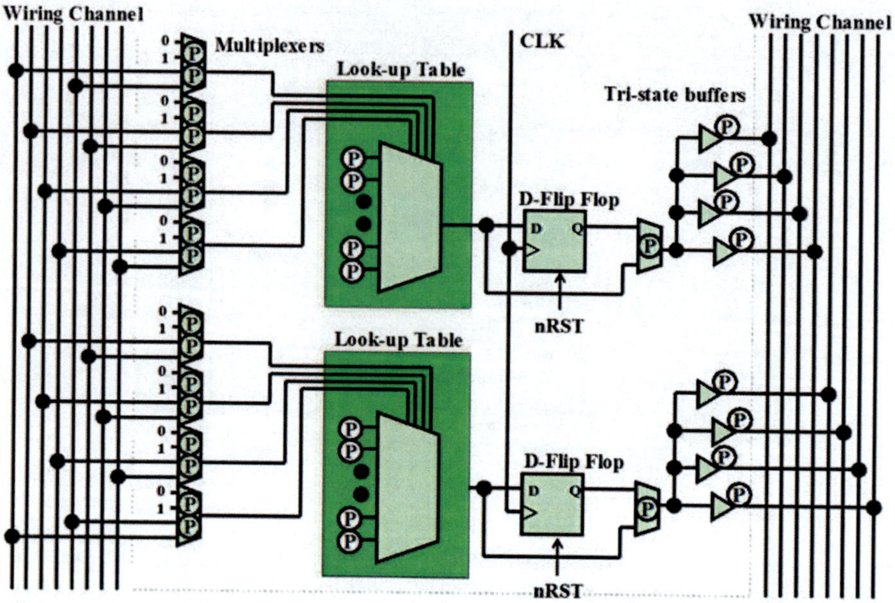

Fig. 7 Block diagram of an optically reconfigurable logic block (ORLB) (Copyright © 2006 Japanese Journal of Applied Physics [53], Copyright © 2005, International Conference on Solid State Devices and Materials [54], INTECH [55])

metal layer is used for guarding transistors from light irradiation; the other three layers were used for wiring. The gate array was designed using Design Compiler (Synopsys Inc.) as a logic synthesis tool and Apollo (Synopsys Inc.) as a place and route tool. The ORGA-VLSI chip consists of 1,508 optically reconfigurable logic blocks (ORLB), 1,589 optically reconfigurable switching matrices (ORSM), and 272 optically reconfigurable I/O bits (ORIOB). One wiring channel has eight wires.

Optically Reconfigurable Logic Block

A block diagram of an ORLB of the ORGA-VLSI chip is presented in Fig. 7. Each ORLB consists mainly of 2 four-input one-output LUTs and 2 delay-type flip-flops, similar to FPGAs. The LUTs are used for implementing Boolean functions. A combinational circuit and sequential circuit can be implemented on it, as in FPGAs. In all, 58 photodiodes are used for programming an ORLB. The ORLB can be reconfigured perfectly in parallel. The CAD layout is depicted in Fig. 8. This is a standard cell-based design.

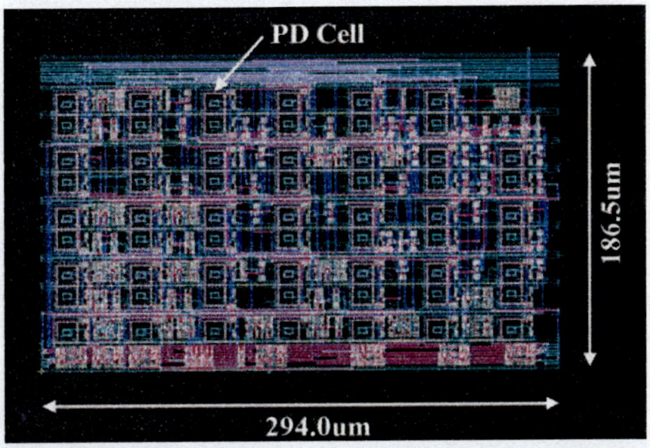

Fig. 8 CAD layout of the optically reconfigurable logic block (ORLB) (Copyright © 2006 Japanese Journal of Applied Physics [53], Copyright © 2005, International Conference on Solid State Devices and Materials [54], INTECH [55])

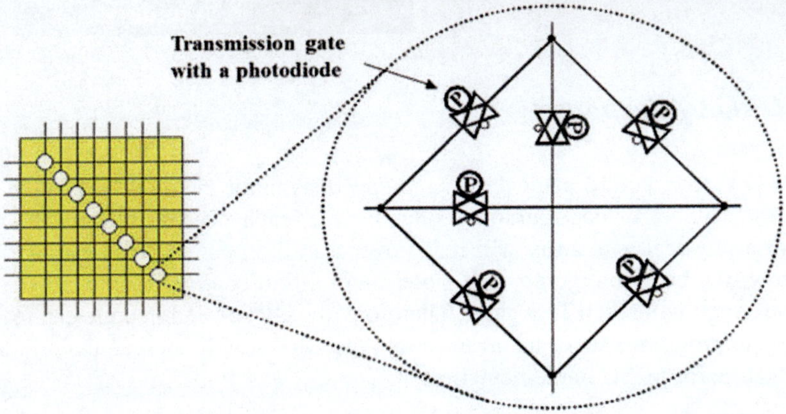

Fig. 9 Block diagram of an ORSM (Copyright © 2006 Japanese Journal of Applied Physics [53], Copyright © 2005, International Conference on Solid State Devices and Materials [54], INTECH [55])

Optically Reconfigurable Switching Matrix

Similarly, ORSMs can be reconfigured optically. A block diagram of the ORSM, portrayed in Fig. 9, shows that its basic construction is the same as that used by Xilinx Inc. Four-directional switching matrices with 48 transmission gates were implemented in the gate array. Each transmission gate can be regarded as a bi-directional switch. A photodiode connected to each transmission gate controls whether the transmission gate is closed or not. The CAD layout is portrayed in Fig. 10.

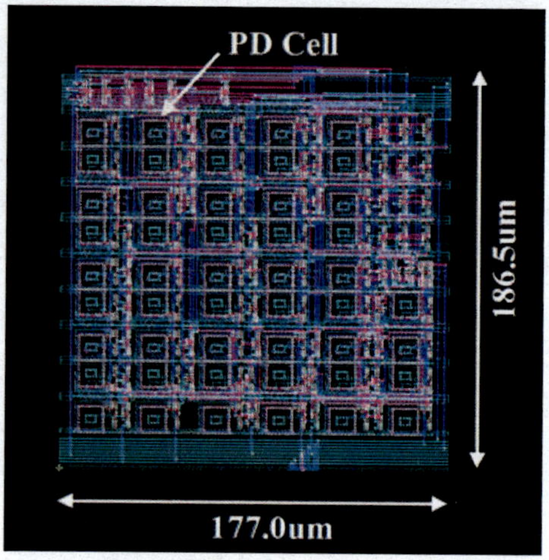

Fig. 10 CAD layout of an ORSM (Copyright © 2006 Japanese Journal of Applied Physics [53], Copyright © 2005, International Conference on Solid State Devices and Materials [54], INTECH [55])

3.3 ORGA Advantages

The ORGA architecture achieves a high-speed dynamic reconfiguration. In the future, the number of configuration contexts will reach over a million. The gate count of a physical gate array will reach over a million gates. Since the physical gate array will be reconfigured with more than million configuration contexts, the total gate count will reach Tera gates. Therefore, the ORGA architecture can realize a large gate programmable gate array so that the ORGA will become the preferred infrastructure for MISC implementation.

4 Mono-Instruction Set Computer

4.1 Concept

A MISC processor represents an instruction of an ALU in a conventional RISC processor. Although each ALU in an RISC processor has many instructions, an MISC processor includes only a single instruction. Changing of instructions is done by reconfiguring its hardware or a programmable gate array, although the hardware of an RISC processor is valid and its instruction change is done by software. To realize various instructions just like a conventional RISC processor, various MISC processors are prepared along with registers. For instance, one MISC has an adder instruction, one MISC processor has a subtractor instruction, one MISC has a multiplier instruction, one MISC processor has a divider instruction, and so on. Moreover, one multiplier-MISC processor might be capable of multiplying two 8-bit numbers, yielding a 16-bit result, while another multiplier-MISC might be

Fig. 11 Example of a large parallel operation of MISC implementation. Each MISC processor is implemented onto a programmable gate array along with data registers

capable of multiplying two 16-bit numbers, yielding a 32-bit result. Consequently, a wide variety of possible MISC processors might be used. Fundamentally, every MISC processor is designed as a single-step operation, although the operation clock frequencies of the respective MISC processors mutually differ. A large advantage of the MISC implementation is that a large parallel operation is possible because the simple architecture of each MISC can be implemented on a small implementation area. Therefore, numerous MISC processors can be implemented onto a programmable gate array. Under a practical MISC implementation, in addition to numerous MISC processors, many data registers are implemented as well as currently available RISC processor, as shown in Fig. 11. A programmable gate array on which MISC processors are implemented is reconfigured dynamically while the values of the data registers are constantly retained without depending on dynamic reconfiguration procedures.

One example of an MISC processor's sequential operation "D0 = D0 × D1 + D2 × D3" is shown in Fig. 12. The example of the MISC processor shows one of many MISC processors implemented on a programmable gate array, as shown in Fig. 11. Here, there are a multiplier MISC, an adder MISC, and four data registers: D0, D1, D2, and D3. First, a multiplier MISC is configured onto a gate array. At that configuration, two inputs of the MISC are connected to the outputs of registers D0 and D1. The output of the MISC is connected to the input of the register D0. Then, a multiplication operation on the MISC is executed in one clock cycle and the multiplication result is stored on the D0 register. Next, the gate array is reconfigured to the second multiplier MISC, leaving data on register D0. In this reconfiguration, two inputs of the MISC are connected to the outputs of registers D2 and D3 and the output of the MISC is connected to the

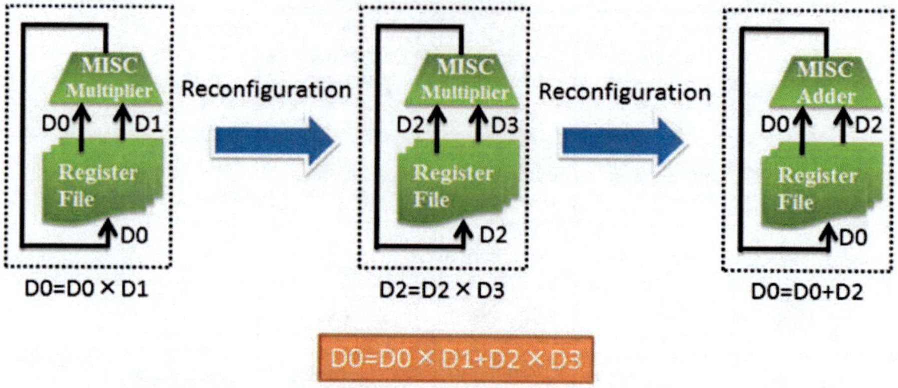

Fig. 12 Example of a sequential operation of a single MISC

input of the register D2. Then, the second multiplication operation on the MISC is executed in one clock cycle and the multiplication result is stored on the register D2. Subsequently, the gate array is reconfigured to the third adder-MISC, maintaining data on all the registers. The adder operation "D0 = D0 + D2" is executed on the MISC. Finally, the additional result is stored on the register D0 so that all operations are completed. In this manner, in the MISC implementation, a software operation is executed by reconfiguring a programmable gate array frequently. Here, in the MISC implementation, a configuration procedure and a gate array operation must be executed as perfectly parallel. When the condition is satisfied, since a configuration procedure can be treated for all purposes as a background job, the reconfiguration overhead time is beyond consideration.

As described above, under the MISC implementation, a necessary MISC is implemented dynamically at the necessary time. The instruction change is based on dynamic reconfiguration. Consequently, the performance of each MISC can be improved. Such high-performance MISC implementation merely requires a high-speed clock-by-clock reconfigurable device, just like an ORGA. Currently, the hardware infrastructure based on ORGAs is mature.

4.2 *Experimental Result of MISC Implementation*

Some results of MISC implementation have been reported to date [25–27]. Here, one example is introduced for discussion of the performance. As described in the paper [27], since ORGA logic synthesis tools and place and route tools are in a development phase [68], a Cyclone II 2C70 FPGA on a DE2-70 Board (Altera Corp.) was used to estimate the MISC implementation. Of course, although the Cyclone II FPGA does not support dynamic reconfiguration of its programmable gate array, it can be considered that the experimental result is obtained using a future

Table 2 FPGA implementation results of 11 kinds of Mono-Instruction Set Computers (MISCs) [27]

Processor type (MISC/RISC)	Logic elements	Clock frequency [MHz]	Total performance ratio (MISC/RISC)
32-bit Adder MISC	99	64.65	203.2
32-bit Subtractor MISC	99	66.05	207.6
32-bit Multiplier MISC	527	46.93	27.7
32-bit Divider MISC	1,145	8.19	2.2
32-bit AND MISC	64	420.17	2,042.4
32-bit OR MISC	64	420.17	2,042.4
32-bit EXOR MISC	64	420.17	2,042.4
32-bit Inverter MISC	64	420.17	2,042.4
Barrel shifter MISC (left, zero)	248	200.52	251.5
Barrel shifter MISC (right, sign)	245	216.08	274.3
Barrel shifter MISC (right, zero)	246	181.69	229.8
32-bit Conventional RISC soft-core processor	2,523	8.11	1

The last line shows a conventional RISC soft-core processor including all instructions of the above 11 kinds of MISC processors, which is a comparison target under the same condition. Both the MISC and RISC processors were implemented onto the same Cyclone II 2C70 FPGA

fine-grained MCFPGA or an ORGA that can support dynamic reconfiguration. The Cyclone II 2C70 FPGA includes 68,416 logic elements, 250 M4K Block RAMs (BRAMs), 150 embedded multipliers, 4 PLLs, and 622 user I/O pins. In this experiment, embedded multipliers and BRAMs were not used. Only logic elements were used for implementing MISC processors. Each logic element consists of a 4-input LUT and a delay type flip-flop. The implemented results of 11 kinds of MISC processors are presented in Table 2. The MISC processors were designed using a logic synthesis and place and route tool (Quartus II Web Edition, ver. 9.0). The maximum clock frequency and the resource usage of each MISC processor shown in Table 2 were reported by the Quartus II Web Edition software.

The first line in Table 2 shows a 32-bit adder MISC processor with a single 32-bit adder function. The 32-bit adder MISC was implemented on 99 logic elements. At that time, the maximum clock frequency was 64.65 MHz. The second line shows a 32-bit subtractor MISC processor with a single 32-bit subtractor function. The 32-bit subtractor MISC has consumed 99 logic elements. The maximum clock frequency was reported as 66.05 MHz. The third and fourth lines in Table 2, respectively, show a 32-bit multiplier MISC and a 32-bit divider MISC. The 32-bit multiplier MISC consumed 527 logic elements while the 32-bit divider MISC used 1,145 logic elements. At that time, the timing report of the Quartus II tool presented that the maximum clock frequencies of the 32-bit multiplier MISC and the 32-bit divider MISC are, respectively, 46.93 and 8.19 MHz. The fifth to eighth lines show 32-bit logical function MISC processors with an AND operation, an OR operation, an EXOR operation, and an inverter operation, respectively. All the 32-bit logical function MISC processors consume 64 logic elements. In addition, the maximum clock frequencies of the 32-bit logical function MISC processors are the same 420.17 MHz. In addition, three 32-bit barrel shifter MISC processors were estimated as shown in the ninth to the eleventh lines of Table 2. The 32-bit barrel shifter MISC processors used 245–248 logic elements. The maximum clock frequencies were 181.69–216.08 MHz. These are sample designs of MISC implementation.

Here, to compare the performance of the MISCs with that of conventional RISC processors, a 32-bit original soft-core RISC processor including all the 11 instructions of the above 11 kinds of MISC processors were designed. The 32-bit soft-core RISC processor has a single ALU that includes all MISC functions: a 32-bit adder function, a 32-bit multiplier, a 32-bit subtractor, a 32-bit divider, four kinds of 32-bit logical functions, and three kinds of 32-bit barrel shifters. The 32-bit soft-core RISC processor was implemented on the same Cyclone II 2C70 FPGA, as shown in the last line of Table 2, as well as MISC processors. The logic synthesis, place, and route were also executed using the same Quartus II Web Edition Software described above. In this case, the maximum clock frequency of the 32-bit soft-core RISC processor was reported as 8.11 MHz. In addition, the resource usage was 2,523 logic elements.

Comparing the a 32-bit adder MISC processor with the 32-bit original soft-core RISC processor, the clock frequency of the 32-bit adder MISC processor is 7.97 times faster than that of the soft-core RISC processor. Moreover, since the implementation area of the 32-bit adder MISC processor is 99 logic elements, if the same number of logic elements as 2,523 logic elements of the soft-core RISC processor are used, then 25.5 MISC processors can be implemented onto the area and can be executed in parallel so that 25.5 times better performance can be achieved. Consequently, since the performance of the MISC processor can be estimated as the product of the clock frequency ratio of the MISC processor to the soft-core RISC processor and the number of implementable modules, the total performance can be estimated as 203.2 times better than that of the conventional RISC architecture of the soft-core RISC processor. In the case of a 32-bit subtractor MISC processor, the clock frequency is 8.14 times higher than that of the soft-core RISC processor. Moreover, given the same number of logic elements as the soft-core RISC processor, 25.5 32-bit subtractor MISC processors can be implemented onto the area and can be executed in parallel so that 25.5 times better performance can be achieved. Finally, the total performance can be estimated as 207.6 times better than that of the soft-core RISC processor with a conventional RISC architecture. For a 32-bit multiplier MISC processor, the clock frequency is 5.79 times higher than that of the soft-core RISC processor; 4.8 multiplier MISC processors can be implemented in the same area as that of the soft-core RISC processor. Finally, the total performance can be estimated as 27.7 times better than that of a conventional soft-core RISC processor. Also, for the 32-bit divider MISC processor, although the clock frequency is similar to that of the RISC processor, 2.2 divider MISC processors can be implemented onto the same area as the soft-core RISC processor. Thereby, 2.2 times higher performance can be achieved. For logical-function MISC processors of an AND operation, an OR operation, an EXOR operation, and an inverter operation, the MISC performance increases are extremely high. The clock frequency is about 51.8 times higher than that of the soft-core RISC processor. Moreover, under the same conditions as those for the soft-core RISC processor, 39.4 logical-function MISC processors can be implemented and can be executed in parallel, thereby achieving 39.4 times higher performance. Finally, total performance can be 2,042.4 times better than that of the conventional soft-core RISC processor. In addition, the 32-bit barrel shifter MISC processors used 245248 logic elements. The maximum clock frequencies were 181.69–216.08 MHz, which is about 22.4–26.6 times faster than that of the soft-core RISC processor. Under the same conditions as those of the soft-core RISC processor, 10.2–10.3 barrel shifter MISC processors can be implemented onto the same area and can be executed in parallel so that 10.2–10.3 times higher performance can be achieved. Finally, its total performance can be estimated as 229.8–274.3 times better than that of a conventional soft-core RISC processor.

As described above, the MISC performance is 2–2,000 times better than that of a conventional soft-core RISC processor which takes static implementation. Of course, since the Cyclone II 2C70 FPGA used in the experiment cannot be reconfigured dynamically, long idle times occur between MISC executions. For that reason, the MISC implementation has been heretofore regarded as an impractical idea. However, high-speed dynamically reconfigurable devices are available now. The high-speed dynamically reconfigurable devices resolve the long idle time. An ORGA is a major candidate device. ORGAs have been developed to support a non-overhead clock-by-clock nanosecond-order reconfiguration. The infrastructure for MISC implementation is being put into place.

4.3 More Practical Implementation of MISC Processors

This section presents a discussion of a more practical implementation of MISC processors. The same FPGA implementation examples of 11 kinds of MISCs as those presented in Table 2 are shown in Table 3. All the conditions are identical to those for Table 2. However, each MISC implementation area is uniformly partitioned to a tile of 66 logic elements, as shown in Fig. 13a. For example, a 32-bit adder MISC can be implemented onto two tiles of 132 logic elements, a 32-bit AND MISC can be implemented onto a single tile of 66 logic elements, and a 32-bit Multiplier MISC consumes eight tiles of 528 logic elements. Each partition has a set of registers in addition to the 66 logic elements to store calculation results. In the case of Fig. 13a, 10 logical function MISC processors with a 32-bit AND operation, a 32-bit OR operation, a 32-bit EXOR operation, or a 32-bit inverter operation are implemented so that a parallel computation can be executed based on the 10 logical function MISC processors. When 32-bit Adder MISC processors are necessary, two tiles are used for implementing each 32-bit Adder MISC processor. An example is shown in Fig. 13b. In this case, four adder MISC processors, an AND MISC

Table 3 The same FPGA implementation examples of 11 kinds of MISCs as those presented in Table 2

Processor type (MISC/RISC)	Logic elements	Clock frequency [MHz]	Total performance ratio (MISC/RISC)
32-bit Adder MISC	132(99)	57.1(1/7)	134.6
32-bit Subtracter MISC	132(99)	57.1(1/7)	134.6
32-bit Multiplier MISC	528(527)	44.4(1/9)	26.2
32-bit Divider MISC	1,188(1,145)	8(1/50)	2.1
32-bit AND MISC	66(64)	400(1)	1,885.4
32-bit OR MISC	66(64)	400(1)	1,885.4
32-bit EXOR MISC	66(64)	400(1)	1,885.4
32-bit Inverter MISC	66(64)	400(1)	1,885.4
Barrel shifter MISC (left, zero)	264(248)	200(1/2)	235.7
Barrel shifter MISC (right, sign)	264(245)	200(1/2)	235.7
Barrel shifter MISC (right, zero)	264(246)	133.3(1/3)	235.7
32-bit Conventional RISC soft-core processor	2,523	8.11	1

All conditions are the same as those in Table 2. However, each MISC implementation area is partitioned into 66 logic elements. Each partition has a set of registers in addition to the 66 logic elements. A base clock, defined as 400 MHz, is divided into various suitable clock frequencies which are distributed for MISC processors. This is a more practical implementation example

processor, and an OR MISC processor are implemented. Moreover, when a 32-bit multiplier MISC processor must be implemented, the multiplier MISC processor is implemented onto 8 tiles of 528 logic elements, as shown in Fig. 13c. The remaining two tiles, for example, can include an EXOR MISC processor and an OR MISC processor.

Since the operation clock frequencies of the respective MISC processors mutually differ, as shown in Table 2, the treatment of the different clock distribution must be discussed. Here, it is assumed that the base clock frequency is defined as 400 MHz. Perhaps the base clock is distributed to all registers and/or flip-flops and the latching timing of the registers and flip-flops is controlled as the multiple-clock-period. For example, the latching timing of the 32-bit adder MISC arrives after 7 clocks under the 400-MHz base clock, whereas the latching timing of the 32-bit multiplier MISC arrives after 9 clocks. Under the practical implementation, because multiple clock period must be used, the net performance of MISC implementation is decreased slightly, as shown in Table 3. However, even under a practical situation, the MISC implementation performance is sufficiently higher than that of the conventional RISC processor. The average performance of the above practical MISC implementation can be estimated as 800 times faster than that of the 32-bit soft-core conventional RISC processor if all the MISC processors described above are used uniformly. Consequently, high-speed dynamic reconfiguration can open a new computation paradigm.

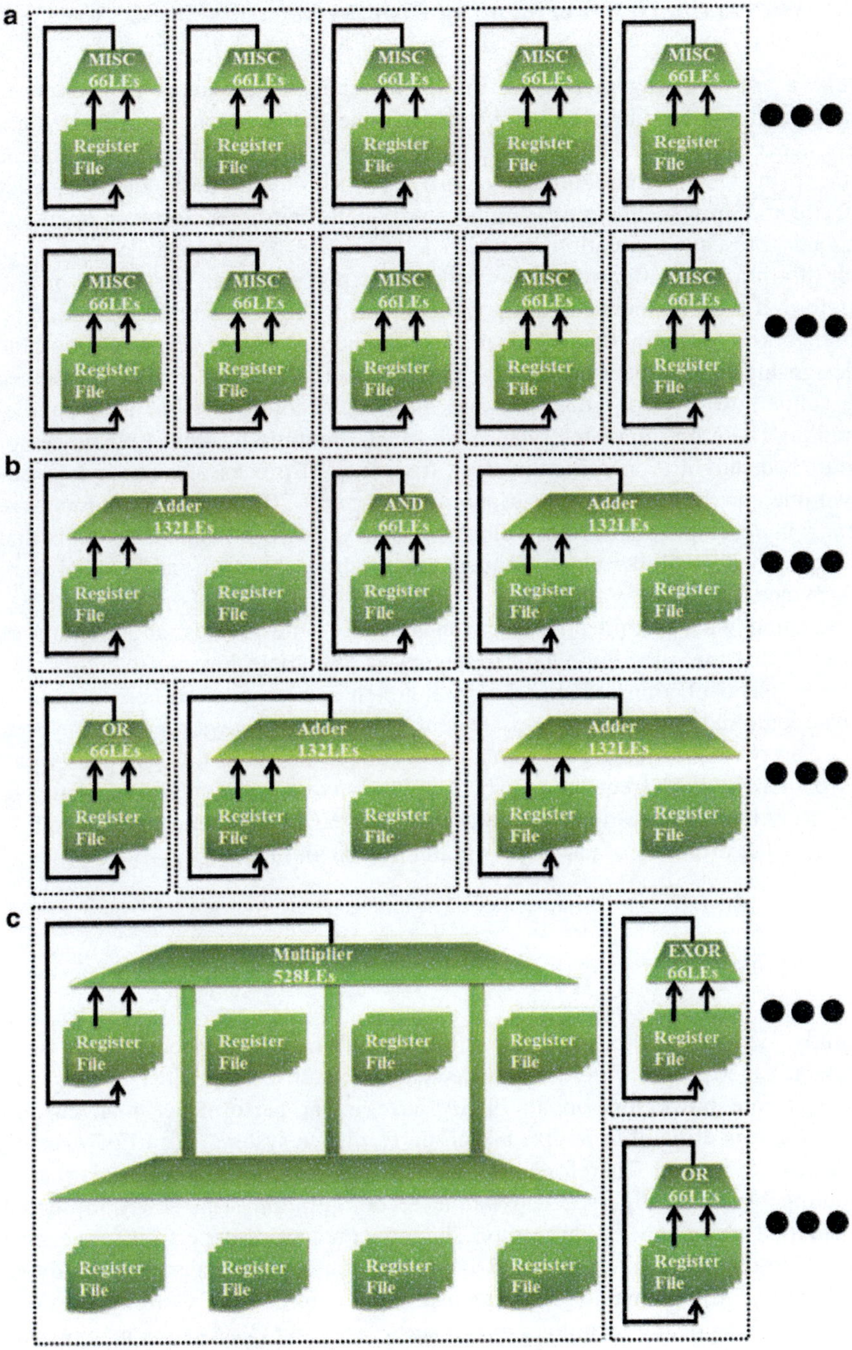

Fig. 13 Three examples of MISC implementation

4.4 What is the Best-Performing Processor?

Here, we are assuming a situation in which our team has many inexperienced programmers with low-level skills. We also assume that the team must design high-performance software. In such a case, what type of processor do you recommend for the team? One might recommend a processor with the highest clock frequency for the team. Software always has numerous branch conditions. Software operation frequently presents a situation in which a successive operation must wait for a result of a prior operation to judge whether a branch condition is satisfied or not. Therefore, if programmers can design software as a single sequential operation on a single-core processor, then even an inexperienced programmer would be able to design high-performance software. Invariably, it is difficult for inexperienced programmers to produce good software on a multi-core processor as a parallel operation. Therefore, until Intel canceled 4-GHz Pentium 4, almost all processor vendors had sought to increase the clock frequency of processors. The most well-known one among them was the Alpha processor [69, 70]. Now, all vendors have given up increasing the processor's clock frequency. However, there is no doubt that a processor with a higher clock frequency is the best choice to increase software performance if a choice is allowed.

Compared with custom processors on an ASIC, the greatest shortcoming of FPGAs is that the operation clock frequency is extremely low. To overcome the operation speed of a custom processor, although a large parallel computation is always required, in this case, a transcendent technique is always needed. However, if a gate array can be reconfigured dynamically and if the MISC implementation can be used, then the clock frequency can be increased so that even complicated software using many branch operations can be accelerated easily. Dynamic reconfiguration can reduce the difficulty of parallel programming on an FPGA.

5 Conclusion

Currently, demand is increasing for implementation of all systems including a processor, a peripheral circuit, and a dedicated circuit onto an FPGA. However, since soft-core processors on an FPGA have lower performance than custom processors, it is difficult to realize a high-performance system on an FPGA using a soft-core processor. Therefore, this chapter has introduced high-performance computing based on high-speed dynamic reconfiguration. The newly presented architecture is called a MISC processor. The soft-core processor performance can be improved dramatically if a high-speed dynamic reconfiguration is exploited fully on a fine-grained programmable gate array just like that of FPGAs. Currently, various high-speed dynamically reconfigurable devices including ORGAs are mature. The author expects that the MISC implementation will be useful for various applications in the future.

References

1. Altera Corporation, Altera Devices, http://www.altera.com
2. Xilinx Inc., Xilinx Product Data Sheets, http://www.xilinx.com
3. J. Perez Acle, M.S. Reorda, M. Violante, Implementing a safe embedded computing system in SRAM-based FPGAs using IP cores: a case study based on the Altera NIOS-II soft processor, in *IEEE Second Latin American Symposium on Circuits and Systems*, 2011, pp. 1–5
4. O.A. Al Rayahi, M.A.S. Khalid, UWindsor Nios II: a soft-core processor for design space exploration, in *IEEE International Conference on Electro/Information Technology*, 2009, pp. 451–457
5. R.H. Klenke, Experiences Using the Xilinx Microblaze softcore processor and uCLinux in computer engineering capstone senior design projects, in *IEEE International Conference on Microelectronic Systems Education*, 2007, pp. 123–124
6. A. Klimm, O. Sander, J. Becker, A MicroBlaze specific co-processor for real-time hyperelliptic curve cryptography on Xilinx FPGAs, in *IEEE International Symposium on Parallel & Distributed Processing*, 2009, pp. 1–8
7. D. Deleganes, J. Douglas, B. Kommandur, M. Patyra, Designing a 3 GHz, 130 nm, IntelR PentiumR 4 processor, in *Symposium on VLSI Circuits Digest of Technical Papers*, 2002, pp. 130–133
8. R. Ali, R. Radhakrishnan, G. Kochhar, J. Hsieh, O. Celebioglu, K. Chadalavada, R. Rajagopalan, Evaluating performance of BLAST on Intel Xeon and Itanium2 processors, in *IEEE International Workshop on Workload Characterization*, 2004, pp. 81–88
9. T.G. Mattson, R. Van der Wijngaart, M. Frumkin, Programming the Intel 80-core network-on-a-chip Terascale processor, in *International Conference for High Performance Computing, Networking, Storage and Analysis*, 2008, pp. 1–11
10. G. Yeung, P. Hoxey, P. Prabhat, Y.K. Chong, D. O'Driscoll, C. Hawkins, Low power memory implementation for a GHz+ Dual Core ARM Cortex A9 processor on a high-K metal gate 32nm low power process, in *International Symposium on VLSI Design, Automation and Test*, 2011, pp. 1–4
11. T. Neagoe, E. Karjala, L. Banica, Why ARM processors are the best choice for embedded low-power applications? in *IEEE 16th International Symposium for Design and Technology in Electronic Packaging (SIITME)*, 2010, pp. 253–258
12. D. Zhou, F.P. Preparata, S.M. Kang, Interconnection delay in very high-speed VLSI. IEEE Trans. Circ. Syst. **38**(7), 779–790 (1991)
13. Y. Wang, L. Wang, H. Ruonan, T. Jiarong, A delay model for SRAM-Based FPGA interconnections, in *IEEE International Midwest Symposium on Circuits and Systems*, vol. 2, 2006, pp. 79–83
14. G.R. Allen, G.M. Swift, G. Miller, Upset characterization and test methodology of the PowerPC405 hard-core processor embedded in Xilinx field programmable gate arrays, in *IEEE Radiation Effects Data Workshop*, vol. 0, 2007, pp. 167–171
15. Altera Corporation, Cyclone V FPGAs, http://www.altera.com/devices/fpga/cyclone-v-fpgas/cyv-index.jsp
16. Y. Pizhou, L. Chaodong, A RISC CPU IP core, in *International Conference on Anti-counterfeiting, Security and Identification*, 2008, pp. 356–359
17. J. Goodacre, A.N. Sloss, Parallelism and the ARM instruction set architecture. Computer **38**(7) 42–50 (2005)
18. T. Jamil, RISC versus CISC. IEEE Potentials **14**(3), 13–16 (1995)
19. D.B. Tolley, Analysis of CISC versus RISC microprocessors for FDDI network interfaces, in *Conference on Local Computer Networks*, 1991, pp. 485–493
20. M.F. Smith, B.E. Luff, Automatic assembler source translation from the Z80 to the MC6809. IEEE Micro **4**(2), 3–9 (1984)
21. M. Breternitz, J.P. Shen, Architecture synthesis of high-performance application-specific processors, in *ACM/IEEE Design Automation Conference*, 1990, pp. 542–548

22. A. Sengupta, R. Sedaghat, Z. Zhipeng, Hardware efficient design of speed optimized power stringent application specific processor, in *International Conference on Microelectronics (ICM)*, 2009, pp. 173–176
23. D. Sheldon, R. Kumar, R. Lysecky, F. Vahid, D. Tullsen, Application-specific customization of parameterized fpga soft-core processors, in *IEEE/ACM International Conference on Computer-Aided Design*, 2006, pp. 261–268
24. T. Good, M. Benaissa, Very small FPGA application-specific instruction processor for AES. IEEE Trans. Circ. Syst. I **53**(7), 1477–1486 (2006)
25. M. Watanabe, F. Kobayashi, Optically reconfigurable gate arrays vs. ASICs, in *IEEE Asia Pacific Conference on Circuits and Systems*, 2006, pp. 1166–1169
26. F. Kobayashi, Y. Morikawa, M. Watanabe, MISC: mono instruction-set computer based on dynamic reconfiguration – A 6502 Perspective, in *International Conference on Engineering of Reconfigurable Systems and Algorithms*, 2008, pp. 222–228
27. Y. Nihira, M, Watanabe, Mono-instruction computer on a dynamically reconfigurable gate array, in *Workshop on Synthesis and System Integration of Mixed Information Technologies*, 2012, pp. 66–70
28. T.P. Chueng, Z.M. Yusoff, A.Z. Sha'ameri, Implementation of pipelined data encryption standard (DES) using Altera CPLD, in *TENCON*, vol. 3, 2000, pp. 17–21
29. P. Sniatala, J. Pierzchlewski, A. Handkiewicz, B. Nowakowski, CPLD based development board for mixed signal chip testing, in *International Conference on Mixed Design of Integrated Circuits and Systems*, 2007, pp. 492–495
30. Lattice Semiconductor Corporation, CPLD Devices, http://www.latticesemi.com/
31. Actel Corp., RTAS FPGA, http://www.actel.com/
32. Actel Corp., Aerospace and RadTolerant FPGAs, http://www.actel.com/techdocs/ds/milaero.aspx
33. R.J. Nejad, P.A. Rickey, K. Konadu, W.J. Stapor, P.T. McDonald, W. Heidergott, Radiation Characterization of a Hardened 0.22 μm anti-fuse field programmable gate array. IEEE Trans. Nucl. Sci. Part 1 **53**(6), 3525–3531 (2006)
34. J. Becker, M. Hubner, G. Hettich, R. Constapel, J. Eisenmann, J. Luka, Dynamic and partial FPGA exploitation. Proc. IEEE **95**(2), 438–452 (2007)
35. E.J. Mcdonald, Runtime FPGA partial reconfiguration. IEEE Aero. Electron. Syst. Mag. **23**(7), 10–15 (2008)
36. H. Nakano, T. Shindo, T. Kazami, M. Motomura, Development of dynamically reconfigurable processor LSI. NEC Tech. J. (Jpn.) **56**(4), 99–102 (2003)
37. H. Amano, Y. Hasegawa, S. Tsutsumi, T. Nakamura, T. Nishimura, V. Tanbunheng, A. Parimala, T. Sano, M. Kato, in *IEEE Asian Solid-State Circuits Conference*, 2007, pp. 384–387
38. N. Miyamoto, T. Ohmi, A 1.6mm^2 4,096 logic elements multi-context FPGA core in 90nm CMOS, in *IEEE Asian Solid-State Circuits Conference*, 2008, pp. 89–92
39. M. Hariyama, S. Ogata, M. Kameyama, Y. Morita, Design of a multi-context FPGA using a floating-gate-MOS functional pass-gate, in *Asian Solid-State Circuits Conference*, 2005, pp. 421–424
40. A. Dehon, Dynamically programmable gate arrays: a step toward increased computational density, in *Fourth Canadian Workshop on Field Programmable Devices*, 1996, pp. 47–54
41. S.M. Scalera, J.R.Vazquez, The design and implementation of a context switching FPGA, in *IEEE Symposium on FPGAs for Custom Computing Machines*, 1998, pp. 78–85
42. S. Trimberger et al., A time–multiplexed FPGA, in *FCCM*, 1997, pp. 22–28
43. D. Jones, D.M. Lewis, A time–multiplexed FPGA architecture for logic emulation, in *Custom Integrated Circuits Conference*, 1995, pp. 495–498
44. J. Mumbru, G. Panotopoulos, D. Psaltis, X. An, F. Mok, S. Ay, S. Barna, E. Fossum, Optically programmable gate array, in *SPIE of Optics in Computing 2000*, vol. 4089, 2000, pp. 763–771
45. J. Mumbru, G. Zhou, X. An, W. Liu, G. Panotopoulos, F. Mok, D. Psaltis, Optical memory for computing and information processing, in *SPIE on Algorithms, Devices, and Systems for Optical Information Processing III*, vol. 3804, 1999, pp. 14–24

46. J. Mumbru, G. Zhou, S. Ay, X. An, G. Panotopoulos, F. Mok, D. Psaltis, Optically reconfigurable processors, in *SPIE Critical Review 1999 Euro-American Workshop on Optoelectronic Information Processing*, vol. 74, 1999, pp. 265–288
47. M. Nakajima, M. Watanabe, A four-context optically differential reconfigurable gate array. J. Lightwave Tech. **27**(20), 4460–4470 (2009)
48. M. Miyano, M. Watanabe, F. Kobayashi, Optically differential reconfigurable gate array. Electron. Comput. Jpn. Part II, **90**(11), 132–139 (2007)
49. M. Watanabe, T. Shiki, F. Kobayashi, Scaling prospect of optically differential reconfigurable gate array VLSIs. Analog Integr. Circ. Signal Process. **60**(1–2), 137–143 (2009)
50. M. Nakajima, M. Watanabe, A 13.75 ns holographic reconfiguration of an optically differential reconfigurable gate array, in *International Conference on Intelligent Information Hiding and Multimedia Signal Processing*, 2009, pp. 852–855
51. Y. Yamaji, M. Watanabe, A 144-configuration context MEMS optically reconfigurable gate array, in *IEEE International SOC Conference, CD-ROM*, 2011
52. H. Morita, M. Watanabe, Microelectromechanical configuration of an optically reconfigurable gate array. IEEE J. Quant. Electron. **46**(9), 1288–1294 (2010)
53. M. Watanabe, F. Kobayashi, Dynamic optically reconfigurable gate array. Jpn. J. Appl. Phys. **45**(4B), 3510–3515 (2006)
54. M. Watanabe, F. Kobayashi, A 51,272-gate-count dynamic optically reconfigurable gate array in a standard 0.35um CMOS technology, in *International Conference on Solid State Devices and Materials*, 2005, pp. 336–337
55. M. Watanabe, A dynamically reconfigurable device, in *Advances in Solid State Circuit Technologies*, Chap. 3, ed. by P.K Chu (INTECH, Rijeka, 2010) [ISBN:978-953-307-086-5]
56. M. Nakajima, M. Watanabe, Fast optical reconfiguration of a nine-context DORGA using a speed adjustment control. ACM Trans. Reconfigurable Technol. Syst. **4**(2) (2011)
57. D. Seto, M. Watanabe, A dynamic optically reconfigurable gate array – Perfect emulation. IEEE J. Quant. Electron. **44**(5), 493–500 (2008)
58. D. Seto, M. Nakajima, M. Watanabe, Dynamic optically reconfigurable gate array very large-scale integration with partial reconfiguration capability. Appl. Optic. **49**(36), 6986–6994 (2010)
59. M. Toishi, T. Tanaka, K. Watanabe, K. Betsuyaku, Analysis of photopolymer media of holographic data storage using non-local polymerization driven diffusion model. Jpn. J. Appl. Phys. **46**(6A), 3438–3447 (2007)
60. A. Pu, D. Psaltis, Holographic data storage with 100 bits/μm^2 density, in *Optical Data Storage Topical Meeting Conference Digest*, 1997, pp. 48–49
61. G.W. Burr et al., Volume holographic data storage at an areal density of 100 Gbit/in^2, in *Conference on Lasers and Electro-Optics*, 2000, pp. 188–189
62. N. Yamaguchi, M. Watanabe, Liquid crystal holographic configurations for ORGAs. Appl. Optic. **47**(28), 4692–4700 (2008)
63. Texas Instruments, DLP, http://www.ti.com/
64. Texas Instruments, Discovery 4000, http://www.ti.com/
65. P.J. van Heerden, Theory of optical information storage in solids. Appl. Optic. **2**(4), 393–400 (1963)
66. N. Butt et al., A 0.039 μm^2 high performance eDRAM cell based on 32nm high-K/Metal SOI technology, in *IEEE International Electron Devices Meeting*, 2010, pp. 27.5.1–27.5.4
67. T. Watanabe, M. Watanabe, A 16-laser array for an optically reconfigurable gate array, in *IEEE International Conference on Space Optical Systems and Applications*, 2011, pp. 255–260
68. M. Watanabe, F. Kobayashi, A logic synthesis and place and route environment for ORGAs, in *International Conference on Engineering of Reconfigurable Systems and Algorithms*, 2006, pp. 237–238
69. D.K. Bhavsar, R.A. Davies, Scan Islands – A scan partitioning architecture and its implementation on the Alpha 21364 processor, in *IEEE VLSI Test Symposium*, 2002, pp. 16–21
70. E. McLellan, The Alpha AXP architecture and 21064 processor. IEEE Micro **13**(3), 36–47 (1993)

Part III
Tools and Methodologies

The third and final part of the book presents tools and methodologies in High-Performance Reconfigurable Computing, a crucial part of making FPGAs an efficient economic solution for HPC applications. The first chapter is a contribution on precision and arithmetic concerns in HPRC from de Dinechin and Pasca of Ecole Normale Superieure de Lyon, France, and Altera European Technology Center, UK, respectively. The following three contributions are on design tools, starting with a contribution from Schafer and Wakabayashi of NEC Corporation, Japan, which presents a comparison between a C-based high level synthesis (HLS) approach and an RTL-based approach to the discrete element method (DEM). Following this, Perry et al. from Edinburgh Parallel Computing Centre (EPCC), UK, present a benchmarking exercise of Euroben kernels on the Maxwell FPGA supercomputer using a hand-coded VHDL-based approach and a C-based HLS approach. After that, a contribution from El Araby et al. from the Catholic University of America, USA, IBM Corporation, India, and the George Washington University, USA, presents a review and taxonomy of high-level languages (HLLs) for HPRC and a framework for their analysis, with programmer productivity taking centre stage. Indeed, programmer productivity in HPRC has been identified by the community as a major problem which needs to be addressed. This part ends with a contribution from Pell et al. from Maxeler Technologies, UK, and Imperial College London, UK, which presents an integrated approach to HPRC based on a dataflow software framework, and hardware that is optimised to the application in hand.

Reconfigurable Arithmetic for High-Performance Computing

Florent de Dinechin and Bogdan Pasca

Abstract An often overlooked way to increase the efficiency of HPC on FPGA is to exploit the bit-level flexibility of the target to match the arithmetic to the application. The ideal operator, for each elementary computation, should toggle and transmit just the number of bits required by the application at this point. FPGAs have the potential to get much closer to this ideal than microprocessors. Therefore, reconfigurable computing should systematically investigate non-standard precisions, but also non-standard number systems and non-standard operations which can be implemented efficiently on reconfigurable hardware. This chapter attempts to review these opportunities systematically.

1 Introduction

High-Performance Computing on FPGAs should tailor, as tightly as possible, the arithmetic to the application. An ideally efficient implementation would, for each of its operations, toggle and transmit just the number of bits required by the application at this point. Conventional microprocessors, with their word-level granularity and fixed memory hierarchy, keep us away from this ideal. FPGAs, with their bit-level granularity, have the potential to get much closer.

Therefore, reconfigurable computing should systematically investigate, in an application-specific way, non-standard precisions, but also non-standard number systems and non-standard arithmetic operations. The purpose of this chapter is to review these opportunities.

F. de Dinechin (✉)
École Normale Supérieure de Lyon, 46 allée d'Italie, 69364 Lyon, France
e-mail: Florent.de.Dinechin@ens-lyon.fr

B. Pasca
Altera European Technology Center, High Wycombe, UK
e-mail: bpasca@altera.com

W. Vanderbauwhede and K. Benkrid (eds.), *High-Performance Computing Using FPGAs*, DOI 10.1007/978-1-4614-1791-0_21, © Springer Science+Business Media, LLC 2013

Table 1 Table of acronyms

MSB	Most significant bit
LSB	Least significant bit
ulp	Unit in the last place (weight of the LSB)
HLS	High-level synthesis
DSP	Digital signal processing
DSP blocks	Embedded multiply-and-accumulate resources targeted at DSP
LUT	Look-up table
HRCS	High-radix carry-save

After a brief overview of computer arithmetic and the relevant features of current FPGAs in Sect. 2, we first discuss in Sect. 3 the issues of *precision analysis* (what is the precision required for each computation point?) and *arithmetic efficiency* (do I need to compute this bit?) in the FPGA context. We then review several approaches to application-specific operator design: *operator specialization* in Sect. 4, *operator fusion* in Sect. 5, and *exotic, non-standard operators* in Sect. 6. Section 7 discusses the application-specific *performance tuning* of all these operators. Finally, Sect. 8 concludes by listing the open issues and challenges in reconfigurable arithmetic.

The systematic study of FPGA-specific arithmetic is also the object of the FloPoCo project (http://flopoco.gforge.inria.fr/). FloPoCo offers open-source implementations of most of the FPGA-specific operators presented in this chapter, and more. It is therefore a good way for the interested reader to explore in more depth the opportunities of FPGA-specific arithmetic.

2 Generalities

Computer arithmetic deals with the representations of numbers in a computer, and with the implementation of basic operations on these numbers. A good introduction on these topics is the textbooks by Ercegovac and Lang [36] and Parhami [59].

In this chapter we will focus on the number systems prevalent in HPC: integer/-fixed point, and floating-point. However, many other number representation systems exist, have been studied on FPGAs, and have proven relevant in some situations. Here are a few examples.

- For integer, redundant versions of the classical position system enable faster addition. These will be demonstrated in the sequel.
- The residue number system (RNS) [59] represents an integer by a set of residues modulo a set of relatively prime numbers. Both addition and multiplication can be computed in parallel over the residues, but comparisons and division are very expensive.
- The logarithm number system (LNS) represents a real number as the value of its logarithm, itself represented in a fixed-point format with e integer bits and f

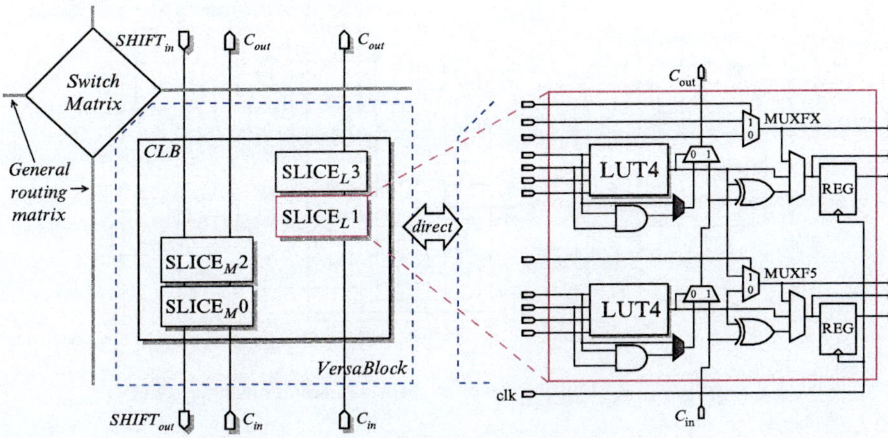

Fig. 1 Schematic overview of the logic blocks in the Virtex 4. More recent devices are similar, with up to 6 inputs to the LUTs

fractional bits. The range and precision of such a format are comparable to those of a floating-point format with e bits of exponent and f bits of fraction. This system offers high-speed and high-accuracy multiplication, division and square root, but expensive addition and subtraction [6, 21].

Current FPGAs support classical binary arithmetic extremely well. Addition is supported in the logic fabric, while the embedded DSP blocks support both addition and multiplication.

They also support floating-point arithmetic reasonably well. Indeed, a floating-point format is designed in such a way that the implementation of most operators in this format reduces to the corresponding binary integer operations, shifts, and leading zero counting.

Let us now review the features of current FPGAs that are relevant to arithmetic design.

2.1 Logic Fabric

Figures 1 and 2 provide a schematic overview of the logic fabric of recent FPGAs from the two main FPGA vendors. The features of interest are the following.

2.1.1 Look-Up Tables

The logic fabric is based on look-up tables with α inputs and one output, with $\alpha = 4,\ldots,6$ for the FPGAs currently on the market, the most recent FPGAs having the

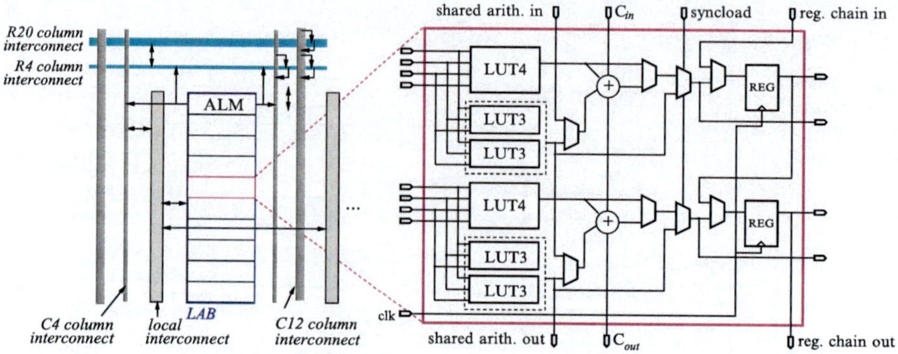

Fig. 2 Schematic overview of the logic blocks of recent Altera devices (Stratix II to IV)

largest α. These LUTs may be combined to form larger LUTs (for instance, the MUXF5 multiplexer visible in Fig. 1 serves this purpose). Conversely, they may be split into smaller LUTs, as is apparent in Fig. 2, where two LUT3 may be combined into an LUT4, and two LUT4 into an LUT5.

As far as arithmetic is concerned, this LUT-based structure means that algorithms relying on the tabulation of 2^{α} values have very efficient implementations in FPGAs. Examples of such algorithms include multiplication or division by a constant (see Sect. 4.1) and function evaluation (see Sect. 6.2).

2.1.2 Fast Carry Propagation

All recent FPGA architectures provide a fast connexion between neighbouring cells in a column, dedicated to carry propagation. This connexion is fast in comparison with the general programmable routing which is slowed down by all the switches enabling this programmability. Compared to classical (VLSI oriented) hardware arithmetic, this considerably changes the rules of the game. For instance, most of the literature regarding fast integer adders is irrelevant on FPGAs for additions smaller than 32 bits: the simple carry-ripple addition exploiting the fast-carry lines is faster, and consumes fewer resources, than the "fast adders" of the literature. Even for larger additions, the optimal solutions on FPGAs are not obtained by blindly applying the classical techniques, but by revisiting them with these new rules [32, 58, 64].

Fast carries are available on both Altera and Xilinx devices, but the detailed structure differs. Both device families allow one to merge an addition with some computations performed in the LUT. Altera devices are designed in such a way to enable the implementation of a 3-operand adder in one ALM level (see Fig. 2).

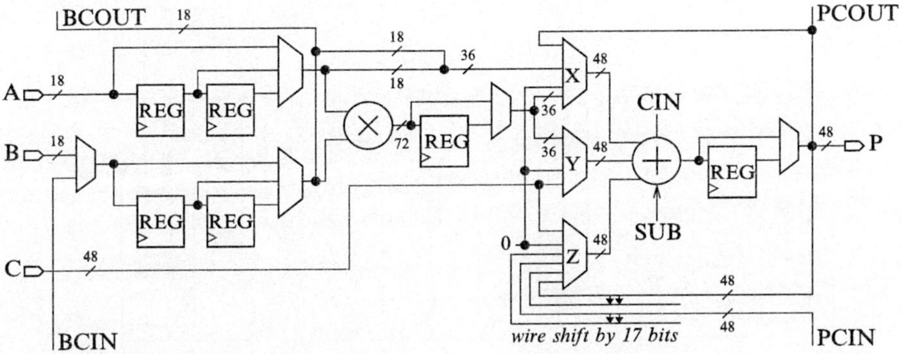

Fig. 3 Simplified overview of the Xilinx DSP48

2.1.3 DSP Blocks

Embedded multipliers (18 × 18-bit signed) first appeared in Xilinx VirtexII devices in 2000 and were complemented by a DSP-oriented adder network in the Altera Stratix in 2003.

DSP blocks not only enhance the performance of DSP applications—and, as we will see, any application using multiplication—they also make this performance more predictable as well.

Xilinx DSP Blocks

A simplified overview of the DSP48 block of Virtex-4 devices is depicted in Fig. 3. It consists of one 18 × 18-bit two's complement multiplier followed by a 48-bit sign-extended adder/subtracter or accumulator unit. The multiplier outputs two subproducts aligned on 36-bits. A 3-input adder unit can be used to add three external inputs, or the two sub-products and a third addend. The latter can be an accumulator (hence the feedback path) or an external input, coming either from global routing or from a neighboring DSP via a dedicated cascading line (PCIN). In this case this input may be shifted by 17 bits. This enables associating DSP blocks to compute large multiplications. In this case unsigned multiplications are needed, so the sign bit is not used, hence the value of 17.

These DSP blocks also feature internal registers (up to four levels) which can be used to pipeline them to high frequencies.

Virtex-5/-6/-7 feature similar DSP blocks (DSP48E), the main difference being a larger (18 × 25-bit, signed) multiplier. In addition the adder/accumulator unit can now perform several other operations such as logic operations or pattern detection. Virtex-6 and later add pre-multiplication adders within the DSP slice.

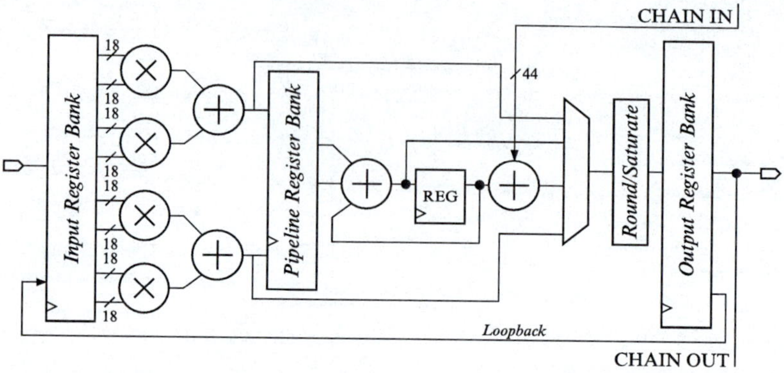

Fig. 4 Simplified overview of the StratixII DSP block, Stratix-III/-IV half-DSP block

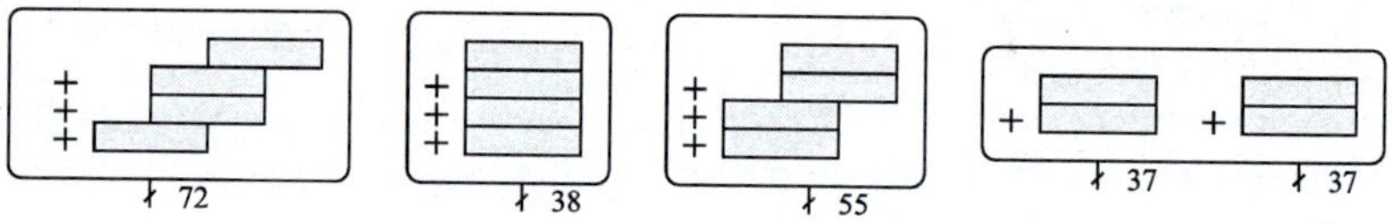

Fig. 5 Main configurations of Stratix DSP. Leftmost can be used to compute a 36×36 bit product, rightmost to compute the product of complex numbers

Altera DSP Blocks

The Altera DSP blocks have a much larger granularity than the Xilinx ones. On StratixII-IV devices (Fig. 4) the DSP block consists of four 18×18 bit (signed or unsigned) multipliers and an adder tree with several possible configurations, represented in Fig. 5. Stratix-III/-IV calls such DSPs half-DSPs, and pack two of them in a DSP block. In these devices, the limiting factor in terms of configurations (preventing us, for instance, to use them as 4 fully independent multipliers) is the number of I/Os to the DSP block. The variable precision DSP block in the StratixV devices is radically different: it is optimized for 27×27-bit or 18×36-bit, and a 36-bit multiplier is implemented in two adjacent blocks. Additionally, all DSPs allow various sum-of-two/four modes for increased versatility. Here also, neighbouring DSP blocks can be cascaded, internal registers allow high-frequency pipelining, and a loopback path enables accumulation. These cascading chains reduce resource consumption, but also latency: a sum-of-two 27-bit multipliers can be clocked at nominal DSP speed in just 2 cycles.

When designing operators for these devices, it is useful to account for these different features and try to fully exploit them. The full details can be found in the vendor documentation.

2.1.4 Embedded Memories

Modern FPGAs also include small and fast on-chip embedded memories. In Xilinx Virtex4 the embedded memory size is 18 Kbits, and 36 Kbits for Virtex5/6. The blocks support various configurations from 16 K × 1-bit to 512 × 36-bit (1 K for Virtex5/6).

Altera FPGAs offer blocks of different sizes. StratixII has three kinds of memory blocks: M512 (512-bit), M4K (4 Kb) and M-RAM (512 Kb); StratixIII-IV have a new family of memory blocks: MLAB (640b ROM/320b RAM), M9K (9 Kbit, up to 256 × 36-bit) and M144K (144 Kbits, up to 2 K × 72-bit); StratixV has MLAB and M20K (20 Kbits, up to 512 × 40-bit).

In both families, these memories can be dual-ported, sometimes with restrictions.

2.2 *Floating-Point Formats for Reconfigurable Computing*

A floating-point (FP) number x is composed of a sign bit S, an exponent field E on w_E bits, and a significand fraction F on w_F bits. It is usually mandated that the significand fraction has a 1 at its MSB: this ensures both uniqueness of representation and maximum accuracy in the case of a rounded result. Floating-point has been standardized in the IEEE-754 standard, updated in 2008 [40]. This standard defines common formats, the most usual being a 32-bit (the sign bit, 8 exponent bits, 23 significand bits) and a 64-bit format (1+12+53). It precisely specifies the basic operations, in particular the rounding behaviour. It also defines exceptional numbers: two signed infinities, two signed zeroes, subnormal numbers for a smooth underflow to zero, and NaN (Not a Number). These exceptional numbers are encoded in the extremal values of the exponent.

This standard was designed for processor implementations and makes perfect sense there. However, for FPGAs, many things can be reconsidered. Firstly, a designer should not restrict himself to the 32-bit and 64-bit formats of IEEE-754: he should aim at optimizing both exponent and significand size for the application at hand. The floating-point operators should be fully parameterized to support this.

Secondly, the IEEE-754 encodings were designed to make the most out of a fixed number of bits. In particular, exceptional cases are encoded in the two extremal values of the exponent. However, managing these encodings has a cost in terms of performance and resource consumption [35]. In an FPGA, this encoding/decoding logic can be saved if the exceptional cases are encoded in two additional bits. This is the choice made by FloPoCo and other floating-point libraries. A small additional benefit is that this choice frees the two extremal exponent values, slightly extending the range of the numbers.

Finally, we choose not to support subnormal numbers support, with flushing to zero instead. This is the most controversial issue, as subnormals bring with them important properties such as $(x-y=0) \iff (x=y)$, which is not true for

Fig. 6 The FloPoCo floating-point format

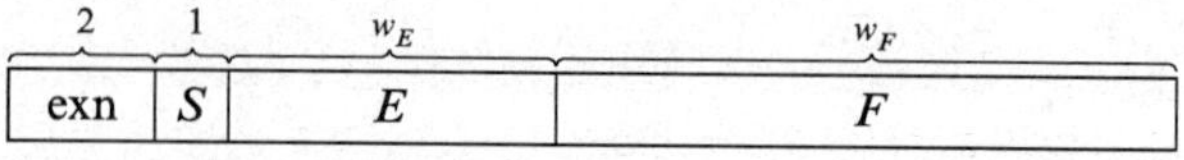

FP numbers close to zero if subnormals are not supported. However the cost of supporting subnormals is quite high, as they require specific shifters and leading-one detectors [35]. Besides, one may argue that adding one bit of exponent brings in all the subnormal numbers, and more, at a fraction of the cost: subnormals are less relevant if the format is fully parameterized. We believe there hasn't been a clear case for subnormal support in FPGA computing yet.

To sum up, Fig. 6 depicts a FloPoCo number, whose value (always normalized) is

$$x = (-1)^S \times 1.F \times 2^{E-E_0} \quad \text{with } E_0 = 2^{w_E-1} - 1.$$

E_0 is called the exponent bias. This representation of signed exponents (taken from the IEEE-754 standard) is prefered over two's complement, because it brings a useful property: positive floating-point numbers are ordered according to the lexicographic order of their binary representation (exponent and significand).

3 Arithmetic Efficiency and Precision Analysis

When implementing a given computation on an FPGA, the goal is usually to obtain an efficient design, be it to maximize performance, minimize the cost of the FPGA chip able to implement the computation, minimize the power consumption, etc. This quest for efficiency has many aspects (parallelism, operator sharing, pipeline balancing, input/output throughputs, FIFO sizes, etc). Here, we focus on an often understated issue, which is fairly specific to numerical computation on FPGAs: *arithmetic efficiency*. A design is arithmetic-efficient if the size of each operator is as small as possible, considering the accuracy requirements of the application. Ideally, no bit should be flipped, no bit should be transferred that is not relevant to the final result.

Arithmetic efficiency is a relatively new concern, because it is less of an issue for classical programming: microprocessors offer a limited choice of registers and operators. The programmer must use 8-, 16-, 32- or 64-bit integer arithmetic, or 32- or 64-bit floating-point. This is often very inefficient. For instance, both standard floating-point formats are vastly overkill for most parts of most applications. In a processor, as soon as you are computing accurately enough, you are very probably computing much too accurately.

In an FPGA, there are more opportunities to compute just right, to the granularity of the bit. Arithmetic efficiency not only saves logic resources, but it also saves routing resources. Finally, it also conserves power, all the more as there is typically more activity on the least significant bits.

Arithmetic efficiency is obtained by bit-width optimization, which in turn requires precision analysis. These issues have been the subject of much research, see, for instance, [47, 57, 65, 66] and references therein.

Range and precision analysis can be formulated as follows: given a computation (expressed as a piece of code or as an abstract circuit), label each of the intermediate variables or signals with information about its range and its accuracy. The range is typically expressed as an interval, for instance variable V lies in the interval $[-17, 42]$. In a fixed-point context, we may deduce from the range of a signal the value of its most significant bit (MSB) which will prevent the occurrence of any overflow. In a floating-point context, the range entails the maximum exponent that the format must accommodate to avoid overflows. In both contexts, accurate determination of the ranges enables us to set these parameters just right.

To compute the range, some information must be provided about the range of the inputs—by default it may be defined by their fixed-point or floating-point format. Then, there are two main methods for computing the ranges of all the signals: dynamic analysis and static analysis.

Dynamic methods are based on simulations. They perform several runs using different inputs, chosen in a more or less clever way. The minimum and maximum values taken by a signal over these runs provides an attainable range. However, there is no guarantee in general that the variable will not take a value out of this range in a different run. These methods are in principle unsafe, although confidence can be attained by very large numbers of runs, but then these methods become very compute-intensive, especially if the input space is large.

Static analysis methods propagate the range information from the inputs through the computation, using variants of interval analysis (IA) [54]. IA provides range intervals that cover all the possible runs and therefore is safe. However, it often overestimates these ranges, leading to bits at the MSB or exponent bits that will never be useful to actual computations. This ill-effect is essentially due to correlations between variables and can be avoided by algebraic rewriting [33] (manual or automated), or higher-order variants of interval arithmetic such as affine arithmetic [47], polynomial arithmetic [12] or Taylor models. In case of loops, these methods must look for a fix point [66]. A general technique in this case is abstract interpretation [18].

Bit-width minimization techniques reduce the size of the data, hence reduce the size and power consumption of all the operators computing on these data. However, there are also less frequent, but more radical operator optimization opportunities. The remainder of this chapter reviews them.

4 Operator Specialization

Operator specialization consists in optimizing the structure of an operator when the context provides some static (compile-time) property on its inputs that can be usefully exploited. This is best explained with some examples.

First, an operator with a constant operand can often be optimized somehow:

- Even in software, it is well known that cubing or extracting a square root is simpler than using the *pow* function x^y.
- For hardware or FPGAs, multiplication by a constant has been extensively studied (although its complexity in the general case is still an open question). There exist several competing constant multiplication techniques, with different relevance domains: they are reviewed in Sect. 4.1.
- One of us has worked recently on the division by a small integer constant [24].
- However, on FPGA technology, there seems to be little to win on addition with a constant operand, except in trivial cases.

It is also possible to specialize an operator thanks to more subtle relationships between its inputs. Here are two examples which will be expanded in Sect. 5.3:

- In terms of bit flipping, squaring is roughly twice cheaper than multiplying.
- If two numbers have the same sign, their floating-point addition is cheaper to implement than a standard addition: the cancellation case (which costs one large leading-zero counter and shifter) never happens [49].

Finally, many functions, even unary ones, can be optimized if their input is statically known to lie within a certain range. Here are some examples.

- If a floating-point number is known to lie in $[-\pi, \pi]$, its sine is much cheaper to evaluate than in the general case (no argument reduction) [22].
- If the range of the input to an elementary function is small enough, a low-degree polynomial approximation may suffice.

Finally, an operator may have its accuracy degraded, as long as the demand of the application is matched. The most spectacular example is truncated multipliers: sacrificing the accuracy of the least significant bit saves almost half the area of a floating-point multiplier [8, 67]. Of course, in the FPGA context, the loss of precision can be recovered by adding one bit to the mantissa, which has a much lower cost.

The remainder of this section focuses on specializations of the multiplication, but designers on FPGAs should keep in mind this opportunity for many other operations.

4.1 Multiplication and Division by a Constant

Multiplication by constants has received much attention in the literature, especially as many digital signal processing algorithms can be expressed as products by constant matrices [13, 52, 62, 72]. There are two main families of algorithms. Shift-and-add algorithms start from the construction of a standard multiplier and simplify it, while LUT-based algorithm tabulate sub-products in LUTs and are thus more specific to FPGAs.

4.1.1 Shift and Add Algorithms

Let C be a positive integer constant, written in binary on k bits:

$$C = \sum_{i=0}^{k} c_i 2^i \quad \text{with } c_i \in \{0,1\}.$$

Let X be a p-bit integer. The product is written $CX = \sum_{i=0}^{k} 2^i c_i X$, and by only considering the non-zero c_i, it is expressed as a sum of $2^i X$. For instance, $17X = X + 2^4 X$. In the following, we will note this using the shift operator $\ll$, which has higher priority than $+$ and $-$. For instance $17X = X + X \ll 4$.

If we allow the digits of the constant to be negative ($c_i \in \{-1,0,1\}$) we obtain a redundant representation, for instance $15 = 01111 = 1000\bar{1}$ ($16-1$ written in signed binary). Among the representations of a given constant C, we may pick up one that minimizes the number of non-zero bits, hence of additions/subtractions. The well-known *canonical signed digits* recoding (or CSD, also called Booth recoding [36]) guarantees that at most $k/2$ bits are non-zero, and in average $k/3$.

The CSD recoding of a constant may be directly translated into an architecture with one addition per non-zero bit, for instance $221X = 100\bar{1}00\bar{1}01_2 X = X \ll 8 + (-X \ll 5 + (-X \ll 2 + X))$. With this right-to-left parenthesing, all the additions are actually of the same size (the size of X): in an addition $X \ll s + P$, the s lower bits of the result are those of P and do not need to participate to the addition.

For large constants, a binary tree adder structure can be constructed out of the CSD recoding of the constant as follows: non-zero bits are first grouped by 2, then by 4, etc. For instance, $221X = (X \ll 8 - X \ll 5) + (-X \ll 2 + X)$. Shifts may also be reparenthesized: $221X = (X \ll 3 - X) \ll 5 + (-X \ll 2 + X)$. After doing this, the leaves of the tree are now multiplications by small constants: $3X, 5X, 7X, 9X, \ldots$ Such a smaller multiple will appear many times in a larger constant, but it may be computed only once: thus the tree is now a DAG (direct acyclic graph), and the number of additions is reduced. A larger example is shown in Fig. 7. This new parenthesing reduces the critical path: for k non-zero bits, it is now of $\lceil \log_2 k \rceil$ additions instead of k in the previous linear architecture. However, additions in this DAG are larger and larger.

This simple DAG construction is the current choice in FloPoCo, but finding the optimal DAG is still an open problem. There is a wide body of literature on constant multiplication, minimizing the number of additions [9, 19, 38, 69, 72], and, for hardware, also minimizing the total size of these adders (hence the logic consumption in an FPGA) [1,19,37,38]. It has been shown that the number of adders in constant multiplication problem is sub-linear in the number of non-zero bits [34]. Exhaustive exploration techniques [19, 38, 69] lead to less than 4 additions for any constant of size smaller than 12 bits, and less than 5 additions for sizes smaller than 19 bits. They become impractical beyond these sizes, and heuristics have to be used. Lefèvre's algorithm [48] looks for maximal repeating bit patterns (in direct or complemented form) in the CSD representation of the constant, then proceeds

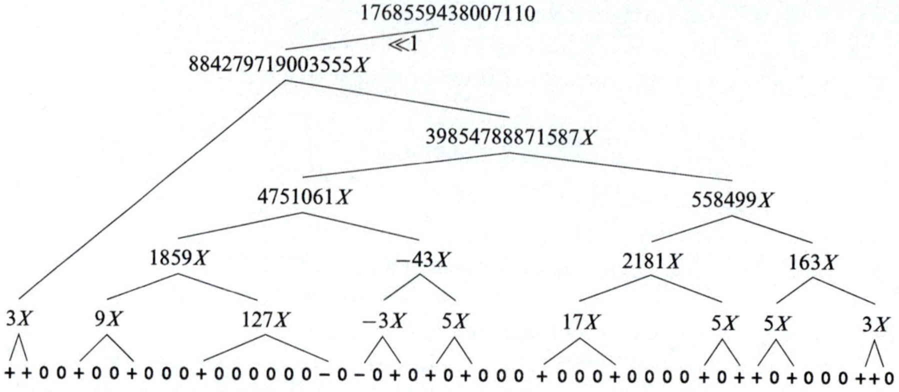

Fig. 7 Binary DAG architecture for a multiplication by 1768559438007110 (the 50 first bits of the mantissa of π)

recursively on these patterns. Experimentally, the number of additions, on randomly generated constants of k bits, grows as $O(k^{0.85})$. However, this algorithm does not currently try to minimize the total size of the adders [14], contrary to Gustafsson et al. [38].

All the previous dealt with multiplication by an integer constant. Multiplying by a real constant (in a fixed-point or floating-point context) raises the additional issue of first approximating this constant by a fixed-point number. Gustafsson and Qureshi suggested to represent a real constant on more than the required number of bits, if it leads to a shift-and-add architecture with fewer additions [37]. This idea was exploited analytically for rational constants, which have a periodic binary representation [23].

4.1.2 Table-Based Techniques

On most FPGAs, the basic logic element is the look-up-table, a small memory addressed by α bits. The KCM algorithm (which probably means "constant (K) Coefficient Multiplier"), due to Chapman [15] and further studied by Wirthlin [76] is an efficient way to use these LUTs to implement a multiplication by an integer constant.

This algorithm, described in Fig. 8, consists in breaking down the binary decomposition of an n-bit integer X into chunks of α bits. This is written as $X = \sum_{i=0}^{\lceil \frac{n}{\alpha} \rceil - 1} X_i.2^{\alpha i}$, where $X_i \in \{0, \ldots, 2^\alpha - 1\}$.

The product of X by an m-bit integer constant C becomes $CX = \sum_{i=0}^{\lceil \frac{n}{\alpha} \rceil} CX_i.2^{-\alpha i}$. We have a sum of (shifted) products CX_i, each of which is an $m+\alpha$ integer. The KCM trick is to read these CX_i from a table of pre-computed values T_i, indexed by X_i, before summing them.

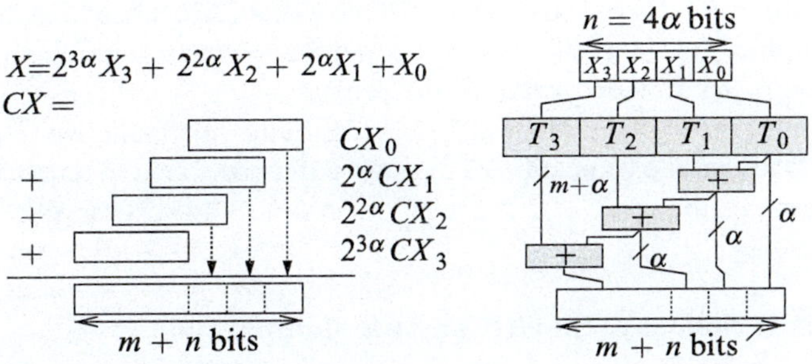

Fig. 8 The KCM LUT-based method (integer × integer)

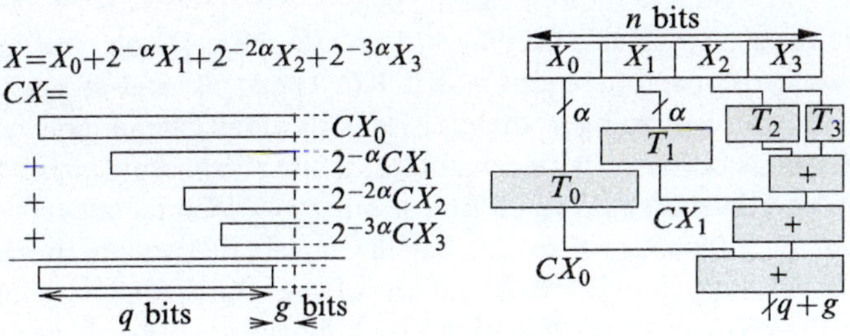

Fig. 9 The KCM LUT-based method (real × fixed-point)

The cost of each table is one FPGA LUT per output bit. The lowest-area way of computing the sum is to use a rake of $\lceil \frac{n}{\alpha} \rceil$ in sequence, as shown in Fig. 8: here again, each adder is of size $m+\alpha$, because the lower bits of a product CX_i can be output directly. If the constant is large, an adder tree will have a shorter latency at a slightly larger area cost. The area is always very predictible and, contrary to the shift-and-add methods, almost independent on the value of the constant (still, some optimizations in the tables will be found by logic optimizers).

There are many possible variations on the KCM idea.

- As all the tables contain the same data, a sequential version can be designed.
- This algorithm is easy to adapt to signed numbers in two's complement.
- Wirthlin [76] showed that if we split the input in chunks of $\alpha - 1$ bits, then one row of LUT can integrate both the table and the corresponding adder, and still exploit the fast-carry logic of Xilinx circuits: this reduces the overall area. Altera FPGAs don't need this trick thanks to their embedded full adder (see Fig. 2).
- It can be adapted to fixed-point input and, more interesting, to an arbitrary real constant C, for instance $\log(2)$ in [31] or FFT twiddle factors in [29]. Figure 9 describes this case. Without loss of generality, we assume a fixed-point input in $[0,1)$: it is now written on n bits as $X = \sum_{i=0}^{\lceil \frac{n}{\alpha} \rceil - 1} X_i . 2^{-\alpha i}$ where $X_i \in \{0,\ldots,2^{\alpha}-1\}$. Each product CX_i now has an infinite number of bits. Assume we want a q-bit result with $q \geq n$. We tabulate in LUTs each product $2^{i\alpha}CX_i$ on

just the required precision, so that its LSB has value $2^{-g}u$ where u is the ulp (unit in the last place) of the result, and g is a number of guard bits. Each table may hold the correctly rounded value of the product of E_i by the *real* value of C to this precision, so entails an error of 2^{-g-1} ulp. In the first table, we actually store $CX_0 + u/2$, so that the truncation of the sum will correspond to a rounding of the product. Finally, the value of g is chosen to ensure 1-ulp accuracy.

4.1.3 Other Variations of Single-Constant Multiplication

Most algorithms can be extended to a floating-point version. As the point of the constant doesn't float, the main question is whether normalization and rounding can be simpler than in a generic multiplication [14].

For simple rational constants such as $1/3$ or $7/5$, the periodicity of their binary representations leads to optimizations both in KCM and shift-and-add methods [23]. The special case corresponding to the division by a small integer constant is quite useful: Integer division by 3 (with remainder) is used in the exponent processing for cube root, and division by 5 is useful for binary to decimal conversion. Fixed-point division by 3 (actually 6 or 24, but the power of two doesn't add to the complexity) enables efficient implementations of sine and cosine based on parallel evaluation of their Taylor series. Floating-point division by 3 is used in the Jacobi stencil algorithm. In addition to techniques considering division by a constant as the multiplication by the inverse [23], a specific LUT-based method can be derived from the division algorithm [24].

4.1.4 Multiple Constant Multiplication

Some signal-processing transforms, in particular finite impulse response (FIR) filters, need a given signal to be multiplied by several constants. This allows further optimizations: it is now possible to share sub-constants (such as the intermediate nodes of Fig. 7) between several constant multipliers. Many heuristics have been proposed for this multiple constant multiplication (MCM) problem [1,13,52,62,72].

A technique called Distributed Arithmetic, which predates FPGA [74], can be considered a generalization of the KCM technique to the MCM problem.

4.1.5 Choosing the Best Approach in a Given Context

To sum up, there is plenty of choice in terms of constant multiplication or division in an FPGA. Table 2 describes the techniques implemented in the FloPoCo tool at the time of writing. This is work in progress.

As a rule of thumb, for small inputs, KCM should be preferred, and for simple constants, shift-and-add should be preferred. In some cases the choice is obvious: for

Table 2 Constant multiplication and division algorithms in FloPoCo 2.3.1

Format	Integer (keep all bits)	Fixed-point (keep higher bits)	Floating-point
Shift-and-add	IntConstMult [14]		FPConstMult [14]
(rational constants)			FPConstMultRational [23]
LUT-based	IntIntKCM [15, 76]	FixRealKCM [29, 31]	FPRealKCM
Division-based	IntConstDiv [24]		FPConstDiv [24]

instance, to evaluate a floating-point exponential, we have to multiply an exponent (a small integer) by $\log(2)$, and we need many more bits on the result: this is a case for KCM, as we would need to consider many bits of the constant. In most usual cases, however, the final choice should probably be done on a trial-and-error basis.

4.2 Squaring

If one computes, using the pen-and-paper algorithm learnt at school, the square of a large number, one will observe that each of the digit-by-digit products is computed twice. This holds also in binary: formally, we have

$$X^2 = \left(\sum_{i=0}^{n-1} 2^i x_i\right)^2 = \sum_{i=0}^{n-1} 2^{2i} x_i + \sum_{0<i<j<n} 2^{i+j+1} x_i x_j$$

and we have a sum of roughly $n^2/2$ partial products, versus n^2 for a standard n-bit multiplication. This is directly useful if the squarer is implemented as LUTs. In addition, a similar property holds for a splitting of the input into several subwords:

$$(2^k X_1 + X_0)^2 = 2^{2k} X_1^2 + 2 \cdot 2^k X_1 X_0 + X_0^2 \tag{1}$$

$$\begin{aligned}(2^{2k} X_2 + 2^k X_1 + X_0)^2 = {} & 2^{4k} X_2^2 + 2^{2k} X_1^2 + X_0^2 \\ & + 2 \cdot 2^{3k} X_2 X_1 \\ & + 2 \cdot 2^{2k} X_2 X_0 \\ & + 2^k X_1 X_0 \end{aligned} \tag{2}$$

Computing each square or product of the above equation in a DSP block yields a reduction of the DSP count from 4 to 3, or from 9 to 6. Besides, this time, it comes at no arithmetic overhead. Some of the additions can be computed in the DSP blocks, too. This has been studied in detail in [25].

Squaring is a specific case of powering, i.e. computing x^p for a constant p. Ad-hoc, truncated powering units have been used for function evaluation [20]. These are based on LUTs and should be reevaluated in the context of DSP blocks.

5 Operator Fusion

Operator fusion consists in building an atomic operator for a non-trivial mathematical expression, or a set of such expressions. The recipe is here to consider the mathematical expression as a whole and to optimize each operator in the context of the whole expression. The opportunities for operator fusion are unlimited, and the purpose of this section is simply to provide a few examples which are useful enough to be provided in an operator generator such as FloPoCo.

5.1 *Floating-Point Sum-and-Difference*

In many situations, the most pervasive of which is probably the Fast Fourier transform (FFT), one needs to compute the sum and the difference of the same two values. In floating-point, addition or subtraction consists in the following steps [56]:

- Alignment of the significands using a shifter, the shift distance being the exponent difference;
- Effective sum or difference (in fixed-point);
- In case of effective subtraction leading to a cancellation, leading zero count (LZC) and normalization shift, using a second shifter;
- Final normalization and rounding.

We may observe that several redundancies exist if we compute in parallel the sum and the difference of the same values:

- The exponent difference and alignment logic is shared by the two operations.
- The cancellation case will appear at most once, since only one of the operations will be an effective subtraction, so only one LZC and one normalization shifter is needed.

Therefore the additional cost of the second operation, with respect to a classical floating-point adder, is only its effective addition/subtraction, and its final normalization and rounding logic. Numerically, a combined sum-and-difference operator needs about one third more logic than a standard adder and has the same latency.

5.2 *Block Floating-Point*

Looking back at the FFT, it is essentially based on multiplication by constants, and the previous sum-and-difference operations. In a floating-point FFT, operator fusion can be pushed a bit further, using a technique called block floating-point [41], first used in the 1950s, when floating point arithmetic was implemented in software, and more recently applied to FPGAs [3, 5]. It consists in an initial alignment of all the input significands to the largest one, which brings them all to the same exponent (hence the phrase "block floating point"). After this alignment, all the computations (multiplications by constants and accumulation) can be performed in fixed point, with a single normalization at the end. Another option, if the architecture implements only one FFT stage and the FFT loops on it, is to perform the normalization of all the values to the largest (in magnitude) of the stage.

Compared with the same computation using standard floating-point operators, this approach saves all the shifts and most of the normalization logic in the intermediate results. The argument is that the information lost in the initial shifts would have been lost in later shifts anyway. However, a typical block floating-point implementation will accumulate the dot product in a fixed-point format slightly larger than the input significands, thus ensuring a better accuracy than that achieved using standard operators.

Block floating-point techniques can be applied to many signal processing transforms involving the product of a signal vector by a constant vector. As it eventually converts the problem to a fixed-point one, the techniques for MCM listed in Sect. 4.1.4 can be used.

5.3 *Floating-Point Sum of Squares*

We conclude this section with the example of a large fused operator that combines several of the FPGA-specific optimizations discussed so far. The datapath described in Fig. 10 inputs three floating-point numbers X, Y and Z, and outputs a floating-point value for $X^2 + Y^2 + Z^2$. Compared to a more naive datapath built out of standard adders and multiplier, it implements several optimizations:

- It uses squarers instead of multipliers, as suggested in Sect. 4.2. These can even be truncated squarers.
- As squares are positive, it can dispose of the leading-zero counters and shifters that, in standard floating-point additions, manage the possible cancellation in case of subtraction [49].
- It saves all the intermediate normalizations and rounding.
- It computes the three squares in parallel and feeds them to a three-operand adder (which is no more expensive than a two-operand adder in Altera devices) instead of computing the two additions in sequence.
- It extends the fixed-point datapath width by $g = 3$ guard bits that ensure that the result is always last-bit accurate, where a combination of standard operators would lead to up to 2.5 ulps of error. This is the value of g for a sum of three squares, but it can be matched to any number of squares to add, as long as this number is known statically.

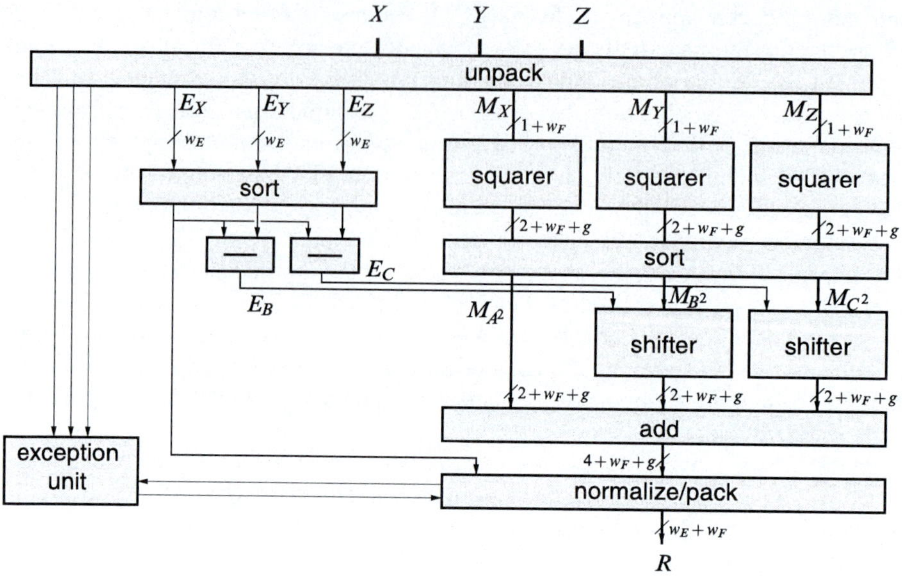

Fig. 10 A floating-point sum-of-squares (for w_E bits of exponent and w_F bits of significand)

- It reflects the symmetry of the mathematical expression $X^2 + Y^2 + Z^2$, contrary to a composition of floating-point operators which computes $(X^2 + Y^2) + Z^2$, leading to slightly different results if X and Z are permuted.

Compared to a naive assembly of three floating-point multipliers and two floating-point adders, the specific architecture of Fig. 10 thus significantly reduces logic count, DSP block count and latency, while being more accurate than the naive datapath. For instance, for double-precision inputs and outputs on Virtex-4, slice count is reduced from 4,480 to 1,845, DSP count is reduced from 27 to 18, and latency is reduced from 46 to 16 cycles, for a frequency of 362 MHz (post-synthesis) which is nominal on this FPGA.

5.4 *Towards Compiler-Level Operator Fusion*

Langhammer proposed an optimizing floating-point datapath compiler [46] that:

- Detects clusters of similar operations and uses a fused operator for the entire cluster;
- Detects dependent operations and fuses the operators by removing or simplifying the normalization, rounding steps and alignment steps of the next operation.

To ensure high accuracy in spite of these simplifications, the compiler relies on additional accuracy provided for free by the DSP blocks. The special floating-point formats used target accuracy "soft spots" for recent Altera DSP blocks (StratixII-IV) which is 36 bits. For instance, in single-precision (24 mantissa bits) the adders use an extended, non-normalized mantissa of up to 31 bits which, when followed by a multiplier stage uses the 36-bit multiplier mode on the 31-bit operands. For this stage as well, an extended mantissa allows for late normalizations while preserving accuracy. The optimizations proposed by Langhammer are available in Altera's DSP Builder Advanced tool [60].

6 Exotic Operators

This section presents in detail three examples of operators that are not present in processors, which gives a performance advantage to FPGAs. There are many more examples, from elementary functions to operators for cryptography.

6.1 Accumulation

Summing many independent terms is a very common operation: scalar products, matrix–vector and matrix–matrix products are defined as sums of products, as are most digital filters. Numerical integration usually consists in adding many elementary contributions. Monte-Carlo simulations also involve sums of many independent terms.

Depending on the fixed/floating-point arithmetic used and the operand count there are several optimization opportunities.

When having to sum a fixed, relatively small number of terms arriving in parallel, one may use adder trees. Fixed-point adder trees benefit from adder support in the FPGA fabric (ternary adder trees can be built on Altera FPGAs). If the precision is large, adders can be pipelined [32] and tessellated [60] for reducing latency and resources (Fig. 11). Floating-point adder trees for positive data may use a dedicated fused operator similar to the one in Fig. 10 for the sum-of-squares. Otherwise, one may rely on the techniques presented by Langhammer for datapath fusion which, depending on the operator count combine clustering and delayed normalizations [46].

For an arbitrary number of summands arriving sequentially, one needs an accumulator, conceptually described in Fig. 12. A fixed-point accumulators may be built out of a binary adder with a feedback loop. This allows good performances for moderate-size formats: as a rule of thumb, a 32-bit accumulator can run at the FPGA nominal frequency (note also that a larger *hard* accumulator is available in modern DSP blocks). If the addition is too wide for the ripple-carry propagation to take place in one clock cycle, a redundant carry-save representation can be used for

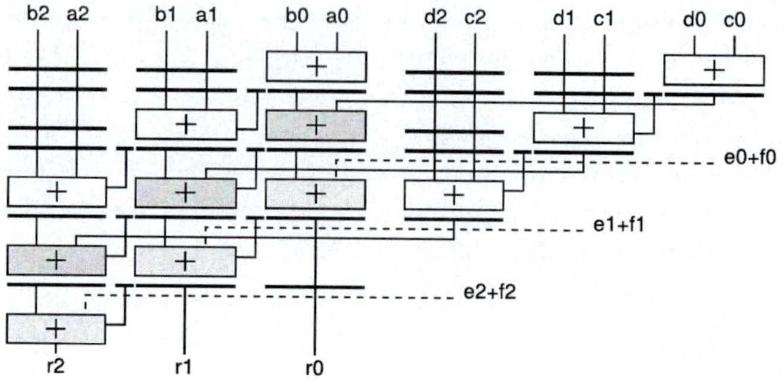

Fig. 11 Fixed-point accumulation for small operand count based on a tessellated adder tree

Fig. 12 An accumulator

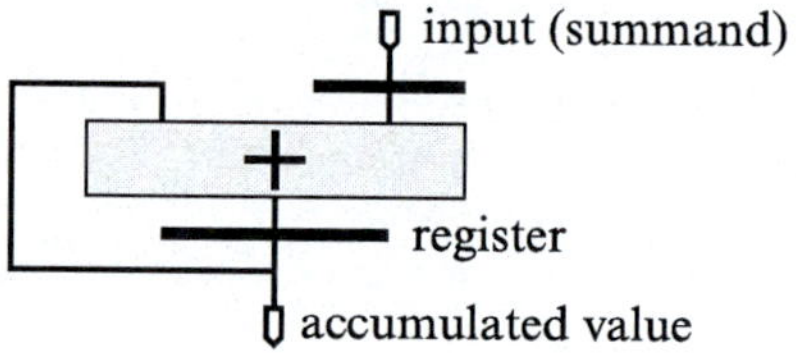

the accumulator. In FPGAs, thanks to fast carry circuitry, a high-radix carry save (HRCS), breaking the carry propagation typically every 32 bits, has a very low area overhead.

Building an efficient accumulator around a floating-point adder is more involved. The problem is that FP adders have long latencies: typically $l = 3$ cycles in a processor, up to tens of cycles in an FPGA. This long latency means that an accumulator based on an FP adder will either add one number every l cycles or compute l independent sub-sums which then have to be added together somehow. One special case is large matrix operations [10, 78], when l parallel accumulations can be interleaved. Many programs can be restructured to expose such sub-sum parallelism [2].

In the general case, using a classical floating point adder of latency l as the adder of Fig. 12, one is left with l independent sub-sums. The log-sum technique adds them using $\lceil \log_2 l \rceil$ adders and intermediate registers [39, 68]. Sun and Zambreno suggest that l can be reduced by having two parallel accumulator memories, one for positive addends and one for negative addends: this way, the cancellation detection and shift can be avoided in the initial floating-point accumulator. This, however, becomes inaccurate for large accumulations whose result is small [68].

Additionally, an accumulator built around a floating-point adder is inefficient, because the significand of the accumulator has to be shifted, sometimes twice (first to align both operands and then to normalize the result). These shifts are in the critical path of the loop. Luo and Martonosi suggested to perform the alignment in two steps, the finest part outside of the loop, and only a coarse alignment inside [50]. Bachir and David have investigated several other strategies to build a single-cycle accumulator, with pipelined shift logic before, and pipelined normalization logic after [7]. This approach was suggested in earlier work by Kulisch, targetting microprocessor floating-point units. Kulisch advocated the concept of an exact accumulator as "the fifth floating-point operation". Such an accumulator is based on a very wide internal register, covering the full floating-point range [43, 44], and accessed using a two-step alignment. One problem with this approach is that in some situations (long carry propagation), the accumulator requires several cycles. This means that the incoming data must be stalled, requiring more complex control. This is also the case in [50].

For FPGA-accelerated HPC, one critics to all previous approaches to universal accumulators is that they are generally overkill: they don't compute just-right for the application. Let us now consider how to build an accumulator of floating-point numbers which is tailored to the numerics of an application. Specifically, we want to ensure that it never overflows and that it eventually provides a result that is as accurate as the application requires. Moreover, it is also designed around a single-cycle accumulator. We present this one [28] in detail as it exhibits many of the techniques used in previously mentioned works.

The accumulator holds the accumulation result in fixed-point format which allows removing any alignments from the loop's critical path. It is depicted in Fig. 13. Single-cycle accumulation at arbitrary frequency is ensured by using an HRCS accumulator if needed.

The bottom part of Fig. 13 presents a component which converts the fixed point accumulator back to floating-point. It makes sense to consider this as a separate component, beause this conversion may be performed in software if the running value of the accumulation is not needed (e.g. in numerical integration applications). In other situations (e.g. matrix–vector product), several accumulators can be scheduled to share a common post-normalization unit. In this unit, the carry-propagation box converts the result into non-redundant format in the case when HCRS is used.

The parameters of the accumulator are explained with the help of Fig. 14:

- MSB_A is the position of the most-significant bit (MSB) of the accumulator. If the maximal expected running sum is smaller than 2^{MSB_A}, no overflow ever occurs.
- LSB_A is the position of the least-significant bit of the accumulator and determines the final accumulation accuracy.
- $MaxMSB_X$ is the maximum expected position of the MSB of a summand. $MaxMSB_X$ may be equal to MSB_A, but very often one is able to tell that each summand is much smaller in magnitude than the final sum. In this case, providing $MaxMSB_X < MSB_A$ will save hardware in the input shifter.

This parameters must be set up in an application-dependent way by considering the numerics of the application to be solved. In many cases, this is easy, because gross overestimation has a moderate impact: taking a margin of three orders of magnitude on MSB_A, for instance, adds only 10 bits to the accumulator size [28].

Fig. 13 The proposed accumulator (*top*) and post-normalisation unit (*bottom*)

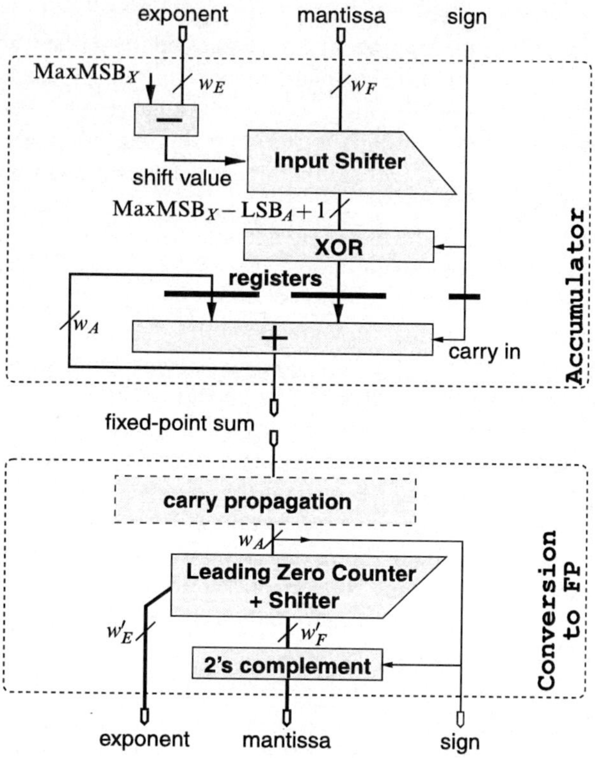

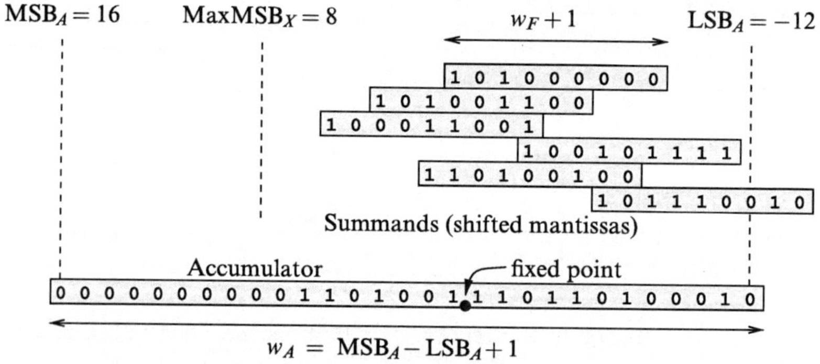

Fig. 14 Accumulation of floating-point numbers into a large fixed-point accumulator

6.2 Generic Polynomial Approximation

Polynomial approximation is an invaluable tool for implementing fixed-point functions (which are also the basis of many floating-point ones) in hardware. Given a function $f(x)$ and an input domain, polynomial approximation starts by finding a polynomial $P(x)$ which approximates $f(x)$. There are several methods for obtaining these polynomials including: the Taylor and Chebyshev series, or the Remez algorithm, a numerical routine that under certain conditions converges to the *minimax* polynomial (the polynomial which minimizes the maximum error between f and P).

There is a strong dependency between the size of the input interval, the polynomial degree and the approximation accuracy: a higher degree polynomial increases accuracy but degrades implementation performance or cost. *Piecewise* polynomial approximation splits the input range into subintervals and uses a different polynomial p_i for each subinterval. This scalable range reduction technique allows reaching an arbitrary accuracy for fixed polynomial degree d. A uniform segmentation scheme, where all subintervals have the same size, has the advantage that interval decoding the is straightforward, just using he leading bits of x. Non-uniform range reduction schemes like the power-of-two segmentation [16] have slightly more complex decoding requirements but can enable more efficient implementation of some functions.

Given a polynomial, there are many possible ways to evaluate it. The HOTBM method [20] uses the developed form $p(y) = a_0 + a_1y + a_2y^2 + \cdots + a_dy^d$ and attempts to tabulate as much of the computation as possible. This leads to a short-latency architecture since each of the a_iy^i may be evaluated in parallel and added thanks to an adder tree, but at a high hardware cost. Conversly, the Horner evaluation scheme minimizes the number of operations, at the expense of latency: $p(y) = a_0 + y \times (a_1 + y \times (a_2 + \cdots . + y \times a_d) \ldots)$ [30]. Between these two extremes, intermediate schemes can be explored. For large degrees, the polynomial may be decomposed into an odd and an even part: $p(y) = p_e(y^2) + y \times p_o(y^2)$. The two sub-polynomial may be evaluated in parallel, so this scheme has a shorter latency than Horner, at the expense of the precomputation of x^2 and a slightly degraded accuracy. Many variations on this idea, e.g. the Estrin scheme, exist [55]. A polynomial may also be refactored to trade multiplications for more additions [42], but this idea is mostly incompatible with range reduction.

When implementing an approximation of f in hardware, there are several error sources which summed-up ($\varepsilon_{\text{total}}$) determine the final implementation accuracy. For arithmetic efficiency, we aim at *faithful rounding*, which means that $\varepsilon_{\text{total}}$ must be smaller than the weight of the LSB of the result, noted u. This error is decomposed as follows: $\varepsilon_{\text{total}} = \varepsilon_{\text{approx}} + \varepsilon_{\text{eval}} + \varepsilon_{\text{finalround}}$ where:

- $\varepsilon_{\text{approx}}$ is the approximation error, the maximum absolute difference between any of the polynomials p_i and the function over its interval. The open-source Sollya tool [17] offers the state of the art for both polynomial approximation and a safe computation of $\varepsilon_{\text{approx}}$.
- $\varepsilon_{\text{eval}}$ is the total of all rounding and truncation errors committed during the evaluation. These can be made arbitrarily small by adding g *guard bits* to the LSB of the datapath.
- $\varepsilon_{\text{finalround}}$ is the error corresponding rounding off the guard bits from the evaluated polynomial to obtain a result in the target format. It is bounded by $u/2$.

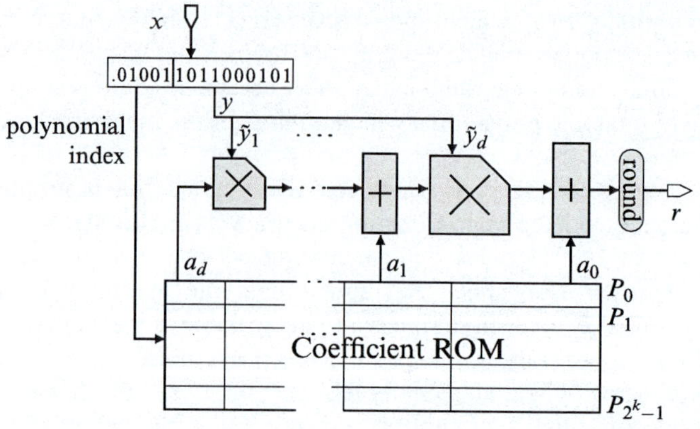

Fig. 15 Function evaluation using an Horner evaluation datapath computing just right

Given that $\varepsilon_{\text{finalround}}$ has a fixed bound $(u/2)$, the aim is to balance the approximation and evaluation error such that the final error remains smaller than u. One idea is to look for polynomials such that $\varepsilon_{\text{approx}} < u/4$. Then, the remaining error budget allocated to the evaluation error: $\varepsilon_{\text{eval}} < u/2 - \varepsilon_{\text{approx}}$.

FloPoCo implements this process (more details in [30]) and builds the architecture depicted in Fig. 15. The datapath is optimized to compute just right at each point, truncating all the intermediate results to the bare minimum and using truncated multipliers [8, 75].

6.3 Putting It All Together: A Floating-Point Exponential

We conclude this section by presenting in Fig. 16 a large operator that combines many of the techniques reviewed so far:

- A fixed-point datapath, surrounded by shifts and normalizations,
- Constant multiplications by $\log(2)$ and $1/\log(2)$,
- Tabulation of pre-computed values in the e^A box,
- Polynomial approximation for the $e^Z - Z - 1$ box,
- Truncated multipliers, and in general computing just right everywhere.

The full details can be found in [31].

Roughly speaking, this operator consumes an amount of resource comparable to a floating-point adder and a floating-point multiplier together. It may be fully pipelined to the nominal frequency of an FPGA, and its throughput, in terms of exponentials computed per second, is about ten times the throughput of the best (software) implementations in microprocessors. In comparison, the throughput of

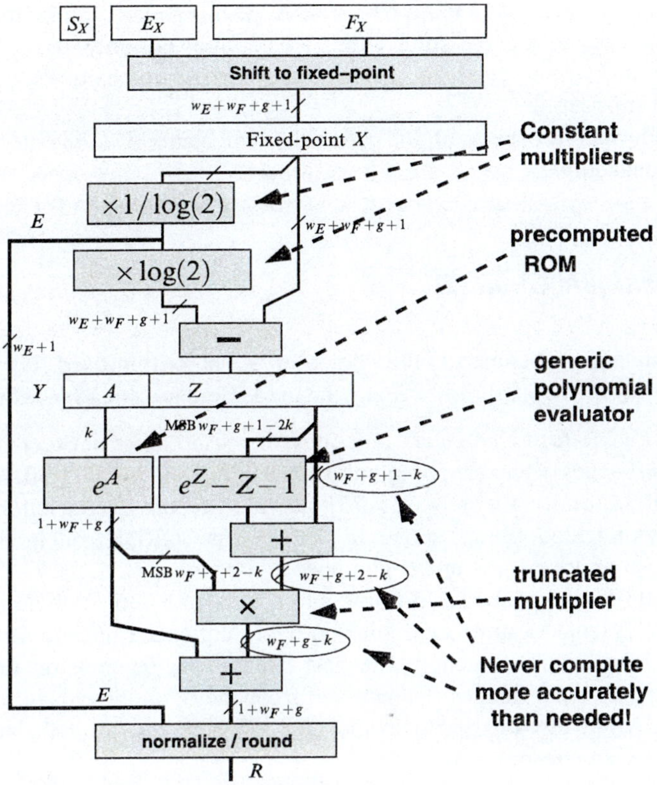

Fig. 16 Architecture of a floating-point exponential operator

floating points adders and multipliers is ten times *less* than the corresponding (hardware) processor implementation. This illustrates the potential of exotic operators in FPGAs.

7 Operator Performance Tuning

Designing an arithmetic operator involves many trade-offs, most often between performance and resource consumption. The architectures of functionally identical operators in microprocessors targetting different markets can can widely differ: compare, for instance, two functionally identical, standard fused multiply-and-add (FMA) operators published in the same conference, one for high-end processors [11], the other for embedded processors [51]. However, for a given processor, the architecture is fixed and the programmer has to live with it.

In FPGAs, we again have more freedom: A given operator can be tuned to the performance needs of a given application. This applies not only to all the FPGA-specific operators we surveyed, but also to classical, standard operators such as plain addition and subtraction.

Let us review a few aspects of this variability which an FPGA operator library or generator must address.

7.1 *Algorithmic Choices*

The most fundamental choice is the choice of the algorithm used. For the same function, widely different algorithms may be used. Here are but a few examples.

- For many algebraic or elementary functions, there is a choice between multiplier-based approaches such as polynomial approximation [16, 20, 30, 61, 70] or Newton–Raphson iterations [45, 55, 73], and digit-recurrence techniques, based essentially on table look-ups and additions, such as CORDIC and its derivatives for exponential, logarithm, and trigonometric functions [4, 55, 63, 71, 77], or the SRT family of algorithms for division and square root [36]. Polynomials have lower latency but consume DSP blocks, while digit-recurrence consume only logic but have a larger latency. The best choice here depends on the format, on the required performance (latency and frequency), on the capabilities of the target FPGA, and also on the global allocation of resources within the application (are DSP a scarce resource or not?).
- Many algorithms replace expensive parts of the computations with tables of pre-computed values. With their huge internal memory bandwidth, FPGAs are good candidates for this. For instance, multiplication modulo some constant (a basic operator for RNS arithmetic or some cryptography applications) can be computed out of the formula $X \times Y \mod n = ((X+Y)^2 - (X-Y)^2)/4 \mod n$, where the squares modulo n can be tabulated (this is a 1-input table, whereas tabulating directly the product modulo n would require a 2-input table of quadratically larger size). Precomputed values are systematically used for elementary functions, for instance the previous exponential, for single-precision, can be built out of one 18-Kbits dual-port memory (holding both boxes e^A and $e^Z - Z - 1$ of Fig. 16) and one 18×18 multiplier [31]. They are also the essence of the multipartite [27] and HOTBM [20] generic function approximation methods. Such methods typically offer a trade-off between computation logic, table size, and performance. Their implementation should expose this trade-off, because the optimal choice will often be application dependent.
- In several operators, such as addition or logarithm, the normalization of the result requires a leading-zero counter. This can be replaced with a leading-zero anticipator (LZA) which runs in parallel of the significand datapath, thus reducing latency [56].

- In floating-point addition, besides the previous LZA, several algorithmic tricks reduce the latency at the expense of area. A dual-path adder implements a separate datapath dedicated to cancellation cases, thus reducing the critical path of the main datapath.
- The Karatsuba technique can be used to reduce DSP consumption of large multiplications at the expense of more additions [25].

7.2 Sequential Versus Parallel Implementation

Many arithmetic algorithms are sequential in nature: they can be implemented either as a sequential operator requiring n cycles on hardware of size n with a throughput of one result every n cycle, or alternatively as a pipelined operator requiring n cycles on hardware of size $n \times s$ with a throughput of one result per cycle. Classical examples are SRT division or square root [36] and CORDIC [4].

Multiplication belongs to this class, too, but with the advent of DSP blocks the granularity has increased. For instance, using DSP blocks with 17×17-bit multipliers, a multiplication of 68×68 (where $68 = 4 \times 17$) can be implemented as either a sequential process consuming 4 DSP blocks with a throughput of one result every 4 cycles, or a fully pipelined operator with a throughput of 1 result per cycle, but consuming 16 DSP blocks.

7.3 Pipelining Tuning

Finally, any combinatorial operator may be pipelined to an arbitrary depth, exposing a trade-off between frequency, latency, and area. FPGAs offer plenty of registers for this: there is one register bit after each LUT, and many others within DSP blocks and embedded memories. Using these is in principle for free: going from a combinatorial to a deeply pipelined implementation essentially means using otherwise unused resources. However, a deeper pipeline will need more registers for data synchronization, and put more pressure on routing.

FloPoCo inputs a target frequency and attempts to pipeline its operators for this frequency [26]. Such frequency-directed pipelining is, in principle, compositional: one can build a large pipeline operating at frequency f out of sub-components operating themselves at frequency f.

8 Open Issues and Challenges

We have reviewed many opportunities of FPGA-specific arithmetic, and there are many more waiting to be discovered. We believe that exploiting these opportunities is a key ingredient of successful HPC on FPGA. The main challenge is now probably to put this diversity in the hands of programmers, so that they can exploit it in a productive way, without having to become arithmetic experts themselves. This section explores this issue and is concluded with a review of possible FPGA enhancements that would improve their arithmetic support.

8.1 Operator Specialization and Fusion in High-Level Synthesis Flows

In the HLS context, many classical optimizations performed by usual standard compilers should be systematically generalized to take into account opportunities of operator specialization and fusion. Let us take just one example. State-of-the-art compilers will consider replacing $A+A$ by $2A$, because this is an optimization that is worth investigating in software: the compiler balances using one instruction, or another. HLS tools are expected to inherit this optimization. Now consider replacing $A*A$ by A^2: this is syntactically similar, and it also consists in replacing one operator with another. But it is interesting only on FPGAs, where squaring is cheaper. Therefore, it is an optimization that we have to add to HLS tools.

Conversely, we didn't dare describe doubling as a specialization of addition, or $A-A=0$ as a specialization of subtraction: it would have seemed too obvious. However they are, and they illustrate that operator specialization should be considered one aspect of compiler optimization and injected in classical optimization problems such as constant propagation and removal, subexpression sharing, strength reduction, and others.

There is one more subtlety here. In classical compilers, algebraic rewriting (for optimization) is often prevented by the numerical discrepancies it would entail (different rounding, or possibly different overflow behaviour, etc). For instance, $(x*x)/x$ should not be simplified into x because it raises a NaN for $x=0$. In HLS tools for FPGAs, it will be legal to perform this simplification, at the very minor cost of "specializing" the resulting x to raise a NaN for $x=0$. This is possible also in software, of course, but at a comparatively larger cost. Another example is overflow behaviour for fixed-point datapath: The opportunity of enlarging the datapath locally (by one bit or two) to absorb possible overflows may enable more opportunities of algebraic rewriting.

However, as often in compilation, optimizations based on operator fusion and specialization may conflict with other optimizations, in particular operator sharing.

8.2 Towards Meta-Operators

We have presented two families of arithmetic cores that are too large to be provided as libraries: multiplication by a constant in Sect. 4.1 (there is an infinite number of possible constants) and function evaluator in Sect. 6.2 (there is an even larger number of possible functions). Such arithmetic cores can only be produced by generators, i.e. programs that input the specification of the operator, and output some structural description of the operator. Such generators were introduced very early by FPGA vendors (with Xilinx LogiCore and Altera MegaWizard). The shift from libraries to generators in turns opens many opportunities in terms of flexibility, parameterization, automation, testing, etc. [26], even to operators that could be provided as a library.

Looking forward, one challenge is now to push this transition one level up, to programming languages and compilers. Programming languages are still, for the most part, based on the library paradigm. We still describe *how* to compute and not *what* to compute. Ideally, the "how" should be compiled out of the "what", using operators generated on demand, and optimized to compute just right.

8.3 What Hardware Support for HPC on FPGA?

We end this chapter with some prospective thoughts on FPGA architecture: how could FPGAs be enhanced to better support arithmetic efficiency? This is a very difficult question as the answer is, of course, very application dependent.

In general, the support of fixed-point is excellent. The combination of fast carries for addition, DSP blocks for multiplication, and LUTs or embedded memories for tabulating precomputed values covers most of the needs. The granularity of the hard multiplications could be smaller: we could gain arithmetic efficiency if we could use a 18×18 DSP block as four independent 9×9 multipliers, for instance. However, such flexibility would double the number of I/O to the DSP block, which has a cost: arithmetic efficiency is but one aspect of the overall chip efficiency.

Floating point support is also fairly good. In general, a floating-point architecture is built out of a fixed-point computation on the significand, surrounded by shifts and leading zero counting for significand alignment and normalization. A straightforward idea could be to enhance the FPGA fabric with hard shifter and LZC blocks, just like hard DSP blocks. However, such blocks are more difficult to compose into larger units than DSP blocks. For the shifts, a better idea, investigated by Moctar et al. [53] would be to perform them in the reconfigurable routing network: it is based on multiplexers whose control signal comes from a configuration bit. Enabling some of these multiplexers to optionally take their control signal from another wire would enable cheaper shifts.

It has been argued that FPGAs should be enhanced with complete hard floating-point units. Current high-end graphical processing units (GPUs) are paved with such units, and indeed this solution is extremely powerful for a large class of floating-point computing tasks. However, there has also been several articles lately showing that FPGAs can outperform these GPUs on various applications thanks to their better flexibility. We therefore believe that floating-point in FPGAs should remain flexible and arithmetic-efficient, and that any hardware enhancement should preserve this flexibility, the real advantage of FPGA-based computing.

Acknowledgments Some of the work presented here has been supported by ENS-Lyon, INRIA, CNRS, Universit Claude Bernard Lyon, the French Agence Nationale de la Recherche (projects EVA-Flo and TCHATER), Altera, Adacsys and Kalray.

References

1. L. Aksoy, E. Costa, P. Flores, J. Monteiro, Exact and approximate algorithms for the optimization of area and delay in multiple constant multiplications. IEEE Trans. Comp.-Aided Des. Integrated Circ. Syst. **27**(6), 1013–1026 (2008)
2. C. Alias, B. Pasca, A. Plesco, Automatic generation of FPGA-specific pipelined accelerators, in *Applied Reconfigurable Computing* (2010) http://www.springer.com/computer/communication+networks/book/978-3-642-12132-6
3. Altera: FFT/IFFT block floating point scaling. Application Note 404 (2005)
4. R. Andraka, A survey of CORDIC algorithms for FPGA based computers, in *Field Programmable Gate Arrays* (ACM, New York, 1998), pp. 191–200
5. R. Andraka, Hybrid floating point technique yields 1.2 gigasample per second 32 to 2048 point floating point FFT in a single FPGA, in *High Performance Embedded Computing Workshop* (2006) http://www.andraka.com/papers.htm
6. M. Arnold, S. Collange, A real/complex logarithmic number system ALU. IEEE Trans. Comp. **60**(2), 202–213 (2011)
7. T.O. Bachir, J.P. David, Performing floating-point accumulation on a modern FPGA in single and double precision, in *Field-Programmable Custom Computing Machines* (IEEE, New York, 2010), pp. 105–108
8. S. Banescu, F. de Dinechin, B. Pasca, R. Tudoran, Multipliers for floating-point double precision and beyond on FPGAs. ACM SIGARCH Comp. Architect. News **38**, 73–79 (2010)
9. R. Bernstein, Multiplication by integer constants. Software Pract. Ex. **16**(7), 641–652 (1986)
10. M.R. Bodnar, J.R. Humphrey, P.F. Curt, J.P. Durbano, D.W. Prather, Floating-point accumulation circuit for matrix applications, in *Field-Programmable Custom Computing Machines* (IEEE, New York, 2006), pp. 303–304
11. M. Boersma, M. Kröner, C. Layer, P. Leber, S.M. Müller, K. Schelm, The POWER7 binary floating-point unit, in *Symposium on Computer Arithmetic* (IEEE, New York, 2011)
12. D. Boland, G. Constantinides, Bounding variable values and round-off effects using Handelman representations. Trans. Comp.-Aided Des. Integrated Circ. Syst. **30**(11), 1691–1704 (2011)
13. N. Boullis, A. Tisserand, Some optimizations of hardware multiplication by constant matrices. IEEE Trans. Comp. **54**(10), 1271–1282 (2005)
14. N. Brisebarre, F. de Dinechin, J.M. Muller, Integer and floating-point constant multipliers for FPGAs, in *Application-specific Systems, Architectures and Processors* (IEEE, New York, 2008), pp. 239–244
15. K. Chapman, Fast integer multipliers fit in FPGAs (EDN 1993 design idea winner).EDN Magazine (1994)
16. R.C.C. Cheung, D.U. Lee, W. Luk, J.D. Villasenor, Hardware generation of arbitrary random number distributions from uniform distributions via the inversion method. IEEE Trans. Very Large Scale Integrat. Syst. **15**(8), 952–962 (2007)
17. S. Chevillard, J. Harrison, M. Joldes, C. Lauter, Efficient and accurate computation of upper bounds of approximation errors. Theor. Comp. Sci. **412**(16), 1523–1543 (2011)
18. P. Cousot, R. Cousot, Abstract interpretation: a unified lattice model for static analysis of programs by construction or approximation of fixpoints, in *Principles of Programming Languages* (ACM, New York, 1977), pp. 238–252
19. A. Dempster, M. Macleod, Constant integer multiplication using minimum adders. Circ. Dev. Syst. **141**(5), 407–413 (1994)

20. J. Detrey, F. de Dinechin, Table-based polynomials for fast hardware function evaluation, in *Application-specific Systems, Architectures and Processors* (IEEE, New York, 2005), pp. 328–333
21. J. Detrey, F. de Dinechin, A tool for unbiased comparison between logarithmic and floating-point arithmetic. J. VLSI Signal Process. **49**(1), 161–175 (2007)
22. J. Detrey, F. de Dinechin, Floating-point trigonometric functions for FPGAs, in *Field Programmable Logic and Applications* (IEEE, New York, 2007), pp. 29–34
23. F. de Dinechin, Multiplication by rational constants. IEEE Trans. Circ. Syst. II **52**(2), pp. 8–102 (2012)
24. F. de Dinechin, L.S. Didier, Table-based division by small integer constants, in *Applied Reconfigurable Computing* (2012), pp. 53–63 http://www.springer.com/computer/communication+networks/book/978-3-642-28364-2
25. F. de Dinechin, B. Pasca, Large multipliers with fewer DSP blocks, in *Field Programmable Logic and Applications* (IEEE, New York, 2009)
26. F. de Dinechin, B. Pasca, Designing custom arithmetic data paths with FloPoCo. IEEE Des. Test Comp. **28**(4), 18–27 (2011)
27. F. de Dinechin, A. Tisserand, Multipartite table methods. IEEE Trans. Comp. **54**(3), 319–330 (2005)
28. F. de Dinechin, B. Pasca, O. Creţ, R. Tudoran, An FPGA-specific approach to floating-point accumulation and sum-of-products, in *Field-Programmable Technology* (IEEE, New York, 2008), pp. 33–40
29. F. de Dinechin, H. Takeugming, J.M. Tanguy, A 128-tap complex FIR filter processing 20 giga-samples/s in a single FPGA, in *44th Asilomar Conference on Signals, Systems & Computers* IEEE, New York (2010) http://www.ieee.org/conferences_events/conferences/conferencedetails/index.html?Conf_ID=18339
30. F. de Dinechin, M. Joldes, B. Pasca, Automatic generation of polynomial-based hardware architectures for function evaluation, in *Application-specific Systems, Architectures and Processors* (IEEE, New York, 2010)
31. F. de Dinechin, B. Pasca, Floating-point exponential functions for DSP-enabled FPGAs, in *Field-Programmable Technology* (IEEE, New York, 2010)
32. F. de Dinechin, H.D. Nguyen, B. Pasca, Pipelined FPGA adders, in *Field Programmable Logic and Applications* (IEEE, New York, 2010)
33. F. de Dinechin, C. Lauter, G. Melquiond, Certifying the floating-point implementation of an elementary function using Gappa. IEEE Trans. Comp. **60**(2), 242–253 (2011)
34. V. Dimitrov, L. Imbert, A. Zakaluzny, Multiplication by a constant is sublinear, in *18th Symposium on Computer Arithmetic* (IEEE, New York, 2007), pp. 261–268
35. P. Echeverría, M. López-Vallejo, Customizing floating-point units for FPGAs: Area-performance-standard trade-offs. Microprocessors Microsyst. **35**(6), 535–546 (2011)
36. M.D. Ercegovac, T. Lang, *Digital Arithmetic* (Morgan Kaufmann, Los Altos, 2004)
37. O. Gustafsson, F. Qureshi, Addition aware quantization for low complexity and high precision constant multiplication. IEEE Signal Process. Lett. **17**(2), 173–176 (2010)
38. O. Gustafsson, A.G. Dempster, K. Johansson, M.D. Macleod, Simplified design of constant coefficient multipliers. Circ. Syst. Signal Process. **25**(2), 225–251 (2006)
39. M. Huang, D. Andrews, Modular design of fully pipelined accumulators, in *Field-Programmable Technology* IEEE (2010), pp. 118–125 http://ieeexplore.ieee.org/xpl/mostRecentIssue.jsp?punumber=5677390
40. IEEE standard for floating-point arithmetic. IEEE 754-2008, also ISO/IEC/IEEE 60559:2011 (2008) http://ieeexplore.ieee.org/xpl/articleDetails.jsp?arnumber=4610935
41. K. Kalliojarvi, J. Astola, Roundoff errors in block-floating-point systems. IEEE Trans. Signal Process. **44**(4), 783–790 (1996)
42. D. Knuth, *The Art of Computer Programming: Seminumerical Algorithms*, vol. 2, 3rd edn. (Addison Wesley, Reading, 1997)
43. U. Kulisch, Circuitry for generating scalar products and sums of floating point numbers with maximum accuracy. United States Patent 4622650 (1986)

44. U.W. Kulisch, *Advanced Arithmetic for the Digital Computer: Design of Arithmetic Units* (Springer, Berlin, 2002)
45. M. Langhammer, Foundation of FPGA acceleration, in *Fourth Annual Reconfigurable Systems Summer Institut* (2008)
46. M. Langhammer, T. VanCourt, FPGA floating point datapath compiler. Field-Program. Custom Comput. Mach. **17**, 259–262 (2009)
47. D. Lee, A. Gaffar, R. Cheung, O. Mencer, W. Luk, G. Constantinides, Accuracy-guaranteed bit-width optimization. Trans. Comp.-Aided Des. Integrated Circ. Syst. **25**(10), 1990–2000 (2006)
48. V. Lefèvre, Multiplication by an integer constant. Tech. Rep. RR1999-06, Laboratoire de l'Informatique du Parallélisme, Lyon, France (1999)
49. J. Liang, R. Tessier, O. Mencer, Floating point unit generation and evaluation for FPGAs, in *Field-Programmable Custom Computing Machines* (IEEE, New York, 2003)
50. Z. Luo, M. Martonosi, Accelerating pipelined integer and floating-point accumulations in configurable hardware with delayed addition techniques. IEEE Trans. Comp. **49**, 208–218 (2000)
51. D.R. Lutz, Fused multiply-add microarchitecture comprising separate early-normalizing multiply and add pipelines, in *Symposium on Computer Arithmetic* (IEEE, New York, 2011), pp. 123–128
52. M. Mehendale, S.D. Sherlekar, G. Venkatesh, Synthesis of multiplier-less FIR filters with minimum number of additions, in *Computer-Aided Design* IEEE/ACM (1995), pp. 668–671 http://ieeexplore.ieee.org/xpl/mostRecentIssue.jsp?punumber=3472
53. Y.O.M. Moctar, N. George, H. Parandeh-Afshar, P. Ienne, G.G. Lemieux, P. Brisk, Reducing the cost of floating-point mantissa alignment and normalization in FPGAs, in *Field Programmable Gate Arrays* (ACM, New York, 2012), pp. 255–264
54. R.E. Moore, *Interval Analysis* (Prentice Hall, Englewood Cliffs, 1966)
55. J.M. Muller, *Elementary Functions, Algorithms and Implementation*, 2nd edn. (Birkhäuser, Boston, 2006)
56. J.M. Muller, N. Brisebarre, F. de Dinechin, C.P. Jeannerod, V. Lefèvre, G. Melquiond, N. Revol, D. Stehlé, S. Torres, *Handbook of Floating-Point Arithmetic* (Birkhäuser, Boston, 2010)
57. A. Nayak, M. Haldar, A. Choudhary, P. Banerjee, Precision and error analysis of MATLAB applications during automated hardware synthesis for FPGAs, in *Design, Automation and Test in Europe* (IEEE, New York, 2001), pp. 722–728
58. H.D. Nguyen, B. Pasca, T. B. Preußer, FPGA-specific arithmetic optimizations of short-latency adders, in *Field Programmable Logic and Applications* (IEEE, New York, 2010)
59. B. Parhami, *Computer Arithmetic: Algorithms and Hardware Designs*, 2nd edn. (Oxford University Press, London, 2010)
60. S. Perry, Model based design needs high level synthesis: a collection of high level synthesis techniques to improve productivity and quality of results for model based electronic design, in *Conference on Design, Automation and Test in Europe* IEEE (2009), pp. 1202–1207 http://ieeexplore.ieee.org/xpl/mostRecentIssue.jsp?punumber=4926138
61. J.A. Piñeiro, J.D. Bruguera, High-speed double-precision computation of reciprocal, division, square root, and inverse square root. IEEE Trans. Comp. **51**(12), 1377–1388 (2002)
62. M. Potkonjak, M. Srivastava, A. Chandrakasan, Efficient substitution of multiple constant multiplications by shifts and additions using iterative pairwise matching, in *Design Automation Conference* (1994), pp. 189–194
63. R. Pottathuparambil, R. Sass, A parallel/vectorized double-precision exponential core to accelerate computational science applications, in *Field programmable gate arrays* (ACM, New York, 2009), pp. 285–285
64. T.B. Preußer, R.G. Spallek, Mapping basic prefix computations to fast carry-chain structures, in *Field Programmable Logic and Applications* (IEEE, New York, 2009), pp. 604–608
65. R. Rocher, D. Menard, N. Herve, O. Sentieys, Fixed-point configurable hardware components. EURASIP J. Embedded Syst. (2006) http://jes.eurasipjournals.com/content/2006/1/023197

66. O. Sarbishei, K. Radecka, Z. Zilic, Analytical optimization of bit-widths in fixed-point LTI systems. IEEE Trans. Comp.-Aided Des. Integrated Circ. Syst. **31**(3), 343–355 (2012)
67. M.J. Schulte, K.E. Wires, J.E. Stine, Variable-correction truncated floating point multipliers, in *Asilomar Conference on Signals, Circuits and Systems* IEEE, New York (2000), pp. 1344–1348 http://www.ieee.org/conferences_events/conferences/conferencedetails/index.html?Conf_ID=18339
68. S. Sun, J. Zambreno, A floating-point accumulator for FPGA-based high performance computing applications, in *Field-Programmable Technology*, IEEE, (2009), pp. 493–499
69. J. Thong, N. Nicolici, An optimal and practical approach to single constant multiplication. IEEE Trans. Comp.-Aided Des. Integrated Circ. Syst. **30**(9), 1373–1386 (2011)
70. A. Tisserand, High-performance hardware operators for polynomial evaluation. Int. J. High Perform. Syst. Architect. **1**, 14–23 (2007)
71. J. Volder, The CORDIC computing technique. IRE Trans. Electron. Comp. **EC-8**(3), 330–334 (1959)
72. Y. Voronenko, M. Püschel, Multiplierless multiple constant multiplication. ACM Trans. Algorithms Article 11, **3**(2) (2007) pp. 1–38 http://dl.acm.org/citation.cfm?id=1240234&dl=ACM&coll=DL&CFID=267988711&CFTOKEN=12528398
73. X. Wang, S. Braganza, M. Leeser, Advanced components in the variable precision floating-point library, in *Field-Programmable Custom Computing Machines* (IEEE Computer Society, Silver Spring, 2006), pp. 249–258
74. S. White, Applications of Distributed Arithmetic to Digital Signal Processing: A Tutorial Review. IEEE ASSP Magazine, **6**(3), pp. 4–19 (1989) http://ieeexplore.ieee.org/xpl/login.jsp?tp=&arnumber=29648&url=http%3A%2F%2Fieeexplore.ieee.org%2Fxpls%2Fabs_all.jsp%3Farnumber%3D29648
75. K.E. Wires, M.J. Schulte, D. McCarley, FPGA resource reduction through truncated multiplication, in *Field Programmable Logic and Applications* (Springer, Berlin, 2001), pp. 574–583
76. M. Wirthlin, Constant coefficient multiplication using look-up tables. J. VLSI Signal Process. **36**(1), 7–15 (2004)
77. Xilinx: LogiCORE IP CORDIC v4.0 (2011) xilinx.com
78. L. Zhuo, V.K. Prasanna, High performance linear algebra operations on reconfigurable systems, in *Supercomputing* (ACM/IEEE, New York, 2005)

Acceleration of the Discrete Element Method: From RTL to C-Based Design

Benjamin Carrion Schafer and Kazutoshi Wakabayashi

Abstract Field programmable gate arrays (FPGAs) have been extensively used to accelerate numerical intensive applications as they provide an excellent platform to exploit low- and high-level parallelism. Most of these computational intensive applications are given in high level languages, e.g. C or C++. A direct path from these descriptions to RTL is therefore highly desirable. Many attempts have been made to provide this direct synthesis path. Many years ago we implemented a multi-FPGA scalable custom hardware architectures to accelerate the discrete element method (DEM) in VHDL (Schafer et al. Comput. Struct. 82(20–21), 1707–1718, 2004) and also tried to use high level synthesis (HLS) tools available at that time, unsuccessfully due to their limitations and bad quality of results (QoR). We have re-implemented this custom hardware architecture in C using a state-of-the-art HLS tool and show in this chapter that it is now possible to design and verify entire systems in C achieving comparable QoR to hand-coded RTL. A step by step guide of how to create C-based FPGA designs with results comparable to that of the original hand-coded RTL architecture is presented as well as different benefits that come along behavioural synthesis, including design turn around time (TAT) reduction, design space exploration and high-speed cycle-accurate model generators for full chip verification.

1 Introduction

Reconfigurable computing is based around the use of field programmable gate arrays (FPGAs) to form co-processors that can be configured to provide custom hardware accelerators. The types of problem that can benefit from reconfigurable computing are established by the properties of the FPGA. In general, FPGAs are

B.C. Schafer (✉) • K. Wakabayashi
System IP Core Department, Central Research Laboratory, NEC Corporation, Kawasaki, Japan
e-mail: schafer@jhu.edu; wakabayashik@bq.jp.nec.com

W. Vanderbauwhede and K. Benkrid (eds.), *High-Performance Computing Using FPGAs*,
DOI 10.1007/978-1-4614-1791-0_22, © Springer Science+Business Media, LLC 2013

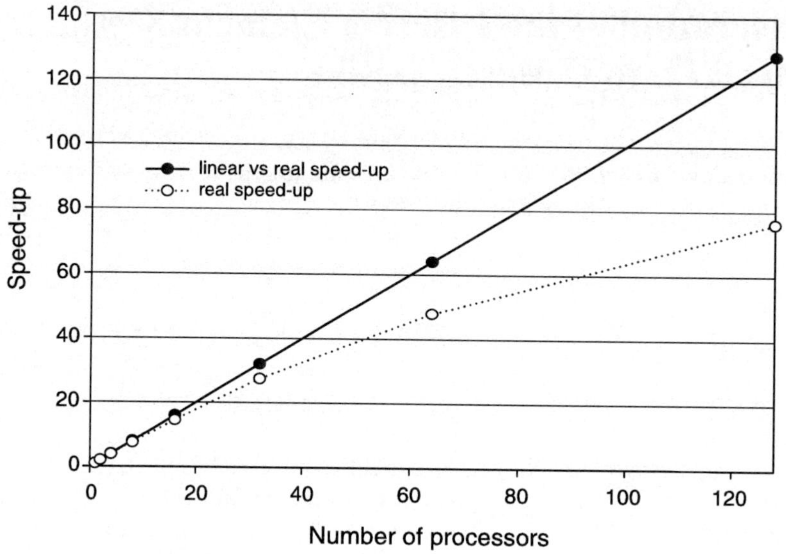

Fig. 1 Alter technologies transputer speed up [1]

good at tasks that use short word length integer or fixed-point data and exhibit a high degree of parallelism. For the right type of application, such a reconfigurable computer can rival and exceed expensive parallel computers that are normally used to accelerate computationally expensive algorithms. FPGAs thus open a new window to low cost hardware acceleration.

Conventional parallel computers suffer from poor system efficiency when solving the discrete element method (DEM), which means that they give a relatively disappointing speed-up due to communication and synchronization penalties as well as load balancing problems. The DEM has properties that suggest that it may be suitable for acceleration using FPGAs: it exhibits an enormous degree of parallelism and, as found in the current work, can be processed using short word length arithmetic.

In order to make it possible to simulate present day large-scale DEM problems, parallel processor systems are used. Having multiple processors working in parallel should accelerate the simulation time considerably, but the load balancing problems and synchronization and communication overheads will prevent these systems from achieving linear speed-ups.

Previous attempts to parallelize the DEM on different multiprocessor platforms are presented in chronological order, so that they can also be evaluated in terms of computational resources available at that time.

The parallel processing lab at the Colorado School of Mines was one of the first to parallelize the DEM using a parallel computer from Alter Technologies, which has 64 T805 processors [1] (see Fig. 1).

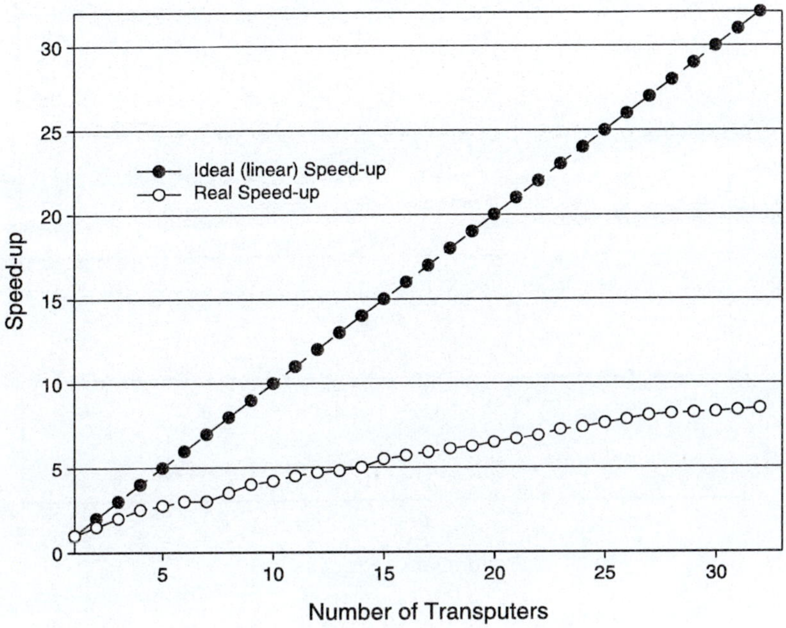

Fig. 2 Cray T3D massively parallel computer [2]

One of the most important choices when mapping the DEM onto a multi-processor machine is the way in which the domain is decomposed, and how each part is assigned to the single processors. This means that different processors handle the different geometric areas. The obvious choice is to divide the domain into rows, columns or a grid. In this case the authors decided to split the domain into vertical columns in order to give good load balance, as the only external force applied to these simulations was gravity. A big advantage of the domain decomposition method is that each processor runs a code that is only a minor modification of the serial version. The only notable changes that are needed are in the set up stage, where the domain has to be split among the different processors, and during simulation, where a new step is needed in order to exchange information between processors.

The department of mathematics of the École Polytechnique of Lausanne also designed a parallel version of the DEM running on a Cray T3D massively parallel computer [2] (see Fig. 2). The CSIRO Mathematical and Information Sciences performed another parallel implementation of the DEM, this time on a Swiss-T0-Dual machine. The system was implemented on a Swiss-T0-Dual machine. This is a cluster computer system consisting of eight Digital Alpha 21164 dual-processor boxes [3] (see Fig. 3).

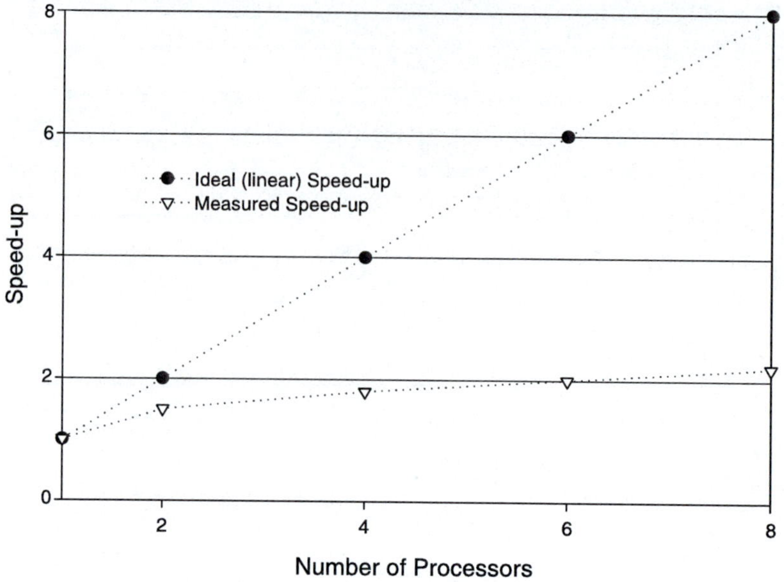

Fig. 3 Swiss-T0-dual machine [3]

All multiprocessor platforms surveyed suffered the same problems: synchronization and communication overheads between processors, as well as poor load balancing between processors, made the speed-up less than linear. Novel computational approaches have to be considered in order to find a more efficient way to accelerate the DEM at an affordable price while obtaining the computational results within a reasonable real time.

The DEM has properties that suggest that it may be suitable for acceleration using FPGAs: (a) It exhibits an enormous degree of parallelism and (b) it is an explicit self-correcting algorithm. It is therefore tempting to examine how well the DEM would map into an FPGA.

2 The Discrete Element Method

Granular materials are formed of distinct particles, which displace independently from one another and interact with each other only at the contact points [4]. The discrete nature of the granular materials leads to a complex behaviour under conditions of loading and unloading. In order to simulate the behaviour of a granular material, a suitable model has to be developed first. A process that happens in nature in a few seconds may take hours or even days to be simulated on a computer.

Cundall and Strack introduced the DEM in 1971 [4, 5]. This numerical method considers every particle as a separate entity. The interaction force, acceleration and movement of each particle are calculated individually at each time step. The assumptions underlying the method are only correct if no disturbance can travel beyond the immediate neighbours of a particle within one time step. This is due to the explicit nature of the method. This generally means that the time step must be limited to a very small value (of the order of milliseconds for the stiffness and density of a typical material, though using scaled stiffness or density can change its value), making the DEM extremely computationally expensive, since many time steps are needed if the dynamic behaviour of the system is required to be modelled accurately.

As computing power increases, so does the number of applications that can be modelled reasonably using the DEM. An even higher growth is expected during this decade as computing power keeps growing, and as the method starts to be used in order to model entire engineering structures (such as dams and tunnels), built of particles bonded together to represent solid material. It is further suggested that continuum methods will be replaced by particle approaches in the future [6], as these capture the behaviour of localized cracks much better than the continuum approach, and a suitable stress–strain law for the material may not exist or the law may be excessively complicated with many obscure parameters.

The main drawback to the application of particle methods to large-scale problems is that their very high computational demands limit the size of system that can be simulated within a feasible timescale. Also, time must be spent calibrating the laws by which the micro-structure affects the overall macro-structure behaviour.

A simulation starts by assuming some initial configuration of particle positions, and then computes which of the particles are touching. The simulation then proceeds by stepping in time, applying the sequence of operations in at each step. The force between two particles can be calculated from the strength of the contact between them. The resultant force on a particle is the vector sum of the forces exerted by each of its neighbours. Once the resultant force on each particle has been computed, it is simple to compute the acceleration, the velocity and the position increment for each particle. Finally, the list of which particles are in contact must be re-computed as shown in Fig. 4. The force interaction, acceleration and movement of each particle are calculated individually at each time step. The assumptions underlying the method are only correct if no disturbance can travel beyond the immediate neighbours of a particle within one time step. This generally means that the time step must be limited to a very small value.

The next sections of this chapter will be devoted to the detailed description of a dedicated FPGA-based DEM HW accelerator. The first part will focus on the architecture itself implemented in RTL (VHDL). The second part of the chapter will describe how the same architecture was re-implemented directly from the original ANSI-C description using a high level synthesis (HLS) tool. The last section highlights the differences between the RTL and the C-based implementation in terms of quality of results (QoR) of results, development time and verification effort required.

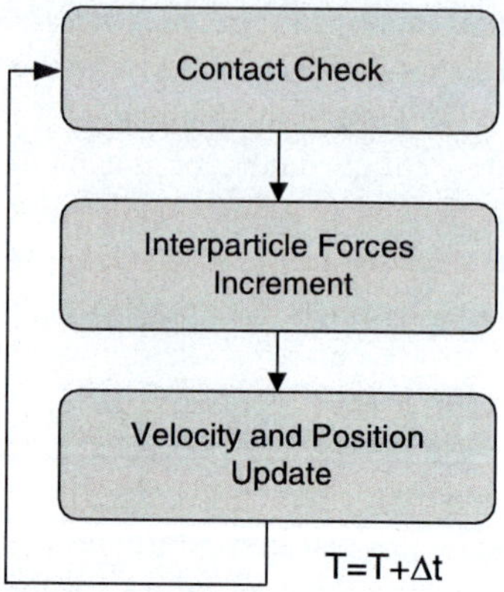

Fig. 4 DEM flow chart

3 Rt-Level DEM Hardware Acceleration

In order to accelerate the DEM as much as possible low- and high-level parallelism needs to be exploited. Figure 5 shows a block diagram of the hardware implementation. It consists of six main units:

- A *contact check unit*, which identifies the particles in contact
- A *force update unit*, which updates the inter-particle forces
- A *movement update unit*, which calculates the particles' new velocities and coordinates
- A *control unit*, which synchronizes all the units and generates all the control and address signals
- An *interface unit* to read and write data to and from the external memory
- A *write back* unit to write the results of the arithmetic units back to the internal FPGA memory

The block RAM of the FPGA is used to hold the data required to describe each particle. This includes position, velocity, angular momentum, identity of neighbours and the force that it is experiencing. Data is read from and written to the internal FPGA memory at a clock speed four times greater than that of the forces update units in order to keep its pipelines fully loaded (on each clock cycle of the force unit, it needs to read and write data of two particles simultaneously). This architecture has shown to reach speed-ups of up to $\times 30$ compared to a state-of-the-art PC and has also shown very good scalability properties allowing to fully overlap computation and communication. The architectural details of this design can be found in [7–9]. Summarizing briefly the results, the design was implemented on a PC containing a reconfigurable computing plug-in card. The FPGA card was a Celoxica RC1000-PP PCI containing a single Xilinx Virtex 2000E-8 and 4 banks of 2 MB of RAM. Due to limitations of hardware resources, only five contact check units could be instantiated in parallel.

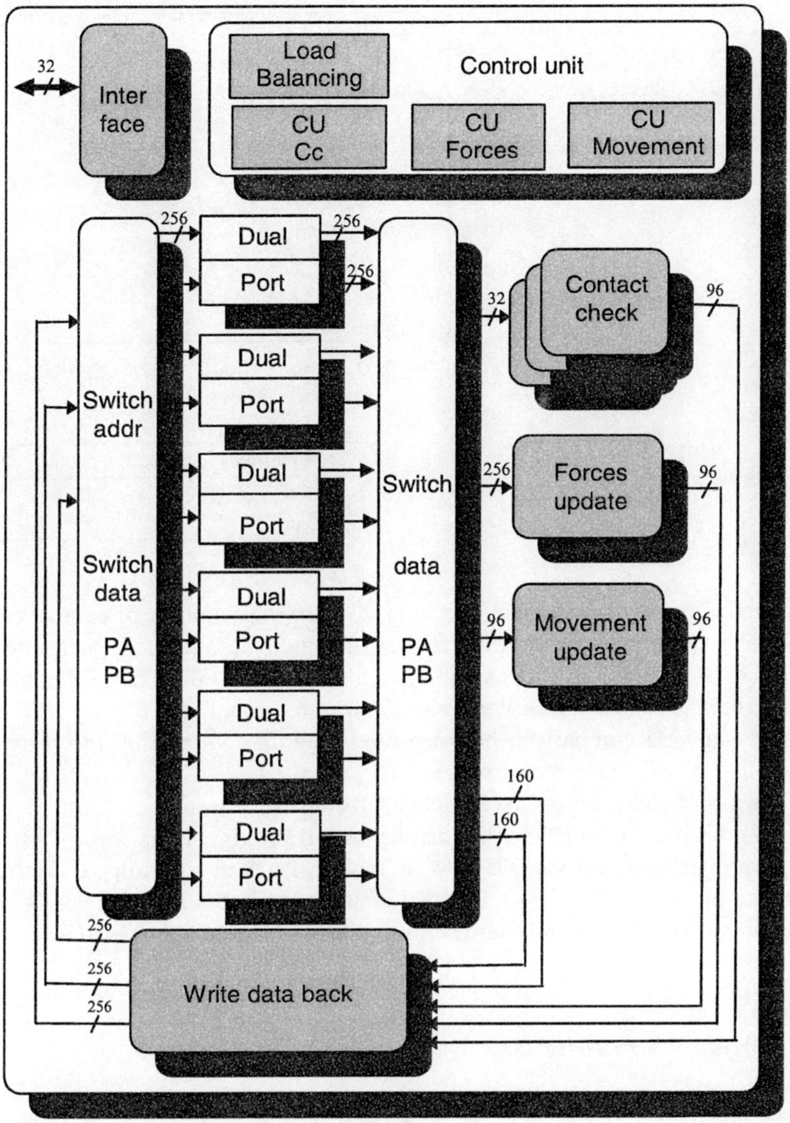

Fig. 5 DEM FPGA implementation block diagram

In order to have the system running at its maximal efficiency there must be as many contact check units as needed to make the time for force update t(fu) equal to the time for contact checking t(cc). Figure 6 illustrates the number of contact check units needed for this condition to be true (for a specific domain depth Y and ball diameter d). The straight line t(fu) is the number of clock cycles required to perform force update on all particles. The curves t(cc) show the number of clock

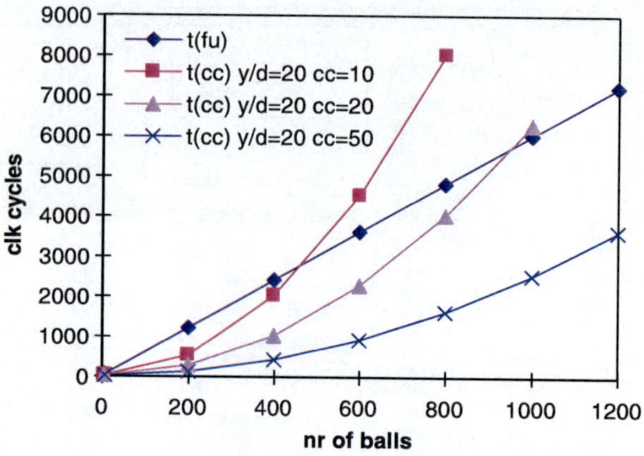

Fig. 6 Contact check units needed for "ideal" load balancing (t(cc) contact check time=t(fu)force update time)

Table 1 Comparisons of speed up obtained by HW DEM

No. of particles	1,500	2,000	2,500
Speed $up_{measured}$	27.4	32.1	29.2
Speed up_{ideal}	48.7	81	100.3

cycles needed to perform contact checking, for varying numbers of contact check units. Where a t(cc) line crosses the t(fu) line, this indicates ideal load balancing for that number of particles. So, for example, a simulation of 1,000 particles has almost ideal load balancing when 20 contact check units are instantiated.

Table 1 shows a comparison between the speed-up achieved by the hardware simulation as compared to the software.

For our simulations, only five contact check units were used due to limitations of hardware resource on the FPGA. As can be seen from Fig. 6, this was insufficient to give a proper load balance. The row in Table 1, labelled speed-up_{ideal} indicates an estimate of the degree of speed up that could be expected if a sufficiently large number of contact check units could be instantiated to give an ideal load balance.

3.1 Low-Level Parallelism

The implementation details of each of the HW design units will be described in detail in this subsection.

3.1.1 Contact Check

For each particle, a "contact list" is formed, which contains references to each of the particles with which it makes contact. In order to detect if two particles are in contact the following equation has to be solved:

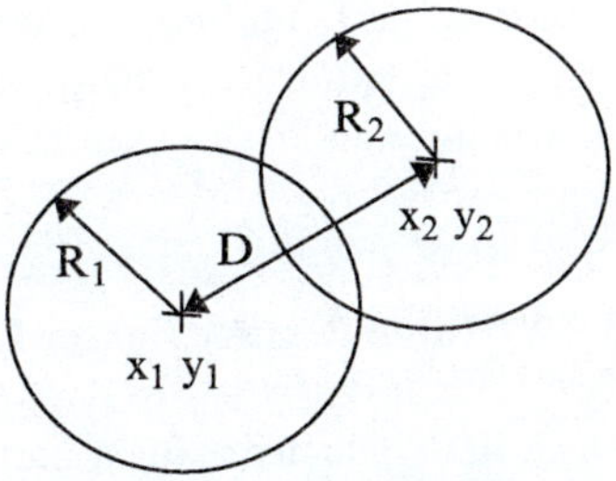

Fig. 7 Balls in contact example

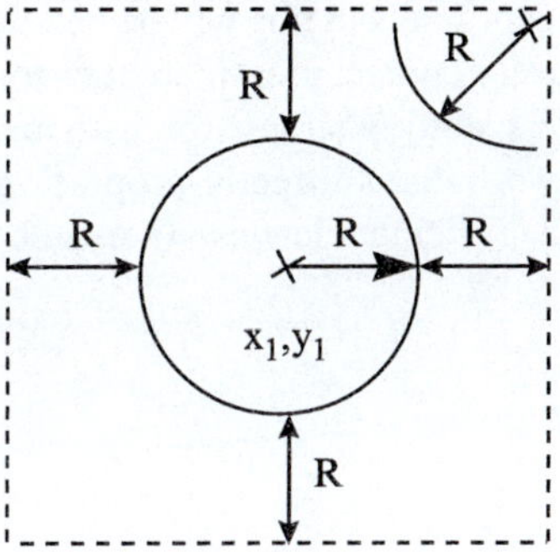

Fig. 8 Neighbour check model

$$\Delta n = R_1 + R_2 - \sqrt{(x_1 - x_2)^2 + (y_1 - y_2)^2} \geq 0 \tag{1}$$

where x_i y_i are the co-ordinates of each particle's centre and R_1 and R_2 are the respective radii (see Fig. 7); the repulsive force between the particles is directly proportional to this overlap. If the condition of Eq. (1) is true, the addresses of the two particles are added to each others' adjacency list. For this investigation, all particles are assumed to have the same radius R. Under this circumstance, simple geometry shows that for a 2-D simulation, the maximum number of contacts that each ball can have is six. This means that contact information can be represented by a very simple data structure, in which each particle has six memory slots allocated to hold the identities of the particles potentially in contact.

If there are N particles within a region of the DEM, then the number of contact checks that must be performed is N^2. The square roots and multiplications used in Eq. (1) are very expensive to perform in FPGA hardware, with the implication that a full contact check would be prohibitively expensive.

In order to reduce the number of operations needed to perform the contact checking, instead of checking for true contacts, our design only checks which particles are within each others' bounding boxes, and this acts as a filter before the actual contact check. Under some circumstances, this means that a pair of particles will be classified as neighbours even though they are not truly in contact (see Fig. 8). This causes no real problem, since it is detected and correctly handled by the force increment unit.

Using the bounding box method to perform contact checking makes this unit very cheap in terms of hardware resources, as it requires only two additions, two subtractions and four comparisons.

3.1.2 Inter-Particle Forces Increment

Once the contact list for a particle has been established, the total force acting on it can be determined. This will require a full solution of Eq. (1) for each contact identified, but this will be needed to be performed only a maximum of 6N times. For every contact identified between two particles, the resulting force is calculated. For this study, a simple force–displacement law is adopted: the resulting force between two balls is directly proportional to the indentation between the balls. The force displacement law used for each ball is as follows:

$$F_{xi} = k_{\mathrm{n}} \Delta n_{xi} \tag{2}$$

$$F_{yi} = k_{\mathrm{s}} \Delta n_{yi} \tag{3}$$

$$M_i = F_{\mathrm{s}i} R \tag{4}$$

where k_i is the stiffness (subscript n for normal and s for shear), n_{xi} and n_{yi} are, respectively, the x and y components of the current ball indentation against particle i, F_{xi} and Fy_i are the components of the force caused by the interaction with ball i, M_i is the moment acting on the current ball due to the ith ball, F_{si} is the shear force acting on the current ball due to the ith ball and R is the ball's radius. The index i runs from the first to the last ball on the present ball's adjacency list, so the resultant force on a ball is the vector sum of the forces caused by each contact with its neighbours.

The force update unit, which does require the computation of the terms in Eq. (1), requires a large amount of hardware. It also operates at a comparatively low clock speed of 7.5 MHz in contrast to the contact check unit which works at the full system clock speed of 30 MHz on the Celoxica RC1000 with a Virtex 2000E-8 FPGA. Re-synthesizing this design on a state-of-the-art FPGA (Virtex4 XC4VLX200) achieved clock speeds of 30 and 120 MHz, respectively.

Figure 9 shows the internal structure of this unit, where each column represents one pipeline stage.

It can be seen that there are three main paths in this structure: one that calculates the forces in the x direction, another that calculates the forces in the y direction, and another shorter path, which computes the terms in Eq. (1) (without doing the square root, which is expensive but unnecessary),to check if the particles are in contact.

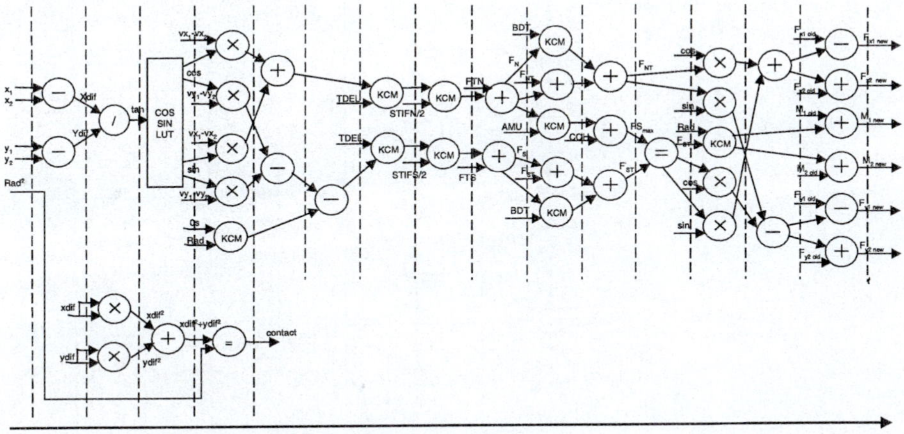

Fig. 9 Forces update unit internal structure

3.1.3 Velocity and Position Update

Once the resultant force on each ball has been calculated, these forces are used to find new accelerations using Newton's second law. In this study, it is assumed that the masses of all the balls are identical. These accelerations are integrated to obtain the velocities in the x and y directions and the angular velocity.

The new coordinates can be found by adding the original coordinates to the incremental displacement obtained by integrating the calculated velocities. The position update unit has an intermediate level of hardware complexity, and operates at the same speed as the force update unit, which is four times slower than the system clock, in order to have its pipeline fully loaded, achieving one new result per clock cycle.

It consists of three pipelines in parallel (see Fig. 10). The first computes x and v_x, the second computes y and v_y and the third computes θ and $\dot{\theta}$.

3.1.4 Write Back Unit

The purpose of this unit is to merge the data for each particle that emerges from the arithmetic units. Each particle is represented by a 256-bit word, but only certain bits of this word are updated by each of the different units. For example, if a new contact list is generated, only the memory locations of the old contact list are overwritten, and the rest of the old data is preserved.

Fig. 10 Velocity and position update unit internal structure

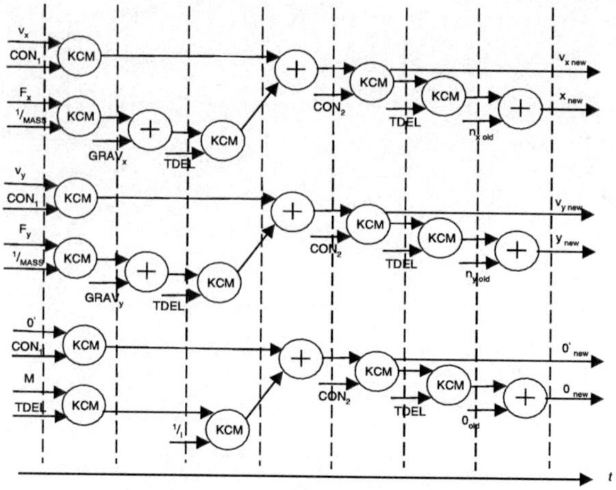

Table 2 Hardware requirements

Contact checking	Force update	Movement update
2 adders	23 adders	8 adders
2 subtractions	10 multipliers	15 KCMs
	8 KCMs	
	1 divider	
	1 Look Up Table (LUT)	

3.1.5 Interface Unit

The interface unit reads and writes data from and to the FPGA's internal memory when instructed to do so by the control unit in the low-level parallelism implementation, it is only used twice in each analysis. Once at the beginning, it is used to read in the new data, and once at the end when the calculations have finished in order to write the data back to the external memory.

3.1.6 Hardware Requirements

The hardware requirements for each of the main functional units are shown in Table 2. Constant coefficient multipliers (KCMs) require much less hardware resource than multipliers that allow both inputs to vary. Having balls of the same radius facilitates the widespread use of KCMs.

The contact checking unit is very simple, requiring little hardware resources and capable of operation at high clock speeds. The force update unit, which does require the computation of the terms in Fig. 9, requires a large amount of hardware. The movement update unit has an intermediate level of hardware complexity.

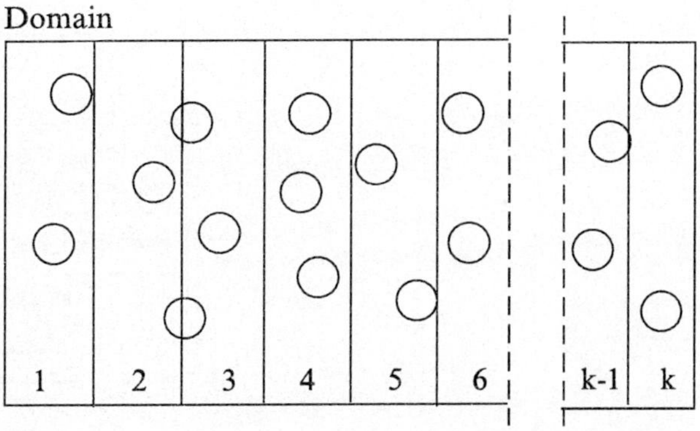

Fig. 11 Domain decomposition

3.2 *High-Level Parallelism*

High-level parallelism exploitation involves having all processing units working in parallel and overlapping computation and communication in order to keep all units at full workload as much as possible.

In order to allow the computational units to operate in parallel, the domain is decomposed into k vertical columnar sub-domains, as shown in Fig. 11. Each particle belongs to a particular cell, and for most particles contact checking and force updating need only be performed against the other particles within the same cell. For the small number of particles that are close to the boundary between two cells, more complicated arrangements are necessary.

The architecture divides the internal block RAM of the FPGA into six dual port RAMs in order to allow data to be read and written to the memory simultaneously. At any given time, six of the columnar cells shown in Fig. 11 are stored within the FPGA and undergo processing. The RAM contains two 256 bit entries for each particle within that cell consisting of 16 bit entries for x, y, θ, v_x, v_y, $\theta\prime$, F_x, F_y, M, a type flag, and the reference of up to six neighbouring particles and another to hold the normal and shear forces for every contact.

The control units generate the necessary control signals to synchronize data between the blocks and to steer the data output from the RAMs through the switch array to the inputs of the appropriate computation unit. The control units also generate the addresses to read and write data back to the internal and external memory. The FPGA must hold six columns at any given time, rather than three because a particle close to the boundary may be in contact not only with particles from its own column, but also from an adjacent column. This situation is handled by an auxiliary memory located within the control unit that handles inter-cell boundaries. Secondly, a particle close to a boundary may transition from one column to another during coordinate update. For such a particle, after the results of the co-ordinate update are written back to the RAM, the particle would have the correct coordinates, but its data would have been stored in the wrong block of RAM.

Fig. 12 DEM tasks duration

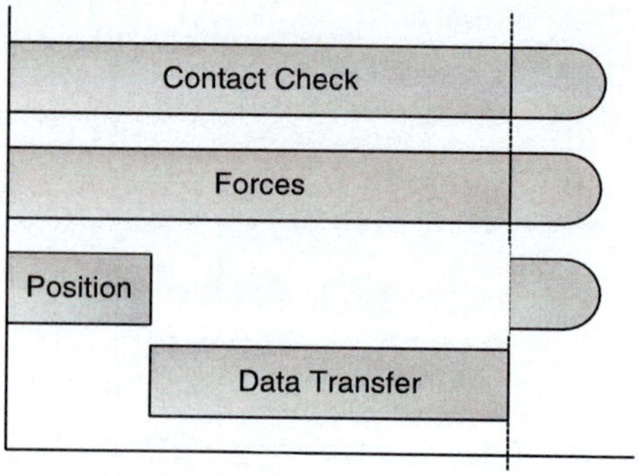

A third complication is that as simulation progresses, particles will move between columns, and some columns may become heavily populated, while others are sparsely populated. It is then necessary to move the cell boundaries, thus expanding some cells and contracting others. This is needed in order to provide good load balancing, and also to prevent overflow of the block RAMs. Movement of cell boundaries is fairly simple. The control unit monitors how many particles are held in each block RAM. When the number falls below a minimum threshold or rises above a maximum, the boundary is moved by a distance *R* so as to expand or contract the cell. When the boundary moves, a number of cells will find that their data is stored in the wrong column of RAM, but this will be automatically detected and corrected by the mechanisms described earlier for handling particles close to boundaries.

Using the procedures described above, the transition of particles from one cell to another is handled without causing any loss of performance. Also, the cell size is adaptively optimized so that good load balancing is always achieved.

With contact checking, force updating and co-ordinate updating being performed in parallel, load balancing problems will inevitably appear, since the overall system speed will be limited by the speed of the slowest of the three units. Figure 12 shows the task graph of the main computational tasks. The coordinate check is the most time consuming, but requires very simple hardware and can operate at high clock speed. In order to improve the load balance, several contact check units are instantiated, and operate in parallel. The number of contact check units to be used is a parameter of the design, which can be easily changed.

The next section describes the HLS tool used to re-implemented the previously described HW architecture directly from the DEM ANSI-C's SW description.

4 CyberWorkBench: Behavioural Synthesis and Verification

NEC has been developing a C-based behavioural synthesis called CyberWorkBench (CWB) since the late 1980s [10] and C-based verification tools, e.g. formal verification and simulation around CWB during the last 10 years [11]. All these tools are integrated into an IDE, where designers execute these tools upon the C-source code.

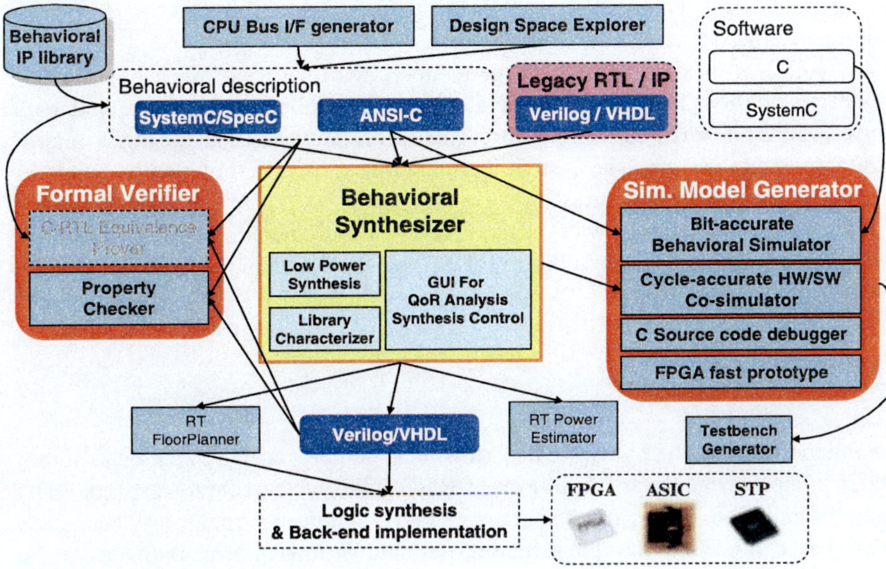

Fig. 13 CyberWorkBench overview

Figure 13 shows a block diagram of the main tools that compose CWB. CWB takes ANSI-C (with or without HW extensions called "BDL", or "Cyber-C") [12], or SystemC and synthesizes it into synthesizable RTL doing traditional HLS given a set of design constraints such as clock frequencies, number and kind of functional units and memories. The only restrictions that apply to the ANSI-C accepted as input are the use of constructs that do not have a direct translation into HW, for example dynamic memory allocation and recursion. Pointers are accepted if they can be resolved statically. Also, the SystemC accepted is only the SystemC synthesizable subset.

Usually legacy RTL or RTL IPs are handled as a black box, but if necessary, this RTL can also be fed to the behavioural synthesizer as an input. The behavioural synthesizer can then insert extra registers to speed up the original RTL and generate new RTL of smaller delay. It can also generate a SystemC cycle-accurate simulation model. The behavioural synthesis can therefore be considered as a Verilog, VHDL, C, C++ and SystemC unification step.

The "Library Characterizer" generates delay and area information of the functional units and memories for a particular ASIC technology or FPGA family. This is important because the synthesizer needs to know the exact delay and area of each basic operation in order to schedule the input description correctly, inserting registers at the right place to meet the target frequency. A Behavioural IP library, called "Cyberware", is also included in the synthesis environment. Cyberware works similarly to Xilinx's CoreGenerator or Altera's Megafunction, but with fully parametrizable IPs described at the C level making them much more flexible.

The designer can generate different RTLs with different area vs. performance trade-offs setting some synthesis parameters.

A "QoR" synthesis report of the generated circuit shows a quick overview of the design quality. The report file includes area, number of states, critical path delay, number of wires and routability. This information is used for quick micro-architectural exploration. The system architecture explorer automatically generates different hardware architectures based on the preferences and constraints entered by the user (area, latency, power) at the C level. A more detailed description of the advantages of design space exploration (DSE) can be found in the next sections.

4.1 Verification Flow

The functionality of the hardware described in C can be verified at the behavioural level, while performance and timing are verified at the cycle-accurate level (or RTL) through simulation. Debugging the generated RTL is however not an easy task since C variables can be shared in the same register, and various optimizations can make the automatically generated RTL not easy to understand. CWB therefore provides a behavioural C source code debugger linking directly the unsynthesized C code with a cycle-accurate model for timing verification.

Figure 14 shows an overview flow diagram of the different level of abstraction on which the different model generators. When using HLS, the first step designers need to take is to manually refine the original SW description in order to make it synthesizable. Some of the typical constructs that are not supported in HLS are dynamic memory allocation and recursion. At this stage the designer will also refine the data types in order to obtain the smallest and most efficient HW design possible. For this purpose most of the HLS tools extend the C syntax providing their own data types. For data type refinement verification a behavioural model generator is provided, which creates a C++ program that models the original behavioural description that has been manually refined. This model generator creates not only a C++ program that models the data types, but also a testbench that allows the re-use of input vectors used in the SW simulation and compare the results of the SW simulation with the results obtained during the behavioural simulation.

After HLS a cycle-accurate simulation can be performed for timing verification. For this purpose a cycle-accurate model generator that creates fast SystemC simulation models is provided. These SystemC models mimic the behaviour of the RTL cycle accurately and are created after the HLS scheduling phase takes place (an intermediate result of HLS). Again a testbench that allows the re-use the untimed SW inputs and outputs is created. Valid signals for each input and output ports are created automatically, because now the timing of when data needs to be read and when valid data is written out needs to be made visible to the testbench. This verification method is called "dynamic equivalence checking", because it is possible to fully re-use untimed input and output test vectors during a timed simulation. The last step in the verification flow is to verify the final RTL created by the HLS.

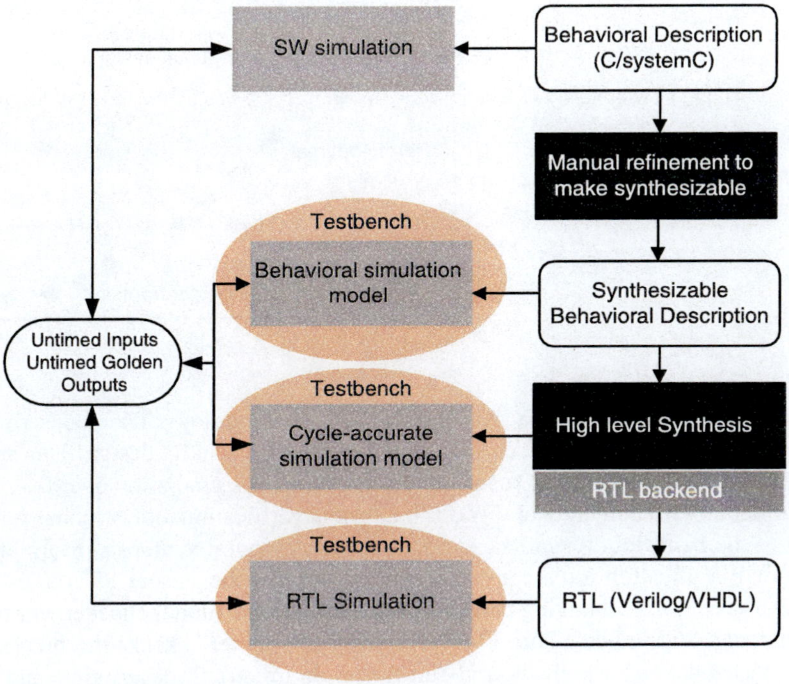

Fig. 14 High level synthesis verification flow overview

For this purpose an RTL testbench generator is provided that creates a testbench that can again re-use the untimed original input test pattern and compare the generated outputs with the original untimed golden outputs.

These different model generators combined with the RTL testbench generator allow designs to be verified at different design stages without having to resort to slow RTL simulations. The model generators have been proved to be consistently faster than RTL by a factor of $\times 1{,}000$ for the behavioural model and $\times 100$ faster for the cycle-accurate model [11].

After verifying each hardware module, the entire SoC is simulated in order to analyse the performance and/or to find inter-modules problems such as low performance through bus collision, or inconsistent bit orders between modules. Since such entire chip performance simulation is extremely slow in RTL-based HW–SW co-simulation, the cycle-accurate SystemC simulation models can be used which can run up to $100\times$ faster than RTL models. The simulator allows designers to simulate and debug both hardware and software at the C source code level at the same time. If any performance problems are found, designers can change the hardware–software partitioning or algorithm directly at the C level, and can then repeat the entire chip simulation. This flow implies a much smaller and therefore

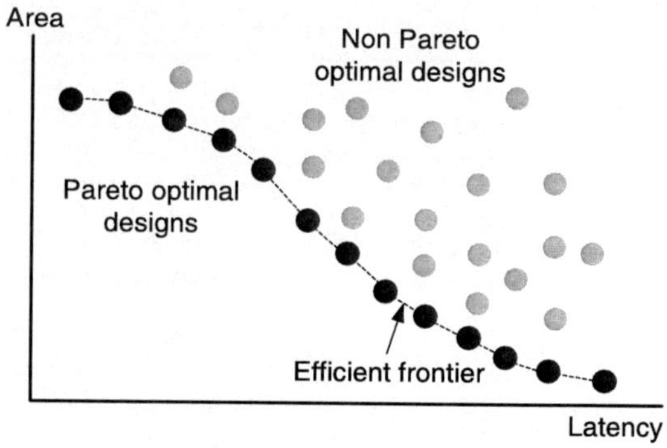

Fig. 15 Design space exploration result example

faster re-design cycle than in a conventional RTL methodology. The C description is the only initial and final SoC description language of the entire design. This entire chip simulation can be further accelerated using an FPGA emulation board.

Another important feature of CWB is the formal verification tool, which is tightly linked to the behavioural synthesizer. With the behavioural synthesis information, the formal verification tools can handle larger circuits than usual RTL tools and have C-source level debugging capability even though the model checker works on the generated RTL model. The "C-RTL equivalence prover" checks the functional equivalence between a behavioural (un-timed or timed) C description and the generated RTL, using information of the optimizations performed such as loop unrolling, loop merge and array expansion performed by the behavioural synthesis. Without such information, the equivalence check is almost impossible for large circuits. Designers can specify assertions or properties at the behavioural C level, similar to our cycle-accurate simulator. Such behavioural level properties/assertions are converted into RTL ones automatically and are passed to our RTL model checker.

4.2 Design Space Exploration

Raising the level of abstraction has a distinct advantage over traditional RTL design approaches. Multiple designs can be easily and quickly generated for the behavioural code, while RTL designs require major rework in order to modify the underlying architecture. Moreover higher levels of abstraction combined with HLS allow the architectural trade-off exploration of the behavioural description. The main objective in DSE is to find optimal implementations with respect to several, often conflicting, objectives. These optimal implementations are called Pareto-optimal designs. The objective is to find all the designs at the efficient frontier (also called Pareto front, see Fig. 15. The tradeoffs can easily be explored within this set rather than considering the entire design space, which would be impractical and irrelevant to the designer. The main problem in DSE is its exponential nature.

Heuristics have been developed to reduce runtime, at the expense of finding less Pareto-optimal designs and some that are not really Pareto-optimal i.e. [13–15]. These designs are called non-dominated designs as they dominate the exploration result obtained so far, but not the overall exploration space

The design options that can be explored during HLS are: (a) Global synthesis options. These are options that apply to the entire design e.g. scheduling mode and speculation (b) Local synthesis directives. These are synthesis directives that are normally manually specified by the user directly at the source code. These directives are in the form of pragmas that the HLS tool processes and in turn synthesizes the instrumented source code accordingly. Some of these include loops, arrays and functions. For example an array specified in C can be synthesized as a memory block or registers and a loop can be unrolled completely, partially or folded. (c) Lastly the maximum number of functional units (FUs) that the synthesizer can instantiate could also be explored. This will limit the amount of parallelism that the scheduler can extract and therefore the amount of resource sharing.

All three exploration methods are orthogonal to each other and should therefore be explored in combination in order to obtain the best possible Pareto-front.

We have developed an automatic design space explorer to find these dominating designs as quick as possible. The next sections describe the results of the C-based HW design including the results of the explorer for the main processes.

5 C-Based DEM Hardware Acceleration

The DEM RTL hardware implementation was re-implemented from the original software description given in ANSI C and synthesized using CWB. The next sections show the result of each of the main processes re-synthesized in C and compare the QoR with respect to the manual RTL design as well as the design turn around time (TAT) including verification.

The starting point for the C-based design is the original SW algorithm used to profile the code in order to determine which parts need to be mapped into dedicated HW and which part would still be executed in SW. The next steps involve re-writing the C code to make it synthesizable. This mainly involves: (1) re-writing non-synthesizable constructs, e.g. mallocs, recursion and pointers that cannot be statically resolved and (2) modifying the data types in order to convert the floating point values into fixed point. Because this had already been done at the RTL HW implementation, the data types and sizes where exactly the same in both implementations.

Once the input description has been made synthesizable, the next step is to set up the HLS tool in order to synthesize each process. This involves specifying the target frequency, i.e. 30 MHz (VirtexE) or 120 MHz (Virtex4), which is the same target frequency used for the RTL implementation and specifying the delay libraries for the target FPGA. In this case Xilinx Virtex4. CWB comes with a library characterizer that can re-generate this delay library for any FPGA family. These libraries are important because the HLS tool schedules as many operations in each clock cycle (given by the target frequency) as possible depending on the logic operations' delays. More information about how HLS works can be found at [16] and a good review of the evolution of HLS at [17]. Once the overall system has been set up we can proceed synthesizing each process separately. It should be remembered that HLS is a single process synthesis and does not perform global optimizations (intra-processes optimizations). Nevertheless the user can specify arrays as outside memories and the HLS tool will created the arbitration logic for the memory.

C-based design has shown to generate comparable results to RTL for some sort of applications [18, 19]. Moreover in order to design and verify complete SoC designs HLS tools need to be expanded to deal with system synthesis issues (ESL Synthesis), e.g. bus generator, top module generator, dealing with hierarchical designs. CWB has the advantage to include these types of system level capabilities making it easier and faster to create entire SoCs [20].

The next subsections will describe the results of the C-based synthesis for the contact check unit, the force and the position update units and the control unit. The control unit implementation is of special interest because it is a "control-intensive" application that traditionally could not be synthesized effectively in C. We will illustrate how this is done here.

5.1 *Contact Check*

In order to replicate the behaviour of the RTL implementation each of the processes is synthesized in automatic pipeline mode. CWB provides three different scheduling modes: (1) Manual, where the user manually times the C code inserting a timing descriptor directly at the source code (see Section 5.4). (2) Automatic, where the tool automatically schedules the input description. (3) Pipeline mode, which creates fully pipelined designs. In this case (pipelined mode) the user needs to specify the data initiation interval (DII). Here the DII=1, meaning that new data is arriving every clock cycle to the pipeline. Figure 16 shows the schematic view of the synthesized circuit. It has a latency of 1 cycle and consumes 120 slice LUTs.

One of the advantages of HLS is that it is very easy to perform DSE by changing the functional unit (FU) constraint file and/or synthesis directives. We execute the automatic DSE explorer of CWB in order to analyse the area vs. performance trade-offs in the contact check unit. The explorer only reports the dominating designs (ideally Pareto-optimal designs should be reported, but due to the size of the design space, Pareto-optimality can normally not be guaranteed). Figure 17 shows the result of the DSE. It can be observed that two designs with very different area vs. latency characteristics have been created. The first replicates the behaviour of the RTL implementation by generating a design with a latency of 1 clock cycle, while the second has a latency of 2 cycles, but requires only 99 LUTs vs. the 120 LUTs required by the first design.

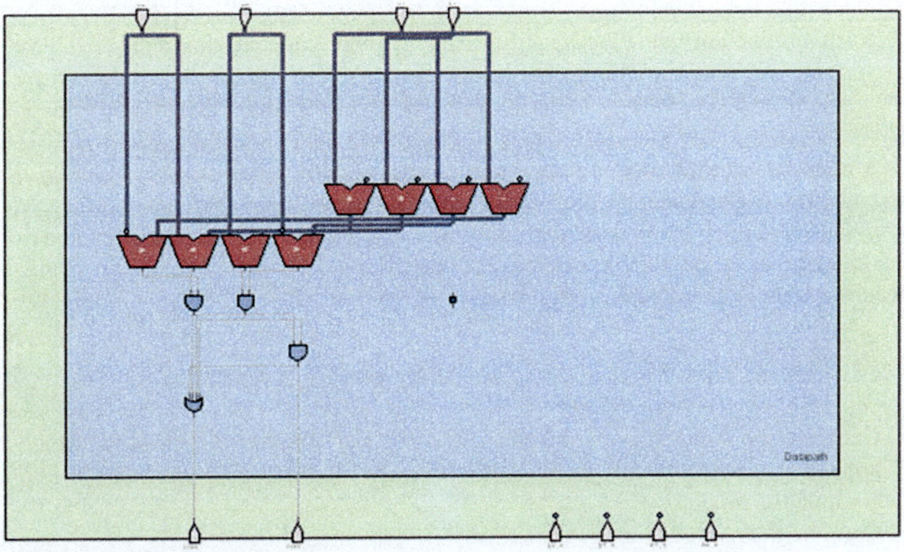

Fig. 16 Contact check unit schematic view

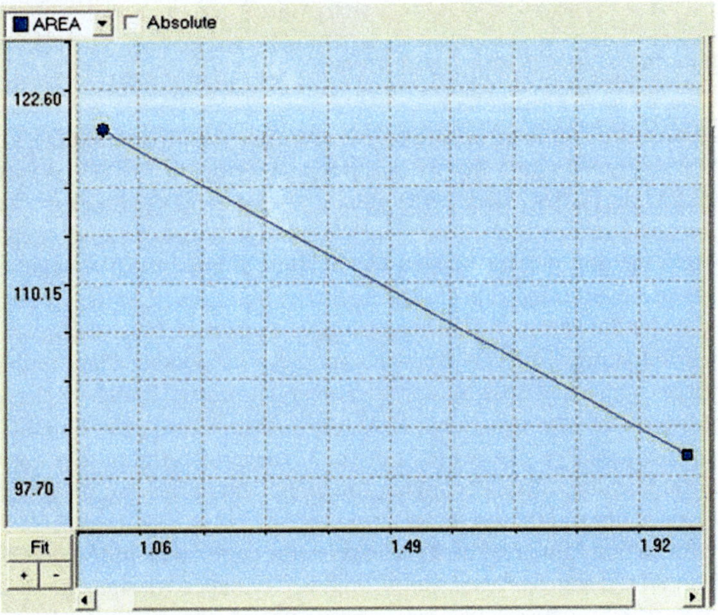

Fig. 17 Contact check DSE result

Once the design has been synthesized the next step is to verify that it is performing the correct functionality. This can be done at the RTL level by performing an RTL simulation, but in order to speed the verification process up, it is recommended to verify the design using the cycle-accurate model generator in CWB.

The cycle-accurate model generator uses the scheduled output of the HLS tool and creates a SystemC program that mimics the behaviour of the RTL cycle accurately. On average it can run around 100× faster than an RTL simulation due to the lack of details that this models has compared to RTL. The flow to use the cycle-accurate model generated is as follows:

- Synthesize the design.
- Generate the input simuli files for the design under test (DUT), in this case the contact check generation unit.
- Generate the cycle-accurate model. This will also create the testbench, which applies a new test-vector to the DUT when needed and also the make file to compile the mode.
- Compile the model using g++ and lastly.
- Execute the binary. The binary generates a vcd file while being executed so that the user can verify the result of the simulation in a wave form viewer.

5.2 *Inter-Particle Forces Increment*

The same procedure as described previously was followed to synthesize the inter-particle forces increment unit. In this case the number of slice LUTs consumed by this unit is 3,251 and 1,024 BlockRAM memory bits and 5 DSP blocks.

As in the previous section we can run the automatic design space explorer in order to obtain the dominating designs for this unit. Figure 19 shows the result of the DSE, where the left-hand side of the schematic shows the FSM that generates the control signals for the data path shown on the right-hand side. It can be observed that many more dominating results are found compared to the contact check unit, mainly due to the larger number of possible combinations. The design in Fig. 18 (largest and fastest) corresponds to the equivalent RTL implementation. Again the DSE shows one of the big advantages of C-based design. A range of different designs can be obtained automatically without changing the input description, by only modifying the FU constraint file and the synthesis attributes.

This approach is impractical in RTL because the time taken to re-code a design is prohibitive. RTL designs are normally "high performance" designs because these are easier to verify. So-called lower performance designs normally involve resource sharing, which make the design much harder to verify as the designer needs to trace all the control signals in the circuit and verify that the data is steered correctly through the data path. Therefore HW designers normally prefer replicating FUs as much as possible to avoid resource sharing, which normally leads to faster designs,

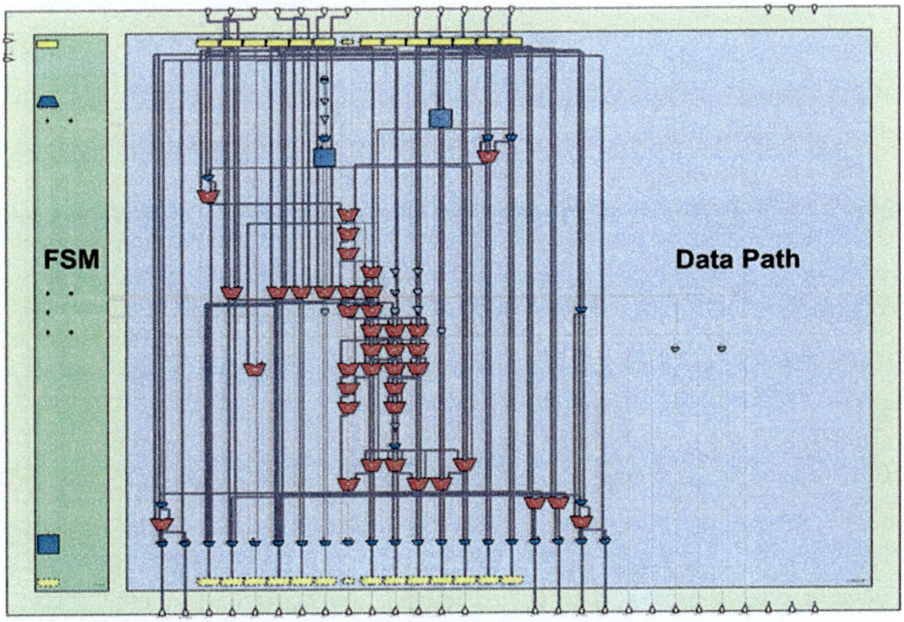

Fig. 18 Inter-particle forces update unit schematic view

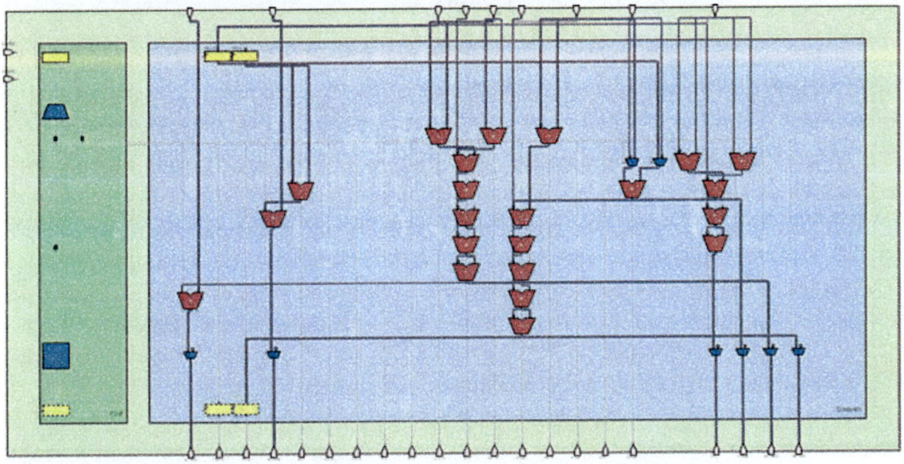

Fig. 19 Forces update unit DSE result

but also larger. It should be also noted that for four-input LUT FPGAs resource sharing normally does not lead to smaller circuits because muxes are extremely expensive in terms of area in FPGAs. For new type of FPGAs this does no hold anymore as shown in [21], because the mux can be included in the larger 6/8-input LUT in combination with the FU.

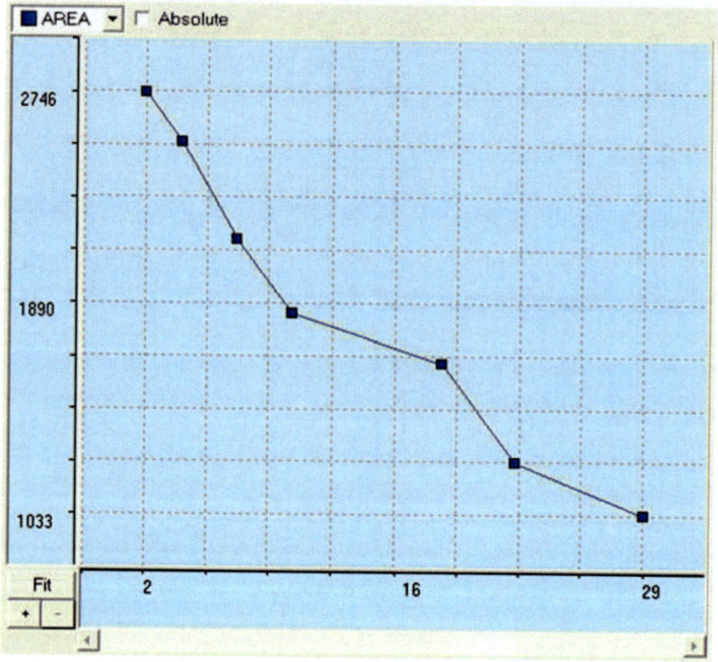

Fig. 20 Velocity and position update unit schematic view

5.3 *Velocity and Position Update*

The velocity and position update unit takes as inputs the coordinates of pairs of particles, their speeds and forces and momentum and computes their new position and velocities. The schematic view of the synthesized circuit can be observed in Fig. 20, following the same steps as for the contact check and forces update unit.

As in the previous processes the automatic design space explorer is executed in order to investigate the area vs. performance trade-off curve of this process, shown in Fig. 21. Again the first design at the graph (largest) corresponds to the equivalent RTL circuit manually generated. It consumes 420 slice LUTs.

5.4 *Control Unit*

The control unit is one of the most important parts of the design as it creates all the control signals to steer the data between the different processing units, reading and writing the data to the internal memory at the correct time. Traditionally HLS has not been able to deal effectively with control dominated circuits and circuits that

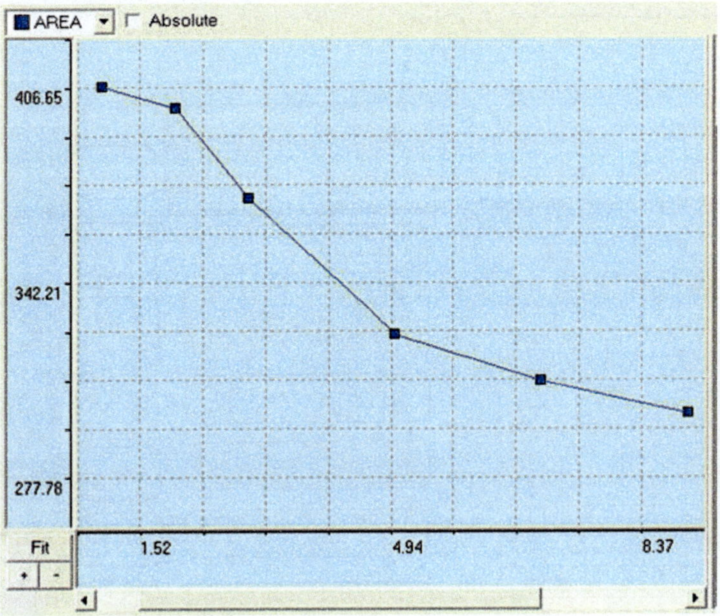

Fig. 21 Position and velocity update unit DSE result

require timing information, e.g. bus controllers. CWB is able to synthesize these types of circuit using its manual scheduling mode, where the user can manually "time" the C-code inserting a timing descriptor ("$").

With this approach the user can describe explicit state machines that are then synthesized into RTL. Figure 22 shows an example of manual scheduling mode. In this example an LCD needs to be controlled with a dedicated controller. When implementing these types of circuits a waveform is normally given as a specification in order to understand the protocol that needs to be followed for creating an interface. CWB's manual scheduling mode allows the user to manually set clock boundaries directly at the C source code. Each step in the figure corresponds to one clock cycle. For example signal "e" (LCD clock) is reset in the first clock cycle and then set to "1" in the next cycle. With this approach a complex state machine can be explicitly specified and synthesized into RTL.

This is the approach followed to synthesize the control unit for the DEM accelerator. Three independent state machines are described. The first state machine performs the contact checking of the particles, the second updates the forces between particles and the third updates the velocity and position of the particles. Three state machines are needed in order to execute these steps concurrently.

CWB's manual scheduling mode does not allow the architectural exploration as the number of FUs and synthesis attributes are optimized for this particular FSM. The three FSM consume a total of 780 slices LUTs.

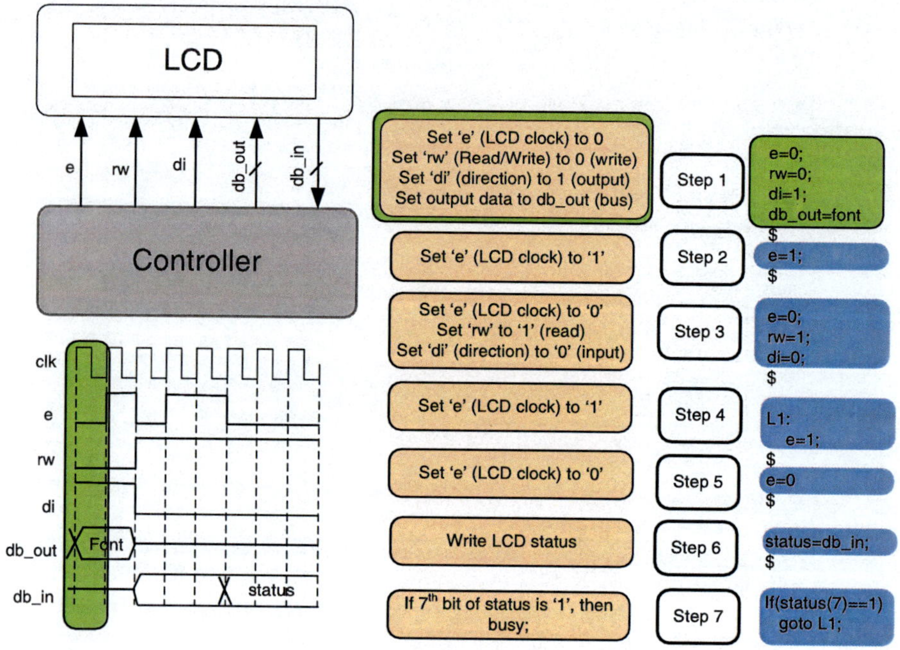

Fig. 22 Manual scheduling mode LCD controller example

6 C-Based vs. RTL-Based Design

It is not easy to compare the RTL vs. the C-based design implementations because the comparison will never be fully fair. Nevertheless there are some objective criteria that can be helpful when comparing both methodologies. Firstly, the QoR of the RTL vs. the QoR of the C-based implementation. In both cases we set the constraint of DII=1 and the latency as a do not care constraint with a target frequency of 120 MHz (30 MHz for the computational units). The DII=1 in the C-based design correspond to the highest performance design in the DSE trade-off curves shown in the previous sections. We can then compare the area size of both implementations, as their performances are identical (same frequency and same DII). Table 3 shows the results of both methodologies after executing Xilinx's Place and Route (ISE 13) in terms of number of LUTs, DSPs and BlockRAM. The target FPGA was in both cases Xilinx Virtex4 XC4VLX200.

It can be observed that the C-based design is slightly larger (~2%) than the hand-coded RTL design. This is mainly due to the numerous low level optimizations coded at the RT-level. It should be noted that the C-based design also makes use of Xilinx's CoreGenerators for the KCMs and Divider.

The second criteria that we consider when comparing both methodologies is the implementation duration, including the verification of each of the processes. Figure 23 shows that TAT difference between the two approaches.It can be observed

Table 3 RTL vs. C-based design result comparison

	RTL design			C-based design		
Unit	LUTs	DSPs	BlockRAM	LUTs	DSPs	BlockRAM
Contact check	120	0	0	120	0	0
Forces update	3,175	5	1,024	3,251	5	1,024
Position update	417	0	0	420	0	0
Control unit	765	0	0	780	0	0
Total	**4,477**	**5**	**1,024**	**4,571**	**5**	**1,024**
Δ %				**2.06%**	**0%**	**0%**

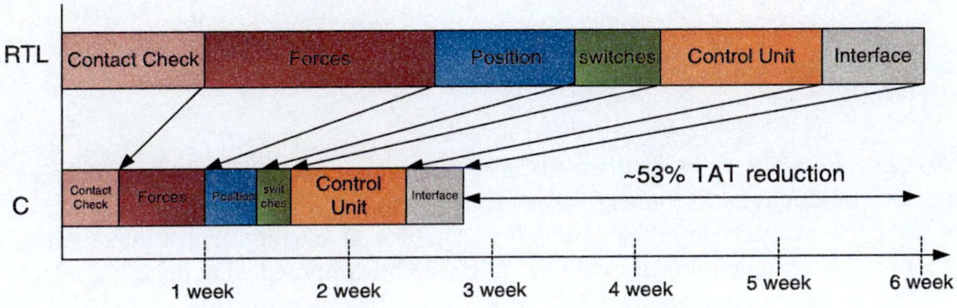

Fig. 23 Turn around time (TAT) comparison between RTL and c-based design

that the C-based design took less than 50% of time to develop. It should be noted that the C-based design was implemented after the RTL implementation, having therefore some advantages, e.g. the data type refinement was done at the RTL implementation. Nevertheless the difference is significant.

Lastly the verification methodology of both methods is also presented. The RTL implementation does not allow the simulation of the entire system due to the simulation speed constraint. In order to verify the functionality of the system different parts are in turn simulated together. The design is partitioned into three main parts for verification. The first simulates the control unit for the contact check unit with the memory and the input and output switches to verify that the contact check information is correctly generated. The second simulation includes the forces update control unit, the memory, switches and forces update unit in order to verify that the forces are correctly update and the last one the position update unit with its control unit.

In the case of the C-based design a fast cycle-accurate model could be generated for the entire design allowing the simulation of the entire system. This permits the verification of the complex interaction between all the concurrent parts being executed. CWB provides a top module generator that interconnects all processes together. A cycle-accurate model can then be generated for this top processes and the functionality be observed in the VCD file dumped out by the model.

The advantage of this approach is enormous compared to the RTL design as many bugs appear at the system integration part. RTL design can mostly only capture this problem when the design is being prototyped directly on the FPGA, which is normally very time-consuming. Being able to catch these bugs at the C-level allows much faster system verification. Another big advantage is the re-use of the original SW test vectors. The model generator also generates a testbench that can read the untimed test vectors and apply these to the DUT at the correct cycle. The testbench can than also compare the golden outputs created during the SW algorithmic development with the outputs generated during the cycle-accurate simulation as the model's testbench knows when a new valid output is being generated. This type of verification is called "transaction level verification" and saves users having to re-time the test vectors each time the latency/timing of the DUT changes.

7 Conclusions

C-based has some very distinct advantages compared to RTL design. It helps improve productivity by raising the level of abstraction and allows the generation of fast cycle-accurate simulation models that in turn enable the verification of entire systems. Also it opens an easy path to DSE, which is impractical at the RT level.

This chapter compares an RTL vs. a C-based implementation of an FPGA-based DEM custom hardware accelerator. We show that C-based design reduces that TAT by over 50%, achieving RTL of similar quality to hand-coded RTL ($\sim$2%) and allowing the verification of the complete system instead of relying on partial simulations and then having to rely on design prototyping.

Acknowledgements The authors would like to acknowledge the work of everyone at the EDA R&D Center, Central Research Laboratories at NEC Corporation, NEC Information Systems Ltd and NEC-HCL-ST for all their work developing CyberWorkBench.

References

1. A.I. Hustrulid, "Parallel implementation of the discrete element method", Colorado school of mines, USA (1996)
2. J.A. Ferrez, D. Mueller, T.M. Liebling, "Parallel Implementation of a distinct element method for granular media simulation on the Cray T3D." EPFL Supercomputing review -SCR No 08, Lausanne, Nov. 1996
3. R. Gruber, Y. Dubois-Pelerin, "Swiss-Tx: first experiences on the T0 system". EPFL Supercomputing Review, SCR 10, 19–23 (1998)
4. P.A. Cundall, O.D.L. Strack, "A discrete numerical model for granular assemblies". Geotechnique **29**, 1–8 (1979)
5. P.A. Cundall, O.D.L. Strack, "The Distinct Element Method as a tool for research in Granular Media", Report to the National Science Foundation Concerning NSF Grant ENG76–20711, Appendix 2, pp 20–21, University of Minnesota, November 19785
6. P.A. Cundall, "A discontinuous future for numerical modelling in geomechanics?", in *Proceedings of the Institution of Civil Engineering, Geotechnical Engineering* **149**, January 2001, Issue 1 Pages 41–476

7. B. Carrion Schafer, S.F. Quigley, A.H.C. Chan, "Acceleration of the discrete element method on a reconfigurable computer". Comput. Struct. **82** (20–21), 1707–1718 (2004)
8. B. Carrion Schafer, S.F Quigley, A.H.C. Chan, "Scalable implementation of the discrete element method on a reconfigurable computing platform", in *12th International Conference on Field Programmable Logic and Applications (FPL)*, Montpellier, 2002, Springer-Verlag
9. B. Carrion Schafer, S.F. Quigley, A.H.C. Chan, "Analysis and implementation of the discrete element method using a dedicated highly parallel architecture in reconfigurable computing", in *IEEE Symposium on Field- Programmable Custom Computing*
10. K. Wakabayashi, "Cyber: High Level Synthesis System from Software into ASIC" (Kluwer Academic, Dordecht, 1991), pp. 127–151
11. K. Wakabayashi, T. Okamoto, "C-based SoC design flow and EDA tools: An ASICand system vendor perspective." IEEE Trans. Comput. Aided Des. Integrated Circ. Syst. **19**(12), 1507–1522 (2000)
12. N. Kobayashi, K. Wakabayashi, H. Tanaka, N. Shinohara, T. Kanoh, "Design experiences with high-level synthesis system Cyber I and behavioral description language BDL," in *Proc. of Asia Pacific Conf on Hardware Description Languages*, Oct 1994
13. B. Carrion Schafer, K. Wakabayashi, "Design space exploration acceleration through operation clustering". IEEE Trans. Comput. Aided Des. Integrated Circ. Syst. (TCAD) **29**(1), 153–157 (2010)
14. C. Haubelt, J. Teich, "Accelerating design space exploration", in *International Conference on ASIC*, 79–84, 2003
15. V. Krishnana, S. Katkoori, A genetic algorithm for the design space exploration of datapaths during high–level synthesis. IEEE Trans. Evol. Comput. **10**(3), 213–229 (2006)
16. D. D. Gajski, "High Level Synthesis: An Introduction to Chip and System Design" (Kluwer Academic, Dordecht, 1992), ISBN 0792391942
17. J. Cong, B. Liu, S. Neuendorffer, J. Noguera, K. Vissers, Z. Zhang, "High-level synthesis for FPGAs: from prototyping to deployment". IEEE Trans. Comput. Aided Des. (TCAD) **30**(4), 473–491 (2011)
18. B. Carrion Schafer, K. Wakabayashi, "Design of complex image processing systems in ESL", ASPDAC, Taiwan, pp. 809–814, 2010
19. S. Morioka, B. Carrion Schafer, K. Wakabayashi, "Complex security engine design in high level synthesis", MPSoC, Savannah, USA, August 2009
20. B. Carrion Schafer, "Complete C-Based SoC design: Is it possible?", MPSoC, Gifu, Japan, June 2010
21. S. Hadjis, A. Canis, J.H. Anderson, J. Choi, K. Nam, S. Brown, T. Czajkowski, "Impact of FPGA architecture on resource sharing in high-level synthesis," in *ACM/SIGDA International Symposium on Field Programmable Gate Arrays*, to be held at Monterey, CA, February 2012
22. K. Wakabayashi, B. Carrion Schafer, in "High-Level Synthesis from Algorithm to Digital Circuit", **XVI**, Chap. 7, ed. by P. Coussy, A. Morawiec (Springer, New York, 2008), ISBN: 978-1-4020-8587-1

Optimising Euroben Kernels on Maxwell

James Perry, Mark Parsons, and Paul Graham

Abstract The ability to run certain common numerical kernels fast is valuable for many applications and fields of scientific research. The University of Edinburgh investigated the possibility of using FPGA devices to accelerate four kernels from the Euroben benchmark suite: dense matrix multiplication, sparse matrix-by-vector multiplication, fast Fourier transform, and random number generation. Each kernel was ported using both the Harwest C compiler and hand-coded VHDL and for each port the performance gain and porting effort were evaluated.

Although all of the kernels ran faster on the FPGA than on the CPU used for comparison, the level of hardware expertise required to port them was high even when using the Harwest compiler. Furthermore, many of the FPGA ports gave only a modest performance improvement over the much more maintainable C implementation, especially when the time taken to copy input and output data across the relatively slow PCI bus to the FPGAs was taken into account. However for certain use cases, for example when a large quantity of random numbers is required or when low power consumption is critical, FPGAs could be a good choice for running this type of kernel.

1 Overview

This chapter describes the work undertaken by EPCC, University of Edinburgh in 2009–2010, to port four kernels from the Euroben benchmark suite to the Maxwell FPGA supercomputer, using both hand-coded VHDL and a C-to-gates compiler (Harwest Compile Environment). Performance and programmability of the two approaches was compared, using the original C code running on an Intel Xeon CPU as a reference.

J. Perry • M. Parsons • P. Graham (✉)
EPCC, University of Edinburgh, Edinburgh, United Kingdom
e-mail: j.perry@epcc.ed.ac.uk

W. Vanderbauwhede and K. Benkrid (eds.), *High-Performance Computing Using FPGAs*,
DOI 10.1007/978-1-4614-1791-0_23,

2 Background

2.1 Prace Prototypes

Work Package 8 of the PRACE (Partnership for Advanced Computing in Europe) project [1] involved evaluating several novel, mostly accelerator-based, computer architectures by porting simple benchmark kernels to them. Four kernels from the Euroben suite [2] were selected—dense matrix-by-matrix multiplication, sparse matrix-by-vector multiplication, fast Fourier transform (FFT) and random number generation. These are very common operations in a multitude of real-world scientific applications and were therefore good candidates for evaluating the performance of next generation systems.

2.2 The Maxwell System

EPCC's FPGA supercomputer, Maxwell, was completed in early 2007 as part of the FPGA High Performance Computing Alliance initiative [3]. It consists of an IBM Blade Center cluster, containing 32 2.8 GHz Intel Xeon nodes, each with 2 Xilinx Virtex-4 accelerator boards attached. Half of these boards are AlphaData boards [4], each of which has a Virtex-4 FX 100 FPGA, and 1 GB of DDR II SDRAM, split into four independent banks. The other half are Nallatech boards [5], each with a Virtex-4 LX 160 FPGA and 512 MB of DRAM.

An interesting feature of the Maxwell system is the direct interconnect between the FPGA boards; the accelerators are connected into a two-dimensional 8-by-8 torus formation using RocketIO cables. This allows fast communication between the FPGAs without having to go through the host CPUs and relatively slow PCI buses and Ethernet links. However, this was not used in the work described here.

2.3 Software Tools

The Xilinx toolchain (ISE 9.1i) [6] was used to generate bitstreams for the FPGAs. Another Xilinx tool, Coregen [7], was used to create building blocks for the VHDL versions of the kernels. Coregen can create optimised, parameterised versions of common digital circuits to avoid the need to program them from scratch in VHDL or Verilog. In this work, various floating-point adders, multipliers and FIFOs (first-in-first-out buffers) of appropriate sizes were generated using Coregen.

In addition, a C-to-VHDL compiler was selected so that its performance and ease of use could be compared to the traditional method of writing VHDL code directly. The Harwest Compilation Environment (HCE) [8] from Ylichron was chosen for this purpose. In the early stages of this project, a compiler backend targeting the AlphaData ADMXRC4FX boards in Maxwell was written by Ylichron.

2.4 The Euroben Kernels

The four kernels selected from the Euroben suite were:

- Dense matrix-by-matrix multiplication
- Sparse matrix-by-vector multiplication
- Fast Fourier transform
- Random number generation

All of these are common operations within HPC codes and are therefore interesting for testing the performance and programmability of new systems. They are also diverse enough to be able to highlight the strengths and weaknesses of different architectures and tools. The original versions of all of these codes used double precision arithmetic; however, this was not possible for all the FPGA ports, as the Harwest compiler did not support double precision at the time.

The main goals of the work were as follows:

- To evaluate the performance of the Maxwell system and compare it to the other PRACE prototype systems, as well as to the original C code running on a traditional processor
- To evaluate the performance and usability of the HCE, in particular how it compared to hand coding VHDL

3 Porting the Kernels

3.1 Matrix Multiplication

The matrix multiplication algorithm used was the benchmark called *mod2am* from the Euroben benchmark suite. It is a simple double precision matrix-by-matrix multiply operation and the Euroben test code exercises it with various square matrix sizes up to 1,000 elements. The algorithm in C (with the loop unroll optimisation of the original removed for clarity, and assuming square matrices) is:

```
for (i = 0; i < n; i++) {
     for (k = 0; k < n; k++) {
          t = 0.0;
          for (j = 0; j < n; j++) {
               t = t + a[i][j] * b[j][k];
            }
            c[i][k] = t;
        }
}
```

where *a* and *b* are the input matrices and *c* is the result of multiplying them together.

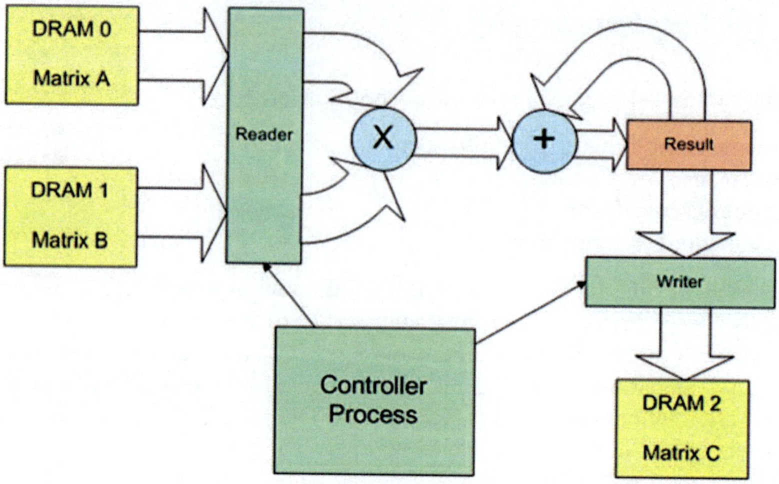

Fig. 1 Block diagram of initial VHDL implementation of matrix multiplier. The DRAM blocks represent memories external to the FPGA, while the Result buffer is an internal register. The circular blocks are arithmetic units, in this case generated by Coregen, and the remainder of the blocks are VHDL logic, usually in the form of a process

3.1.1 VHDL Port

3.2 *Initial Implementation*

The initial implementation made use of one floating point multiplier and one floating point adder, as well as some logic to fetch the inputs and store the results at the right times. DRAM banks 0 and 1 were each used to hold one input matrix, with the result of the multiplication being stored to bank 2. A block diagram is shown in Fig. 1.

The reader process fetches values from the input matrices at the appropriate times, iterating over j, k and i as for the C version. The multiplier and adder form a pipeline; effectively one multiplication and one addition can be run per clock cycle once the pipeline is filled, but there is a latency of 10 cycles associated with the multiplier and 16 cycles for the adder, so there is a lag of 26 cycles in total between the values entering the multiplier and the corresponding result emerging from the adder. The result of the addition is fed back into the adder as an input so that the products of the input matrix elements are accumulated. When the inner loop (over j) completes, the writer process writes an element to matrix C and the accumulation is zeroed ready for the next loop. The controller process keeps track of the stage the whole operation is at and sends out control signals to the other components at the right times.

There are a few complications not shown in the block diagram. Firstly, because of the 16 cycle latency of the adder, it would be inefficient to simply accumulate the result in the normal order, as this would mean waiting 16 cycles for the result of the previous addition before feeding it back into the adder for the next one. To overcome this, 16 accumulations are run concurrently, allowing full utilisation of the adder. So the first calculation done is for $k = 0$, $i = 0$, $j = 0$; the second is for $k = 0$, $i = 1$, $j = 0$; next $k = 0$, $i = 2$, $j = 0$; $k = 0$, $i = 3$, $j = 0$; and so on up to $k = 0$, $i = 15$, $j = 0$; and by this time the result of the first addition is available, so $k = 0$, $i = 0$, $j = 1$ can be processed.

Secondly, the memories are not always able to deliver data at a steady rate, so the whole pipeline must keep track of which data items are valid and which are not. This is achieved by associating a 10-bit shift register with the multiplier and a 16-bit one with the adder. On each clock cycle that valid input values are read from DRAMs 0 and 1, a 1 bit is shifted into the left-hand side of the multiplier's shift register, otherwise a 0 bit is shifted in. Both registers are shifted right one place per clock cycle, and the bit shifted out of the multiplier's register is shifted into the adder's register. In this way, the 1 bits track where the valid data items are, and the bit being shifted out of the adder's register on each clock cycle indicates whether the current result from the adder is valid or not.

Because of this, it is not sufficient to keep the adder result for one cycle to be fed back in to the accumulation. It might have to be kept more than one cycle, and during this time additional values might "queue up" behind it. So a FIFO (first-in-first-out buffer) is used to hold as many accumulation results as necessary in the right order before they are fed back into the accumulator.

Unfortunately the performance of this design turned out to be very poor, taking about 300 s to multiply two $1{,}024 \times 1{,}024$ matrices—a factor of 100 slower than the host CPU running the original code. This was mainly due to the access pattern of the input data in DRAM—the DRAMs respond extremely slowly to random accesses of single values, which is what the reader process was doing. Additionally, because of the FPGA's much slower clock speed as compared to the host CPU, it would have to perform many floating point operations per clock cycle to have a chance of running faster (the host performs about one every five clock cycles and is clocked roughly 20 times faster than the FPGA!).

3.3 *Second Design: Fast But Size-Restricted*

In the second implementation, block RAM (very fast on-chip memory) was used to store the input matrices, and a network of eight multipliers and eight adders to perform several multiply-adds per clock cycle (see Fig. 2 for details).

Unlike the DRAMs, the block RAMs are capable of sustaining a read every clock cycle, even with a completely random access pattern. They also have dual ports, which means two locations in each block RAM can be read independently per clock cycle. In this design, each input matrix is split across four banks of block RAM, so that eight values can be read from each matrix every clock cycle. (The split is done based on the low two bits of the j co-ordinate; because the A and B values that are

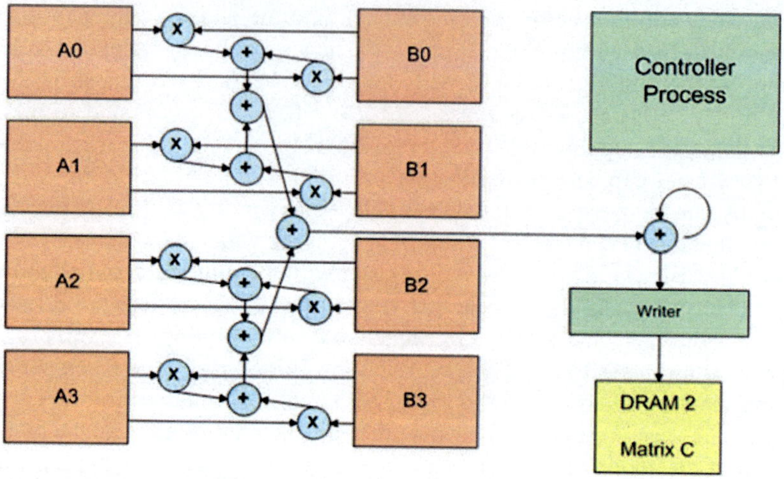

Fig. 2 Block diagram of faster VHDL matrix multiplier

multiplied together always have the same *j* co-ordinate, this ensures that the values that need to be multiplied are always available from corresponding banks.)

The values read from block RAM go into a hierarchical pipeline of multipliers and adders that, once the pipeline is filled, compute eight products and add them to the accumulation every clock cycle. Because of the block RAM's consistent access time, there is no need for the shift registers employed in the first design to keep track of valid data: the pipeline is always full of valid data. There is also no need for the FIFO to hold a queue of previous results to feed back into the final adder, although due to the adder's latency it is still necessary to perform 16 accumulations at the same time to achieve optimum throughput.

The resulting values from the accumulator are written to DRAM by exactly the same writer component as in the initial design. Because the source matrices are accessed many more times than the destination, the use of DRAM rather than block RAM here does not cause a bottleneck. The controller process generates read addresses for the block RAMs, tells the writer component when a value is ready to be written, and resets the accumulator when required.

This design is fast, multiplying a 256×256 matrix roughly four times faster than the C code. However, it has a major drawback: because of the limited size of block RAM available, it cannot work with matrices larger than 256×256. The Virtex-4 FX100 has approximately 6 Mbit of block RAM, but depending on the configuration required and the amount taken up by common code such as the AlphaData DRAM interfaces, less than this may be usable in practice. It was possible to instantiate up to eight banks of 16k floating point numbers, enough for two 256×256 source matrices, before running out of usable block RAM.

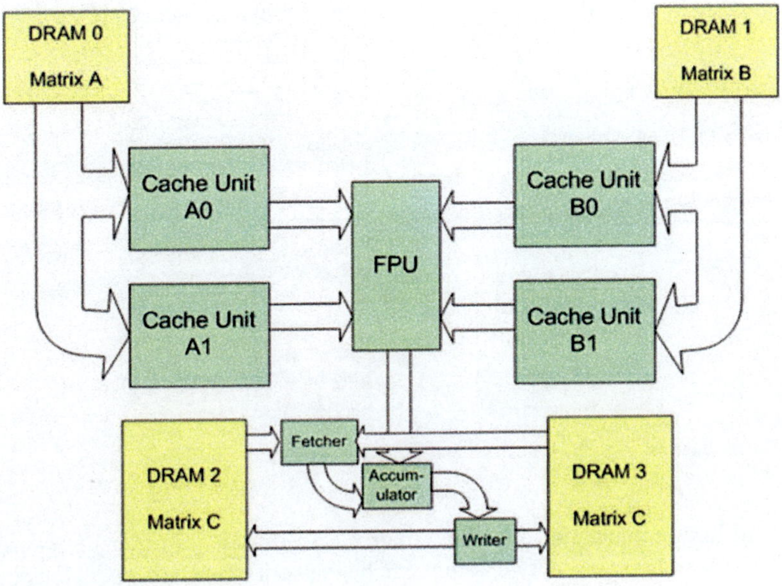

Fig. 3 Block diagram of final VHDL implementation of matrix multiplier

3.4 *The Final Design*

By employing a more intelligent algorithm and caching parts of the source matrices in block RAM, it is possible to build a multiplier capable of multiplying larger ($1{,}024 \times 1{,}024$) matrices almost as fast as the second design. The high level structure of this is shown in Fig. 3.

This design takes advantage of the fact that the values being multiplied together from matrices A and B are always from the same j index (j defines the column of matrix A and the row of matrix B from which data is taken). The problem is divided into 32 blocks in the j dimension, each of which is computed separately. This involves fetching 32 columns of matrix A and 32 rows of matrix B into the caches and working on them, adding the results to previous partial results which are streamed from DRAM bank 2 to bank 3 or vice versa.

There are four identical cache units, two for each of the input matrices. Each one holds 32 complete rows or columns of data. At any time, one pair of cache units (A0 and B0 or A1 and B1) is active (i.e. its contents are being used in calculations) while the other is being refilled (fetching the next 32k values from DRAM).

The internal structure of one of the cache units is shown in Fig. 4.

It consists of four block RAMs, each able to hold 8,192 32-bit values, and a refiller process that fetches the next 32,768 values from DRAM into the block RAMs when a refill operation is triggered. (Note: so that the DRAM is always accessed sequentially, which is much faster than random access to individual elements, matrix

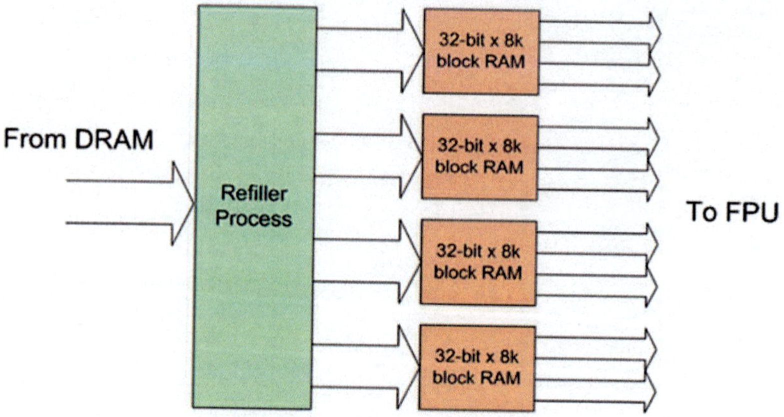

Fig. 4 Block diagram of cache unit from matrix multiplier

A is stored in column-major order and matrix B in row-major order.) Because the block RAMs are dual ported, two values can be read out of each one per clock cycle, allowing each cache unit to feed eight floating point values into the floating point unit every clock cycle.

The floating-point unit at the centre is identical to the structure of multipliers and adders in the second design; in each clock cycle eight pairs of numbers flow in are multiplied together, and the multiplication results are summed. The result of this summation is added to an accumulator. As in the first two designs, 16 accumulations are in progress simultaneously to overcome the latency of the floating point adder in the accumulator.

Because the caches only hold 32 rows/columns of the input matrices at a time, and because eight values from each are processed in one go, only four accumulations are performed before the partial result is ready to be written to DRAM. On the first iteration, the accumulator is initialised with zeroes before each operation and the results are simply written to DRAM 2. On the second iteration, the accumulator is initialised with the partial results from DRAM 2 and the new results (with the second block of inputs now processed) are written to DRAM 3. This continues on subsequent iterations, the partial results being streamed from one DRAM bank to the other and having the latest data added to them in the process. After 32 iterations, DRAM 3 contains the final complete result of the multiplication.

Two components, the fetcher and the writer, handle this streaming of data. The writer component is quite simple and similar to the one in the previous implementations. It simply writes values out to the destination DRAM until the whole $1{,}024 \times 1{,}024$ partial result matrix has been written. The fetcher is slightly more complex. It fetches values from the source DRAM in bursts of length 128 and stores them in a FIFO ready for the accumulator to access. The timing of this is rather sensitive and the burst reads have to be initiated at the right times to ensure that the FIFO is never either empty or overflowing. This is the responsibility of the main controller process which co-ordinates the data flow of the entire calculation.

The clock speed had to be dropped from 140 to 115 MHz before the DRAM accesses were fully reliable; therefore this design is not quite as fast as the second one. However, it still achieves an approximate 3× speed-up over the software version.

3.5 Converting to Double Precision

All the work described so far was applicable only to matrices of single precision floating point values; although the original Euroben code is double precision, single precision operations result in a much simpler bitstream which compiles faster, so doing all the initial development in single precision allowed for a faster development cycle.

Converting a VHDL design from single to double precision generally entails at least:

- Replacing any floating point arithmetic blocks with their double precision equivalents
- Widening any signals and registers that carry floating point values from 32 to 64 bits

In the case of the matrix multiplication, some additional changes were also required. Because the double precision floating point multiplier had a different latency from the single precision one, the pipeline timings had to be changed. It also turned out that it was no longer possible to use multipliers which took full advantage of the DSP (digital signal processor) blocks in the Virtex-4, as there were not enough DSP blocks for eight double precision multipliers. The component settings were changed in Coregen to use fewer DSP blocks and more generic logic instead.

Most significantly, because double precision values take up twice as much storage space as single precision, it was only possible to cache 16 rows or columns of each input matrix instead of 32. This in turn made the timings required for the streaming of the result matrix more demanding, as a partial result value was required on average every two cycles instead of every four. However, by increasing the size of the FIFO, fetching values in bursts of length 256 instead of 128, pre-fetching the first two bursts before the main loop across each block of the matrix, and reducing the clock speed slightly from 115 to 110 MHz, it was possible to meet this requirement. Performance was 1,700 MFlops, around three times faster than the host code.

3.5.1 HCE Port

Getting a port of the kernel up and running correctly using HCE was quite easy. It took approximately 2.5 days' effort to create the initial port; this compares well with the 5 days taken to produce a working VHDL version. Performance was poor with the initial naïve port, only 14 MFlops.

However, with Ylichron's help, an optimised version was produced. Blocks of the matrix were cached in fast internal memory, the main loop was unrolled 64 times, and a highly pipelined tree-structured adder was used to accumulate results. The code was annotated with the HCE directives *pipeline*, *split*, *unroll* and *combinational* in order to give the compiler hints about how to optimise the kernel. Resulting performance was very good, at 2.8 GFlops (single precision only as HCE did not support double), faster than the hand-coded VHDL (single precision) version at 1.8 GFlops.

Even taking into account the hand tuning and compiler directives, the code was very much more compact and easy to write and maintain than the VHDL version (543 lines of code compared to 1,252 in VHDL). However, compilation times were much longer for the HCE kernel.

HCE code is standard C code which will compile and run with any C compiler, but with directives applied to certain constructs (typically loops, functions or individual lines of code) to improve performance when using the HCE tools. An example function is shown below:

```
/*#HWST combinational*/
unsigned short ResolveAddressWriteC(unsigned short
int i, unsigned short int j2)
{
       return (i * 4 * N + j2);
}
```

The *combinational* directive instructs the compiler to implement the function in combinational, rather than sequential, logic if possible.

3.6 Sparse Matrix-by-Vector Kernel

The sparse matrix-by-vector kernel used was the **mod2as** kernel from Euroben. Again it is a very simple kernel expressed in a few lines of C code:

```
for (i = 0; i < nrows - 1; i++) {
     outvec[i] = 0.0;
     for (j = rowp[i]; j < rowp[i+1]; j++) {
          outvec[i] += mat-vals[j]*invec[indx[j]];
     }
}
outvec[nrows-1] = 0.0;
for (j = rowp[nrows-1]; j < nelmts; j++){
     outvec[nrows-1] += mat-vals[j]*invec[indx[j]];
}
```

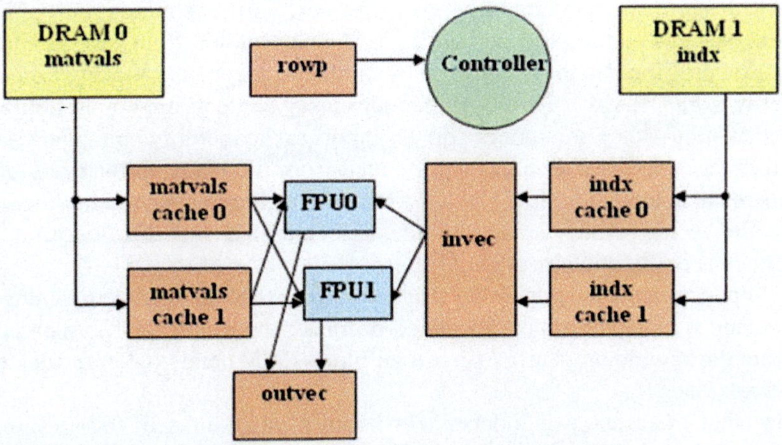

Fig. 5 Block diagram of VHDL implementation of sparse matrix multiplier

invec and *outvec* are the input and output vectors. *nrows* is the number of rows in the matrix, which is always square, and also the number of elements in each vector. *nelmts* is the total number of non-zero elements in the matrix. *matvals* contains the non-zero values of the matrix and *indx* maps them to their column positions. *rowp* contains the starting index of each row in the matrix.

The original kernel is double precision; however, only a single precision VHDL port was created. This is because the main purpose of the VHDL port was for comparison with the Harwest port, which does not support double precision, so the effort required to convert to double precision would not have been worthwhile.

3.6.1 VHDL Port

Obviously the sparse-matrix-by-vector kernel has some commonalities with the matrix-by-matrix kernel already ported: the arithmetic operation of multiplying and accumulating is identical and the same strategy of having two alternating caches for subsections of large input data arrays can be used. However, the differences in where the source data comes from, and especially the unpredictable number of elements in each matrix row, make this kernel significantly more complicated to implement efficiently in hardware.

The VHDL was implemented with a maximum matrix size of 2,047 square in mind, as the original Euroben test data set goes up to a maximum of a 2,000 square matrix with 300,000 non-zero elements. This means that the input and output vectors and the row pointers array are small enough to fit comfortably into the internal block RAM—only the matrix values and column index arrays must be placed in DRAM.

The architecture of the first implementation is shown in Fig. 5.

DRAM bank 0 is used to store the matrix values. As in the matrix multiplier, at any particular time one of the *matvals* caches is being refilled from DRAM and the other is providing input to the calculation. A similar arrangement is used to cache the index array from DRAM 1. However, the index array contents are not used directly in the calculation—they are used as address inputs to the memory holding the input vector data. So that several values can be read from the input vector every clock cycle, there are actually multiple block RAMs holding duplicate copies of the input data. As the vector is fairly small and does not change during the algorithm, this arrangement is easily implemented.

The output from the calculation is written to another block RAM containing the output vector. A main controller process co-ordinates the whole algorithm based on input from the row pointers array (in a final block RAM) and communicates with the host processor.

Notice that there are two independent floating point units in this design, as distinct from the single unit in the matrix multiplier design. The FPUs are of the same structure (a tree of floating point multipliers and adders) but instead of a single tree taking eight pairs of inputs, there are two separate trees each taking four pairs of inputs. This is because with a large FPU, very small row sizes could result in inefficient use of resources. For example, imagine a sparse matrix with only three elements in each row. With an eight input FPU, this would result in only three inputs being used and five sitting idle the whole time. With twin four input FPUs, this inefficiency is reduced considerably—each FPU can process rows independently so only one input of each is idle.

Partly because of the twin FPUs and partly because of the variable row lengths, the controller process has a much more complicated task than in the matrix-by-matrix multiplier. Recall that each FPU is processing 16 accumulations at once (due to the 16 cycle latency on floating point addition); with 2 FPUs there are now 32 rows being processed concurrently and each row could be a different length. The controller process has to keep track of the status of each of the 32 "row slots", feeding in the correct input data and being ready to start a new row as soon as a slot becomes vacant.

To free up the controller from also having to keep track of writing the results to the output vector, each FPU contains a shift register that is initialised with the destination index when the inputs are written. This index is shifted down as the data makes its way through the pipeline and at the end is used to write the final FPU output to the right place in the *outvec* array. Because the *outvec* block RAM is dual ported, writes from both FPUs on the same clock cycle are not a problem.

To simplify the hardware as much as possible, some of the higher level processing is done by the host CPU—specifically, converting the row pointer array to a more useful form and dividing the data up into blocks that fit within the caches. Instead of passing the simple *rowp* array to the hardware, the host first converts it into a structure containing a source address for *matvals* and *indx*, a destination address for *outvec*, and a length element for each row. This saves the hardware from having to perform these calculations, which are slightly complicated but unlikely to be a bottleneck.

At start up, the host divides the input data into blocks that fit within the block RAM caches. It still copies the whole data set to the FPGA board memory in one go for efficiency, but at the end of processing each block the control process waits for the host to tell it the start address and length of the next block before proceeding. Again this is unlikely to become a bottleneck and saves having to implement hardware to perform operations better suited to software.

But despite all the efficiency measures described, this implementation ran slower than the original software. This was due to the fact that each element of the largest input data array, *matvals*, is only used once in this kernel—unlike the matrix-by-matrix multiply, this algorithm is memory bound.

3.6.1.1 A Faster Version

Although memory bound codes are not normally good candidates for FPGA acceleration, it is sometimes possible to improve their speed by making intelligent use of the multiple memory banks available on the FPGA boards (the ADMXRC4FX board has four independent banks of DRAM that can be accessed simultaneously). To take advantage of this, the VHDL was rewritten so that the *matvals* array was split across DRAM banks 0 and 1 (with even elements in one bank and odd elements in the other), and the *indx* array similarly split across banks 2 and 3. This doubled the memory bandwidth available when refilling the caches and halved the time taken for a cache refill. This gave the algorithm a modest speed advantage over the software (622 MFlops/s versus 489 MFlops/s on the host), although this advantage is lost if the time taken to transfer the input data to the FPGA board is taken into account.

3.6.2 HCE Port

As with the matrix multiplier kernel, producing a working port of the sparse matrix code to HCE was straightforward and took only about 1.5 days, compared with 10 days for the VHDL version. However, the initial version was again very slow, running at only 2.9 MFlops (compared to 489 MFlops on the host CPU and 622 MFlops for the VHDL implementation). Again this was optimised by unrolling the main loop, making a pipelined adder, and using HCE compiler directives to provide further optimisation hints. This gave a 12-fold speed up over the original version, but at 34.6 MFlops, performance was still poor.

The sparse matrix kernel is inherently more difficult to optimise than the dense one; it is memory bound and therefore highly sensitive to anything that may cause sub-optimal memory throughput. It also involves unpredictable data sizes, which can vary greatly from one row of the matrix to the next, and this does not fit well with HCE's optimisations, which mostly work best on large, regularly sized data.

3.7 Random Number Generator

The random number generator kernel used was the Euroben kernel **mod2h**. It generates 64-bit double precision random numbers from five 64-bit integer state variables (z1–z5). L1 to L5 are constants used to mask off only certain bits of the state variables and NORM is a constant that scales a 64-bit integer down to the range -0.5 to +0.5:

```
double rand64(void)
{
        unsigned long long b;

        b = (((z1 << 1) ^ z1) >> 53);
        z1 = (((z1 & L1) << 10) ^ b);
        b = (((z2 << 24) ^ z2) >> 50);
        z2 = (((z2 & L2) << 5) ^ b);
        b = (((z3 << 3) ^ z3) >> 23);
        z3 = (((z3 & L3) << 29) ^ b);
        b = (((z4 << 5) ^ z4) >> 24);
        z4 = (((z4 & L4) << 23) ^ b);
        b = (((z5 << 3) ^ z5) >> 33);
        z5 = (((z5 & L5) << 8) ^ b);
        return((z1 ^ z2 ^ z3 ^ z4 ^ z5)*NORM + 0.5);
}
```

The state variable updates are a mixture of bit shifts, logical ands and exclusive-ORs. The final line generates the result by XOR-ing all the state variables together and converting them to a floating point number in the range 0.0–1.0.

3.7.1 VHDL Port

This algorithm was ideal for FPGA implementation and the very high performance achieved reflects this. The state variable transforms and the final exclusive-OR can be implemented almost trivially in VHDL, as a component which takes the five variables as inputs and generates the updated state variables and the 64-bit integer result of the exclusive-OR. The bit shifts in fact require no resources at all in hardware—they just affect where each bit is routed to.

Several instances of this transform component can be chained together in order to generate multiple values in the sequence in a single clock cycle. Our double precision implementation chains together three instances, generating three values per clock cycle (we also implemented a single precision port which chains together six instances). If multiple independent sequences of numbers were to be generated instead, each with its own state variables, it is likely that performance would be higher still.

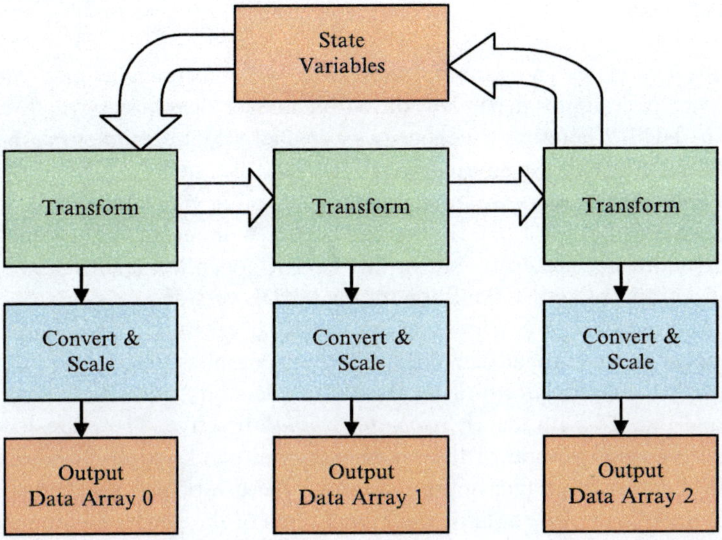

Fig. 6 Block diagram of VHDL implementation of random number generator

All that remains outside the lightweight transform component is the final conversion to a floating-point number in the range 0.0–1.0. This could be done using a standard integer-to-float conversion followed by a multiplication and an addition as it is in the software; however, a more efficient way is possible in VHDL. The addition of 0.5 can be achieved by simply flipping the most significant bit of the integer before the conversion, and the floating point conversion and scaling can be combined into one custom operation, generating the correct IEEE double precision bit pattern from the input. This custom conversion component also completes in one clock cycle, after which the result can be written to RAM. This is a good demonstration of an optimisation that goes beyond what could reasonably be done in software.

The architecture of the whole algorithm is shown in Fig. 6.

This implementation can run at a clock frequency of up to 150 MHz, and it generates three random numbers per clock cycle (six in the single precision version). It can therefore generate 450 million double precision (or 900 million single precision) numbers per second—over 30 times more than the host processor. However, when the time taken to transfer the output back to the host is taken into account, the PCI bus becomes a severe bottleneck and the overall speed drops to slower than the software. Clearly for this optimised implementation to be of use, the numbers would have to be consumed by another process on the FPGA.

3.7.2 HCE Port

Getting this kernel up and running with HCE was more challenging than the previous two; it is highly dependent on 64-bit integer operations, which are not available in HCE. Therefore a version was created which decomposes the state transformation into multiple 32-bit operations. As HCE also did not support double precision, the final conversion from integer to floating point was performed in host code for this port. As before, the initial port without any platform-specific optimisations did not perform well; in this case it generated 2.76 million values per second, compared with 450 million for the VHDL version.

Optimisation of this kernel, however, was fairly successful. In order to make the compiler generate a combinatorial block that could complete in a single clock cycle for the state variable transform (as in the VHDL version), this was moved into a function and flagged with the HCE *combinational* directive. The outer loop was then unrolled so that several of these combinatorial blocks could be instantiated in parallel. Another important optimisation was the use of two separate memory buffers for the outputs, one holding the high 32 bits of the generated numbers, the other the low 32 bits. This allowed each number to be written to memory in a single clock cycle, rather than needing two sequential writes to the same memory buffer, requiring at least two clock cycles.

Performance of the optimised HCE version of the kernel reached 217 million values per second, slower than the VHDL version but still around 15 times faster than the original code running on the Xeon.

3.8 Fast Fourier Transform

The third kernel studied was a standard FFT algorithm, **mod2f** from Euroben. In its original version it is a radix 4 FFT, but the results should be the same as if coming from any FFT algorithm so the VHDL port is a simple radix 2 implementation. The original operates on double precision complex numbers; the hardware version uses single precision complex numbers.

3.8.1 VHDL Port

The central FFT "butterfly" operation can be turned into a very efficient pipeline capable of performing one butterfly every clock cycle. It consists of one complex multiplier, a complex adder and a complex subtractor (which translates to four real multipliers, three adders and three subtractors in total) as well as some shift registers to delay the even input while the odd input is multiplied by the twiddle factor (see Fig. 7).

Two instances of this butterfly component were used as the core of the algorithm. Block RAMs hold the input and output and another block RAM holds the twiddle

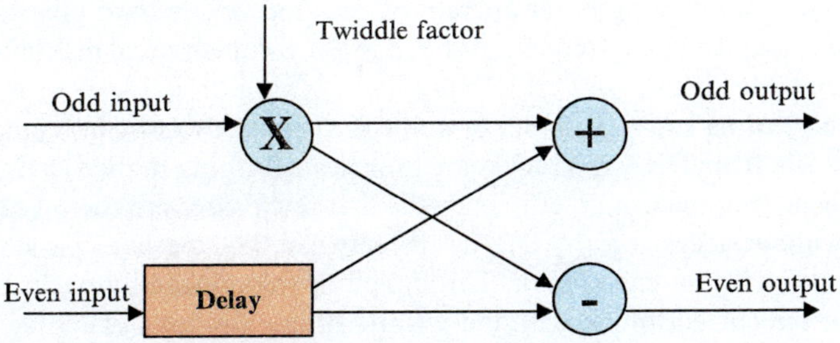

Fig. 7 Diagram of FFT "butterfly" block from VHDL implementation

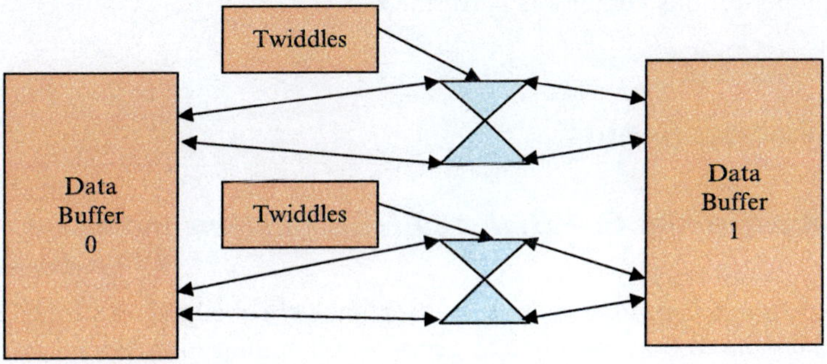

Fig. 8 Block diagram of VHDL implementation of FFT

factors. In fact there are two copies of the twiddle factors, one for each butterfly. To save space, only the real part of the twiddle factor is stored and the imaginary part is obtained by taking the real part from 90° out of phase and where appropriate negating it. On the FPGA where fast memory is constrained but computation is cheap, this is an advantage.

The architecture of the FFT implementation is shown in Fig. 8.

The data flow is bidirectional because the algorithm consists of multiple passes over the data. In even passes data flows from buffer 0 to buffer 1, and in odd passes from 1 to 0. A control process co-ordinates the data flow and generates addresses for the data buffers and the twiddle factor arrays. The address generation, which involves bit manipulation, is implemented easily and efficiently in hardware, without the nested loops used in software to do the same thing.

Each data buffer actually consists of two dual port block RAMs, allowing four reads or writes per clock cycle. The data is partitioned in such a way that there will always be two accesses to each block RAM on each clock cycle. The memory locations are 64 bits wide so that an entire single precision complex number can be stored in each one.

FFT sizes up to 32,768 points can be run entirely within the internal block RAM. The speed is about 2,280 MFlops/s, approximately six times faster than the software running on the Xeon. The pipelined butterflies are responsible for most of this advantage—with 20 floating point operations per clock cycle (two pipelines of 10 operations each) and a clock frequency of 115 MHz, the theoretical maximum speed would be 2,300 MFlops/s.

Larger transforms (up to 1,048,576 point is supported) must be computed by copying the data from DRAM a block at a time, transforming it, then saving it back to DRAM again. For maximum efficiency the first few passes are run on each block in one go without accessing the DRAM in between, but the later passes have to be run one at a time on each block. Throughput is maximised by overlapping the transform of the current block with the writing of the previous block back to two banks of DRAM and fetching the next block from the other two banks of DRAM. The FPGA is about three times faster than the software for larger data sizes.

No HCE port of this kernel was performed.

4 Summary of Results

4.1 Comparison of C, VHDL and HCE Implementations

The timings from the original codes in this section were obtained on a 2.8 GHz Intel Xeon, running the code as compiled by gcc with full optimisation. For completeness the reference performance values were also obtained on a dual socket Nehalem-EP system using the Intel compiler with full optimisation as well as the Intel Math Kernel Library (MKL) are listed in the tables below. In general the FPGA performance is better than that of the Xeon (which is about the same age) but does not compare well to the Nehalem system. However, it should be noted that (a) the energy efficiency is likely to be higher (see Section 4.15), and (b) FPGAs generally perform better with integer codes than with floating point.

As well as the performance of each kernel implementation (measured in megaflops per second for most of them, but values per second for the random number generator), several other items of interest have been recorded. Estimated implementation effort in days and lines of code are given to provide some idea of the relative complexity of each implementation. For the FPGA implementations, device utilisation (the percentage of the FPGA's logic blocks required for the kernel) and the maximum speed at which the FPGA could be clocked for reliable operation are also given. A low device utilisation indicates that there is space remaining on the FPGA that could possibly be used for further parallelising the kernel or for accelerating an additional part of a larger code as well as the kernel. A low clock speed suggests that power consumption is likely to be low.

For most of the kernels, both "on-board" and "off-board" timings are given; in these cases, the on-board timing measures only the actual computation time and does not take account of the time necessary for transferring data between the host system and the FPGA board memory. The off-board timing is an inclusive

Table 1 mod2am performance results

Algorithm version	Speed (MFlop/s)	Implementation effort (day)	Lines of code	Device utilisation (%)	Clock speed (MHz)
Original (SP)	587	0.5	38	N/A	N/A
Reference platform (1,024 × 1,024) (DP)	63,000	∼ 0.5	277	N/A	2,530
Harwest (on-board)	2,859	10	543	93	67
Harwest (off-board)	2,463	10	543	93	67
VHDL (on-board)	1,846	16	1,252	39	115
VHDL (off-board)	1,081	16	1,252	39	115

measurement that does take the data transfers into account. As can be seen from the differences between these figures, transferring data across the PCI bus can be a major bottleneck for accelerator-based kernel implementations.

4.1.1 Mod2am

The VHDL version timings are taken from multiplying two 1,024 × 1,024 matrices together, as it is easier to work with powers of two in hardware implementations (a version able to work with non-power of two sizes would be more complex but probably just as fast). It performs 16 floating-point operations per clock cycle and runs approximately three times faster than the Xeon, dropping to about twice as fast when the time taken to transfer data across the PCI bus is taken into account Table 1.

The Harwest port uses a similar strategy of caching subsections of the matrices in block RAM and using a pipelined tree of floating-point operations; however, the Harwest port pipelines 63 floating-point operations rather than the 16 in the VHDL version and achieves better performance. The code is very much altered from the original C version and contains unrolled loops, pipelined functions and HCE-specific code and directives.

4.1.2 Mod2as

The sparse matrix-by-vector multiplication kernel (spmxv) is in some ways similar to the matrix-by-matrix multiplication (mxm). However, while the mxm performs a large number of operations on each item of input data and can therefore make good use of cache memory, the spmxv operation uses each element of the matrix only once. This means that it is memory bound and an FPGA is unlikely to have much advantage over a conventional processor. A modest speed up over the gcc-generated code on the host CPU was achieved with the VHDL version, probably due to the advantage gained by fetching from multiple external memory banks simultaneously. Additionally, the time taken to transfer the input data over the PCI bus to the FPGA

Table 2 mod2as performance results

Algorithm version	Speed (MFlop/s)	Implementation effort (day)	Lines of code	Device utilisation (%)	Clock speed (MHz)
Original (SP)	489	0.5	23	N/A	N/A
Reference platform single core (DP)	1,392	~ 0.5	300	N/A	2,530
Reference platform MKL (DP)	3,067	~ 0.5	300	N/A	2,530
HCE (on-board)	34.6	4	132	55	67
HCE (off-board)	4.9	4	132	55	67
VHDL (on-board)	622	10	2,490	43	110
VHDL (off-board)	6.2	10	2,490	43	110

device is a major bottleneck. When this is taken into account the performance of the VHDL version is impaired by a factor of 100 Table 2.

The HCE version of the kernel uses a similar structure to the VHDL version but scope for optimisation is more limited as many of HCE's parallelisation directives only work effectively on relatively large data blocks of predetermined size, which does not apply well to this kernel. Although it was possible to speed up the original port 12-fold by pipelining eight multiply and add operations, the HCE version is still much slower than the host or VHDL. It is likely that this could be improved further given more time.

4.1.3 Mod2h

The VHDL version of the algorithm is capable of generating 900 million single precision values per second (six per clock cycle with the design clocked at 150 MHz). This algorithm, consisting primarily of bit shifts and exclusive or operations, was particularly amenable to FPGA acceleration. The main transform of the random number generator can be expressed elegantly in VHDL and it was also possible to eliminate the final floating-point scaling from the VHDL implementation as it was just a bit shift, replacing it with a custom integer-to-float conversion which completes in one clock cycle instead. However, when the time taken to transfer the results back to the host is taken into account, the VHDL version's performance is dramatically reduced by a factor of more than 100, making it slower than the software kernel. The bandwidth over the PCI bus appears to be limited to approximately 32 MB/s (Table 3).

Porting to Harwest was slightly complicated as the original kernel makes extensive use of 64-bit integer operations, which Harwest does not support. The algorithm had to be rewritten and tested using multiple 32-bit operations to achieve the same result. Because Harwest does not support double precision floating-point, this port passes back the unconverted 64-bit integer results so that the host can perform the final conversion to floating-point in full precision. By employing similar

Table 3 mod2h performance results

Algorithm version	Speed (millions of values per second)	Implementation effort (day)	Lines of code	Device utilisation (%)	Clock speed (MHz)
Original (SP)	14.4	1	37	N/A	N/A
Reference platform single core (DP)	47.6	~ 0.25	37	N/A	2,530
Harwest (on-chip, int)	217.6	6	171	26	80
Harwest (off-chip, int)	4.0	6	171	26	80
VHDL (on-chip)	900.0	6	690	54	150
VHDL (off-chip)	8.1	6	690	54	150

optimisation strategies to those used in the VHDL version (using HCE-specific directives to turn the main transform into a combinatorial operation, and running three separate instances of it in parallel), numbers can be generated at a rate of 217 million per second, much faster than the host, though slower than the hand-coded VHDL. However, this is subject to the same PCI bus bandwidth problem, so when the time taken to transfer the results back to the host is included, the speed again falls to 4 million values per second.

4.1.4 Mod2f

When running on data in its internal memory, the FPGA reaches speeds approximately six times faster than the original gcc-generated code running on the host CPU. But comparing the FPGA-based results to the reference platform performance one clearly can see that the latest generation of general purpose processors do provide FFT performance values which are quite comparable to the values obtained on FPGAs. The original kernel is a radix 4 FFT. The VHDL version is a simple radix 2 implementation but gives identical results. When its pipeline is filled, it can complete two FFT butterflies (each one comprising 10 individual floating-point additions, subtractions and multiplications) per clock cycle. At 115 MHz this gives a sustained speed of almost 2.3 GFlop/s.

Table 4 shows the maximum performance achieved by each implementation. As the FFT performance is more sensitive to the size of the data set than that of the other kernels, performance for each data set size is given in Fig. 9.

The performance of the VHDL version varies much more than that of the host; with faster computation, factors such as memory bandwidth, PCI bus bandwidth and start-up overhead become more significant. Looking at the on-board performance, the VHDL version is fairly slow for a 256 point transform—this data size is small enough that start-up overheads for the algorithm and for each pass become noticeable. As the size increases, performance improves and reaches its peak for the 32,768 point transform, which the FPGA can perform around six times faster than the gcc-code run on the host CPU. This is the largest transform which can

Table 4 mod2f performance results

Algorithm version	Max speed (MFlop/s)	Implementation effort (day)	Lines of code	Device utilisation (%)	Clock speed (MHz)
Original (DP)	481	N/A	420	N/A	N/A
Original (SP)	465	0.5	420	N/A	N/A
Reference platform (DP)	2,778	~ 2	398	N/A	2,530
VHDL (on-board)	2,284	11	2,119	50	115
VHDL (off-board)	1,177	11	2,119	50	115

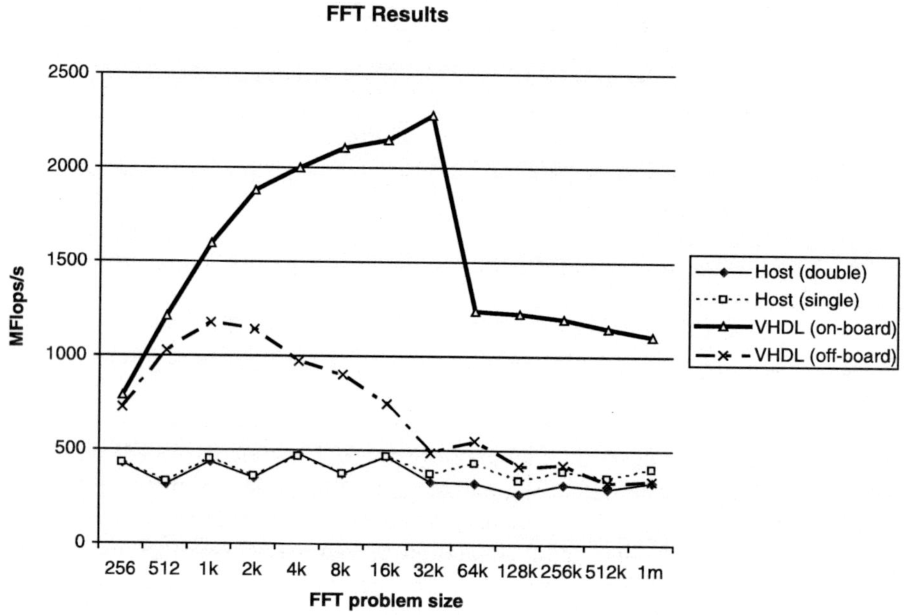

Fig. 9 Performance of mod2f for different problem sizes

be performed entirely within the FPGA's internal block RAM and does not require any transfers to or from DRAM during the algorithm. For larger sizes, performance tails off as the rate at which data can be read from or written to the DRAM banks becomes more and more significant. The FPGA is still faster than the host by a factor of 3–4, probably because reading and writing four separate external memory banks simultaneously gives it an advantage. When the time taken to transfer data to and from the FPGA is taken into account (off-board column in the table), much of the VHDL version's speed advantage is lost. Unfortunately there was insufficient time to complete a Harwest port of this kernel.

4.1.5 Notes on Power Consumption

Low power consumption is a major advantage of FPGAs so it is worth considering the PRACE kernel implementations in this context. Unfortunately the Maxwell system does not provide a facility for measuring the actual power consumed by the accelerator boards, so only a rough estimate of power usage based on data available online is possible.

Virtex-4 devices are much more power efficient than those of previous FPGA generations. Information released by Xilinx suggests that a large Virtex-4 FPGA (in this case an LX device, but the FX family used on the AlphaData boards in Maxwell is likely to be similar) running a demanding bitstream with most of the slices occupied at 200 MHz is likely to consume around 2–3 W. Given that the implementations of the Euroben kernels all run at 150 MHz or less and mostly occupy a lower percentage of the logic slices, it seems reasonable to assume that they also will not consume more than 3 W. Assuming 3 W power consumption, the Flops-per-Watt value for the Euroben kernels reaches almost 1,000 MFlops/W (for the matrix multiply kernel which is the fastest).

This compares very favourably to contemporary conventional CPUs; the thermal design power for the single core Xeons in Maxwell is 77 W, and the actual power consumed by one of these under high load is likely to be at least 50 W. Assuming 50 W power consumption, and taking the highest megaflop figure achieved in our testing (587 MFlops for the dense matrix multiplication), the Xeon is performing only 11.7 MFlops/W.

5 Conclusions

The PRACE work on Maxwell demonstrates a number of points about FPGA acceleration.

- FPGAs can be much faster than contemporary CPUs for certain tasks (e.g. the Euroben random number generator is 30 times faster on the FPGA compared to the Xeon) but do not provide significant benefits for all codes (e.g. the sparse-matrix-by-vector multiplication kernel is only slightly faster on the FPGA).
- C-to-gates tools (e.g. HCE) can be performance-competitive with hand-coded VHDL in some circumstances. They can also remove a lot of the most tedious and error-prone elements from the porting work.
- Hand tuning is still essential in order to get good performance from FPGAs, whatever language they are programmed in.
- FPGA power consumption, including flops-per-watt metric, can be very good due to low clock speeds.
- Most current accelerator designs (including FPGA boards and GPUs) suffer from data transfer bottleneck between accelerator and host (e.g. PCI bus), which can badly affect some codes.

FPGA-based accelerators can be a valuable aid to improving application performance, and particularly performance-per-watt. However, there are still programmability challenges; while modern C-to-gates tools can sometimes match the performance of hand-coded designs at a lower development cost, there is a lack of standardisation and code still tends to be very platform- and compiler-specific. A standard for C-to-gates languages, perhaps based on OpenCL, would be a great step forward in this area.

References

1. PRACE: Partnership for Advanced Computing in Europe. http://www.prace-project.eu/. Accessed 28th May 2012
2. Home Page of the EuroBen Benchmark. http://www.hpcresearch.nl/euroben/index.php. Accessed 28th May 2012
3. The FHPCA. http://www.fhpca.org. Accessed 28th May 2012
4. AlphaData ADM-XRC-4FX product page. http://www.alphadata.co.uk/products.php?product=ADM-XRC-4FX. Accessed 28th May 2012
5. Nallatech H101 product overview. http://www.nallatech.com/H101-PCIXM/overview.html. Accessed 28th May 2012
6. Xilinx ISE Design Suite. http://www.xilinx.com/products/design-tools/ise-design-suite/. Accessed 28th May 2012
7. Xilinx Core Generator System. http://www.xilinx.com/tools/coregen.htm. Accessed 28th May 2012
8. Marongiu A, Palazzari P *The HARWEST Compiling Environment: Accessing the FPGA World through ANSI-C Programs* (Cray User Group publications, 2008) https://cug.org/5-publications/proceedings_attendee_lists/2008CD/S08_Proceedings/pages/Authors/01--5Monday/Palazzari-Monday4C/Palazzari-Monday4C-paper.pdf. Accessed 28th May 2012

Assessing Productivity of High-Level Design Methodologies for High-Performance Reconfigurable Computers

Esam El-Araby, Saumil G. Merchant, and Tarek El-Ghazawi

Abstract In spite of their potential to provide substantial performance improvements over traditional supercomputers, high-performance reconfigurable computers (HPRCs) and their broad acceptance have been hindered by productivity challenges. These challenges arise from increased design complexity, a wide array of custom design languages and tools, and often overblown sales literature. Therefore, it is essential to review, evaluate, and assess the productivity of this technology. This chapter presents a review and taxonomy of high-level languages (HLLs) for HPRCs and a framework for the comparative analysis of their features. It also introduces new metrics and an assessment model based on computational effort. The proposed concepts are inspired by Newton's equations of motion and the notion of work and power in an abstract multidimensional space of design specifications. The metrics are devised to highlight two aspects of the design process: the total time-to-solution and the efficient utilization of user and computing resources at discrete time steps along the development path. The study involves analytical and experimental evaluations demonstrating the applicability of the proposed model.

E. El-Araby (✉)
Electrical Engineering and Computer Science, The Catholic University of America, Pangborn Hall Room #314A, 620 Michigan Ave., NE, Washington, DC 20064, USA
e-mail: aly@cua.edu

S.G. Merchant
IBM Corporation, India B148, 6th Floor, Embassy Golf Links Block D, Domlur Kormangala Intermediate Ring Rd, Bangalore, KA 560071, India
e-mail: smerchant@in.ibm.com

T. El-Ghazawi
The George Washington University, 624D Academic Center, 801 22nd Street, NW, Washington, DC 20052, USA
e-mail: tarek@gwu.edu

W. Vanderbauwhede and K. Benkrid (eds.), *High-Performance Computing Using FPGAs*, DOI 10.1007/978-1-4614-1791-0_24, © Springer Science+Business Media, LLC 2013

1 Introduction

Programmability challenges with high-performance reconfigurable computers (HPRCs) have hindered their widespread acceptance among the supercomputing community. Application development on these systems typically requires software and hardware programming expertise for which design paradigms and tools have been traditionally separate. The standard way of describing software is using high-level languages (HLLs), such as C, C++, or Fortran, whereas, hardware is typically designed using hardware description languages (HDLs), such as VHDL and Verilog. Fragmented design flow and the need for expertise in parallel software and hardware design are major productivity hurdles facing HPRCs. To bridge the productivity gap several HLL tools such as Xilinx Forge, Handel-C, Impulse-C, Mitrion-C, and Bluespec have been proposed which attempt to abstract underlying hardware design details and streamline the disparate design flows. These tools often tradeoff performance for programmability. Dataflow design tools, based on the graphical user interface, e.g., DSPLogic, seem to offer an interesting compromise between HLLs and HDLs. These languages offer a trade-off between a shorter development time and a performance overhead imposed by HLLs.

Streamlining hardware description using HLLs typically used in software programming, or at least using dataflow languages, is a major and distinctive feature of HPRCs that potentially allows domain scientists to develop entire applications without relying on hardware designers. However, an HLL compiler for HPRCs must combine the capabilities of tools for traditional microprocessor compilation and tools for computer-aided design with FPGAs. It must also extend these two separate set of tools with capabilities for mutual synchronization and data transfer between microprocessors and reconfigurable processor subsystems [1, 2]. The problems are further escalated by lack of standard interfaces and architectural diversity in reconfigurable computing subsystems. Moreover, the range of tool choices and puffed up sales literature make it hard to comprehend real differences.

This chapter aims to present a framework and a mathematical model to compare and contrast different HLL languages and their features. For this a detailed review and a taxonomy of the existing design languages for HPRCs is provided. The model and the metrics to evaluate HLLs are inspired from the principles of Newton's equations of motion and the notions of work and power in an abstract multidimensional space of design specifications. They highlight two distinct aspects of the development process: (a) the total time-to-solution and (b) the efficient utilization of user and computing resources. We believe that this enables a comprehensive evaluation of the design languages. The experimental study presented includes HLLs from the imperative and the dataflow programming paradigms, showcasing the wide applicability of our methodology. In brief, the major contributions of this chapter are as follows: (a) A detailed review and taxonomy of HLL languages; (b) new evaluation metrics to emphasize the total time to solution as well as the resource usage efficiency of an HLL tool/language along the development path; and (c) an analytical framework for comparative analyses of HLL languages.

The remainder of this chapter is organized as follows. Section 2 presents a detailed review and taxonomy of HLL design tools for HPRCs. Section 3 presents related work. Section 4 introduces the model and the framework to evaluate and compare HLLs. Section 5 presents an experimental study that uses the proposed framework to evaluate HLLs from imperative and dataflow programming styles. Finally, Sect. 6 concludes the chapter.

2 HLL Review and Taxonomy

To understand and evaluate the different language attributes, taxonomy is imperative. This section provides a thorough review and taxonomy of available HLL tools in research and commercial literature. Table 1 shows a list of HLLs reviewed. Some of the listed languages are text-based, either C-, Fortran-, Java-, or Matlab-like, others are graphical-based.

Our review revealed that various vendors provide not only a HLL, but also a complete development environment that may integrate with the tools of the basic development flow, see Fig. 1 Users can develop their applications using either the standard HDL flow or a suite of higher-level languages such as C and C++ or the Xilinx System Generator for DSP package.

Impulse Accelerated Technologies, for example, provides a C-based development kit, Impulse-C, that allows the users to program their applications in standard C with the aid of a library of functions for describing parallel processes, partitioning the application into software and hardware parts, and simulating and instrumenting the application. The Impulse-C programming model provides stream-based communication between processes. The kit provides a compiler that generates HDL code for synthesis from the hardware parts of the application targeting different HPRC platforms such as Cray-XD1 and/or SGI Altix/RASC [3,4] Impulse-C represents a class of imperative languages with syntax based strongly on ANSI C [5]. The language is extended to address specific hardware concepts such as

Table 1 Reviewed HLLS

1	Impulse-C	11	SA-C	21	JHDL	31	C2Verilog
2	Handel-C	12	Trident Compiler	22	Galadriel & Nenya	32	Bach C
3	Mitrion-C	13	CHiMPS	23	Viva	33	SpecC
4	Dime-C	14	System-Studio	24	Ptolemy II	34	Ocapi
5	System-C	15	Convergen SC	25	SysGen	35	HardwareC
6	Catapult-C	16	Transmogrifier-C	26	RC Toolbox	36	Cones
7	Carte-C	17	SPARK	27	CoreFire	37	BDL
8	Streams-C	18	Brass	28	DSS	38	Cocentric
9	AccelChip	19	DeFacto	29	Forge	39	C2H
10	NAPA-C	20	MATCH	30	CASH	40	Blue Spec

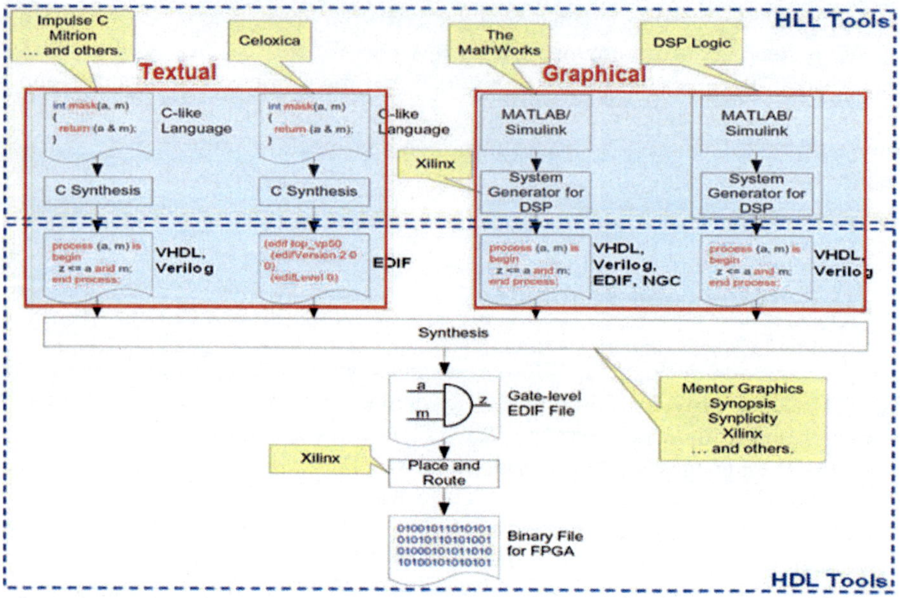

Fig. 1 Development flow of high-level tools [3]

communicating sequential processes (CSP) and streams. Existing VHDL designs may also be incorporated and called from the Impulse-C code as external functions.

Mentor Graphics, formerly Celoxica and later Agility DS, provides a C-based hardware design language, Handel-C, and the DK Design Suite that can be customized for specific HPRC platforms such as the Cray-XD1 and/or SGI Altix/RASC systems [3, 4]. DK Design Suite unifies system verification,hardware/software codesign, and Handel-C synthesis in a GUI-based development environment. The customization includes a Platform Support Library (PSL), which includes Handel-C interfaces to the underlying HPRC platform [6].

Mitrionics provides a C-based hardware design language, Mitrion-C, and an abstract machine, the Mitrion Virtual Processor (MVP), which allows users to develop portable high-level code for FPGA applications. The Mitrion architecture uses a data-driven representation of the program, which the tools map onto programmable logic. Mitrion-C is an ANSI C-based functional language. Mitrion-C programming language is an implicitly parallel programming language with syntax similar to C. The language centers on parallelism and data dependencies. In contrast, traditional languages are sequential and center on order-of-execution. In Mitrion-C there is no order-of-execution; any operation may be executed as soon as its data dependencies are fulfilled. Mitrion-C is a Single-Assignment language (variables may only be assigned once in a scope) in order to prevent variables from having different values within the same scope. Software written in the Mitrion-C programming language is compiled into a configuration of the MVP. The MVP is a fine-grain, massively parallel, reconfigurable soft-core processor [7].

BlueSpec provides BlueSpec system verilog (BSV) language high-level synthesis tools which produce systemC executables and verilog RTL models for RTL synthesis and/or simulation with third party tools. The input to these tools are models, transactors, test benches, and implementations written in BSV. BSV is a high-level functional hardware description programming language and is a synthesizable subset of system verilog. It enables some powerful features emanating from parameterization, atomic transactions, synthesizable transactors, and synthesizable verification. It targets both ASIC and FPGA design flows by integrating into popular design flows such as that of Cadence, Synopsys, Mentor Graphics, and Magma [8].

The Xilinx System Generator (SysGen) for DSP tool uses a somewhat different approach for designing digital signal processing (DSP) blocks. It integrates with the MATLAB and Simulink packages from MathWorks to allow users to design the algorithmic block of their applications in the MATLAB GUI environment [9].

DSPlogic provides the Rapid Reconfigurable Computing Development Kit (RC Toolbox), which integrates with MATLAB and Simulink from MathWorks and with Xilinx tools. It allows users to design, verify, and build the FPGA logic for DSP applications entirely from within the MATLAB/Simulink environment. The tool also includes the Reconfigurable Computing I/O (RCIO) library, which provides a portable application programming interface for communications between the software application that runs on the host processor(s) and the attached FPGAs. DSPLogic RC Toolbox is a combined graphical and text-based programming environment for HPRC application development. Blocks from the DSPlogic RC blockset and Xilinx System Generator are used to create a data flow diagram. Existing VHDL designs may also be incorporated using System Generator's HDL co-simulation capabilities [10].

The Carte-C (Carte-Fortran) development environment is somewhat different from the above mentioned flows in the sense that it is tightly integrated with SRC systems. It is a C-based (Fortran-based) environment that allows users to program their applications in standard C (Fortran) with the aid of a library of pre-synthesized hardware functions. There are two types of application source files to be compiled, one that targets the microprocessor and another that targets the reconfigurable processor. Since users often wish to extend the built-in set of operators, the compiler allows users to integrate their own VHDL/Verilog macros. The environment also provides a means for debugging and verification [11].

2.1 HLL Tool Taxonomy

After reviewing the literature of HLLs, we recognized the need for a taxonomy of their programming models that would provide a useful means for the characterization of the differences among them. Figure 2 shows our taxonomy of the programming models of the different HLLs. The programming model can be defined as the hardware abstract view presented to the programmer by the programming

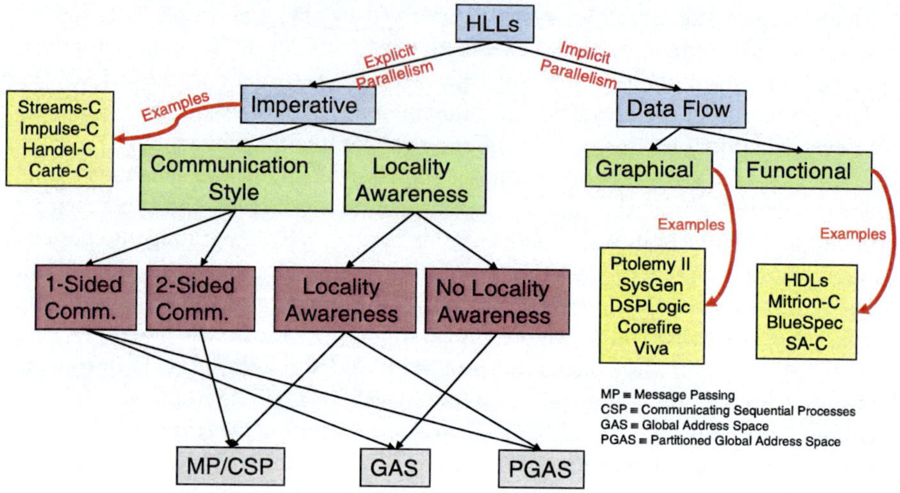

Fig. 2 Taxonomy of HLL programming models

tool. Thus, a programming model defines which parts of the hardware architecture will become visible to the programmer and be under his/her direct control.

In general, HLLs can be categorized as either imperative or dataflow programming paradigms. This is mainly dependent on how parallelism is expressed and/or extracted. In imperative paradigms parallelism is nonnative and expressed in an explicit manner which is solely the user's responsibility. In other words, in imperative languages, everything is sequential unless otherwise stated. On the other hand, in dataflow paradigms parallelism is native and expressed in an implicit manner. In other words, in dataflow languages, everything is parallel unless otherwise stated.

Dataflow paradigms, as shown in Fig. 2, can be further divided into two subcategories, namely functional and graphical paradigms. Functional languages are basically the text-based versions of dataflow paradigms as opposed to the graphical version. In general, functional HLLs are characterized as being single-assignment languages. In light of this, HDLs, i.e., VHDL and/or Verilog, fall under this category. Examples of this family are Mitrion-C, BlueSpec, SA-C, etc. Examples of graphical dataflow languages include SysGen, DSPLogic, CoreFire, Viva, etc.

Imperative paradigms, on the other hand, include languages such as Streams-C, Impulse-C, Handel-C, Carte-C, etc. However, because parallelism is nonnative, locality awareness and communication style, from an HPC perspective, become issues that are difficult to express in imperative paradigms. These issues are resolved by either the introduction of new extensions to the language, as in the case of Handel-C, or by the insertion of compiler directives/pragmas, as in the case of Impulse-C and Carte-C. In addition, concepts such as Message Passing (MP), Communication Sequential Processes (CSP), Global Address Space (GAS), and Partitioned Global Address Space (PGAS), are commonly found among the plethora of languages that fall under imperative paradigms. Figure 2 shows how those languages deal with communication issues as well as with locality awareness. For example, languages such as Streams-C and Impulse-C provide a two-sided, i.e., send-and-receive, communication style and also show an awareness of data locality through the MP/CSP model. The MP/CSP model is also supported by Carte-C and Handel-C.

3 Related Works

The objective of this chapter is to formalize a statistical framework to evaluate various HLL features/attributes to characterize and compare different high-level design languages/methodologies. To achieve this objective our approach is based on leveraging previous work and concepts that were introduced, and proved useful to us, in similar investigations. For example, Holland et al. [12] reviewed some C-based HLLs and highlighted some of the differences among them. Similarly, Edwards [13] discussed the challenges of synthesizing hardware from C-like languages. Finally, our previous work [14, 15] provided a formal and empirical comparative analysis of HLLs along with experimental work conducted on Cray-XD1. The work presented here significantly extends the model in [14, 15] and leverages some of its terminology and concepts. It enables evaluation of language features as well as experimental metrics, both of which are termed as attributes in the model. Furthermore, in our investigation the elimination of biasing effects has been formalized based on statistical analysis and validation.

In our study we considered productivity as one of the evaluation metrics. The definition and model of productivity have been widely discussed in the literature. For example, Sterling [16] in his "Special Theory of Productivity" defines it as utility divided by time, where utility being the useful work, e.g., operations. While Snir and Bader [17] defined productivity of a system as the time dependent utility of the answers it produces divided by the total lifetime cost. Kennedy et al. [18] definition is tailored towards tool expressiveness and efficiency. Abstract expressiveness determines the programming tool development power and computational efficiency is the programming tool execution efficiency. Kepner [19, 20] combines all of the above ideas into a single synthesis formulation. In addition, Numrich [21] presents his generic performance metric based on computational action. He examines work as it evolves in time and computes computational action as the integral of the work function over time.

Our work leverages and builds off concepts from Numrich's research on performance and productivity metrics based on the principle of computational least action [21–25]. The metrics proposed here emphasize the rate of this computational work/effort. We call this rate as the work progress rate. Computational work or effort is the work done by the user–tool combination in traversing an abstract specification space. As it will be shown later, the productivity metric emphasizing the total time-to-solution is a special case of this new metric.

4 HLL Evaluation Framework

4.1 Formalizing the Framework

We start our formalization by visualizing the different HLLs as being observations of a space of attributes. These attributes can be either qualitative language features such as support for pointers and debugging capability, or experimental metrics such as development time and productivity. Therefore, in this multidimensional space of attributes each language can be considered as a single point or as an observation in that space. For example, Fig. 3 shows a case where this space has been hypothetically reduced to a two-dimensional space of, collectively, two orthogonal sets of attributes, namely software features and hardware features. Therefore, the evaluation of the different HLLs can be simply performed by comparing the different components along the dimensions of that space, e.g., feature coverage and/or feature loss. Figure 3 shows a pictorial comparison of two hypothetical HLLs with different degrees of feature coverage and/or loss.

It is worth mentioning that in developing our framework we needed to minimize certain biasing effects associated with small sample sizes. For example, biasing effects may include previous user experience and knowledge of a specific language and/or design methodology. Therefore, we formalized our framework by instrumenting a normalization mechanism through which each user is required to provide three different observations of the same trial (application). In other words, each user develops the same application in three different implementations (languages). The first implementation is performed using the language under consideration, while the other two are performed using both pure software, e.g., in C/C++, and pure

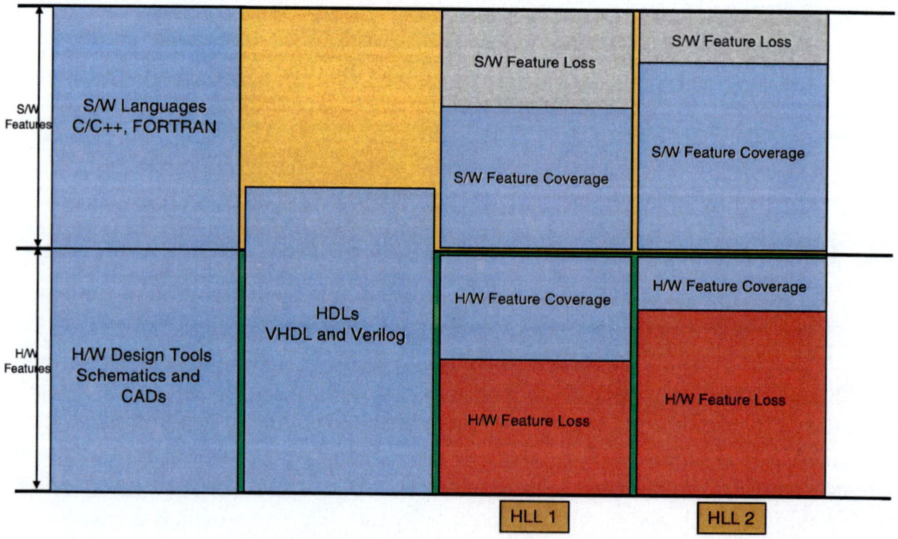

Fig. 3 HLLs as observations in the feature space

a

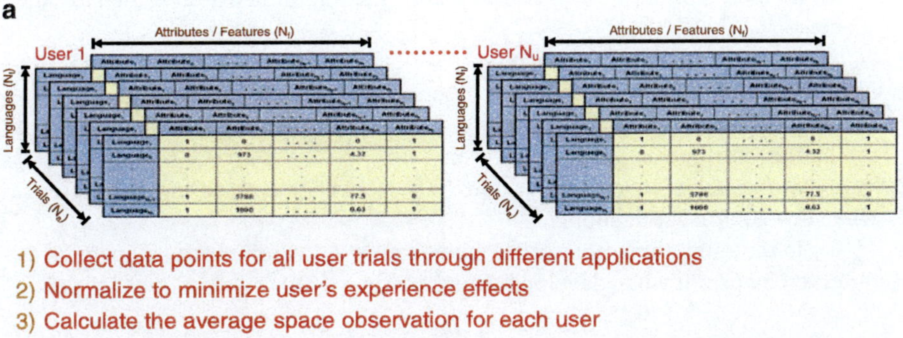

b

Attributes / Features (N_f)

Languages (N_l)

	Attribute$_1$	Attribute$_2$	....	Attribute $_{N_f-1}$	Attribute $_{N_f}$
Language$_1$	1	0	....	0	1
Language$_2$	0	973	....	4.32	1
.	.	.	.	.	.
Language $_{N_l-1}$	1	5780	....	77.5	0
Language $_{N_l}$	1	1000	....	0.63	1

4) Calculate the average space observation for all users

$$s_{l,f} = \frac{1}{N_u}\sum_{u=1}^{N_u} s_{u,l,f} \Rightarrow s_{l,f} = \frac{1}{N_u N_a}\sum_{u=1}^{N_u}\sum_{a=1}^{N_a}\left(\frac{v_{a,u,l,f} - v_{a,u,f}^{\min}}{v_{a,u,f}^{\max} - v_{a,u,f}^{\min}}\right)$$

$$\textit{where} \left\{ a \in [1, N_a]\ ,\ u \in [1, N_u]\ ,\ l \in [1, N_l]\ ,\ f \in [1, N_f] \right\},\ \textit{and}$$

$$\left\{ v_{a,u,f}^{\min} = \min_{l=1}^{N_l}\left(v_{a,u,l,f}\right),\ v_{a,u,f}^{\max} = \max_{l=1}^{N_l}\left(v_{a,u,l,f}\right) \right\}$$

Fig. 4 Formalizing the scoring mechanism. (**a**) Preliminary attribute/feature matrices. (**b**) Final attribute/feature matrix

hardware, e.g., in VHDL/Verilog. These two other observations (implementations and/or languages) serve as reference points that can minimize the biasing effects of user experience and knowledge of a given language. In other words, each user observation for a certain trial is normalized to his/her own experience making the observations more language-specific rather than being user-specific measurements. Figure 4 summarizes the steps involved in our formal methodology for conducting the experimental work. Figure 4a shows the preliminary formulation of attribute matrices leading to the final attribute matrix shown in Fig. 4b. Each row in the final attributes matrix represents an observation (language) projected in the multidimensional space of attributes/features.

Based on the above discussion, we introduce the following notations in order to quantify our concepts:

- N_f is the total number of attributes/features, i.e., the dimension of the attribute space.
- N_l is the total number of languages.
- N_u is the total number of independent users involved in the experiments.
- N_a is the total number of applications developed by each user, i.e., the number of trials of the attribute space for each user.
- $v_{a,u,l,f}$ is the value of attribute/feature f for language l as observed by user u when developing application a.
- $v_{a,u,l,f}^{\min}$ is the minimum value of attribute/feature f for reference language l_0 as observed by user u when developing application a

$$v_{a,u,f}^{\min} = \min_{l=1}^{N_l} \left(v_{a,u,l,f} \right) \tag{1}$$

- $v_{a,u,l,f}^{\max}$ is the maximum value of attribute/feature f for reference language l_1 as observed by user u when developing application a

$$v_{a,u,f}^{\max} = \max_{l=1}^{N_l} \left(v_{a,u,l,f} \right) \tag{2}$$

- $s_{a,u,l,f}$ is the normalized value of attribute/feature f for language l, with respect to reference language l_0 and language l_1, as observed by user u when developing application a

$$s_{a,u,l,f} = \frac{v_{a,u,l,f} - v_{a,u,f}^{\min}}{v_{a,u,f}^{\max} - v_{a,u,f}^{\min}} \tag{3}$$

- $s_{u,l,f}$ is the average normalized value of attribute/feature f for language l, with respect to reference language l_0 and language l_1, as observed by user u across all trials (applications), i.e., the average space observation for each user

$$s_{u,l,f} = \frac{1}{N_a} \sum_{a=1}^{N_a} s_{a,u,l,f} = \frac{1}{N_a} \sum_{a=1}^{N_a} \left(\frac{v_{a,u,l,f} - v_{a,u,f}^{\min}}{v_{a,u,f}^{\max} - v_{a,u,f}^{\min}} \right) \tag{4}$$

- $s_{l,f}$ is the average normalized value of attribute/feature f for language l, with respect to reference language l_0 and language l_1, as observed by all users across all trials (applications), i.e., the average space observation for all users

$$s_{l,f} = \frac{1}{N_u} \sum_{u=1}^{N_u} s_{u,l,f}$$

$$\Rightarrow s_{l,f} = \frac{1}{N_u N_a} \sum_{u=1}^{N_u} \sum_{a=1}^{N_a} \left(\frac{v_{a,u,l,f} - v_{a,u,f}^{\min}}{v_{a,u,f}^{\max} - v_{a,u,f}^{\min}} \right) \tag{5}$$

where

$$\{a \in [1,N_a],\ u \in [1,N_u],\ l \in [1,N_l],\ f \in [1,N_f]\}$$

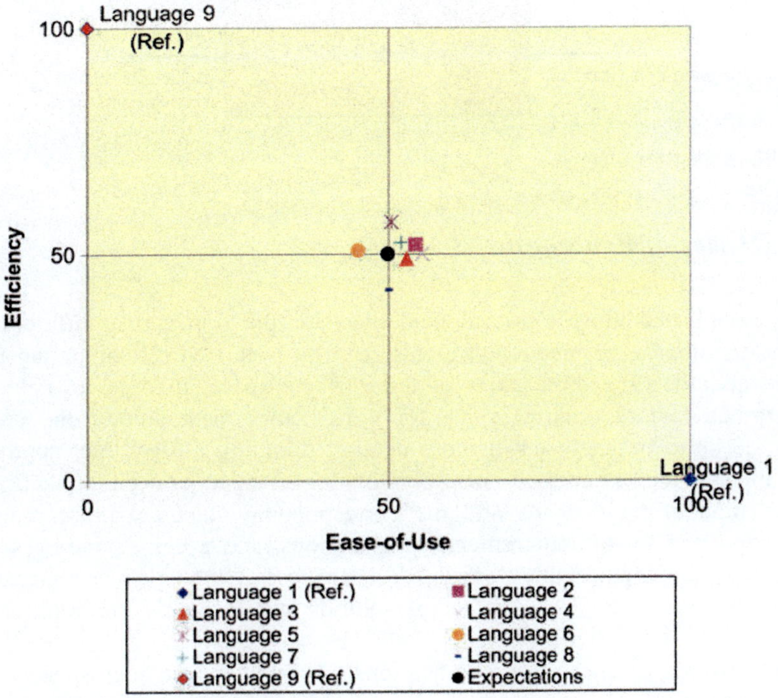

Fig. 5 Validation of the formal framework

4.2 Validating the Framework

We statistically validated the fairness of our framework by applying our formulation to the evaluation metrics (attributes), i.e., ease-of-use and efficiency, as proposed and defined in [14]. This validation was performed by the replacement of all entries in the preliminary attribute matrices, i.e., attribute value $v_{a,u,l,f}$, with independent and identically distributed random variables. The random variables were uniformly distributed. After applying our framework through Eqs. (1)–(5), we compared the theoretical expected values of the metrics of evaluation with the observed values. We found those to be almost identical. Furthermore, the variance of the data points was measured to be minimum and consistent with the theoretical expectations, see Fig. 5. In other words, the different hypothetical languages (observations) in the attribute space were closely clustered with minimum relative dispersion around the expected value. This proved to us the fairness of the proposed framework as well as the minimization of biasing effects towards certain languages over others. Figure 5 shows our findings with this respect. One can also note the relative placement of the reference languages as two extremes.

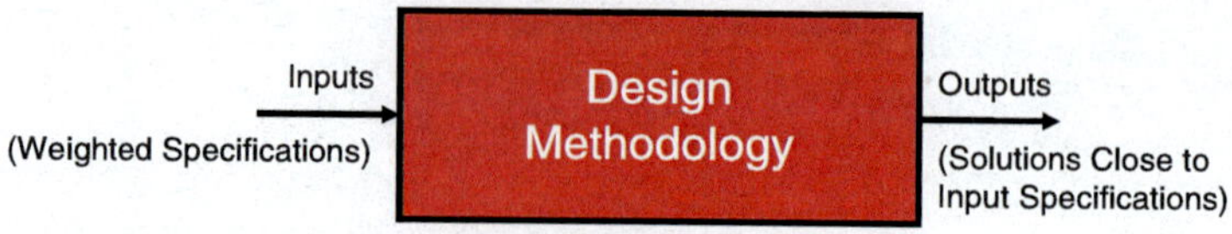

Fig. 6 Black box representation of design methodologies

4.3 Metrics of Evaluation

Having established those top-level guidelines for the framework, different design methodologies, as mentioned earlier, can be evaluated by comparing their components along the dimensions of the attribute space. In order to calculate each attribute/feature (metric) value for a particular methodology, the design process needs to be analyzed in more details. Adopting a black box approach, design methodologies and their corresponding tools/languages need a quantitative representation of their inputs and the corresponding outcome. Inputs can be represented as a set of requirements/specifications, and a corresponding set of preferences/weights adjustable along the design process. Outputs can be represented as a set of solutions where the target solution would be the preferred input specifications, see Fig. 6.

In other words, given a set of specifications, $S = \{$power, resources, speed, etc.$\}$, with their allowable ranges, $S_{min} = \{$minimum power, minimum resources, minimum speed, etc.$\}$ and $S_{max} = \{$maximum power, maximum resources, maximum speed, etc.$\}$, and also given their corresponding weights or preference $W = \{$power weight, resources weight, speed weight, etc.$\}$, the goal is to achieve a target set of specifications, $S_{target} = \{$target power, target resource utilization, target frequency, etc.$\}$. More formally, this can be represented as a multidimensional space of specifications whose basis can be described by the following vector representation:

$$\vec{S} = \begin{bmatrix} s_1 \\ s_2 \\ \cdots \\ s_N \end{bmatrix}, \overrightarrow{S_{\min}} = \begin{bmatrix} s_{1_{\min}} \\ s_{2_{\min}} \\ \cdots \\ s_{N_{\min}} \end{bmatrix}, \overrightarrow{S_{\max}} = \begin{bmatrix} s_{1_{\max}} \\ s_{2_{\max}} \\ \cdots \\ s_{N_{\max}} \end{bmatrix}, \overrightarrow{S_{\text{target}}} = \begin{bmatrix} s_{1_{\text{target}}} \\ s_{2_{\text{target}}} \\ \cdots \\ s_{N_{\text{target}}} \end{bmatrix},$$

$$\vec{W} = \begin{bmatrix} w_1 \\ w_2 \\ \cdots \\ w_N \end{bmatrix} \tag{6}$$

where

$N \equiv$ Dimensionality of the specification (solution) space

Due to the different scales and units for each component of the specification vector, a normalized and unitless representation of the space is desirable. Therefore, a mapping function is needed to establish the correspondence relation between the original space and the normalized space. The mapping is done such that more desirable solutions always have higher coordinates in the normalized space, see Eq. (7). The design problem becomes now a search process for the target solution. For specifications that need to be maximized, e.g., frequency, search in the normalized space moves in the same direction as in the original space

$$x_i = \left\{ \begin{array}{ll} w_i \times \frac{s_i - s_{i_{\min}}}{s_{i_{\max}} - s_{i_{\min}}}, & s_i \text{ to be maximized} \\ w_i \times \frac{s_{i_{\max}} - s_i}{s_{i_{\max}} - s_{i_{\min}}}, & s_i \text{ to be minimized} \end{array} \right\}, i = 1,2,\ldots,N$$

$$\Delta x_i = \left\{ \begin{array}{ll} w_i \times \frac{\Delta s_i}{s_{i_{\max}} - s_{i_{\min}}}, & s_i \text{ to be maximized} \\ w_i \times \frac{-\Delta s_i}{s_{i_{\max}} - s_{i_{\min}}}, & s_i \text{ to be minimized} \end{array} \right\}, \; i = 1,2,\ldots,N$$

$$\overrightarrow{X} = \begin{bmatrix} x_1 \\ x_2 \\ \cdots \\ x_N \end{bmatrix}, \; \overrightarrow{X_{\text{optimal}}} = \begin{bmatrix} x1_{\text{optimal}} \\ x2_{\text{optimal}} \\ \cdots \\ xN_{\text{optimal}} \end{bmatrix} = \begin{bmatrix} w_1 \\ w_2 \\ \cdots \\ w_N \end{bmatrix} = \overrightarrow{W}, \; \overrightarrow{X_{\text{target}}} = \begin{bmatrix} x1_{\text{target}} \\ x2_{\text{target}} \\ \cdots \\ xN_{\text{target}} \end{bmatrix} \tag{7}$$

where

$\overrightarrow{X} \equiv$ Solution position vector in the normalized space

For specifications that need to be minimized, e.g., area and/or power, search in the normalized space moves in the opposite direction to that in the original space, see Fig. 7. It can be seen in Fig. 7 that the most desirable solution, i.e., optimal solution with respect to the given range of specifications, is located at the positive extreme of the normalized space.

Based on this representation, design methodologies/tools can be evaluated by analyzing the time evolution of the search path towards the target solution. Different methodologies differ in the selection of the search path, i.e., location of candidate solutions in the search space. They also differ in the manner through which the search path is being time sampled, i.e., total number of iterations and/or the candidate solutions, M, along the search path. In other words, any given design methodology can be characterized by the instantaneous progress of the search path as a function of time. Under this representation, several useful quantities can be defined and used in studying the design process. More specifically, the target vector, $\boldsymbol{T_k}$, at time sample (iteration) k, can be defined as the distance from a given candidate solution, $\boldsymbol{X_k}$, to the target solution, $\boldsymbol{X_{\text{target}}}$. $\boldsymbol{T_k}$ measures the closeness of a given candidate solution to the final target solution. Also, displacement vector, $\boldsymbol{D_k}$, at time sample (iteration) k, which is the shift between consecutive candidates in the solution space measures the distance traversed along the search path from a candidate solution to the following one. Moreover, sensitivity, λ_k, at time sample (iteration) k, defined as the shift between consecutive candidate solutions in the direction of the target solution, i.e., the projection of $\boldsymbol{D_k}$ onto $\boldsymbol{T_k}$, represents a measure of the convergence towards the target solution. φ_k is the angle of projection or the angle of deviation of the tool away from the target. Additionally, the velocity

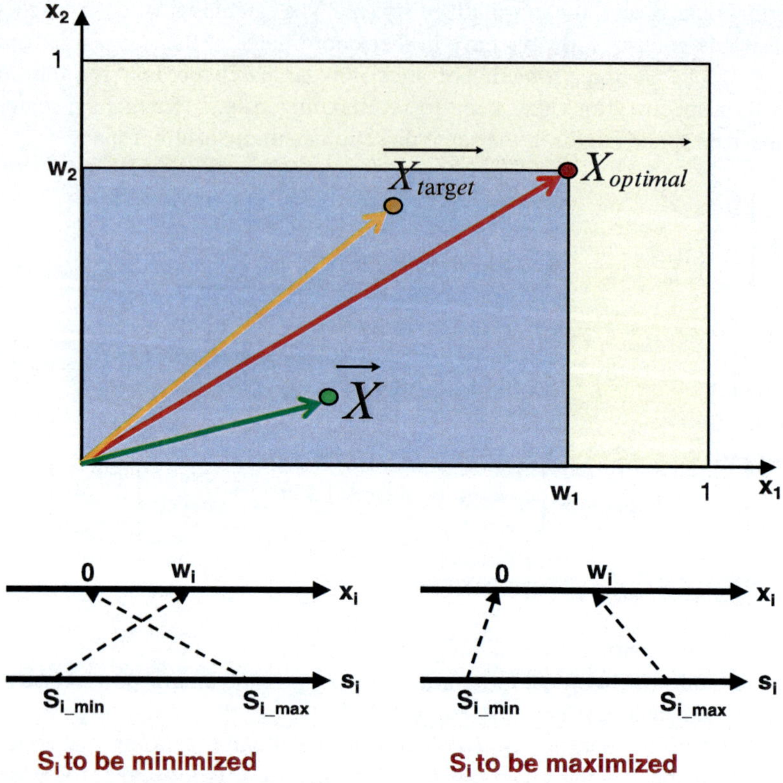

Fig. 7 Space for the design problem (solution search-process)

vector, υ_k, at time sample (iteration) k, represents the instantaneous traversal speed along the search path from a candidate solution to the following one. The definitions of these quantities are given in Eq. (8) and illustrated in Fig. 8

$$\begin{Bmatrix} \overrightarrow{X_k} = \overrightarrow{X(t_k^{;})},\ \overrightarrow{D_k} = \overrightarrow{D(t_k^{;})} \\ \overrightarrow{T_k} = \overrightarrow{T(t_k^{;})},\ \overrightarrow{\upsilon_k} = \overrightarrow{\upsilon(t_k^{;})},\ \lambda_k = \lambda(t_k^{;}) \end{Bmatrix}$$

$$\left.\begin{array}{l} \overrightarrow{T_k} = \overrightarrow{X_{\text{target}}} - \overrightarrow{X_k} \\ \overrightarrow{D_k} = \overrightarrow{X_{k+1}} - \overrightarrow{X_k} \\ \lambda_k = \dfrac{\overrightarrow{D_k} \cdot \overrightarrow{T_k}}{\left\|\overrightarrow{T_k}\right\|} = \dfrac{\overrightarrow{D_k} \cdot \overrightarrow{T_k}}{T_k} = D_k \cos(\varphi_k) \end{array}\right\} \begin{array}{l} ,\ 0 \le k < M \\ ,\ \text{and}\ \overrightarrow{X_0} = \overrightarrow{0} \end{array} \tag{8}$$

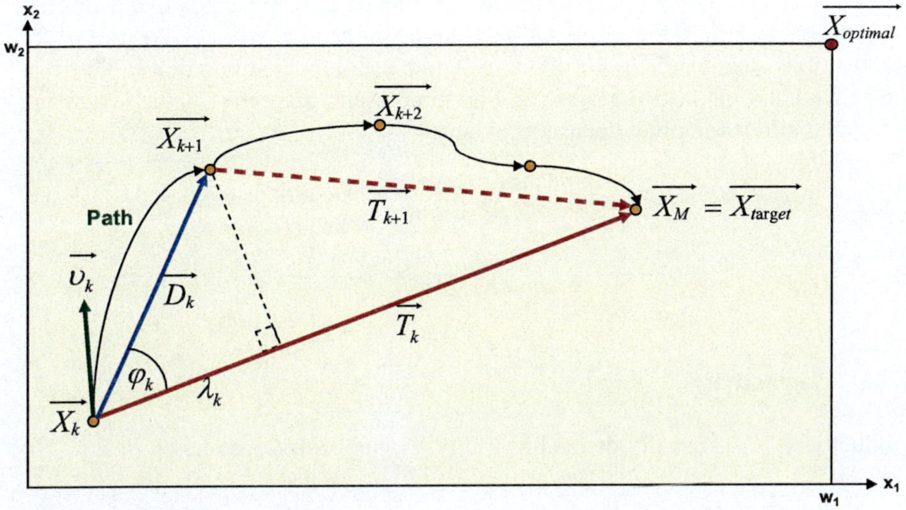

Fig. 8 Time evolution of the search path

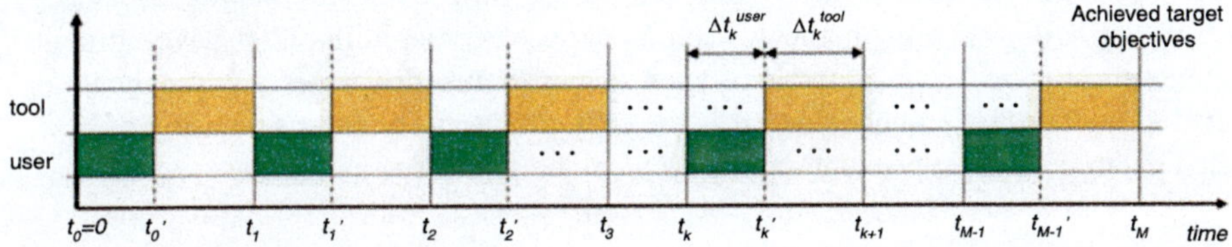

Fig. 9 Activities of the design process

where

$$\vec{A} \cdot \vec{B} \equiv \text{Dot product of vectors } \vec{A} \text{ and } \vec{B}$$

$$\left\|\vec{A}\right\| \equiv A \equiv \text{Length of vector } \vec{A}$$

Finally, it is essential to consider two activities that are typically associated with the design process, namely the user activity and the tool activity. These two activities are disjoint and mutually exclusive in time (consecutive activities that do not occur simultaneously), see Fig. 9, however, they are dependent in a causal (cause-and-effect) relationship. The design process typically starts with a user activity, i.e., initial design given to the tool, and ends with a tool activity after which the target solution at minimum is reached, see Fig. 8. The design process alternates (iterates) between the two exclusive activities, see Fig. 9. The user activity can be viewed as corrective to the tool deviations away from the target solution. Therefore, if we assume that t_M is the total development time, t_k represents the time spent after k iterations, Δt_k is the time period between two consecutive iterations, the expression given by Eq. (9) can be used to describe the timeline of the user and tool activities. It can be seen in Fig. 9 that Δt_k is the summation of the time spent by the user Δt_k^{user} and the time spent by the tool Δt_k^{tool} to generate the outputs of iteration k. Note that the shaded and un-shaded regions in Fig. 9 represent, respectively, the active and idle time periods for either the user or the tool

$$t_{k+1} = t_k + \Delta t_k = t_k + \left(\Delta t_k^{\text{user}} + \Delta t_k^{\text{tool}}\right) = t'_k + \Delta t_k^{\text{tool}},\ 0 \le k < M \tag{9}$$

where

$$t'_k = t_k + \Delta t_k^{\text{user}}$$

4.3.1 Productivity

Productivity, Ψ, is usually defined as utility, U, per cost, C, as shown in Eq. (10) [16–23]

$$\Psi = \frac{U}{C} \tag{10}$$

Utility is typically a function of achieved design objectives such as performance, power, and area, etc. Cost is the development cost expressed in either development time or proportional time equivalents such as man-hours, dollars, etc. When the total development time, t_M, is considered as the cost, productivity gives a measure of how fast a desired solution was obtained which can be expressed as follows:

$$\Psi = \frac{U}{C} = \frac{\left\|\overrightarrow{T_0}\right\|}{t_M} = \frac{\left\|\overrightarrow{X_{\text{target}}}\right\|}{t_M} \equiv \left\|\overrightarrow{v_{\text{target}}}\right\| \tag{11}$$

where $\boldsymbol{T_0}$ is the target vector at time sample (iteration) 0. In other words, $\boldsymbol{T_0}$ is the initial distance to the target solution, $\boldsymbol{X_{\text{target}}}$, see Fig. 8.Relative productivity of two methodologies is the ratio of individual productivities. Thus for two methodologies attaining the same design objectives the relative productivity is the inverse ratio of their development costs, see Eq. (12)

$$\Psi_{1/2} = \frac{\Psi_1}{\Psi_2} = \frac{C_2}{C_1} = \frac{t_{M_2}}{t_{M_1}} \tag{12}$$

Also, based on this metric two methodologies achieving the same design objectives with the same development costs have the same productivity. Although this is a logical inference, the metric has no notion of the user effort, tool maturity, or how efficiently the resources of the development process interact are being used or interact. Two tools could have taken two completely different paths in the objective search space to reach the target in the same amount of time, and have different levels of user involvement, design iterations, tool algorithm complexities, and compute resource requirements. The productivity metric based on time-to-solution cannot capture these facets of development process. Hence evaluating methodologies based on solely the productivity metric is incomplete at best. To address this we present a new metric, i.e., *work progress rate*, as defined below.

4.3.2 Work Progress Rate

The purpose of work progress rate metric is to capture the efficiency of resource utilization, by the user and the tool combined, at discrete time steps along the search path progressing towards the target specifications. The metric draws from the principles of classical mechanics based on Newton's laws of motion. It evaluates the computational effort exerted by the user and the tool in moving towards the target specifications along the search path in the specification space. Based on the fact that the user and the tool computational efforts are mutually time exclusive activities as discussed earlier, the instantaneous computational effort, E_k, in any given time period Δt_k of iteration k, can be expressed as a vector with two components. The two components are the user effort, E_k^{user}, and the tool effort, E_k^{tool}, as shown in Eq. (13)

$$\vec{E_k} = \begin{bmatrix} E_k^{\text{user}} \\ E_k^{\text{tool}} \end{bmatrix} \tag{13}$$

The instantaneous work progress rate, Γ_k, in any given time period Δt_k is the rate of exerting computational effort in order to move along the search path in that time period. It is also a vector with two components, the user work progress rate, Γ_k^{user}, and the tool work progress rate, Γ_k^{tool}, as shown in Eqs. (14a) and (14b)

$$\vec{\Gamma_k} \equiv \begin{bmatrix} \Gamma_k^{\text{user}} \\ \Gamma_k^{\text{tool}} \end{bmatrix} \equiv \frac{\mathrm{d}\vec{E_k}}{\mathrm{d}t_k} \cong \frac{\Delta\vec{E_k}}{\Delta t_k}$$

$$\Delta\vec{E_k} \equiv \frac{\partial\vec{E_k}}{\partial t_k^{\text{user}}}\Delta t_k^{\text{user}} + \frac{\partial\vec{E_k}}{\partial t_k^{\text{tool}}}\Delta t_k^{\text{tool}}$$

$$\Delta\vec{E_k} = \frac{\partial}{\partial t_k^{\text{user}}}\begin{bmatrix} E_k^{\text{user}} \\ E_k^{\text{tool}} \end{bmatrix}\Delta t_k^{\text{user}} + \frac{\partial}{\partial t_k^{\text{tool}}}\begin{bmatrix} E_k^{\text{user}} \\ E_k^{\text{tool}} \end{bmatrix}\Delta t_k^{\text{tool}} \tag{14a}$$

$$\Delta\vec{E_k} = \begin{bmatrix} \frac{\mathrm{d}E_k^{\text{user}}}{\mathrm{d}t_k^{\text{user}}} \\ 0 \end{bmatrix}\Delta t_k^{\text{user}} + \begin{bmatrix} 0 \\ \frac{\mathrm{d}E_k^{\text{tool}}}{\mathrm{d}t_k^{\text{tool}}} \end{bmatrix}\Delta t_k^{\text{tool}} = \begin{bmatrix} \frac{\mathrm{d}E_k^{\text{user}}}{\mathrm{d}t_k^{\text{user}}}\Delta t_k^{\text{user}} \\ \frac{\mathrm{d}E_k^{\text{tool}}}{\mathrm{d}t_k^{\text{tool}}}\Delta t_k^{\text{tool}} \end{bmatrix}$$

$$\Rightarrow \Gamma_k^{\text{user}} = \frac{\mathrm{d}E_k^{\text{user}}}{\mathrm{d}t_k^{\text{user}}} \cong \frac{E_k^{\text{user}}}{\Delta t_k^{\text{user}}}, \text{ and } \Gamma_k^{\text{tool}} = \frac{\mathrm{d}E_k^{\text{tool}}}{\mathrm{d}t_k^{\text{tool}}} \cong \frac{E_k^{\text{tool}}}{\Delta t_k^{\text{tool}}} \tag{14b}$$

The overall work progress rate, Γ, can be defined as the statistical average of the instantaneous rates across all iterations along the search path in the specifications space. This can be described as shown in Eq. (15)

$$\vec{\Gamma} \equiv \frac{1}{M}\sum_{k=0}^{M-1}\vec{\Gamma_k} = \frac{1}{M}\sum_{k=0}^{M-1}\left(\begin{bmatrix} \frac{E_k^{\text{user}}}{\Delta t_k^{\text{user}}} \\ \frac{E_k^{\text{tool}}}{\Delta t_k^{\text{tool}}} \end{bmatrix}\right)$$

$$\Rightarrow \vec{\Gamma} \equiv \begin{bmatrix} \Gamma^{\text{user}} \\ \Gamma^{\text{tool}} \end{bmatrix} = \begin{bmatrix} \frac{1}{M}\sum_{k=0}^{M-1}\left(\frac{E_k^{\text{user}}}{\Delta t_k^{\text{user}}}\right) \\ \frac{1}{M}\sum_{k=0}^{M-1}\left(\frac{E_k^{\text{tool}}}{\Delta t_k^{\text{tool}}}\right) \end{bmatrix} \tag{15}$$

Based on Eq. (15), two methodologies achieving the same design objectives, in equal amount of time, along different paths and across different number of iterations can have different work progress rates even though their productivities as given by Eq. (11) is the same. This depends on the user and tool computational efforts along the search path. Thus, the work progress rate metric allows us to compare two methodologies not only based on how fast the target objectives were achieved, but how efficiently the resources were used along the way to achieve the target objectives.

It is necessary at this point to analyze and model the computational efforts exerted by both the user and the tool. In our model we leverage the concept of computational force as introduced and defined by Numrich [24, 25]. We will also define the computational effort as the work, as defined in classical mechanics, done by a force field to move an object between two positions in a given space.

In our model, the user, at any given time period Δt_k of iteration k, expends effort at the beginning of the iteration to maximize the displacement in the specifications space towards the target solution. Hence the computational effort expended by the user can be modeled by the work done by the user to push the tool to move from a given position (candidate solution), $\boldsymbol{X}_k$, in the specifications space to the next position, $\boldsymbol{X}_{k+1}$. Because the target solution is always known to the user at any point of time, the user's computational force is assumed to be pointing towards the target position. In other words, the computational force, $\boldsymbol{F}_k^{\text{user}}$, applied by the user is an impact force used to give the initial push to the tool to move in the specifications space and is aligned with the direction of the current target vector, $\boldsymbol{T}_k$. This is shown in Fig. 10 and expressed in Eq. (16). Similarly, the computational effort expended by the tool is the work done by the tool to cause the displacement $\boldsymbol{D}_k$ in the specifications space, see Fig. 10 and Eq. (17). The reader is reminded of the fact that the user and tool computational efforts, although being mutually dependent in a causal (cause-and-effect) relationship, are time exclusive activities and hence their computational forces cannot be used simultaneously to describe the dynamics of motion in the solution space, see Eqs. (16) and (17). More specifically,

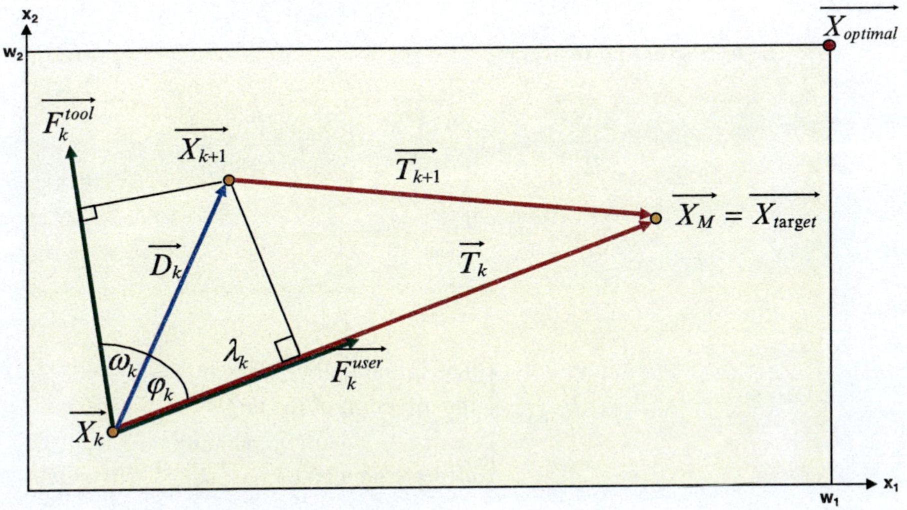

Fig. 10 User and tool computational forces

the user,during the time interval Δt_k^{user}, exerts an effort only through his/her own computational force, $\boldsymbol{F_k^{\text{user}}}$, in the direction of the target vector, $\boldsymbol{T_k}$. After that time interval the tool receives an initial impact velocity, $\boldsymbol{\upsilon_{0k}^{\text{tool}}}$, from the user and starts searching the space for a duration of Δt_k^{tool}. It is assumed that the user starts his activity from rest, i.e., zero velocity, and continues his activity until he reaches a final velocity, $\boldsymbol{\upsilon_k^{\text{user}}}$, after a time interval of Δt_k^{user}. It is also assumed that the user final velocity is transferred to the tool as an initial velocity, $\boldsymbol{\upsilon_{0k}^{\text{tool}}}$, in a perfectly inelastic collision/impact. These dynamics are described through the equations of motion (18) and (19)

$$E_k^{\text{user}} = \overrightarrow{F_k^{\text{user}}} \cdot \overrightarrow{T_k} = F_k^{\text{user}} T_k \tag{16}$$

$$E_k^{\text{tool}} = \overrightarrow{F_k^{\text{tool}}} \cdot \overrightarrow{D_k} = F_k^{\text{tool}} D_k \cos \omega_k \tag{17}$$

$$\overrightarrow{F_k^{\text{user}}} \equiv m^{\text{user}} \frac{\mathrm{d}\overrightarrow{\upsilon_k^{\text{user}}}}{\mathrm{d}t_k^{\text{user}}} \cong m^{\text{user}} \frac{\Delta \overrightarrow{\upsilon_k^{\text{user}}}}{\Delta t_k^{\text{user}}} = m^{\text{user}} \frac{\overrightarrow{\upsilon_k^{\text{user}}} - \overrightarrow{\upsilon_{0k}^{\text{user}}}}{\Delta t_k^{\text{user}}}$$

$$\overrightarrow{F_k^{\text{user}}} = m^{\text{user}} \frac{\overrightarrow{\upsilon_k^{\text{user}}}}{\Delta t_k^{\text{user}}}$$

$$\Rightarrow \overrightarrow{\upsilon_k^{\text{user}}} = \frac{\overrightarrow{F_k^{\text{user}}}}{m^{\text{user}}} \Delta t_k^{\text{user}} \equiv \frac{\overrightarrow{T_k}}{\Delta t_k^{\text{user}}} \tag{18}$$

where

$m^{\text{user}} \equiv$ The user's inertial resistance

$$\overrightarrow{F_k^{\text{tool}}} \equiv m^{\text{tool}} \frac{d\overrightarrow{\upsilon_k^{\text{tool}}}}{dt_k^{\text{tool}}} \cong m^{\text{tool}} \frac{\Delta\overrightarrow{\upsilon_k^{\text{tool}}}}{\Delta t_k^{\text{tool}}} = m^{\text{tool}} \frac{\overrightarrow{\upsilon_k^{\text{tool}}} - \overrightarrow{\upsilon_{0k}^{\text{tool}}}}{\Delta t_k^{\text{tool}}}$$

$$\Rightarrow \overrightarrow{\upsilon_k^{\text{tool}}} = \overrightarrow{\upsilon_{0k}^{\text{tool}}} + \frac{\overrightarrow{F_k^{\text{tool}}}}{m^{\text{tool}}} \Delta t_k^{\text{tool}} \tag{19}$$

where

$$\overrightarrow{\upsilon_{0k}^{\text{tool}}} = \overrightarrow{\upsilon_k^{\text{user}}} = \frac{\overrightarrow{T_k}}{\Delta t_k^{\text{user}}} \equiv \begin{cases} \text{The initial tool velocity in} \\ \text{the direction of the target} \\ \text{vector } \overrightarrow{T_k} \text{ duetocomputational} \\ \text{effort expended by the user} \end{cases}$$

$m^{\text{tool}} \equiv$ The tool's inertial resistance

The instantaneous work progress rate can now be given as

$$\overrightarrow{\Gamma_k} \equiv \begin{bmatrix} \Gamma_k^{\text{user}} \\ \Gamma_k^{\text{tool}} \end{bmatrix} = \begin{bmatrix} \frac{E_k^{\text{user}}}{\Delta t_k^{\text{user}}} \\ \frac{E_k^{\text{tool}}}{\Delta t_k^{\text{tool}}} \end{bmatrix} = \begin{bmatrix} \frac{\overrightarrow{F_k^{\text{user}}} \cdot \overrightarrow{T_k}}{\Delta t_k^{\text{user}}} \\ \frac{\overrightarrow{F_k^{\text{tool}}} \cdot \overrightarrow{D_k}}{\Delta t_k^{\text{tool}}} \end{bmatrix} = \begin{bmatrix} \overrightarrow{F_k^{\text{user}}} \cdot \overrightarrow{\upsilon_k^{\text{user}}} \\ \overrightarrow{F_k^{\text{tool}}} \cdot \overrightarrow{\upsilon_k^{\text{tool}}} \end{bmatrix} \tag{20}$$

It is worth mentioning that the user final velocity, $\boldsymbol{\upsilon_k^{\text{user}}}$, and hence the initial tool velocity, $\boldsymbol{\upsilon_{0k}^{\text{tool}}}$, are proportional to the user's experience. For a given iteration target, $\boldsymbol{T_k}$, advanced users spend less time, Δt_k^{user}, to debug and redevelop/modify their designs than novice users before they start their tools, see Eqs. (18) and (19). On the other hand, least experienced users spend long periods of times, i.e., $\Delta t_k^{\text{user}} \to \infty$, with almost no guidance or corrective efforts to the tool, i.e., $\boldsymbol{\upsilon_{0k}^{\text{tool}}} = \boldsymbol{\upsilon_k^{\text{user}}} = 0$ and $\boldsymbol{\Gamma_k^{\text{user}}} = 0$, see Eqs. (19) and (20).

As mentioned earlier, the user effort is always corrective to the tool effort. Therefore, it is desirable at this point to investigate the divergent behavior of the tool from that of the user's away from the target solution. In other words, it is essential to calculate the initial direction, ω_k, of the tool computational force with respect to the final displacement vector, $\boldsymbol{D_k}$, see Fig. 10. ω_k represents the angle with which the tool starts searching the space. Using Eqs. (16)–(19) and the geometrical properties shown in Fig. 10, the following expression can be derived for calculating ω_k:

$$\cos(\omega_k) = \frac{1 - \left(\frac{\lambda_k}{D_k}\right)^2 \left(\frac{\upsilon_{0k}^{\text{tool}} \Delta t_k^{\text{tool}}}{\lambda_k}\right)}{\sqrt{1 + \left(\frac{\lambda_k}{D_k}\right)^2 \left(\frac{\upsilon_{0k}^{\text{tool}} \Delta t_k^{\text{tool}}}{\lambda_k}\right) \left(\frac{\upsilon_{0k}^{\text{tool}} \Delta t_k^{\text{tool}}}{\lambda_k} - 2\right)}} \tag{21}$$

where

$$\frac{\lambda_k}{D_k} = \cos(\varphi_k)$$

Based on the proposed model, important characteristics of the development process can be understood by considering some special cases of Eq. (21). For example, at any design iteration k, when there is no initial push to the tool by the user, i.e., $\boldsymbol{\upsilon}_{\boldsymbol{0k}}^{\mathbf{tool}} = 0$ and $\Gamma_k^{\text{user}} = 0$, due to his/her lack of experience putting together a good initial design/development, the tool is left on its own searching the design space with $\omega_k = 0$, see Eq. (22a) and Fig. 10. This results in a random displacement, $\boldsymbol{D_k}$, the magnitude and direction of which are purely dependent on the computational force of the tool and on how efficient the tool is. On the other extreme, advanced users tend to give the tool the maximum guidance and/or corrective efforts, i.e., $\boldsymbol{\upsilon}_{\boldsymbol{k}}^{\mathbf{user}} = \boldsymbol{\upsilon}_{\boldsymbol{0k}}^{\mathbf{tool}} \rightarrow \infty$, resulting in a pure resistive behavior of the tool opposing the user effort, i.e., $\omega_k = \pi - \varphi_k$, see Eq. (22b) and Fig. 10. Consider the case of a highly experienced (ideal)user, i.e., $\upsilon_k^{\text{user}} = \upsilon_{0k}^{\text{tool}} \rightarrow \infty$, using an inefficient tool. Although the user gives the maximum guidance (best possible initial design) to the tool, the target solution might not be reached in one iteration or even in a short time. This is a result of the opposing behavior of the tool to the user effort, i.e., $\omega_k = \pi - \varphi_k$. It is also worth mentioning that more efficient tools remain to resist (oppose) the ideal user effort but with less resistive force, see Eq. (19),resulting in reaching the target solution in possibly one iteration and/or shorter time but never in zero time. Additionally, if the sensitivity λ_k, i.e., the distance traveled by the tool towards the target solution, is solely due to the user's first push, i.e., $(\boldsymbol{\upsilon}_{\boldsymbol{0k}}^{\mathbf{tool}} \Delta t_k^{\text{tool}})$, then the tool has not performed any useful work and its computational force has been orthogonal, i.e., $\omega_k = \pi/2 - \varphi_k$, to the target vector, see Eq. (22c) and Fig. 10

$$\lim_{\upsilon_{0k}^{\text{tool}} \rightarrow 0} (\omega_k) = 0 \tag{22a}$$

$$\lim_{\upsilon_{0k}^{\text{tool}} \rightarrow \infty} (\omega_k) = \pi - \varphi_k \tag{22b}$$

$$\lim_{\frac{\upsilon_{0k}^{\text{tool}} \Delta t_k^{\text{tool}}}{\lambda_k} \rightarrow 1} (\omega_k) = \frac{\pi}{2} - \varphi_k \tag{22c}$$

5 Experimental Evaluation

The framework was applied to evaluate the support for qualitative language features of a subset of the HLLs from our review list. A number of useful language features are considered including behavioral hierarchy, structural hierarchy, supported architectures, degree of parallelism, locality exploration, portability, and dynamic memory allocation. The quantifying process is a simple scoring system that ranks the degree of support of each HLL tool for the specific qualitative language feature.

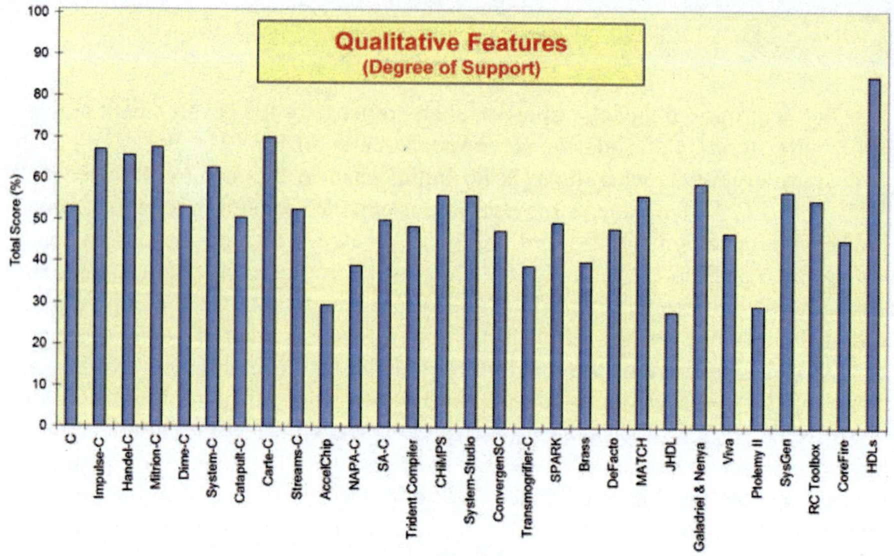

Fig. 11 Degree of support for qualitative language features

The scoring mechanism allows weighting, see Eq. (23), to rate specific features higher according to evaluator preferences. In our case we used equal weights since we consider each of those features equally with no particular preference. Figure 11 plots the final scores for the languages evaluated

$$\text{Score}_l = 100 \times \left(\frac{\sum_{f=1}^{N_f} s_{l,f} \cdot w_f}{\sum_{f=1}^{N_f} s_f^{\max} \cdot w_f} \right), \text{ where } l \in [1, N_l] \tag{23}$$

$$f \in [1, N_f], \; s_f^{\max} = \max_{l=1}^{N_l} (s_{l,f}), \text{ and } s_{l,f} \in [0, s_f^{\max}]$$

For our experimental evaluation we selected the HLL tools that scored the highest in this scoring mechanism. We also considered representative high-level tools that were selected to represent imperative, functional, and graphical programming. The availability of these tools for experimentation was also a factor in our selection. The HLL tools selected for our experimental evaluation are listed in Table 2.

Four workloads were selected for implementation using the selected HLL tools. The first workload is a simple pass-through implementation that reads input from the host microprocessor and sends it back unmodified. The purpose of this simple application is to measure the overhead caused by each tool on the FPGA with respect to the area utilization and also to measure the maximum clocking rates reached by

Table 2 Experimental results

	Tool	Development time (h)	Frequency (MHz)	Area (% utilization)	Normalized frequency (X1)	Normalized area (X2)
1	C	3	0	100	0	0
2	Impulse-C	6	125	21.25	0.625	0.7875
3	Handel-C	7.5	200	20.25	1	0.7975
4	Carte-C	6.5	100	19.5	0.5	0.805
5	Mitrion-C	10.75	100	22.75	0.5	0.7725
6	SysGen	9	200	19	1	0.81
7	RC Toolbox	9	200	19	1	0.81
8	HDLs	15.25	200	16.75	1	0.8325

each tool in the simplest of applications. This will give an initial and basic idea of the performance of each tool. The second application implemented is a discrete wavelet transform (DWT) [26, 27]. The third and fourth applications implemented are the data encryption standard (DES) and the DES breaking algorithms [28]. DWT and DES were selected as representative workloads of communication-intensive applications while DES breaker is a computational-intensive workload.

In conducting our experiments, five students and two faculty members, with different levels of experience and backgrounds in computer science, computer engineering, and electrical engineering were selected. As mentioned earlier, we developed our framework such that certain biasing effects associated with small sample sizes are minimized. For example, biasing effects may include previous user experience and knowledge of a specific language and/or design methodology. Therefore, we formalized our framework by instrumenting a normalization mechanism through which each user is required to provide three different observations of the same trial (application). In other words, each user develops the same application in three different design paradigms (languages). The first of which is performed using the language under consideration, while the other two are performed using a pure software approach, e.g., in C/C++, and a pure hardware approach, e.g., in VHDL/Verilog. These two other observations serve as reference points that can minimize the biasing effects of previous user programming experience which can range from software-centric programming experience to hardware-centric programming experience. In other words, in our experiments each user observation for a certain trial is normalized to his/her own experience making the observations more language-specific rather than being user-specific measurements.

5.1 *Results*

Table 2 shows the final attribute matrix for the four workloads implemented using the selected HLL tools. The reference implementations in HDLs and C are also included. The specifications space is assumed to be two-dimensional with area and

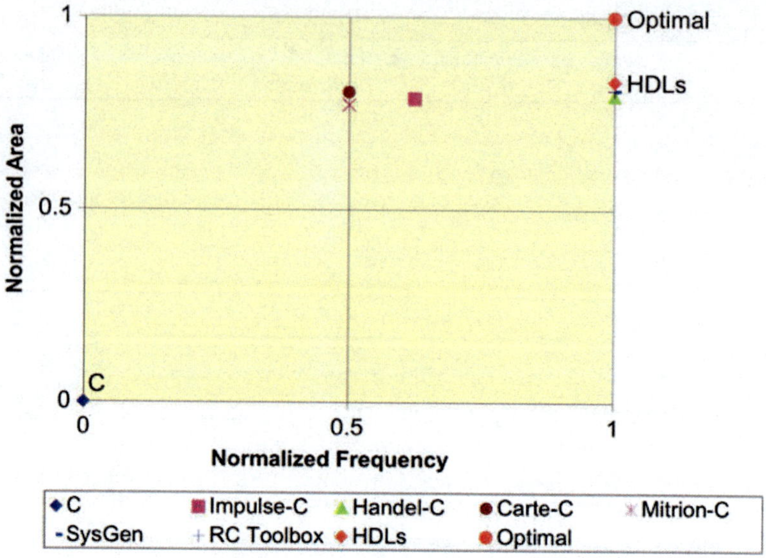

Fig. 12 HLL results plotted onto the specification space

frequency as the specifications. Figure 12 shows the frequency and area utilization achieved by each tool plotted on the normalized specification space. On the assumption that each tool generated the result in a single iteration, .the work progress rate for each tool is equal to the productivity, see Eqs. (11) and (15). Figure 13 presents the results projected onto the attribute space while Fig. 14 plots the productivity of each tool.

We may observe from the experimental results that imperative approaches proved to be the easiest to use while performing reasonably and comparably with standard HDL approaches. On the other hand, dataflow approaches, both functional and graphical, proved to achieve the high utility but were not as easy to use as imperative counterparts. Pure functional approaches proved to be the most difficult to use among the three approaches. Moreover, HDL approaches achieve the highest utility (close to optimal with this respect) but at the expense of being the most difficult to use for application developers. These observations are captured in Figs. 12 and 13.

6 Conclusions

The work reported in this chapter presents a review and a comprehensive taxonomy of HLL languages for HPRCs. It also presents new metrics and a framework for comparative evaluation of the HLLs. The concepts and methodology are inspired from the principles of Newtonian mechanics, which are applied notionally to the

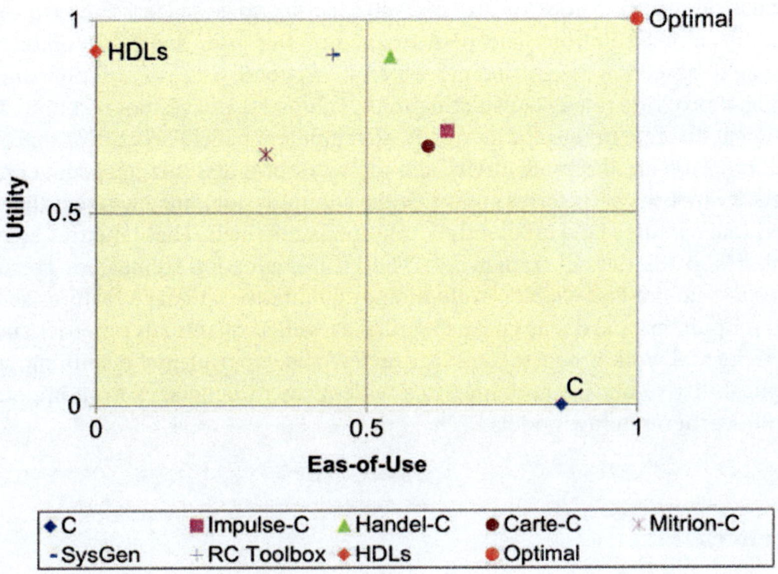

Fig. 13 HLL results plotted onto the attribute space

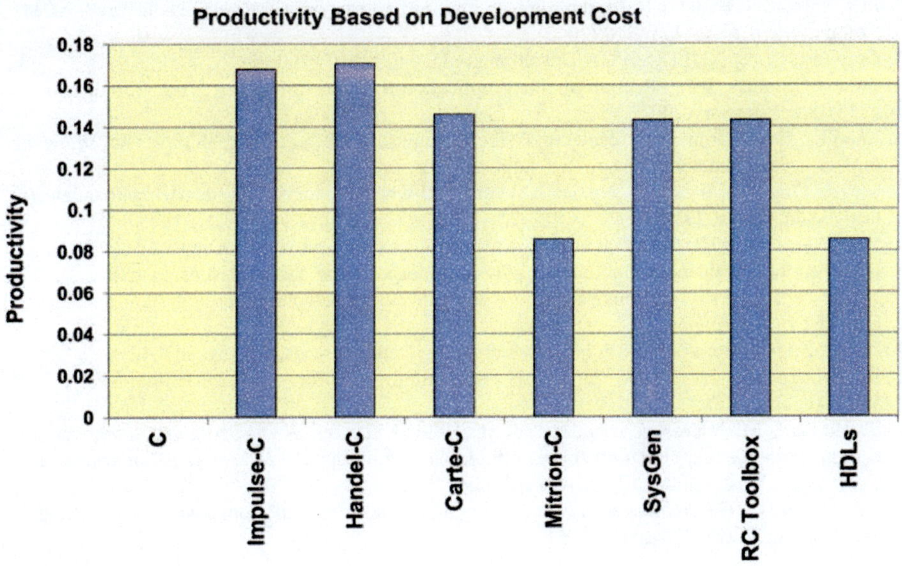

Fig. 14 Productivity of evaluated HLL tools

movement of user and tool in an abstract specifications space as they progress towards the target solution. The performance of this user and tool combination is evaluated based on two principle criteria: (a) total time to solution and (b) incremental progress rate encapsulating the combined user and tool resource usage efficiency at discrete time steps along the development path. The metrics that focus on each criterion are the productivity and the work progress rate, respectively. The productivity metric characterizes how fast the solution is obtained whereas the work progress rate captures how efficiently a solution is obtained. These metrics are used as attributes in the overall framework. The HLL evaluation framework presented provides a structure that enables evaluation of qualitative language features such as support for pointers and debugging capability, as well as quantitative metrics such as productivity and work progress rate. Our review and experimental results showcase the applicability of our methodology to a wide-array of languages from imperative to dataflow programming models.

References

1. W. Luk, N. Shirazi, P.Y.K. Cheung, "Compilation tools for run-time reconfigurable designs", in *IEEE Symposium on Field-Programmable Custom Computing Machines, FCCM* 1997, pp. 56–65
2. K. Compton, S. Hauck, "Reconfigurable computing: a survey of systems and software", ACM Comput. Surv. **34**(2), 171–210 (2002)
3. Cray Inc., "Cray XD1TM FPGA Development (S-6400–14)", 2006
4. Silicon Graphics, Inc., "Reconfigurable Application-Specific Computing User's Guide (007–4718–005)", January 2007
5. Impulse C – "Impulse Accelerated Technologies" web site available at http://www.impulseaccelerated.com/, last visited August 2012
6. Mentor Graphics, Inc., web site available at http://www.mentor.com/products/fpga/handel-c/, last visited August 2012
7. Mitrionics web site available at http://www.mitrionics.com, last visited August 2012
8. Bluespec, Inc., website available at http://www.bluespec.com/, last visited August 2012
9. Xilinx Inc., web site available at http://www.xilinx.com/ise/optional_prod/system_generator.htm, last visited August 2012
10. DSPLogic web site available at http://www.dsplogic.com, last visited August 2012
11. SRC Computers, Inc., "SRC Carte™ C Programming Environment v2.2 Guide (SRC-007–18)", August 2006
12. B. Holland, M. Vacas, V. Aggarwal, R. DeVille, I. Troxel, A.D. George, "Survey of C-based application mapping tools for reconfigurable computing", in *2005 MAPLD International Conference*, Washington, DC, USA, September, 2005
13. S.A. Edwards, "The challenges of synthesizing hardware from C-like languages", IEEE Design Test Comput. **23**(5), 375–386 (2006)
14. E. El-Araby, M. Taher, M. Abouellail, T. El-Ghazawi, G.B. Newby, "Comparative analysis of high level programming for reconfigurable computers: methodology and empirical study", in *III Southern Conference on Programmable Logic (SPL2007)*, Mar del Plata, Argentina, February, 2007
15. E. El-Araby, P. Nosum, T. El-Ghazawi, "Productivity of high-level languages on reconfigurable computers: an HPC perspective", in *IEEE International Conference on Field-Programmable Technology (FPT 2007)*, Japan, December, 2007

16. T. Sterling, "Productivity metrics and models for high performance computing", Int. J. High Perform. Comput. Appl. **18**, 433–440 (2004)
17. M. Snir, D.A. Bader, "A framework for measuring supercomputer productivity", Int. J. High Perform. Comput. Appl. **18**, 417–432 (2004)
18. K. Kennedy, C. Koelbel, R. Schreiber, "Defining and measuring the productivity of programming languages", Int. J. High Perform. Comput. Appl. **18**, 441–448 (2004)
19. J. Kepner, "HPC productivity: an overarching view", Int. J. High Perform. Comput. Appl. **18**, 393–397 (2004)
20. J. Kepner, "High performance computing productivity model synthesis",Int. J. High Perform. Comput. Appl. **18**, 505–516 (2004)
21. R.W. Numrich, "Performance metrics based on computational action",Int. J. High Perform. Comput. Appl. **18**(4), 449–458 (2004)
22. R.W. Numrich, "A metric space for computer programs and the principle of computational least action", J. Supercomput. **43**(3), 281–298 (2008)
23. R.W. Numrich, L. Hochstein, V. Basili, "A metric space for productivity measurement in software development", in *Proceedings of the Second International Workshop on Software Engineering for High Performance Computing System Applications (SE-HPCS'05)*, St. Louis, Missouri, 15 May 2005
24. R.W. Numrich, "Computational force: a unifying concept for scalability analysis", Adv. Parallel Comput. **15** (2008). ISSN 0927–5452, ISBN 978-1-58603-796-3 (IOS Press)
25. R.W. Numrich, "Computational force, mass and energy", Int. J. Mod. Phys. C **8**(3), 437–457 (1997)
26. E. El-Araby, M. Taher, T. El-Ghazawi, J. Le Moigne, "Remote sensing and high performance reconfigurable computing systems", in *High Performance Computing in Remote Sensing*, vol 16, ed. by A.J. Plaza, C.I. Chang, *CRC Computer & Information Science Series* (Chapman & Hall, New York, 2007), pp. 496. ISBN: 9781584886624, ISBN 10: 1584886625
27. E. El-Araby, T. El-Ghazawi, J. Le Moigne, K. Gaj, "Wavelet spectral dimension reduction of hyperspectral imagery on a reconfigurable computer", in *IEEE FPT 2004*, Brisbane, Australia, December, 2004
28. O.D. Fidanci, H. Diab, T. El-Ghazawi, K. Gaj, N. Alexandridis, "Implementation trade-offs of triple DES in the SRC-6E reconfigurable computing environment", in *Proc. MAPLD* 2002

Maximum Performance Computing with Dataflow Engines

Oliver Pell, Oskar Mencer, Kuen Hung Tsoi, and Wayne Luk

Abstract Maximum Performance Computing (MPC) means striving to deliver the maximum possible performance within a space and/or power budget. The essence of the method is to start with a particular application and develop an appropriate computer by iterating between algorithm optimization and machine optimization, essentially, cross-optimizing across the layers of abstraction from mathematics to logic gates. An MPC system pairs fast scalar processors with dataflow engines which can be emulated on FPGAs. In this chapter we outline the general approach, and describe in detail example hardware architecture, programming model and tools. We also discuss additional issues that arise at the cluster level, and describe a detailed case study of applying MPC to Reverse Time Migration, a computational geophysics algorithm widely used in the oil industry.

1 Introduction

Since the introduction of computers into routine scientific work in the middle of the last century, their role in the scientific endeavor has changed dramatically from mere auxiliary tools of numerical computation into an essential element of scientific discovery. The first hint pointing at the forthcoming change was seen as early as in 1953, when Fermi, Pasta and Ulam conducted the first fundamentally important numerical experiment [2]. Fermi and his colleagues used digital computers to simulate the expected ergodicity in a system of coupled nonlinear oscillators, a phenomenon that was beyond analytical description. Their results not only had a

O. Pell (✉) • O. Mencer
Maxeler Technologies, 1 Down Place, London W6 9JH, UK
e-mail: oliver@maxeler.com

K.H. Tsoi • W. Luk
Imperial College London, London SW7 2AZ, UK
e-mail: wl@doc.ic.ac.uk

W. Vanderbauwhede and K. Benkrid (eds.), *High-Performance Computing Using FPGAs*, DOI 10.1007/978-1-4614-1791-0_25,

profound influence on the theory of chaos but also influenced the way scientists think when tackling basic theoretical problems. The numerical experiment approach of Enrico Fermi and his colleagues, while revolutionary at its time, has since become a mainstream method of modern theoretical investigations in such diverse fields as fluid dynamics, molecular electronic structure and nuclear dynamics, material science, etc. In fact, the role of numerical experiments has grown so much that today intricate analytical techniques are routinely used to interpret numerical findings much in the same way as they have been used to interpret "real" experiments.

Tremendous advances in the application of numerical techniques in science have been driven by progress in computer technology. The available computational power and computer memory resources define the scope of the scientific problems that can be addressed, as well as the achievable accuracy level of theoretical modeling and thus also the reliability of the scientific prediction. Modern science strives to address more and more challenging problems on a larger and larger scale using accurate theoretical models. Achieving such challenging goals in the coming years will require development of powerful computational technology at new levels of performance. It will not be practical to build HPC systems capable of processing at this level by simply scaling existing CPU technology [13].

Maxeler dataflow computing has been shown to lead to orders of magnitude lower power consumption and lower data center space needs than conventional CPU systems and has been deployed by companies ranging from JP Morgan for financial analytics to Chevron for oil exploration. One obvious question is whether dataflow fits a wide range of applications. We have seen excellent results for applications ranging from large-scale complex Monte Carlo simulations and irregular financial tree-based partial differential equation (PDE) solvers[16] to 3D finite difference (FD) and finite element (FE) solvers [6]. In one extreme case [12] a 3U Maxeler dataflow node delivered equivalent compute performance to over 1,800 high-end x86 control flow cores for a finite difference application with over 100 GB of state that needed to be iterated across time steps of the simulation.

In this chapter we describe the basis of our multi-disciplinary dataflow computing approach to computing and discuss a case study in detail.

2 What is Dataflow Computing?

Maxeler's dataflow computing paradigm is orthogonally different to computing with conventional CPUs (control flow cores), as shown in Fig. 1 and represents an evolution of concepts of dataflow computers [1] and systolic array processors [4] developed in the 1970s and 1980s.

At its heart, a control flow processor contains a latency critical loop. Data is read from memory into the processor core, where operations are performed and the results are written back to memory. Modern processors contain many levels of caching, forwarding and prediction logic to improve the efficiency of this paradigm; however, the model is inherently sequential with performance limited by the speed at which data can move around this loop.

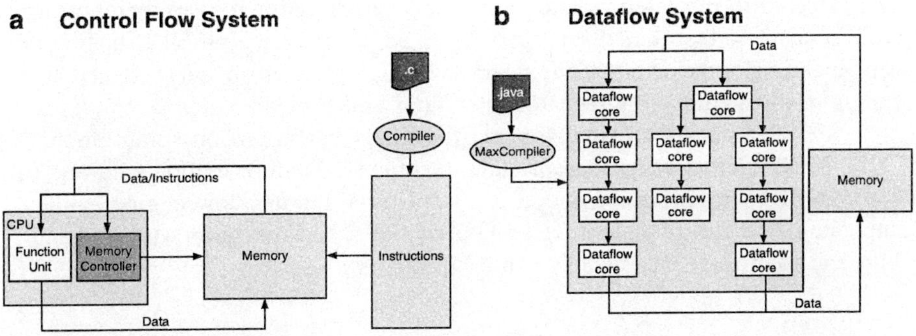

Fig. 1 Computing with a control flow core compared to dataflow cores

A dataflow engine (DFE) operates differently. Data is *streamed* from memory onto the chip where operations are performed and data is forwarded directly from one functional unit ("dataflow core") to another as the results are needed, without ever being written to the off-chip memory until the chain of processing is complete. Each dataflow core computes only a single type of operation (for example an addition or multiplication) and is thus simple so thousands can fit on one chip with every dataflow core computing simultaneously. Unlike in the control flow core where operations are computed at different points in time on the same functional units ("computing in time"), the complete dataflow computation is laid out spatially on the chip ("computing in space"). Dependencies in the dataflow are resolved statically at compile time, and because there are no new dependencies present at run-time the whole dataflow engine can be deeply pipelined (1,000–10,000 stages). Every stage of the pipeline computes in parallel with the dataflow architecture maintaining a throughput of one result per cycle. This dataflow engine structure can then be emulated on large FPGAs. One analogy for moving from control flow to dataflow is the Ford car manufacturing model, where expensive highly skilled craftsman (control flow cores) are replaced by a factory line, moving cars through a sea of single-skill workers (dataflow cores).

The dataflow engine structure itself represents the computation thus there is no need for instructions per se; they are replaced by operation executing units. Because there are no instructions there is no need for instruction decode logic. A static dataflow machine doesn't require techniques like branch prediction (since there aren't any branches) or out-of-order scheduling (parallelism is explicit). And since data is always available on chip for as long as it is needed, general purpose caches are not needed with the minimum amount of buffering memory automatically utilized as necessary. This saves silicon area and power budget. By eliminating these extraneous functions, the full resources of the chip are dedicated to performing computation.

In summary, the advantage of the dataflow engine is based on the elimination of sequentially iterated instruction streams and reduction in memory accesses for both instructions and data. Instead of using indexed register files and hierarchical caching memory, the intermediate values and variables for the computation are stored in sparsely distributed registers which are close to the physical operators. Deep pipelining of the dataflow engine provides high parallelism and sustained high throughput when computing repeated operations on large data volumes, but means that dataflow engines are less optimized for single operations on small amounts of data. Thus it is typically appropriate to combine a dataflow engine with one or more control flow CPU cores and at a system level, the dataflow engine handles computation of the large scale streaming operations in an application while dynamic events and control are managed by control flow core(s).

3 Building Dataflow Computers

A dataflow computing system is usually a heterogeneous system including a number of dataflow processors and one or more conventional CPUs. The CPUs are required for system level tasks like file storage, networking and process management. They may also share the computation workload which is best mapped to the control-flow computing paradigm. There are various schemes for combining the two radically different processing devices and allowing them to collaborate in a system.

3.1 Integrating DFEs and CPUs

3.1.1 Roles in the System

The two types of computing devices may play different roles in a computing system. One possible scheme is to form a master-slave relationship. In this scenario, the CPUs perform all management tasks and initiate the computations. The DFEs passively accept commands and data from the master CPUs and perform the required computation as instructed. This scheme has a number of advantages including the ease of system construction, the readily available technologies and the ability to reuse existing systems by adding DFEs as extensions.

On the other hand, DFEs and CPUs can play the same role in the system. In this scheme, both are able to actively access all system resources with the same priorities. This creates the possibility of DFEs requiring and instructing CPUs for specific operations. Due to the increased accessible resources and the more active role in control, the DFEs may perform better in some applications. The challenge of this scheme is usually in designing the interface between the DFEs and the CPUs, at both physical level and programming level. Also, the complicated interface and active role require more controlling logic in the DFEs. This may defeat the original purpose of having only streamlined data processing in the DFEs.

Recent developments in cloud computing and virtualization technologies provide a new direction for incorporating DFEs and CPUs. The new schemes is to allow the DFE clusters working together with CPU clusters and providing platform as a service (PaaS) to the CPU-centric servers. This isolation between DFEs and CPUs enables more flexibilities in system cost, hardware deployment, cluster management and application development. Since the approach is relatively new, there is no existing standard and very few adaptations of utilizing DFEs in PaaS. However, we are expecting to see more dataflow computing systems deployed in this scheme in the future.

3.1.2 Access to Memory

In high performance computation, efficient access to external memory is usually one of the most critical factors in system performance. By the law of diminishing returns, optimizing the internal architecture of CPUs or dataflow processors will become insignificant when memory access occupies a large part of the application execution time. Different schemes of memory access in dataflow computing systems will impact the performance of the specific implementations.

It is possible to provide a unified flat memory addressing space for both CPUs and DFEs. In this arrangement, any processing unit can access any memory location directly, regardless of the physical hierarchy of the memory system. This is the most flexible scheme for application programmers. However, it has practical issues such as data coherence across heterogeneous computing devices and memory bus contention. These issues limit the achievable performance of the applications.

A common practice of maximizing memory bandwidth is to associate each DFE with multiple banks of local memory. This direct coupling between memory and DFE creates an explicit divide in the memory hierarchy of the dataflow computing system, which can provide maximum performance potential but requires extra efforts from application developers to manage the memory hierarchy explicitly within the applications. Although the physical connections of this divided memory hierarchy are rigid, abstractions at software level can simplify the task of transferring data between DFEs and CPUs.

3.1.3 Coupling

Another representative characteristic of a dataflow computing system is the scheme of coupling between CPUs and DFEs. Figure 2 illustrates four types of common interconnection schemes.

The tightest scheme is to place both the dataflow processor and the control flow processor on the same silicon as shown in Fig. 2a. The dataflow processor communicates directly with the CPU core through a dedicated low latency interface. In this configuration, the dataflow processor is often activated by executing special instructions in the CPU core. With the highest communication efficiency, this scheme can enable beneficial offload of small operations; however, this scheme has the disadvantage of requiring modification of the CPU internal architecture

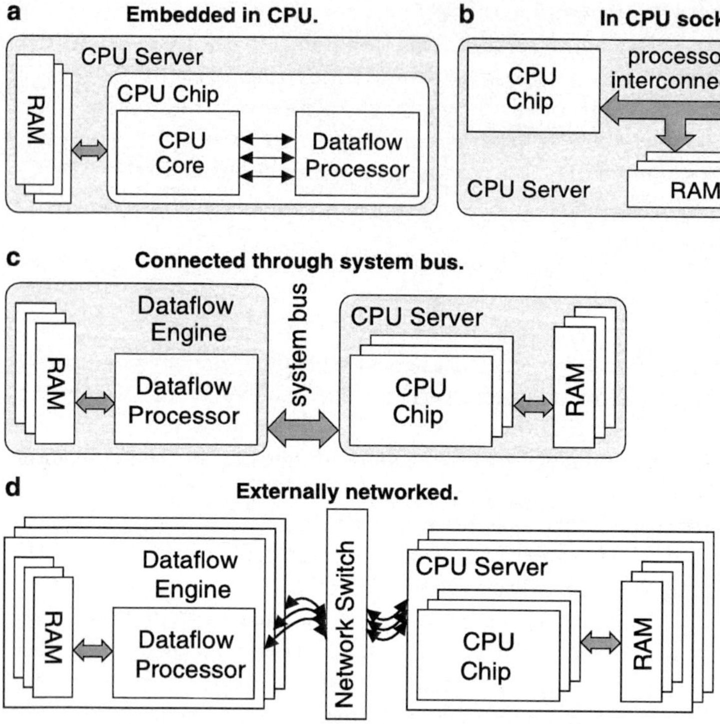

Fig. 2 Interconnection schemes for CPUs and DFEs

which is often not feasible. In addition, since the dataflow processor accesses system memory through the CPU core, many of the advantages of dataflow computation for computation on large data volumes cannot be easily achieved in this scheme.

Another tightly coupled scheme is to place the dataflow processors on the CPU sockets side-by-side with the CPUs as shown in Fig. 2b. The dataflow processors act as conventional CPUs in a multi-processor system, communicating through a standard processor specific interconnect such as HyperTransport or QuickPath Interconnect. However, the dataflow processor must now incorporate a full featured CPU interface which consumes the DFE's resources and also limits the number of supported DFEs in the system.

Another scheme is to connect the DFEs through a system bus (e.g. PCI Express) in the CPU server as shown in Fig. 2c. The interface to the system bus can be implemented within the dataflow processors or in separate components in the DFEs. This interfacing logic introduces overheads in terms of transmission latency and silicon area. Modern system bus architectures have sufficient bandwidth and robust communication protocols for a few DFEs interfacing to a CPU. However, the scalability of this approach is limited by both the maximum number of devices supported by the bus controller logically and the maximum number of devices supported by the server system physically.

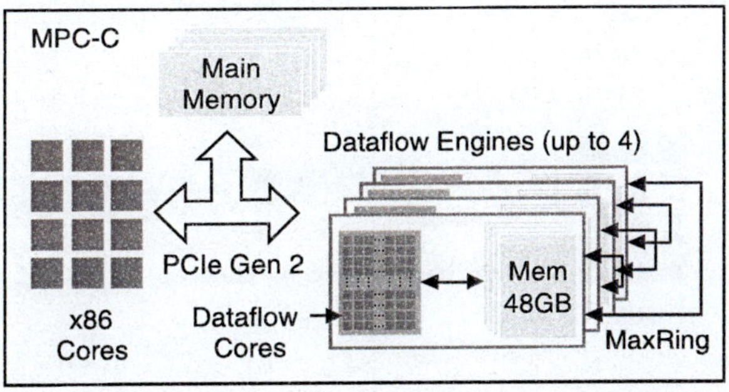

Fig. 3 *MPC-C* Series Architecture. A single node contains both x86 CPUs and dataflow engines connected via PCI Express

Figure 2d shows a more loosely coupled scheme where a cluster of DFEs is connected to a cluster of CPU servers through an inter-node network such as Ethernet or Infiniband. This scheme sacrifices communication latency for system level flexibility. The external switching network usually has higher latency than the internal connections in the previous schemes; however, this overhead may not be significant if the communication is balanced with the computation. Using this architecture, a cluster of DFEs can potentially be shared by all the CPU servers on the network, improving efficiency in the use of the DFE resources and facilitating the PaaS scheme discussed earlier.

3.2 Real World Examples: MPC-C and MPC-X

Here we present real world computing systems which implement the two most practically useful DFE-CPU coupling schemes: the Maxeler *MPC-C* series and *MPC-X* series. The MPC-C series (shown in Fig. 3) compute nodes contain x86 CPUs directly attached to DFEs via PCI Express, while the MPC-X series (shown in Fig. 4) are stand-alone dataflow nodes that connect to CPU-only nodes in a system via an Infiniband network. DFEs within a node have an additional high-bandwidth, low-latency direct interconnect between the dataflow processors called *MaxRing* which is utilized for fast communication between neighbouring DFEs.

The MPC-C series allows a stand-alone deployment of dataflow technology with a fixed combination of tightly coupled CPUs and dataflow engines. The MPC-X series allows for a heterogeneous system with a tailored balance of different compute technologies, since the balance of CPUs to DFEs is flexible at run-time. For example, if an application runs on a fixed number of CPU cores and continuously utilizes one or more DFEs, an MPC-C series platform could be the best fit; while if the application has several stages (some of which use the DFEs and some of which do not) or might require a varying number of DFEs depending on the particular problem it is running the MPC-X series may be more appropriate.

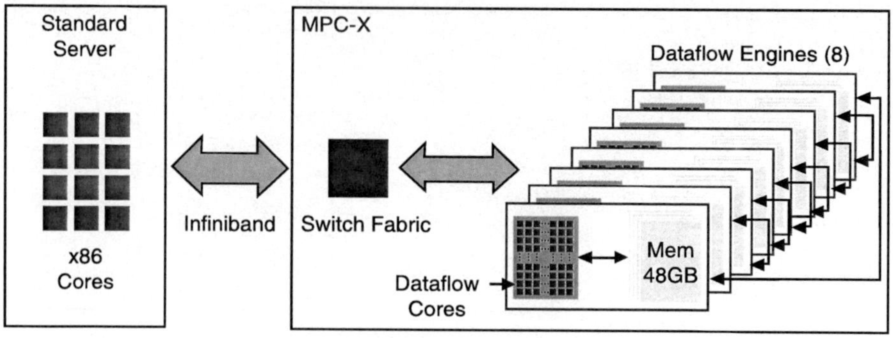

Fig. 4 *MPC-X* Series Architecture. Multiple CPU nodes can connect to *MPC-X* nodes and utilize the dataflow engines. The number of dataflow engines used by each CPU node varies dynamically at run-time

The common part in these two configurations is the individual Maxeler dataflow engine hardware. The *MAX3* engine is the current state-of-the-art with up to 48 GB of DDR3 memory and a Xilinx Virtex-6 FPGA providing the configurable logic fabric. Multiple MAX3 engines can work together on a single application, sharing data over their high-speed local MaxRing connection.

Maxeler's dataflow systems are designed to integrate into production server environments, supporting standard operating systems and management tools. Multiple applications can make use of the MAX3 DFEs within a system. Maxeler's management software (*MaxelerOS*) coordinates resource use, scheduling and data movement within the dataflow compute environment. Application programmers can concentrate on improving their application for maximum performance without worrying about low-level resource management.

4 Maximum Performance Computing Approach

It is widely accepted that modeling and measurement of application characteristics can be used to optimize the design of a cluster, for example in terms of relative resources dedicated to computation, memory and interconnect. The same approach can be extended to every level in a computer system, developing dedicated, problem-specific computer architectures to address different domains of scientific problems in the most efficient way, achieving the maximum performance.

In fact, to deliver maximum application performance, we espouse a multi-disciplinary approach to scientific computing, where a team of natural and computer scientists and engineers work together to *codesign* a system all the way from the formulation of the computational problem down to design of the best possible computer architecture for its solution. The essence of this approach is to optimize the scientific algorithm to match the capabilities of the computer at the same

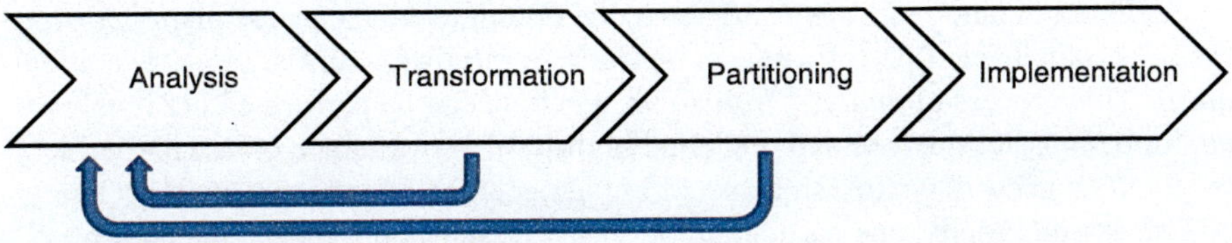

Fig. 5 Process for accelerating applications with dataflow computing

time as optimizing the computer to match the requirements of the algorithm. Dataflow computing provides a key part of the solution by combining flexibility with maximum performance and high efficiency.

Figure 5 shows the process for maximizing application performance with dataflow computing.

The first step is application analysis, the main objective of which is to establish an understanding of the application, the algorithms used and the potential performance bottlenecks. Application analysis should cover all the layers involved in a computational problem, from the mathematics and algorithms to the low level computer science and electrical engineering aspects. This includes considerations such as:

- Type and regularity of computation
- Ratio of computation and memory access
- Balance between local computation and network communication
- Balance between computation and disk I/O
- Trade-offs between (re)computation and storage

All of these can have a large impact on the final system performance—even if an application has a highly optimized computational kernel this will have no impact if an application is limited by the speed it can read/write data to disk. The second step in the process is to *transform* the application to create a structure that is suitable for acceleration. This includes algorithm-level transformations (for example: is there an alternative algorithm for achieving the same result?) as well as code transformations and data layout or representation transformations. The analysis and transformation steps are those that require the largest degree of multidisciplinary codesign cooperation. Natural scientists with expertise in the problem domain can collaborate with computer scientists and engineers to make sure the full range of possibilities are explored.

In the third step, we *partition* the application between the different resources in the computing system. There are two key things to be partitioned: code and data. For program code, we consider multiple *code partitioning options* which determine where each operation runs—on the conventional CPUs, or on the dataflow engines. We orthogonally consider *data access plans*—the location of each significant data array used in the computation. For the dataflow engines, data can either be present in the large (tens of gigabytes) off-chip memory, or the smaller on-chip memory (typically a few megabytes). For the CPUs, data can reside on disk, in main memory or in one of several levels of on-chip cache.

Each combination of transformation, code option and data access plan will provide a certain level of performance in return for a certain amount of implementation effort. This process of *analyze, transform, partition* can be performed iteratively as additional possibilities are explored, and high-level performance modeling be used to quickly narrow down to a small set of implementation options that are most likely be successful. Finally, one particular option can be selected for implementation.

A case study of this approach can be seen in Sect. 7.

5 Programming with MaxCompiler

Maxeler's dataflow computers are programmed in standard C/C++/FORTRAN and Java combined with Maxeler's language extensions to drive dataflow generation.

Figure 6 shows the logical architecture of a dataflow computing system. The programmer creates a dataflow application with three parts:

- A CPU application, typically written in C/C++ or FORTRAN, which runs on the conventional CPUs and controls the system
- One or more dataflow kernels, written in Java
- A manager, written in Java

Kernels are hardware data-paths implementing the arithmetic and logical computations needed within an algorithm. The Manager describes the logic which orchestrates data flow between Kernels and off-chip I/O in the form of streams. By using a streaming model for off-chip I/O to CPU interconnect (e.g. PCI Express, Infiniband), MaxRing and DRAM memory, Managers guide high utilization of available bandwidth in off-chip communication channels.

Separating computation and communication into Kernels and a Manager is beneficial as it enables logic for Kernels to be deeply pipelined without running into

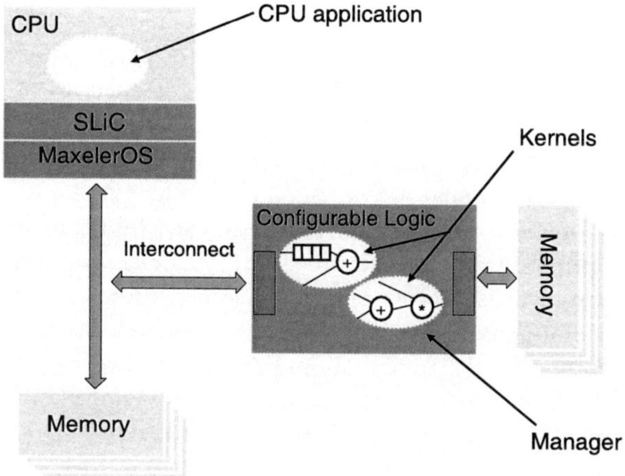

Fig. 6 Logical architecture of a dataflow computing system with one CPU and one dataflow engine

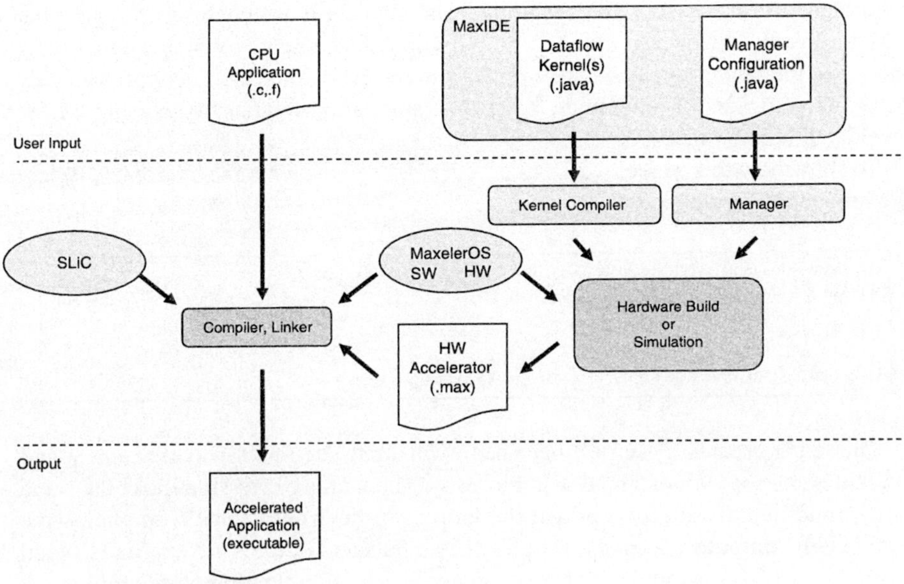

Fig. 7 Compilation flow of a MaxCompiler application

the synchronization problems which plague parallel programming on control flow cores. High performance in a dataflow solution is achieved through a combination of deep-pipelining and exploiting both inter- and intra-Kernel parallelism. The number of pipelines within a Kernel and number of parallel Kernels is limited only by the parallelism inherent in the application and the size of the dataflow engine.

Figure 7 shows the compilation flow of a typical MaxCompiler application. The programmer describes both the Kernels and Manager in Java with Maxeler language extensions (which we call *MaxJ*). The Java is a meta-program that describes the structure of the dataflow engine that should be created. When the Java program is executed the chip configuration file requested by the user is generated. This chip configuration file (the `.max` file) contains the bitstream to configure the chip, meta-information about the configuration and also CPU function calls enabling it to be easily integrated into the full CPU application. The CPU code is compiled as normal and linked with the `.max` file and the *SLiC* (Simple Live CPU) interface library using the standard linker.

The result of compilation is a single application executable that contains all the binary code to run on both the conventional CPUs and the dataflow engines in a system. When the CPU program makes a function call which uses a dataflow engine the engine will automatically be configured to run that application.

Compiling the dataflow engine into hardware can take some time (typically several hours up to a day for a large program) so during development it is often more appropriate to use *simulation*. The user's MaxCompiler program can be compiled directly to a high-level simulation running in native CPU code executed on a standard processor. The resulting simulated DFE is much slower than the real hardware; however, it can be compiled in a few minutes which dramatically improves the debug cycle. MaxCompiler's high-level simulation is approximately two to three orders of magnitude faster than simulating the hardware using a low-level HDL simulator.

To show how this model works in practice, let's consider some examples. Below is a snippet of Kernel code:

```
HWVar x = io.input("x", hwFloat(8, 24));
HWVar y = io.input("y", hwFloat(8, 24));

HWVar o = x + x * y;

io.output("o", o, hwFloat(8,24));
```

This code creates a Kernel that takes two input streams (arrays) x and y and computes $x + x \times y$ for each index in the array. The maths expression looks the same as it would in software, except that the loop over the array values is implicit—the Kernel will compute $x_i + x_i \times y_i$ for all i. The inputs and outputs are explicitly typed as *hwFloat(8,24)* meaning they are IEEE single precision floating point numbers.

A more complex Kernel might need to make decisions about what values to output: for example, we could output either x^2 or y^2 depending on which was larger. In software, we would represent this with an *if* statement. In our Kernel language we are creating operations instantiated spatially on a chip—so it is not possible to choose different branches of execution. However, it is possible to compute multiple possibilities and select between them. We do this using the ? operator, for example:

```
HWVar x2 = x * x;
HWVar y2 = y * y;

HWVar o = x2 > y2 ? x2:y2;
```

Another operation we might want to do is to access values from other positions in the stream. Consider a simple 3-point moving average computation:

$$y_i = (x_{i-1} + x_i + x_{i+1})/3 \tag{1}$$

To compute this, we need to access not just the value of x at location i but also at $i-1$ and $i+1$. Because the x array has become a stream, rather than indexing into the array (as we might in software), our Kernel program uses *stream offsets* relative to the current position to access the previous and next elements it needs for the calculation:

```
HWVar prev = stream.offset(x, -1);
HWVar next = stream.offset(x, +1);
HWVar o = (prev + x + next) / 3;
```

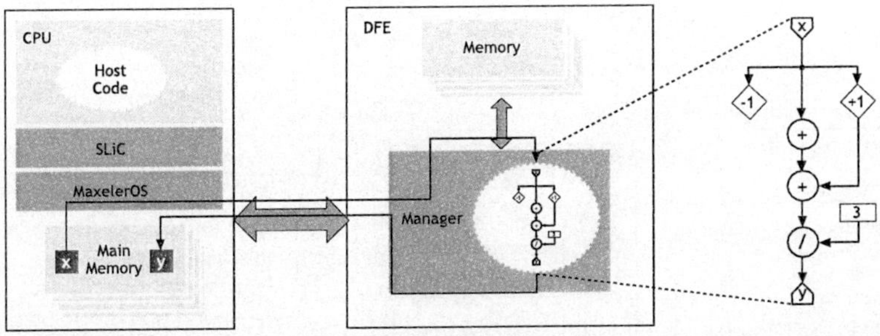

CPU Code (.c)

```c
#include "AVG.max"

float x[N], y[N];
...
AVG(x, y, N);
```

Manager (.java)

```java
Manager m = new Manager("AVG");
Kernel k = new AverageKernel();

m.setKernel(k);
m.setIO(
   link("x", CPU),
   link("y", CPU),
m.addMode(modeDefault());
m.build();
```

Kernel (.java)

```java
HWVar x = io.input("x", hwFloat(8, 24));
HWVar prev = stream.offset(x, -1);
HWVar next = stream.offset(x, 1);

HWVar sum = prev+x+next;
HWVar result = sum/3;

io.output("y", result, hwFloat(8, 24));
```

Fig. 8 The CPU code, Manager code and Kernel code for a simple 3-point moving average calculation, showing the data transfer that occurs during the computation

Once we have a Kernel, we can integrate it into a Manager to create a full dataflow engine. Figure 8 shows how this can be achieved with the moving average Kernel.

The center section of Fig. 8 shows the small Java program we use to configure a Manager that includes the moving average Kernel and connects it to the CPU. The left section shows the CPU software program. The compilation of the manager and kernel generates the SLiC interface function `AVG(x,y,N)` which the CPU program calls to run the computation. When the function is called, the CPU will connect to a dataflow engine (automatically configuring the chip if necessary) and stream N items of the x array to the engine to perform the calculation, receiving N items back and storing them into array y. The data will stream from memory, over the CPU to DFE interconnect, through the Kernel and back into memory automatically.

6 Cluster-Level Considerations

The dataflow computing paradigm and the dataflow processor have been previously proposed and studied. However only now they are starting to appear in production systems, since most of the research results achieved in academic environments cannot be applied to existing data center infrastructure. These previous studies usually considered dataflow processors as a stand-alone component attached to a

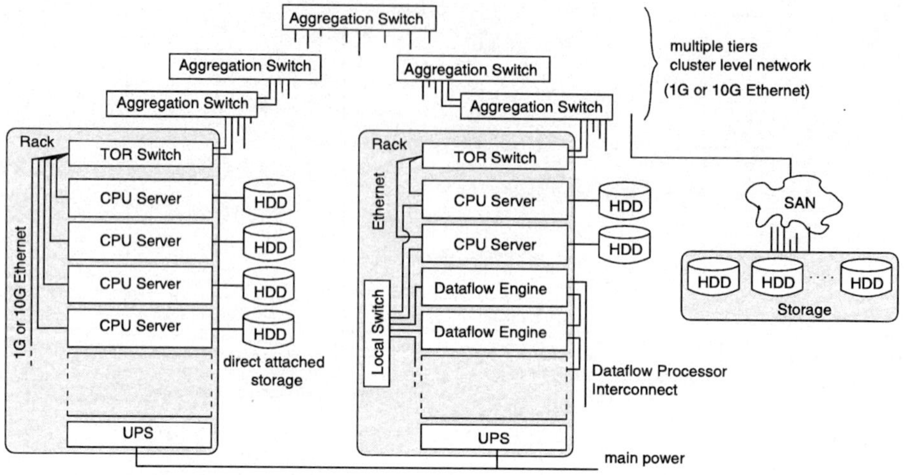

Fig. 9 Rack configuration in data center

single node but in reality, the system must be scaled to a cluster of nodes with multiple CPUs and DFEs. If done improperly, the advantages of employing this high performance hardware can be overridden by the inefficiency of ad hoc integration approaches. In this section we briefly describe some of the issues that must be tackled when building a cluster to deliver maximum application performance with conventional CPU and dataflow compute nodes.

6.1 Designing a Cluster

In modern data centers or supercomputing centers, computing facilities are usually grouped into physical containers called *racks*. This is a physical partition of the resources for ease of management. Usually, the computing nodes and other appliances are mounted on the racks' frame. A common configuration of a multi-rack computing system in a data center is shown in Fig. 9. The computing capabilities of a rack are limited by the physical space, the power supply, cooling capacity, network capacity and the access to storage.

6.1.1 Compute

Typical data center racks have 40–50 "rack units" of space, capable of holding 40–50 dense 1U compute servers (so named because they take up 1 rack unit of space). However, electric power supplied to a rack is typically in the range of a few kilowatts up to a few tens of kilowatts, which limits the total computing capability within the rack. For a common dual socket 1U CPU server, power consumption under load can easily be in the region of 300–600 W. If each machine consumes 500 W, 40 such machines will consume 20 KW and more importantly will require a matching amount of cooling, since most of this power consumption is converted into heat.

When comparing a dataflow computing rack to a conventional rack it is possible to consider comparisons in terms of space, power, or compute performance. Adding dataflow engines to a compute rack has only a small impact on power consumption, since the power consumed by a single DFE is typically in the region of 50–100 W. If a dataflow node runs 30 times faster than a conventional CPU node and consumes 25% more power, the overall improvement in energy per result is 24 times. To deliver results in the same time (i.e. fixed performance), the cluster can be 30 times smaller, with 24 times lower power consumption. Or utilizing a fixed power budget to the fullest (fixed power consumption), a cluster could be built that is 80% of the size of the CPU cluster but still delivers 24 times the performance. At fixed size, the cluster would consume 25% more power, but deliver 30 times the performance. The precise balance between these considerations can be determined based on user demands.

As we discussed earlier, there are tasks or applications which require more CPU-centric computation while others can be highly optimized for DFEs. So it is sometimes not obvious how the power or space budget, constrained within a rack, should be allocated between the CPUs and the DFEs. The solution is limited if the DFEs are hosted within the CPU servers and the ratio of CPUs to DFEs is rigidly defined in deployment stage, although this can be optimized to the best balance given the expected particular application demands. With the help of a loosely coupled system such as the MPC-X dataflow nodes, higher flexibility is provided, particularly for clusters running more than one type of application.

6.1.2 Memory

Each node in a cluster will typically have its own local memory to store working data. In general, if nodes have large enough memories remote communication can be minimized, however applications with large memory requirements may not fit into the memory of one computation node and will instead be split across nodes. Mechanisms such as message passing interface (*MPI*) can be used to create a distributed program where each node operates on part of the overall data volume. Other applications may be parallelizable across nodes with minimal communication—for example Monte Carlo simulations can often generate results for each simulation independently and then combine results at the end of the computation.

6.1.3 Disk Storage

In addition to main memory, disk storage is a significant part of most systems. Clusters may have a centralized data storage, such as the storage area network

(SAN) in Fig. 9 which is where all nodes fetch data from and store results to. Individual nodes may also have local storage, which may be more efficient for some applications.

A key consideration is the storage use-case: shared or exclusive. Nodes require shared storage for loading global input data, writing final output data, or communicating large files between nodes. Exclusive storage is often required for intermediate results or caching. In contrast, the storage resources as described above can be centralized or distributed. Centralized storage can be a good fit for storing shared data, provided the centralized resource can provide sufficient performance to satisfy the demands of all the compute nodes. Distributed storage is a good fit for exclusive use by one node, since it is local to the node and the provision of the storage resource scales as the number of compute nodes is increased.

Distributed storage can also be used to provide shared access; however, data synchronization will be difficult if any node requires access to the global data set. For these applications, data can be partitioned in a way that the nodes spend most of the time computing with a local data set while a remote request is made if the needed data are not in local scope.

Various hybrid schemes exist to try to provide the performance benefits of distributed storage with the ease-of-programming and data coherency benefits of shared storage. One example of such schemes is parallel file systems such as Lustre [9], which stripe data across many file servers in parallel. In these schemes, bottlenecks around individual file servers or disks can be minimized; however, care must be taken that the network itself does not then overly limit performance.

Ultimately, the critical factor in the performance of a storage system is how well it supports the applications that need to run on it. However at a lower level, the performance of a storage system can be modeled as a trade-off between cost and three main characteristics:

1. Capacity
2. Throughput for linear access (MB/s)
3. IOPS (Input/output operations per second)

Streaming throughput is more interesting when accessing large files, while IOPS are more critical when reading/writing many small files. Different underlying storage technologies can offer alternative points in this space. For applications where streaming access is most important, conventional hard drives often provide the best capacity and price/performance. On the other hand, if IOPS are critical SSDs provide dramatically better performance, but at reduced capacity and higher cost.

6.1.4 Network

As shown in Fig. 9, there may be multiple networks with multiple levels connecting the nodes in a cluster system. Networks can be measured in terms of latency between two points (which can be approximated by the number of hops in the network) and bandwidth, in terms of both aggregate bandwidth (how many nodes can send messages at the same time) and bisection bandwidth (aggregate bandwidth at a particular cut in the network).

Typically high performance network traffic in current clusters uses networks such as Gigabit Ethernet, 10 Gigabit Ethernet or Infiniband. In many clusters, the nodes within a rack communicate through top-of-rack (TOR) switches. These TOR switches then provide up-links to the wider network (other racks). This tree architecture can be provisioned as a "fat-tree" [5] which maintains full non-blocking bandwidth between any points on the network, or as a under-provisioned "thinner" structure where full bandwidth is only available between local nodes and bandwidth to the wider network must be shared. The level of "fatness" at different points in the network tree determines the amount of bandwidth available between two points.

Clusters will often have a separate management network as well as the main network, which provides access to each node to provide remote access, load configurations, power on/off, etc. This management network is generally not bandwidth sensitive and can be built with a thin tree structure, often of 10/100 Mbit or 1 Gbit Ethernet links.

In addition to the global network(s), local networks may be present within small parts of the system. One example of this is a SAN, where storage nodes may communicate with themselves over a dedicated interconnect such as Fibrechannel. Another example is a dedicated DFE/CPU network where CPU nodes communicate with local dataflow nodes over faster links than are provided globally.

Global networks need not be tree-based, and other network topologies can provide substantially superior performance if they are a good fit to the application. For example, a torus network where nodes are connected directly to their neighbors, provides a good fit for clusters where most communication is expected to be between neighboring nodes (e.g. exchanging MPI messages).

DFEs can also have their own dedicated networks; for example Maxeler nodes contain internal DFE torus networks (MaxRing) which connect each DFE to its neighbors. This provides an extra layer of tightly integrated and low latency networking in the cluster.

6.2 Cluster Management

Software support for cluster level management is also important for adopting dataflow computing in a large scale deployment. The main tasks of a cluster management system include monitoring status of the cluster nodes, scheduling the compute jobs and providing a control interface between users and the cluster. Although CPU-centric cluster management software is widely adopted in production systems, most such software packages are not designed to manage clusters with dataflow computing elements.

Dataflow computing puts two significant new constraints on cluster management systems. Firstly, they must be able to support the allocation and scheduling of DFE resources, which are different to CPUs; and secondly they must support nodes that compute substantially faster.

Resource allocation, sharing and scheduling operates differently for DFEs to CPUs. A CPU server is intrinsically able to support multiple processes in a time sharing style with the help of the operating system. But DFEs do not usually support time sharing for multiple applications in this way—partly because many of the performance benefits come from keeping data on-chip and it would thus take a very long time to swap a process off the DFE compared to the CPU.

Maxeler's MaxelerOS management software provides resource allocation and sharing for DFEs, but under strict user control. Programmers can choose to access a *group* of dataflow engines and submit individual processing operations to that group. Each operation is guaranteed to complete atomically, without interruption by another process, but at the group level individual DFEs are shared between many nodes and processing operations are queued until DFEs are available for them.

Dataflow engine groups can also grow or shrink automatically depending on demand, something particularly useful when there is more than one application using the DFE resources. For example, if there are some number of CPU nodes accessing an MPC-X series system with 8 dataflow engines, and applications A and B are running, then the 8 engines should balance between those applications i.e. if application A needs one third as much DFE run-time as application B, then 2 DFEs might be allocated to A, and 6 to B.

Because this resource sharing requires programmer knowledge at the application level, it also requires scheduling DFEs as a new type of cluster resource in the cluster management software. One possibility is to simply allocate blocks of DFEs to particular application jobs, and then allow them to share DFEs for any use within that block. Another alternative is that in systems with a fixed DFE to CPU mapping, such as the MPC-C series, DFEs can be allocated automatically as part of the allocation of a CPU node.

To facilitate the management of dataflow nodes, a cluster system must also be able to monitor various characteristics of the DFEs. These include health, ownership, temperature and utilization. Maxeler provides the *maxtop* and *maxstatuscheck* monitoring utilities which return status information about DFEs within a particular system; then this output can be integrated into cluster-level systems; however, the management system must be extensible to absorb this and translate it into useful information for the management process.

On the performance side, cluster schedulers are often built to schedule relatively long-running "batch" processes, where any scheduling overhead is minimal compared to the process run-time. If a dataflow accelerated process performs the same computation 100 times faster, scheduling overhead can suddenly become the major cost in a cluster. One solution to this is to modify the application level to increase the amount of work performed per scheduling unit; another is to modify the cluster scheduler to improve performance for scheduling small jobs.

6.3 Resiliency

When dataflow computing is adopted for a cluster, the computational performance and energy efficiency is usually the main focus. However, like any other machines, dataflow computing nodes may fail during operation for various reasons. A conventional CPU server may fail for reasons such as disk failure, power failure, memory error or failures within the processor while the reasons for a DFE to fail depend on the underlying technologies employed in the hardware. For FPGA-based DFEs, memory errors are the main source of failure. The memory can be either the internal SRAM storage for configuration and status or the external DRAM storage for application data storage.

The reliability of a cluster can be measured by the mean-time-between-failures (MTBF). The MTBF of a cluster is directly related to the number of nodes in the cluster as shown in the following equation.

$$\mathrm{MTBF_{cluster}} = \frac{\mathrm{MTBF_{node}}}{\mathrm{NumNodes}} \tag{2}$$

This means that clusters will tend to experience failures more often as the number of nodes is increased. Thus, an intrinsic advantage of a dataflow accelerated cluster is that by utilizing a smaller number of nodes to perform the same amount of computation, the rate of failure can be reduced if DFE nodes have a similar individual MTBF to CPU nodes.

In addition, unlike CPU servers, DFE nodes may not have direct attached hard disk storage as shown in Fig. 9. Instead, the engines make use of large amount of on-board DRAM. Statistic results show that, for a single node, 76% of the failures are due to the hard drivers or the RAID controller while only 5% are accounted for by DRAM [15], so a diskless DFE node can be more reliable than a disk-based CPU server.

On the other side, DFE nodes often have more memory than CPU nodes, which makes them more likely to see memory errors. However most errors in DRAM in large-scale deployments are correctable [14]. In CPU servers, ECC memory is used to handle errors in DRAM. Using ECC memory, 1-bit errors can be corrected and 2-bit errors can be detected in each 64-bit memory word. This provides a uniform memory error protection across all applications running on the CPU server. On the other hand, a DFE can be customized with different levels of protection according to the practical requirements of the application. For example, a video streaming application may require less protection than a cryptographic application. The customization includes different type of error correction and detection schemes, different protection granularities in memory word size and the option of embedding the protection logic in application kernels.

7 Case Study: Reverse Time Migration

Reverse time migration (RTM) is a geoscience algorithm used in complex geologies to give detailed subsurface images based on seismic survey data [17]. It is widely used in oil and gas exploration.

The concept behind RTM is relatively simple. We start with a known earth model. This earth model might be simply acoustic velocity but could also be anisotropic, elastic, or even visco-elastic. Scientists conduct two modeling experiments simultaneously through the earth model. Both attempt to simulate the seismic experiment conducted in the field—one from the source's perspective and one from the receiver's perspective. To simulate the wave propagation, we start from the acoustic wave equation (where u is pressure and v is velocity):

$$\frac{\partial^2 u}{\partial t^2} = \mathbf{v}^2 \left(\frac{\partial^2 \mathbf{u}}{\partial \mathbf{x}^2} + \frac{\partial^2 \mathbf{u}}{\partial \mathbf{y}^2} + \frac{\partial^2 \mathbf{u}}{\partial \mathbf{z}^2} \right) \tag{3}$$

We can use a Taylor expansion to approximate these derivatives and implement this as a convolution filter. Based on this, Fig. 10 shows pseudo-code for a simple wave propagator. In this case, the computation requires the storage of two pressure volumes for the current and previous state of the wavefield and these are used to compute the next state. The vv volume contains v^2, the earth model. Modeling more complex physics will typically require more earth model parameters and more state volumes to be stored, as well as greater computation. For example a tilted transverse isotropic (TTI) anisotropic modeling algorithm could require up to 6 parameter volumes: velocity, density, two anisotropy parameters and two angles.

We utilize the wave propagator as the computational kernel of RTM, which can then be implemented using pseudo-code as shown in Fig. 11. The source experiment involves injecting our estimated source wavelet into the earth and propagating it from $t = 0$ to our maximum recording time $t_{\max}$, creating a 4D source field $s(x,y,z,t)$. For the receiver data, we inject and propagate our recorded data starting from $t_{\max}$ to t_0, creating a similar 4D volume $r(x,y,z,t)$.

A reflection is inferred where the energy propagated from the source and receiver is located at the same space–time point, thus the final image is the summation of correlating the source and receiver wavefield at every time and every shot.

```
curr = zeros()
prev = zeros()
for t = 0 to tmax:
  for i = 0 to X*Y*Z-1:
    l = convolve(curr[i], stencil)
    next[i] = 2 * curr[i] - prev[i]
              + vv[i] * l
    next[i] += stimulus[t, i]
  apply_boundary_condition(curr, next)
  swap(prev, curr, next)
```

Fig. 10 Pseudo-code for a finite difference wave propagator

```
final_image = zeros()
for i = 0 to nshots:
        shot_image(i) = zeros()
        s = propagate_source(i, 0, tmax, source_wavelet)
        r = propagate_receiver(i, tmax, 0, receiver_data)
        for t = 0 to tmax:
                shot_image += s(t) * r(t)
        final_image += shot_image
```

Fig. 11 Pseudo-code for reverse time migration

$$i(x,y,z) = \sum_{\text{shots}} \sum_{t=0}^{t_{\max}} s(x,y,z,t)r(x,y,z,t) \qquad (4)$$

7.1 Analysis

To apply the *Maximum Performance Computing* approach to the RTM application we must consider the different possible bottlenecks that affect performance.

On the computation side, the most significant cost is the wave propagation calculation. The cross-correlation between source and receiver fields can also be significant but contains many fewer operations than the modeling calculation.

On the memory/storage side, the main cost is the requirement to store a 4D volume for the source or receiver; however, the 3D current/previous pressure fields used for the wave propagation calculation can also be an issue.

The explicit time marching scheme used for wave propagation described above has an operation count proportional to the number time steps *nt* times the size of the domain—$O(nx \times ny \times nz \times nt)$. In order to obtain a stable and non-dispersive solution, we are constrained in choosing our sampling in time and space. If we use too large time steps or sample our medium too coarsely for a given velocity or frequency in our data, we can run into problems with stability, dispersion, and/or accuracy. Our sampling is controlled by the minimum and maximum velocity in the media, the maximum frequency we want propagated, and the order of accuracy of the derivative approximations. Figure 12 shows the impact of wave frequency on the computation and memory costs. For 70 Hz RTM, a data volume can easily be over 10 bn points (40 gigabytes per array, if single precision floating point is used for storage).

7.2 Transformations

We can consider a number of transformations to maximize the performance potential of RTM.

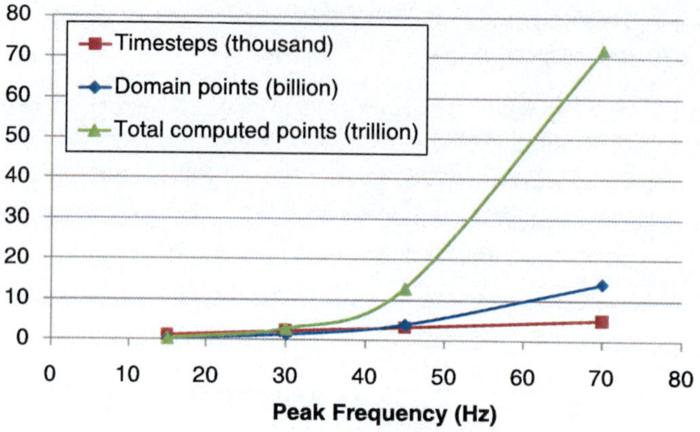

Fig. 12 Impact of maximum frequency on computation and memory cost for modeling wave propagation

7.2.1 Modeling Kernel

Assuming we have selected a particular physics model, the main opportunity for algorithmic transformation in the modeling kernel is to select an optimized convolution stencil for the computation of the derivatives.

By using higher approximations to the space and time derivatives (larger stencils), we can use a coarser sampling in time and space with the trade-off of a larger number of operations per sample. This is generally a beneficial trade-off, particularly for the spatial derivatives, because we trade off linear growth in our derivative approximation with cubic growth in the number of samples.

Another consideration is the shape of stencil. Typically stencils are "star" shaped as shown in Fig. 13a, with each leg computing an x, y or z derivative; however, an alternative is to use a "cube" shaped stencil as shown in Fig. 13b. The cube requires approximately 20% more arithmetic operations than the star for the same order derivative approximation but only requires one third of the on-chip buffering memory when implemented as a dataflow architecture since it is more compact. For implementations on CPU, the cube stencil has similar or worse performance to the star. However when used in a dataflow architecture the lower buffering memory requirements can improve performance in some cases [3].

This performance improvement comes about from considering the different axes of parallelization that are possible for the finite difference kernel within the dataflow engine. The first level is *pipeline parallelism*—that all stages of the pipeline will be computing simultaneously. However if this does not utilize all the silicon resources available it makes sense to utilize additional parallelism to further increase performance.

The second level of parallelization possible is *multi-pipeline* parallelism—the instantiation of multiple copies of the computation pipeline beside eachother on the

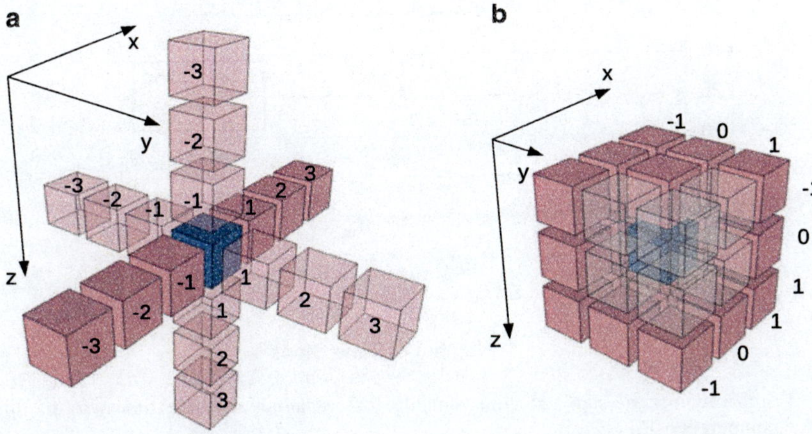

A 7-pt star stencil

A 3x3x3 cube stencil

Fig. 13 Two alternative stencil shapes providing 6th order in space accuracy

chip, processing multiple data items per cycle. Multiple dataflow pipelines can share the same on-chip buffering memory so this is an efficient method of parallelization; however, the memory bandwidth required increases linearly as the number of *pipes* is increased.

An alternative, third level of parallelization is *multi-timestep* computation. In this case, multiple timesteps are *cascaded* on the chip, without intermediate data being written back to memory. This means that the amount of computation performed in parallel can be increased by lengthening the pipeline without increasing memory bandwidth requirements; however, this does increase on-chip buffering memory requirements. As buffering memory becomes a scarce resource, the overall efficiency of the computation will drop since some data must be read/written from memory multiple times.

Figure 14 shows the performance impact of cascading multiple timesteps with different stencils for a simple physics model. Star stencil performance begins to drop off after a small number of cascade steps due to the limitations of on-chip buffering, while cube stencil performance continues to scale.

7.2.2 Data Management

RTM requires storage of at least one 4D data volume that is typically larger than it is practical to hold in main memory (e.g. several terabytes). If this data volume is stored on disk, disk speed could potentially limit performance. At the transformation stage, we should consider alternative schemes that could eliminate this bottleneck.

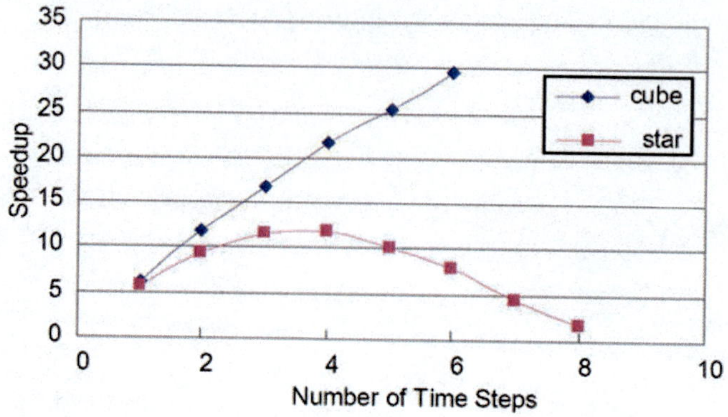

Fig. 14 Comparing use of star and cube stencils for cascading multiple timesteps in finite difference computations[3]

Broadly, we can consider:

1. Ensure the 4D volume fits into main memory.
2. Store the 4D volume to a fast disk system.
3. Recompute parts of the volume as necessary.

Option 1 above is possible for either low frequencies with small amounts of data, or by provisioning large amounts of memory in a system. For example, if the problem is split across multiple MPC-X dataflow systems, each of which has nearly 400 GB of RAM, it is practical to store the full 4D volume in memory for many useful sizes.

Option 2 requires that the disk performance can keep up with the compute performance. One way to achieve this is to have compute nodes with many high-performance disks and pair them with a relatively small number of dataflow engines. Compression can also be utilized to reduce the data volume that must be written to disk at the cost of extra computation.

Option 3 is an explicit trade-off between storage and computation and may be most interesting in terms of future scalability since we expect compute speed to grow faster than storage speed as time goes forward. There are a number of different possible schemes to trade computation for storage, but one of the most effective is to use the property of the wave equation that it is *reversible*.

The requirement to store the 4D volume arises because the source and receiver experiments are being performed in opposite time directions, so the first imaging step cannot be begun until the source wavefield has been fully propagated to $t = t_{\max}$. An alternative to storing $s(nx, ny, nz, nt)$ is to compute the source wavefield until $t_{\max}$ without storing the intermediate data, then to reverse the time direction and recompute the source wave back from $t_{\max}$ to t_0 in parallel to the receiver propagation. This approach requires approximately 50% more computation but removes the need to store a 4D data volume entirely.

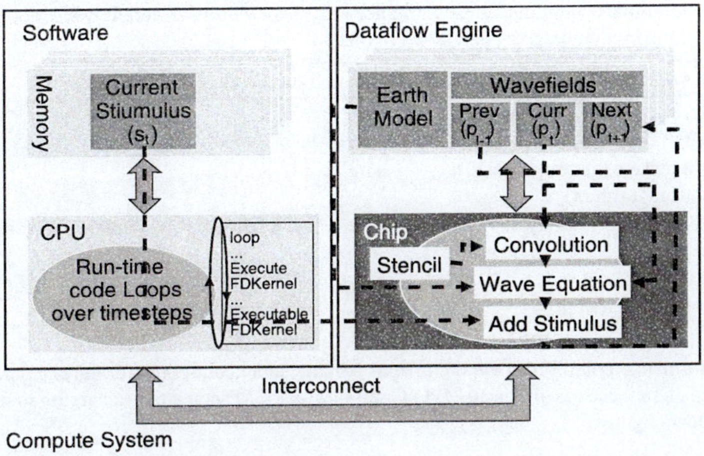

Fig. 15 Partitioning computation and data for the wave propagation calculation

7.3 Partitioning

The main wave propagation calculation is a good fit for the dataflow engines in the system and it is relatively straightforward to assign that computation to the DFEs. Similarly, program initialization and set-up and control at run-time can easily remain on the CPU since it is irregular and not performance critical. Figure 15 shows how this partitioning operates.

The imaging calculation may be executed on either the CPUs or the DFEs. If a recomputation scheme is chosen from Sect. 7.2.2, then all computation data will be present in DFE memory at run-time so it may make sense for the DFE to perform the imaging calculation as well even though it is not a significant part of the run-time. Alternatively, if the data is being stored to the CPU memory or disk, it may make more sense for the CPU to perform the imaging calculation in parallel as the DFE computes the wave propagation.

For parallelizing across multiple dataflow engines, we select to duplicate the full computation on each engine and split the problem domain across them with each DFE computing on a sub-domain of the whole. At the boundaries of each DFE's region, it will need to communicate data elements with neighboring DFEs and this can be done efficiently using the dedicated *MaxRing* interconnect.

7.4 Implementation

To implement the RTM, we create a software program to run on the conventional CPUs (or modify an existing program) and program the dataflow engines with the wave propagation (and possibly imaging) computation. We can implement the

Table 1 Summary of published performance results for various different wave modeling computations implemented as dataflow engines

Physics model	Compared hardware	Equiv. cores	Ref.
Isotropic, variable velocity and density (1bn pt)	1 core vs. MAX2	240	[11]
Vertical transverse isotropic (VTI) anisotropy (1bn pt)	1 core vs. MAX2	56	[8]
Isotropic, variable velocity (14bn pt)	32 cores[a] vs. 8 MAX2	~ 2000	[12]
Tilted transverse isotropic (TTI) anisotropy (0.5bn pt)	8 cores vs. 4 MAX3	~ 500	[7]

[a]Parallelization using MPI over Infiniband

For performance comparison we quote approximate equivalent cores for the accelerated node, which in these cases was an MPC-C series with 8 CPU cores plus a varying number of dataflow engines

DFE part using MaxCompiler, or using Maxeler *MaxGenFD* [10]. MaxGenFD is a high-level domain-specific language compiler which takes a higher-level input description in terms of pressure fields, convolution stencils and basic arithmetic operations and automatically generates the MaxCompiler Kernels and Manager.

MaxGenFD handles the complexities facing any finite difference implementation such as managing very large data sets, boundary conditions and domain decomposition across multiple DFEs with halo exchange. In addition, the compiler removes the need for the geoscience programmer to perform hardware-specific optimizations such as customizing data-types and generating optimized convolution stencil descriptions for hardware.

At the cluster level, ideally each shot that is computed runs on a single node. Depending on the amount of data required for that shot, multiple DFEs may be needed. If a scheme to store the 4D wavefield volume is used, each compute node should have relatively few DFEs and a large number of fast hard disks to store the temporary volumes. These 4D volumes should not be stored on a centralized file server.

If a recomputation scheme is used, little or no disks are required for a node, but the number of DFEs should be increased to maximize the compute density. A shared file server or SAN can usually supply the input data for each shot computation. Output data can either be written to the shared file server for each shot, or accumulated on each node and then only aggregated data written back to the shared file server.

A variety of RTM implementations on Maxeler hardware have been reported. Table 1 shows a summary of published performance results for a variety of real-world geoscience applications with a range of physics model complexity. Performance on real applications is dependent on not only just arithmetic and memory bandwidth but also the balance of different types of data and the data access pattern. The real applications have been implemented with a range of different CPUs and DFEs but consistently show that a single dataflow engine can be equivalent to dozens of conventional CPUs.

For large-scale isotropic modeling [12], a single dataflow engine equates to the performance of 100–200 CPU cores, partly because the CPU performance is limited by communication overhead to decompose the large domain, while the dataflow implementation operates within a single node using MaxRing. For VTI modeling [8] in contrast, the algorithm maps less well to the dataflow architecture and each DFE is equivalent to about 25 CPU cores. In this instance the VTI propagator is very limited by memory bandwidth despite the efficient memory use of the dataflow paradigm and so is unable to fully exploit the arithmetic capabilities of the chip.

8 Conclusion

Maximum Performance Computing strives to deliver the maximum possible performance within a space and/or power budget. Dataflow computing is a powerful computational paradigm for supporting this goal, maximizing the efficient use of silicon area and delivering significantly better performance than conventional CPUs for a wide range of problems in scientific computing and further afield.

One of the key advantages of dataflow computing is constructing a computer system around the flow of data in an application and inherently minimizing data movement rather than purely considering the computation. As silicon technology continues to advance in the future, the cost of computation will decrease with Moore's Law, while the cost of moving data and accessing memory will become increasingly dominant. This implies that the advantages of the dataflow computing approach will become more and more significant in the future.

References

1. J.B. Dennis, Data flow supercomputers. Computer **13**(11), 48–56 (1980)
2. E. Fermi, J. Pasta, S. Ulam, Studies of nonlinear problems. Tech. Rep. Document LA-1940 (1955)
3. H. Fu, R.G. Clapp, O. Mencer, O. Pell, Accelerating 3D convolution using streaming architectures on FPGAs. SEG Expanded Abstracts **28** (2009)
4. H.T. Kung, Why systolic architectures? Computer **15** (1982)
5. C.E. Leiserson, Fat-trees: universal networks for hardware-efficient supercomputing. IEEE Trans. Comp. **34** (1985)
6. O. Lindtjorn, R.G. Clapp, O. Pell, O. Mencer, M.J. Flynn, H. Fu, Beyond traditional microprocessors for geoscience high-performance computing applications. IEEE Micro **31**(2), 41–49 (2011)
7. O. Lindtjorn, R.G. Clapp, O. Pell, O. Mencer, M.J. Flynn, H. Fu, Beyond traditional microprocessors for geoscience high-performance computing applications. IEEE Micro (2011)
8. W. Liu et al., Anisotropic reverse-time migration using co-processors. SEG Expanded Abstracts **28** (2009)
9. Lustre, http://www.lustre.org. Accessed 26th Jaunary 2013
10. Maxeler Technologies: MaxGenFD white paper. Tech. rep. (2010)

11. T. Nemeth, J. Stefani, W. Liu, R. Dimond, O. Pell, R. Ergas, An implementation of the acoustic wave equation on FPGAs. SEG Expanded Abstracts **27** (2008)
12. D. Oriato, O. Pell, C. Andreoletti, N. Bienati, FD modeling beyond 70hz with FPGA acceleration, in *SEG 2010 HPC Workshop*, Denver (2010)
13. O. Pell, O. Mencer, Surviving the end of frequency scaling with reconfigurable dataflow computing. SIGARCH Comput. Archit. News **39**(4), 60–65 (2011)
14. B. Schroeder, E. Pinheiro, W.D. Weber, DRAM errors in the wild: a large-scale field study, in *Proceedings of eleventh International Joint Conference on Measurement and Modeling of Computer Systems* (SIGMETRICS '09) (ACM, New York, USA, 2009), pp. 193–204. DOI 10.1145/1555349.1555372, http://doi.acm.org/10.1145/1555349.1555372
15. K.V. Vishwanath, N. Nagappan, Characterizing cloud computing hardware reliability, in *Proceedings of the 1st ACM Symposium on Cloud Computing* (SoCC '10) (ACM, New York, USA, 2010), pp. 193–204. DOI 10.1145/1807128.1807161, http://doi.acm.org/10.1145/1807128.1807161
16. S. Weston, J. Spooner, S. Racaniere, O. Mencer, Rapid computation of value and risk for derivatives portfolios. Concurrency Comput.: Pract. Ex. (2011)
17. K. Yoon, C. Chin, S. Suh, L. Lines, S. Hong, 3D reverse-time migration using the acoustic wave equation: an experience with the SEG/EAGE dataset. The Leading Edge **22**, 38–41 (2003)

ERRATUM

Chapter 19
MEMSCALE: Re-architecting Memory Resources for Clusters

Holger Frönin
University of Heidelberg, Germany

Federico Silla • Hector Montaner
Universitat Politècnica de València, Spain

W. Vanderbauwhede and K. Benkrid (eds.), *High-Performance Computing Using FPGAs*, DOI 10.1007/978-1-4614-1791-0_19, pp. 569–604,

---

DOI 10.1007/978-1-4614-1791-0_26

There is an incorrect spelling to the first author's name in chapter 19. The correct spelling is "Holger Fröning".

The online version of the original chapter can be found at
http://dx.doi.org/10.1007/978-1-4614-1791-0_19

Index

W. Vanderbauwhede and K. Benkrid (eds.), *High-Performance Computing Using FPGAs*, DOI 10.1007/978-1-4614-1791-0, © Springer Science+Business Media, LLC 2013

B

C

D

I

W

X